SI UNITS USED IN MECHANICS

Quantity	Unit	SI Symbol
(*Base Units*)		
Length	meter°	m
Mass	kilogram	kg
Time	second	s
(*Derived Units*)		
Acceleration, linear	meter/second^2	m/s^2
Acceleration, angular	radian/second^2	rad/s^2
Area	meter^2	m^2
Density	kilogram/meter^3	kg/m^3
Force	newton	N ($= \text{kg·m/s}^2$)
Frequency	hertz	Hz ($= 1/s$)
Impulse, linear	newton-second	N·s
Impulse, angular	newton-meter-second	N·m·s
Moment of force	newton-meter	N·m
Moment of inertia, area	meter^4	m^4
Moment of inertia, mass	kilogram-meter^2	kg·m^2
Momentum, linear	kilogram-meter/second	kg·m/s ($=$ N·s)
Momentum, angular	kilogram-meter^2/second	kg·m^2/s ($=$ N·m·s)
Power	watt	W ($= $ J/s $=$ N·m/s)
Pressure, stress	pascal	Pa ($= \text{N/m}^2$)
Product of inertia, area	meter^4	m^4
Product of inertia, mass	kilogram-meter^2	kg·m^2
Spring constant	newton/meter	N/m
Velocity, linear	meter/second	m/s
Velocity, angular	radian/second	rad/s
Volume	meter^3	m^3
Work, energy	joule	J ($=$ N·m)
(*Supplementary and Other Acceptable Units*)		
Distance (navigation)	nautical mile	($= 1.852$ km)
Mass	ton (metric)	t ($= 1000$ kg)
Plane angle	degrees (decimal)	°
Plane angle	radian	—
Speed	knot	(1.852 km/h)
Time	day	d
Time	hour	h
Time	minute	min

° Also spelled *metre*.

SI UNIT PREFIXES

Multiplication Factor	Prefix	Symbol
$1\ 000\ 000\ 000\ 000 = 10^{12}$	terra	T
$1\ 000\ 000\ 000 = 10^{9}$	giga	G
$1\ 000\ 000 = 10^{6}$	mega	M
$1\ 000 = 10^{3}$	kilo	k
$100 = 10^{2}$	hecto	h
$10 = 10$	deka	da
$0.1 = 10^{-1}$	deci	d
$0.01 = 10^{-2}$	centi	c
$0.001 = 10^{-3}$	milli	m
$0.000\ 001 = 10^{-6}$	micro	μ
$0.000\ 000\ 001 = 10^{-9}$	nano	n
$0.000\ 000\ 000\ 001 = 10^{-12}$	pico	p

SELECTED RULES FOR WRITING METRIC QUANTITIES

1. (a) Use prefixes to keep numerical values generally between 0.1 and 1000.
   (b) Use of the prefixes hecto, decka, deci, and centi should be generally avoided except for certain areas or volumes where the numbers would be otherwise awkward.
   (c) Use prefixes only in the numerator of unit combinations. The one exception is the base unit kilogram. (*Example:* write kN/m not N/mm; J/kg not mJ/g)
   (d) Avoid double prefixes. (*Example:* write GN not kMN)
2. Unit designations
   (a) Use a dot for multiplication of units. (*Example:* write N·m not Nm)
   (b) Avoid ambiguous double solidus. (*Example:* write N/m^2 not N/m/m)
   (c) Exponents refer to entire unit. (*Example:* mm^2 means (mm)^2)
3. Number grouping
   Use a space rather than a comma to separate numbers in groups of three, counting from the decimal point in both directions. (*Example:* 4 607 321.048 72)
   Space may be omitted for numbers of four digits. (*Example:* 4296 or 0.0476)

ENGINEERING MECHANICS STATICS AND DYNAMICS

J. L. MERIAM

Professor of Mechanical Engineering
California Polytechnic State University

JOHN WILEY & SONS, New York • Chichester • Brisbane • Toronto

This book was printed and bound by Kingsport Press.
It was set in Laurel by York Graphic Services.
The drawings were designed and executed by John Balbalis
with the assistance of the Wiley Illustration Department.

Text design by Jerome Wilke, and Edward A. Butler.

Cover design by Edward A. Butler.

Library of Congress Cataloging in Publication Data

Meriam, James L.
   Engineering mechanics: Statics and Dynamics

   Ed. of 1966 and 1969 published under title: Statics
and dynamics.
   Includes index.
   1. Mechanics, Applied. I. Title.
TA350.M458   1978b   620.1   78-518
ISBN 0-471-01979-8

Printed in the United States of America

10 9 8

FOREWORD

The innovations and contributions of Dr. James L. Meriam to the field of engineering mechanics cannot be overstated. He has undoubtedly had more influence on instruction in mechanics during the last quarter century than any other individual.

No one who commenced the study of engineering mechanics after 1950 can fully appreciate the apprehension and lack of understanding that the average engineer once experienced when facing a problem in mechanics. Professor Meriam did much to bring clarity and comprehension to this subject. His first book on mechanics in 1951 literally reconstructed undergraduate mechanics and became the definitive textbooks for the next decade. His texts were logically organized, easy to read, directed to the average engineering undergraduate and were packed with exciting examples of real-life engineering problems superbly illustrated. These books became the model for other engineering mechanics texts in the 1950s and 1960s.

Dr. Meriam began his work in mechanical engineering at Yale University where he earned his B.E., M. Eng., and Ph.D. degrees. He had early industrial experience with Pratt and Whitney Aircraft and the General Electric Company, which stimulated his first contributions to mechanics in mathematical and experimental stress analysis. In the Second World War he served in the U.S. Coast Guard. These early experiences influenced Professor Meriam in two ways. First, he discovered his profound interest in the practical applications of mechanics and, second, he developed a lifelong interest in ships and the sea.

Dr. Meriam was a member of the faculty of the University of California, Berkeley, for twenty-one years where he served as Professor of Engineering Mechanics, Assistant Dean of Graduate Studies, and Chairman of the Division of Mechanics and Design. In 1963 he became Dean of Engineering at Duke University where he devoted his full energies to the development of its School of Engineering. In 1972 Professor Meriam followed his desire to return to full-time teaching and accepted appointment as Professor of Mechanical Engineering at the California Polytechnic State University. Dr. Meriam has always placed great emphasis on teaching, and this trait has been recognized by his students wherever he has taught. For example, at Berkeley in 1963 he was the first recipient of the Outstanding Faculty Award of Tau Beta Pi, given primarily for excellence in teaching.

The free-body diagram is the foundation for mechanics. This was not a new concept with Dr. Meriam, but the emphasis and rigor with which he developed the free body in mechanics were new and highly successful. They permeate his writing and teaching. He was the first author to show clearly how the method of virtual work in statics can be employed to solve a class of problems largely neglected by previous authors. In dynamics, plane motion became understandable, and in his later editions, three-dimensional kinematics and kinetics received the same treatment. He is credited with original developments in the theory of variable-mass dynamics, which are contained in his *Dynamics, 2nd Edition.* More recently, Professor Meriam has been a leader in promoting the use of SI units, and his *SI Versions* of *Statics* and *Dynamics* published in 1975 were the first mechanics textbooks in SI units in this country.

Professor Meriam's new book promises to meet and even to exceed the high standards that he has set in the past. Without question it contains one of the most outstanding collections of problems yet assembled. This new text is especially designed to assist students in the preliminary stages of each new topic, and then to lead them to more challenging engineering applications as well. The new book will appeal to a wide audience of students, teachers, and engineers and will further extend the author's contributions to mechanics.

Robert F. Steidel
Professor of Mechanical Engineering
University of California, Berkeley

PREFACE
To the Student

As you undertake the study of engineering mechanics, first statics and then dynamics, you will be building a foundation of analytical capability for the solution of a great variety of engineering problems. Modern engineering practice demands a high level of analytical capability, and you will find that your study of mechanics will help you immensely in developing this capacity.

In engineering mechanics we learn to construct and solve mathematical models which describe the effects of force and motion on a variety of structures and machines that are of concern to engineers. In applying our principles of mechanics we formulate these models by incorporating appropriate physical assumptions and mathematical approximations. Both in the formulation and solution of mechanics problems you will have frequent occasion to use your background in plane and solid geometry, scalar and vector algebra, trigonometry, analytic geometry, and calculus. Indeed, you are likely to discover new significance to these mathematical tools as you make them work for you in mechanics.

Your success in mechanics (and throughout engineering) will be highly contingent upon developing a well-disciplined method of attack from hypothesis to conclusion in which the applicable principles are applied rigorously. From many years of experience as a teacher and an engineer I know the importance of developing the ability to represent one's work in a clear, logical, and concise manner. Mechanics is an excellent place in which to develop these habits of logical thinking and effective communication.

ENGINEERING MECHANICS contains a large number of sample problems in which the solutions are presented in detail. Also included in these examples are helpful observations with mention made of common errors and pitfalls to be avoided. In addition, the book contains a large selection of simple, introductory problems and problems of intermediate difficulty to help you gain initial confidence and understanding of each new topic. Also included are many problems which illustrate significant and contemporary engineering situations to stimulate your interest and help develop your appreciation for the many applications of mechanics in engineering.

I am pleased to extend my encouragement to you as a student of mechanics. I hope this book will provide both help and stimulation as you develop your background in engineering.

J. L. Meriam

Santa Barbara, California
January 1978

PREFACE
To the Instructor

The primary purpose of the study of engineering mechanics is to develop capacity to predict the effects of force and motion in the course of carrying out the creative design function of engineering. Successful prediction requires more than a mere knowledge of the physical and mathematical principles of mechanics. Also required is the ability to visualize physical configurations in terms of real materials, actual constraints, and practical limitations which govern the behavior of machines and structures. One of our primary objectives in teaching mechanics should be to help the student develop this ability to visualize, which is so vital to problem formulation. Indeed, the construction of a meaningful mathematical model is often a more important experience than its solution. Maximum progress is made when the principles and their limitations are learned together within the context of engineering application.

Courses in mechanics are often regarded by students as a difficult requirement and frequently as an uninteresting academic hurdle as well. The difficulty stems from the extent to which reasoning from fundamentals, as distinguished from rote learning, is required. The disinterest which is frequently felt is due primarily to the extent to which mechanics is presented as an academic discipline largely lacking in engineering purpose and challenge. This attitude is traceable to the frequent tendency in the presentation of mechanics to use problems mainly as a vehicle to illustrate theory rather than to develop theory for the purpose of solving problems. When the first view is allowed to predominate, problems tend to become overly idealized and unrelated to engineering with the result that the exercise becomes dull, academic, and uninteresting. This approach deprives the student of much of the valuable experience in formulating problems and thus of discovering the need for and meaning of theory. The second view provides by far the stronger motive for learning theory and leads to a better balance between theory and application. The crucial role of interest and purpose in providing the strongest possible motive for learning cannot be overemphasized. Further, we should stress the view that, at best, theory can only approximate the real world of mechanics rather than the view that the real world approximates the theory. This difference in philosophy is indeed basic and distinguishes the *engineering* of mechanics from the *science* of mechanics.

During the past twenty years there has been a strong trend in engineering education to increase the extent and level of theory in the engineering-science courses. Nowhere has this trend been felt more than in mechanics courses. To the extent that students are prepared to handle the accelerated treatment, the trend is beneficial. There is evidence and justifiable concern, however, that a significant disparity has more recently appeared between coverage and comprehension. Among the contributing factors there are three trends which we should note. First, emphasis on the geometric and physical meanings of prerequisite mathematics appears to have diminished. Second, there has been a significant reduction and even elimination of instruction in graphics which in the past served to enhance the visualization and representation of mechanics problems. Third, in advancing the mathematical level of our treatment of mechanics there has been a tendency to allow the notational manipulation of vector operations to mask or replace geometric visualization. Mechanics is inherently a subject which depends on geometric and physical perception, and we should increase our efforts to develop this ability.

One of our responsibilities as teachers of mechanics is to use the mathematics which is most appropriate for the problem at hand. The use of vector notation for one-dimensional problems is usually trivial; for two-dimensional problems it is often optional; but for three-dimensional problems it is quite essential. As we introduce vector operations in two-dimensional problems, it is especially important that their geometric meaning be emphasized. A vector equation is brought to life by a sketch of the corresponding vector polygon, which often discloses through its geometry the shortest solution. There are, of course, many mechanics problems where the complexity of variable interdependence is beyond normal powers of visualization and physical perception, and reliance on analysis is essential. This fact notwithstanding, our students become better engineers when their abilities to perceive, visualize, and represent are developed to the fullest.

As teachers of engineering mechanics we have the strongest obligation to the engineering profession to set reasonable standards of performance and to uphold them. In addition, we have a serious responsibility to encourage our students to think for themselves. Too much help with details that the student should be reasonably able to handle from prerequisite subjects can be as bad as too little help and can easily condition him to becoming overly dependent on others rather than to exercise his own initiative and ability. Also, when mechanics is subdivided into an excessive number of small compartments, each with detailed and repetitious instructions, the student can have difficulty seeing the 'forest' for the 'trees' and, consequently, fail to perceive the unity of mechanics and the far-reaching applicability of its few basic principles and methods.

ENGINEERING MECHANICS is written with the foregoing philosophy in mind. It is intended primarily for the first engineering course in mechanics, generally taught in the second year of study. The book omits a number of the more advanced topics contained in the author's more extensive treatments, *Statics* and *Dynamics, 2nd Edition* and *SI Version,* and is designed especially to facilitate self-study. To this end a major feature of the book is a greatly expanded treatment of sample problems which are presented in a single-page format for more convenient study. In addition to presenting the solution in detail, each sample problem also contains comments and cautions keyed to salient points in the solution and printed in colored type. These comments alert students to common pitfalls and should provide a valuable aid to their self-study efforts.

ENGINEERING MECHANICS contains 168 sample problems and 1820 unsolved problems from which a wide choice of assignments can be made. Of these "problems over 50 percent are totally new and the balance selected from the author's *Statics* and *Dynamics, 2nd Edition* and *SI Version.* In recognition of the need for the predominant emphasis on SI units, there are two problems in SI units for every one in U.S. customary units. This apportionment between the two sets of units permits classroom use anywhere from a 50-50 emphasis to a 100 percent SI treatment. Each problem set begins with relatively simple, uncomplicated problems to help the student gain confidence with the new topic. Many practical problems and examples of interesting engineering situations drawn from a wide range of applications are represented in the problem collection. Simple numerical values have been used throughout, however, so as not to complicate the solutions and divert attention from the principles. The problems are arranged generally in order of increasing difficulty, and the answers to a majority of them are given. The more difficult problems are identified by a ▶ mark or the mark ▶ and may often be used to provide a comprehensive classroom experience when solved by the instructor. All numerical solutions have been carried out and checked with an electronic calculator without rounding intermediate values. Consequently, the final answers should be correct to within the number of significant figures cited. The author is confident that the book is exceptionally free from error.

ENGINEERING MECHANICS is written in a style which is both concise and friendly. The major emphasis is focused on basic principles and methods rather than on a multitude of special cases. Strong effort has been made to show both the cohesiveness of the relatively few fundamental ideas and the great variety of problems which these few ideas will solve.

Volume 1, Statics. In Chapter 2 the properties of forces, moments, couples, and resultants are developed so that the student may proceed directly to the equilibrium of noncurrent force systems in Chapter 3 without belaboring unnecessarily the relatively trivial problem of the equilibrium of concurrent forces acting on a particle. In both Chapters 2 and 3 analysis of two-dimensional problems is presented before three-dimensional problems are treated. The vast majority of students acquire a greater physical insight and understanding of mechanics by first gaining confidence in two-dimensional analysis before coping with the third dimension.

Application of equilibrium principles to simple trusses and to frames and machines is presented in Chapter 4 with primary attention given to two-dimensional systems. A sufficient number of three-dimensional examples are included, however, to enable the student to exercise his more general vector tools of analysis.

The concepts and categories of distributed forces are introduced at the beginning of Chapter 5 with the balance of the chapter divided into two main sections. Section A treats centroids and mass centers where detailed examples are presented to help the student master his early applications of calculus to physical and geometrical problems. Section B includes the special topics of beams, flexible cables, and fluid forces which may be omitted without loss of continuity of basic concepts.

Chapter 6 on friction is divided into Section A on the phenomenon of dry friction, and Section B on selected machine applications. Although Section B may be omitted if time is limited, this material does provide a valuable experience for the student in dealing with distributed forces.

Chapter 7 presents a consolidated introduction to virtual work with application limited to single-degree-of-freedom systems. Special emphasis is placed on the advantage of the virtual-work and energy method for interconnected systems and stability determination. Virtual work provides an excellent opportunity to convince the student of the power of mathematical analysis in mechanics.

Volume 2, Dynamics. The logical division between particle dynamics and rigid-body dynamics, with each part treating the kinematics prior to the kinetics, has been followed in Chapters 2 and 3. This arrangement greatly facilitates a more thorough and rapid excursion in rigid-body dynamics with the prior benefit of a comprehensive introduction to particle dynamics.

Chapter 3 on particle kinetics focuses on the three basic methods, force-mass-acceleration, work-energy, and impulse-momentum. The special topics of central-force motion, impact, relative motion, and vibrations are grouped together in Chapter 4 on special applications and serve as optional material to be assigned according to instructor preference and available time. With this arrangement the attention of the student is focused more strongly on the three basic approaches to kinetics which are developed in the single chapter. Vibrations, once treated as a major topic in dynamics, is now more frequently covered in other courses. Consequently, the treatment of vibrations has been limited to a single article which, however, is sufficient to introduce the formulation and solution of the equation for free and forced linear oscillation.

Chapter 5 on systems of particles is a generalization of the principles of motion for a single particle. The chapter also includes the topics of steady mass flow and variable mass which may be considered as optional material depending on the time available.

In Chapter 6 on the kinematics of rigid bodies in plane motion, emphasis is placed jointly on the geometry and algebra of vector solutions to relative-velocity and relative-acceleration equations. Again, this dual approach serves the purpose of re-enforcing the meaning of vector mathematics.

In Chapter 7 on the kinetics of rigid bodies in plane motion each basic motion is separately identified and solved. Strong dependence is placed on forming the direct equivalence between the actual forces and their $m\bar{a}$ and $\bar{I}\alpha$ resultants. In this way the versatility of the moment principle is emphasized, and the student is encouraged to think directly in terms of resultant dynamics effects.

Chapter 8, which may be treated as optional, provides a basic introduction to three-dimensional dynamics which is sufficient to solve many of the more common space-motion problems. For students who later pursue more advanced work in dynamics, Chapter 8 will provide a solid foundation. Gyroscopic motion with steady precession is treated in two ways. The first approach makes use of the analogy between the relation of force and linear-momentum vectors and the relation of moment and angular-momentum vectors. With this treatment the student can understand the gyroscopic phenomenon of steady precession and handle most of the engineering problems on gyros without a detailed study of three-dimensional dynamics. The second approach makes use of the more general momentum equations for three-dimensional rotation where all components of momentum are accounted for.

Moments and products of inertia of areas are presented in Appendix A (Statics). This topic helps to bridge the subjects of statics and solid mechanics. Moments and products of inertia of mass are included in Appendix A (Dynamics). Appendix B contains a summary review of selected topics of elementary mathematics that the student should be prepared to use in mechanics.

It is a pleasure for me to recognize again the continuing contribution of Dr. A. L. Hale of the Bell Telephone Laboratories for his invaluable suggestions and careful checking of the manuscript. The critical reviews of Professor Andrew Pytel of The Pennsylvania State University and Professors Kenneth Schneider and John Biddle of the California State Polytechnic University have also been of great assistance and are acknowledged with gratitude. In addition, appreciation is expressed to Professor J. M. Henderson of the University of California, Davis, for helpful comments and suggestions of selected problems. Contribution by the staff of John Wiley & Sons during the planning and production of the book reflects a high degree of professional competence and is duly recognized. The support of the California Polytechnic State University in granting me a leave of absence in which to prepare this book is likewise acknowledged. Finally, I acknowledge the patience and forebearance of my wife, Julia, during the many hours required to prepare this manuscript.

J. L. Meriam

Santa Barbara, California
January 1978

CONTENTS

STATICS

DYNAMICS

PART I DYNAMICS OF PARTICLES

VOLUME 1
STATICS

INTRODUCTION TO STATICS

1

1/1 MECHANICS. Mechanics is that physical science which deals with the state of rest or motion of bodies under the action of forces. No one subject plays a greater role in engineering analysis than does mechanics. The early history of this subject is synonymous with the very beginnings of engineering. Modern research and development in the fields of vibrations, stability and strength of structures and machines, rocket and spacecraft design, automatic control, engine performance, fluid flow, electrical machines and apparatus, and molecular, atomic, and subatomic behavior are highly dependent upon the basic principles of mechanics. A thorough understanding of this subject is an absolute prerequisite for work in these and many other fields.

Mechanics is the oldest of the physical sciences. The earliest recorded writings in this field are those of Archimedes (287–212 B.C.) which concern the principle of the lever and the principle of buoyancy. Substantial progress awaited the formulation of the laws of vector combination of forces by Stevinus (1548–1620), who also formulated most of the principles of statics. The first investigation of a dynamic problem is credited to Galileo (1564–1642) in connection with his experiments with falling stones. The accurate formulation of the laws of motion, as well as the law of gravitation, was made by Newton (1642–1727), who also conceived the idea of the infinitesimal in mathematical analysis. Substantial contributions to the development of mechanics were also made by da Vinci, Varignon, D'Alembert, Lagrange, Laplace, and others.

The principles of mechanics as a science embody the rigor of mathematics upon which they are highly dependent. On the other hand, the purpose of engineering mechanics is the application of these principles to the solution of practical problems. In this book we shall be concerned both with the rigorous development of principles and their application. The basic principles of mechanics are relatively few in number, but they have exceedingly wide application, and the methods employed in mechanics will carry over into many fields of engineering endeavor.

The subject of mechanics is logically divided into two parts, *statics*, which concerns the equilibrium of bodies under the action of forces, and *dynamics,* which concerns the motion of bodies. ENGINEERING MECHANICS is divided into these two parts, Volume 1 *Statics* and Volume 2 *Dynamics.*

1/2 **BASIC CONCEPTS.** Certain concepts and definitions are basic to the study of mechanics, and they should be understood at the outset.

Space is the geometric region occupied by bodies whose positions are described by linear and angular measurements relative to a coordinate system. For three-dimensional problems our space will require three independent coordinates. For two-dimensional problems only two coordinates will be required.

Time is the measure of the succession of events and is a basic quantity in dynamics. Time is not directly involved in the analysis of statics problems.

Mass is a measure of the inertia of a body, which is its resistance to a change of motion. Of more importance to us in statics, mass is also the property of every body by which it experiences mutual attraction to other bodies.

Force is the action of one body on another. A force tends to move a body in the direction of its action. The action of a force is characterized by its *magnitude,* by the *direction* of its action, and by its *point of application.* Force is a vector quantity, and its properties are discussed in detail in Chapter 2.

Particle. A body of negligible dimensions is called a particle. In the mathematical sense a particle is a body whose dimensions approach zero so that it may be analyzed as a point mass. Frequently a particle is chosen as a differential element of a body. Also, when the dimensions of a body are irrelevant to the description of its position or the action of forces applied to it, the body may be treated as a particle.

Rigid body. A body is considered rigid when the relative movements between its parts are negligible for the purpose at hand. For instance, the calculation of the tension in the cable which supports the boom of a mobile crane under load is essentially unaffected by the small internal strains (deformations) in the structural members of the boom. For the purpose, then, of determining the external forces which act on the boom, we may treat it as a rigid body. Statics deals primarily with the calculation of external forces which act on rigid bodies which are in equilibrium. To determine the internal stresses and strains, the deformation characteristics of the material of the boom would have to be analyzed. This type of analysis belongs in the study of the mechanics of deformable bodies which comes after the study of statics.

1/3 **SCALARS AND VECTORS.** In statics we deal with two kinds of quantities, scalars and vectors. Scalar quantities are those with which a magnitude alone is associated. Examples of scalar quantities in mechanics are time, volume, density, speed, energy, and mass. Vector quantities, on the other hand, possess direction as well as magnitude and must obey the parallelogram law of addition as described in this article. Examples of vectors are displacement, velocity, acceleration, force, moment, and momentum.

Physical quantities that are vectors fall into one of three classifications, free, sliding, or fixed.

A *free vector* is one whose action is not confined to or associated with a unique line in space. For example, if a body moves without rotation, then the movement or displacement of any point in the body may be taken as a vector, and this vector will describe equally well the direction and magnitude of the displacement of every point in the body. Hence we may represent the displacement of such a body by a free vector.

A *sliding vector* is one for which a unique line in space must be maintained along which the quantity acts. When we deal with the external action of a force on a rigid body, the force may be applied at any point along its line of action without changing its effect on the body as a whole° and hence may be considered a sliding vector.

A *fixed vector* is one for which a unique point of application is specified, and therefore the vector occupies a particular position in space. The action of a force on a deformable or nonrigid body must be specified by a fixed vector at the point of application of the force. In this problem the forces and deformations internal to the body will be dependent on the point of application of the force as well as its magnitude and line of action.

A vector quantity **V** is represented by a line segment, Fig. 1/1, having the direction of the vector and having an arrowhead to indicate the sense. The length of the directed line segment represents to some convenient scale the magnitude |**V**| of the vector and is written with lightface italic type *V*. In scalar equations and frequently on diagrams where only the magnitude of a vector is labeled the symbol will appear in lightface italic type. Boldface type is used for vector quantities whenever the directional aspect of the vector is a part of its mathematical representation. When writing vector equations we must preserve the mathematical distinction between vectors and scalars. It is recommended that in all handwritten work a distinguishing mark be used for each vector quantity, such as an underline, $\underline{V}$, or an arrow over the symbol, $\vec{V}$, to take the place of boldface type in print. The direction of the vector **V** may be measured by an angle θ from some known reference direction as indicated. The negative of **V** is a vector −**V** directed in the sense opposite to **V** as shown.

Figure 1/1

° This is the so-called *principle of transmissibility*, which is discussed in Art. 2/2.

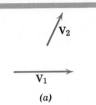

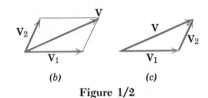

(a)

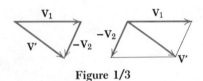

(b)   (c)

Figure 1/2

Figure 1/3

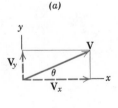

(a)

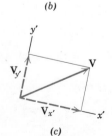

(b)

(c)

Figure 1/4

In addition to possessing the properties of magnitude and direction, vectors must also obey the parallelogram law of combination. This law requires that two vectors V_1 and V_2, treated as free vectors, Fig. 1/2a, may be replaced by their equivalent V which is the diagonal of the parallelogram formed by V_1 and V_2 as its two sides, as shown in Fig. 1/2b. This combination or vector sum is represented by the vector equation

$$\mathbf{V} = \mathbf{V}_1 + \mathbf{V}_2$$

where the plus sign used in conjunction with the vector quantities (boldface type) means *vector* and not *scalar* addition. The scalar sum of the magnitudes of the two vectors is written in the usual way as $V_1 + V_2$, and it is clear from the geometry of the parallelogram that $V \neq V_1 + V_2$.

The two vectors V_1 and V_2, again treated as free vectors, may also be added head-to-tail by the triangle law, as shown in Fig. 1/2c, to obtain the identical vector sum V. We see from the diagram that the order of addition of the vectors does not affect their sum, so that $V_1 + V_2 = V_2 + V_1$.

The difference $V_1 - V_2$ between the two vectors is easily obtained by adding $-V_2$ to V_1 as shown in Fig. 1/3 where either the triangle or parallelogram procedure may be used. The difference V' between the two vectors is expressed by the vector equation

$$\mathbf{V'} = \mathbf{V}_1 - \mathbf{V}_2$$

where the minus sign is used to denote *vector subtraction*.

Any two or more vectors whose sum equals a certain vector V are said to be the *components* of that vector. Hence the vectors V_1 and V_2 in Fig. 1/4a are the components of V in the directions 1 and 2, respectively. It is usually most convenient to deal with vector components that are mutually perpendicular, and these are called *rectangular components*. The vectors V_x and V_y in Fig. 1/4b are the x- and y-components, respectively, of V. Likewise, in Fig. 1/4c, $V_{x'}$ and $V_{y'}$ are the x'- and y'-components of V. When expressed in rectangular components, the direction of the vector with respect to, say, the x-axis is clearly specified by

$$\theta = \tan^{-1} \frac{V_y}{V_x}$$

For some problems, particularly three-dimensional ones, it is convenient to express the rectangular components of V, Fig. 1/5, in terms of unit vectors $\mathbf{i}, \mathbf{j}, \mathbf{k}$, which are vectors in the x-, y-, and z-directions, respectively, with magnitudes of unity. The vector sum of the components is written

$$\boxed{V = iV_x + jV_y + kV_z}$$

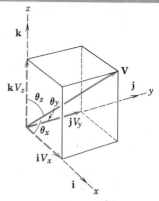

When we substitute the *direction cosines l, m,* and *n* of **V** given by

$$l = \cos\theta_x \qquad m = \cos\theta_y \qquad n = \cos\theta_z$$

we may write the magnitudes of the components of **V** as

with
$$\boxed{\begin{array}{ccc} V_x = lV & V_y = mV & V_z = nV \\ & V^2 = V_x^2 + V_y^2 + V_z^2 & \end{array}}$$

Figure 1/5

Note also that $l^2 + m^2 + n^2 = 1$.

1/4 NEWTON'S LAWS. Sir Isaac Newton was the first to state correctly the basic laws governing the motion of a particle and to
demonstrate their validity.[°] Slightly reworded to use modern terminology, these laws are:

    Law I. A particle remains at rest or continues to move in a
straight line with a uniform velocity if there is no unbalanced force
acting on it.

    Law II. The acceleration of a particle is proportional to the
resultant force acting on it and is in the direction of this force.

    Law III. The forces of action and reaction between interacting
bodies are equal in magnitude, opposite in direction, and collinear.

The correctness of these laws has been verified by innumerable
accurate physical measurements. Newton's second law forms the
basis for most of the analysis in dynamics. As applied to a particle of
mass *m* it may be stated as

$$\boxed{F = ma} \tag{1/1}$$

where **F** is the resultant force acting on the particle and **a** is the
resulting acceleration. This equation is a *vector* equation since the
direction of **F** must agree with the direction of **a** in addition to the
equality in magnitudes of **F** and *m***a**. Newton's first law contains
the principle of the equilibrium of forces, which is the main topic of
concern in statics. Actually this law is a consequence of the second
law, since there is no acceleration when the force is zero, and the
particle either is at rest or is moving with a constant velocity. The
first law adds nothing new to the description of motion but is included here since it was a part of Newton's classical statements.

[°] Newton's original formulations may be found in the translation of his *Principia*
(1687) revised by F. Cajori, University of California Press, 1934.

The third law is basic to our understanding of force. It states that forces always occur in pairs of equal and opposite forces. Thus the downward force exerted on the desk by the pencil is accompanied by an upward force of equal magnitude exerted on the pencil by the desk. This principle holds for all forces, variable or constant, regardless of their source and holds at every instant of time during which the forces are applied. Lack of careful attention to this basic law is the cause of frequent error by the beginner. In analyzing bodies under the action of forces it is absolutely necessary to be clear about which of the pair of forces is being considered. It is necessary first of all to *isolate* the body under consideration and then to consider only the one force of the pair which acts *on* the body in question.

1/5 UNITS. The International System of Units, abbreviated SI (from the French, Système International d'Unités), has been accepted throughout the world and in the United States is rapidly replacing the U.S. customary or British system of units. During the years of transition engineers will need to be familiar with both systems, although their primary use will be for the SI metric system. In this book both SI and U.S. units are used, although the principal emphasis is on SI units.

The SI system is an absolute system of units based on the quantities of length, time, and mass with the units of force derived from Eq. 1/1. The main characteristics of SI units are set forth inside the front cover of the book along with the numerical conversions between U.S. customary and SI units. In addition, charts which give the approximate conversions between selected quantities in the two systems appear inside the back cover of the book for convenient reference. In statics we are primarily concerned with the units of length and force, with mass being involved only when we compute gravitational force, as explained in the next article. In SI units the unit of force is the newton (symbol N) which, by definition, is that force required to give a one-kilogram mass an acceleration of one meter per second squared. The kilogram is a unit of mass and *not* a unit of force. In the MKS gravitational system, which has been in use for years in many countries, the kilogram, like the pound in the U.S. customary or British system, has been used both as a unit of mass and a unit of force. It is necessary to guard against this practice when using the SI system.

Primary standards for the measurements of mass, length, and time have been established by international agreement and are as follows:

Mass. The kilogram is defined as the mass of a certain platinum-iridium cylinder which is kept at the International Bureau of Weights and Measures near Paris, France. An accurate copy of this cylinder is kept at the National Bureau of Standards in the United States and serves as the standard of mass for this country.

Length. The meter, originally defined as one ten millionth of the distance from the pole to the equator along the meridian through Paris, was later defined as the length of a certain platinum-iridium bar kept at the International Bureau of Weights and Measures. The difficulty of accessibility and accuracy of reproduction of measurement prompted the adoption of a more accurate and reproducible standard of length for the meter, which is now defined as 1 650 763.73 wavelengths of a certain radiation of the krypton-86 atom.

Time. The second was originally defined as the fraction 1/(86 400) of the mean solar day. Irregularities in the Earth's rotation have led to difficulties with this definition, and a more accurate and reproducible standard has been adopted. The second is now defined as the duration of 9 192 631 770 periods of the radiation of a certain state of the cesium-133 atom.

Clearly, for most engineering work, and for our purpose in studying mechanics, the accuracy of these standards is considerably beyond our needs.

In the U.S. customary or British system the unit of force is the pound (symbol lb). The standard pound is the force required to give a one-pound mass an acceleration of 32.1740 ft/sec^2, which is the standard value of gravitational acceleration g at sea level and at a latitude of 45°. Other units of force in use with the U.S. customary system are the *kilopound* (kip), which equals 1000 lb and the *ton*, which equals 2000 lb. In SI units the corresponding standard value of g is 9.806 65 m/s^2. By applying Eq. 1/1 to free-fall conditions for the standard one-pound mass, we find that one pound of force is equivalent to (0.453 592 37)(9.806 65) = 4.4482 newtons of force. To three significant figures, then, we have the equivalences

$$1 \text{ lb} = 4.45 \text{ N} \qquad \text{or} \qquad 1 \text{ N} = 0.225 \text{ lb}$$

The unit of mass in the U.S. customary or British system is derived from Eq. 1/1 and is the mass which has an acceleration of 1 ft/sec^2 when acted upon by a 1-lb force. This mass is called a *slug* and equals 32.1740 pounds mass. Thus from Eq. 1/1

$$\text{force (lb)} = \text{mass (slugs)} \times \text{acceleration (ft/sec}^2)$$

From the gravitational experiment where W is the weight we derive the mass m from

$$m \text{ (slugs)} = \frac{W \text{ (lb)}}{g \text{ (ft/sec}^2)}$$

In Fig. 1/6 are depicted examples of force, mass, and length in the two systems of units to aid in visualizing their relative magnitudes.

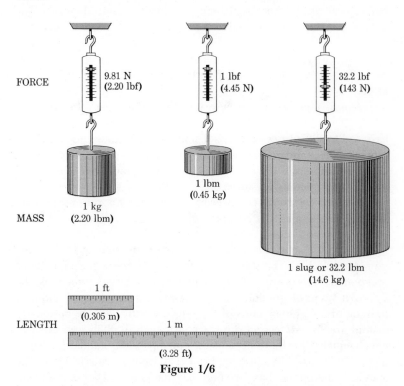

FORCE

9.81 N
(2.20 lbf)

1 lbf
(4.45 N)

32.2 lbf
(143 N)

1 lbm
(0.45 kg)

1 kg
(2.20 lbm)

MASS

1 slug or 32.2 lbm
(14.6 kg)

1 ft

(0.305 m)

LENGTH   1 m

(3.28 ft)

Figure 1/6

1/6 LAW OF GRAVITATION.

In statics as well as dynamics we have frequent need to compute the weight of (gravitational force acting on) a body. This computation depends on the law of gravitation, which was also formulated by Newton. The law of gravitation is expressed by the equation

$$F = K\,\frac{m_1 m_2}{r^2} \tag{1/2}$$

where F = the mutual force of attraction between two particles
K = a universal constant known as the constant of gravitation
m_1, m_2 = the masses of the two particles
r = the distance between the centers of the particles

The mutual forces F obey the law of action and reaction, since they are equal and opposite and are directed along the line joining the centers of the particles. By experiment the gravitational constant is found to be $K = 6.673(10^{-11})\text{m}^3/(\text{kg}\cdot\text{s}^2)$. Gravitational forces exist between every pair of bodies. On the surface of the earth the only gravitational force of appreciable magnitude is the force due to the earth's attraction. Thus each of two iron spheres 100 mm in diameter is attracted to the earth with a gravitational force of 37.9 N, which is called its weight. On the other hand the force of mutual attraction between the spheres if they are just touching is 0.000 000 099 4 N. This force is clearly negligible compared with the earth's attraction of 37.9 N, and consequently the gravitational attraction of the earth is the only gravitational force of any appreciable magnitude which need be considered for most engineering experiments conducted on the earth's surface.

The gravitational attraction of the earth on a body is known as the *weight* of the body. This force exists whether the body is at rest or in motion. Since this attraction is a force, the weight of a body should be expressed in newtons (N) in SI units. Unfortunately in common practice the mass unit kilogram (kg) has been used extensively as a measure of weight. When expressed in kilograms the word "weight" technically means mass. To avoid confusion the word "weight" in this book shall be restricted to mean the force of gravitational attraction, and it will always be expressed in newtons.

For a body of mass m on the surface of the earth the earth's gravitational attraction on the body as specified by Eq. 1/2 may be calculated from the results of the simple gravitational experiment. If the gravitational force or weight has a magnitude W, then, since the body falls with an acceleration g, Eq. 1/1 gives

$$W = mg \qquad (1/3)$$

The weight W will be in newtons (N) when m is in kilograms (kg) and g is in meters per second squared (m/s^2). The standard value $g = 9.81$ m/s^2 will be sufficiently accurate for our calculations in statics. The corresponding value of g in U.S. customary or British units is 32.2 ft/sec^2.

The true weight (gravitational attraction) and the apparent weight (as measured by a spring scale) are slightly different. The difference, which is due to the rotation of the earth, is quite small and will be neglected. This effect will be discussed in *Vol. 2 Dynamics*.

1/7 ACCURACY, LIMITS, AND APPROXIMATIONS. The number of significant figures shown in an answer should be no greater than the number of figures which can be justified by the accuracy of the given data. Hence the cross-sectional area of a square bar whose side, 24 mm, say, was measured to the nearest half millimeter should be written as 580 mm^2 and not as 576 mm^2, as would be indicated if the numbers were multiplied out.

When calculations involve small differences in large quantities, greater accuracy in the data is required to achieve a given accuracy in the results. Hence it is necessary to know the numbers 4.2503 and 4.2391 to an accuracy of five significant figures in order that their difference 0.0112 be expressed to three-figure accuracy. It is often difficult in somewhat lengthy computations to know at the outset the number of significant figures needed in the original data to ensure a certain accuracy in the answer. Accuracy to three significant figures is considered satisfactory for the majority of engineering calculations.

The *order* of differential quantities is the subject of frequent misunderstanding. Higher-order differentials may always be neglected compared with lower-order differentials when the mathematical limit is approached. As an example the element of volume ΔV of a right circular cone of altitude h and base radius r may be taken to be a circular slice a distance x from the vertex and of thickness Δx. It can be verified that the complete expression for the volume of the element may be written as

$$\Delta V = \frac{\pi r^2}{h^2}[x^2\,\Delta x + x(\Delta x)^2 + \tfrac{1}{3}(\Delta x)^3]$$

It should be recognized that, when passing to the limit in going from ΔV to dV and from Δx to dx, the terms in $(\Delta x)^2$ and $(\Delta x)^3$ drop out, leaving merely

$$dV = \frac{\pi r^2}{h^2}x^2\,dx$$

which gives an exact expression when integrated.

When dealing with small angles we can usually make use of simplifying assumptions. Consider the right triangle of Fig. 1/7 where the angle θ, expressed in radians, is relatively small. With the hypotenuse taken as unity, we see from the geometry of the figure that the arc length $1 \times \theta$ and $\sin \theta$ are very nearly the same. Also $\cos \theta$ is close to unity. Furthermore, $\sin \theta$ and $\tan \theta$ have almost the same values. Thus for small angles we may write

$$\sin \theta \approx \tan \theta \approx \theta \qquad \cos \theta \approx 1$$

These approximations amount to retaining only the first terms in the series expansions for these three functions. As an example of these approximations, for an angle of $1°$

$$1° = 0.017\,453 \text{ rad}$$
$$\sin 1° = 0.017\,452$$
$$\tan 1° = 0.017\,455$$
$$\cos 1° = 0.999\,848$$

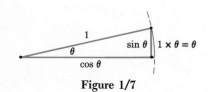

Figure 1/7

If a closer approximation is desired, the first two terms may be retained, and they are

$$\sin \theta = \theta - \theta^3/6 \qquad \tan \theta = \theta + \theta^3/3 \qquad \cos \theta = 1 - \theta^2/2$$

The error in replacing the sine by the angle for $1°$ is only 0.005 percent. For $5°$ the error is 0.13 percent, and for $10°$ the error is still only 0.51 percent. As the angle θ approaches zero, it should now be clear that the following relations are true in the mathematical limit:

$$\sin d\theta = \tan d\theta = d\theta$$

$$\cos d\theta = 1$$

The angle $d\theta$ is, of course, expressed in radian measure.

1/8 DESCRIPTION OF STATICS PROBLEMS. The study of statics is directed toward the quantitative description of forces that act on engineering structures in equilibrium. Mathematics establishes the relations between the various quantities involved and makes it possible for us to predict effects from these relations. A dual thought process is required in formulating this description. It is necessary to think in terms of the physical situation and in terms of the corresponding mathematical description. Analysis of every problem will require the repeated transition of thought between the physical and the mathematical. Without question, one of the most common problems encountered by the student is the difficulty in making this transition of thought freely. We should recognize that the mathematical formulation of a physical problem represents an ideal limiting description, or model, which approximates but never quite matches the actual physical situation.

When constructing the idealized mathematical model for a given engineering problem certain approximations will always be involved. Some of these approximations may be mathematical, whereas others will be physical. For instance, it is often necessary for us to neglect small distances, angles, or forces compared with large distances, angles, or forces. A force which is actually distributed over a small area of the body upon which it acts may be considered a concentrated force if the dimensions of the area involved are small compared with other pertinent dimensions. The weight of a steel cable per unit length may be neglected if the tension in the cable is many times greater than its total weight, whereas the cable weight may not be neglected if the problem calls for a determination of the deflection or sag of a suspended cable due to its weight. Thus the degree of assumption involved depends on what information is desired and on the accuracy required. We must be constantly alert to the various assumptions called for in the formulation of real problems. The ability to understand and make use of the appropriate assumptions in the formulation and solution of engineering problems is certainly one of the most important characteristics of a successful engineer. One of the major aims of this book is to provide a maximum of opportunity to develop this ability through the formulation and analysis of many practical problems involving the principles of statics.

Graphics is an important analytical tool which serves us in three capacities. First, it makes possible the representation of a physical system on paper by means of a sketch or diagram. Geometrical representation is vital to physical interpretation and aids greatly in visualizing the three-dimensional aspects of many problems. Second, graphics often affords a means of solving physical relations where a direct mathematical solution would be awkward or difficult. Graphical solutions not only provide us with a practical means for obtaining results, but they also aid greatly in making the transition of thought between the physical situation and the mathematical expression because both are represented simultaneously. A third use of graphics is in the display of results in charts or graphs which become a valuable aid to representation.

An effective method of attack on statics problems, as in all engineering problems, is essential. The development of good habits in formulating problems and in representing their solutions will prove to be an invaluable asset. Each solution should proceed with a logical sequence of steps from hypothesis to conclusion, and its representation should include a clear statement of the following parts, each clearly identified:

1. Given data
2. Results desired
3. Necessary diagrams
4. Calculations
5. Answers and conclusions

In addition it is well to incorporate a series of checks on the calculations at intermediate points in the solution. We should observe reasonableness of numerical magnitudes, and the accuracy and dimensional homogeneity of terms should be frequently checked. It is also important that all work be neat and orderly. Careless solutions that cannot be easily read by others are of little or no value. The discipline involved in adherence to good form will in itself be an invaluable aid to the development of the abilities for formulation and analysis. Many problems that at first may seem difficult and complicated become clear and straightforward once they are begun with a logical and disciplined method of attack.

The subject of statics is based on surprisingly few fundamental concepts and involves mainly the application of these basic relations to a variety of situations. In this application the *method* of analysis is all-important. In solving a problem it is essential that the laws which apply be carefully fixed in mind and that we apply these principles literally and exactly. In applying the principles which define the requirements for forces acting on a body it is essential that we *isolate* the body in question from all other bodies so that complete and accurate account of all forces which act on this body may be taken. This *isolation* should exist mentally as well as be represented on paper. The diagram of such an isolated body with the representation of *all* external forces acting on it is called a *free-body diagram*. It has long been established that the *free-body-diagram method* is the key to the understanding of mechanics. This is so because the *isolation* of a body is the tool by which *cause* and *effect* are clearly separated and by which our attention to the literal application of a principle is accurately focused. The technique of drawing free-body diagrams is covered in Chapter 3 where they are first used.

In applying the laws of statics, we may use numerical values of the quantities directly in proceeding toward the solution, or we may use algebraic symbols to represent the quantities involved and leave the answer as a formula. With numerical substitution the magnitude of each quantity expressed in its particular units is evident at each stage of the calculation. This approach offers advantage when the practical significance of the magnitude of each term is important. The symbolic solution, however, has several advantages over the numerical solution. First, the abbreviation achieved by the use of symbols aids in focusing our attention on the connection between the physical situation and its related mathematical description. Second, a symbolic solution allows us to make a dimensional check at every step, whereas dimensional homogeneity may be lost when numerical values only are used. Third, we may use a symbolic solution repeatedly for obtaining answers to the same problem when different sets and sizes of units are used. Facility with both forms of solution is essential, and ample practice with each should be sought in the problem work.

The student will find that solutions to the problems of statics may be obtained in one of three ways. First, we may utilize a direct mathematical solution by hand calculation where answers appear either as algebraic symbols or as numerical results. The large majority of problems come under this category. Second, we may approximate the results of certain problems by graphical solutions. Third, the modern digital computer is of particular advantage where a large number of equations or repeated data are involved in numerical form. Students who have ready access to digital computation facilities may wish to solve a selected few problems by this means. In order to reduce computation time in the problem work the data for most of the problems are given in simple numbers. The choice of the most expedient method of solution is an important aspect of the experience to be gained from the problem work.

FORCE SYSTEMS

2

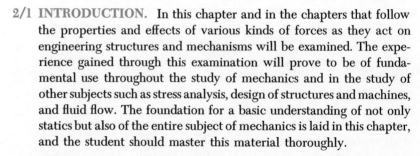

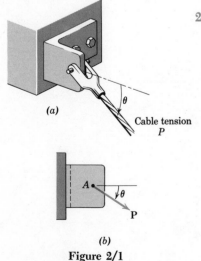

(a)

Cable tension
P

(b)

Figure 2/1

2/1 INTRODUCTION. In this chapter and in the chapters that follow the properties and effects of various kinds of forces as they act on engineering structures and mechanisms will be examined. The experience gained through this examination will prove to be of fundamental use throughout the study of mechanics and in the study of other subjects such as stress analysis, design of structures and machines, and fluid flow. The foundation for a basic understanding of not only statics but also of the entire subject of mechanics is laid in this chapter, and the student should master this material thoroughly.

2/2 FORCE. Before dealing with a group or *system* of forces it is necessary for us to examine the properties of a single force in some detail. A force has been defined as the action of one body on another. We find that force is a vector quantity, since its effect depends on the direction as well as on the magnitude of the action and since forces may be combined according to the parallelogram law of vector combination. The action of the cable tension on the bracket in Fig. 2/1a is represented in Fig. 2/1b by the force vector **P** of magnitude P. The effect of this action on the bracket will depend on P, the angle θ, and the location of the point of application A. Changing any one of these three specifications will alter the effect on the bracket, as could be detected, for instance, by the force in one of the bolts which secure the bracket to the base or the internal stress and strain in the material of the bracket at any point. Thus the complete specification of the action of a force must include its *magnitude, direction,* and *point of application,* in which case it is treated as a fixed vector.

Force is applied either by direct mechanical contact or by remote action. Gravitational and magnetic forces are applied by remote action. All other forces are applied through direct physical contact.

The action of a force on a body can be separated into two effects, *external* and *internal*. For the bracket of Fig. 2/1 the effects of **P** external to the bracket are the reactions or forces (not shown) exerted on the bracket by the foundation and bolts because of the action of **P**. Forces external to a body are then of two kinds, *applied* forces and *reactive* forces. The effects of **P** internal to the bracket are the resulting internal stresses and strains distributed throughout the material of the bracket. The relation between internal forces and internal strains involves the material properties of the body and is studied in strength of materials, elasticity, and plasticity.

In dealing with the mechanics of rigid bodies, where concern is given only to the net *external* effects of forces, experience shows us that it is not necessary to restrict the action of an applied force to a given point. Hence the force **P** acting on the rigid plate in Fig. 2/2 may be applied at *A* or at *B* or at any other point on its action line, and the net external effects of **P** on the bracket will not change. The external effects are the force exerted on the plate by the bearing support at *O* and the force exerted on the plate by the roller support at *C*. This conclusion is described by the *principle of transmissibility*, which states that a force may be applied at any point on its given line of action without altering the resultant effects of the force *external* to the *rigid* body on which it acts. When only the resultant external effects of a force are to be investigated, the force may be treated as a *sliding* vector, and it is necessary and sufficient to specify the *magnitude, direction,* and *line of action* of the force. Since this book deals essentially with the mechanics of rigid bodies, we will treat almost all forces as sliding vectors for the rigid body on which they act.

Forces may be either *concentrated* or *distributed*. Actually every contact force is applied over a finite area and is therefore a distributed force. When the dimensions of the area are very small compared with the other dimensions of the body, we may consider the force to be concentrated at a point with negligible loss of accuracy. Force may be distributed over an area, as in the case of mechanical contact, or it may be distributed over a volume when gravity or magnetic force is acting. The *weight* of a body is the force of gravitational attraction distributed over its volume and may be taken as a concentrated force acting through the center of gravity. The position of the center of gravity is frequently obvious from considerations of symmetry. If the position is not obvious, then a separate calculation, explained in Chapter 5, will be necessary to locate the center of gravity.

A force may be measured either by comparison with other known forces, using a mechanical balance, or by the calibrated movement of

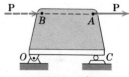

Figure 2/2

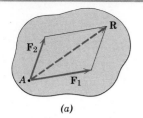

(a)

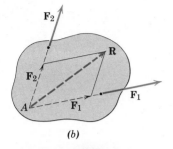

(b)

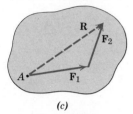

(c)

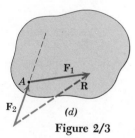

(d)

Figure 2/3

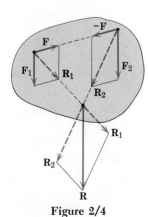

Figure 2/4

an elastic element. All such comparisons or calibrations have as their basis a primary standard. The standard unit of force in SI units is the newton (N) and in the U.S. customary system is the pound (lb), as defined in Art. 1/5.

The characteristic of a force expressed by Newton's third law must be carefully observed. The *action* of a force is always accompanied by an *equal* and *opposite reaction*. It is essential for us to fix clearly in mind which force of the pair is being considered. The answer is always clear when the body in question is *isolated* and the force exerted *on* that body (not *by* the body) is represented. It is very easy to make a careless mistake and consider the wrong force of the pair unless we distinguish carefully between action and reaction.

Two forces $\mathbf{F}_1$ and $\mathbf{F}_2$ that are concurrent may be added by the parallelogram law in their common plane to obtain their sum or *resultant* $\mathbf{R}$ as shown in Fig. 2/3*a*. If the two concurrent forces lie in the same plane but are applied at two different points as in Fig. 2/3*b*, by the principle of transmissibility we may move them along their lines of action and complete their vector sum $\mathbf{R}$ at the point of concurrency. The resultant $\mathbf{R}$ may replace $\mathbf{F}_1$ and $\mathbf{F}_2$ without altering the external effects on the body upon which they act. The triangle law may also be used to obtain $\mathbf{R}$, but it will require moving the line of action of one of the forces as shown in Fig. 2/3*c*. In Fig. 2/3*d* the same two forces are added, and although the correct magnitude and direction of $\mathbf{R}$ are preserved, we lose the correct line of action, since $\mathbf{R}$ obtained in this way does not pass through A. This type of combination should be avoided. Mathematically the sum of the two forces may be written by the vector equation

$$\mathbf{R} = \mathbf{F}_1 + \mathbf{F}_2$$

In addition to the need for combining forces to obtain their resultant, we often have occasion to replace a force by its *components* which act in two specified directions. Thus the force $\mathbf{R}$ in Fig. 2/3*a* may be replaced by or *resolved* into two components $\mathbf{F}_1$ and $\mathbf{F}_2$ with these specified directions merely by completing the parallelogram as shown to obtain the magnitudes of $\mathbf{F}_1$ and $\mathbf{F}_2$.

A special case of addition is presented when the two forces $\mathbf{F}_1$ and $\mathbf{F}_2$ are parallel, Fig. 2/4. They may be combined by first adding two equal, opposite, and collinear forces $\mathbf{F}$ and $-\mathbf{F}$ of convenient magnitude which taken together produce no external effect on the body. Adding $\mathbf{F}_1$ and $\mathbf{F}$ to produce $\mathbf{R}_1$ and combining with the sum $\mathbf{R}_2$ of $\mathbf{F}_2$ and $-\mathbf{F}$ yield the resultant $\mathbf{R}$ correct in magnitude, direction, and line of action. This procedure is also useful in obtaining a graphical combination of two forces that are almost parallel and hence have a point of concurrency which is remote and inconvenient.

It is usually helpful to master the analysis of force systems in two dimensions before undertaking three-dimensional analysis. To this end the remainder of the chapter is subdivided into these two categories. However, for students who have a good command of vector analysis, these sections may be studied simultaneously if preferred.

SECTION A. TWO-DIMENSIONAL FORCE SYSTEMS

2/3 RECTANGULAR COMPONENTS. The most common two-dimensional resolution of a force $\mathbf{F}$ is resolution into *rectangular components* $\mathbf{F}_x$ and $\mathbf{F}_y$ as shown in Fig. 2/5. It should be immediately evident from the figure that

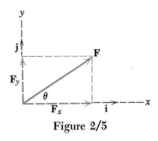

$$
\boxed{
\begin{array}{ll}
F_x = F \cos \theta & F = \sqrt{F_x{}^2 + F_y{}^2} \\
F_y = F \sin \theta & \theta = \tan^{-1} \dfrac{F_y}{F_x}
\end{array}
} \tag{2/1}
$$

Figure 2/5

where F is the magnitude of $\mathbf{F}$ and where F_x and F_y are the magnitudes of $\mathbf{F}_x$ and $\mathbf{F}_y$. If we introduce unit vectors $\mathbf{i}$ and $\mathbf{j}$ in the x- and y-directions as shown in Fig. 2/5, we may write the vector equation

$$
\boxed{\mathbf{F} = \mathbf{F}_x + \mathbf{F}_y = \mathbf{i}F_x + \mathbf{j}F_y} \tag{2/2}
$$

To eliminate any ambiguity it is desirable to show the components of a force in dotted lines, as in Fig. 2/5, and the force in a full line, or vice versa. With either of these conventions it will always be clear that a force and its components are being represented and not three separate forces as would be implied by three solid-line vectors.

Actual problems do not come with reference axes, so their assignment is a matter of arbitrary convenience, and the choice is frequently up to the student. The logical choice is usually indicated by the manner in which the geometry of the problem is specified. When the principal dimensions of a body are given in the horizontal and vertical directions, for example, then assignment of reference axes in these directions is generally convenient. However, dimensions are not always given in horizontal and vertical directions, angles need not be measured counterclockwise from the x-axis, and the origin of coordinates need not be on the line of action of a force. Therefore, it is essential that we be able to determine the correct components of a force no matter how the axes are oriented or how the angles are measured. Figure 2/6 suggests a few typical examples of resolution situations in two dimensions, the results of which should be readily apparent. Thus it is seen that memorization of Eqs. 2/1 is not a substitute for an understanding of the parallelogram law and for the correct projection of a vector onto a reference axis. A neatly drawn sketch always helps to clarify the geometry and avoid error.

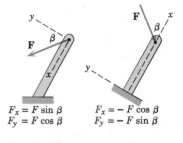

$$
\begin{array}{ll}
F_x = F \sin \beta & F_x = -F \cos \beta \\
F_y = F \cos \beta & F_y = -F \sin \beta
\end{array}
$$

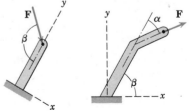

$$
\begin{array}{ll}
F_x = F \sin(\pi - \beta) & F_x = F \cos(\beta - \alpha) \\
F_y = -F \cos(\pi - \beta) & F_y = F \sin(\beta - \alpha)
\end{array}
$$

Figure 2/6

Sample Problem 2/1

The 100-N force **F** is applied to the fixed bracket as shown. Determine the magnitudes of the rectangular components of **F** in (1) the x- and y-directions and (2) the x'- and y'-directions. Also (3) find the components of **F** in the x'- and y-directions.

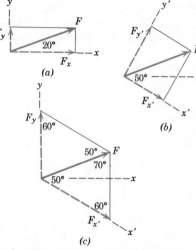

Solution. Part (1). The x- and y-components of **F** are shown in Fig. a and are

$$F_x = F \cos \theta_x = 100 \cos 20° = 94.0 \text{ N}$$

$$F_y = F \cos \theta_y = 100 \cos 70° = 34.2 \text{ N} \qquad Ans.$$

Part (2). The x'- and y'-components of **F** are the projections on these axes as shown in Fig. b and are

$$F_{x'} = F \cos \theta_{x'} = 100 \cos 50° = 64.3 \text{ N}$$

$$F_{y'} = F \cos \theta_{y'} = 100 \cos 40° = 76.6 \text{ N} \qquad Ans.$$

Part (3). The components of **F** in the x'- and y-directions are non-rectangular and are obtained by completing the parallelogram as shown in ① Fig. c. The components may be calculated by the law of sines which gives

$$\frac{F_{x'}}{\sin 70°} = \frac{F}{\sin 60°} \qquad F_{x'} = \frac{0.940}{0.866} 100 = 108.5 \text{ N}$$

$$\frac{F_y}{\sin 50°} = \frac{F}{\sin 60°} \qquad F_y = \frac{0.766}{0.866} 100 = 88.5 \text{ N} \qquad Ans.$$

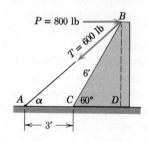

① Obtain $F_{x'}$ and F_y graphically and compare your results with the calculated values.

Sample Problem 2/2

Combine the two forces **P** and **T**, which act on the fixed structure at B, into a single force **R**.

① **Solution.** The parallelogram addition of **T** and **P** is shown in the figure. We must first find the angle α before R can be found by the law of cosines. From the given figure

$$\tan \alpha = \frac{\overline{BD}}{\overline{AD}} = \frac{6 \sin 60°}{3 + 6 \cos 60°} = 0.866, \quad \alpha = 40.9°$$

The law of cosines applied to the parallelogram of vectors gives

$$R^2 = 600^2 + 800^2 - 2(600)(800) \cos 40.9° = 274\,300$$

$$R = 524 \text{ lb} \qquad Ans.$$

The angle θ gives the direction of **R**, and from the law of sines

$$\frac{600}{\sin \theta} = \frac{524}{\sin 40.9°}, \quad \sin \theta = 0.750, \quad \theta = 48.6° \qquad Ans.$$

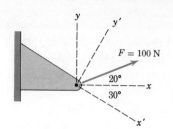

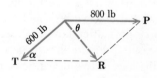

① Note carefully the repositioning of P to permit parallelogram addition at B.

PROBLEMS

2/1 Calculate the x- and y-components of the force $\mathbf{P}$ acting on the structural member if the magnitude of the force is 20 kN.

Ans. $P_x = -19.70$ kN, $P_y = 3.47$ kN

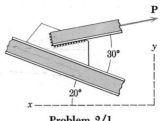

Problem 2/1

2/2 Find the rectangular component of the 600-lb tension $\mathbf{T}$ of Sample Problem 2/2 along the line BC.

2/3 When the load L is 7 m from the pivot at C, the tension $\mathbf{T}$ in the cable has a magnitude of 15 kN. Express $\mathbf{T}$ as a vector using the unit vectors $\mathbf{i}$ and $\mathbf{j}$.

Ans. $\mathbf{T} = 12.86\mathbf{i} + 7.72\mathbf{j}$ kN

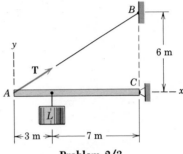

Problem 2/3

2/4 A cable exerts a force $\mathbf{F}$ on the bracket of the structural member to which it is attached. If the magnitude of the x-component of $\mathbf{F}$ is 900 N, calculate the y-component and the magnitude of $\mathbf{F}$.

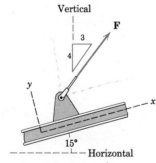

Problem 2/4

2/5 The turnbuckle C is tightened until the light cable which is stretched across the frame is under a tension of 900 N. Compute the x- and y-components of the force exerted *on* the frame at B *by* the cable.

Ans. $F_x = -879$ N, $F_y = -195.2$ N

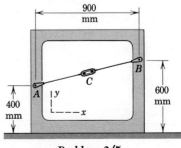

Problem 2/5

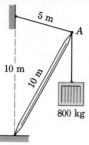

Problem 2/6

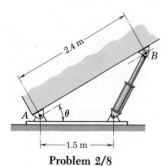

Problem 2/7

Problem 2/8

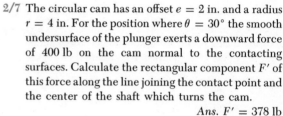

Problem 2/9

2/6 The tension **T** in the vertical cable equals the weight of the crate. Calculate the components T_t and T_n along and normal to the boom, respectively, of the force **T** applied to the boom at A by the crate.

2/7 The circular cam has an offset $e = 2$ in. and a radius $r = 4$ in. For the position where $\theta = 30°$ the smooth undersurface of the plunger exerts a downward force of 400 lb on the cam normal to the contacting surfaces. Calculate the rectangular component F' of this force along the line joining the contact point and the center of the shaft which turns the cam.

Ans. $F' = 378$ lb

2/8 The hydraulic cylinder exerts a force of 40 kN in the direction of its shaft against the load that it is hoisting. Determine the components F_n and F_t normal and tangent to AB for the position $\theta = 30°$.

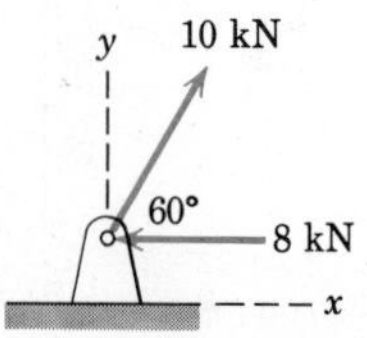

2/9 Calculate the magnitude of the single force **R** which is equivalent to the two forces shown. Also calculate the angle θ measured counterclockwise from the positive x-axis to **R**.

Ans. $R = 9.17$ kN, $\theta = 109.1°$

2/10 Solve Problem 2/9 graphically.

2/11 At what angle θ must the 400-lb force be applied in order that the resultant **R** of the two forces will have a magnitude of 1000 lb? For this condition what will be the angle β between **R** and the horizontal?

Ans. $\theta = 51.3°$, $\beta = 18.2°$

400 lb

700 lb

Problem 2/11

2/12 The 10-kN vertical force is to be replaced by two forces, $\mathbf{F}_1$ directed along the 45° line *a-a* and $\mathbf{F}_2$ which has a magnitude of 8 kN. Calculate the magnitude of $\mathbf{F}_1$ and the counterclockwise angle θ made by $\mathbf{F}_2$ with the *x*-axis.

Ans. $F_1 = 10.81$ kN with $\theta = 17.1°$,
or $F_1 = 3.33$ kN with $\theta = 72.9°$

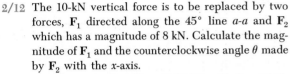

Problem 2/12

2/13 Solve Prob. 2/12 graphically.

2/14 The rigid member *ABC* is supported by the pin *A* and the hinged link *D* and is subjected to a force **F** at *C*. Can it be concluded from the principle of transmissibility that the force exerted by the pin at *A* on member *ABC* would be the same if **F** were applied either at *D* or at *E* rather than at *C*?

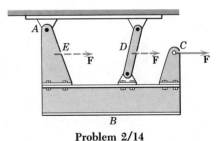

Problem 2/14

2/15 Resolve the 4-kN force into two components, one along *AB* and the other along *BC*.

Ans. $F_{AB} = 2.93$ kN, $F_{BC} = 3.59$ kN

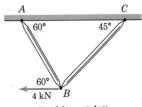

Problem 2/15

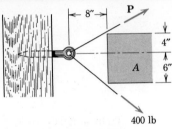

Problem 2/16

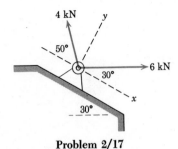

Problem 2/17

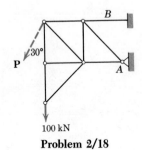

100 kN

Problem 2/18

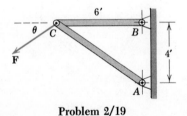

Problem 2/19

2/16 It is desired to remove the spike from the timber by applying force along its horizontal axis. An obstruction A prevents direct access, so that two forces, one 400 lb and the other $\mathbf{P}$, are applied by cables as shown. Compute the magnitude of $\mathbf{P}$ necessary to ensure tension T along the axis of the spike. Also find T.

2/17 Replace the 6-kN and 4-kN forces by a single equivalent force $\mathbf{R}$ expressed as a vector using unit vectors in the x- and y-directions. Compute the angle θ made by $\mathbf{R}$ with the x-axis.

Ans. $\mathbf{R} = 2.63\mathbf{i} + 6.06\mathbf{j}$ kN, $\theta = 66.6°$

2/18 The resultant of the 100-kN load and the accompanying tension T in member B passes through point A and results in a certain force on the pin which supports the truss at this point. If the 100-kN load is replaced by a force $\mathbf{P}$ applied along the dotted line shown, determine the magnitude of $\mathbf{P}$ which will result in the same effect on the pin at A as when the 100-kN force was applied. Specify the corresponding increment ΔT in the tension of member B. All interior angles of the truss are either 45° or 90°. (*Hint:* The resultant of P and the new T would be the same vector as the resultant of the 100-kN force and the original T.)

2/19 At what maximum angle θ should the force $\mathbf{F}$ be directed so that the magnitude of its component along CA does not exceed 80 percent of the magnitude of its component along BC?   *Ans.* $\theta = 53.0°$

2/20 The force $\mathbf{F} = -40\mathbf{i} + 60\mathbf{j}$ lb is to be replaced by two forces, one along the y-axis and one in the horizontal h-direction, which produce the same combined effect as $\mathbf{F}$. Determine the magnitudes of these components.

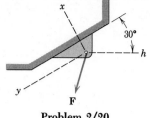

Problem 2/20

2/21 When resolved into rectangular components along the x- and y-axes, the force $\mathbf{P}$ has an x-component of 500 N. For resolution along the x'- and y'-axes, $P_{x'} = 800$ N. Express $\mathbf{P}$ as a vector using unit vectors with the x-y axes and calculate the y'-component of $\mathbf{P}$. (With the aid of the geometry of the vector figure, carry out the solution without involving simultaneous solutions.)

$Ans.$ $\mathbf{P} = 500\mathbf{i} + 965\mathbf{j}$ N, $P_{y'} = 736$ N

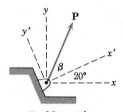

Problem 2/21

2/22 Solve Prob. 2/21 graphically.

2/23 It is known that the resultant of the two forces passes through point A. Determine the magnitude of $\mathbf{P}$.

$Ans.$ $P = 675$ N

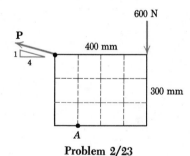

Problem 2/23

2/24 The two forces which act on the rigid frame are to be replaced by a single equivalent force $\mathbf{R}$ applied at point A. Determine the magnitude of $\mathbf{R}$ and the distance x to point A. Solve graphically or algebraically.

$Ans.$ $R = 898$ lb, $x = 4.05$ ft

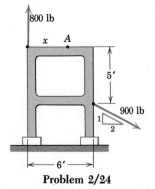

Problem 2/24

2/4 MOMENT. In addition to the tendency to move a body in the direction of its application, a force also tends to rotate the body about any axis which does not intersect the line of action of the force and which is not parallel to it. This tendency is known as the *moment* **M** of the force about the given axis. The moment of a force is also frequently referred to as *torque*.

Figure 2/7*a* shows a two-dimensional body acted upon by a force **F** in its plane. The magnitude of the moment or tendency of the force to rotate the body about the axis *O-O* normal to the plane of the body is, clearly, proportional both to the magnitude of the force and to the *moment arm d*, which is the perpendicular distance from the axis to the line of action of the force. Therefore the magnitude of the moment is defined as

$$M = Fd \qquad (2/3)$$

The moment is a vector **M** perpendicular to the plane of the body. The sense of **M** depends on the direction in which **F** tends to rotate the body. The right-hand rule, Fig. 2/7*b*, is used to identify this sense, and the moment of **F** about *O-O* may be represented as a vector pointing in the direction of the thumb with the fingers curled in the direction of the tendency to rotate. The moment **M** obeys all the rules of vector combination and may be considered a sliding vector with a line of action coinciding with the moment axis. The basic units of moment in SI units are newton-meters (N·m) and in the U.S. customary system are pound-feet (lb-ft).

When dealing with forces all of which act in a given plane we customarily speak of the moment about a point. Actually the moment with respect to an axis normal to the plane and passing through the point is implied. Thus the moment of force **F** about point *O* in Fig. 2/7*c* has the magnitude $M_O = Fd$ and is counterclockwise. Vector representation of moments for coplanar forces is unnecessary, since the vectors are either out from the paper (counterclockwise) or

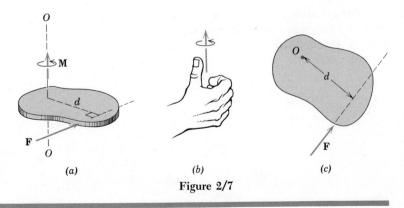

(a)   *(b)*   *(c)*

Figure 2/7

into the paper (clockwise). Since the addition of parallel free vectors may be accomplished with *scalar* algebra, the moment directions may be accounted for by using a plus sign (+) for counterclockwise moments and a minus sign (−) for clockwise moments, or vice versa. It is necessary only for us to be consistent within a given problem in using either sign convention.

Principle of moments. One of the most important principles of mechanics is *Varignon's theorem,* or the *principle of moments,* which for coplanar forces states that the moment of a force about any point is equal to the sum of the moments of the components of the force about the same point. To prove this statement we consider a force *R* and two equivalent components *P* and *Q* acting at point *A*, Fig. 2/8.

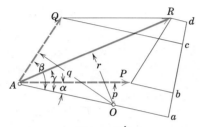

Figure 2/8

Point *O* is selected arbitrarily as the moment center. Construct the line *AO* and project the three vectors onto the normal to this line. Also construct the moment arms *p*, *q*, *r* of the three forces to point *O* and designate the angles of the vectors to the line *AO* by α, β, γ as shown in the figure. Since the parallelogram whose sides are *P* and *Q* requires that $\overline{ac} = \overline{bd}$, it is evident that

$$\overline{ad} = \overline{ab} + \overline{bd} = \overline{ab} + \overline{ac}$$

or

$$R \sin \gamma = P \sin \alpha + Q \sin \beta$$

Multiplying by the distance $\overline{AO}$ and substituting the values of *p*, *q*, *r* give

$$Rr = Pp + Qq$$

which proves that the moment of a force about any point equals the sum of the moments of its two components about the same point. Varignon's theorem need not be restricted to the case of only two components but applies equally well to three or more, since it is always possible by direct combination to reduce the number of components to two for which the theorem was proved.

Sample Problem 2/3

Calculate the moment about the base point O of the 600-N force in four different ways.

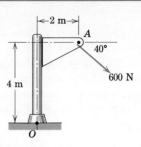

Solution. (*I*) The moment arm to the 600-N force is

$$d = 4 \cos 40° + 2 \sin 40° = 4.35 \text{ m}$$

① By $M = Fd$ the moment is clockwise and has the magnitude

$$M_O = 600(4.35) = 2610 \text{ N·m} \qquad Ans.$$

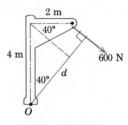

(*II*) Replace the force by its rectangular components at A

$$F_1 = 600 \cos 40° = 460 \text{ N}, \quad F_2 = 600 \sin 40° = 386 \text{ N}$$

By Varignon's theorem, then, the moment becomes

② $$M_O = 460(4) + 386(2) = 2610 \text{ N·m} \qquad Ans.$$

① The required geometry here and in similar problems should not cause difficulty if the sketch is carefully drawn.

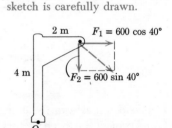

(*III*) By the principle of transmissibility move the 600-N force along its line of action to point B, which eliminates the moment of the component F_2. The moment arm of F_1 becomes

$$d_1 = 4 + 2 \tan 40° = 5.68 \text{ m}$$

and the moment is

$$M_O = 460(5.68) = 2610 \text{ N·m} \qquad Ans.$$

② This procedure is frequently the shortest approach.

③ (*IV*) Moving the force to point C eliminates the moment of the component F_1. The moment arm of F_2 becomes

$$d_2 = 2 + 4 \text{ ctn } 40° = 6.77 \text{ m}$$

and the moment is

$$M_O = 386(6.77) = 2610 \text{ N·m} \qquad Ans.$$

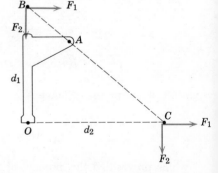

③ The fact that points B and C are not on the body proper should not cause concern, as the mathematical calculation of the moment of a force does not require that the force be on the body.

PROBLEMS

2/25 In Sample Problem 2/3 calculate the magnitude of the smallest force **P** which can be applied at point A and produce the same moment about O as that produced by the 600-N force. Find the corresponding angle θ between **P** and the horizontal.

Ans. $P = 584$ N, $\theta = 26.6°$

2/26 The moment of the force **P** about point A is 30 N·m. Calculate the magnitude of **P**. The plate upon which the force acts is divided into 0.1-m squares.

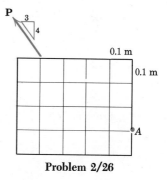

Problem 2/26

2/27 The rectangular plate is made up of 1-m squares as shown. A 10-kN force is applied at point A in the direction shown. Calculate the moment M_B of the force about point B.   *Ans.* $M_B = 27.7$ kN·m clockwise

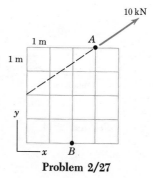

Problem 2/27

2/28 Compute the moment of the 400-N force about point O.

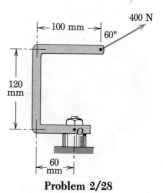

Problem 2/28

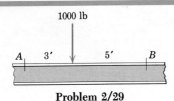

Problem 2/29

2/29 It is desired to replace the 1000-lb force acting on the beam by two downward forces, R_A at A and R_B at B, which will not alter the overall external effect on the beam. Calculate these forces.

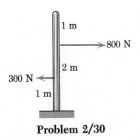

Problem 2/30

2/30 Determine the distance y down from the top of the pole at which a single horizontal force $\mathbf{P}$ must be applied in order to duplicate the external effects of the two given forces.   *Ans.* $y = 0.4$ m

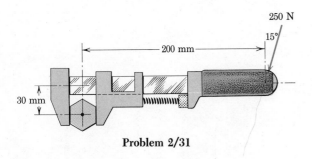

Problem 2/31

2/31 Calculate the moment of the 250-N force on the handle of the monkey wrench about the center of the bolt.

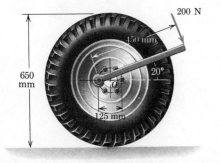

Problem 2/32

2/32 A force of 200 N is applied to the end of the wrench to tighten a flange bolt which holds the wheel to the axle. Determine the moment M produced by this force about the center O of the wheel for the position of the wrench shown.   *Ans.* $M = 78.3$ N·m

2/33 Calculate the moment about point *A* exerted by the 40,000-lb force supported by the hoisting cable of the tractor crane for the position shown.

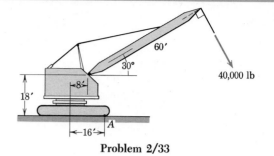

Problem 2/33

2/34 In raising the flagpole from the position shown, the tension *T* in the cable must supply a moment about *O* of 72 kN·m. Determine *T*. *Ans. T* = 8.65 kN

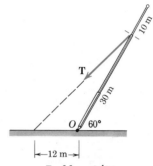

Problem 2/34

2/35 Determine the angle θ which will maximize the moment M_O of the 50-lb force about the shaft axis at *O*. Also compute M_O.

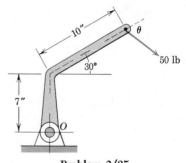

Problem 2/35

2/36 If the combined moment about point *A* of the 50-kN force and the force **P** is zero, determine both graphically and algebraically the magnitude of **P**. The plate upon which the forces act is divided into squares. *Ans. P* = 51.5 kN

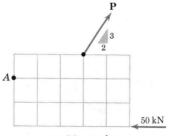

Problem 2/36

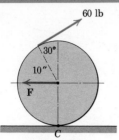

Problem 2/37

2/37 If the combined moment about C of the two forces is zero, determine the magnitude R of their resultant.

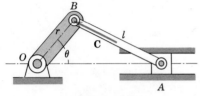

Problem 2/38

2/38 In the slider-crank mechanism shown, the connecting rod AB of length l supports a variable compressive force $\mathbf{C}$. Derive an expression for the moment of $\mathbf{C}$ about the crank axis O in terms of C, r, l, and the variable angle θ.

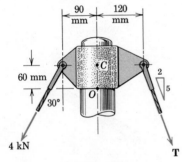

Problem 2/39

2/39 The masthead fitting supports the two forces shown. Determine the magnitude of $\mathbf{T}$ which will cause no bending (zero moment) of the mast at point C.

2/40 For Prob. 2/39 calculate the magnitude of $\mathbf{T}$ so that there will be no bending in the mast at point O.
                               Ans. $T = 3.23$ kN

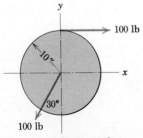

Problem 2/41

2/41 Determine graphically or algebraically the coordinates of the point A on the rim of the wheel about which the combined moment M of the two forces is a maximum. Find M for this point.
      Ans. $A(-8.66$ in., -5 in.$)$, $M = 2000$ lb-in.

2/5 COUPLE. The moment produced by two equal and opposite and noncollinear forces is known as a *couple.* Couples have certain unique properties and have important applications in mechanics.

Consider the action of two equal and opposite forces **F** and −**F** a distance *d* apart, Fig. 2/9*a.* These two forces cannot be combined into a single force, since their sum in every direction is zero. Their effect is entirely to produce a tendency of rotation. The combined moment of the two forces about an axis normal to their plane and passing through any point such as *O* in their plane is the *couple* **M.** It has a magnitude

$$M = F(a + d) - Fa$$

or

$$M = Fd$$

and is in the counterclockwise direction when viewed from above for the case illustrated. Note especially that the magnitude of the couple contains no reference to the dimension *a* which locates the forces with respect to the moment center *O.* It follows that the moment of a couple has the same value for *all* moment centers. We may therefore represent a couple by a *free* vector **M,** as shown in Fig. 2/9*b,* where the direction of **M** is normal to the plane of the couple and the sense of the vector is established by the right-hand convention.

Since the couple vector **M** will always be perpendicular to the plane of the forces which constitute the couple, in two-dimensional analysis we can represent the sense of a couple vector as clockwise or counterclockwise by one of the conventions shown in Fig. 2/9*c.* Later when we deal with couple vectors in three-dimensional problems, we will retain the vector notation for their representation.

A couple is unchanged as long as the magnitude and direction of its vector remain constant. Consequently a given couple will not be altered by changing the values of *F* and *d* as long as their product remains the same. Likewise a couple is not affected by allowing the

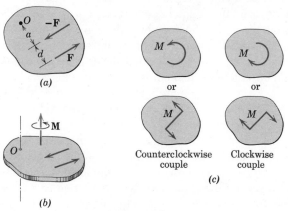

(a)

(b)

M

or

M

or

M M

Counterclockwise   Clockwise
couple   couple

(c)

Figure 2/9

forces to act in any one of parallel planes. Figure 2/10 shows four different configurations of the same couple **M**. In each of the four cases the couple is described by the identical free vector that represents the identical tendencies to rotate the bodies in the direction shown.

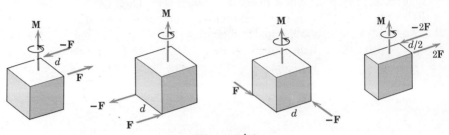

Figure 2/10

The effect of a force acting on a body has been described in terms of the tendency to push or pull the body in the direction of the force and to rotate the body about any axis which does not intersect the line of the force. The representation of this dual effect is often facilitated by replacing the given force by an equal parallel force and a couple to compensate for the change in the moment of the force. This resolution of a force into a force and a couple is illustrated in Fig. 2/11, where

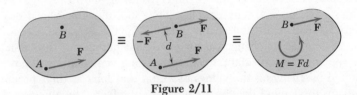

Figure 2/11

the given force **F** acting at point A is replaced by an equal force **F** at some point B and the counterclockwise couple $M = Fd$. The transfer is seen from the middle figure, where the equal and opposite forces **F** and $-$**F** are added at point B without introducing any net external effects on the body. We now see that the original force at A and the equal and opposite one at B constitute the couple $M = Fd$, which is counterclockwise for the sample chosen, as shown in the right-hand part of the figure. Thus we have replaced the original force at A by the same force acting at a different point B and a couple without altering the external effects of the original force on the body. It follows also that a given couple and a force which lies in the plane of the couple (normal to the couple vector) may be combined to produce a single force by reversing the procedure. The resolution of a force into an equivalent force and couple is a step that finds repeated application in mechanics and should be thoroughly mastered.

Sample Problem 2/4

The rigid bracket is subjected to a couple consisting of the 200-N forces. Replace this couple by an equivalent couple consisting of the two forces **P** and −**P** which have a magnitude of 500 N. Find the proper angle θ.

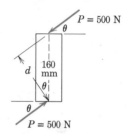

Dimensions in Millimeters

Solution. The given couple is counterclockwise when viewed in the plane of the forces from above, and its magnitude is

$[M = Fd]$ $M = 200(0.200 + 0.100) = 60$ N·m

The forces **P** and −**P** produce a counterclockwise couple

$[M = Fd]$ $M = 500(0.100 + 0.060) \cos \theta$

Equating the two expressions gives

$$60 = 500(0.160) \cos \theta$$

$$\theta = \cos^{-1} \frac{60}{80} = 41.4° \qquad Ans.$$

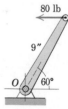

① Note that the only dimensions that are relevant are those that give the perpendicular distances between the forces of the couples.

Sample Problem 2/5

Replace the horizontal 80-lb force acting on the lever by an equivalent system consisting of a force at O and a couple.

Solution. We apply two equal and opposite 80-lb forces at O and identify the counterclockwise couple

$[M = Fd]$ $M = 80(9 \sin 60°) = 624$ lb-in. *Ans.*

Thus the original force is equivalent to the 80-lb force at O and the 624-lb-in. couple as shown in the third of the three equivalent figures.

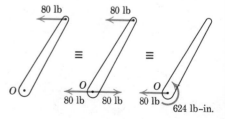

① The reverse of this problem is often encountered, namely, the replacement of a force and a couple by a single force. Proceeding in reverse is the same as replacing the couple by two forces one of which is equal and opposite to the 80-lb force at O. The moment arm to the second force would be $M/F =$

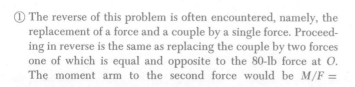

$624/80 = 7.79$ in. which is $9 \sin 60°$, thus determining the line of action of the single resultant force of 80 lb.

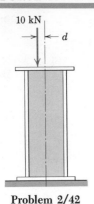

Problem 2/42

PROBLEMS

2/42 The action of the 10-kN force on the steel column can be analyzed by considering it to produce a compression along the centerline and a couple. If the couple has a magnitude of 800 N·m, determine the offset *d*.   *Ans. d* = 80 mm

2/43 In designing the lifting hook we note that the action of the applied force **F** at the critical section of the hook is a pull at *B* and a couple. If the magnitude of the couple is 4000 lb-ft, determine the magnitude of **F**.

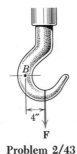

F

Problem 2/43

2/44 Each propeller of the twin-screw ship develops a full-speed thrust *F* of 300 kN. In maneuvering the ship one propeller is turning full speed ahead and the other full speed in reverse. What thrust *P* must each tug exert on the ship to counteract the turning effect of the ship's propellers?   *Ans. P* = 51.4 kN

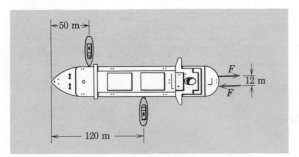

Problem 2/44

2/45 Express the combined moment of the two forces about the *y*-axis and about the *y'*-axis in vector notation using the unit vectors shown.

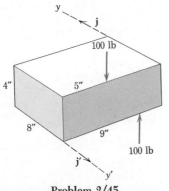

Problem 2/45

2/46 The simple truss supports a load of 40 kN. The vertical wall exerts a horizontal force against the supporting roller at *A*, and the hinged connection at *B* exerts the additional force on the truss required to maintain equilibrium. The 40-kN load and the vertical component of the reaction at *B* constitute a couple that is equal and opposite to the couple due to the two horizontal forces. Calculate the magnitude *B* of the force acting on the pin connection at *B*.
Ans. $B = 75.5$ kN

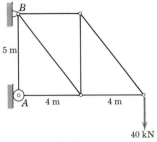

Problem 2/46

2/47 The angle plate is subjected to the two 250-N forces shown. It is desired to replace these forces by an equivalent set consisting of the 200-N force applied at *A* and a second force applied at *B*. Determine the *y*-coordinate of *B*.

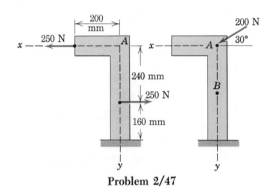

Problem 2/47

2/48 The rear wheel of an accelerating car is acted upon by a friction force *F* of 2.4 kN and a torque on the axle which amounts to a couple *M*. If the force and the couple are replaced by an equivalent force which acts through a point 12 mm directly above the center of the wheel, find *M*. *Ans.* $M = 929$ N·m

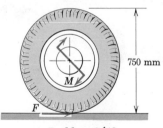

Problem 2/48

2/49 The 300-lb-in. couple is applied to the vertical shaft which is welded to the horizontal rectangular plate. If the couple and 60-lb force are to be replaced by a single equivalent force at *B*, determine the distance *x*.

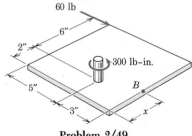

Problem 2/49

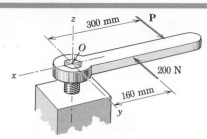

300 mm

P

z

O

x

200 N

160 mm

y

Problem 2/50

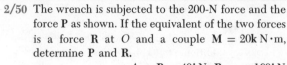

2/50 The wrench is subjected to the 200-N force and the force **P** as shown. If the equivalent of the two forces is a force **R** at O and a couple $\mathbf{M} = 20\mathbf{k} \ \text{N·m}$, determine **P** and **R**.

Ans. $\mathbf{P} = 40\mathbf{j} \ \text{N}, \ \mathbf{R} = -160\mathbf{j} \ \text{N}$

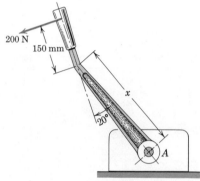

200 N

150 mm

x

20°

A

Problem 2/51

2/51 The control lever is subjected to a clockwise couple of 80 N·m exerted by its shaft at A and is to be designed to operate with a 200-N pull as shown. If the resultant of the couple and the force passes through A, determine the proper dimension x of the lever.

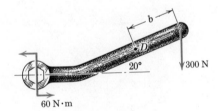

b

D

20°

300 N

60 N·m

Problem 2/52

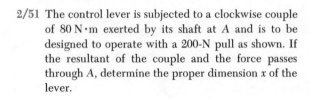

2/52 Replace the couple and force shown by a single force **F** applied at a point D. Locate D by determining the distance b. *Ans.* $b = 213$ mm

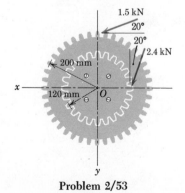

1.5 kN

20°

20°

2.4 kN

200 mm

x

120 mm

O

y

Problem 2/53

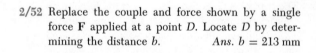

2/53 The figure represents two integral gears subjected to the tooth-contact forces shown. Replace the two forces by an equivalent single force **R** at the rotation axis O and a corresponding couple M. If the gears start from rest under the action of the tooth loads shown, in what direction would rotation occur?

Ans. $R = 3.56$ kN, $\theta_x = 51.1°$
$M = 11.28$ N·m counterclockwise

2/54 The bracket is fastened to the girder by means of the two rivets A and B and supports the 520-lb force. Replace this force by a force acting along the horizontal centerline between the rivets and a couple. Then redistribute this force and couple by replacing them by two forces, one at A and one at B, and thus determine the forces supported by the rivets.

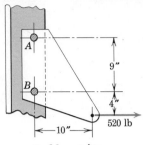

Problem 2/54

2/55 If the force system $\mathbf{F}_1$ and $\mathbf{F}_2$ is to be equivalent to the system of the two 200-N forces, determine θ and the magnitudes of $\mathbf{F}_1$ and $\mathbf{F}_2$.

Ans. $\theta = 26.6°$, $|\mathbf{F}_1| = |\mathbf{F}_2| = F = 335$ N

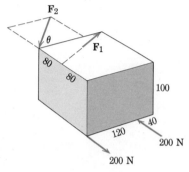

Dimensions in Millimeters

Problem 2/55

2/6 RESULTANTS. In the previous four articles we have developed the properties of force, moment, and couple. With the aid of these descriptions we are now ready to describe the resultant action of a group or *system* of forces. Most problems in mechanics deal with a system of forces, and it is generally necessary to reduce the system to its simplest form in describing its action. The resultant of a system of forces is the simplest force combination that can replace the original forces without altering the external effect of the system on the rigid body to which the forces are applied. The equilibrium of a body is the condition where the resultant of all forces that act on it is zero. When the resultant of all forces on a body is not zero, the acceleration of the body is described by equating the force resultant to the product of the mass and acceleration of the body. Thus the determination of resultants is basic to both statics and dynamics.

The most common type of force system occurs when the forces all act in a single plane, say the x-y plane, as illustrated by the system of three forces F_1, F_2, and F_3 in Fig. 2/12a. The resultant force $\mathbf{R}$ is

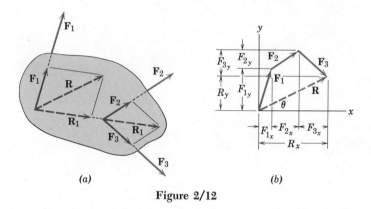

(a)   (b)

Figure 2/12

obtained in magnitude and direction by forming the *force polygon* in the b-part of the figure where the forces are added head-to-tail in any sequence. Thus for any system of coplanar forces we may write

$$\mathbf{R} = \mathbf{F}_1 + \mathbf{F}_2 + \mathbf{F}_3 + \cdots = \Sigma\mathbf{F}$$
$$R_x = \Sigma F_x \qquad R_y = \Sigma F_y \qquad R = \sqrt{(\Sigma F_x)^2 + (\Sigma F_y)^2}$$
$$\theta = \tan^{-1}\frac{R_y}{R_x} = \tan^{-1}\frac{\Sigma F_y}{\Sigma F_x}$$

$$(2/4)$$

Graphically the correct line of action of $\mathbf{R}$ may be obtained by preserving the correct lines of action of the forces and adding them by the parallelogram law as indicated in the a-part of the figure for the case of three forces where the sum $\mathbf{R}_1$ of $\mathbf{F}_2$ and $\mathbf{F}_3$ is added to $\mathbf{F}_1$ to obtain $\mathbf{R}$. In this process the principle of transmissibility has been used.

Algebraically, we may locate the resultant force by using Varignon's moment principle with the selection of some convenient point O as a moment center, Fig. 2/13. Thus the unknown arm d is computed from

$$Rd = \Sigma Fd = F_1 d_1 - F_2 d_2 + F_3 d_3$$

or simply

$$\boxed{Rd = \Sigma M_O} \qquad\qquad (2/5)$$

which is a restatement of Varignon's theorem or the *principle of moments*, which says that the moment of the resultant force about any point O equals the sum of the moments of the forces of the system about that same point. This principle is one of the most widely used of all principles in mechanics.

We see that it requires three equations, $R_x = \Sigma F_x$, $R_y = \Sigma F_y$, and $Rd = \Sigma M_O$, to determine completely the resultant $\mathbf{R}$ of a general coplanar force system. For a concurrent force system, the moment equation about the point of concurrency is automatically satisfied, and only the two force equations are needed to determine $\mathbf{R}$. For a parallel force system, one force equation in the direction of the forces and one moment equation will be sufficient to determine $\mathbf{R}$.

If the resultant force $\mathbf{R}$ for a given force system is zero, the resultant of the system need not be zero as it may be a couple M. In this event $\Sigma M_O = M$, where O is any convenient moment center. The three forces in Fig. 2/14, for instance, have a zero resultant force but have a resultant clockwise couple $M = F_3 d$.

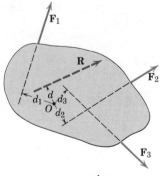

Figure 2/13

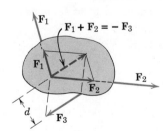

Figure 2/14

Sample Problem 2/6

Determine the resultant of the four forces and one couple that act on the plate shown.

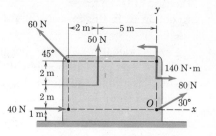

Solution. Point O is selected arbitrarily as a convenient origin of coordinates and moment center. The components R_x and R_y, the magnitude of the resultant $\mathbf{R}$, and the angle θ made by $\mathbf{R}$ with the x-axis become

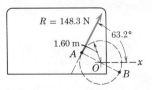

$[R_x = \Sigma F_x]$ $R_x = 40 + 80 \cos 30° - 60 \cos 45° = 66.9$ N

$[R_y = \Sigma F_y]$ $R_y = 50 + 80 \sin 30° + 60 \sin 45° = 132.4$ N

$[R = \sqrt{R_x^2 + R_y^2}]$ $R = \sqrt{(66.9)^2 + (132.4)^2} = 148.3$ N   *Ans.*

$\left[\theta = \tan^{-1}\dfrac{R_y}{R_x}\right]$ $\theta = \tan^{-1}\dfrac{132.4}{66.9} = 63.2°$ *Ans.*

Although the couple has no influence on the magnitude and direction of $\mathbf{R}$, it does influence the moment of the resultant, which will now be determined. The position of the line of action of $\mathbf{R}$ is found from the principle of moments (Varignon's theorem). With O as the moment center, with d as the moment arm of $\mathbf{R}$, and with the counterclockwise sense chosen arbitrarily as positive, this principle requires

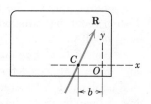

① $[Rd = \Sigma M_O]$ $148.3d = 140 - 50(5) + 60 \cos 45°(4) - 60 \sin 45°(7)$

$d = -1.60$ m

The negative sign indicates that the moment of the resultant is acting in a clockwise rather than counterclockwise sense about O. Hence the resultant may be applied at any point on a line, making an angle of 63.2° with the x-axis and tangent to a circle of 1.6-m radius about O as shown in the figure. The clockwise moment of $\mathbf{R}$ requires that the line of action of $\mathbf{R}$ be tangent at point A and not at point B, as would have been the case if the moment had been acting in a counterclockwise sense.

Alternatively, we may locate $\mathbf{R}$ by finding the point C on, say, the x-axis through which $\mathbf{R}$ must pass. If we assumed initially that C was to the left of O, then a clockwise moment sum would give us

② $[Rd = \Sigma M_O]$ $132.4b = -140 + 50(5) - 60 \cos 45°(4) + 60 \sin 45°(7)$

$b = 1.79$ m

① It is noted that the choice of point O as a moment center eliminated any moments due to the two forces that pass through O. The careful selection of a convenient moment center that eliminates as many terms as possible from the moment equations is an important simplification in mechanics calculations.

② If we had assumed that C was on the positive x-axis, b would have been negative. As a check note that $d = b \sin \theta$, $1.60 = 1.79 \sin 63.2°$. Also observe that we could locate $\mathbf{R}$ by finding the point on the y-axis through which $\mathbf{R}$ passes. In this calculation R_y would not appear.

PROBLEMS

2/56 Determine the resultant **R** of the four forces acting on the gusset plate. Also find the magnitude of **R** and the angle θ_x which the resultant makes with the x-axis.

$$\text{Ans. } \mathbf{R} = 34.9\mathbf{i} + 41.8\mathbf{j} \text{ kN}$$
$$R = 54.5 \text{ kN}$$
$$\theta_x = 50.2°$$

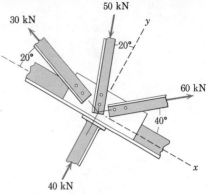

Problem 2/56

2/57 Calculate the magnitude of the tension **T** and the angle θ for which the eye bolt will be under a resultant downward force of 15 kN.

$$\text{Ans. } T = 12.85 \text{ kN}, \theta = 38.9°$$

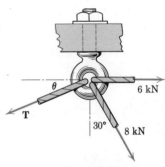

Problem 2/57

2/58 Determine the height h above the base B at which the resultant of the three forces acts.

Problem 2/58

2/59 Where does the resultant of the two forces act?

$$\text{Ans. } 10.70 \text{ m to the left of } A$$

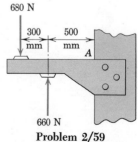

Problem 2/59

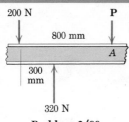

Problem 2/60

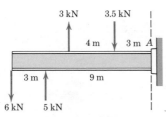

Problem 2/61

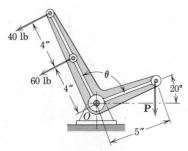

Problem 2/62

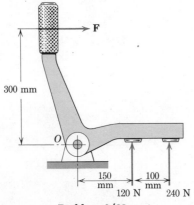

Problem 2/63

2/60 Explain why the resultant of the three parallel forces always passes through point A regardless of the magnitude of **P.**

2/61 Determine the force R that could replace the four forces that act on the cantilever beam and not change the reaction on the end of the beam at the supporting weld at A. Locate R by finding its distance b to the left of A.

Ans. $R = 1.5$ kN down, $b = 11$ m

2/62 In the equilibrium position shown the resultant of the three forces acting on the bell crank passes through the bearing O. Determine the vertical force **P.** Does the result depend on θ?

2/63 Determine the magnitude of the force **F** applied to the handle which will make the resultant of the three forces pass through O.   *Ans.* $F = 260$ N

2/64 The gear reducer shown is acted upon by the two couples, its weight of 200 N, and a vertical force at each of the mountings A and B. If the resultant of this system of two couples and three forces is zero, determine the vertical forces at A and B.

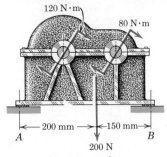

Problem 2/64

2/65 Determine the resultant **R** of the four forces and locate the point A on the upper chord through which **R** must pass.   *Ans.* **R** = 4000**i** − 11,460**j** lb, 6.11 ft to the right of B

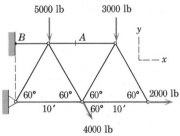

Problem 2/65

2/66 Replace the three forces and couple by an equivalent force **R** at A and a couple M. Specify M and the magnitude of **R**.

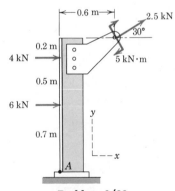

Problem 2/66

2/67 Determine the resultant **R** of the three forces and two couples shown. Find the coordinate x of the point on the x-axis through which **R** passes.
   Ans. **R** = −1.5**i** − 2**j** kN, x = 290 mm

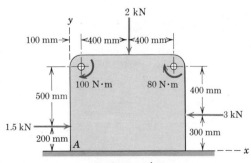

Problem 2/67

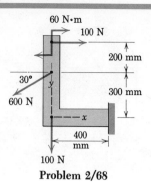

Problem 2/68

2/68 Calculate the y-coordinate of the point on the y-axis through which the resultant of the three forces and couple must pass.   *Ans.* $y = 109$ mm

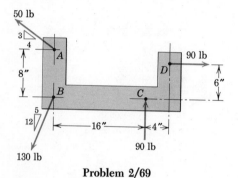

Problem 2/69

2/69 Determine graphically the resultant of the four forces. Check your result by a calculation.

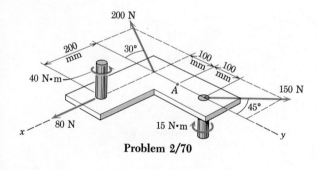

Problem 2/70

2/70 Replace the three forces and two couples acting on the rigid unit by an equivalent single force **R** at point A and a couple M.

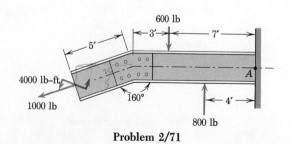

Problem 2/71

2/71 Represent the resultant of the three forces and couple by a force **R** at A and a couple M. Find M and the magnitude of **R**.
      Ans. $M = 8420$ lb-ft counterclockwise, $R = 950$ lb

2/72 The gear and attached V-belt pulley are turning counterclockwise and are subjected to the tooth load of 1600 N and the 800-N and 450-N tensions in the V-belt. Represent the action of these three forces by a resultant force **R** at O and a couple of magnitude M. Is the unit slowing down or speeding up?   *Ans.* **R** $= 930\mathbf{i} + 1666\mathbf{j}$ N, speeding up

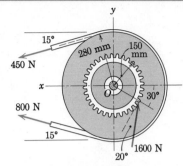

Problem 2/72

▶2/73 The jet plane has a mass m of 30 Mg and is climbing at the 15° angle at constant velocity under the steady thrust T of 86 kN. Since the velocity is constant, the resultant of all forces acting on the aircraft is zero. Calculate the drag D, the lift L, and the force P acting on the stabilizer for the conditions represented.

Ans. $D = 9.83$ kN, $L = 279.6$ kN, $P = 4.69$ kN

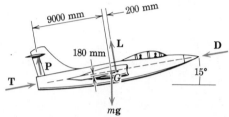

Problem 2/73

SECTION B. THREE-DIMENSIONAL FORCE SYSTEMS

2/7 RECTANGULAR COMPONENTS.   Many problems in mechanics require analysis in three dimensions, and it is often necessary to resolve a force into its three mutually perpendicular components. Thus, the force **F** acting at point O in Fig. 2/15 has the *rectangular components* F_x, F_y, F_z, where

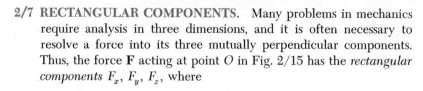

Figure 2/15

$$
\begin{aligned}
F_x &= F \cos \theta_x & F &= \sqrt{F_x^{\,2} + F_y^{\,2} + F_z^{\,2}} \\
F_y &= F \cos \theta_y & \mathbf{F} &= \mathbf{i}F_x + \mathbf{j}F_y + \mathbf{k}F_z \\
F_z &= F \cos \theta_z & \mathbf{F} &= F(\mathbf{i} \cos \theta_x + \mathbf{j} \cos \theta_y + \mathbf{k} \cos \theta_z)
\end{aligned}
\qquad (2/6)
$$

The unit vectors **i**, **j**, **k** are in the x-, y-, and z-directions, respectively. If we introduce the *direction cosines* of **F** which are $l = \cos \theta_x$, $m = \cos \theta_y$, $n = \cos \theta_z$, where $l^2 + m^2 + n^2 = 1$, we may write the force as

$$
\mathbf{F} = F(\mathbf{i}l + \mathbf{j}m + \mathbf{k}n)
\qquad (2/7)
$$

The choice of orientation of the coordinate system is quite arbitrary, with convenience being the primary consideration. But we must use a right-handed set of axes in our three-dimensional work so as to preserve the relative orientations of the axes. When we rotate from the x- to the y-axis, the positive direction for the z-axis in a right-handed system is that of the advancement of a right-handed screw when rotated in the same sense.

Rectangular components of a force **F** (or other vector) may be written alternatively with the aid of the vector operation known as the *dot* or *scalar product*.° By definition, the dot product of two vectors **P** and **Q**, Fig. 2/16a, is the product of their magnitudes times the cosine of the angle α between them and is written

$$
\mathbf{P} \cdot \mathbf{Q} = PQ \cos \alpha
$$

This product may be viewed either as the projection (component)

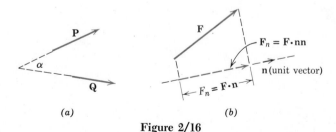

(a)   (b)

Figure 2/16

° See item 6 of section 7 in Appendix B.

$P \cos \alpha$ of $\mathbf{P}$ in the direction of $\mathbf{Q}$ multiplied by Q or as the projection (component) $Q \cos \alpha$ of $\mathbf{Q}$ in the direction of $\mathbf{P}$ multiplied by P. In either case the dot product of the two vectors is a scalar quantity. Thus the component $F_x = F \cos \theta_x$ of the force $\mathbf{F}$ in Fig. 2/15, for instance, may be written as $F_x = \mathbf{F} \cdot \mathbf{i}$ where $\mathbf{i}$ is the unit vector in the x-direction. In more general terms, if $\mathbf{n}$ is a unit vector in a specified direction, the component of $\mathbf{F}$ in the $\mathbf{n}$-direction, Fig. 2/16b, has the magnitude $F_n = \mathbf{F} \cdot \mathbf{n}$. If it is desired to write the component vector in the $\mathbf{n}$-direction as a vector quantity, then its scalar magnitude, expressed by $\mathbf{F} \cdot \mathbf{n}$, must be multiplied by the unit vector $\mathbf{n}$ to give $\mathbf{F}_n = (\mathbf{F} \cdot \mathbf{n})\mathbf{n}$, which may be written merely as $\mathbf{F}_n = \mathbf{F} \cdot \mathbf{nn}$.

If $\mathbf{n}$ has direction cosines α, β, γ, then we may write $\mathbf{n}$ in vector component form like any other vector as

$$\mathbf{n} = \mathbf{i}\alpha + \mathbf{j}\beta + \mathbf{k}\gamma$$

where in this case its magnitude is unity. If $\mathbf{F}$ has the direction cosines l, m, n with respect to reference axes x-y-z, then the component of $\mathbf{F}$ in the $\mathbf{n}$-direction becomes

$$F_n = \mathbf{F} \cdot \mathbf{n} = F(\mathbf{i}l + \mathbf{j}m + \mathbf{k}n) \cdot (\mathbf{i}\alpha + \mathbf{j}\beta + \mathbf{k}\gamma)$$

$$= F(l\alpha + m\beta + n\gamma)$$

since $\mathbf{i} \cdot \mathbf{i} = \mathbf{j} \cdot \mathbf{j} = \mathbf{k} \cdot \mathbf{k} = 1$

and $\mathbf{i} \cdot \mathbf{j} = \mathbf{j} \cdot \mathbf{i} = \mathbf{i} \cdot \mathbf{k} = \mathbf{k} \cdot \mathbf{i} = \mathbf{j} \cdot \mathbf{k} = \mathbf{k} \cdot \mathbf{j} = 0$

If the angle between the force $\mathbf{F}$ and the direction specified by the unit vector $\mathbf{n}$ is θ, then by virtue of the dot-product relationship we have $\mathbf{F} \cdot \mathbf{n} = Fn \cos \theta = F \cos \theta$ where $|\mathbf{n}| = n = 1$. Thus the angle between $\mathbf{F}$ and $\mathbf{n}$ is given by

$$\theta = \cos^{-1} \frac{\mathbf{F} \cdot \mathbf{n}}{F} \tag{2/8}$$

or, in general, the angle between any two vectors $\mathbf{P}$ and $\mathbf{Q}$ is

$$\theta = \cos^{-1} \frac{\mathbf{P} \cdot \mathbf{Q}}{PQ} \tag{2/8a}$$

If a force $\mathbf{F}$ is perpendicular to a line whose direction is specified by the unit vector $\mathbf{n}$, then $\cos \theta = 0$, and $\mathbf{F} \cdot \mathbf{n} = 0$. Note especially that this relationship does not mean that either $\mathbf{F}$ or $\mathbf{n}$ is zero, as would be the case with scalar multiplication where $(A)(B) = 0$ requires that either A or B be zero.

It should be observed that the dot-product relationship applies to nonintersecting vectors as well as to intersecting vectors. Thus the dot product of the nonintersecting vectors $\mathbf{P}$ and $\mathbf{Q}$ in Fig. 2/17 is the projection of $\mathbf{P}'$ on $\mathbf{Q}$, which is $P'Q \cos \alpha = PQ \cos \alpha$ since $\mathbf{P}'$ and $\mathbf{P}$ are the same free vectors.

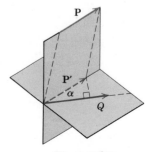

Figure 2/17

48

Sample Problem 2/7

A force $\mathbf{F}$ with a magnitude of 100 N is applied at the origin O of the axes x-y-z as shown. The line of action of $\mathbf{F}$ passes through a point A whose coordinates are 3 m, 4 m, and 5 m. Determine (a) the x-, y-, and z-components of $\mathbf{F}$, (b) the projection of $\mathbf{F}$ on the x-y plane, and (c) the component F_n of $\mathbf{F}$ in the direction of the line O-n which passes through point B as shown.

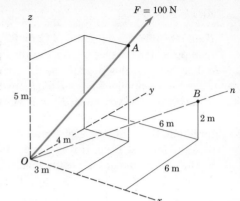

Solution. **Part (a).** The direction cosines of $\mathbf{F}$ are

$$l = \frac{3}{7.071} = 0.424 \qquad m = \frac{4}{7.071} = 0.566 \qquad n = \frac{5}{7.071} = 0.707$$

where the diagonal to point A is $\sqrt{3^2 + 4^2 + 5^2} = \sqrt{50} = 7.071$ m. The components become

$$F_x = Fl = 100(0.424) = 42.4 \text{ N}$$
$$F_y = Fm = 100(0.566) = 56.6 \text{ N}$$
$$F_z = Fn = 100(0.707) = 70.7 \text{ N} \qquad\qquad Ans.$$

Part (b). The cosine of the angle θ_{xy} between $\mathbf{F}$ and the x-y plane is

$$\cos \theta_{xy} = \frac{\sqrt{3^2 + 4^2}}{7.071} = 0.707$$

so that $F_{xy} = F \cos \theta_{xy} = 100(0.707) = 70.7 \text{ N}.$ *Ans.*

Part (c). The direction cosines of a unit vector $\mathbf{n}$ along O-n are

$$\alpha = \beta = \frac{6}{\sqrt{6^2 + 6^2 + 2^2}} = 0.688 \qquad \gamma = \frac{2}{\sqrt{6^2 + 6^2 + 2^2}} = 0.229$$

Thus the component of $\mathbf{F}$ along O-n becomes

$$F_n = \mathbf{F} \cdot \mathbf{n} = 100(0.424\mathbf{i} + 0.566\mathbf{j} + 0.707\mathbf{k}) \cdot (0.688\mathbf{i} + 0.688\mathbf{j} + 0.229\mathbf{k})$$
$$= 100[(0.424)(0.688) + (0.566)(0.688) + (0.707)(0.229)]$$
$$= 84.4 \text{ N} \qquad\qquad Ans.$$

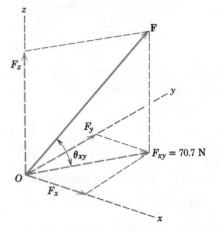

① In this example all components are positive. Be prepared for the case where a direction cosine, and hence the component, is negative.

② The dot product automatically finds the length of the projection or component of $\mathbf{F}$ on line On as shown. To express this component as a vector, we would write $\mathbf{F} \cdot \mathbf{nn}$ or $84.4\mathbf{n}$ N.

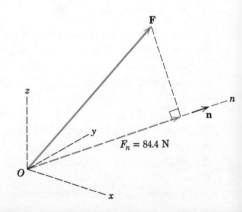

PROBLEMS

2/74 If the *x*-component of **P** equals 60 lb, determine the *y*-component of **P**.   *Ans.* $P_y = 85.9$ lb

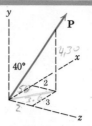

Problem 2/74

2/75 The turnbuckle is tightened until the tension in the cable *AB* equals 1.2 kN. Write the vector expression for the tension **T** as a force acting on the lever.

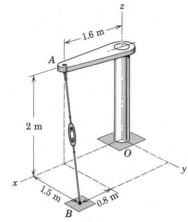

Problem 2/75

2/76 The cable exerts a tension of 2 kN on the fixed bracket at *A*. Write the vector expression for the tension **T**.
     Ans. $\mathbf{T} = 2(-0.920\mathbf{i} + 0.383\mathbf{j} + 0.077\mathbf{k})$ kN

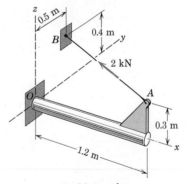

Problem 2/76

2/77 The 140-lb force **F** is directed along the diagonal of the rectangular parallelepiped whose sides are in the ratios 2:3:6. Express **F** as a vector.

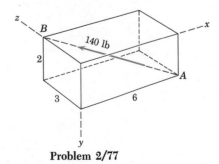

Problem 2/77

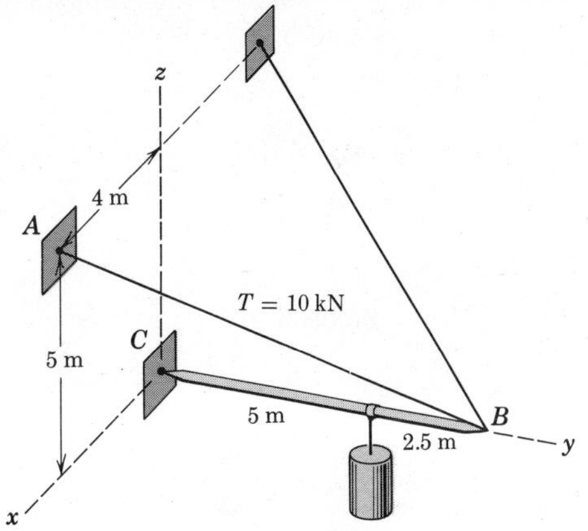

Problem 2/78

2/78 The tension in the supporting cable AB is 10 kN. Express this tension as a force vector $\mathbf{T}$ acting on BC.

Ans. $\mathbf{T} = \frac{10}{\sqrt{389}}(8\mathbf{i} - 15\mathbf{j} + 10\mathbf{k})$ kN

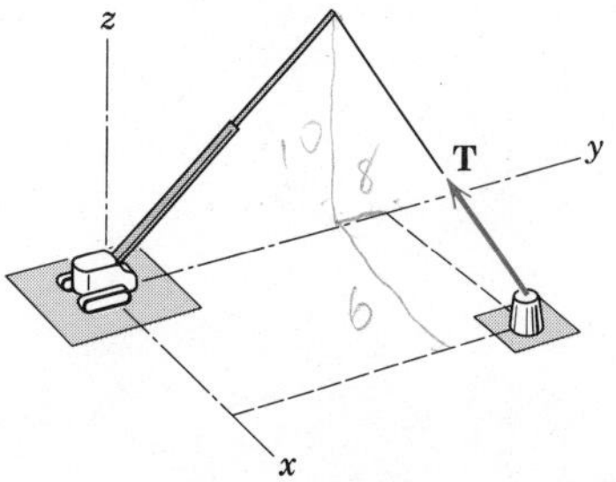

Problem 2/79

2/79 The crane exerts a force $\mathbf{T} = -6\mathbf{i} - 8\mathbf{j} + 10\mathbf{k}$ kN on the concrete pillar to which its cable is attached. Calculate the angles which $\mathbf{T}$ makes with the positive x-axis and with the x-y plane.

2/80 A force $\mathbf{F}$ has a magnitude of 1200 N. If its direction cosine with respect to the x-axis is 0.25 and if the ratio of its x-component to its z-component is 0.6, compute F_y. *Ans.* $F_y = 1049$ N

2/81 The line of action of a force $\mathbf{P}$ makes an angle of 120° with the positive x-axis, and $\mathbf{P}$ has a direction cosine with respect to the z-axis of 0.6. If the y-component of $\mathbf{P}$ is 200 N, write $\mathbf{P}$ as a vector.

2/82 The direction cosines of a certain force vector $\mathbf{F}$ with respect to the x- and y-directions are -0.4 and 0.6, respectively. If $\mathbf{F}\cdot\mathbf{k} = 60$ lb, compute the angle θ_{xy} made by $\mathbf{F}$ with the x-y plane and find the x-component of $\mathbf{F}$.

Ans. $\theta_{xy} = 43.9°$, $F_x = -34.6$ lb

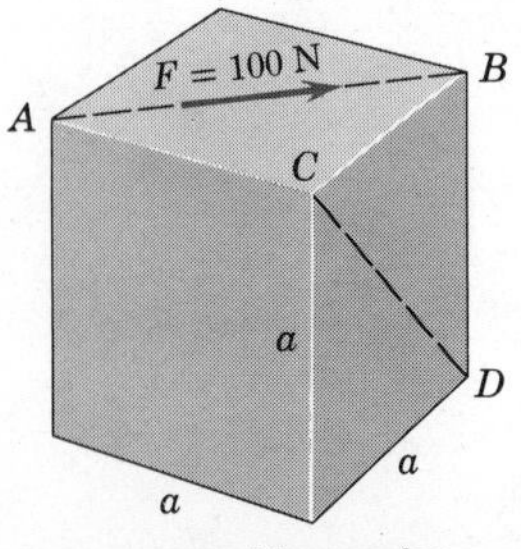

Problem 2/83

2/83 Calculate the magnitude of the projection F_{CD} of the 100-N force on the face diagonal CD of the cube.

2/84 Determine the *x*-coordinate of the small pulley at *C* which will ensure that there is no component of the tension in cable *OC* along *AB*.   *Ans.* $x = 5.33$ ft

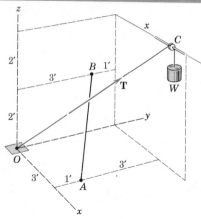

Problem 2/84

2/85 Derive the expression for the component F_{DC} of the force **F** along the line directed from *D* to *C*.

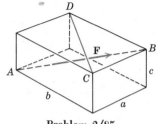

Problem 2/85

2/86 The force **F** has a magnitude of 2 kN and is directed from *A* to *B*. Calculate the component F_{CD} of **F** along *CD*.   *Ans.* $F_{CD} = \sqrt{6/5}$ kN

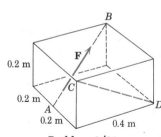

Problem 2/86

▶**2/87** The access door is held in the 30° open position by the chain *AB*. If the tension in the chain is 100 N, determine the component of the tension along the diagonal axis *CD* of the door.   *Ans.* $T_{CD} = 46.0$ N

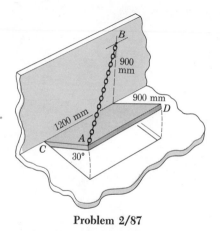

Problem 2/87

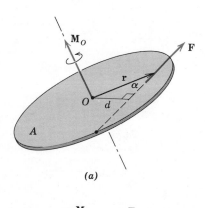

(a)

(b)

Figure 2/18

2/8 MOMENT AND COUPLE. A somewhat more general formulation of the concepts of moment and couple than was used in two-dimensional analysis will now be developed for our use in three-dimensional problems. It is only when we work in three dimensions that the full advantage of vector analysis becomes apparent.

Moment. Consider a force **F** with a given line of action, Fig. 2/18a, and any point O not on this line. Point O and the line of **F** establish a plane A. The moment $\mathbf{M}_O$ of **F** about an axis through O normal to the plane has the magnitude $M_O = Fd$, where d is the perpendicular distance from O to the line of **F**. This moment is also referred to as the moment of **F** about the *point* O. The vector $\mathbf{M}_O$ is normal to the plane and along the axis through O. Both the magnitude and the direction of $\mathbf{M}_O$ may be described by the vector operation known as the *cross* or *vector product* (see item 7 in Art. B7 of Appendix B). A vector **r** is introduced that extends from O to *any* point on the line of action of **F**. By definition, the cross product of **r** and **F** is written **r** × **F** and has the magnitude $(r \sin \alpha)F$, which is the same as Fd, the magnitude of $\mathbf{M}_O$. The correct direction and sense of the moment are established by the right-hand rule, described previously in Arts. 2/4 and 2/5. Thus, with **r** and **F** treated as free vectors, Fig. 2/18b, the thumb points in the direction of $\mathbf{M}_O$ if the fingers of the right hand curl in the direction of rotation from **r** to **F**. Therefore we may write the moment of **F** about the axis through O as

$$\boxed{\mathbf{M}_O = \mathbf{r} \times \mathbf{F}} \qquad (2/9)$$

The order **r** × **F** of the vectors *must* be maintained, since **F** × **r** would produce a vector with a sense opposite to that of $\mathbf{M}_O$ or $\mathbf{F} \times \mathbf{r} = -\mathbf{M}_O$.

The cross-product expression for $\mathbf{M}_O$ may be written in determinant form which is

$$\boxed{\mathbf{M}_O = \begin{vmatrix} \mathbf{i} & \mathbf{j} & \mathbf{k} \\ r_x & r_y & r_z \\ F_x & F_y & F_z \end{vmatrix}} \qquad (2/10)$$

(Again, refer to item 7 in Art. B7 of Appendix B if you are not already familiar with the expansion of the cross product.) The symmetry and order of the terms should be carefully noted, and care must be exercised to ensure the use of a *right-handed* coordinate system upon which the correct evaluation of vector operations depends. Expansion of the determinant gives

$$\mathbf{M}_O = \mathbf{i}(r_y F_z - r_z F_y) + \mathbf{j}(r_z F_x - r_x F_z) + \mathbf{k}(r_x F_y - r_y F_x)$$

To gain more confidence in the cross-product relationship let us observe the three components of the moment of a force about a point as seen from Fig. 2/19, which shows the three components of a force **F** acting at a point *A* located from *O* by the vector **r**. The scalar magnitudes of the moments of these forces about the positive *x*-, *y*-, and *z*-axes through *O* are seen to be

$$M_x = r_y F_z - r_z F_y, \quad M_y = r_z F_x - r_x F_z, \quad M_z = r_x F_y - r_y F_x$$

which agree with the respective terms in the determinant expansion for the cross product **r × F**.

The moment $\mathbf{M}_\lambda$ of **F** about *any* axis λ through *O*, Fig. 2/20, may now be written. If **n** is a unit vector in the λ-direction, then by using the dot product expression for the component of a vector as described in Art. 2/7, the component of $\mathbf{M}_O$ in the direction of λ is merely $\mathbf{M}_O \cdot \mathbf{n}$, which is the scalar magnitude of the moment $\mathbf{M}_\lambda$ of **F** about λ. To obtain the vector expression for the moment of **F** about λ, the magnitude must be multiplied by the directional unit vector **n** to give

$$\boxed{\mathbf{M}_\lambda = (\mathbf{r} \times \mathbf{F} \cdot \mathbf{n})\mathbf{n}} \qquad (2/11)$$

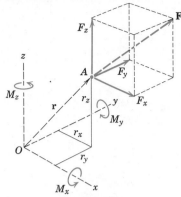

Figure 2/19

where **r × F** replaces $\mathbf{M}_O$. The expression **r × F · n** is known as a *triple scalar product* (see item 8 in Art. B7, Appendix B). It need not be written **(r × F) · n**, since the association **r × (F · n)** would have no meaning because a cross product cannot be formed by a vector and a scalar. The triple scalar product may be represented by the determinant

$$\boxed{|\mathbf{M}_\lambda| = M_\lambda = \begin{vmatrix} r_x & r_y & r_z \\ F_x & F_y & F_z \\ \alpha & \beta & \gamma \end{vmatrix}} \qquad (2/12)$$

where α, β, γ are the direction cosines of the unit vector **n**. This result is easily verified when the indicated operations are carried out.

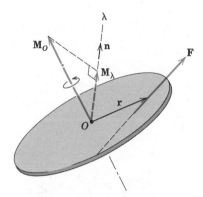

Figure 2/20

Principle of moments. In Art. 2/4 on the moment of a force we introduced the *principle of moments* or *Varignon's theorem* for forces in two dimensions. The principle is equally applicable in three dimensions and is easily proved by applying the distributive law for the sum of vector products. In Fig. 2/21 is shown a system of forces $\mathbf{F}_1$, $\mathbf{F}_2$, $\mathbf{F}_3$, ... concurrent at point *A* whose position vector from point *O* is **r**. The sum of the moments about *O* of the forces of the system is

$$\mathbf{r} \times \mathbf{F}_1 + \mathbf{r} \times \mathbf{F}_2 + \mathbf{r} \times \mathbf{F}_3 + \cdots = \mathbf{r} \times (\mathbf{F}_1 + \mathbf{F}_2 + \mathbf{F}_3 + \cdots) = \mathbf{r} \times \Sigma\mathbf{F}$$

or

$$\boxed{\Sigma\mathbf{M}_O = \mathbf{r} \times \mathbf{R}} \qquad (2/13)$$

Thus the *sum of the moments of a system of concurrent forces about a given point equals the moment of their sum about the same point.* As mentioned in Art. 2/4 this principle finds repeated application in

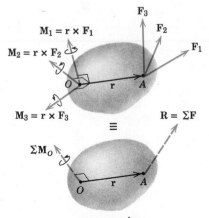

Figure 2/21

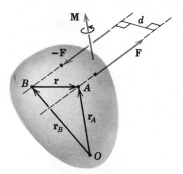

Figure 2/22

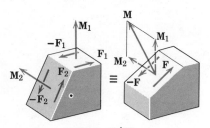

Figure 2/23

mechanics not only for the moments of force vectors but for the moments of other vectors as well.

Couple. The concept of the couple was introduced in Art. 2/5 and is easily extended to three dimensions. Figure 2/22 shows two equal and opposite forces $\mathbf{F}$ and $-\mathbf{F}$ acting on a body. The vector $\mathbf{r}$ joins *any* point B on the line of action of $-\mathbf{F}$ to *any* point A on the line of action of $\mathbf{F}$. Points A and B are located by position vectors $\mathbf{r}_A$ and $\mathbf{r}_B$ from *any* point O. The combined moment of the two forces about O is

$$\mathbf{M} = \mathbf{r}_A \times \mathbf{F} + \mathbf{r}_B \times (-\mathbf{F}) = (\mathbf{r}_A - \mathbf{r}_B) \times \mathbf{F}$$

But $\mathbf{r}_A - \mathbf{r}_B = \mathbf{r}$, so that all reference to the moment center O disappears, and the *moment of the couple* becomes

$$\boxed{\mathbf{M} = \mathbf{r} \times \mathbf{F}} \tag{2/14}$$

Thus the moment of a couple is the *same about all points*. We see that the magnitude of $\mathbf{M}$ is $M = Fd$ where d is the perpendicular distance between the lines of action of the two forces, as described in Art. 2/5.

The moment of a couple is a *free vector,* whereas the moment of a force about a point (which is also the moment about a defined axis through the point) is a *sliding vector* whose direction is along the axis through the point. As in the case of two dimensions the action of the couple is to produce a pure rotation on the body about an axis normal to the plane of the forces which constitute the couple.

Couple vectors obey all of the rules which govern vector quantities. Thus in Fig. 2/23 the couple vector $\mathbf{M}_1$ due to $\mathbf{F}_1$ and $-\mathbf{F}_1$ may be added as shown to the couple vector $\mathbf{M}_2$ due to $\mathbf{F}_2$ and $-\mathbf{F}_2$ to produce the couple $\mathbf{M}$ which, in turn, can be produced by $\mathbf{F}$ and $-\mathbf{F}$.

In Art. 2/5 we learned to replace a force by its equivalent consisting of a force and a couple. We should also be prepared to carry out this replacement in three dimensions. The procedure is represented in Fig. 2/24 where the force $\mathbf{F}$ acting on a rigid body at point A is replaced by an equal force at point B and the couple $\mathbf{M} = \mathbf{r} \times \mathbf{F}$. By adding the equal and opposite forces $\mathbf{F}$ and $-\mathbf{F}$ at B we have the couple composed of $-\mathbf{F}$ and the original $\mathbf{F}$. Thus we see that the couple vector is simply the moment of the original force about the point to which the force is being moved. Again we note that $\mathbf{r}$ is a vector from B to *any* point on the line of action of the original force passing through A.

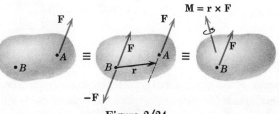

Figure 2/24

Sample Problem   2/8

A tension **T** of magnitude 10 kN is applied to the cable attached to the top A of the rigid mast and secured to the ground at B. Determine the moment M_z of **T** about the z-axis passing through the base O of the mast.

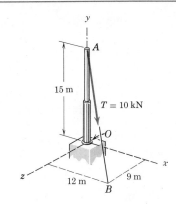

Solution (a).   The required moment may be obtained by finding the component along the z-axis of the moment $\mathbf{M}_O$ of **T** about point O. The vector $\mathbf{M}_O$ is normal to the plane defined by **T** and point O as shown in the accompanying figure. In using Eq. 2/9 to find $\mathbf{M}_O$, the vector **r** is any vector from point O to the line of action of **T**. The simplest choice is the vector from O to A, which is written as $\mathbf{r} = 15\mathbf{j}$ m. The vector expression for **T** requires finding its direction cosines, which are $12/\overline{AB} = 0.566$, $-15/\overline{AB} = -0.707$, and $9/\overline{AB} = 0.424$ where $\overline{AB} = \sqrt{9^2 + 15^2 + 12^2} = \sqrt{450} = 21.2$ m. Therefore we have

$$\mathbf{T} = 10(0.566\mathbf{i} - 0.707\mathbf{j} + 0.424\mathbf{k}) \text{ kN}$$

From Eq. 2/9,

$$[\mathbf{M}_O = \mathbf{r} \times \mathbf{F}] \quad \mathbf{M}_O = 15\mathbf{j} \times 10(0.566\mathbf{i} - 0.707\mathbf{j} + 0.424\mathbf{k})$$

$$= 150(-0.566\mathbf{k} + 0.424\mathbf{i}) \text{ kN} \cdot \text{m}$$

The magnitude M_z of the desired moment is the component of $\mathbf{M}_O$ in the z-direction or $M_z = \mathbf{M}_O \cdot \mathbf{k}$. Therefore

$$M_z = 150(-0.566\mathbf{k} + 0.424\mathbf{i}) \cdot \mathbf{k} = -84.9 \text{ kN} \cdot \text{m} \qquad Ans.$$

The minus sign indicates that the vector $\mathbf{M}_z$ is in the negative z-direction. Expressed as a vector, the moment is $\mathbf{M}_z = -84.9\mathbf{k} \text{ kN} \cdot \text{m}$

Solution (b).   The force of magnitude T is resolved into components T_z and T_{xy} in the x-y plane, which is normal to the moment axis z. Since T_z is parallel to the z-axis, it can exert no moment about this axis. The moment M_z is, then, due only to T_{xy} and is $M_z = T_{xy} d$ where d is the perpendicular distance from T_{xy} to O. The cosine of the angle between T and T_{xy} is $\sqrt{15^2 + 12^2}/\sqrt{15^2 + 12^2 + 9^2} = 0.906$, and therefore

$$T_{xy} = 10(0.906) = 9.06 \text{ kN}$$

The moment arm d equals $\overline{OA}$ multiplied by the sine of the angle between T_{xy} and OA, or

$$d = 15 \frac{12}{\sqrt{12^2 + 15^2}} = 9.37 \text{ m}$$

Hence the moment of **T** about the z-axis is

$$M_z = 9.06(9.37) = 84.9 \text{ kN} \cdot \text{m} \qquad Ans.$$

and is clockwise when viewed in the x-y plane.

Solution (c).   The moment is also easily calculated by resolving T_{xy} into its components T_x and T_y. It is clear that T_y exerts no moment about the z-axis since it passes through it, so that the required moment is due to T_x alone. The direction cosine of **T** with respect to the x-axis is $12/\sqrt{9^2 + 12^2 + 15^2} = 0.566$ so that $T_x = 10(0.566) = 5.66$ kN. Thus

$$M_z = 5.66(15) = 84.9 \text{ kN} \cdot \text{m} \qquad Ans.$$

① We could also use the vector from O to B for **r** and obtain the same result, but using vector OA is simpler.

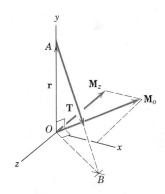

② It is always helpful to accompany your vector operations with a sketch of the vectors so as to retain a clear picture of the geometry of the problem.

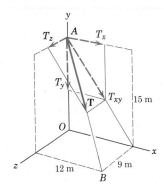

③ Sketch the x-y view of the problem and show d.

Sample Problem 2/9

Determine the magnitude and direction of the couple **M** that will replace the two given couples and still produce the same external effect on the block. Specify the two forces **F** and **−F**, applied in the two faces of the block parallel to the *y-z* plane, that may replace the four given forces. The 30-N forces act parallel to the *y-z* plane.

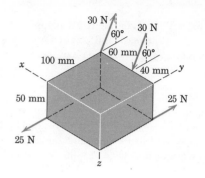

Solution. The couple due to the 30-N forces has the magnitude $M_1 = 30(0.06) = 1.80$ N·m. The direction of M_1 is normal to the plane defined by the two forces, and the sense, shown in the figure, is established by the right-hand convention. The couple due to the 25-N forces has the magnitude $M_2 = 25(0.10) = 2.50$ N·m with the direction and sense shown in the same figure. The two couple vectors combine to give the components

$$M_y = 1.80 \sin 60° = 1.559 \text{ N·m}$$

$$M_z = -2.50 + 1.80 \cos 60° = -1.600 \text{ N·m}$$

① Thus $\quad M = \sqrt{(1.559)^2 + (-1.600)^2} = 2.23$ N·m   *Ans.*

with $\quad \theta = \tan^{-1}\dfrac{1.559}{1.600} = \tan^{-1} 0.974 = 44.3°$ *Ans.*

The forces **F** lie in a plane normal to the couple **M**, and their moment arm as seen from the right-hand figure is 100 mm. Thus each force has the magnitude

$$F = \frac{2.23}{0.10} = 22.3 \text{ N}$$ *Ans.*

and the direction $\theta = 44.3°$.

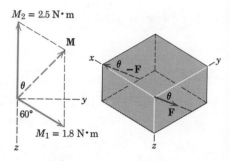

① Bear in mind that the couple vectors are *free* vectors and therefore have no unique lines of action.

Sample Problem 2/10

A force of 40 lb is applied at *A* to the handle of the control lever that is attached to the fixed shaft *OB*. In determining the effect of the force on the shaft at a cross section such as at *O*, the force may be replaced by an equivalent force at *O* and a couple. Describe this couple as a vector **M**.

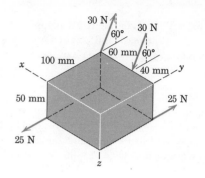

Solution. The couple may be expressed in vector notation as $\mathbf{M} = \mathbf{r} \times \mathbf{F}$ where $\mathbf{r} = \overrightarrow{OA} = 8\mathbf{j} + 5\mathbf{k}$ in. and $\mathbf{F} = -40\mathbf{i}$ lb. Thus

$$\mathbf{M} = (8\mathbf{j} + 5\mathbf{k}) \times (-40\mathbf{i})$$

$$= -200\mathbf{j} + 320\mathbf{k} \text{ lb-in.}$$ *Ans.*

from which the magnitude and direction of **M** may be written. Alternatively we see that moving the 40-lb force through a distance $d = \sqrt{5^2 + 8^2} = 9.43$ in. to a parallel position through *O* requires the addition of a couple **M** whose magnitude is

$$M = Fd = 40(9.43) = 377 \text{ lb-in.}$$ *Ans.*

The couple vector is perpendicular to the plane in which the force is shifted, and its sense is that of the moment of the given force about *O*. The direction of **M** in the *y-z* plane is given by

$$\theta = \tan^{-1}\tfrac{5}{8} = 32.0°$$ *Ans.*

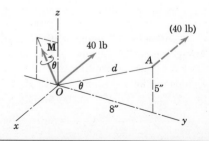

PROBLEMS

2/88 Determine the moment of force **F** about point *A*.

$$Ans.\ \mathbf{M}_A = F(b\mathbf{i} + a\mathbf{j})$$

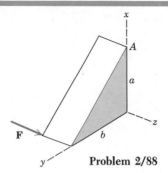

Problem 2/88

2/89 Determine the moment of the force **P** about point *A*.

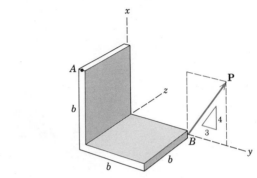

Problem 2/89

2/90 A steel H-beam acts as a column and supports the two vertical loads shown. Replace these forces by a single equivalent force along the vertical centerline of the column and a couple **M**.

$$Ans.\ \mathbf{M} = (-0.450\mathbf{i} + 0.300\mathbf{j})10^6\ \text{lb-in.}$$

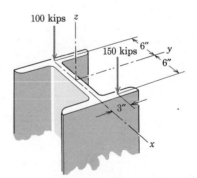

Problem 2/90

2/91 The directions of rotation of the input shaft *A* and output shaft *B* of the worm-gear reducer are indicated by the curved dotted arrows. An input torque (couple) of 80 N·m is applied to shaft *A* in the direction of rotation. The output shaft *B* supplies a torque of 320 N·m to the machine that it drives (not shown). The shaft of the driven machine exerts an equal and opposite reacting torque on the output shaft of the reducer. Determine the resultant **M** of the two couples that act on the reducer unit and calculate the direction cosine of **M** with respect to the *x*-axis.

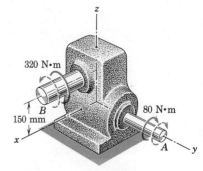

Problem 2/91

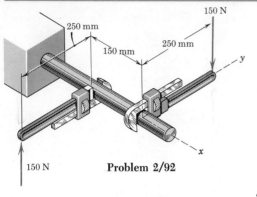

150 N

250 mm

150 mm

250 mm

y

x

150 N

Problem 2/92

2/92 The two forces acting on the handles of the pipe wrenches constitute a couple **M.** Express the couple as a vector.   *Ans.* $\mathbf{M} = -75\mathbf{i} + 22.5\mathbf{j}$ N·m

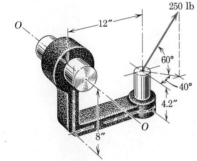

O

250 lb

12″

60°

40°

4.2″

8″

O

Problem 2/93

2/93 Compute the magnitude of the moment of the 250-lb force about the axis *O-O*.

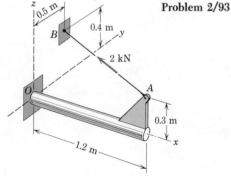

z

0.5 m

B

0.4 m

y

2 kN

A

0.3 m

x

O

1.2 m

Problem 2/94

2/94 Calculate the moment about the *z*-axis of the 2-kN tension in cable *AB* of Prob. 2/76, repeated here.
    Ans. $M_z = 0.920$ kN·m

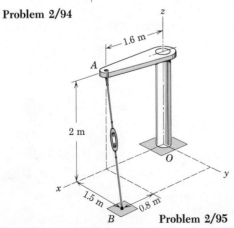

z

1.6 m

A

2 m

O

y

x

1.5 m

0.8 m

B   **Problem 2/95**

2/95 The figure for Prob. 2/75 is repeated here where the cable *AB* exerts a force of 1.2 kN on the lever in the direction from *A* to *B*. Calculate the magnitude of the moment of this force about point *O*.

2/96 In picking up a load from position A a cable tension **T** of 21 kN is developed. Calculate the moment that **T** produces about the base O of the construction crane.   *Ans.* $\mathbf{M}_O = -374\mathbf{i} + 93.5\mathbf{j} - 56.1\mathbf{k}$ kN·m

2/97 If the crane of Prob. 2/96 picks up a load at B rather than at A and develops an initial 21-kN tension T in its cable, determine the moment $\mathbf{M}_O$ of this force about the origin O.

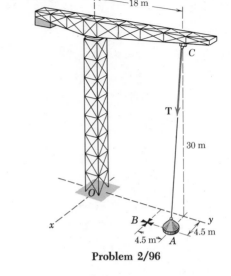

Problem 2/96

2/98 Three couples are formed by the three pairs of equal and opposite forces. Determine the resultant **M** of the three couples.

$$\text{Ans. } \mathbf{M} = -20\mathbf{i} - 6.77\mathbf{j} - 37.2\mathbf{k} \text{ N·m}$$

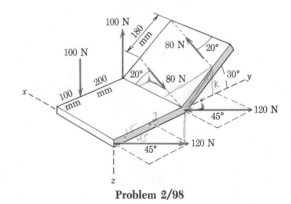

Problem 2/98

2/99 Calculate the magnitude of the moment about the x-axis of the 100-lb force acting on the bracket at A. Also determine the vector expression for the moment of the force about point O without using the cross product.

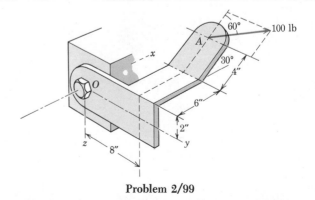

Problem 2/99

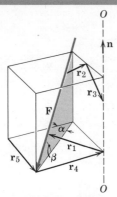

Problem 2/100

2/100 The force **F** exerts a certain moment **M** about the axis *O-O*. Determine which of the expressions cited correctly describes **M**. The vector **n** is a unit vector along *O-O*, and the box is rectangular.

$$\mathbf{r}_1 \times \mathbf{F} \cdot \mathbf{n} \qquad (\mathbf{r}_1 \times \mathbf{F} \cdot \mathbf{n})\mathbf{n}$$

$$(-\mathbf{r}_4 \times \mathbf{F} \cdot \mathbf{n})\mathbf{n} \qquad [\mathbf{F} \times (\mathbf{r}_2 + \mathbf{r}_3) \cdot \mathbf{n}]\mathbf{n}$$

$$(r_5 F \cos \beta \cos \alpha)\mathbf{n}$$

2/101 A 50-kN force with direction cosines proportional to 2, 6, 9 passes through a point *P* whose *x*-, *y*-, *z*-coordinates in millimeters are 3, 2, −5. Determine the moment **M** of the force about a point whose coordinates in millimeters are 2, 2, −3.

Ans. $\mathbf{M} = \frac{50}{11}(12\mathbf{i} - 13\mathbf{j} + 6\mathbf{k})$ N·m

2/102 A 100-lb force passes through the two points *A* and *B* in the sense from *A* to *B*. The *x*-, *y*-, *z*-coordinates of *A* and *B* expressed in inches are −3, −1, 4 and 3, 4, 5, respectively. Calculate the moment **M** of the force about point *C* whose coordinates in inches are 2, −2, 1.

2/103 The special-purpose milling cutter is subjected to the force of 1200 N and a couple of 240 N·m as shown. Determine the moment of this system about point *O*. *Ans.* $\mathbf{M}_O = -260\mathbf{i} + 328\mathbf{j} + 88\mathbf{k}$ N·m

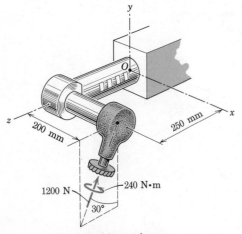

Problem 2/103

104 The rectangular steel plate is tilted about its horizontal edge *AB* and held in the position shown by a cable from corner *C* to point *O* on the ground. The tension in the cable is 20 kN. Calculate the magnitude of the moment of the cable tension about edge *AB*. Work the problem using the approach most appropriate for each of the two ways in which the same problem is specified. (Observe that the choice of method of solution often depends on how the data are given.)

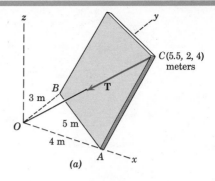

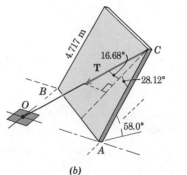

(a)

(b)

Problem 2/104

105 The figure for Prob. 2/84 is repeated here. The tension in the cable equals 200 lb. For the position $x = 5.33$ ft for pulley *C*, calculate the magnitude of the moment of the tension about the axis *AB*.

Ans. M = 141.4 lb-ft

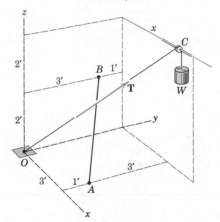

Problem 2/105

106 The figure of Prob. 2/86 is shown again here. If the magnitude of the moment of **F** about line *CD* is 50 N·m, determine the magnitude of **F**.

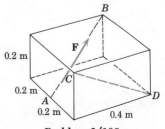

Problem 2/106

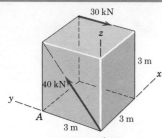

Problem 2/107

2/107 Replace the two forces that act on the 3-m cube by an equivalent single force **F** at *A* and a couple **M**.
Ans. $\mathbf{F} = -1.72\mathbf{j} + 28.28\mathbf{k}$ kN
$\mathbf{M} = 5.15\mathbf{i} - 90\mathbf{k}$ kN·m

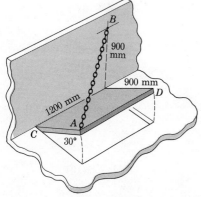

Problem 2/108

▶**2/108** The access door of Prob. 2/87 is shown again here. If the tension in the chain *AB* is 100 N, determine the magnitude *M* of its moment about the hinge axis.   *Ans.* $M = 46.8$ N·m

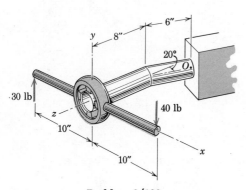

Problem 2/109

▶**2/109** The threading die is screwed onto the end of the fixed pipe which is bent through an angle of 20°. Replace the two forces by an equivalent force at *O* and a couple **M**. Find **M** and calculate the magnitude *M'* of the moment which tends to screw the pipe into the fixed block about its angled axis through *O*.
Ans. $\mathbf{M} = 136\mathbf{i} - 679\mathbf{k}$ lb-in., $M' = 685$ lb-in.

2/9 RESULTANTS. In Art. 2/6 we defined the resultant as the simplest force combination that can replace a given system of forces without altering the external effect on the rigid body upon which the forces act. We found the magnitude and direction of the resultant force for the two-dimensional force system by a vector summation of forces, Eq. 2/4, and we located the line of action of the resultant force by applying the principle of moments, Eq. 2/5. These same principles may be extended to three dimensions.

In the previous article we showed that a force could be moved to a parallel position by adding a corresponding couple. Thus for the system of forces $\mathbf{F}_1, \mathbf{F}_2, \mathbf{F}_3 \ldots$ acting on a rigid body in Fig. 2/25a, we may move each of them in turn to the arbitrary point O provided we also introduce a couple for each force transferred. Thus, for example, we may move force $\mathbf{F}_1$ to O provided we introduce the couple $\mathbf{M}_1 = \mathbf{r}_1 \times \mathbf{F}_1$ where $\mathbf{r}_1$ is a vector from O to any point on the line of action of $\mathbf{F}_1$. When all forces are shifted to O in this manner, we have a system of concurrent forces at O and a system of couple vectors, as represented in the b-part of the figure. The concurrent forces may then be added vectorially to produce a resultant force $\mathbf{R}$, and the couples may also be added to produce a resultant couple $\mathbf{M}$, Fig. 2/25c. The general force system, then, is reduced to

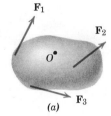
(a)

$$\boxed{\begin{aligned} \mathbf{R} &= \mathbf{F}_1 + \mathbf{F}_2 + \mathbf{F}_3 + \cdots = \Sigma\mathbf{F} \\ \mathbf{M} &= \mathbf{M}_1 + \mathbf{M}_2 + \mathbf{M}_3 + \cdots = \Sigma(\mathbf{r} \times \mathbf{F}) \end{aligned}} \qquad (2/15)$$

The couple vectors are shown through point O, but since they are free vectors, they may be represented in any parallel positions. The magnitudes of the resultants and their components are

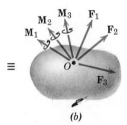
(b)

$$\boxed{\begin{aligned} R_x &= \Sigma F_x \qquad R_y = \Sigma F_y \qquad R_z = \Sigma F_z \\ R &= \sqrt{(\Sigma F_x)^2 + (\Sigma F_y)^2 + (\Sigma F_z)^2} \\ M_x &= \Sigma(\mathbf{r} \times \mathbf{F})_x \qquad M_y = \Sigma(\mathbf{r} \times \mathbf{F})_y \qquad M_z = \Sigma(\mathbf{r} \times \mathbf{F})_z \\ M &= \sqrt{M_x^2 + M_y^2 + M_z^2} \end{aligned}} \qquad (2/16)$$

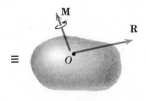
(c)

Figure 2/25

The point O that is selected as the point of concurrency for the forces is arbitrary, and the magnitude and direction of $\mathbf{M}$ will depend on the particular point O selected. The magnitude and direction of $\mathbf{R}$, however, are the same no matter which point is selected. In general any system of forces may be replaced by its resultant force $\mathbf{R}$ and the resultant couple $\mathbf{M}$. In dynamics we usually select the mass center as the reference point, and the change in the linear motion of the body is determined by the resultant force and the change in the angular motion of the body is determined by the resultant couple. In statics the complete equilibrium of a body is specified when the resultant force $\mathbf{R}$ is zero and the resultant couple $\mathbf{M}$ is zero. Thus the determination of resultants is essential in both statics and dynamics.

The resultants for several special force systems will now be noted.

Concurrent forces. When forces are concurrent at a point, only the first of Eqs. 2/15 need to be used since there are no moments about the point of concurrency.

Parallel forces. For a system of parallel forces not all in the same plane the magnitude of the parallel resultant force is simply the algebraic sum of the given forces, and the position of its line of action is easily obtained by applying the two components of the moment principle about axes in a plane normal to the forces.

Coplanar forces. Article 2/6 was devoted to this force system.

Wrench resultant. When the resultant couple vector **M** is parallel to the resultant force **R**, as shown in Fig. 2/26, the resultant is said to be a *wrench*. By definition a wrench is positive if the couple and force vectors point in the same direction and negative if they point in opposite directions. A common example of a positive wrench is found with the application of a screwdriver.

Any general force system may be represented by a wrench applied along a unique line of action. This reduction is illustrated in Fig. 2/27, where the *a*-part of the figure represents for the general force system the resultant force **R** acting at some point *O* and the corresponding resultant couple **M**. Although **M** is a free vector, for convenience we represent it through *O*. In the *b*-part of the figure, **M** is resolved into components $\mathbf{M}_1$ along the direction of **R** and $\mathbf{M}_2$ normal to **R**. In the *c*-part of the figure the couple $\mathbf{M}_2$ is replaced by its equivalent of two forces **R** and −**R** separated a distance $d = M_2/R$ with −**R** applied at *O* to cancel the original **R**. This step leaves the resultant **R**, which acts along a new and unique line of action, and the parallel couple $\mathbf{M}_1$, which is a free vector, as shown in the *d*-part of the figure. Thus the resultants of the original general force system have been transformed into a wrench (positive in this illustration) with its unique axis defined by the new position of **R**. We see from Fig. 2/27 that the axis of the wrench resultant lies in a plane through *O* normal to the plane defined by **R** and **M**. The wrench is the simplest form in which the resultant of a general force system may be expressed. This form of the resultant, however, has limited application, since it is usually more convenient to use as the reference point some point *O* such as the mass center of the body or other convenient origin of coordinates not on the wrench axis.

Positive wrench   Negative wrench

Figure 2/26

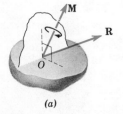
(a)

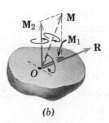
(b)

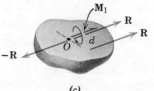
(c)

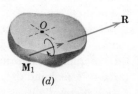
(d)

Figure 2/27

Sample Problem 2/11

Replace the two forces and the negative wrench by a single force **R** applied at A and the corresponding couple **M**.

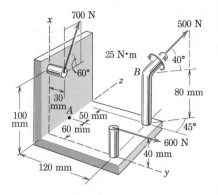

Solution. The resultant force has the components

$[R_x = \Sigma F_x]$ $R_x = 500 \sin 40° + 700 \sin 60° = 928$ N

$[R_y = \Sigma F_y]$ $R_y = 600 + 500 \cos 40° \cos 45° = 871$ N

$[R_z = \Sigma F_z]$ $R_z = 700 \cos 60° + 500 \cos 40° \sin 45° = 621$ N

Thus $\mathbf{R} = 928\mathbf{i} + 871\mathbf{j} + 621\mathbf{k}$ N

and $R = \sqrt{(928)^2 + (871)^2 + (621)^2} = 1416$ N   *Ans.*

The couple to be added as a result of moving the 500-N force is

$[\mathbf{M} = \mathbf{r} \times \mathbf{F}]$ $\mathbf{M}_{500} = (0.08\mathbf{i} + 0.12\mathbf{j} + 0.05\mathbf{k}) \times 500(\mathbf{i} \sin 40°$
$$+ \mathbf{j} \cos 40° \cos 45° + \mathbf{k} \cos 40° \sin 45°)$$

where **r** is the vector from A to B.

The term-by-term, or determinant, expansion gives

$$\mathbf{M}_{500} = 18.95\mathbf{i} - 5.59\mathbf{j} - 16.90\mathbf{k} \text{ N·m}$$

The moment of the 600-N force about A is written by inspection of its x- and z-components, which give

$$\mathbf{M}_{600} = (600)(0.060)\mathbf{i} + (600)(0.040)\mathbf{k}$$
$$= 36.0\mathbf{i} + 24.0\mathbf{k} \text{ N·m}$$

The moment of the 700-N force about A is easily obtained from the moments of the x- and z-components of the force. The result becomes

$$\mathbf{M}_{700} = (700 \cos 60°)(0.030)\mathbf{i} - [(700 \sin 60°)(0.060)$$
$$+ (700 \cos 60°)(0.100)]\mathbf{j} - (700 \sin 60°)(0.030)\mathbf{k}$$
$$= 10.5\mathbf{i} - 71.4\mathbf{j} - 18.19\mathbf{k} \text{ N·m}$$

Also, the couple of the given wrench may be written

$$\mathbf{M}' = 25.0(-\mathbf{i} \sin 40° - \mathbf{j} \cos 40° \cos 45° - \mathbf{k} \cos 40° \sin 45°)$$
$$= -16.07\mathbf{i} - 13.54\mathbf{j} - 13.54\mathbf{k} \text{ N·m}$$

Therefore, the resultant couple upon adding together the **i**-, **j**-, and **k**-terms of the four **M**'s is

$$\mathbf{M} = 49.4\mathbf{i} - 90.5\mathbf{j} - 24.6\mathbf{k} \text{ N·m}$$

and $M = \sqrt{(49.4)^2 + (90.5)^2 + (24.6)^2} = 106.0$ N·m   *Ans.*

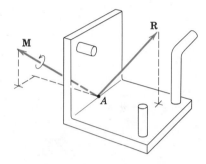

① *Suggestion:* Check the cross-product results by evaluating the moments about A of the components of the 500-N force directly from the sketch.

② For the 600-N and 700-N forces it is easier to obtain the components of their moments about the coordinate directions through A by inspection of the figure than it is to set up the cross-product relations.

③ The 25-N·m couple vector of the *wrench* points in the direction opposite to that of the 500-N force, and we must resolve it into its x-, y-, and z-components to be added to the other couple-vector components.

④ Although the resultant couple vector **M** in the sketch of the resultants is shown through A, we recognize that a couple vector is a free vector and therefore has no specified line of action.

Sample Problem 2/12

Determine the wrench resultant of the three forces acting on the bracket. Calculate the coordinates of the point P in the x-y plane through which the resultant force of the wrench acts. Also find the magnitude of the couple $\mathbf{M}$ of the wrench.

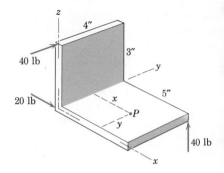

Solution. The direction cosines of the couple $\mathbf{M}$ of the wrench must ① be the same as those of the resultant force $\mathbf{R}$, assuming that the wrench is positive. The resultant force is

$$\mathbf{R} = 20\mathbf{i} + 40\mathbf{j} + 40\mathbf{k} \text{ lb} \qquad R = \sqrt{(20)^2 + (40)^2 + (40)^2} = 60 \text{ lb}$$

and its direction cosines are

$$\cos \theta_x = 20/60 = 1/3, \ \cos \theta_y = 40/60 = 2/3, \ \cos \theta_z = 40/60 = 2/3$$

The moment of the wrench couple must equal the sum of the moments of the given forces about point P through which $\mathbf{R}$ passes. The moments about P of the three forces are

$$(\mathbf{M})_{R_x} = 20y\mathbf{k} \text{ lb-in.}$$

$$(\mathbf{M})_{R_y} = -40(3)\mathbf{i} - 40x\mathbf{k} \text{ lb-in.}$$

$$(\mathbf{M})_{R_z} = 40(4 - y)\mathbf{i} - 40(5 - x)\mathbf{j} \text{ lb-in.}$$

and the total moment is

$$\mathbf{M} = (40 - 40y)\mathbf{i} + (-200 + 40x)\mathbf{j} + (-40x + 20y)\mathbf{k} \text{ lb-in.}$$

The direction cosines of $\mathbf{M}$ are

$$\cos \theta_x = (40 - 40y)/M$$

$$\cos \theta_y = (-200 + 40x)/M$$

$$\cos \theta_z = (-40x + 20y)/M$$

where M is the magnitude of $\mathbf{M}$. Equating the direction cosines of $\mathbf{R}$ and $\mathbf{M}$ gives

$$40 - 40y = \frac{M}{3}$$

$$-200 + 40x = \frac{2M}{3}$$

$$-40x + 20y = \frac{2M}{3}$$

Solution of the three equations gives

$$M = -120 \text{ lb-in.}, \quad x = 3 \text{ in.}, \quad y = 2 \text{ in.} \qquad \textit{Ans.}$$

We see that M turned out to be negative, which means that the couple vector is pointing in the direction opposite to $\mathbf{R}$, which makes the wrench negative.

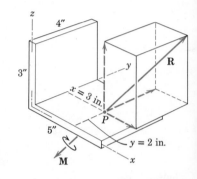

① We shall assume initially that the wrench is positive. If $\mathbf{M}$ turns out to be negative, then the direction of the couple vector is opposite to that of the resultant force.

PROBLEMS

110 The concrete slab supports the six vertical loads shown. Determine the resultant of these forces and the x- and y-coordinates of a point through which it acts.   *Ans.* $x = 7.30$ m, $y = 15.83$ m

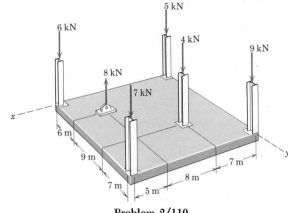

Problem 2/110

111 Determine the x- and y-coordinates of a point through which the resultant of the parallel forces passes.

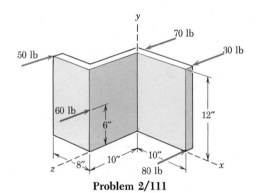

Problem 2/111

112 Determine the angle θ so that the net downward force applied to the fixed eyebolt is 750 lb. Determine the corresponding magnitude of the resultant of the three forces and the direction cosines of the resultant.

Ans. $\theta = 33.3°$, $R = 1021$ lb, $\cos \theta_x = 0.525$,
$\cos \theta_y = 0.430$, $\cos \theta_z = -0.734$

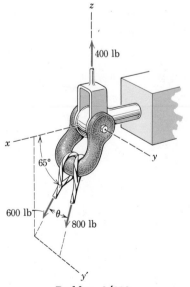

Problem 2/112

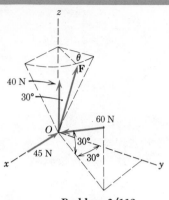

Problem 2/113

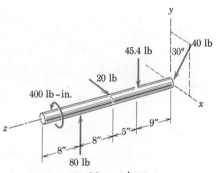

Problem 2/115

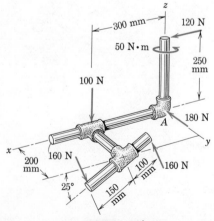

Problem 2/116

2/113 The four forces are concurrent at the origin O of coordinates. If the x-component of their resultant **R** is -50 N and the z-component is 100 N, determine F, θ, and R.

2/114 Use Fig. 2/25c and show that the most general force system may be represented by two nonintersecting forces.

2/115 Determine the resultant of the four forces and couple acting on the shaft.
    Ans. Couple **M** $= -1351\mathbf{i} + 280\mathbf{j} + 400\mathbf{k}$ lb-in.

2/116 Represent the resultant of the force system acting on the pipe assembly by a single force **R** at A and a couple **M**.

117 The four forces are acting along the edges of the 0.8-m cube as shown. Represent the resultant of these forces by a force **R** through point *A* and a couple **M**.

$$Ans. \ \mathbf{R} = 200(\mathbf{j} + \mathbf{k}) \ N$$
$$\mathbf{M} = 560\mathbf{i} - 320\mathbf{j} + 400\mathbf{k} \ N \cdot m$$

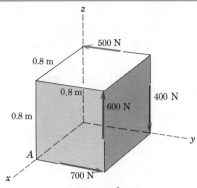

Problem 2/117

118 Replace the two forces of equal magnitude *F* by a single force **R** passing through point *A* and a couple **M**.

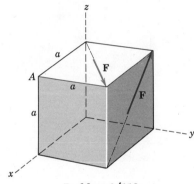

Problem 2/118

119 The combined action of the three forces on the base at *O* may be obtained by establishing their resultant through *O*. Determine the magnitudes of **R** and the accompanying couple **M**.

$$Ans. \ R = 1093 \ lb, \ M = 9734 \ lb\text{-}ft$$

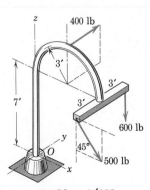

Problem 2/119

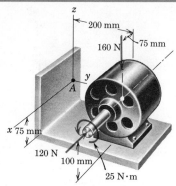

Problem 2/120

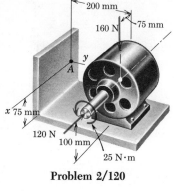

Problem 2/121

2/120 The motor mounted on the bracket is acted upon by its 160-N weight, and its shaft resists the 120-N thrust and 25-N·m couple applied to it. Determine the resultant of the force system shown in terms of a force **R** at *A* and a couple **M**.

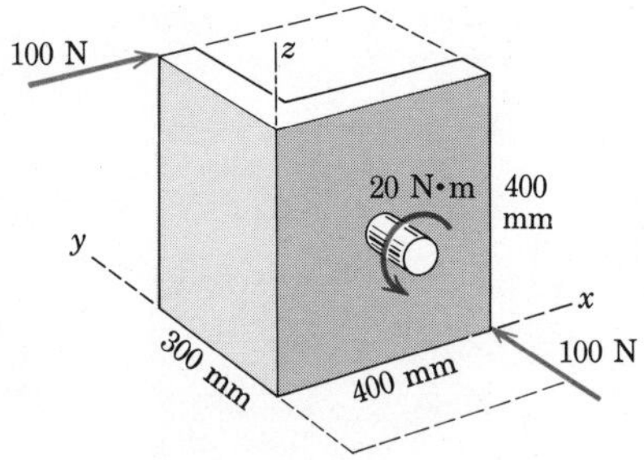

▶**2/121** The resultant of the two forces and couple may be represented by a wrench. Determine the vector expression for the moment **M** of the wrench and find the coordinates of the point *P* in the *x-z* plane through which the resultant force of the wrench passes.
Ans. **M** $= 10(\mathbf{i} + \mathbf{j})$ N·m, $x = 100$ mm, $z = 100$ mm

▶**2/122** Replace the force system acting on the pipe assembly of Prob. 2/116 by a wrench. Find the magnitude of the moment **M** of the wrench and the coordinates of the point *P* in the *x-y* plane through which the resultant force of the wrench passes.
Ans. $M = 95.6$ N·m, $x = 271$ mm, $y = -486$ mm

▶**2/123** The resultant of a general force system may be expressed as a wrench along a unique line of action. For the force system of Prob. 2/120 determine the coordinates of the point *P* which is the intersection of the line of action of the wrench with the *x-y* plane. *Ans.* $x = 56.2$ mm, $y = 100$ mm

EQUILIBRIUM

3

3/1 **INTRODUCTION.** The subject of statics deals primarily with the description of the conditions of force that are both necessary and sufficient to maintain the state of equilibrium of engineering structures. This chapter on equilibrium, therefore, constitutes the most central part of statics and should be mastered thoroughly. We shall make continuous use of the concepts developed in Chapter 2 involving forces, moments, couples, and resultants as we apply the principles of equilibrium. The procedures which we shall develop in Chapter 3 constitute a comprehensive introduction to the approach used in the solution of countless problems in mechanics and in other engineering areas as well. The approach which we shall develop is basic to the successful mastery of statics, and the student is urged to read and study the following articles with special effort and attention to detail.

A body is in equilibrium when the resultant of *all* forces acting on it is zero. Thus the resultant force $\mathbf{R}$ and the resultant couple $\mathbf{M}$ are both zero, and we have the equilibrium equations

$$\mathbf{R} = \Sigma\mathbf{F} = 0 \qquad \mathbf{M} = \Sigma\mathbf{M} = 0 \qquad (3/1)$$

These requirements are both necessary and sufficient conditions for equilibrium.

We shall follow the arrangement used in Chapter 2 and discuss in Section A the equilibrium of two-dimensional force systems and in Section B the equilibrium of three-dimensional force systems.

SECTION A. EQUILIBRIUM IN TWO DIMENSIONS

3/2 **MECHANICAL SYSTEM ISOLATION.** Before we apply Eqs. 3/1 it is essential that we define unambiguously the particular body or mechanical system to be analyzed and represent clearly and completely *all* forces which act *on* the body. The omission of a force or the inclusion of a force which does not act on the body in question will give erroneous results.

A mechanical system is defined as a body or group of bodies that can be isolated from all other bodies. Such a system may be a single body or a combination of connected bodies. The bodies may be rigid or nonrigid. The system may also be a defined fluid mass, liquid or gas, or the system may be a combination of fluids and solids. In statics we direct our attention primarily to a description of the forces that act on rigid bodies at rest, although consideration is also given to the statics of fluids. Once we reach a decision about which body or combination of bodies is to be analyzed, then this body or combination treated as a single body is *isolated* from all surrounding bodies. This isolation is accomplished by means of the *free-body diagram*, which is a diagrammatic representation of the isolated body or combination of bodies considered as a single body, showing all forces applied to it by other bodies that are imagined to be removed. Only after such a diagram has been carefully drawn should the equilibrium equations be written. Because of its critical importance we emphasize here that

> *the free-body diagram is the most important single step in the solution of problems in mechanics.*

Before we attempt to draw free-body diagrams, the mechanical characteristics of force application must be recognized. In Art. 2/2 the basic characteristics of force were described with primary attention focused on the vector properties of force. We noted that forces are applied both by direct physical contact and by remote action and that forces may be either internal or external to the body under consideration. We further observed that the application of external forces is accompanied by reactive forces and that both applied and reactive forces may be either concentrated or distributed. Additionally the principle of transmissibility was introduced which permits the treatment of force as a sliding vector insofar as its external effects on a rigid body are concerned. We will now use these characteristics of force in developing the analytical model of an isolated mechanical system to which the equations of equilibrium will then be applied.

Figure 3/1 shows the common types of force application on mechanical systems for analysis in two dimensions. In each example the force exerted *on* the body to be isolated *by* the body to be removed is indicated. Newton's third law, which notes the existence of an equal and opposite reaction to every action, must be carefully observed.

In example 1 the action of a flexible cable, belt, rope, or chain on the body to which it is attached is depicted. Because of its flexibility a rope or cable is unable to offer any resistance to bending, shear, or compression and therefore exerts a tension force in a direction tangent to the cable at its point of attachment. The force exerted *by* the cable *on* the body to which it is attached is always *away* from the body. When the tension T is large compared with the weight of the

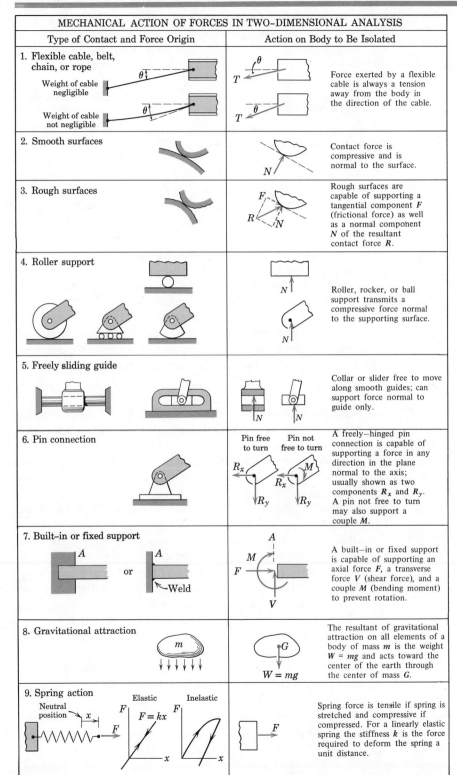

MECHANICAL ACTION OF FORCES IN TWO-DIMENSIONAL ANALYSIS

Type of Contact and Force Origin	Action on Body to Be Isolated
1. Flexible cable, belt, chain, or rope — Weight of cable negligible; Weight of cable not negligible	Force exerted by a flexible cable is always a tension away from the body in the direction of the cable.
2. Smooth surfaces	Contact force is compressive and is normal to the surface.
3. Rough surfaces	Rough surfaces are capable of supporting a tangential component F (frictional force) as well as a normal component N of the resultant contact force R.
4. Roller support	Roller, rocker, or ball support transmits a compressive force normal to the supporting surface.
5. Freely sliding guide	Collar or slider free to move along smooth guides; can support force normal to guide only.
6. Pin connection	A freely-hinged pin connection is capable of supporting a force in any direction in the plane normal to the axis; usually shown as two components R_x and R_y. A pin not free to turn may also support a couple M.
7. Built-in or fixed support	A built-in or fixed support is capable of supporting an axial force F, a transverse force V (shear force), and a couple M (bending moment) to prevent rotation.
8. Gravitational attraction	The resultant of gravitational attraction on all elements of a body of mass m is the weight $W = mg$ and acts toward the center of the earth through the center of mass G.
9. Spring action	Spring force is tensile if spring is stretched and compressive if compressed. For a linearly elastic spring the stiffness k is the force required to deform the spring a unit distance.

Figure 3/1

cable, we may assume that the cable forms a straight line. When the cable weight is not negligible compared with its tension, the sag of the cable becomes important, and the tension in the cable changes direction and magnitude along its length. At its attachment the cable exerts a force tangent to itself.

When the smooth surfaces of two bodies are in contact, as in example 2, the force exerted by one on the other is *normal* to the tangency of the surfaces and is compressive. Although no actual surfaces are perfectly smooth, we are justified in making this assumption for practical purposes in many instances.

When mating surfaces of contacting bodies are rough, example 3, the force of contact may not necessarily be normal to the tangent to the surfaces but may be resolved into a *tangential* or frictional component F and a *normal component N.*

Example 4 illustrates a number of forms of mechanical support which effectively eliminate tangential friction forces, and here the net reaction is normal to the supporting surface.

Example 5 shows the action of a smooth guide on the body it supports. Resistance parallel to the guide is absent.

Example 6 illustrates the action of a pin connection. Such a connection is able to support force in any direction normal to the axis of the pin. We usually represent this action in terms of two rectangular components. If the joint is free to turn about the pin, only the force R can be supported. If the joint is not free to turn, a resisting couple M may also be supported.

Example 7 shows the resultants of the rather complex distribution of force over the cross section of a slender bar or beam at a built-in or fixed support.

One of the most common forces is that due to gravitational attraction, example 8. This force affects all elements of mass of a body and is, therefore, distributed throughout it. The resultant of the gravitational forces on all elements is the weight $W = mg$ of the body, which passes through the center of mass G and is directed toward the center of the earth for earthbound structures. The position of G is frequently obvious from the geometry of the body, particularly where conditions of symmetry exist. When the position is not readily apparent, the location of G must be calculated or determined by experiment. Similar remarks apply to the remote action of magnetic and electric forces. These forces of remote action have the same overall effect on a rigid body as forces of equal magnitude and direction applied by direct external contact.

Example 9 illustrates the action of a linear elastic spring and of a nonlinear spring. The force exerted by a linear spring, in tension or compression, is given by $F = kx$, where k is the stiffness of the spring and x is its deformation measured from the neutral or undeformed position. The linearity of the force-deformation relation tells us that the magnitudes of the tensile and compressive forces in a given spring are equal if the magnitudes of the respective tension and compression deformations are equal.

The student is urged to study these nine conditions and to identify them in the problem work so that the correct free-body diagrams may be drawn. The representations in Fig. 3/1 are *not* free-body diagrams but are merely elements in the construction of free-body diagrams.

The full procedure for drawing a free-body diagram which accomplishes the isolation of the body or system under consideration will now be described.

Construction of free-body diagrams. The following steps are involved.

Step 1. A clear decision is made concerning which body or combination of bodies is to be isolated. The body chosen will usually involve one or more of the desired unknown quantities.

Step 2. The body or combination chosen is next isolated by a diagram that represents its *complete external boundary*.

This boundary defines the isolation of the body from *all* other contacting or attracting bodies, which are considered removed. This step is often the most crucial of all. We should always be certain that we have *completely isolated* the body before proceeding with the next step.

Step 3. All forces that act *on* the isolated body as applied *by* the removed contacting and attracting bodies are next represented in their proper positions on the diagram of the isolated body. A systematic traverse of the entire boundary will disclose all contact forces. Weights, where appreciable, must be included. Known forces should be represented by vector arrows with their proper magnitude, direction, and sense indicated. Unknown forces should be represented by vector arrows with the unknown magnitude or direction indicated by symbol. If the sense of the vector is also unknown, it may be arbitrarily assumed. The calculations will reveal a positive quantity if the correct sense was assumed and a negative magnitude if the incorrect sense was assumed. It is necessary to be *consistent* with the assigned characteristics of unknown forces throughout all of the calculations.

Step 4. The choice of coordinate axes should be indicated directly on the diagram. Pertinent dimensions may also be represented for convenience. Note, however, that the free-body diagram serves the purpose of focusing accurate attention on the action of the external forces, and therefore the diagram should not be cluttered with excessive extraneous information. Force arrows should be clearly distinguished from any other arrows which may appear so that confusion will not result. For this purpose a colored pencil may be used.

When the foregoing four steps are completed, a correct free-body diagram will result, and the way will be clear for a straightforward and successful application of the governing equations, both in statics and in dynamics.

Many students are often tempted to omit from the free-body diagram certain forces that may not appear at first glance to be needed in the calculations. When we yield to this temptation we invite serious error. It is only through *complete* isolation and a systematic representation of *all* external forces that a reliable accounting of the effects of all applied and reactive forces can be made. Very often a force that at first glance may not appear to influence a desired result does indeed have an influence. Hence the only safe procedure is to make certain that all forces whose magnitudes are not negligible appear on the free-body diagram.

The free-body diagram has been explained in some detail because of its great importance in mechanics. The free-body method ensures an accurate definition of a mechanical system and focuses attention on the exact meaning and application of the force laws of statics and dynamics. Indeed the free-body method is so important that students are strongly urged to reread this section several times in conjunction with their study of the sample free-body diagrams shown in Fig. 3/2 and the sample problems which appear at the end of the next article.

Figure 3/2 gives four examples of mechanisms and structures together with their correct free-body diagrams. Dimensions and magnitudes are omitted for clarity. In each case we treat the entire system as a single body, so that the internal forces are not shown. The characteristics of the various types of contact forces illustrated in Fig. 3/1 are included in the four examples as they apply.

In example 1 the truss is composed of structural elements that, taken all together, constitute a rigid framework. Thus we may remove the entire truss from its supporting foundation and treat it as a single rigid body. In addition to the applied external load P, the free-body diagram must include the reactions on the truss at A and B. The rocker at B can support a vertical force only, and this force is transmitted to the structure at B (example 4 of Fig. 3/1). The pin connection at A (example 6 of Fig. 3/1) is capable of supplying both a horizontal and a vertical component of force to the truss. In this relatively simple example it is clear that the vertical component A_y must be directed down to prevent the truss from rotating clockwise about B. Also, the horizontal component A_x will be to the left to keep the truss from moving to the right under the influence of the horizontal component of P. If the total weight of the truss members is appreciable compared with P and the forces at A and B, then the weights of the members must be included on the free-body diagram as external forces.

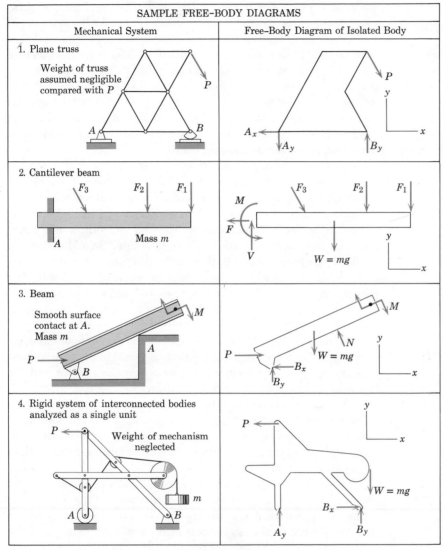

SAMPLE FREE–BODY DIAGRAMS

Mechanical System	Free–Body Diagram of Isolated Body
1. Plane truss Weight of truss assumed negligible compared with P	
2. Cantilever beam Mass m	$W = mg$
3. Beam Smooth surface contact at A. Mass m	$W = mg$
4. Rigid system of interconnected bodies analyzed as a single unit Weight of mechanism neglected	$W = mg$

Figure 3/2

In example 2 the cantilever beam is secured to the wall and subjected to the three applied loads. When we isolate that part of the beam to the right of the section at A, we must include the reactive forces applied to the beam by the wall. The resultants of these reactive forces are shown acting on the section of the beam (example 7 of Fig. 3/1). A vertical force V to counteract the excess of downward applied force is shown, and a tension T to balance the excess of applied force to the right must also be included. Then to prevent the beam from rotating about A a counterclockwise couple M is also required. The weight mg of the beam must also be represented through the mass center (example 8 of Fig. 3/1). The free-body diagram is now complete and shows the beam in equilibrium under the action of six forces and one couple.

In example 3 the weight $W = mg$ is shown acting through the center of mass of the beam, which is assumed known (example 8 of Fig. 3/1). The force exerted by the corner A on the beam is normal to the smooth surface of the beam (example 2 of Fig. 3/1). If the contacting surfaces at the corner were not smooth, a tangential frictional component of force could be developed. In addition to the applied force P and couple M, there is the pin connection at B, which exerts both an x- and a y-component of force on the beam. The positive senses of these components are assigned arbitrarily.

In example 4 the free-body diagram of the entire isolated mechanism discloses three unknown quantities for equilibrium with the given loads mg and P. Any one of many internal configurations for securing the cable leading from the mass m would be possible without affecting the external response of the mechanism as a whole, and this fact is brought out by the free-body diagram.

The positive senses of B_x and B_y in example 3 and B_y in example 4 are assumed on the free-body diagrams, and the correctness of the assumptions would be proved or disproved according to whether the algebraic signs of the terms were plus or minus when the calculations were carried out in the actual problems.

The isolation of the mechanical system under consideration will be recognized as a crucial step in the formulation of the mathematical model. The student is again urged to devote special attention to this step. Before direct use is made of the free-body diagram in the application of the principles of force equilibrium in the next article, some initial practice with the drawing of free-body diagrams is helpful. The problems that follow are designed to provide this practice.

PROBLEMS

3/1 In each of the five following examples, the body to be isolated is shown in the left-hand diagram, and an *incomplete* free-body diagram (FBD) of the isolated body is shown on the right. Add whatever forces are necessary in each case to form a complete free-body diagram. The weights of the bodies are negligible unless otherwise indicated. Dimensions and numerical values are omitted for simplicity.

	Body	Incomplete *FBD*
1. Bell crank supporting mass m with pin support at A	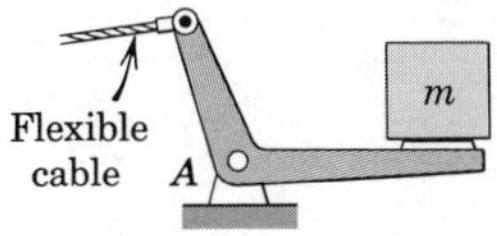	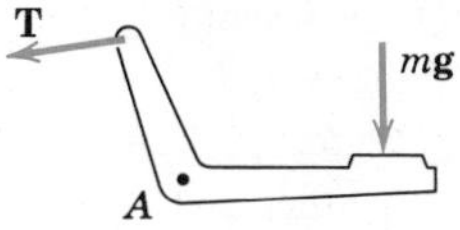
2. Control lever applying torque to shaft at O	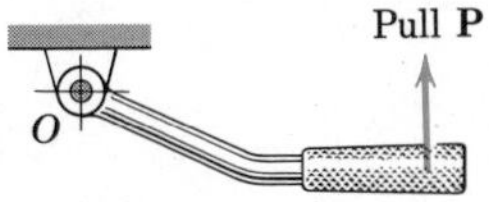	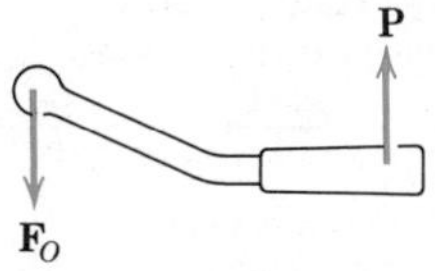
3. Boom OA, of negligible mass compared with mass m. Boom hinged at O and supported by hoisting cable at B.	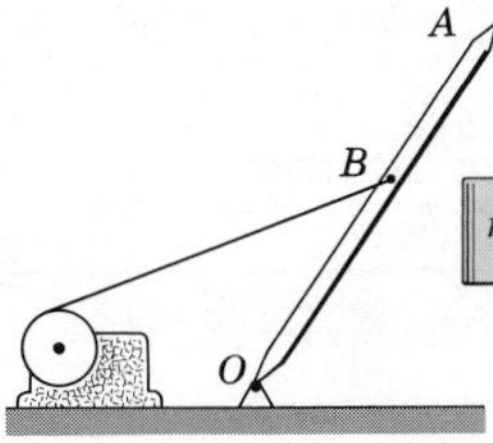	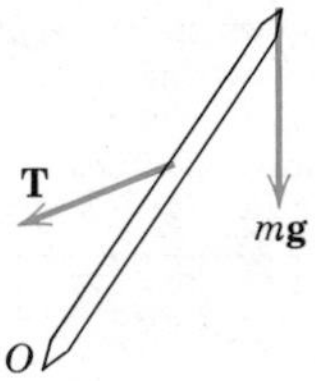
4. Uniform crate of mass m leaning against smooth vertical wall and supported on a rough horizontal surface	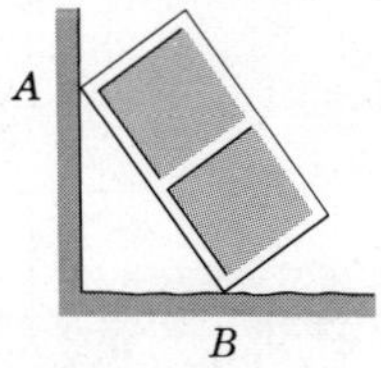	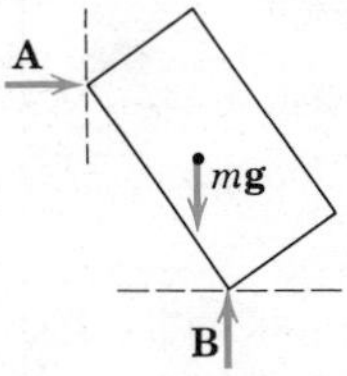
5. Loaded bracket supported by pin connection at A and fixed pin in smooth slot at B	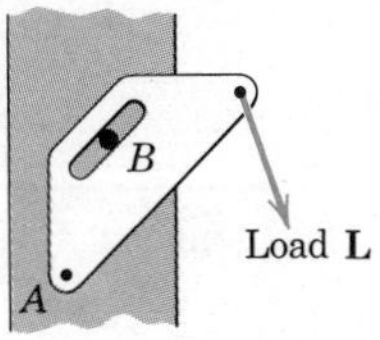	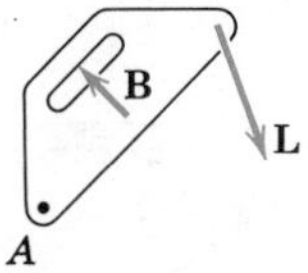

Problem 3/1

3/2 In each of the five following examples, the body to be isolated is shown in the left-hand diagram, and either a *wrong* or an *incomplete* free-body diagram (FBD) is shown on the right. Make whatever changes or additions are necessary in each case to form a correct and complete free-body diagram. The weights of the bodies are negligible unless otherwise indicated. Dimensions and numerical values are omitted for simplicity.

	Body	Wrong or Incomplete *FBD*
1. Lawn roller of mass m being pushed up incline θ		
2. Pry bar lifting body A having smooth horizontal surface. Bar rests on hori–zontal rough surface	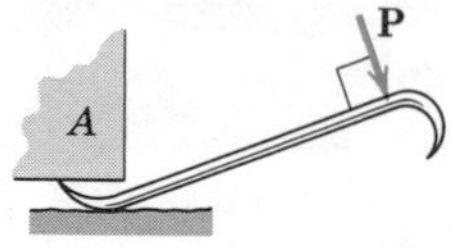	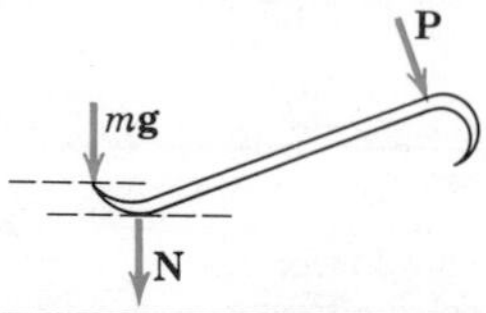
3. Uniform pole of mass m being hoisted into position by winch. Horizontal supporting surface notched to prevent slipping of pole	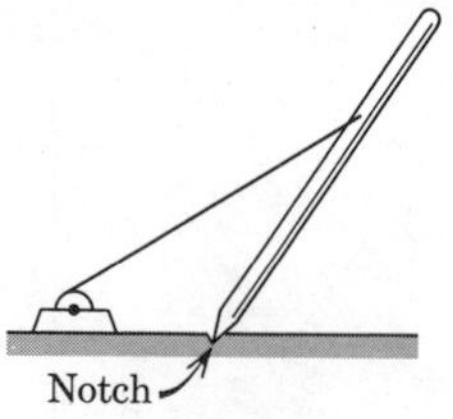	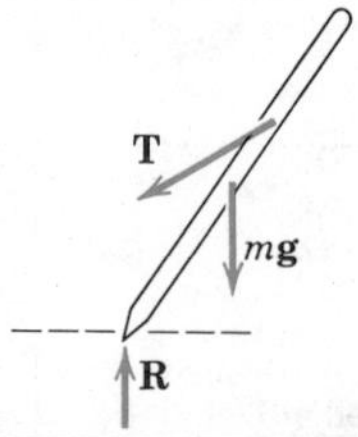
4. Supporting angle bracket for frame. Pin joints	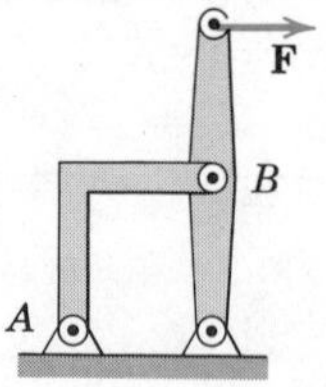	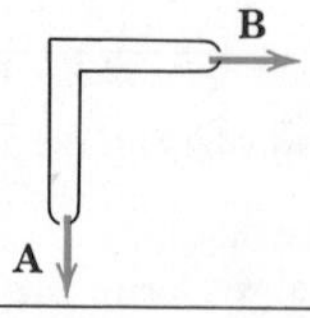
5. Bent rod welded to support at A and subjected to two forces and couple	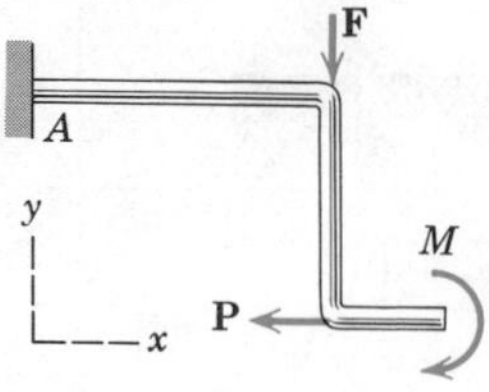	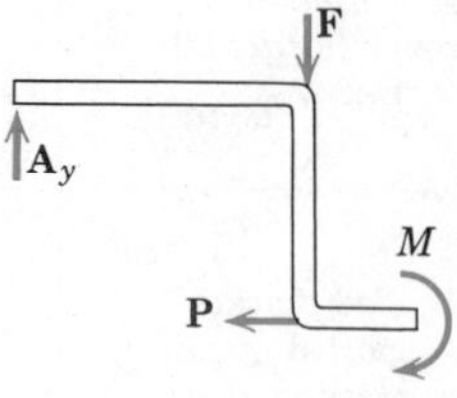

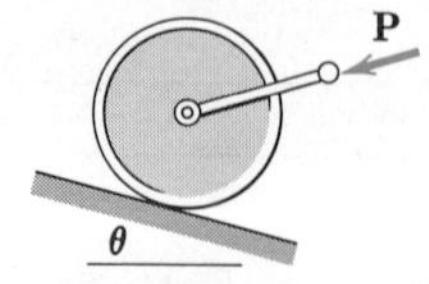

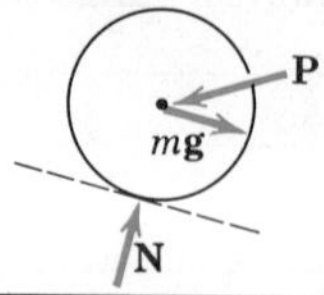

Problem 3/2

3/3 Draw a complete and correct free-body diagram of each of the bodies designated in the statements. The weights of the bodies are significant only if the mass is stated. All forces, known and unknown, should be labeled. (*Note:* The sense of some reaction components cannot always be determined without numerical calculation.)

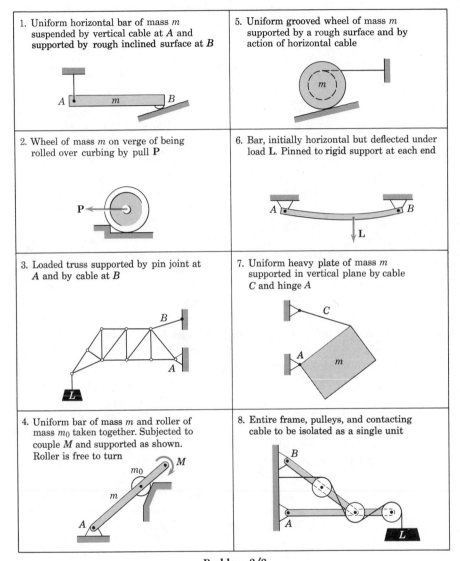

1. Uniform horizontal bar of mass m suspended by vertical cable at A and supported by rough inclined surface at B

2. Wheel of mass m on verge of being rolled over curbing by pull **P**

3. Loaded truss supported by pin joint at A and by cable at B

4. Uniform bar of mass m and roller of mass m_0 taken together. Subjected to couple M and supported as shown. Roller is free to turn

5. Uniform grooved wheel of mass m supported by a rough surface and by action of horizontal cable

6. Bar, initially horizontal but deflected under load **L**. Pinned to rigid support at each end

7. Uniform heavy plate of mass m supported in vertical plane by cable C and hinge A

8. Entire frame, pulleys, and contacting cable to be isolated as a single unit

Problem 3/3

3/3 EQUILIBRIUM CONDITIONS. In Art. 3/1 we defined equilibrium as the condition in which the resultant of all forces acting on a body is zero. Stated in another way, a body is in equilibrium if all forces and moments applied to it are in balance. These requirements are contained in the vector equations of equilibrium, Eqs. 3/1, which in two dimensions may be written in scalar form as

$$\Sigma F_x = 0 \qquad \Sigma F_y = 0 \qquad \Sigma M_O = 0 \qquad\qquad (3/2)$$

The third equation represents the zero sum of the moments of all forces about any point O on or off the body. Equations 3/2 are the necessary and sufficient conditions for complete equilibrium in two dimensions. They are necessary conditions because, if not satisfied, there can be no force or moment balance. They are sufficient since once satisfied there can be no unbalance, and equilibrium is assured.

The equations relating force and acceleration for rigid-body motion are developed in Vol. 2 *Dynamics* from Newton's second law of motion. These equations show that the acceleration of the mass center of a body is proportional to the resultant force $\Sigma\mathbf{F}$ acting on the body. Consequently, if a body moves with constant velocity (zero acceleration), the resultant force on it must be zero, and the body may be treated as in a state of equilibrium.

For complete equilibrium in two dimensions all three of Eqs. 3/2 must hold. However, these conditions are independent requirements, and one may hold without another. Take, for example, a body which slides along a horizontal surface with increasing velocity under the action of applied forces. The force-equilibrium equations will be satisfied in the vertical direction where the acceleration is zero but not in the horizontal direction. Also, a body, such as a flywheel, which rotates about its fixed mass center with increasing angular speed is not in rotational equilibrium, but the two force-equilibrium equations will be satisfied.

(*a*) *Categories of Equilibrium.* Applications of Eqs. 3/2 fall naturally into a number of categories that are easily identified. These categories of force systems acting on bodies in two-dimensional equilibrium are summarized in Fig. 3/3 and are explained further as follows:

Case 1, equilibrium of collinear forces, clearly requires only the one force equation in the direction of the forces (*x*-direction), since all other equations are automatically satisfied.

Case 2, equilibrium of forces that lie in a plane (*x*-*y* plane) and are concurrent at a point O, requires the two force equations only, since the moment sum about O, that is, about a *z*-axis through O, is necessarily zero.

Case 3, equilibrium of parallel forces in a plane, requires the one

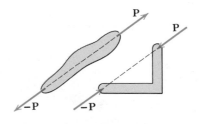

CATEGORIES OF EQUILIBRIUM IN TWO DIMENSIONS

Force System	Free–Body Diagram	Independent Equations
1. Collinear	F_3, F_2, F_1 (x)	$\Sigma F_x = 0$
2. Concurrent at a point	F_1, F_2, F_4, F_3, O (y, x)	$\Sigma F_x = 0$ $\Sigma F_y = 0$
3. Parallel	F_1, F_2, F_3, F_4 (y, x)	$\Sigma F_x = 0$ $\Sigma M_z = 0$
4. General	F_1, F_2, F_3, M, F_4 (y, x)	$\Sigma F_x = 0$ $\Sigma M_z = 0$ $\Sigma F_y = 0$

Figure 3/3

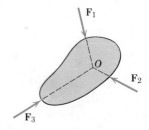

Two–force members

(a)

Figure 3/4

force equation in the direction of the forces (x-direction) and one moment equation about an axis (z-axis) normal to the plane of the forces.

Case 4, equilibrium of a general system of forces in a plane (x-y), requires the two force equations in the plane and one moment equation about an axis (z-axis) normal to the plane.

There are two equilibrium situations that occur frequently to which the student should be alerted. The first situation is the equilibrium of a body under the action of two forces only. Two examples are shown in Fig. 3/4, and we see that for such a *two-force member* the forces must be equal, opposite, and collinear. The shape of the member should not obscure this simple requirement. In the illustrations cited we consider the weights of the members to be negligible compared with the applied forces.

The second situation is the equilibrium of a body under the action of three forces, Fig. 3/5*a*. We see that the lines of action of the three forces must be *concurrent*. If they were not concurrent, then one of the forces would exert a resultant moment about the point of concurrency of the other two, which would violate the requirement of zero moment about every point. The only exception occurs when

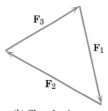

(a) Three–force member

(b) Closed polygon satisfies $\Sigma F = 0$

Figure 3/5

the three forces are parallel. In this case we may consider the point of concurrency to be at infinity. The principle of the concurrency of three forces in equilibrium is of considerable use in carrying out a graphical solution of the force equations. In this case the polygon of forces is drawn and made to close, as shown in Fig. 3/5*b*. Frequently, a body in equilibrium under the action of more than three forces may be reduced to a *three-force member* by a combination of two or more of the known forces.

(*b*) *Alternative Equilibrium Equations.* There are two additional ways in which we may express the general conditions for the equilibrium of forces in two dimensions. For the body shown in Fig. 3/6*a*, if $\Sigma M_A = 0$, then the resultant **R**, if it still exists, cannot be a

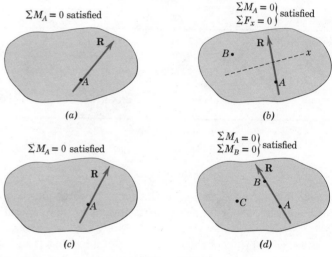

Figure 3/6

couple but must be a force **R** passing through *A*. If now the equation $\Sigma F_x = 0$ holds, where the *x*-direction is perfectly arbitrary, it follows from Fig. 3/6*b* that the resultant force **R**, if it still exists, not only must pass through *A*, but also must be perpendicular to the *x*-direction as shown. Now, if $\Sigma M_B = 0$, where *B* is any point such that the line *AB* is not perpendicular to the *x*-direction, we see that **R** must be zero, and hence the body is in equilibrium. Therefore an alternative set of equilibrium equations is

$$\Sigma F_x = 0 \qquad \Sigma M_A = 0 \qquad \Sigma M_B = 0$$

where the two points *A* and *B* must not lie on a line perpendicular to the *x*-direction.

A third formulation of the conditions of equilibrium may be

made for a coplanar force system. Again, if $\Sigma M_A = 0$ for any body such as that shown in Fig. 3/6c, the resultant, if any, must be a force **R** through A. In addition if $\Sigma M_B = 0$, the resultant, if one still exists, must pass through B as shown in Fig. 3/6d. Such a force cannot exist, however, if $\Sigma M_C = 0$, where C is not collinear with A and B. Hence we may write the equations of equilibrium as

$$\Sigma M_A = 0 \qquad \Sigma M_B = 0 \qquad \Sigma M_C = 0$$

where A, B, and C are any three points not on the same straight line.

When equilibrium equations are written which are not independent, redundant information is obtained, and solution of the equations will yield $0 = 0$. For example, for a general problem in two dimensions with three unknowns, three moment equations written about three points which lie on the same straight line are not independent. Such equations will contain duplicated information, and solution of two of them can at best determine two of the unknowns with the third equation merely verifying the identity $0 = 0$.

(c) *Constraints and Statical Determinacy.* The equilibrium equations developed in this article, once satisfied, are both necessary and sufficient conditions to establish the equilibrium of a body. However, they do not necessarily provide all the information that is required to calculate all the unknown forces that may act on a body in equilibrium. The question of adequacy lies in the characteristics of the constraints to possible movement of the body provided by its supports. By constraint we mean the restriction of movement. In example 4 of Fig. 3/1 the roller, ball, and rocker provide constraint normal to the surface of contact but none tangent to the surface. Hence a tangential force cannot be supported. For the collar and slider of example 5 constraint exists only normal to the guide. In example 6 the fixed-pin connection provides constraint in both directions but offers no resistance to rotation about the pin unless the pin is not free to turn. The fixed support of example 7, however, offers constraint against rotation as well as lateral movement.

If the rocker that supports the truss of example 1 in Fig. 3/2 were replaced by a pin joint, as at A, there would be one additional constraint beyond that required to support an equilibrium configuration without collapse. The three scalar conditions of equilibrium, Eqs. 3/2, would not provide sufficient information to determine all four unknowns, since A_x and B_x could not be separated. These two components of force would be dependent on the deformation of the members of the truss as influenced by their corresponding stiffness properties. The horizontal reactions A_x and B_x would also be dependent on any initial deformation required to fit the dimensions of the structure to those of the foundation between A and B. Again referring to Fig. 3/2, if the pin B in example 3 were not free to turn, the support could transmit a couple to the beam through the pin. Therefore there would be four unknown supporting reactions acting on the beam, namely, the force at A, the two components of force at B, and the couple at B. Consequently the three independent scalar equations of equilibrium would not provide enough information to compute all four unknowns.

A body, or rigid combination of elements considered as a single body, that possesses more external supports or constraints than are necessary to maintain an equilibrium position is called *statically indeterminate*. Supports that can be removed without destroying the equilibrium condition of the body are said to be *redundant*. The number of redundant supporting elements present corresponds to the degree of statical indeterminacy and equals the total number of unknown external forces minus the number of available independent equations of equilibrium. On the other hand, bodies that are supported by the minimum number of constraints necessary to ensure an equilibrium configuration are called *statically determinate*, and for such bodies the equilibrium equations are sufficient to determine the unknown external forces.

The problems on equilibrium included in this article and throughout *Statics* are generally restricted to statically determinate bodies where the constraints are just sufficient to ensure a stable position and where the unknown supporting forces can be completely determined by the available independent equations of equilibrium. The student is alerted at this point by this brief discussion, however, to the fact that we must be aware of the nature of the constraints before we attempt to solve an equilibrium problem. A body will be recognized as statically indeterminate when there are more unknown external reactions than there are available independent equilibrium equations for the force system involved. It is always well to count the number of unknown forces on a given body and to be certain that an equal number of independent equations may be written; otherwise effort may be wasted in attempting an impossible solution with the aid of the equilibrium equations only. Unknowns may be forces, couples, distances, or angles.

In discussing the relationship between constraints and equilibrium, we should look further at the question of the adequacy of constraints. The existence of three constraints for a two-dimensional problem does not always guarantee a stable configuration. Figure 3/7 shows four different types of constraints. In the *a*-part of the figure point A of the rigid body is fixed by the two links and cannot move, and the third link prevents any rotation about A. Thus this body is completely fixed with three adequate (proper) constraints. In the *b*-part of the figure the third link is positioned so that the force transmitted by it passes through point A where the other two constraint forces act. Thus this configuration of constraints can offer no initial resistance to rotation about A, which would occur when exter-

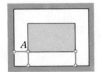

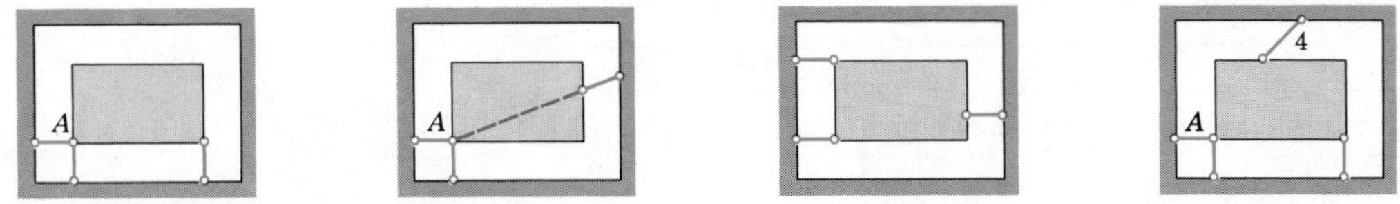

(a) Complete fixity
Adequate constraints

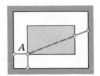

(b) Incomplete fixity
Partial constraints

(c) Incomplete fixity
Partial constraints

(d) Excessive fixity
Redundant constraint

Figure 3/7

nal loads were applied to the body. We conclude, therefore, that this body is incompletely fixed under partial constraints. The configuration in the *c*-part of the figure gives us a similar condition of incomplete fixity since the three parallel links could offer no initial resistance to a small vertical movement of the body as a result of external loads applied to it in this direction. The constraints in these two examples are often termed *improper*. In the *d*-part of Fig. 3/7 we have a condition of complete fixity with link 4 acting as an unnecessary fourth constraint to maintain a fixed position. Link 4, then, is a *redundant* constraint, and the body is statically indeterminate.

As in the four examples of Fig. 3/7 it is generally possible by direct observation to conclude whether the constraints on a body in two-dimensional equilibrium are adequate (proper), partial (improper), or redundant. As indicated previously, the vast majority of problems in this book are statically determinate with adequate (proper) constraints.

The sample problems at the end of the article illustrate the application of free-body diagrams and the equations of equilibrium to typical statics problems. When equilibrium equations are applied, one of the most useful steps is an expeditious choice of reference axes and moment centers. Generally, the best choice of a moment center is one through which as many unknown forces pass as possible. Simultaneous solutions of equilibrium equations are frequently necessary but can be minimized or avoided by a careful choice of reference axes and moment centers.

In the problem work it is always advisable to precede each major application of a principle of mechanics by a symbolic statement of the principle or governing equation which is involved. In the sample problems these statements are set forth in brackets to the left of the calculations and serve as a reminder of the justification for each major step. Also, the recommendations set forth in Art. 1/8 will be valuable to the student particularly at this stage as he begins to form habits of approach to the solution of engineering problems.

Sample Problem 3/1

Determine the magnitudes of the forces **C** and **T** which, along with the other three forces shown, act on the members of the bridge-truss joint.

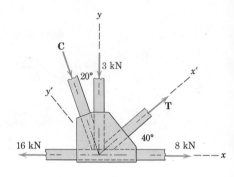

Solution. The given sketch constitutes the free-body diagram of the isolated section of the joint in question and shows the five forces which are in equilibrium.

Solution I (*scalar algebra*). For the x-y axes as shown we have

$$[\Sigma F_x = 0] \qquad 8 + T\cos 40° + C\sin 20° - 16 = 0$$

$$0.766T + 0.342C = 8 \qquad (a)$$

$$[\Sigma F_y = 0] \qquad T\sin 40° - C\cos 20° - 3 = 0$$

$$0.643T - 0.940C = 3 \qquad (b)$$

Simultaneous solution of Eqs. (a) and (b) produces

$$T = 9.09 \text{ kN} \qquad C = 3.03 \text{ kN} \qquad \qquad Ans.$$

Solution II (*scalar algebra*). To avoid a simultaneous solution we may use axes x'-y' with the first summation in the y'-direction to eliminate reference to T. Thus

$$[\Sigma F_{y'} = 0] \quad -C\cos 20° - 3\cos 40° - 8\sin 40° + 16\sin 40° = 0$$

$$C = 3.03 \text{ kN} \qquad \qquad Ans.$$

$$[\Sigma F_{x'} = 0] \quad T + 8\cos 40° - 16\cos 40° - 3\sin 40° - 3.03\sin 20° = 0$$

$$T = 9.09 \text{ kN} \qquad \qquad Ans.$$

Solution III (*vector algebra*). With unit vectors **i** and **j** in the x- and y-directions the zero summation of forces for equilibrium yields the vector equation

$$[\Sigma \mathbf{F} = 0] \quad 8\mathbf{i} + (T\cos 40°)\mathbf{i} + (T\sin 40°)\mathbf{j} - 3\mathbf{j} + (C\sin 20°)\mathbf{i}$$

$$- (C\cos 20°)\mathbf{j} - 16\mathbf{i} = 0$$

Equating the coefficients of the **i**- and **j**-terms to zero gives

$$8 + T\cos 40° + C\sin 20° - 16 = 0$$

$$T\sin 40° - 3 - C\cos 20° = 0$$

which are the same, of course, as Eqs. (a) and (b) which we solved above.

Solution IV (*geometric*). The polygon representing the zero vector sum of the five forces is shown. Equations (a) and (b) are seen immediately to give the projections of the vectors onto the x- and y-directions. Similarly, projections onto the x'- and y'-directions give the alternative equations in Solution II.

A graphical solution is easily obtained. The known vectors are laid off head-to-tail to some convenient scale, and the directions of **T** and **C** are then drawn to close the polygon. The resulting intersection at point P completes the solution, thus enabling us to measure the magnitudes of **T** and **C** directly from the drawing to whatever degree of accuracy we incorporate into the construction.

① The selection of reference axes to facilitate computation is always an important consideration. Alternatively in this example we could take a set of axes along and normal to the direction of **C** and employ a force summation normal to **C** to eliminate it.

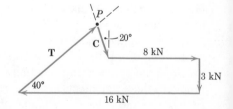

② The known vectors may be added in any order desired, but they must be added before the unknown vectors.

Sample Problem 3/2

Determine the magnitude T of the tension in the supporting cable and the magnitude of the force on the pin at A for the jib crane shown. The beam AB is a standard 0.5-m I-beam with a mass of 95 kg per meter of length.

Algebraic solution. The system is symmetrical about the vertical x-y plane through the center of the beam, so the problem may be analyzed as the equilibrium of a coplanar force system. The free-body diagram of the beam is shown in the figure with the pin reaction at A represented in terms of its two rectangular components. The weight of the beam is $95(10^{-3})(5)9.81 = 4.66$ kN and acts through its center. Note that there are three unknowns A_x, A_y, and T which may be found from the three equations of equilibrium. We begin with a moment equation about A which eliminates two of the three unknowns from the equation. In applying the moment equation about A it is simpler to consider the moments of the x- and y-components of $\mathbf{T}$ than it is to compute the perpendicular distance from $\mathbf{T}$ to A. Hence with the counterclockwise sense as positive we write

$[\Sigma M_A = 0]$ $(T \cos 25°)0.25 + (T \sin 25°)(5 - 0.12)$

$$- 10(5 - 1.5 - 0.12) - 4.66(2.5 - 0.12) = 0$$

from which

$$T = 19.61 \text{ kN} \qquad Ans.$$

Equating the sum of forces in the x- and y-directions to zero gives

$[\Sigma F_x = 0]$ $A_x - 19.61 \cos 25° = 0,$ $A_x = 17.77$ kN

$[\Sigma F_y = 0]$ $A_y + 19.61 \sin 25° - 4.66 - 10 = 0,$ $A_y = 6.37$ kN

$[A = \sqrt{A_x^2 + A_y^2}]$ $A = \sqrt{(17.77)^2 + (6.37)^2},$ $A = 18.88$ kN   *Ans.*

Graphical solution. The principle that three forces in equilibrium must be concurrent is utilized for a graphical solution by combining the two known vertical forces of 4.66 and 10 kN into a single 14.66-kN force located as shown on the modified free-body diagram of the beam in the lower figure. The position of this resultant load may easily be determined graphically or algebraically. The intersection of the 14.66-kN force with the line of action of the unknown tension $\mathbf{T}$ defines the point of concurrency O through which the pin reaction $\mathbf{A}$ must pass. The unknown magnitudes of $\mathbf{T}$ and $\mathbf{A}$ may now be found by adding the forces head-to-tail to form the closed equilibrium polygon of forces, thus satisfying their zero vector sum. After the known vertical load is laid off to a convenient scale, as shown in the lower part of the figure, a line representing the given direction of the tension $\mathbf{T}$ is drawn through the tip of the 14.7-kN vector. Likewise a line representing the direction of the pin reaction $\mathbf{A}$, determined from the concurrency established with the free-body diagram, is drawn through the tail of the 14.7-kN vector. The intersection of the lines representing vectors $\mathbf{T}$ and $\mathbf{A}$ establishes the magnitudes T and A which are necessary to make the vector sum of the forces equal to zero. These magnitudes may be scaled directly from the diagram. The x- and y-components of $\mathbf{A}$ may be constructed on the force polygon if desired.

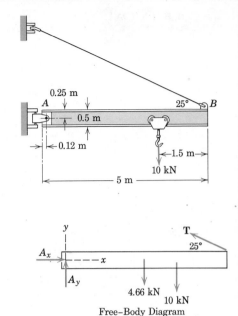

Free–Body Diagram

① The justification for this step is, of course, Varignon's theorem, the principle of moments, explained in Art. 2/4. Be prepared to take full advantage of this principle frequently.

② The calculation of moments in two-dimensional problems is generally handled more simply by scalar algebra than by the vector cross-product $\mathbf{r} \times \mathbf{F}$. In three dimensions, as we shall see later, the reverse is often the case.

③ The direction of the force at A could be easily calculated if desired. However, in designing the pin A or in checking its strength it is the magnitude only of the force which matters.

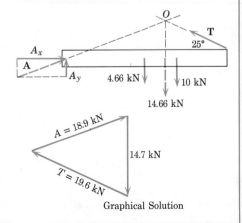

Graphical Solution

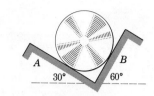

Problem 3/4

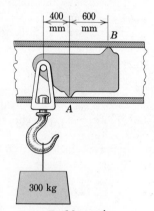

Problem 3/5

Problem 3/6

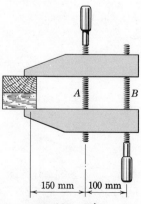

Problem 3/7

PROBLEMS

3/4 The homogeneous cylinder has a mass of 40 kg and rests on smooth surfaces A and B which are inclined 30° and 60°, respectively, from the horizontal. Determine the contact forces at A and B.

Ans. $A = 340$ N, $B = 196$ N

3/5 Determine the magnitude P of the horizontal force which the gardener must exert on the light handle of the 200-lb lawn roller to hold it in place on the 10° slope. Also determine the force exerted by the ground on the roller. The center of gravity of the roller is at its center O.

3/6 To facilitate shifting the position of a lifting hook when it is not under load, the sliding hanger shown is used. The projections at A and B engage the flanges of a box beam when a load is supported, and the hook projects through a horizontal slot in the beam. Compute the forces at A and B when the hook supports a 300-kg mass.

Ans. $A = 4.91$ kN, $B = 1.96$ kN

3/7 If the screw B of the wood clamp is tightened so that the two blocks are under a compression of 500 N, determine the force in screw A. (*Note:* The force supported by each screw may be taken in the direction of the screw.)

3/8 A spring scale indicates a tension T in the right-hand cable of the pulley-cable system which supports the body of mass m. Express m in terms of T. Neglect the mass of the pulleys. *Ans.* $m = 8T/g$

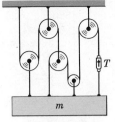

Problem 3/8

3/9 The uniform 18-ft beam weighs 500 lb and is loaded in the vertical plane by the parallel forces shown. Calculate the reactions at the supports A and B.

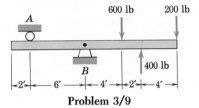

Problem 3/9

3/10 Calculate the pull P on the rope which the man must exert in order to suspend the 200-kg crate in the deflected position shown. Choose your reference axis so as to solve for P in one equation without involving the tension in the upper rope.
  Ans. $P = 871$ N

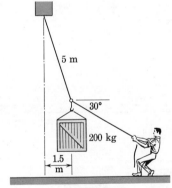

Problem 3/10

3/11 Solve for the pull P and the tension T in the upper rope of Prob. 3/10 graphically.

3/12 Compute the magnitudes T_1 and T_2 of the tensions in the two cables attached to the 100-kg crate.
  Ans. $T_1 = 1067$ N, $T_2 = 1730$ N

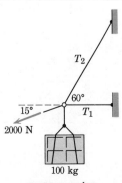

Problem 3/12

Problem 3/13

3/13 Determine the force P that the 80-kg worker must exert on the rope in order to support himself in the bosun's chair. What force R does the man exert on the seat of the chair?

Ans. $P = 157$ N, $R = 628$ N

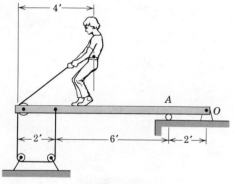

Problem 3/14

3/14 To test the deflection of the uniform 200-lb beam the 120-lb boy exerts a pull of 40 lb on the rope rigged as shown. Compute the force supported by the pin at the hinge O. *Ans. $F_O = 820$ lb*

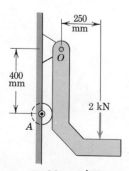

Problem 3/15

3/15 The resistance of the bracket to bending is tested under the 2-kN load. Compute the force on the roller at A and the total force supported by the pin at O.

3/16 The T-frame weighs 500 lb with mass center at G. Calculate the total force supported by the pin at O after the 600-lb force is applied to the cable.

Ans. $O = 465$ lb

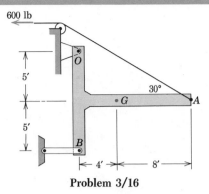

Problem 3/16

3/17 A jet airplane having a mass of 8 Mg is flying horizontally at a constant speed of 1000 km/h under a thrust of 16 kN from its turbojet engines. If the pilot increases the fuel rate to give a thrust of 20 kN and noses the plane upward to maintain a constant 1000-km/h air speed, determine the angle θ made by the new line of flight with the horizontal. Note that the air resistance in the line of flight at the particular altitude involved is a function of air speed.

Ans. $\theta = 2.92°$

3/18 If the mass of the boom is negligible compared with that of the load L, find the force F on the ball joint at A and show that the magnitude F is constant for all values of θ. Determine the limiting value of T as θ approaches $90°$.

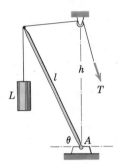

Problem 3/18

3/19 The hinged member is used to activate a latching device for securing a large trailer body to its undercarriage. If a tension $T = 80$ lb is required in the horizontal control rod to trip the plunger against which the member acts at C, estimate the force supported by the pin at A by means of a free-hand vector solution.

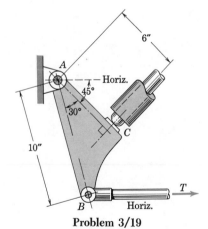

Problem 3/19

3/20 Calculate the value of the couple M required to roll the 40-kg wheel up the incline. Also determine the contact force R at A. The surface of the incline is sufficiently rough to prevent slipping.

Ans. $M = 47.1$ N·m, $R = 393$ N

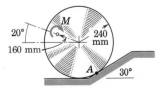

Problem 3/20

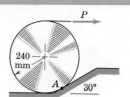

Problem 3/21

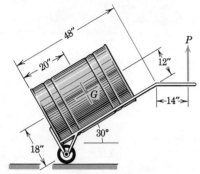

Problem 3/22

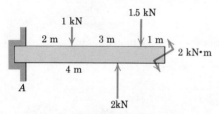

Problem 3/23

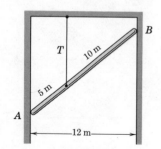

Problem 3/24

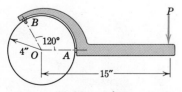

Problem 3/25

3/21 Replace the couple M of Prob. 3/20 by the horizontal force P and compute its value required to roll the 40-kg wheel up the incline. Also determine the contact force R at A. No slipping occurs.

Ans. $R = 406$ N

3/22 The oil drum weighs 620 lb when full and has a mass center at G. Calculate the vertical force P required to maintain equilibrium of the drum and dolly in the position shown. The weight of the dolly may be neglected compared with that of the drum.

3/23 The uniform cantilever beam has a mass of 50 kg per meter of length and supports the couple and three forces shown. Isolate the beam to the right of section A and calculate the moment (couple) M and shear (vertical force) V exerted by the wall support on the beam at this section.

Ans. $V = 3.44$ kN up
$M = 8.33$ kN·m counterclockwise

3/24 The uniform 15-m pole has a mass of 150 kg and is supported by its smooth ends against the vertical walls and by the tension T in the vertical cable. Compute the reactions at A and B.

3/25 The hook wrench or pin spanner is used to turn shafts and collars. If a moment of 60 lb-ft is required to turn the 8-in.-diameter collar about its center O under the action of the applied force P, determine the contact force R on the smooth surface at A. Engagement of the pin at B may be considered to occur at the periphery of the collar.

Ans. $R = 236$ lb

3/26 A drum of 600-mm radius with mass center G at its
geometric center has a total mass of 1600 kg and
rests on a cradle consisting of two long rollers of
240-mm diameter. The rollers rest on a smooth
horizontal surface and are prevented from separating by horizontal links, one on each end of the
rollers as shown. Calculate the tension T in each
link and the reaction R between the drum and each
roller.   *Ans. $T = 3.79$ kN, $R = 10.91$ kN*

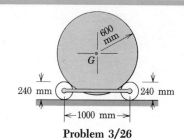

Problem 3/26

3/27 The rigid truss is subjected to the four loads shown
and is supported by the hinge at A and the link BC.
Neglect the weight of the truss and compute the
magnitude of the force supported by the pin at A.
Why is the direction of the force at A parallel to BC
with the given loading?

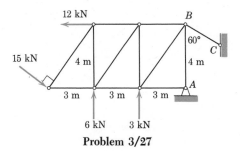

Problem 3/27

3/28 Magnetic tape under a tension of 10.0 N at D passes
around the guide pulleys and through the erasing
head at C at constant speed. As a result of a small
amount of friction in the bearings of the pulleys, the
tape at E is under a tension of 11.0 N. Determine
the tension T in the supporting spring at B. The
plate is horizontal and is mounted on a precision
needle bearing at A.   *Ans. $T = 10.6$ N*

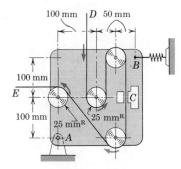

Problem 3/28

3/29 Calculate the magnitude of the force supported by
the pin at A under the action of the 250-lb-in.
couple applied to the light bracket.

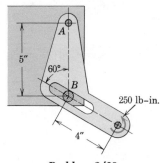

Problem 3/29

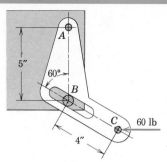

Problem 3/30

3/30 The applied couple in Prob. 3/29 is replaced by the 60-lb horizontal force shown here. Calculate the magnitude of the force supported by the pin at A which holds the light bracket. *Ans.* $A = 147.5$ lb

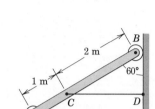

Problem 3/31

3/31 The uniform 50-kg bar with end rollers is held in the equilibrium position by the horizontal cord CD as shown. Determine the tension T in the cord by writing only one equation of equilibrium. Then obtain the forces at A and B by inspection.

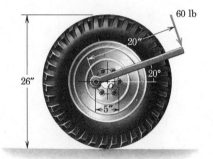

Problem 3/32

3/32 When the 60-lb force is applied to the tire wrench as shown, the wheel is prevented from turning because of the friction force between it and the ground. By using only one equation of equilibrium determine the horizontal component O_x of the reaction exerted on the wheel by the fixed bearing at O. Analyze the wrench and wheel as a single body. *Ans.* $O_x = 102$ lb

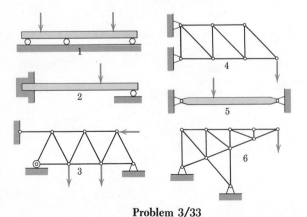

Problem 3/33

3/33 Each of the structures is statically indeterminate. Describe at least one modification in the supports for each case which would make the structure statically determinate.

3/34 The figure shows a series of rectangular plates and their constraints, all confined to the plane of representation. The plates could be subjected to various known loads applied in the plane of the plate. Identify the plates that belong to each of the following categories.

   (A) Complete fixity with minimum number of adequate constraints
   (B) Partial fixity with inadequate constraints
   (C) Complete fixity with redundant constraints
   (D) Partial fixity with redundant constraint

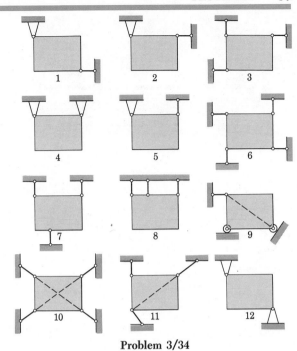

Problem 3/34

3/35 The uniform bar with end rollers weighs 60 lb and is supported by the horizontal and vertical surfaces and by the wire *AC*. Calculate the tension *T* in the wire and the reactions against the rollers at *A* and *B*.
   Ans. T = 60.2 lb, *A* = 15 lb, *B* = 40 lb

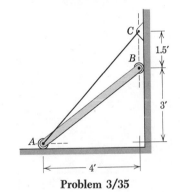

Problem 3/35

3/36 The uniform concrete slab shown in edge view has a mass of 25 Mg and is being hoisted slowly into a vertical position by the tension *P* in the hoisting cable. For the position where $\theta = 60°$ calculate the tension *T* in the horizontal anchor cable by using only one equation of equilibrium.

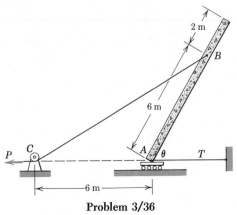

Problem 3/36

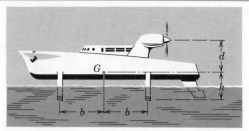

Problem 3/37

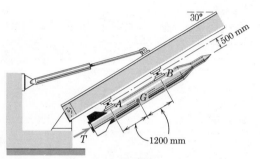

Problem 3/38

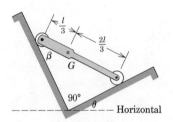

Problem 3/39

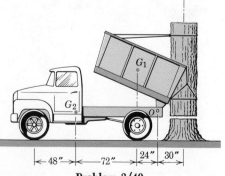

Problem 3/40

3/37 An experimental boat is equipped with four hydrofoils, two on each side as shown. The boat has a total mass m with center of mass at G. Thrust is provided with the air screw. The ratio of lift to drag for each foil is n. Lift is the vertical force supported by each foil, and drag is the horizontal resistance to motion through the water. For a given thrust T of the propeller write the expression for the drag D on each of the two forward foils.

$$\text{Ans. } D = \frac{1}{4n}\left(mg + T\frac{d+h}{b}\right)$$

3/38 The supporting slippers, which hold the missile to the launcher, slide in T-slots in the guiding rail. In a static test of a missile the slipper at A is clamped securely to the rail whereas the slipper at B is not clamped. If the missile has a mass of 1500 kg with center of mass at G and if the static thrust T is 20 kN, determine the force supported by the pin that connects the missile to the slipper at B.

3/39 The bar and its end rollers have a mass center at G and are placed on the inclined surfaces shown. For a given value of θ determine the one angle β for which the bar will be in an unstable equilibrium position. Also indicate two additional values of β for stable positions.　　　$\text{Ans. } \beta = \tan^{-1}(\frac{1}{2}\tan\theta)$

3/40 The dump truck is used to lift a cut section of trunk from the stump of a large tree. The section of trunk is measured and calculated to weigh 1200 lb. The body of the rotatable dump weighs 600 lb, and its center of mass G_1 is directly over the rear wheels in the position for lifting. Calculate the necessary torque M applied to the dump through its shaft at O to make the lift. Also calculate the corresponding forces under the front and rear pairs of wheels of the truck. The truck weighs 6200 lb exclusive of the dump, and its center of mass is at G_2.

3/41 The 300-mm-radius wheel has a mass of 60 kg with center of mass at G and rests on a rough horizontal surface. Under the action of the couple $M = 50$ N·m applied at A the wheel bears against the roller at B, which prevents it from rolling further. Calculate the magnitude of the total force exerted on the wheel by the horizontal surface at C.

Ans. $F_C = 343$ N

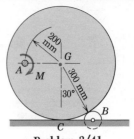

Problem 3/41

3/42 The specially built trailer is used to transport and erect a missile to its vertical launch position. The trailer body and missile have a combined mass of 6.20 Mg with center of mass at G. The unit is tilted into position by two hydraulic cylinders, one on each side of the trailer. Compute the compressive force C in each piston rod of the cylinders for the position where the axis AB of the cylinder is perpendicular to the longitudinal axis of the trailer and missile.

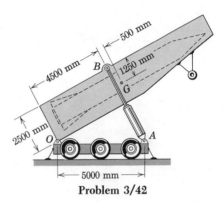

Problem 3/42

3/43 Determine the magnitude F of the total force supported by the hinge at O for the missile transport described in Prob. 3/42. A graphical solution is suggested.

Ans. $F \approx 26$ kN

3/44 The portable floor crane in the automotive shop is lifting a 420-lb engine. For the position shown compute the magnitude of the force supported by the pin at C and the oil pressure p against the 3.20-in.-diameter piston of the hydraulic-cylinder unit AB.

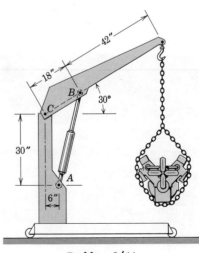

Problem 3/44

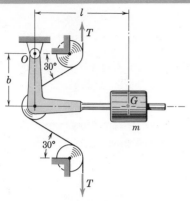

Problem 3/45

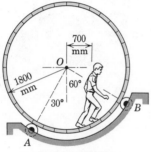

Problem 3/46

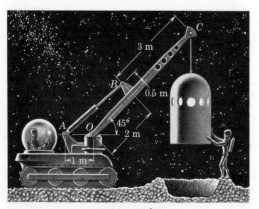

Problem 3/47

3/45 Determine the dimension l if the mass m maintains a specified tension T in the belt for the position shown. Neglect the mass of the arm and central pulley compared with m. Also find the magnitude of the force R supported by the pin at O.

Ans. $R = \sqrt{3T^2 + (mg)^2}$

3/46 The uniform 400-kg drum is mounted on a line of rollers at A and a line of rollers at B. An 80-kg man moves slowly a distance of 700 mm from the vertical centerline before the drum begins to rotate. All rollers are perfectly free to rotate except one of them at B which must overcome appreciable friction in its bearing. Calculate the friction force F exerted by that one roller tangent to the drum and find the magnitude R of the force exerted by all rollers at A on the drum for this condition.

3/47 The lunar shelter is fabricated from aluminum and has a mass of 240 kg. The boom OC of the lunar crane has a mass of 48 kg, and its center of mass is at its midlength. Determine the force F in the hydraulic cylinder AB of the crane in the position shown as it is placing the shelter in position on the lunar surface. Recall that gravitational attraction on the moon is $\frac{1}{6}$ that on the earth.   *Ans.* $F = 2340$ N

3/48 The crane is hoisting a 4.20-Mg bulldozer. The mass center of the 2-Mg boom *OA* is at its midlength. Calculate the tension *T* in the cable where it attaches at *B* and the magnitude of the force supported by the hinge at *O* for equilibrium in the 60° position. Neglect the width of the boom.

Ans. T = 61.5 kN, O = 99.2 kN

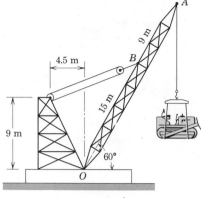

Problem 3/48

3/49 The weight of the rigid truss *ABCDE* is small compared with that of the 12-Mg mass which it supports. Compute the force in the horizontal link at *B* and the magnitude of the force supported by the pin joint at *A*.

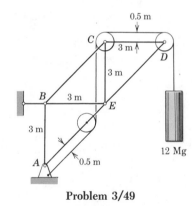

Problem 3/49

3/50 The member *OBC* and sheave at *C* together weigh 500 lb, with a combined center of mass at *G*. Calculate the magnitude of the force supported by the pin connection at *O* when the 300-lb load is applied. The collar at *A* can provide support in the horizontal direction only. *Ans. O = 1345 lb*

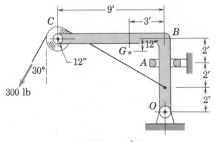

Problem 3/50

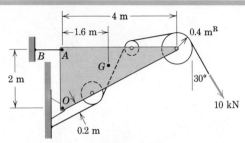

Problem 3/51

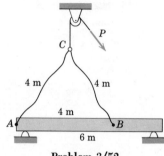

Problem 3/52

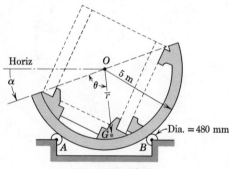

Problem 3/53

3/51 The triangular frame with pulleys has a total mass of 200 kg with mass center at G. Calculate the magnitude of the force supported by the pin at O when the 10-kN force is applied to the cable. (*Hint:* In using the moment equation take advantage of the replacement of a force by a force and a couple.)

Ans. $O = 34.3$ kN

▶**3/52** The uniform beam has an overall length of 6 m and a mass of 300 kg. The force P applied to the hoisting cable is slowly increased to raise the ring C, the two 4-m ropes AC and BC, and the beam. Compute the tensions in the ropes at A and B when the beam is clear of its supports and the force P is equal to the weight of the beam.

Ans. $T_A = 816$ N, $T_B = 2449$ N

▶**3/53** A special jig for turning large concrete pipe sections (shown dotted) consists of an 80-Mg sector mounted on a line of rollers at A and a line of rollers at B. One of the rollers at B is a gear which meshes with a ring of gear teeth on the sector so as to turn the sector about its geometric center O. When $\alpha = 0$, a counterclockwise torque of 2460 N·m must be applied to the gear at B to keep the assembly from rotating. When $\alpha = 30°$, a clockwise torque of 4680 N·m is required to prevent rotation. Locate the mass center G of the jig by calculating $\bar{r}$ and θ.

Ans. $\bar{r} = 367$ mm, $\theta = 79.8°$

SECTION B. EQUILIBRIUM IN THREE DIMENSIONS

3/4 EQUILIBRIUM CONDITIONS. We now extend our principles and methods developed for two-dimensional equilibrium to the case of three-dimensional equilibrium. In Art. 3/1 the general conditions for the equilibrium of a body were stated in Eqs. 3/1 which require that the resultant force and resultant couple on a body in equilibrium must be zero. These two vector equations of equilibrium and their scalar components may be written as

$$
\begin{aligned}
\Sigma \mathbf{F} = \mathbf{0} \quad &\text{or} \quad
\begin{cases}
\Sigma F_x = 0 \\
\Sigma F_y = 0 \\
\Sigma F_z = 0
\end{cases} \\[2ex]
\Sigma \mathbf{M} = \mathbf{0} \quad &\text{or} \quad
\begin{cases}
\Sigma M_x = 0 \\
\Sigma M_y = 0 \\
\Sigma M_z = 0
\end{cases}
\end{aligned}
\tag{3/3}
$$

The first three scalar equations state that there is no resultant force acting on a body in equilibrium in any of the three coordinate directions. The second three scalar equations express the further equilibrium requirement that there is no resultant moment acting on the body about any of the coordinate axes or about axes parallel to the coordinate axes. These six equations are both necessary and sufficient conditons for complete equilibrium. The reference axes may be chosen arbitrarily as a matter of convenience, the only restriction being that a right-handed coordinate system must be used with vector notation.

The six scalar relationships of Eqs. 3/3 are independent conditions since any of them may be valid without the others. For example, for a car which accelerates on a straight and level road in the x-direction, Newton's second law tells us that the resultant force on the car equals its mass times its acceleration. Hence, $\Sigma F_x \neq 0$ but the remaining force-equilibrium equations are satisfied since all other acceleration components are zero. Similarly, if the flywheel of the engine of the accelerating car is rotating with increasing angular speed about the x-axis, it is not in rotational equilibrium about this axis. Hence $\Sigma M_x \neq 0$ along with $\Sigma F_x \neq 0$, but the remaining four equilibrium equations would be satisfied for mass-center axes.

In applying the vector form of Eqs. 3/3 we first express each of the forces in terms of the coordinate unit vectors $\mathbf{i}$, $\mathbf{j}$, and $\mathbf{k}$. For the first equation, $\Sigma \mathbf{F} = \mathbf{0}$, the vector sum will be zero only if the coefficients of $\mathbf{i}$, $\mathbf{j}$, and $\mathbf{k}$ in the expression are, respectively, zero. These three sums when each set equal to zero yield precisely the three scalar equations of equilibrium, $\Sigma F_x = 0$, $\Sigma F_y = 0$, $\Sigma F_z = 0$.

For the second equation, $\Sigma \mathbf{M} = \mathbf{0}$, where the moment sum may be taken about any convenient point O, we express the moment of each force as the cross product $\mathbf{r} \times \mathbf{F}$ where $\mathbf{r}$ is the position vector from O to any point on the line of action of the force $\mathbf{F}$. Thus $\Sigma \mathbf{M} = \Sigma \mathbf{r} \times \mathbf{F} = \mathbf{0}$. The coefficients of $\mathbf{i}$, $\mathbf{j}$, and $\mathbf{k}$ in the resulting moment equation when set equal to zero, respectively, produce exactly the three scalar moment equations $\Sigma M_x = 0$, $\Sigma M_y = 0$, and $\Sigma M_z = 0$.

(*a*) *Free-Body Diagrams.* The summations in Eqs. 3/3 include the effects of *all* forces on the body under consideration. We learned in the previous article that the free-body diagram is the only reliable method for disclosing all forces and moments which should be included in our equilibrium equations. In three dimensions the free-body diagram serves the same essential purpose as it does in two dimensions and should *always* be drawn. We have our choice either of drawing a pictorial view of the isolated body with all external forces represented or of drawing the orthogonal projections of the free-body diagram. Both representations will be illustrated in the sample problems at the end of the article.

The correct representation of forces on the free-body diagram requires a knowledge of the characteristics of contacting surfaces. These characteristics were set forth in Fig. 3/1 for two-dimensional problems, and their extension to three-dimensional problems is represented in Fig. 3/8 for the most common situations of force transmission. The representations in both Figs. 3/1 and 3/8 will be used in three-dimensional analysis.

(*b*) *Categories of Equilibrium.* Application of Eqs. 3/3 falls into four categories which we can easily identify with the aid of Fig. 3/9 (pg. 106).

Case 1, equilibrium of forces that are concurrent at a point O, requires all three force equations but no moment equations since the moment of the forces about any axis through O is automatically zero.

Case 2, equilibrium of forces that are concurrent with a line, requires all equations except the moment equation about that line, which is automatically satisfied.

Case 3, equilibrium of parallel forces, requires only one force equation in the direction of the forces (x-direction as shown) and two moment equations about axes (y and z) that are normal to the direction of the forces.

Case 4, equilibrium of a general system of forces, requires all three force equations and all three moment equations.

The observations contained in these statements are generally quite evident when a given problem is being solved.

(*c*) *Constraints and Statical Determinacy.* The six scalar relations of Eqs. 3/3, although necessary and sufficient conditions to

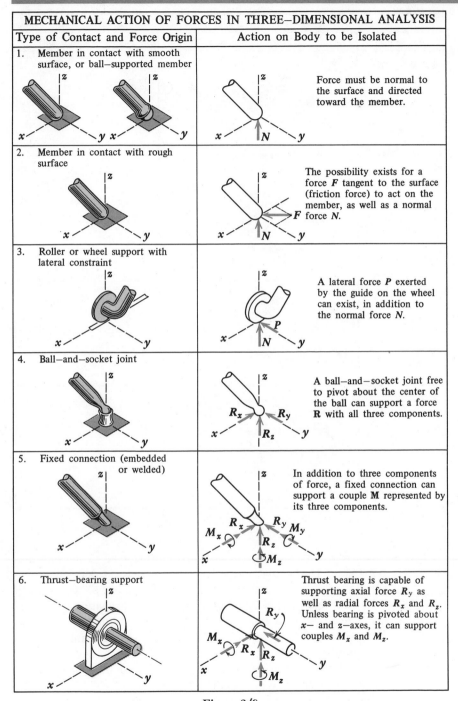

MECHANICAL ACTION OF FORCES IN THREE–DIMENSIONAL ANALYSIS	
Type of Contact and Force Origin	Action on Body to be Isolated
1. Member in contact with smooth surface, or ball–supported member	Force must be normal to the surface and directed toward the member.
2. Member in contact with rough surface	The possibility exists for a force F tangent to the surface (friction force) to act on the member, as well as a normal force N.
3. Roller or wheel support with lateral constraint	A lateral force P exerted by the guide on the wheel can exist, in addition to the normal force N.
4. Ball–and–socket joint	A ball–and–socket joint free to pivot about the center of the ball can support a force R with all three components.
5. Fixed connection (embedded or welded)	In addition to three components of force, a fixed connection can support a couple M represented by its three components.
6. Thrust–bearing support	Thrust bearing is capable of supporting axial force R_y as well as radial forces R_x and R_z. Unless bearing is pivoted about x– and z–axes, it can support couples M_x and M_z.

Figure 3/8

CATEGORIES OF EQUILIBRIUM IN THREE DIMENSIONS

Force System	Free–Body Diagram	Independent Equations
1. Concurrent at a point	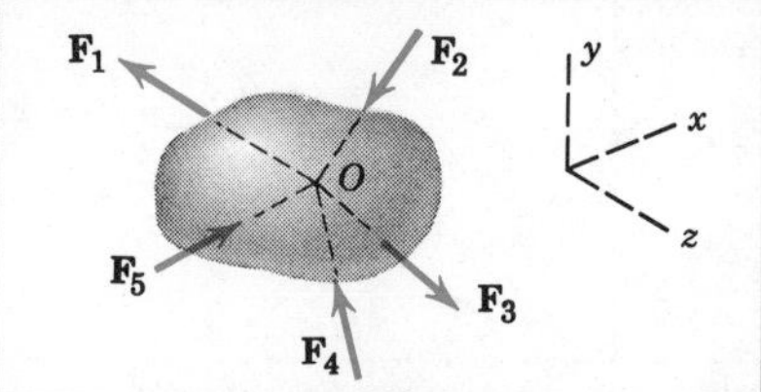	$\Sigma F_x = 0$ $\Sigma F_y = 0$ $\Sigma F_z = 0$
2. Concurrent with a line	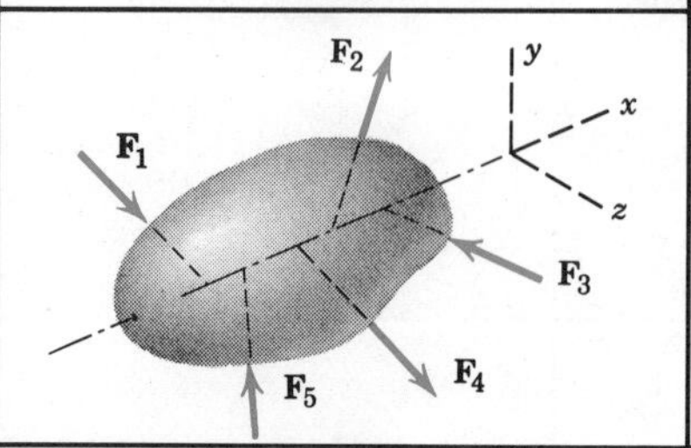	$\Sigma F_x = 0$ $\Sigma M_y = 0$ $\Sigma F_y = 0$ $\Sigma M_z = 0$ $\Sigma F_z = 0$
3. Parallel	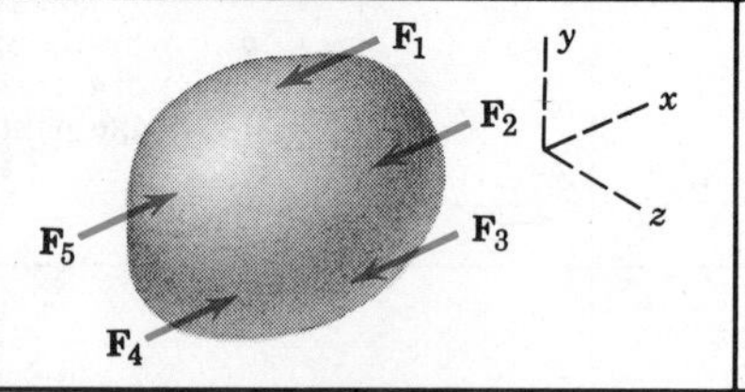	$\Sigma F_x = 0$ $\Sigma M_y = 0$ $\Sigma M_z = 0$
4. General	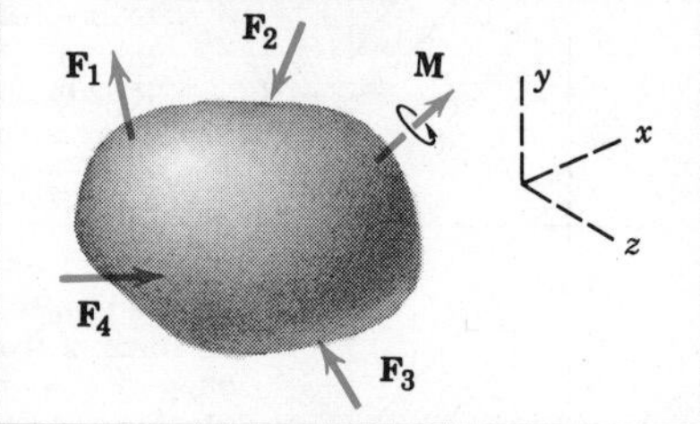	$\Sigma F_x = 0$ $\Sigma M_x = 0$ $\Sigma F_y = 0$ $\Sigma M_y = 0$ $\Sigma F_z = 0$ $\Sigma M_z = 0$

Figure 3/9

establish equilibrium, do not necessarily provide all of the information required to calculate the unknown forces acting in a three-dimensional equilibrium situation. Again, as we found to be the case with two dimensions, the question of adequacy of information lies in the characteristics of the constraints provided by the supports. An analytical criterion for determining the adequacy of constraints is available but is beyond the scope of this treatment.° In Fig. 3/10, however, we cite four examples of constraint conditions to alert the reader to the problem. In the *a*-part of the figure is shown a rigid body whose corner *A* is completely fixed by the links 1, 2, and 3. Links 4, 5, and 6 prevent rotations about the axes of links 1, 2, and 3, respectively, so

° See the author's *Statics, 2nd Edition* or *SI Version*, Art. 16.

that the body is completely fixed and the constraints are said to be adequate. The *b*-part of Fig. 3/10 shows the same number of constraints, but we see that they provide no resistance to a moment which might be applied about axis *AE*. Here the body is incompletely fixed and only partially constrained. Similarly, in Fig. 3/10*c* the constraints provide no resistance to an unbalanced force in the *y*-direction, so here also is a case of incomplete fixity with partial constraints. In Fig. 3/10*d*, if a seventh constraining link were imposed on a system of six constraints placed properly for complete fixity, more supports would be provided than would be necessary to establish the equilibrium position, and link 7 would be *redundant*. The body would then be *statically indeterminate* with such a seventh link in place. With only a few exceptions the supporting constraints for rigid bodies in equilibrium in this book are adequate and the bodies are statically determinate.

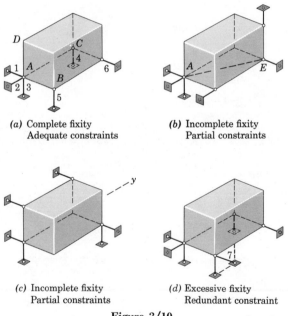

(*a*) Complete fixity
    Adequate constraints

(*b*) Incomplete fixity
    Partial constraints

(*c*) Incomplete fixity
    Partial constraints

(*d*) Excessive fixity
    Redundant constraint

Figure 3/10

Sample Problem 3/3

The uniform 7-m steel shaft has a mass of 200 kg and is supported by a ball-and-socket joint at A in the horizontal floor. The ball end B rests against the smooth vertical walls as shown. Compute the forces exerted by the walls and the floor on the ends of the shaft.

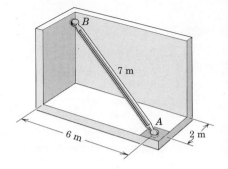

Solution. The free-body diagram of the shaft is first drawn where the contact forces acting on the shaft at B are shown normal to the wall surfaces. In addition to the weight $W = mg = 200(9.81) = 1962$ N, the force exerted by the floor on the ball joint at A is represented by its x-, y-, and z-components. These components are shown in their correct physical sense as should be evident from the requirement that A is held in place. The vertical position of B is found from $7 = \sqrt{2^2 + 6^2 + h^2}$, $h = 3$ m. Right-handed coordinate axes are conveniently assigned as shown.

Vector solution. We will use A as a moment center to eliminate reference to the forces at A. The position vectors needed to compute the moments about A are

$$\mathbf{r}_{AG} = -1\mathbf{i} - 3\mathbf{j} + 1.5\mathbf{k} \text{ m} \quad \text{and} \quad \mathbf{r}_{AB} = -2\mathbf{i} - 6\mathbf{j} + 3\mathbf{k} \text{ m}$$

where the mass center G is located halfway between A and B for the uniform shaft.

The vector moment equation gives

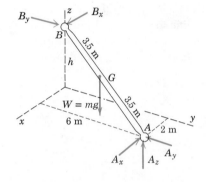

$$[\Sigma \mathbf{M}_A = 0] \quad \mathbf{r}_{AB} \times (\mathbf{B}_x + \mathbf{B}_y) + \mathbf{r}_{AG} \times \mathbf{W} = 0$$

$$(-2\mathbf{i} - 6\mathbf{j} + 3\mathbf{k}) \times (i B_x + j B_y) + (-\mathbf{i} - 3\mathbf{j} + 1.5\mathbf{k}) \times (-1962\mathbf{k}) = 0$$

$$\begin{vmatrix} \mathbf{i} & \mathbf{j} & \mathbf{k} \\ -2 & -6 & 3 \\ B_x & B_y & 0 \end{vmatrix} + \begin{vmatrix} \mathbf{i} & \mathbf{j} & \mathbf{k} \\ -1 & -3 & 1.5 \\ 0 & 0 & -1962 \end{vmatrix} = 0$$

$$(-3B_y + 5886)\mathbf{i} + (3B_x - 1962)\mathbf{j} + (-2B_y + 6B_x)\mathbf{k} = 0$$

Equating the coefficients of $\mathbf{i}$, $\mathbf{j}$, and $\mathbf{k}$ to zero and solving give

$$B_x = 654 \text{ N} \quad \text{and} \quad B_y = 1962 \text{ N} \qquad \textit{Ans.}$$

The forces at A are easily determined by

$$[\Sigma \mathbf{F} = 0] \quad (654 - A_x)\mathbf{i} + (1962 - A_y)\mathbf{j} + (-1962 + A_z)\mathbf{k} = 0$$

and

$$A_x = 654 \text{ N} \quad A_y = 1962 \text{ N} \quad A_z = 1962 \text{ N}$$

Finally

$$A = \sqrt{A_x^2 + A_y^2 + A_z^2}$$

$$= \sqrt{(654)^2 + (1962)^2 + (1962)^2} = 2851 \text{ N} \qquad \textit{Ans.}$$

Scalar solution. If we evaluate the scalar moment equations about axes through A parallel, respectively, to the x- and y-axes, we may write

$$[\Sigma M_{A_x} = 0] \quad 1962(3) - 3B_y = 0 \quad B_y = 1962 \text{ N}$$

$$[\Sigma M_{A_y} = 0] \quad -1962(1) + 3B_x = 0 \quad B_x = 654 \text{ N}$$

The force equations give, simply,

$$[\Sigma F_x = 0] \quad -A_x + 654 = 0 \quad A_x = 654 \text{ N}$$

$$[\Sigma F_y = 0] \quad -A_y + 1962 = 0 \quad A_y = 1962 \text{ N}$$

$$[\Sigma F_z = 0] \quad A_z - 1962 = 0 \quad A_z = 1962 \text{ N}$$

① We could, of course, assign all of the unknown components of force in the positive mathematical sense in which case A_x and A_y would turn out to be negative upon computation. The free-body diagram describes the physical situation, so it is generally preferable to show the forces in their correct physical senses wherever possible.

② Note that the third equation $-2B_y + 6B_x = 0$ merely checks the results of the first two equations. This result could be anticipated from the fact that an equilibrium system of forces concurrent with a line requires only two moment equations (Case 2 under *Categories of equilibrium*).

③ We observe that a moment sum about an axis through A parallel to the z-axis merely gives us $6B_x - 2B_y = 0$, which serves only as a check as noted previously. Alternatively we could have first obtained A_z from $\Sigma F_z = 0$ and then taken our moment equations about axes through B to obtain A_x and A_y.

Sample Problem 3/4

A 200-N force is applied to the handle of the hoist in the direction shown. The bearing A supports the thrust (force in the direction of the shaft axis) while bearing B supports only radial load (load normal to the shaft axis). Determine the mass m which can be supported and the total radial force exerted on the shaft by each bearing.

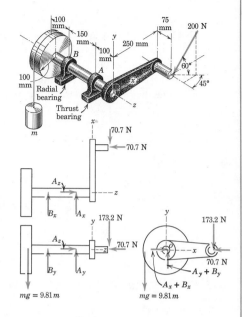

Solution. The system is clearly three-dimensional with no lines or planes of symmetry, and therefore the problem must be analyzed as a general space system of forces. A scalar solution is used here to illustrate this approach, although a solution using vector notation would also be satisfactory. The free-body diagram of the shaft, lever, and drum considered a single body could be shown by a space view if desired but is

① represented here by its three orthogonal projections.

The 200-N force is resolved into its three components, and each of the three views shows two of these components. The correct directions of A_x and B_x may be seen by inspection by observing that the line of action of the resultant of the two 70.7-N forces passes between A and B. The correct sense of the forces A_y and B_y cannot be determined until the magnitudes of the moments are obtained, so they are arbitrarily assigned. The x-y projection of the bearing forces is shown in terms of the sums of the unknown x- and y-components. The addition of A_z and the weight $W = mg$ completes the free-body diagrams. It should be noted that the three views represent three two-dimensional problems related by the corresponding components of the forces.

① If the standard three views of orthographic projection are not entirely familiar, then review and practice them. Visualize the three views as the images of the body projected onto the front, top, and end surfaces of a clear plastic box placed over and aligned with the body.

② From the x-y projection

$$[\Sigma M_O = 0] \qquad 100(9.81m) - 250(173.2) = 0, \qquad m = 44.1 \text{ kg} \qquad Ans.$$

② We could have started with the x-z projection rather than with the x-y projection.

From the x-z projection

$$[\Sigma M_A = 0] \qquad 150 B_x + 175(70.7) - 250(70.7) = 0, \qquad B_x = 35.4 \text{ N}$$

$$[\Sigma F_x = 0] \qquad A_x + 35.4 - 70.7 = 0, \qquad A_x = 35.3 \text{ N}$$

③ The y-z view gives

$$[\Sigma M_A = 0] \qquad 150 B_y + 175(173.2) - 250(44.1)(9.81) = 0, \qquad B_y = 520 \text{ N}$$

$$[\Sigma F_y = 0] \qquad A_y + 520 - 173.2 - (44.1)(9.81) = 0, \qquad A_y = 86.8 \text{ N}$$

$$[\Sigma F_z = 0] \qquad A_z = 70.7 \text{ N}$$

③ The y-z view could have followed immediately after the x-y view since the determination of A_y and B_y may be made after mg is found.

The total radial forces on the bearings become

$$[A_r = \sqrt{A_x{}^2 + A_y{}^2}] \qquad A_r = \sqrt{(35.3)^2 + (86.8)^2} = 93.5 \text{ N} \qquad Ans.$$

$$[B = \sqrt{B_x{}^2 + B_y{}^2}] \qquad B = \sqrt{(35.4)^2 + (520)^2} = 521 \text{ N} \qquad Ans.$$

Sample Problem 3/5

The welded tubular frame is secured to the horizontal x-y plane by a ball-and-socket joint at A and receives support from the loose-fitting ring at B. Under the action of the 2-kN load, rotation about a line from A to B is prevented by the cable CD, and the frame is stable in the position shown. Neglect the weight of the frame compared with the applied load and determine the tension T in the cable, the reaction at the ring, and the reaction components at A.

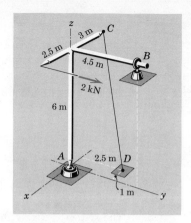

Solution. The system is clearly three-dimensional with no lines or planes of symmetry, and therefore the problem must be analyzed as a general space system of forces. The free-body diagram is drawn, where the ring reaction is shown in terms of its two components. All unknowns except $\mathbf{T}$ may be eliminated by a moment sum about the line AB. The

① direction of AB is specified by the unit vector $\mathbf{n} = \dfrac{1}{\sqrt{6^2 + 4.5^2}}(4.5\mathbf{j} + 6\mathbf{k})$

$= \frac{1}{5}(3\mathbf{j} + 4\mathbf{k})$. The moment of $\mathbf{T}$ about AB is the component in the direction of AB of the vector moment about the point A and equals

② $\mathbf{r}_1 \times \mathbf{T} \cdot \mathbf{n}$. Similarly the moment of the applied load F about AB is $\mathbf{r}_2 \times \mathbf{F} \cdot \mathbf{n}$. With $\overrightarrow{CD} = \sqrt{46.25}$ m the vector expressions for $\mathbf{T}$, $\mathbf{F}$, $\mathbf{r}_1$, and $\mathbf{r}_2$ are

$$\mathbf{T} = \frac{T}{\sqrt{46.25}}(2\mathbf{i} + 2.5\mathbf{j} - 6\mathbf{k}), \qquad \mathbf{F} = 2\mathbf{j} \text{ kN}$$

$$\mathbf{r}_1 = -\mathbf{i} + 2.5\mathbf{j} \text{ m}, \qquad \mathbf{r}_2 = 2.5\mathbf{i} + 6\mathbf{k} \text{ m}$$

The moment equation now becomes

$$[\Sigma M_{AB} = 0] \qquad (-\mathbf{i} + 2.5\mathbf{j}) \times \frac{T}{\sqrt{46.25}}(2\mathbf{i} + 2.5\mathbf{j} - 6\mathbf{k}) \cdot \tfrac{1}{5}(3\mathbf{j} + 4\mathbf{k})$$

$$+ (2.5\mathbf{i} + 6\mathbf{k}) \times (2\mathbf{j}) \cdot \tfrac{1}{5}(3\mathbf{j} + 4\mathbf{k}) = 0$$

Completion of the vector operations gives

$$-\frac{48T}{\sqrt{46.25}} + 20 = 0, \qquad T = 2.83 \text{ kN} \qquad \textit{Ans.}$$

and the components of T become,

$$T_x = 0.833 \text{ kN} \qquad T_y = 1.042 \text{ kN} \qquad T_z = -2.50 \text{ kN}$$

We may find the remaining unknowns by moment and force summations as follows.

$[\Sigma M_z = 0] \quad 2(2.5) - 4.5B_x - 1.042(3) = 0 \qquad B_x = 0.417 \text{ kN} \qquad \textit{Ans.}$

$[\Sigma M_x = 0] \quad 4.5B_z - 2(6) - 1.042(6) = 0 \qquad B_z = 4.06 \text{ kN} \qquad \textit{Ans.}$

$[\Sigma F_x = 0] \quad A_x + 0.417 + 0.833 = 0 \qquad A_x = -1.250 \text{ kN} \qquad \textit{Ans.}$

③ $[\Sigma F_y = 0] \quad A_y + 2 + 1.042 = 0 \qquad A_y = -3.04 \text{ kN} \qquad \textit{Ans.}$

$[\Sigma F_z = 0] \quad A_z + 4.06 - 2.50 = 0 \qquad A_z = -1.56 \text{ kN} \qquad \textit{Ans.}$

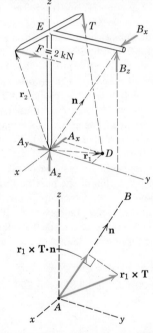

① Recall that the vector $\mathbf{r}$ in the expression $\mathbf{r} \times \mathbf{F}$ for the moment of a force is a vector from the moment center to *any* point on the line of action of the force. Instead of $\mathbf{r}_1$ an equally simple choice would be the vector $\overrightarrow{AC}$.

② The advantage of using vector notation in this problem is the freedom to take moments directly about any axis. In this problem this freedom permits the choice of an axis that eliminates five of the unknowns.

③ The negative signs with the A-components indicate that they are in the opposite direction to those shown on the diagram.

PROBLEMS

54 As a check on the balance of the aircraft each of the three wheels is run onto a scale, and the force readings are $A = 22.0$ kN, $B = 22.4$ kN, $C = 3.43$ kN. Calculate the x-y coordinates of the mass center of the plane through which the total weight of the airplane acts.

Ans. $\bar{x} = -0.0201$ m, $\bar{y} = 0.215$ m

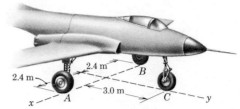

Problem 3/54

55 The two steel I-beams, each weighing 28 lb per foot of length, are welded together at right angles and lifted by the vertical cables so that the beams remain in a horizontal plane. Compute the tension in each of the cables A, B, and C.

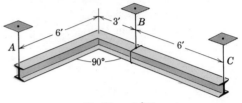

Problem 3/55

56 The mass center of the 30-kg door is in the center of the panel. If the weight of the door is supported entirely by the lower hinge A, calculate the magnitude of the total force supported by the hinge at B.

Ans. $B = 190.2$ N

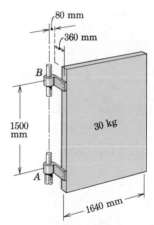

Problem 3/56

57 The industrial door is a uniform rectangular panel weighing 1200 lb and rolls along the fixed rail D on its hanger-mounted wheels A and B. The door is maintained in a vertical plane by the floor-mounted guide roller C which bears against the bottom edge. For the position shown compute the horizontal side thrust on each of the wheels A and B.

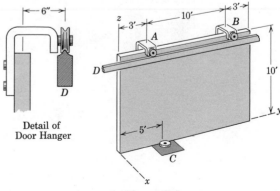

Detail of
Door Hanger

Problem 3/57

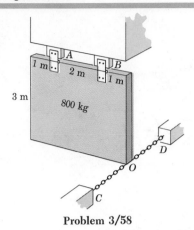

Problem 3/58

3/58 The large rectangular sign has a mass of 800 kg with center of mass in the center of the rectangle. The sign is prevented from swinging by the light chains *OC* and *OD*. The two chains together are slightly longer than the distance from *C* to *D* so that only one of them is taut at any time. Calculate the magnitude of the total force supported by the hinge at *A* when a horizontal wind exerts a force of 2 kN normal to the sign and acting at its center.

Ans. $A = 4200$ N

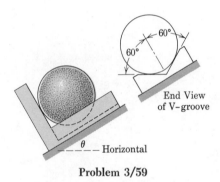

Problem 3/59

3/59 The smooth homogeneous sphere rests in the 120° groove and bears against the end plate which is normal to the direction of the groove. Determine the angle θ, measured from the horizontal, for which the reaction on each side of the groove equals the force supported by the end plate.

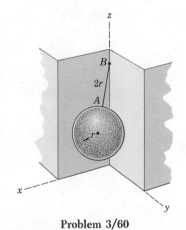

Problem 3/60

3/60 A smooth homogeneous sphere of mass *m* and radius *r* is suspended by a wire *AB* of length 2*r* from point *B* on the line of intersection of the two vertical walls at right angles to one another. Determine the reaction *R* of each wall against the sphere.

Ans. $R = mg/\sqrt{7}$

/61 A high-voltage power line is suspended from a transmission tower by the projecting framework shown. If the tension in the power line is 3 kN, calculate the tension T in link AD and the compression C in links AB and AC.

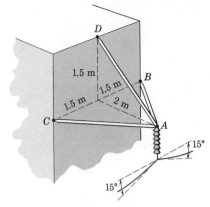

Problem 3/61

/62 The center of mass of the 120-kg workbench is on a vertical line through the center of the square top. A 400-N downward force is applied to the handle of the pipe wrench used to twist the pipe in its flange. If legs A and C are slightly shorter than legs B and D, calculate the forces supported by the three legs which contact the floor. Assume contact takes place under the outside corners of the legs and neglect the mass of the wrench, pipe, and vise compared with the mass of the table.

        Ans. $B = 522$ N, $C = 367$ N, $D = 689$ N

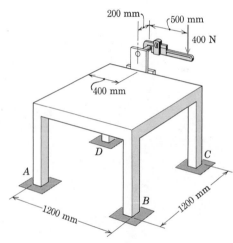

Problem 3/62

/63 The uniform ventilator door has a mass of 200 kg and is hinged at the corners A and B of its upper edge. The door is held open in a horizontal position by the wire from C to point D in the vertical wall. Compute the tension in the wire and the forces normal to the hinge axis supported by the hinge pins at A and B.

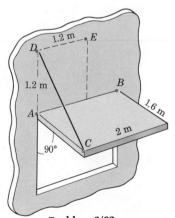

Problem 3/63

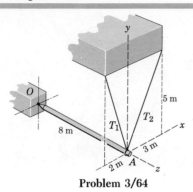

Problem 3/64

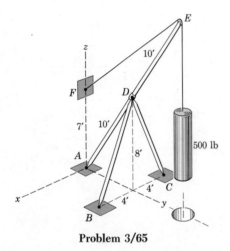

Problem 3/65

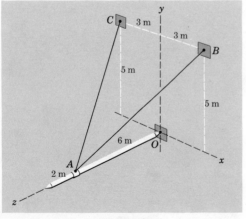

Problem 3/66

3/64 The uniform horizontal boom has a mass of 160 kg and is supported by the two cables in the vertical *x-y* plane and by the ball-and-socket joint at *O*. Calculate the tensions T_1 and T_2 in the cables.

Ans. $T_1 = 507$ N, $T_2 = 366$ N

3/65 The three poles are erected as shown and support the 500-lb cylinder prior to lowering it into the hole. The cable attached to the cylinder passes over a small pulley at *E* and is secured at point *F*. The connections at the ends of the poles may be treated as ball-and-socket joints, and the weight of the poles is small compared with the loads they support. Calculate the compression *P* in each of the equal legs *BD* and *CD* and find the magnitude of the total force at *A*.

3/66 The uniform horizontal boom has a mass of 240 kg and is supported by the two cables anchored at *B* and *C* and by the ball-and-socket joint at *O*. Calculate the tension *T* in cable *AC*.   *Ans.* $T = 1310$ N

67 A uniform steel ring 60 in. in diameter and weighing
600 lb is lifted by the three cables, each 50 in. long,
attached at points A, B, and C as shown. Compute
the tension in each cable.

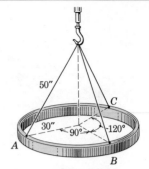

Problem 3/67

68 The square steel plate has a mass of 1800 kg with
mass center at its center G. Calculate the tension in
each of the three cables with which the plate is
lifted while remaining horizontal.

$$Ans. \ T_A = T_B = 5.41 \ \text{kN}, \ T_C = 9.87 \ \text{kN}$$

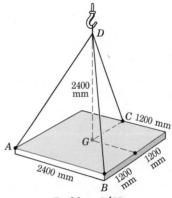

Problem 3/68

69 Determine the force in each member of the tripod.
Each of the three members is secured at its ends by
ball-and-socket connections and is capable of supporting tension or compression. The weights of the
members may be neglected.

$$Ans. \ A = 2.041 \ \text{kN tension}$$
$$B = 0.861 \ \text{kN compression}$$
$$C = 1.269 \ \text{kN compression}$$

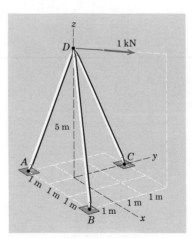

Problem 3/69

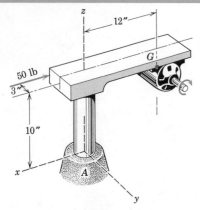

Problem 3/70

3/70 The rigid unit of post, bracket, and motor weighs 60 lb with its mass center at G located 12 in. from the vertical centerline of the post. The post is welded to the fixed base at A. The motor, which drives a machine through a flexible shaft, turns in the direction indicated and delivers a torque of 1500 lb-in. In addition, a 50-lb force is applied to the bracket as shown. Determine the vector expressions for the total force **R** and moment **M** applied to the post at A by the supporting base. (*Caution:* Be careful to assign the torque (couple) which acts on the motor shaft in its correct sense consistent with Newton's third law.)

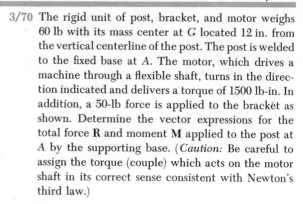

Problem 3/71

3/71 A torque of 20 N·m is applied to the flexible shaft as shown. After the initial angular twist has taken place, one revolution of the input end will be accompanied by one revolution of the output end, and the output torque will equal the input torque. The shaft housing is mounted in a self-aligning support at B which is capable of exerting a single force normal to the shaft axis. Support A, on the other hand, is a loose-fitting sleeve and can support the shaft housing in either of the two ways shown in the separate sectional views. Insofar as equilibrium is concerned the shaft housing may be treated as a rigid body for the position in which it is bent. Determine the forces exerted by the supports at B, C, and D on the housing. Neglect the weight of the shaft. (*Caution:* Observe Newton's third law at the output end regarding the torque applied *to* the shaft.)   *Ans.* $\mathbf{B} = \mathbf{D} = 50\mathbf{k}$ N, $\mathbf{C} = -100\mathbf{k}$ N

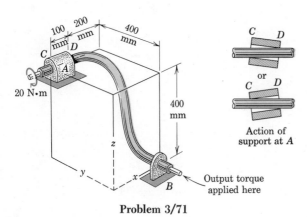

Problem 3/72

3/72 The uniform door weighs 50 lb and is hinged at A and B with hinge A carrying the entire weight. A 40-lb horizontal pull parallel to the suface of the door is applied to the handle of a claw hammer to remove a nail. The door is prevented from opening further by a stop at C which exerts a force at the lower edge of the door normal to its surface. Determine the magnitude of the horizontal force supported by the hinge at B.

73 Gear *C* drives the V-belt pulley *D* at a constant speed. For the belt tensions shown calculate the gear-tooth force *P* and the magnitudes of the total forces supported by the bearings at *A* and *B*.
     Ans. $P = 70.9$ N, $A = 83.3$ N, $B = 208$ N

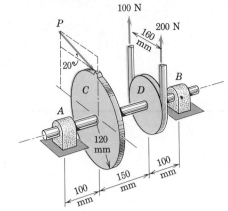

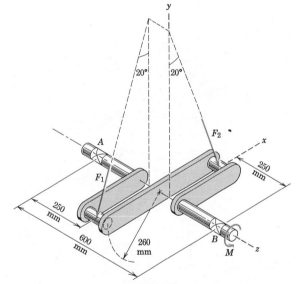

Problem 3/73

74 The crankshaft for a two-cylinder compressor is mounted in bearings at *A* and *B*. For the position shown the forces exerted by the connecting rods on the crank pins are $F_1 = 16$ kN and $F_2 = 8$ kN in the directions indicated. For an equilibrium condition of the crankshaft calculate the *x*- and *y*-components of the total forces supported by the bearings at *A* and *B* and the torque *M* applied to the shaft.

Problem 3/74

75 If the ventilator door of Prob. 3/63 is held in position by a wire from *C* to *E* rather than from *C* to *D*, compute the tension in the wire and the forces normal to the hinge axis supported by the hinge pins.   *Ans.* $T = 1907$ N, $A = 523$ N, $B = 1256$ N

76 If the weight of the mast is negligible compared with the applied 30-kN load, determine the two cable tensions T_1 and T_2 and the force *A* acting at the ball joint at *A*.

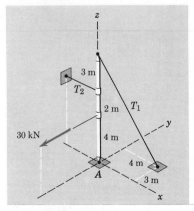

Problem 3/76

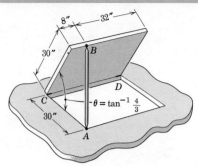

Problem 3/77

3/77 The uniform 30- by 40-in. trap door weighs 200 lb and is propped open by the light strut AB at the angle $\theta = \tan^{-1}(4/3)$. Calculate the compression F_B in the strut and the force supported by the hinge D normal to the hinge axis. Assume that the hinges act at the extreme ends of the lower edge.

> *Ans.* $F_B = 70$ lb, $D_n = 101$ lb

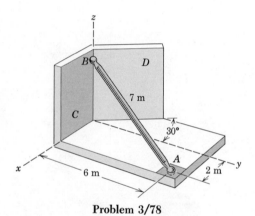

Problem 3/78

3/78 One of the vertical walls supporting end B of the 200-kg uniform shaft of Sample Problem 3/3 is turned through a 30° angle as shown here. End A is still supported by the ball-and-socket connection in the horizontal x-y plane. Calculate the magnitudes of the forces $\mathbf{P}$ and $\mathbf{R}$ exerted on the ball end B of the shaft by the vertical walls C and D, respectively.

> *Ans.* $P = 1584$ N, $R = 755$ N

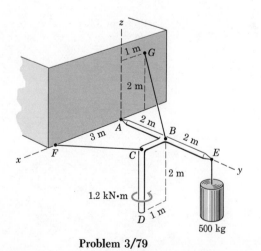

Problem 3/79

3/79 The light rigid frame is secured to the vertical wall by a ball-and-socket joint at A and by cables BG and CF. In addition to supporting the 500-kg load, the frame supports a 1.2-kN·m couple applied to the vertical leg CD. Compute the y-component of the force supported by the ball-and-socket joint at A. (*Suggestion:* Use one vector equation which eliminates reference to all unknowns other than A_y.)

3/80 The 9-m steel boom has a mass of 600 kg with center of mass at midlength. It is supported by a ball-and-socket joint at A and the two cables under tensions T_1 and T_2. The cable which supports the 2000-kg load leads through a sheave at B and is secured to the vertical x-y plane at F. Calculate the magnitude of the tension T_1. (*Hint:* Write a moment equation which eliminates all unknowns except T_1.)

Ans. $T_1 = 19.76$ kN

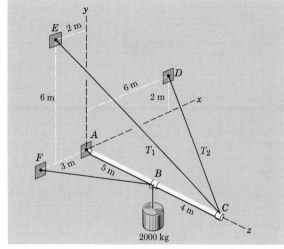

Problem 3/80

3/81 The welded tubular frame of negligible mass is supported by a ball-and-socket joint at A, by a smooth ring at B, and by the wire from the strut at C to the fixed point D. The frame is loaded by the 300-lb ball that is welded to the horizontal strut. Calculate the force exerted on the frame by the fixed ring at B. (*Hint:* The problem may be solved without involving the force at A.)

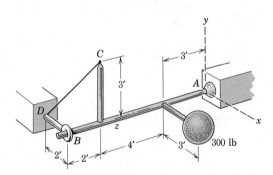

Problem 3/81

3/82 The rigid member ABC is attached to the vertical x-y surface by a ball-and-socket joint at A and is supported by the cables BE and CD. The mass of the member may be neglected compared with that of the 5-Mg load which it supports. There is one position of D along the horizontal slot through which the cable must be passed and secured for the member to maintain the position shown. Find x.

Ans. $x = 3.75$ m

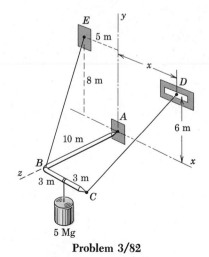

Problem 3/82

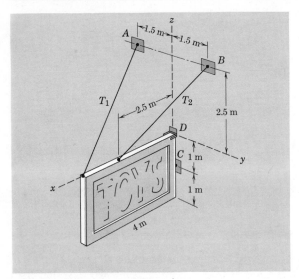

Problem 3/83

▶3/83 A rectangular sign over a store has a mass of 100 kg, with the center of mass in the center of the rectangle. The support against the wall at point C may be treated as a ball-and-socket joint. At corner D support is provided in the y-direction only. Calculate the tensions T_1 and T_2 in the supporting wires, the total force supported at C, and the lateral force R supported at D.

Ans. $T_1 = 347$ N $T_2 = 431$ N
$R = 63.1$ N $C = 768$ N

3/5 PROBLEM FORMULATION AND REVIEW. In Chapter 3 we have applied our knowledge of the properties of forces, moments, and couples studied in Chapter 2 to solve problems in the equilibrium of rigid bodies. Each such body in complete equilibrium is characterized by the requirement that the vector resultant of all forces acting on it is zero ($\Sigma \mathbf{F} = \mathbf{0}$) and the vector resultant of all moments on the body about a point (or axis) is also zero ($\Sigma \mathbf{M} = \mathbf{0}$). We are guided in all of our solutions by these two requirements which are easily comprehended physically.

As is often the case it is not the theory but its application which presents the difficulty. The crucial steps in applying our principles of equilibrium should be quite familiar by now. They are:

1. Make an unequivocal decision as to which body in equilibrium is to be analyzed.
2. Isolate the body in question from all contacting bodies by drawing its *free-body diagram* upon which *all* forces that act on the isolated body from external sources are represented.
3. Observe the principle of action and reaction (Newton's third law) when assigning the sense of each force.
4. Label all forces, known and unknown.
5. Choose and label reference axes, always using a right-handed set for three-dimensional analysis.

These five steps should become automatic, and following them consistently before beginning the equilibrium calculation is the very best assurance of a correct solution.

When solving an equilibrium problem we should first check to see that the body is statically determinate. If there are more supports than are necessary to hold the body in place, the body is statically indeterminate, and the equations of equilibrium by themselves will not enable us to solve for all of the external reactions. In applying the equations of equilibrium, we choose scalar algebra, vector algebra, or graphical analysis according to both preference and experience, with vector algebra being particularly useful for many three-dimensional problems.

One of our most useful procedures is to simplify the algebra of a solution by the choice of a moment axis that eliminates as many unknowns as possible or by the choice of a direction for a force summation that avoids reference to certain unknowns. A few moments of thought to take advantage of these simplifications can save appreciable time and effort.

The principles and methods which are covered in Chapters 2 and 3 constitute the most basic part of statics. They lay the foundation for what follows not only in statics but in dynamics as well.

REVIEW PROBLEMS

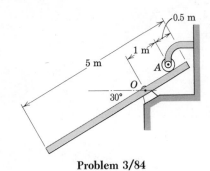

Problem 3/84

3/84 The uniform 5-m bar with a mass of 100 kg is hinged at *O* and prevented from rotating in the vertical plane beyond the 30° position by the fixed roller at *A*. Calculate the magnitude of the total force supported by the pin at *O*.   *Ans. O* = 1769 N

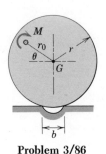

Problem 3/85

3/85 The jaws of the open-ended wrench are smooth case-hardened steel and, for the given thickness of the wrench, can withstand up to 7.5 kN of concentrated force against the corner of a hexagonal hardened bolt without surface damage. Calculate the maximum force *P* at a lever arm of 300 mm that can safely be applied to the wrench. Assume that there is a very slight clearance between the head of the bolt and the jaws of the wrench.

Ans. P = 0.689 kN

Problem 3/86

3/86 A wheel of mass *m* and radius *r* with its mass center *G* at the geometric center rests in a small depression of width *b*. Determine the minimum couple *M* applied to the wheel at the location shown in order to roll the wheel out of the depression. Assume no slipping occurs. What is the influence of r_0, *r*, and θ?

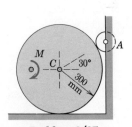

Problem 3/87

3/87 The 100-kg wheel rests on a rough surface and bears against the roller *A* when the couple *M* is applied to it. If *M* = 60 N·m with the wheel not slipping, compute the reaction on the roller *A*.

Ans. F_A = 231 N

3/88 A large symmetrical drum for drying sand is operated by the geared motor drive shown. If the mass of
the sand is 750 kg and an average gear-tooth force of
2.6 kN is supplied by the motor pinion A to the
drum gear normal to the contacting surfaces at B,
calculate the average offset $\bar{x}$ of the center of mass
G of the sand from the vertical centerline. Neglect
all friction in the supporting rollers.

Ans. $\bar{x} = 199$ mm

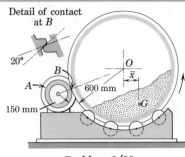

Detail of contact at B

Problem 3/88

3/89 The roller-band device consists of two rollers, each
of radius r, encircled by a flexible band of negligible
thickness and subjected to the two tensions T. Write
the expression for the contact force R between the
band and the flat supporting surfaces at A and B.
The action is in the horizontal plane, so that the
weights of the rollers and band are not involved.

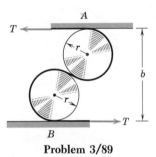

Problem 3/89

3/90 The device shown in section can support the load L
at various heights by resetting the pawl C in another
tooth at the desired height on the fixed vertical
column D. Determine the distance b at which the
load should be positioned in order for the two rollers
A and B to support equal forces. The weight of the
device is negligible compared with L.

Ans. $b = 10.33$ in.

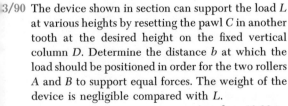
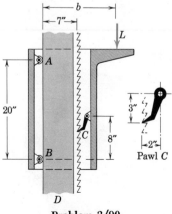

Problem 3/90

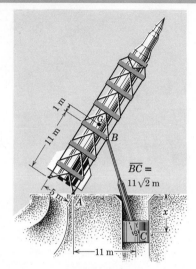

Problem 3/91

3/91 The elevating structure for a rocket test stand and the rocket within it have a combined mass of 635 Mg with center of mass at G. For the position in which $x = 5$ m determine the equilibrium value of the force on the hinge axis A. Solve graphically.

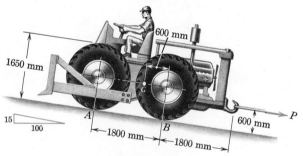

Problem 3/92

3/92 The rubber-tired tractor shown has a mass of 13.5 Mg with center of mass at G and is used for pushing or pulling heavy loads. Determine the load P which the tractor can pull at a constant speed of 5 km/h up the 15-percent grade if the driving force exerted by the ground on each of its four wheels is 80 percent of the normal force under that wheel. Also find the total normal reaction N_B under the rear pair of wheels at B.

Ans. $P = 85$ kN, $N_B = 125$ kN

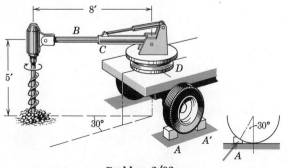

Problem 3/93

3/93 The power unit of the post-hole digger supplies a torque of 4000 lb-in. to the auger. The arm B is free to slide in the supporting sleeve C but is not free to rotate about the horizontal axis of C. If the unit is free to swivel about the vertical axis of the mount D, determine the force exerted against the right rear wheel by the block A (or A') which prevents the unbraked truck from rolling. (*Hint:* View the system from above.)   *Ans. $A' = 41.7$ lb*

3/94 The uniform steel plate of mass m is triangular in shape and is supported in the horizontal plane by three vertical wires from its vertices. Show that the tension in each wire is always $mg/3$ regardless of the shape of the triangle.

Problem 3/94

3/95 A vertical force P on the foot pedal of the bell crank is required to produce a tension T of 400 N in the vertical control rod. Determine the corresponding bearing reactions at A and B.

Ans. $A = 184$ N, $B = 424$ N

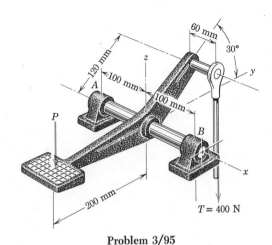

Problem 3/95

3/96 One of the three landing pads for the Mars Viking lander is shown in the figure with its approximate dimensions. The mass of the lander is 600 kg. Compute the force in each leg when the lander is resting on a horizontal surface on Mars. (Assume equal support by the pads and consult Table C2 in Appendix C as needed.)

Ans. $F_{CD} = 1046$ N compression
$F_{AC} = F_{CB} = 240$ N tension

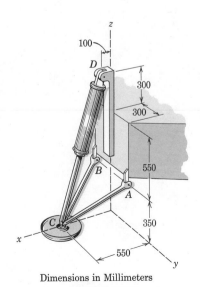

Dimensions in Millimeters
Problem 3/96

Problem 3/97

3/97 Three identical steel balls each of mass m are placed in the cylindrical ring which rests on a horizontal surface and whose height is slightly greater than the radius of the balls. The diameter of the ring is such that the balls are virtually touching one another. A fourth identical ball is then placed on top of the three balls. Determine the force P exerted by the ring on each of the three lower balls.

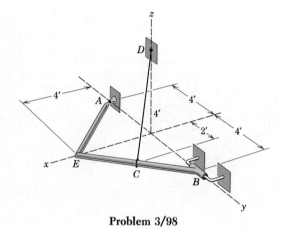

Problem 3/98

3/98 Each of the two legs of the welded frame weighs 100 lb. A wire from C to D prevents the frame from rotating out of the horizontal plane about an axis through its bearing at B and its ball-and-socket joint at A. Calculate the tension T in the wire and the magnitude of the total force supported by the connection at A.   *Ans. $T = 245$ lb, $A = 122.5$ lb*

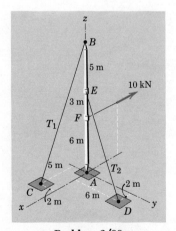

Problem 3/99

3/99 The mast has a mass of 300 kg and is supported by a ball-and-socket joint at A. Calculate the tension T_1 if the horizontal 10-kN force is applied at F.
                                        Ans. $T_1 = 11.34$ kN

STRUCTURES

4

4/1 INTRODUCTION. In Chapter 3 we focused our attention on the equilibrium of a single rigid body or upon a system of connected members which, when taken as a whole, could be treated as a single rigid body. In these problems we first drew a free-body diagram of this single body showing all forces external to the isolated body and then we applied the force and moment equations of equilibrium. In Chapter 4 our attention is directed toward the determination of the forces internal to a structure, that is, forces of action and reaction between the connected members. An engineering structure is any connected system of members built to support or transfer forces and to withstand safely the loads applied to it. In the force analysis of structures it is necessary to dismember the structure and to analyze separate free-body diagrams of individual members or combinations of members in order to determine the forces internal to the structure. This analysis calls for very careful observance of Newton's third law, which states that each action is accompanied by an equal and opposite reaction.

In Chapter 4 we shall analyze the internal forces acting in several types of structures, namely, trusses, frames, and machines. In this treatment we shall consider only *statically determinate* structures, that is, structures which do not have more supporting constraints than are necessary to maintain an equilibrium configuration. Thus, as we have already seen, the equations of equilibrium are adequate to determine all unknown reactions.

The student who has mastered the basic procedure developed in Chapter 3 of defining unambiguously the body under consideration by constructing a correct free-body diagram will have little difficulty with the analysis of statically determinate structures. The analysis of trusses, frames and machines, and beams under concentrated loads constitutes a straightforward application of the material developed in the previous two chapters.

4/2 PLANE TRUSSES. A framework composed of members joined at their ends to form a rigid structure is known as a truss. Bridges, roof supports, derricks, and other such structures are common examples of trusses. Structural members used are I-beams, channels, angles, bars, and special shapes which are fastened together at their ends by welding, riveted connections, or large bolts or pins. When the members of the truss lie essentially in a single plane, the truss is known as a *plane truss*. Plane trusses, such as those used for bridges, are commonly designed in pairs with one truss panel placed on each side of the bridge and connected together by cross beams that support the roadway and transfer the applied loads to the truss members. Several examples of commonly used trusses that can be analyzed as plane trusses are shown in Fig. 4/1.

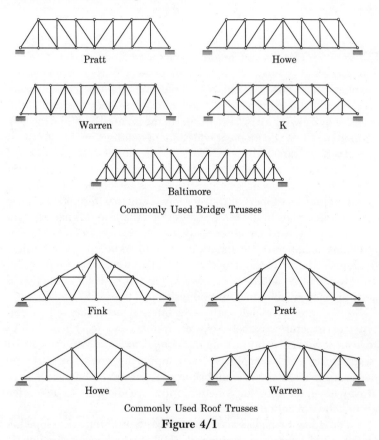

Commonly Used Bridge Trusses

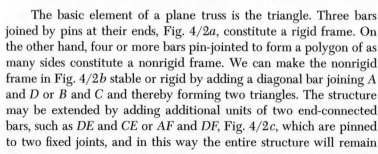

Commonly Used Roof Trusses

Figure 4/1

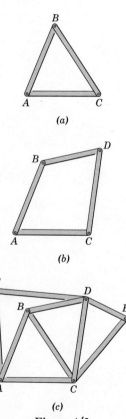

(a)

(b)

(c)

Figure 4/2

The basic element of a plane truss is the triangle. Three bars joined by pins at their ends, Fig. 4/2a, constitute a rigid frame. On the other hand, four or more bars pin-jointed to form a polygon of as many sides constitute a nonrigid frame. We can make the nonrigid frame in Fig. 4/2b stable or rigid by adding a diagonal bar joining A and D or B and C and thereby forming two triangles. The structure may be extended by adding additional units of two end-connected bars, such as DE and CE or AF and DF, Fig. 4/2c, which are pinned to two fixed joints, and in this way the entire structure will remain rigid. The term rigid is used in the sense of noncollapsible and also in the sense that deformation of the members due to induced internal strains is negligible.

Structures that are built from a basic triangle in the manner described are known as *simple trusses*. When more members are present than are needed to prevent collapse, the truss is statically indeterminate. A statically indeterminate truss cannot be analyzed by the equations of equilibrium alone. Additional members or supports that are not necessary for maintaining the equilibrium position are called *redundant*.

The design of a truss involves the determination of the forces in the various members and the selection of appropriate sizes and structural shapes to withstand the forces. Several assumptions are made in the force analysis of simple trusses. First, we assume all members to be *two-force members*. A two-force member is one in equilibrium under the action of two forces only, as defined in general terms with Fig. 3/4 in Art. 3/3. For trusses each member is a straight link joining the two points of application of force. The two forces are applied at the ends of the member and are necessarily equal, opposite, and *collinear* for equilibrium. We see that the member may be in tension or compression, as shown in Fig. 4/3. When we represent the equilibrium of a portion of a two-force member, the tension T or compression C acting on the cut section is the same for all sections. We assume here that the weight of the member is small compared with the force it supports. If it is not, or if the small effect of the weight is to be accounted for, the weight W of the member, if uniform, may be assumed to be replaced by two forces, each $W/2$, acting at each end of the member. These forces, in effect, are treated as loads externally applied to the pin connections. Accounting for the weight of a member in this way gives the correct result for the average tension or compression along the member but will not account for the effect of bending of the member.

When welded or riveted connections are used to join structural members, the assumption of a pin-jointed connection is usually satisfactory if the centerlines of the members are concurrent at the joint as in Fig. 4/4.

We also assume in the analysis of simple trusses that all external forces are applied at the pin connections. This condition is satisfied in most trusses. In bridge trusses the deck is usually laid on cross beams that are supported at the joints.

Provision for expansion and contraction due to temperature changes and for deformations resulting from applied loads is usually made at one of the supports for large trusses. A roller, rocker, or some kind of slip joint is provided. Trusses and frames wherein such provision is not made are statically indeterminate, as explained in Art. 3/3.

Two methods for the force analysis of simple trusses will be

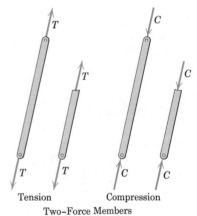

Tension   Compression
Two-Force Members
Figure 4/3

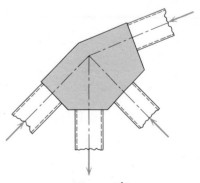

Figure 4/4

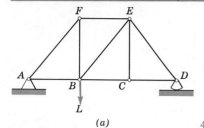

(a)

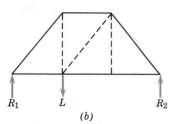

(b)

Figure 4/5

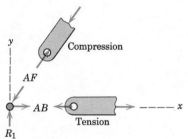

Figure 4/6

given, and reference will be made to the simple truss shown in Fig. 4/5*a* for each of the two methods. The free-body diagram of the truss as a whole is shown in Fig. 4/5*b*. The external reactions are usually determined by computation from the equilibrium equations applied to the truss as a whole before proceeding with the force analysis of the remainder of the truss.

4/3 METHOD OF JOINTS. This method for finding the forces in the members of a simple truss consists of satisfying the conditions of equilibrium for the forces acting on the connecting pin of each joint. The method therefore deals with the equilibrium of concurrent forces, and only two independent equilibrium equations are involved. We begin the analysis with any joint where at least one known load exists and where not more than two unknown forces are present. Solution may be started with the pin at the left end, and its free-body diagram is shown in Fig. 4/6. With the joints indicated by letters, we may designate the force in each member by the two letters defining the ends of the member. The proper directions of the forces should be evident for this simple case by inspection. The free-body diagrams of portions of members AF and AB are also shown to indicate clearly the mechanism of the action and reaction. The member AB actually makes contact on the left side of the pin, although the force AB is drawn from the right side and is shown acting away from the pin. Thus, if we consistently draw the force arrows on the *same* side of the pin as the member, then tension (such as AB) will always be indicated by an arrow *away* from the pin, and compression (such as AF) will always be indicated by an arrow *toward* the pin. The magnitude of AF is obtained from the equation $\Sigma F_y = 0$ and AB is then found from $\Sigma F_x = 0$.

Joint F must be analyzed next, since it now contains only two unknowns, EF and BF. Joints B, C, E, and D are subsequently analyzed in that order. The free-body diagram of each joint and its corresponding force polygon which represents graphically the two equilibrium conditions $\Sigma F_x = 0$ and $\Sigma F_y = 0$ are shown in Fig. 4/7. The numbers indicate the order in which the joints are analyzed. We note that, when joint D is finally reached, the computed reaction R_2 must be in equilibrium with the forces in members CD and ED, determined previously from the two neighboring joints. This requirement will provide a check on the correctness of our work. We should also note that isolation of joint C quickly discloses the fact that the force in CE is zero when the equation $\Sigma F_y = 0$ is applied. The force in this member would not be zero, of course, if an external vertical load were applied at C.

It is often convenient to indicate the tension T and compression C of the various members directly on the original truss diagram by drawing arrows away from the pins for tension and toward the pins

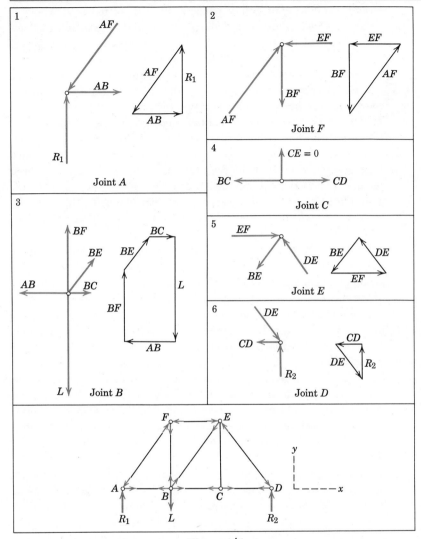

Figure 4/7

for compression. This designation is illustrated at the bottom of Fig. 4/7.

In some instances it is not possible to assign initially the correct direction of one or both of the unknown forces acting on a given pin. In this event we may make an arbitrary assignment. A negative value from the computation indicates that the assumed direction is incorrect.

If a simple truss has more external supports than are necessary to ensure a stable equilibrium configuration, the truss as a whole is statically indeterminate, and the extra supports constitute *external* redundancy. If the truss has more internal members than are necessary to prevent collapse, then the extra members constitute *internal* redundancy and the truss is statically indeterminate. For a truss that is statically determinate externally, there is a definite relation between the number of its members and the number of its joints necessary for internal stability without redundancy. Since we can specify the equilibrium of each joint by two scalar force equations, there are in all $2j$ such equations for a simple truss with j joints. For the entire truss composed of m two-force members with a maximum of three unknown support reactions, there are in all $m + 3$ unknowns. Thus, for a simple plane truss composed of triangular elements, the equation $m + 3 = 2j$ will be satisfied if the truss is statically determinate internally.

This relation is a necessary condition for stability but it is not a sufficient condition, since one or more of the m members can be arranged in such a way as not to contribute to a stable configuration of the entire truss. If $m + 3 > 2j$, there are more members than there are independent equations, and the truss is statically indeterminate internally with redundant members present. If $m + 3 < 2j$, there is a deficiency of internal members, and the truss is unstable and will collapse under load.

The force polygon for each joint, shown in Fig. 4/7, may be constructed graphically to obtain the unknown forces in the members as an alternative to or as a check on the algebraic calculations using the force equations of equilibrium. If a consistent sequence around each joint, clockwise, for example, has been used for the addition of the forces, we may superpose these force polygons on one another to form a composite graphical figure known as the *Maxwell diagram.*[*] The force and its sense may be obtained directly from the diagram. The student who is interested in structures may wish to experiment with this construction and to consult other books dealing more completely with structural analysis for a more detailed description of the Maxwell diagram.

Special conditions. We draw attention to several special conditions which occur frequently in the analysis of simple trusses. When two collinear members are under compression, as indicated in Fig. 4/8a, it is necessary to add a third member to maintain alignment of the two members and prevent buckling. We see very quickly from a force summation in the y-direction that the force F_3 in the third member must be zero and from the x-direction that $F_1 = F_2$. This conclusion holds regardless of the angle θ and, of course, holds if the collinear members are in tension. If an external force with a compo-

---

[*] The method was published by James Clerk Maxwell in 1864.

nent in the y-direction were applied to the joint, then, of course, F_3 would no longer be zero.

When two noncollinear members are joined as shown in Fig. 4/8b, then in the absence of an externally applied load at this joint, the forces in both members must be zero as we see from the two force summations.

When two pairs of collinear members are joined as shown in Fig. 4/8c, the forces in each pair must be equal and opposite. This conclusion follows from the force summations indicated in the figure.

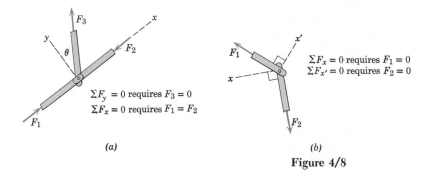

(a)   (b)   (c)

Figure 4/8

Truss panels are frequently cross-braced as shown in Fig. 4/9a. Such a panel is statically indeterminate if each brace is capable of supporting either tension or compression. However, when the braces are flexible members incapable of supporting compression, as are cables, then only the tension member acts and the other member is disregarded. It is usually evident from the asymmetry of the loading how the panel will deflect. If the deflection is as indicated in Fig. 4/9b, then member AB should be retained and CD disregarded. When this choice cannot be made by inspection, we may make an arbitrary selection of the member to be retained. If the assumed tension turns out to be positive upon calculation, then the choice was correct. If the assumed tension force turns out to be negative, then the opposite member must be retained and the calculation redone.

The simultaneous solution of the equations for two unknown forces at a joint may be avoided by a careful choice of reference axes. Thus for the joint indicated schematically in Fig. 4/10 where L is known and F_1 and F_2 are unknown, a force summation in the x-direction eliminates reference to F_1 and a force summation in the x'-direction eliminates reference to F_2. When the angles involved are not easily found, then a simultaneous solution of the equations using one set of reference directions for both unknowns may be preferable.

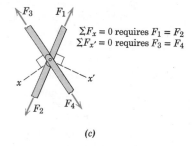

(a)   (b)

Figure 4/9

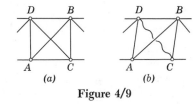

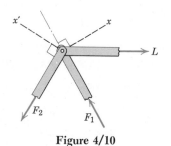

Figure 4/10

Sample Problem 4/1

Compute the force in each member of the loaded cantilever truss by the method of joints.

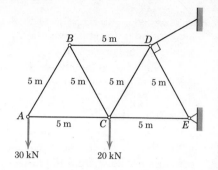

Solution. If it were not desired to calculate the external reactions at D and E, the analysis for a cantilever truss could begin with the joint at the loaded end. However, this truss will be analyzed completely, so the first step will be to compute the external forces at D and E from the free-body diagram of the truss as a whole. The equations of equilibrium give

$[\Sigma M_E = 0]$ $5T - 20(5) - 30(10) = 0$ $T = 80.0$ kN

$[\Sigma F_x = 0]$ $80.0 \cos 30° - E_x = 0$ $E_x = 69.3$ kN

$[\Sigma F_y = 0]$ $80.0 \sin 30° + E_y - 20 - 30 = 0$ $E_y = 10.0$ kN

Next we draw free-body diagrams showing the forces acting on each of the connecting pins. The correctness of the assigned directions of the forces is verified when each joint is considered in sequence. There should be no question about the correct direction of the forces on joint A. Equilibrium requires

$[\Sigma F_y = 0]$ $0.866AB - 30 = 0$ $AB = 34.64$ kN T *Ans.*

$[\Sigma F_x = 0]$ $AC - 0.5(34.64) = 0$ $AC = 17.32$ kN C *Ans.*

① where T stands for tension and C stands for compression.

Joint B must be analyzed next, since there are more than two unknown forces on joint C. The force BC must provide an upward component, in which case BD must balance the force to the left. Again the forces are obtained from

$[\Sigma F_y = 0]$ $0.866BC - 0.866(34.64) = 0$ $BC = 34.64$ kN C *Ans.*

$[\Sigma F_x = 0]$ $BD - 2(0.5)(34.64) = 0$ $BD = 34.64$ kN T *Ans.*

Joint C now contains only two unknowns, and these are found in the same way as before:

$[\Sigma F_y = 0]$ $0.866CD - 0.866(34.64) - 20 = 0$

$CD = 57.74$ kN T *Ans.*

$[\Sigma F_x = 0]$ $CE - 17.32 - 0.5(34.64) - 0.5(57.74) = 0$

$CE = 63.51$ kN C *Ans.*

Finally, from joint E there results

$[\Sigma F_y = 0]$ $0.866DE = 10.00$ $DE = 11.55$ kN   *Ans.*

and the equation $\Sigma F_x = 0$ checks.

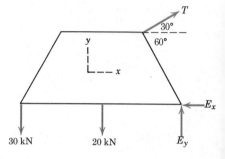

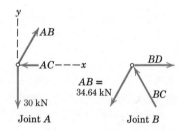

① Note that we draw the force arrow on the same side of the joint as the member which exerts the force. In this way tension (arrow away from the joint) is distinguished from compression (arrow toward the joint).

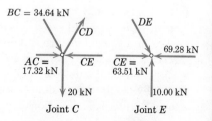

PROBLEMS

(Solve the following problems by the method of joints. Neglect the weights of the members compared with the forces they support unless otherwise indicated.)

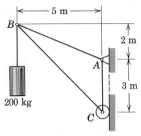

Problem 4/1

4/1 Calculate the force in each member of the loaded truss.

Ans. AB = 3.52 kN T
BC = 4.62 kN C
AC = 3.27 kN T

4/2 Calculate the force in each member of the truss. (1 kip = 1000 lb)

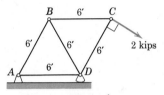

Problem 4/2

4/3 Calculate the force in member *CF* of the loaded truss.

Ans. CF = 3.33 kN C

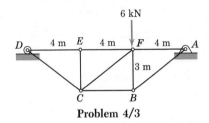

Problem 4/3

4/4 Calculate the force in each member of the loaded truss. All triangles are isosceles.

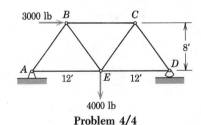

Problem 4/4

4/5 Calculate the forces in members *CG* and *CF* for the truss shown.

Ans. CG = 2.24 kN T
CF = 1.00 kN C

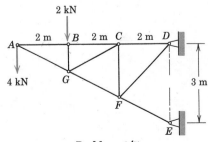

Problem 4/5

4/6 If the 2-kN force acting on the truss of Prob. 4/5 were removed, identify by inspection those members in which the forces are zero. On the other hand, if the 2-kN force were applied at *G* instead of *B*, would there be any zero-force members?

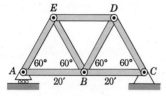

Problem 4/7

4/7 Each member of the truss is a uniform 20-ft bar weighing 400 lb. Calculate the average tension or compression in each member due to the weights of the members.

$$Ans. \ AE = CD = 2000/\sqrt{3} \text{ lb } C$$
$$AB = BC = 1000/\sqrt{3} \text{ lb } T$$
$$BE = BD = 800/\sqrt{3} \text{ lb } T$$
$$DE = 1400/\sqrt{3} \text{ lb } C$$

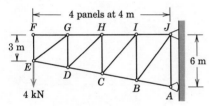

Problem 4/8

4/8 Calculate the forces in members *FG*, *EG*, and *GD* for the loaded cantilever truss.

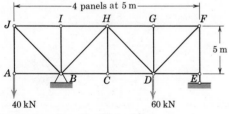

Problem 4/9

4/9 Calculate the forces in members *JB* and *BH* for the loaded truss. $Ans. \ JB = 56.6$ kN C
$$BH = 47.1 \text{ kN } C$$

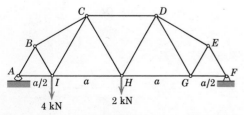

Problem 4/10

4/10 Determine the forces in members *BI*, *CI*, and *HI* for the loaded truss. All angles are 30°, 60°, or 90°.

4/11 A snow load transfers the forces shown to the upper joints of a Pratt roof truss. Neglect any horizontal reactions at the supports and compute the forces in members *BH*, *BC*, and *CH*.

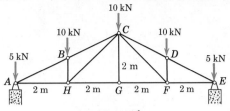

Problem 4/11

4/12 Calculate the forces induced in members *GH* and *ED* for the crane truss when it lifts an 1800-kg car.

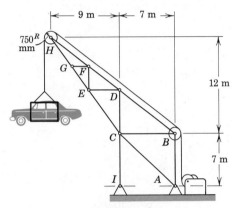

Problem 4/12

4/13 The signboard truss is designed to support a horizontal wind load of 800 lb. A separate analysis shows that $\frac{5}{8}$ of this force is transmitted to the center connection at *C* and the rest is equally divided between *D* and *B*. Calculate the forces in members *BE* and *BC*.   *Ans. BE* = 559 lb *T*, *BC* = 300 lb *T*

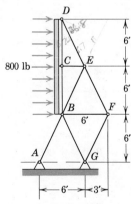

Problem 4/13

4/14 Calculate the forces in members *CF*, *BF*, *BG*, and *FG* for the simple crane truss.

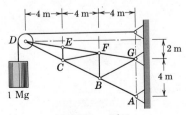

Problem 4/14

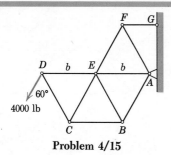

Problem 4/15

4/15 Calculate the forces in all members of the loaded truss supported by the horizontal link *FG* and the hinge at *A*. All interior angles are 60°.

Ans. $AB = CB = DC = 4000$ lb C
$BE = CE = DE = 4000$ lb T, $AE = 0$
$EF = 8000$ lb T, $AF = 8000$ lb C

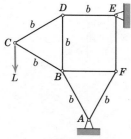

Problem 4/16

4/16 Show that the truss is statically determinate and determine the forces in members *BD* and *BF* in terms of the applied load *L*.

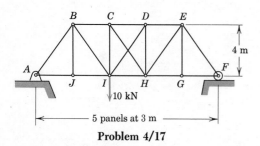

Problem 4/17

4/17 Calculate the forces in members *AB*, *BJ*, *BI*, and *CI*. Members *CH* and *DI* are cables that are capable of supporting tension only.

Ans. $AB = 7.5$ kN C, $BJ = CI = 0$, $BI = 7.5$ kN T

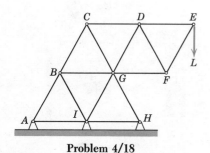

Problem 4/18

4/18 By inspection designate those members of the truss that cause the structure to be statically indeterminate.

4/19 Each of the loaded trusses has supporting constraints which are statically indeterminate. List all members of each truss whose forces are not affected by the indeterminacy of the supports and that may be computed directly by using only the equations of equilibrium. Assume that the loading and dimensions of the trusses are known.

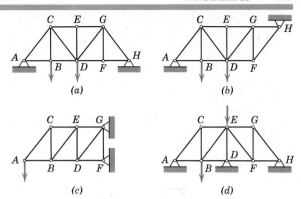

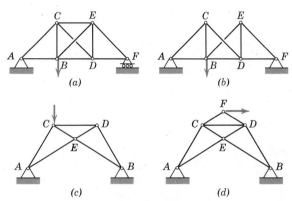

Problem 4/19

4/20 Verify the fact that each of the trusses contains one or more elements of redundancy, and propose two separate changes, either one of which would remove the redundancy and produce complete statical determinacy. All members can support compression as well as tension.

Problem 4/20

4/21 The movable gantry is used to erect and prepare a 500-Mg rocket for firing. The primary structure of the gantry is approximated by the symmetrical plane truss shown, which is statically indeterminate. As the gantry is positioning a 60-Mg section of the rocket suspended from A, strain gage measurements indicate a compressive force of 50 kN in member AB and a tensile force of 120 kN in member CD due to the 60-Mg load. Calculate the corresponding forces in members BF and EF.

Ans. $BF = 188.4$ kN C, $EF = 120$ kN T

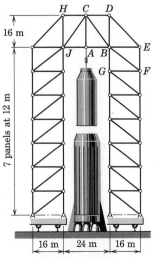

Problem 4/21

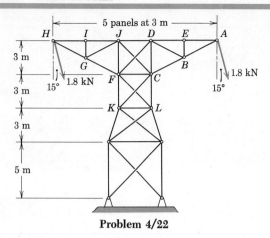

Problem 4/22

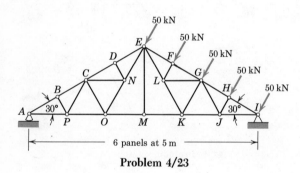

Problem 4/23

▶4/22 The tower for a transmission line is modeled by the truss shown. The crossed members in the center sections of the truss may be assumed capable of supporting tension only. For the loads of 1.8 kN applied in the vertical plane, compute the forces induced in members *AB*, *DB*, and *CD*.

Ans. AB = 3.89 kN *C, DB* = 0, *CD* = 0.93 kN *C*

▶4/23 Find the forces in members *EF*, *KL*, and *GL* for the Fink truss shown. (*Hint:* Note that the forces in *BP*, *PC*, *DN*, etc., are zero.)

Ans. EF = 202 kN *C*
KL = 100 kN *T*
GL = 50.0 kN *T*

4/4   METHOD OF SECTIONS.   In the previous article on the analysis of plane trusses by the method of joints we took advantage of only two of the three equilibrium equations, since the procedures involve concurrent forces at each joint. We may take advantage of the third or moment equation of equilibrium by selecting an entire section of the truss for the free body in equilibrium under the action of a nonconcurrent system of forces. This *method of sections* has the basic advantage that the force in almost any desired member may be found directly from an analysis of a section which has cut that member. Thus it is not necessary to proceed with the calculation from joint to joint until the member in question has been reached. In choosing a section of the truss we note that, in general, not more than three members whose forces are unknown may be cut, since there are only three available equilibrium relations which are independent.

The method of sections will now be illustrated for the truss in Fig. 4/5, which was used in the explanation of the previous method. The truss is shown again in Fig. 4/11*a* for ready reference. The external reactions are first computed as before, considering the truss as a whole. Now let us determine the force in the member *BE* for example. An imaginary section, indicated by the dotted line, is passed through the truss, cutting it into two parts, Fig. 4/11*b*. This section has cut three members whose forces are initially unknown. In order for the portion of the truss on each side of the section to remain in equilibrium it is necessary to apply to each cut member the force that was exerted on it by the member cut away. These forces, either tensile or compressive, will always be in the directions of the respective members for simple trusses composed of two-force members. The left-hand section is in equilibrium under the action of the applied load *L*, the end reaction R_1, and the three forces exerted on the cut members by the right-hand section which has been removed. We may usually draw the forces with their proper senses by a visual approximation of the equilibrium requirements. Thus in balancing the moments about point *B* for the left-hand section, the force *EF* is clearly to the left, which makes it compressive, since it acts toward the cut section of member *EF*. The load *L* is greater than the reaction R_1, so

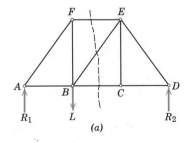
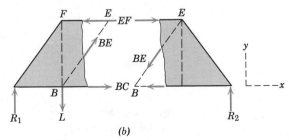

(a) *(b)*

Figure 4/11

that the force *BE* must be up and to the right to supply the needed upward component for vertical equilibrium. Force *BE* is therefore tensile, since it acts away from the cut section. With the approximate magnitudes of R_1 and L in mind we see that the balance of moments about point *E* requires that *BC* be to the right. A casual glance at the truss should lead to the same conclusion when it is realized that the lower horizontal member will stretch under the tension caused by bending. The equation of moments about joint *B* eliminates three forces from the relation, and *EF* may be determined directly. The force *BE* is calculated from the equilibrium equation for the *y*-direction. Finally, we determine *BC* by balancing moments about point *E*. In this way each of the three unknowns has been determined independently of the other two.

The right-hand section of the truss, Fig. 4/11*b*, is in equilibrium under the action of R_2 and the same three forces in the cut members applied in the directions opposite to those for the left section. The proper sense for the horizontal forces may easily be seen from the balance of moments about points *B* and *E*.

We may use either section of a truss for the calculations, but the one involving the smaller number of forces will usually yield the simpler solution.

It is essential to understand that in the method of sections an entire portion of the truss is considered a single body in equilibrium. Thus the forces in members internal to the section are not involved in the analysis of the section as a whole. In order to clarify the free body and the forces acting externally on it, the section is preferably passed through the members and not the joints.

The moment equations are used to great advantage in the method of sections, and a moment center, either on or off the section, through which as many forces pass as possible should be chosen. It is not always possible to assign an unknown force in the proper sense when the free-body diagram of a section is initially drawn. With an arbitrary assignment made, a positive answer will verify the assumed sense and a negative result will indicate that the force is in the sense opposite to that assumed. Any system of notation desired may be used, although usually it is found convenient to letter the joints and designate a member and its force by the two letters defining the ends of the member.

An alternative notation preferred by some is to assign all unknown forces arbitrarily as positive in the tension direction (away from the section) and let the algebraic sign of the answer distinguish between tension and compression. Thus a plus sign would signify tension and a minus sign compression. On the other hand the advantage of assigning forces in their correct sense on the free-body diagram of a section wherever possible is that it emphasizes the physical action of the forces more directly and is preferred in this treatment.

Sample Problem 4/2

Calculate the forces induced in members *KL*, *CL*, and *CB* by the 20-ton load on the cantilever truss.

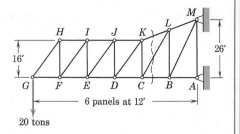

Solution. Although the vertical components of the reactions at *A* and *M* are statically indeterminate with the two fixed supports, all members other than *AM* are statically determinate. We may pass a section directly through members *KL*, *CL*, and *CB* and analyze the portion of the truss to the left of this section as a statically determinate rigid body.

The free-body diagram of the portion of the truss to the left of the section is shown. A moment sum about *L* quickly verifies the assignment of *CB* as compression, and a moment sum about *C* quickly discloses that *KL* is in tension. The direction of *CL* is not quite so obvious until we observe that *KL* and *CB* intersect at a point *P* to the right of *G*. A moment sum about *P* eliminates reference to *KL* and *CB* and shows that *CL* must be compressive to balance the moment of the 20-ton force about *P*. With these considerations in mind the solution becomes straightforward, as we now see how to solve for each of the three unknowns independently of the other two.

Summing moments about *L* requires the moment arm $BL = 16 + (26 - 16)/2 = 21$ ft. Thus

$$[\Sigma M_L = 0] \qquad 20(5)(12) - CB(21) = 0 \qquad CB = 57.1 \text{ tons } C \qquad Ans.$$

Next we take moments about *C* which requires a calculation of $\cos \theta$. From the given dimensions we see $\theta = \tan^{-1}(5/12)$ so that $\cos \theta = 12/13$. Therefore

$$[\Sigma M_C = 0] \qquad 20(4)(12) - \tfrac{12}{13}KL(16) = 0 \qquad KL = 65.0 \text{ tons } T \qquad Ans.$$

Finally we may find *CL* by a moment sum about *P* whose distance from *C* is given by $PC/16 = 24/(26 - 16)$ or $PC = 38.4$ ft. We also need β which is given by $\beta = \tan^{-1}(\overline{CB}/\overline{BL}) = \tan^{-1}(12/21) = 29.7°$ and $\cos \beta = 0.868$. We now have

$$[\Sigma M_P = 0] \qquad 20(48 - 38.4) - CL(0.868)(38.4) = 0$$
$$CL = 5.76 \text{ tons } C \qquad Ans.$$

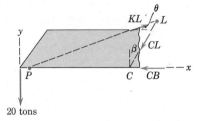

① We note that analysis by the method of joints would necessitate working with eight joints in order to calculate the three forces in question. Thus the method of sections offers a considerable advantage in this case.

② We could have started with moments about *C* or *P* just as well.

③ We could also have determined *CL* by a force summation in either the *x*- or *y*-direction.

Sample Problem 4/3

Calculate the force in member DJ of the Howe roof truss illustrated. Neglect any horizontal components of force at the supports.

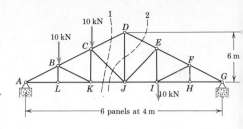

Solution. It is not possible to pass a section through DJ without cutting four members whose forces are unknown. Although three of these cut by section 2 are concurrent at J and therefore the moment equation about J could be used to obtain DE, the force in DJ cannot be obtained from the remaining two equilibrium principles. It is necessary to consider first the adjacent section 1 before considering section 2.

The free-body diagram for section 1 is drawn and includes the reaction of 18.3 kN at A, which is previously calculated from the equilibrium of the truss as a whole. In assigning the proper directions for the forces acting on the three cut members we see that a balance of moments about A eliminates the effects of CD and JK and clearly requires that CJ be up and to the left. A balance of moments about C eliminates the effect of the three forces concurrent at C and indicates that JK must be to the right to supply sufficient counterclockwise moment. Again it should be fairly obvious that the lower chord is under tension because of the bending tendency of the truss. Although it should also be apparent that the top chord is under compression, for purposes of illustration the force in CD ① will be arbitrarily assigned as tension.

By the analysis of section 1, CJ is obtained from

$$[\Sigma M_A = 0] \qquad 0.707\,CJ(12) - 10(4) - 10(8) = 0 \qquad CJ = 14.1 \text{ kN } C$$

In this equation the moment of CJ is calculated by considering its horizontal and vertical components acting at point J. Equilibrium of moments about J requires

$$[\Sigma M_J = 0] \qquad 0.894\,CD(6) + 18.3(12) - 10(4) - 10(8) = 0$$

$$CD = -18.6 \text{ kN}$$

The moment of CD about J is calculated here by considering its two ② components as acting through D. The minus sign indicates that CD was assigned in the wrong direction.

Hence $\qquad\qquad\qquad\qquad CD = 18.6 \text{ kN } C \qquad\qquad\qquad\qquad$ *Ans.*

From the free-body diagram of section 2, which now includes the known value of CJ, a balance of moments about G is seen to eliminate DE and JK. Thus

$$[\Sigma M_G = 0] \qquad 12\,DJ + 10(16) + 10(20) - 18.3(24) - 14.1(0.707)(12) = 0$$

$$DJ = 16.6 \text{ kN } T \qquad\qquad\qquad\qquad \text{*Ans.*}$$

Again the moment of CJ is determined from its components considered to be acting at J. The answer for DJ is positive, so that the assumed tensile direction is correct. An analysis of the joint D alone also verifies this conclusion.

In choosing a section it is always important to match the number of ③ unknowns with the number of independent equilibrium equations which may be applied.

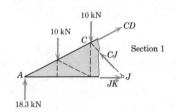

① There is no harm in assigning one or more of the forces in the wrong direction as long as the calculations are consistent with the assumption. A negative answer will show the need for reversing the direction of the force.

② If desired, the direction of CD may be changed on the free-body diagram and the algebraic sign of CD reversed in the calculations, or else the work may be left as it stands with a note stating the proper direction.

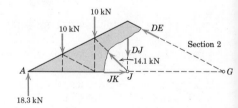

③ Observe that a section through members CD, DJ, and DE could be taken that would cut only three unknown members. However, since the forces in these three members are all concurrent at D, a moment equation about D would yield no information about them. The remaining two force equations would not be sufficient to solve for the three unknowns.

PROBLEMS

(Solve the following problems by the method of sections. Neglect the weight of the members compared with the forces they support.)

4/24 Calculate the forces in members *AB*, *BF*, and *EF* in the loaded truss.

   Ans. AB = 8 kN *T, BF* = 2 kN *C, EF* = $4\sqrt{5}$ kN *C*

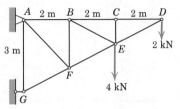

Problem 4/24

4/25 Calculate the forces in members *ED* and *EB* in the loaded truss composed of equilateral triangles.

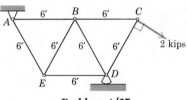

Problem 4/25

4/26 Determine the force in member *CF* in terms of the applied load *L*. All interior angles are 60°.

   Ans. CF = $2L/\sqrt{3}$, *C*

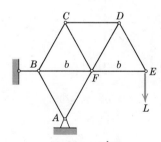

Problem 4/26

4/27 Calculate the forces in members *CD*, *BC*, and *CG* in the loaded truss composed of equilateral triangles.

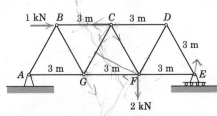

Problem 4/27

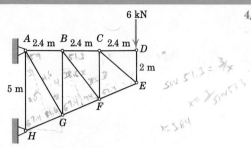

Problem 4/28

4/28 Calculate the force in members *AB*, *BG*, and *GF*. Solve for each force from an equilibrium equation which contains that force as the only unknown.

Ans. AB = 7.2 kN *T*, *BG* = 3 kN *C*,
GF = 7.8 kN *C*

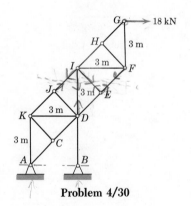

Problem 4/29

4/29 Assume that the cross braces in the bridge truss are flexible members incapable of supporting compression. Calculate the force in member *DE* for the loading condition shown.

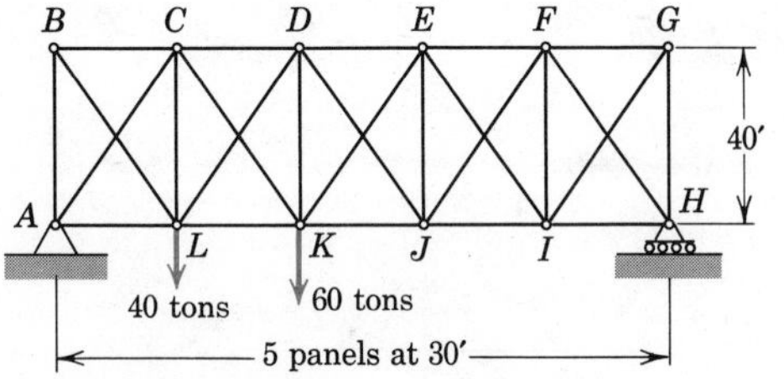

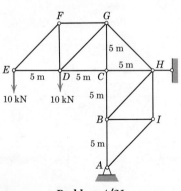

Problem 4/30

4/30 Calculate the forces in members *DI*, *DE*, and *EI* for the loaded truss shown.

Ans. DI = 18 kN *C*, *DE* = 25.5 kN *C*,
EI = 0

4/31 Calculate the forces in members *CH*, *CB*, and *GH* for the cantilevered truss. Solve for each force from a moment equation which contains that force as the only unknown.

Problem 4/31

4/32 Compute the forces in members *BC*, *CI*, and *HI* for the truss of Prob. 4/10 repeated here. Solve for each force from an equilibrium equation which contains that force as the only unknown.

Ans. $BC = 4.33$ kN *C*, $CI = 2.12$ kN *T*

$HI = 2.69$ kN *T*

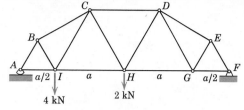

Problem 4/32

4/33 The roof truss is composed of 30°–60° right triangles and is loaded as shown. Compute the forces in members *BH* and *HG*.

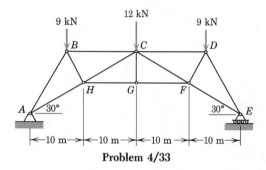

Problem 4/33

4/34 A crane is modeled by the simple truss shown. Compute the forces in members *DE*, *DG*, and *HG* under the load of the 2-ton tractor.

Ans. $DE = 2$ tons *T*, $DG = 4.24$ tons *T*

$HG = 5$ tons *C*

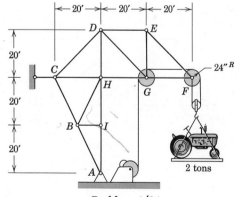

Problem 4/34

4/35 Compute the forces in members *CH*, *CD*, and *HI* for the crane truss of Prob. 4/34.

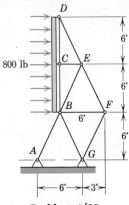

Problem 4/36

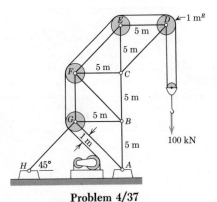

Problem 4/37

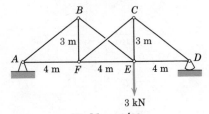

Problem 4/38

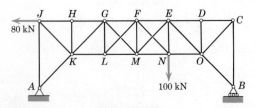

Problem 4/39

4/36 Solve for the forces in members *BG* and *BF* of the signboard truss of Prob. 4/13 repeated here. The resultant of the 800-lb wind load passes through *C*.

Ans. $BG = 400\sqrt{5}$ lb *C*, $BF = 800$ lb *T*

4/37 Calculate the forces in members *FC* and *FB* due to the 100-kN load on the crane truss.

4/38 Each of the members *BE* and *FC* is capable of supporting compression as well as tension. Compute the forces in members *CF*, *BE*, and *EF*.

Ans. $BE = 1.667$ kN *C*, $FC = 3.33$ kN *C*

$EF = 4$ kN *T*

4/39 The truss shown is composed of 45° right triangles. The crossed members in the center two panels are slender tie rods incapable of supporting compression. Retain the two rods which are under tension and compute the magnitudes of their tensions. Also find the force in member *MN*.

4/40 The crane truss is secured to the fixed supports at *A* and *K*, and its winch *W* is locked in position while supporting the 2000-lb tank. Identify any statically indeterminate members and calculate the force in member *HG*.   *Ans. HG* = 8121 lb *C*

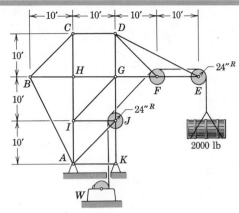

Problem 4/40

4/41 The transmission-line truss of Prob. 4/22 is shown again here. Assume that the crossed members are capable of supporting tension only and compute the force in member *FC* under the action of the loading shown.

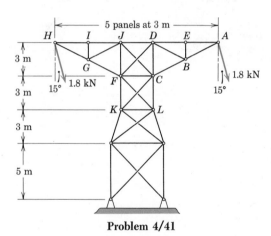

Problem 4/41

4/42 Find the force in member *JQ* for the Baltimore truss where all angles are 30°, 60°, 90°, or 120°.
                                  Ans. JQ = 57.7 kN *C*

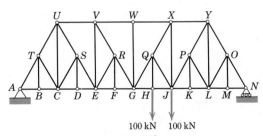

Problem 4/42

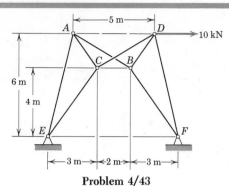

Problem 4/43

▶4/43 The hinged frames *ACE* and *DFB* are connected by two hinged bars, *AB* and *CD*, which cross without being connected. Compute the force in *AB*.

Ans. $AB = 3.78$ kN C

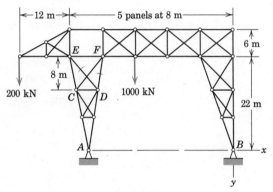

Problem 4/44

▶4/44 In the traveling bridge crane shown all crossed members are slender tie rods incapable of supporting compression. Determine the forces in members *DF* and *EF* and find the horizontal reaction on the truss at *A*. Show that if *CF* = 0, *DE* = 0 also.

Ans. $DF = 768$ kN C
$EF = 364$ kN C
$A_x = 101$ kN to the right

4/5 SPACE TRUSSES. A space truss is the three-dimensional counterpart of the plane truss described in the two previous articles. The idealized space truss consists of rigid links connected at their ends by ball-and-socket joints. We saw that a triangle of pin-connected bars forms the basic noncollapsible unit for the plane truss. A space truss, on the other hand, requires six bars joined at their ends to form the edges of a tetrahedron for the basic noncollapsible unit. In Fig. 4/12*a* the two bars *AD* and *BD* joined at *D* require a third support *CD* to keep the triangle *ADB* from rotating about *AB*. In Fig. 4/12*b* the supporting base is replaced by three more bars *AB, BC,* and *AC* to form a tetrahedron not dependent on the foundation for its own rigidity. We may form a new rigid unit to the structure with three additional concurrent bars whose ends are attached to three fixed joints on the existing structure. Thus in Fig. 4/12*c* the bars *AF, BF,* and *CF* are attached to the foundation and therefore fix point *F* in space. Likewise point *H* is fixed in space by the bars *AH, DH,* and *CH.* The three additional bars *CG, FG,* and *HG* are attached to the three fixed points *C, F,* and *H* and therefore fix *G* in space. Point *E* is similarly established. We see now that the structure is entirely rigid. The two applied loads shown will result in forces in all of the members.

　　Ideally there must be point support, such as represented by a ball-and-socket joint, for the connections of a space truss so that there will be no bending in the members. Again, as in riveted and welded connections for plane trusses, if the centerlines of joined members intersect at a point, we may justify the assumption of two-force members under simple tension and compression.

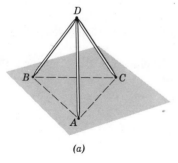

(a)

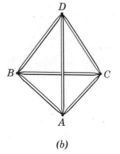

(b)

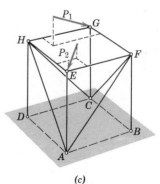

(c)

Figure 4/12

For a space truss which is supported externally in such a way that it is statically determinate as an entire unit, a relationship exists between the number of its joints and the number of its members necessary for internal stability without redundancy. Since the equilibrium of each joint is specified by three scalar force equations, there are in all $3j$ such equations for a simple space truss with j joints. For the entire truss composed of m members there are m unknowns plus six unknown support reactions in the general case of a statically determinate space structure. Thus, for a simple space truss composed of tetrahedral elements, the equation $m + 6 = 3j$ will be satisfied if the truss is statically determinate internally. Again, as in the case of the plane truss, this relation is a necessary condition for stability, but it is not a sufficient condition, since one or more of the m members can be arranged in such a way as not to contribute to a stable configuration of the entire truss. If $m + 6 > 3j$, there are more members than there are independent equations, and the truss is statically indeterminate internally with redundant members present. If $m + 6 < 3j$, there is a deficiency of internal members, and the truss is unstable and subject to collapse under load. The foregoing relationship between the number of joints and the number of members for a space truss is very helpful in the preliminary design for such a truss, since the configuration is not nearly so obvious as in the case of a plane truss, where the geometry for statical determinacy is generally quite apparent.

The method of joints developed in Art. 4/3 for plane trusses may be extended directly to space trusses by satisfying the complete vector equation

$$\Sigma \mathbf{F} = \mathbf{0}$$

for each joint. It is necessary to start at some joint where at least one known force acts and not more than three unknown forces are present. Adjacent joints upon which not more than three unknown forces act may be analyzed in turn.

The method of sections developed in the previous article may also be applied to space trusses. The two vector equations

$$\Sigma \mathbf{F} = \mathbf{0} \quad \text{and} \quad \Sigma \mathbf{M} = \mathbf{0}$$

must be satisfied for any section of the truss, where the zero moment sum will hold for all moment axes. Since the two vector equations are equivalent to six scalar equations, we conclude that a section should in general not be passed through more than six members whose forces are unknown. The method of sections for space trusses is not widely used, however, because a moment axis can seldom be found which eliminates all but one unknown as in the case of plane trusses.

Vector notation for expressing the terms in the force and moment equations for space trusses is of considerable advantage and is used in the sample problem that follows.

Sample Problem 4/4

The space truss consists of the rigid tetrahedron *ABCD* anchored by a ball-and-socket connection at *A* and prevented from any rotation about the *x*-, *y*-, or *z*-axes by the respective links 1, 2, and 3. The load *L* is applied to joint *E*, which is rigidly fixed to the tetrahedron by the three additional links. Solve for the forces in the members at joint *E* and indicate the procedure for the solution of the forces in the remaining members of the truss.

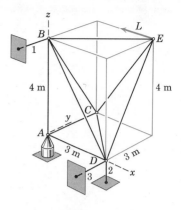

Solution. We note first that the truss is supported with six properly placed constraints, which are the three at *A* and the links 1, 2, and 3. Also, with $m = 9$ members and $j = 5$ joints the condition $m + 6 = 3j$ for a sufficiency of members to provide a noncollapsible structure is satisfied.

The external reactions at *A*, *B*, and *D* can be calculated easily as a first step, although their values will be determined from the solution of all forces on each of the joints in succession.

We must start with a joint upon which at least one known force and not more than three unknown forces act, which in this case is joint *E*. The free-body diagram of joint *E* is shown with all force vectors arbitrarily but consistently indicated in their positive tension directions (away from the joint). The vector expressions for the three unknown forces are

$$\mathbf{F}_{EB} = \frac{F_{EB}}{\sqrt{2}}(-\mathbf{i} - \mathbf{j}), \quad \mathbf{F}_{EC} = \frac{F_{EC}}{5}(-3\mathbf{i} - 4\mathbf{k}), \quad \mathbf{F}_{ED} = \frac{F_{ED}}{5}(-3\mathbf{j} - 4\mathbf{k})$$

Equilibrium of joint *E* requires

$$[\Sigma\mathbf{F} = 0] \qquad \mathbf{L} + \mathbf{F}_{EB} + \mathbf{F}_{EC} + \mathbf{F}_{ED} = 0 \qquad \text{or}$$

$$-L\mathbf{i} + \frac{F_{EB}}{\sqrt{2}}(-\mathbf{i} - \mathbf{j}) + \frac{F_{EC}}{5}(-3\mathbf{i} - 4\mathbf{k}) + \frac{F_{ED}}{5}(-3\mathbf{j} - 4\mathbf{k}) = 0$$

Rearranging terms gives

$$\left(-L - \frac{F_{EB}}{\sqrt{2}} - \frac{3F_{EC}}{5}\right)\mathbf{i} + \left(-\frac{F_{EB}}{\sqrt{2}} - \frac{3F_{ED}}{5}\right)\mathbf{j} + \left(-\frac{4F_{EC}}{5} - \frac{4F_{ED}}{5}\right)\mathbf{k} = 0$$

Equating the coefficients of the **i**-, **j**-, and **k**-terms to zero gives the three equations

$$\frac{F_{EB}}{\sqrt{2}} + \frac{3F_{EC}}{5} = -L \qquad \frac{F_{EB}}{\sqrt{2}} + \frac{3F_{ED}}{5} = 0 \qquad F_{EC} + F_{ED} = 0$$

Solving the equations gives us

$$F_{EB} = -L/\sqrt{2} \qquad F_{EC} = -5L/6 \qquad F_{ED} = 5L/6 \qquad \textit{Ans.}$$

Thus we conclude that F_{EB} and F_{EC} are compressive forces and F_{ED} is tension.

Unless we have computed the external reactions first, we must next analyze joint *C* with the known value of F_{EC} and the three unknowns F_{CB}, F_{CA}, and F_{CD}. The procedure is identical with that used for joint *E*. Joints *B*, *D*, and *A* are then analyzed in the same way and in that order, which limits the unknowns to three for each joint. The external reactions computed from these analyses must, of course, agree with the values which can be determined initially from an analysis of the truss as a whole.

In a more general case, if there is no loaded joint which has as few as three unknown forces acting, it is necessary to compute the external reactions first and begin the analysis at one of the external reaction joints upon which no more than three unknown forces act.

① *Suggestion:* Draw a free-body diagram of the truss as a whole and verify that the external forces acting on the truss are $\mathbf{A}_x = L\mathbf{i}$, $\mathbf{A}_y = L\mathbf{j}$, $\mathbf{A}_z = (4L/3)\mathbf{k}$, $\mathbf{B}_y = 0$, $\mathbf{D}_y = -L\mathbf{j}$, $\mathbf{D}_z = -(4L/3)\mathbf{k}$.

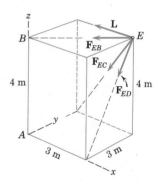

② A negative numerical value for a force indicates compression.

PROBLEMS

(In the following problems use plus for tension and minus for compression.)

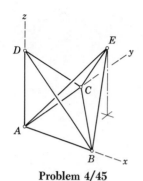

Problem 4/45

4/45 The space truss is built upon the triangular base ABC. The locations of joints D and E are established by the links shown. Show that this configuration is internally stable. Also replace link AE by a different link that will preserve the rigidity of the truss.

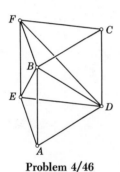

Problem 4/46

4/46 The prismatic space truss has a horizontal base ADE and a parallel top face BCF in the shape of equal equilateral triangles which are connected by three equal vertical legs and braced by three diagonal members as shown. Show that this truss represents a stable configuration.

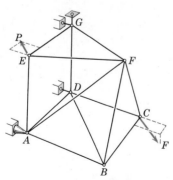

Problem 4/47

4/47 The space truss is shown in an intermediate stage of design. The external constraints indicated are sufficient to maintain external equilibrium. How many additional members are needed to prevent internal instability and where can they be placed?

4/48 For the space truss of Sample Problem 4/4 use the result $F_{EC} = -5L/6$ and compute the forces in members CB, CA, and CD.

Ans. $F_{CD} = L/\sqrt{2}$, $F_{CB} = 5L/6$, $F_{CA} = -L$

4/49 The space truss in the form of a tetrahedron is supported by ball-and-socket connections at its base points A and B and is prevented from rotating about AB by the vertical tie bar CD. After noting the vertical components of the reactions under the symmetrical truss at A and B, draw a free-body diagram of the triangular configuration of links BDE and determine the x-component of the force exerted by the foundation on the truss at B.

Ans. $B_x = P$

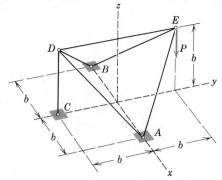

Problem 4/49

4/50 The rectangular space truss 16 m in height is erected on a horizontal square base 12 m on a side. Guy wires are attached to the structure at E and G as shown and are tightened until the tension T in each wire is 9 kN. Calculate the force F in each of the diagonal members. Ans. $F = -3.72$ kN

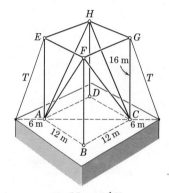

Problem 4/50

4/51 The tetrahedral space truss has a horizontal base ABC in the form of an isosceles triangle and legs AD, BD, and CD which support the mass m from point D. Each vertex of the base is suspended by a vertical wire from overhead supports. Calculate the force induced in members AC and AB.

Ans. $F_{AC} = -\dfrac{5mg}{54}$, $F_{AB} = -\dfrac{4mg}{27}$

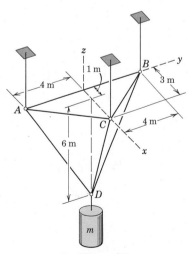

Problem 4/51

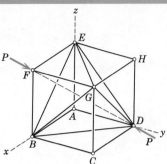

Problem 4/52

▶4/52 A space truss is constructed in the form of a cube with six diagonal members shown. Verify that the truss is internally stable. If the truss is subjected to the compressive forces P applied at F and D along the diagonal FD, determine the forces in members FE and EG.   *Ans.* $F_{FE} = -P/\sqrt{3}$, $F_{EG} = P/\sqrt{6}$

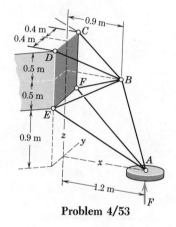

Problem 4/53

▶4/53 Each of the landing struts for a lunar spacecraft is a space truss symmetrical about the vertical x-z plane as shown. For a landing force of $F = 2.2$ kN, calculate the corresponding force in member BE. The assumption of static equilibrium for the truss is permissible if the mass of the truss is very small. Assume equal loads in the symmetrically placed members.   *Ans.* $F_{BE} = 1.620$ kN

4/6 FRAMES AND MACHINES. Structures composed of joined members any one of which has more than two forces acting on it fall in the categories of frames or machines. Frames are structures which are designed to support applied loads and are usually fixed in position. Machines are structures which contain moving parts and are designed to transmit forces or couples from input values to output values.

Since frames and machines contain multiforce members (three or more forces), these forces in general will *not* be in the directions of the members. Therefore we cannot analyze these structures by the methods developed in Arts. 4/3, 4/4, and 4/5 for simple trusses composed of two-force members where the forces are in the directions of the members.

In the previous chapter the equilibrium of multiforce bodies was discussed and illustrated, but we focused attention on the equilibrium of a *single* rigid body. In this present article attention is focused on the equilibrium of *interconnected* rigid bodies which contain multiforce members. Although most such bodies may be analyzed as two-dimensional systems, there are numerous examples of frames and machines that are three-dimensional.

The forces acting on each member of a connected system are found by isolating the member with a free-body diagram and applying the established equations of equilibrium. The *principle of action and reaction* must be carefully observed when we represent the forces of interaction on the separate free-body diagrams. If the structure contains more members or supports than are necessary to prevent collapse, then, as in the case of trusses, the problem is statically indeterminate, and the principles of equilibrium, although necessary, are not sufficient for solution. Although many frames and machines are statically indeterminate, we will consider in this article only those that are statically determinate.

If the frame or machine constitutes a rigid unit by itself when removed from its supports, as is the A-frame in Fig. 4/13a, the analysis is best begun by establishing all the forces external to the

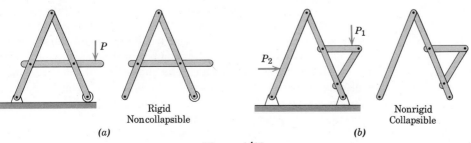

Rigid
Noncollapsible

(a)

Nonrigid
Collapsible

(b)

Figure 4/13

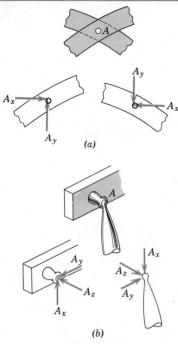

Figure 4/14

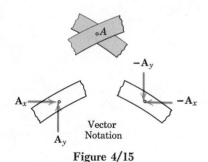

Vector
Notation

Figure 4/15

structure considered as a single rigid body. We then dismember the structure and consider the equilibrium of each part separately. The equilibrium equations for the several parts will be related through the terms involving the forces of interaction. If the structure is not a rigid unit by itself but depends on its external supports for rigidity, as illustrated in Fig. 4/13b, then the calculation of the external support reactions cannot be completed until the structure is dismembered and the individual parts are analyzed.

In most cases we find that the analysis of frames and machines is facilitated by representing the forces in terms of their rectangular components. This is particularly so when the dimensions of the parts are given in mutually perpendicular directions. The advantage of this representation is that the calculation of moment arms is accordingly simplified. In some three-dimensional problems, particularly when moments are evaluated about axes that are not parallel to the coordinate axes, we find the use of vector notation is an advantage.

It is not always possible to assign every force or its components in the proper sense when drawing the free-body diagrams, and it becomes necessary for us to make an arbitrary assignment. In any event it is *absolutely necessary* that a force be *consistently* represented on the diagrams for interacting bodies which involve the force in question. Thus for two bodies connected by the pin A, Fig. 4/14a, the components when separated must be consistently represented in the *opposite* directions. For a ball-and-socket connection between members of a space frame, we must apply the action-and-reaction principle to all three components as shown in Fig. 4/14b. The assigned directions may prove to be wrong when the algebraic signs of the components are determined upon calculation. If A_x, for instance, should turn out to be negative, it is actually acting in the direction opposite to that originally represented. Accordingly it would be necessary for us to reverse the direction of the force on *both* members and to reverse the sign of its force terms in the equations. Or we may leave the representation as originally made, and the proper sense of the force will be understood from the negative sign. If we choose to use vector notation in labeling the forces, then we must be careful to use a plus sign for an action and a minus sign for the corresponding reaction, as shown in Fig. 4/15.

Finally, situations occasionally arise where it is necessary to solve two or more equations simultaneously in order to separate the unknowns. In most instances, however, we may avoid simultaneous solutions by careful choice of the member or group of members for the free-body diagram and by a careful choice of moment axes which will eliminate undesired terms from the equations. The method of solution described in the foregoing paragraphs is illustrated in the following sample problems.

Sample Problem 4/5

The frame supports the 400-kg load in the manner shown. Neglect the weights of the members compared with the forces induced by the load and compute the horizontal and vertical components of all forces acting on each of the members.

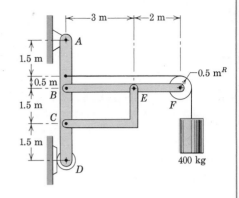

Solution. We observe first that the three supporting members that constitute the frame form a rigid assembly that can be analyzed as a single unit. We also observe that the arrangement of the external supports makes the frame statically determinate.

From the free-body diagram of the entire frame we determine the external reactions. Thus

$$[\Sigma M_A = 0] \qquad 5.5(0.4)(9.81) - 5D = 0 \qquad D = 4.32 \text{ kN}$$

$$[\Sigma F_x = 0] \qquad A_x - 4.32 = 0 \qquad A_x = 4.32 \text{ kN}$$

$$[\Sigma F_y = 0] \qquad A_y - 3.92 = 0 \qquad A_y = 3.92 \text{ kN}$$

Next we dismember the frame and draw a separate free-body diagram of each member. The diagrams are arranged in their approximate relative positions to aid in keeping track of the common forces of interaction. The external reactions just obtained are entered onto the diagram for AD. Other known forces are the 3.92-kN forces exerted by the shaft of the pulley on the member BF, as obtained from the free-body diagram of the pulley. The cable tension of 3.92 kN is also shown acting on AD at its attachment point.

Next, the components of all unknown forces are shown on the diagrams. Here we observe that CE is a two-force member. The force components on CE have equal and opposite reactions, which are shown on BF at E and on AD at C. We may not recognize the actual sense of the components at B at first glance, so they may be arbitrarily but consistently assigned.

The solution may proceed by use of a moment equation about B or E for member BF followed by the two force equations. Thus

$$[\Sigma M_B = 0] \qquad 3.92(5) - \tfrac{1}{2}E_x(3) = 0, \qquad E_x = 13.08 \text{ kN} \qquad Ans.$$

$$[\Sigma F_y = 0] \qquad B_y + 3.92 - 13.08/2 = 0, \qquad B_y = 2.62 \text{ kN} \qquad Ans.$$

$$[\Sigma F_x = 0] \qquad B_x + 3.92 - 13.08 = 0, \qquad B_x = 9.15 \text{ kN} \qquad Ans.$$

Positive numerical values of the unknowns mean that we assumed their directions correctly on the free-body diagrams. The value of $C_x = E_x = 13.08$ kN obtained by inspection of the free-body diagram of CE is now entered onto the diagram for AD, along with the values of B_x and B_y just determined. The equations of equilibrium may now be applied to member AD as a check, since all the forces acting on it have already been computed. The equations give

$$[\Sigma M_C = 0] \qquad 4.32(3.5) + 4.32(1.5) - 3.92(2) - 9.15(1.5) = 0$$

$$[\Sigma F_x = 0] \qquad 4.32 - 13.08 + 9.15 + 3.92 - 4.32 = 0$$

$$[\Sigma F_y = 0] \qquad -13.08/2 + 2.62 + 3.92 = 0$$

① We see that the frame corresponds to the category illustrated in Fig. 4/13a.

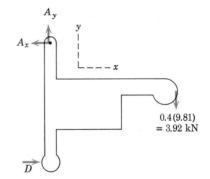

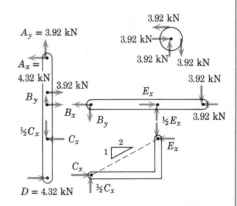

② Without this critical observation the problem cannot be solved. Note especially that the direction of the line joining the two points of application of force, and not the shape of the member, determines the direction of the force and hence the ratio of the force components acting at C and E.

Sample Problem 4/6

Neglect the weight of the frame and compute the forces acting on all of its members.

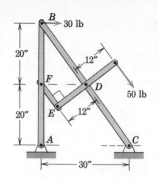

Solution. We note first that the frame is not a rigid unit when removed from its supports since *BDEF* is a movable quadrilateral and not a rigid triangle. Consequently the external reactions cannot be completely determined until the individual members are analyzed. However, we can determine the vertical components of the reactions at *A* and *C* from the free-body diagram of the frame as a whole. Thus

$$[\Sigma M_C = 0] \quad 50(12) + 30(40) - 30A_y = 0 \quad A_y = 60 \text{ lb} \quad \textit{Ans.}$$

$$[\Sigma F_y = 0] \quad C_y - 50(4/5) - 60 = 0 \quad C_y = 100 \text{ lb} \quad \textit{Ans.}$$

Next we dismember the frame and draw the free-body diagram of each part. Since *EF* is a two-force member, the direction of the force at *E* on *ED* and at *F* on *AB* is known. We may assume that the 30-lb force is applied to the pin as a part of member *BC*. There should be no difficulty in assigning the correct directions for forces *E*, *F*, *D*, and *B_x*. The direction of *B_y*, however, may not be assigned by inspection and therefore is arbitrarily shown as downward on *AB* and upward on *BC*.

Member ED. The two unknowns are easily obtained by

$$[\Sigma M_D = 0] \quad 50(12) - 12E = 0 \quad E = 50 \text{ lb} \quad \textit{Ans.}$$

$$[\Sigma F = 0] \quad D - 50 - 50 = 0 \quad D = 100 \text{ lb} \quad \textit{Ans.}$$

Member EF. Clearly *F* is equal and opposite to *E* with the magnitude of 50 lb.

Member AB. Since *F* is now known, we solve for *B_x*, *A_x*, and *B_y* from

$$[\Sigma M_A = 0] \quad 50(3/5)(20) - B_x(40) = 0 \quad B_x = 15 \text{ lb} \quad \textit{Ans.}$$

$$[\Sigma F_x = 0] \quad A_x + 15 - 50(3/5) = 0 \quad A_x = 15 \text{ lb} \quad \textit{Ans.}$$

$$[\Sigma F_y = 0] \quad 50(4/5) - 60 - B_y = 0 \quad B_y = -20 \text{ lb} \quad \textit{Ans.}$$

The minus sign shows that we assigned *B_y* in the wrong direction.

Member BC. The results for *B_x*, *B_y*, and *D* are now transferred to *BC*, and the remaining unknown *C_x* is found from

$$[\Sigma F_x = 0] \quad 30 + 100(3/5) - 15 - C_x = 0 \quad C_x = 75 \text{ lb} \quad \textit{Ans.}$$

We may apply the remaining two equilibrium equations as a check. Thus

$$[\Sigma F_y = 0] \quad 100 + (-20) - 100(4/5) = 0$$

$$[\Sigma M_C = 0] \quad (30 - 15)(40) + (-20)(30) = 0$$

① We see that this frame corresponds to the category illustrated in Fig. 4/13*b*.

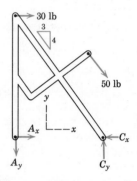

② The directions of *A_x* and *C_x* are not obvious initially and can be assigned arbitrarily to be corrected later if necessary.

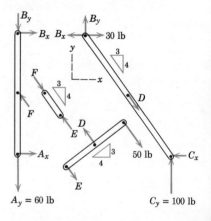

③ Alternatively the 30-lb force could be applied to the pin considered a part of *BA*, with a resulting change in the reaction *B_x*.

④ Alternatively we could have returned to the free-body diagram of the frame as a whole and found *C_x*.

Sample Problem 4/7

The machine shown is an overload protection device which releases the load when it exceeds a predetermined value T. A soft metal shear pin S is inserted in a hole in the lower half and is acted upon by the upper half. When the total force on the pin exceeds its strength, it will break. The two halves then rotate about A under the action of the tensions in BD and CD, as shown in the second sketch, and rollers E and F release the eye bolt. Determine the maximum allowable tension T if the pin S will shear when the total force on it is 800 N. Also compute the corresponding force on the hinge pin A.

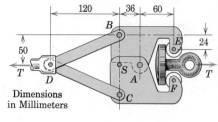

Dimensions in Millimeters

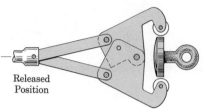

Released Position

Solution. Because of symmetry we analyze only one of the two hinged members. The upper part is chosen, and its free-body diagram along with that for the connection at D is drawn. Because of symmetry the forces at S and A have no x-components. The two-force members BD and CD exert forces of equal magnitude $B = C$ on the connection at D. Equilibrium of the connection gives

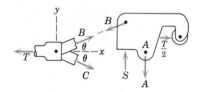

$[\Sigma F_x = 0]$ $B \cos \theta + C \cos \theta - T = 0$, $2B \cos \theta = T$

$B = T/(2 \cos \theta)$

From the free-body diagram of the upper part we express the equilibrium of moments about point A. Substituting $S = 800$ N and the expression for B gives

$[\Sigma M_A = 0]$

$$\frac{T}{2 \cos \theta}(\cos \theta)(50) + \frac{T}{2 \cos \theta}(\sin \theta)(36) - 36(800) - \frac{T}{2}(26) = 0$$

Substituting $\sin \theta / \cos \theta = \tan \theta = 5/12$ and solving for T give

$$T\left(25 + \frac{5(36)}{2(12)} - 13\right) = 28\ 800$$

$T = 1477$ N   or $T = 1.477$ kN   *Ans.*

Finally, equilibrium in the y-direction gives us

$[\Sigma F_y = 0]$ $S - B \sin \theta - A = 0$

$800 - \dfrac{1477}{2(12/13)} \dfrac{5}{13} - A = 0$ $A = 492$ N   *Ans.*

① Symmetry is always useful to recognize. Here it tells us that the forces acting on the two parts behave as mirror images of each other with respect to the x-axis. Thus we cannot have an action on one member in the plus x-direction and its reaction on the other member in the negative x-direction. Consequently the forces at S and A have no x-components.

② Be careful not to forget the moment of the y-component of B. Note that our units here are newton-millimeters.

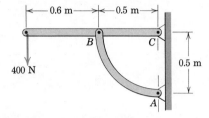

Problem 4/54

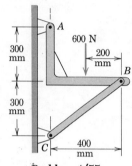

Problem 4/55

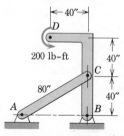

Problem 4/56

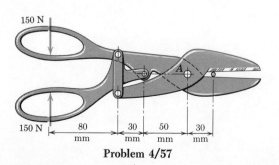

Problem 4/57

PROBLEMS

(Unless otherwise indicated neglect the mass of the various members in the problems that follow.)

4/54 Calculate the magnitude of the force supported by the pin at C for the loaded frame.

Ans. C = 1002 N

4/55 Calculate the magnitude of the force supported by the pin at A which secures the loaded bracket to the wall.

4/56 Calculate the magnitude of the force supported by the pin B of the frame loaded by the 200-lb-ft couple.   *Ans. B* = 69.3 lb

4/57 Replace the 200-lb-ft couple in Prob. 4/56 by a 200-lb downward force applied at point D and compute the magnitude of the force acting at B.

4/58 Compound-lever snips, shown in the figure, are often used in place of regular tinners' snips when large cutting forces are required. For the gripping force of 150 N, what is the cutting force P at a distance of 30 mm along the blade from the pin at A?   *Ans. P* = 1467 N

/59 A small bolt cutter operated by hand for cutting small bolts and rods is shown in the sketch. For a hand grip $P = 150$ N, determine the force Q developed by each jaw on the rod to be cut.

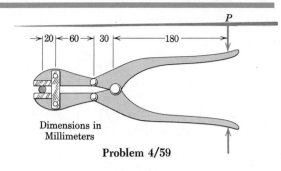

Dimensions in Millimeters

Problem 4/59

/60 A limiting torque wrench is designed with a shear pin B which breaks when the force on it exceeds its strength and, hence, limits the torque which can be applied by the wrench. If the limiting strength of the shear pin is 900 N in double shear (i.e., $V/2 = 450$ N), calculate the limiting torque M. What would be the effect on M of increasing b, all other conditions remaining unchanged?

$$Ans. \ M = 75 \ \text{N} \cdot \text{m}$$
$$M \text{ is decreased as } b \text{ is increased}$$

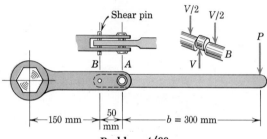

Problem 4/60

/61 Calculate the x- and y-components of the force supported by the pin C for the loaded frame.

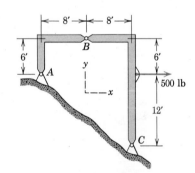

Problem 4/61

/62 Solve for the x- and y-components of the force supported by the hinge at C for the loaded frame.

$$Ans. \ C_x = 5.25 \ \text{kN}, \ C_y = 1 \ \text{kN}$$

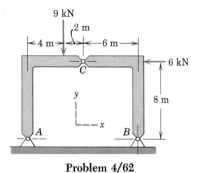

Problem 4/62

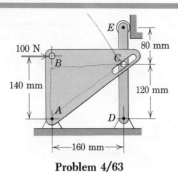

Problem 4/63

4/63 Calculate the magnitude of the force acting on the pin at D. Pin C is fixed in DE and bears against the smooth slot in the triangular plate.

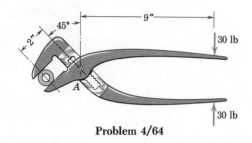

Problem 4/64

4/64 Compute the force supported by the pin at A for the channel-lock pliers under a grip of 30 lb.

Ans. $A = 157.6$ lb

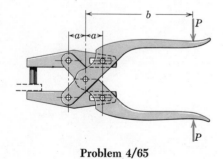

Problem 4/65

4/65 For the paper punch shown find the punching force Q corresponding to a hand grip P.

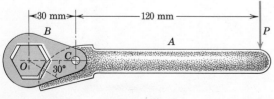

Problem 4/66

4/66 The special box wrench with head B swiveled at C to the handle A will accommodate a range of sizes of hexagonal bolt heads. For the nominal size shown where the center O of the bolt and the pin C are in line with the handle, compute the magnitude of the force supported by the pin at C if $P = 160$ N. Assume the surface of the bolt head to be smooth.

Ans. $C = 1367$ N

/67 The speed reducer consists of the input shaft A and attached pinion B, which drives gear C and its output shaft D with a 2:1 reduction. The center of mass of the 30-kg unit is at G. If a clockwise input torque of 50 N · m is applied to shaft A and if the output shaft D drives a machine at a constant speed, determine the net forces exerted *on* the base flange of the reducer at E and F by the combined action of the bolts and the supporting foundation. (*Caution:* Be careful to apply the reactive torque on the output shaft in the correct direction.)

Ans. $E = 139$ N down, $F = 433$ N up

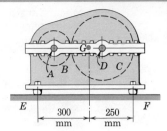

Problem 4/67

/68 The two spur gears A and B drive the bevel gears C and D. For a given torque M_0 express the torque M on the output shaft required to maintain equilibrium.

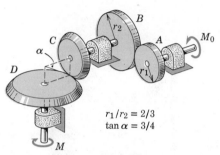

$r_1/r_2 = 2/3$
$\tan \alpha = 3/4$

Problem 4/68

/69 Compute the force acting in the link AB of the lifting tongs which cross without touching.

Ans. $F_{AB} = 1650$ lb tension

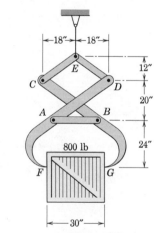

Problem 4/69

/70 Determine the vertical clamping force at E in terms of the force P applied to the handle of the toggle clamp which holds the workpiece F in place.

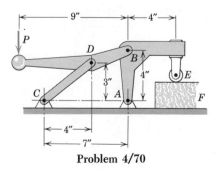

Problem 4/70

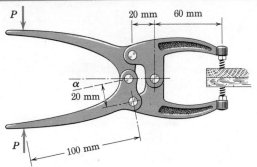

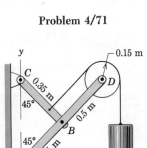

Problem 4/71

4/71 The toggle pliers are used for a variety of clamping purposes. For the handle position given by $\alpha = 10°$ and for a handle grip of $P = 150$ N calculate the clamping force C produced.   *Ans. $C = 1368$ N*

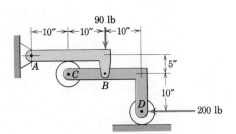

Problem 4/72

4/72 Find the x- and y-components of the pin reaction at B for the loaded frame.

Problem 4/73

4/73 Calculate the magnitude of the force supported by the pin at B for the loaded frame.
                                             Ans. $B = 233$ lb

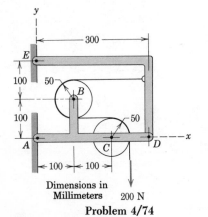

Dimensions in
Millimeters   200 N
Problem 4/74

4/74 Calculate the x- and y-components of all forces acting on each member of the loaded frame.

4/75 Calculate the *x*- and *y*-components of the force supported by the pin at *B*. The cables are securely wrapped around the pulleys which are fastened together.   *Ans. $B_x = 809$ N, $B_y = 785$ N*

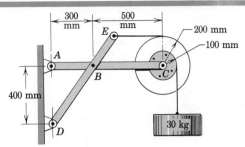

Problem 4/75

4/76 Compute the *x*- and *y*-components of the force acting on the pin *B* connecting the two members of the machine subjected to the 200-N load.
         Ans. $B_x = 287$ N, $B_y = 215$ N

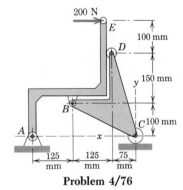

Problem 4/76

4/77 Replace the 200-N force at *E* of Prob. 4/76 by a 50-N·m clockwise couple and compute the *x*- and *y*-components of all forces on each of the two members of the machine.

4/78 Calculate the *x*- and *y*-components of the force supported by the pin at *E* for the loaded frame.
         Ans. $E_x = 5$ lb, $E_y = 5\sqrt{3}$ lb

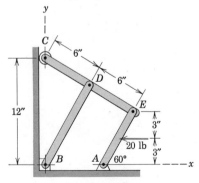

Problem 4/78

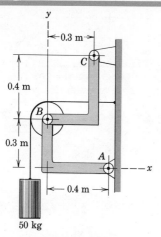

Problem 4/79

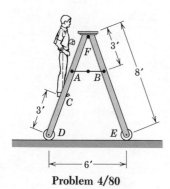

Problem 4/80

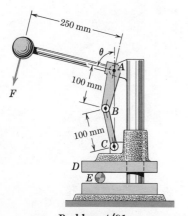

Problem 4/81

4/79 Find the x- and y-components of the forces at A and C of the loaded frame.

4/80 The symmetrical 40-lb stepladder is mounted on wheels so that it may be moved easily. If a 180-lb man stands on the step at C, calculate the tension T in the connecting link AB. The top hinge may be assumed to be on the centerlines of the legs.

Ans. T = 47.2 lb

4/81 The upper jaw D of the toggle press slides with negligible frictional resistance along the fixed vertical column. Calulate the compressive force R exerted on the cylinder E and the force supported by the pin at A if a force $F = 200$ N is applied to the handle at an angle of $\theta = 75°$.

4/82 A double-axle suspension for use on small trucks is shown in the figure. The mass of the central frame F is 40 kg, and the mass of each wheel and attached link is 35 kg with center of mass 680 mm from the vertical centerline. For a load of $L = 12$ kN transmitted to the frame F, compute the total shear force supported by the pin at A.   *Ans. A = 1.75 kN*

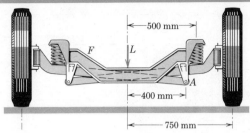

Problem 4/82

4/83 The nose-wheel assembly is raised by the application of a torque M to link BC through the shaft at B. If the arm and wheel AO have a combined weight of 100 lb with center of gravity at G, find the value of M necessary to lift the wheel when D is directly under B at which position angle θ is 30°.

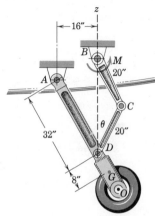

Problem 4/83

4/84 The aircraft landing gear consists of a spring- and hydraulically-loaded piston and cylinder D and the two pivoted links OB and CB. If the gear is moving along the runway at a constant speed with the wheel supporting a stabilized constant load of 24 kN, calculate the total force that the pin at A supports.   *Ans. A = 44.7 kN*

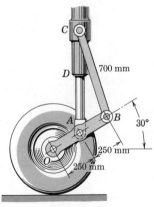

Problem 4/84

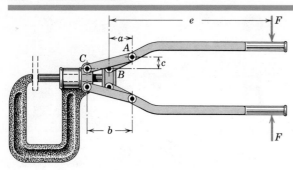
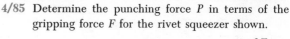

Problem 4/85

4/85 Determine the punching force P in terms of the gripping force F for the rivet squeezer shown.

$$Ans.\ P = \frac{2Fe}{c\left(1 - \dfrac{a}{b}\right)}$$

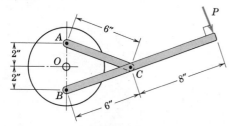

Problem 4/86

4/86 The device shown is an adjustable form of spanner wrench in which the pins at A and B fit into holes in the face of the disk that is to be screwed onto its fixed shaft O. If a torque (moment) of 600 lb-in. about O is required to tighten the disk on the shaft, compute the force supported by each of the pins at A and B when the applied force P has the required value.

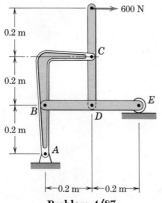

Problem 4/87

4/87 Determine the force supported by the pin at C for the loaded frame.   *Ans.* $C = 2160$ N

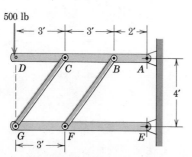

Problem 4/88

4/88 Calculate the magnitude of the force supported by the pin at A for the loaded frame.

4/89 Arm *CE* and the toothed wheel are mounted independently about the same axis at *C*. A torque
$M_0 = 30$ N·m on the shaft of arm *AB* is required to
hold the wheel against a resisting torque *M* acting
on it for the position shown where link *BE* is normal
to *AB* and *CE*. Calculate the magnitude of the
corresponding force acting on end *C* of the arm *EC*.
                                        Ans. C = 908 N

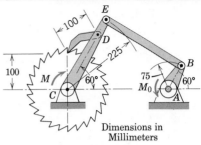

Dimensions in
Millimeters

Problem 4/89

4/90 The portable hoist shown is used to raise construction materials to the roof of a building. The hoist has
a weight of 100 lb with mass center at *G*. If a 180-lb
load is supported by applying the force *F* on the
handle of the drum, calculate the magnitude of the
force supported by the pin at *A*.
                                        Ans. A = 352 lb

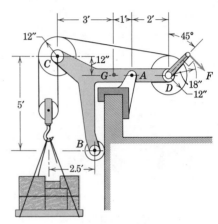

Problem 4/90

4/91 Determine the shearing force *Q* applied to the bar if
a 400-N force is applied to the handle for
$\theta = 30°$. For a given applied force what value of θ
gives the greatest shear?   *Ans. Q* = 13.18 kN

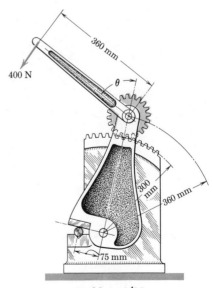

Problem 4/91

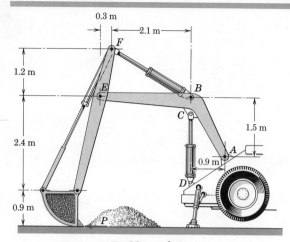

Problem 4/92

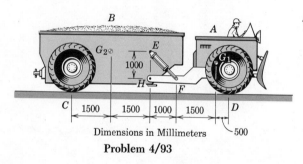

Dimensions in Millimeters
Problem 4/93

Dimensions in
Meters
Problem 4/94

4/92 The backhoe is controlled by the three hydraulic cylinders, and, in the particular position shown, the hoe can apply a horizontal force $P = 10$ kN. Neglect the masses of the members and compute the magnitude of the forces supported by the pins at A and E.   *Ans.* $A = 22.4$ kN, $E = 36.7$ kN

4/93 The power unit A of the tractor-scraper has a mass of 4 Mg with mass center at G_1. The trailer-scraper unit B fully loaded has a mass of 24 Mg with mass center at G_2. Positioning of the scraper is controlled by two hydraulic cylinders EF, one on each side of the machine. Calculate the compression F in each of the cylinders and the magnitude of the force supported by each of the pins at H, one on each side. Assume that the wheels are free to turn so that there are no horizontal components of force under the wheels.   *Ans.* $F = 131.8$ kN, $H = 113.9$ kN

4/94 The figure shows a special rig for erecting vertical sections of a construction tower. The assembly A has a mass of 1.5 Mg and is elevated by the platform B which itself has a mass of 2 Mg. The platform is guided up the fixed vertical column by rollers and is activated by the hydraulic cylinder CD and links EDF and FH. For the particular position shown calculate the force R exerted by the hydraulic cylinder at D and the magnitude of the force supported by the pin at E. Neglect the mass of the cylinder and links.   *Ans.* $R = 59.7$ kN, $E = 40.0$ kN

4/95 The hoisting mechanism for the dump truck is shown in the enlarged view. Determine the compression P in the hydraulic cylinder BE and the magnitude of the force supported by the pin at A for the particular position shown where BA is perpendicular to OAE and link DC is perpendicular to AC. The dump and its load together weigh 20,000 lb with center of mass at G. All dimensions for the indicated geometry are given on the figure.

Ans. $P = 26{,}930$ lb, $A = 14{,}600$ lb

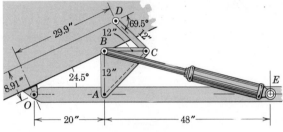

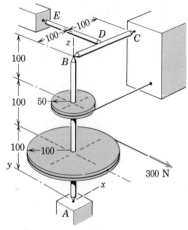

Detail of Hoisting Mechanism

Problem 4/95

4/96 Calculate the x-, y-, and z-components of all forces on each of the three parts of the mechanism. The two disks are rigidly mounted on their shaft, and the larger one is loaded by the 300-N force in the cord wrapped around it and leading in the negative y-direction. The cord around the smaller disk leads in the x-direction and is secured to prevent rotation. Treat all joints as ball-and-socket connections and neglect the weights of all members.

Dimensions in Millimeters

Problem 4/96

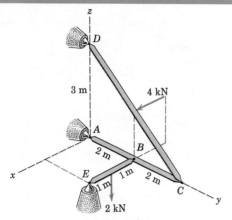

Problem 4/97

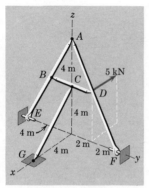

Problem 4/98

4/97 Calculate the x-, y-, and z-components of all forces acting on each member of the light space frame which is loaded as shown.

▶**4/98** The construction A-frame shown in the figure is hinged about the y-axis through points E and F, which cannot offer restraint in the y-direction. The connections at A, B, C, D, and G may be treated as ball-and-socket joints. Link CG is the only two-force member. Compute the x-, y-, and z-components of all forces acting on each member of the frame. The weights of the members may be neglected compared with the loads transmitted.

Ans. $A_x = A_y = B_y = D_y = E_x = F_x = 1.25$ kN
$B_x = B_z = D_x = D_z = E_z = F_z = 2.5$ kN
$C_x = C_z = 5$ kN, $A_z = 0$

4/7 PROBLEM FORMULATION AND REVIEW. In Chapter 4 we have applied the principles of equilibrium to two classes of problems (*a*) simple trusses and (*b*) frames and machines. Basically, no new theory was needed since we merely drew the necessary free-body diagrams and applied our familiar equations of equilibrium. The structures dealt with in Chapter 4, however, have given us the opportunity to develop further our appreciation for the systematic approach to mechanics problems which has been frequently illustrated and emphasized.

The most essential features of the analysis of these two classes of structures are reviewed in the statements which follow.

(*a*) *Simple trusses*

1. Simple trusses are composed of two-force members joined at their ends and capable of supporting tension or compression. Each internal force, therefore, is always in the direction of its member.

2. Simple trusses are built around the basic rigid (noncollapsible) unit of the triangle for plane trusses and the tetrahedron for space trusses. Additional units of a truss are formed by adding new members, two for plane trusses and three for space trusses, attached to existing joints and joined at their ends to form a new joint.

3. The joints for simple trusses are assumed to be pin connections for plane trusses and ball-and-socket connections for space trusses. Thus the joints can transmit force but not moment.

4. External loads on simple trusses are applied only at the joints.

5. Trusses are statically determinate externally when the external constraints are not in excess of those required to maintain an equilibrium position.

6. Trusses are statically determinate internally when constructed in the manner described in item (2) where internal members are not in excess of those required to prevent collapse.

7. The *method of joints* utilizes the force equations of equilibrium for each joint. Analysis must begin at a joint where at least one force is known and not more than two forces are unknown for plane trusses or not more than three forces are unknown for space trusses.

8. The *method of sections* utilizes a free body of an entire section of a truss containing two or more joints and, in general, will involve the equilibrium of a nonconcurrent system of forces. The moment equation of equilibrium is of special value in using the method of sections. In general the forces acting on a section which cuts more than three mem-

bers whose forces are unknown for a plane truss cannot be solved completely since there are only three independent equations of equilibrium.

9. The vector representing a force acting on a joint or a section is drawn on the same side of the joint or section as is the member which transmits the force. With this convention tension is indicated when the force arrow is away from the joint or section, and compression is indicated when the arrow points toward the joint or section.

10. When the two diagonal members which brace a quadrilateral panel are flexible members incapable of supporting compression, only the one in tension is retained, and the panel remains statically determinate.

11. When two joined members under load are collinear and a third member with a different direction is joined with their connection, the force in the third member must be zero, unless an external force is applied at the joint with a component in the direction of the third member.

(b) Frames and machines

1. Frames and machines are multiforce structures, that is, structures which contain one or more members subjected to more than two forces.

2. Frames are structures designed to support loads generally under static conditions. Machines are structures which transform input forces and moments to output forces and moments and generally involve one or more moving parts. Some structures may be classified in either category.

3. Only frames and machines which are statically determinate externally and internally are considered in this treatment.

4. The same procedures for the analysis of frames apply to machines.

5. If a frame or machine as a whole is a rigid (noncollapsible) unit when its external supports are removed, then the analysis is begun by computing the external reactions on the entire unit. If a frame or machine as a whole is a nonrigid (collapsible) unit when its external supports are removed, then the analysis of the external reactions cannot be completed until the structure is dismembered.

6. Forces acting in the internal connections of frames and machines are calculated by dismembering the structure and constructing a separate free-body diagram of each part. The principle of action and reaction must be *strictly* observed; otherwise error will result.

7. The force and moment equations of equilibrium are applied to the members as needed to compute the desired unknowns.

REVIEW PROBLEMS

4/99 A frame used to test the compressive strength of concrete blocks is composed of the four pressure pads, the four compression links, and the three outer links arranged in a square as shown. Express the compressive force C acting on each side of the square concrete block in terms of the applied forces P.

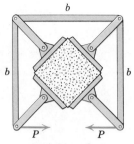

Problem 4/99

4/100 Calculate the forces in members BH, CD, and GD for the truss loaded by the 40- and 60-kN forces.

Ans. $BH = 47.1$ kN C
$CD = 6.7$ kN C
$GD = 0$

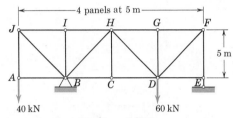

Problem 4/100

4/101 Determine the force in each member of the two trusses that support the 10-kN load at their common pin K.

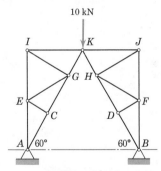

Problem 4/101

4/102 Calculate the force in member BG using a free-body diagram of the rigid member ABC.

Ans. $BG = 1800$ lb C

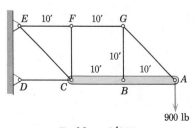

Problem 4/102

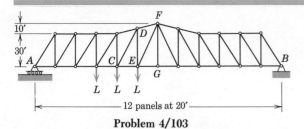

Problem 4/103

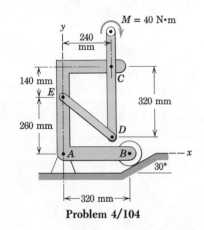

Problem 4/104

4/103 Each of the three loads L on the bridge truss is 20 tons (1 U.S. ton = 2000 lb). Determine the forces in members EG and DF with a minimum of calculation.   *Ans.* $EG = 60$ tons T, $DF = 82.5$ tons C

4/104 Calculate the x- and y-components of the forces acting at A and C for the frame loaded by the couple $M = 40$ N·m.

4/105 An adjustable tow bar connecting the tractor unit H with the landing gear J of a large aircraft is shown in the figure. Adjusting the height of the hook F at the end of the tow bar is accomplished by the hydraulic cylinder CD activated by a small hand pump (not shown). For the nominal position shown of the triangular linkage ABC, calculate the force P supplied by the cylinder to the pin C to position the tow bar. The rig has a total weight of 100 lb and is supported by the tractor hitch at E.

Ans. $P = 60.8$ lb

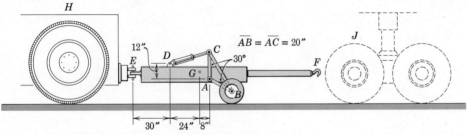

Problem 4/105

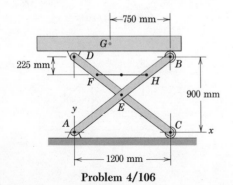

Problem 4/106

4/106 The top of the folding workbench has a mass of 50 kg with mass center at G. Calculate the x- and y-components of the force supported by the pin at E.

107 Solve for the force in member *GL* of the loaded
tower truss. Is the truss statically determinate?

Ans. *GL* = 37.5 kN *T*

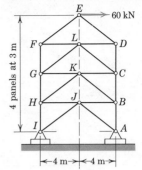

Problem 4/107

108 Determine the force in member *AC* in terms of the
mass *m* supported by the truss. All interior acute
angles are either 30° or 60°. Ans. *AC* = *mg*/3, *C*

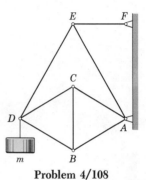

Problem 4/108

109 Verify the fact that each of the loaded trusses shown
is unstable internally (nonrigid), and indicate at
least two alternative ways to insure internal stability (rigidity) for each truss by the addition of one or
more members without introducing redundancy.

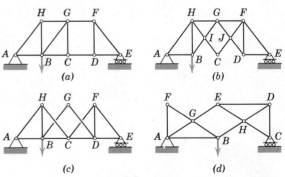

Problem 4/109

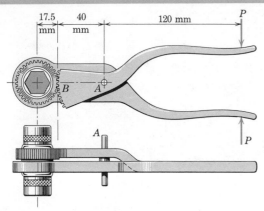

Problem 4/110

4/110 An antitorque wrench is designed for use by a crewman of a spacecraft where he has no stable platform against which to push as he tightens a bolt. The pin A fits into an adjacent hole in the structure which contains the bolt to be turned. Successive oscillations of the gear and handle unit turns the socket in one direction through the action of a ratchet mechanism. The reaction against the pin A provides the "antitorque" characteristic of the tool. For a gripping force of $P = 150$ N determine the torque M transmitted to the bolt and the external reaction R against the pin A normal to the line AB. (One side of the tool is used for tightening and the opposite side for loosening a bolt.)

Ans. $M = 7.88$ N·m, $R = 137.0$ N

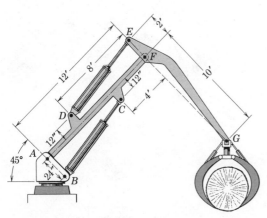

Problem 4/111

4/111 In the special position shown for the log hoist, booms AF and EG are at right angles to one another and AF is perpendicular to AB. If the hoist is handling a log weighing 4800 lb, compute the force supported by the pins at A and D in this one position due to the weight of the log.

Ans. $A = 34,000$ lb, $D = 17,100$ lb

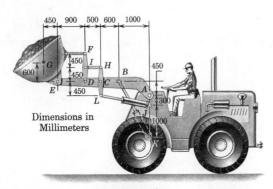

Problem 4/112

4/112 The loader has a capacity of 4 m^3 and is handling shale having a density of 2.6 Mg/m^3. For the particular position shown, where the arm EB is horizontal, find the compressive force in the hydraulic piston rod JL and the total shear force supported by the pin at A. The machine is symmetrical about a central vertical plane in the fore-and-aft direction and has two sets of the linkages shown.

Ans. $L = 52.0$ kN, $A = 247$ kN

113 The A-frame hoist supports the 100-kg load with a force P on the crank handle. Calculate the x- and y-components of all forces acting on each of the three members of the frame for the vertical position of the crank handle.

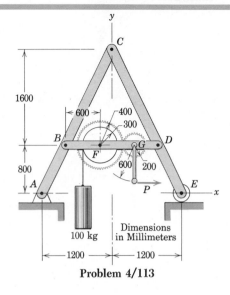

Problem 4/113

114 Compute the magnitude of the force acting on the member BD at D for the space frame loaded by the two forces shown. Each of the connections may be treated as a ball-and-socket joint.

Ans. $D = 4.90$ kN

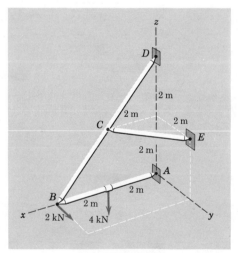

Problem 4/114

DISTRIBUTED FORCES

5

5/1 INTRODUCTION. In the previous chapters we have treated all forces as concentrated along their lines of action and at their points of application. This treatment has provided a reasonable model for the forces with which we have dealt. Actually, however, "concentrated" forces do not exist in the exact sense, since every external force applied mechanically to a body is distributed over a finite contact area however small. The force exerted by the pavement on an automobile tire, for instance, is applied to the tire over its entire area of contact, Fig. 5/1a, which may be appreciable if the tire is soft. When the dimension b of the contact area is negligible compared with the other pertinent dimensions, such as the distance between wheels, then replacement of the actual distributed forces of contact by their resultant R considered a concentrated force raises no question when we are analyzing the forces acting on the car as a whole. Even the force of contact between a hardened steel ball and its race in a loaded ball bearing, Fig. 5/1b, will be applied over a finite contact area the dimensions of which, of course, are extremely small. The forces applied to a two-force member of a truss, Fig. 5/1c, are applied over an actual area of contact of the pin against the hole and internally across the cut section in some such manner as indicated. In these and other similar examples we have no hesitation in treating the forces as concentrated when analyzing their external effects on bodies as a whole.

If, on the other hand, we are interested in finding the distribution of *internal* forces in the material of the body in the neighborhood of the contact location where the internal stresses and strains may be appreciable, then we no longer treat the load as concentrated but would be obliged to take the actual distribution into account. This type of problem requires a knowledge of the properties of the material and belongs in more advanced treatments of the mechanics of materials and the theories of elasticity and plasticity.

When forces are applied over a region whose dimensions are not negligible compared with other pertinent dimensions, then we must account for the actual manner in which the force is distributed by summing up the effects of the distributed force over the entire region. We carry out this process by using the procedures of mathematical

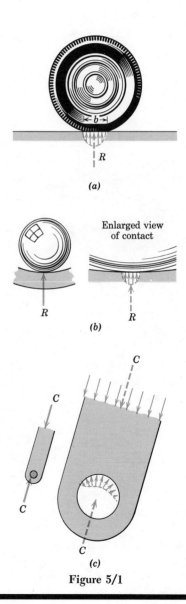

Figure 5/1

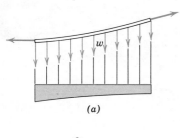

(a)

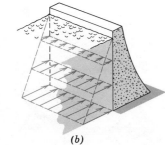

(b)

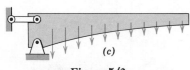

(c)

Figure 5/2

integration. For this purpose we need to know the intensity of the force at any location. There are three categories into which such problems fall.

Line distribution. When a force is distributed along a line, as in the continuous vertical load supported by a suspended cable, Fig. 5/2*a*, the intensity w of the loading is expressed as force per unit length of line, newtons per meter (N/m), or pounds per foot (lb/ft).

Area distribution. When a force is distributed over an area, as with the hydraulic pressure of water against the inner face of a section of dam, Fig. 5/2*b*, the intensity is expressed as force per unit area. This intensity is known as *pressure* for the action of fluid forces and *stress* for the internal distribution of forces in solids. The basic unit for pressure or stress in SI is the newton per square meter (N/m^2), which is also called the *pascal* (Pa). This unit, however, is too small for most applications (6895 Pa = 1 lb/in.²), and the kilopascal (kPa), which equals 10^3 Pa, is more commonly used for fluid pressure and the megapascal, which equals 10^6 Pa, is used for stress. In the U.S. customary system of units both fluid pressure and mechanical stress are commonly expressed in pounds per square inch (lb/in.²).

Volume distribution. A force which is distributed over the volume of a body is known as a *body force*. The most common body force is the force of gravitational attraction which acts on all elements of mass of a body. The determination of the forces on the supports of the heavy cantilevered structure in Fig. 5/2*c*, for example, would require accounting for the distribution of gravitational force throughout the structure. The intensity of gravitational force is the *specific weight* ρg where ρ is the density (mass per unit volume) and g is the acceleration due to gravity. The units for ρg are $(kg/m^3)(m/s^2) = N/m^3$ in SI and lb/ft³ or lb/in.³ in the U.S. customary system.

The body force due to the earth's gravitational attraction (weight) is by far the most commonly encountered distributed force. Section A of the chapter deals with the determination of the point in a body through which the resultant gravitational force acts. Section B of the chapter treats the important special problems of the distributed forces which act on and in beams and flexible cables and distributed forces which fluids exert on exposed surfaces.

SECTION A. CENTERS OF MASS AND CENTROIDS

5/2 CENTER OF MASS. Consider a three-dimensional body of any size, shape, and mass m. If we suspend the body, as shown in Fig. 5/3, from any point such as A, the body will be in equilibrium under the action of the tension in the cord and the resultant W of the gravitational forces acting on all particles of the body. This resultant is clearly collinear with the cord, and it will be assumed that we mark its position by drilling a hypothetical hole of negligible size along its line of action. We repeat the experiment by suspending the body from other points such as B and C, and in each instance we mark the line of action of the resultant force. For all practical purposes these lines of action will be concurrent at a point G, which is known as the *center of gravity* of the body. An exact analysis, however, would take into account the fact that the directions of the gravity forces for the various particles of the body differ slightly because they converge toward the center of attraction of the earth. Also, since the particles are at different distances from the earth, the intensity of the earth's force field is not exactly constant over the body. These considerations lead to the conclusion that the lines of action of the gravity force resultants in the experiments just described will not quite be concurrent, and therefore no unique center of gravity exists in the exact sense. This condition is of no practical importance as long as we deal with bodies whose dimensions are small compared with those of the earth. We therefore assume a uniform and parallel field of force due to the earth's gravitational attraction, and this condition results in the concept of a unique center of gravity.

To determine mathematically the location of the center of gravity of any body, Fig. 5/4a, we apply the *principle of moments* (Varignon's theorem) to the parallel system of gravitational forces to locate its resultant. The moment of the resultant gravitational force W about any axis equals the sum of the moments about the same axis of the gravitational forces dW acting on all particles considered as infinitesimal elements of the body. The resultant of the gravitational forces acting on all elements is the weight of the body and is given by the sum $W = \int dW$. If we apply the moment principle about the y-axis, for example, the moment about this axis of the elemental weight is $x \, dW$, and the sum of these moments for all elements of the body is $\int x \, dW$. This sum of moments must equal $W\bar{x}$, the moment of the sum. With the substitution of $W = mg$ and $dW = g \, dm$, the moment expressions for all three axes become

$$\bar{x} = \frac{\int x \, dm}{m} \qquad \bar{y} = \frac{\int y \, dm}{m} \qquad \bar{z} = \frac{\int z \, dm}{m} \qquad (5/1)$$

The numerator of each expression represents the *sum of the moments*, and the product of m and the corresponding coordinate of G represents the *moment of the sum*. (The third equation results from reorienting the body and axes together so that $\bar{z}$ is horizontal.)

Equations 5/1 may be expressed in vector form with the aid of Fig. 5/4b, where the elemental mass and the position of G are located by their respective position vectors $\mathbf{r} = \mathbf{i}x + \mathbf{j}y + \mathbf{k}z$ and $\bar{\mathbf{r}} = \mathbf{i}\bar{x} + \mathbf{j}\bar{y} + \mathbf{k}\bar{z}$. Thus Eqs. 5/1 are the components of the single vector equation

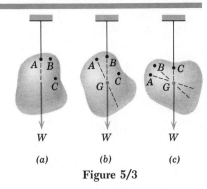

(a)   *(b)*   *(c)*

Figure 5/3

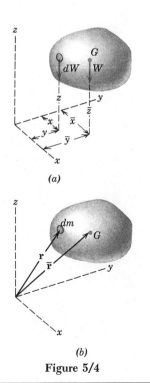

(a)

(b)

Figure 5/4

$$\bar{\mathbf{r}} = \frac{\int \mathbf{r}\, dm}{m} \tag{5/2}$$

The density ρ of a body is its mass per unit volume. Thus the mass of a differential element of volume dV becomes $dm = \rho\, dV$. In the event that ρ is not constant throughout the body but can be expressed as a function of the coordinates of the body, it will be necessary to account for this variation in the calculation of both the numerators and denominators of Eqs. 5/1. We may then write these expressions as

$$\bar{x} = \frac{\int x\rho\, dV}{\int \rho\, dV} \qquad \bar{y} = \frac{\int y\rho\, dV}{\int \rho\, dV} \qquad \bar{z} = \frac{\int z\rho\, dV}{\int \rho\, dV} \tag{5/3}$$

Equations 5/1, 5/2, and 5/3 define the position of the *center of mass*, which is clearly the same point as the center of gravity as long as the gravity field is treated as uniform and parallel. It is meaningless for us to speak of the center of gravity of a body that is removed from the earth's gravitational field, since no gravitational forces would act on the body. It would, however, still possess its unique center of mass. For the most part we will make reference henceforth to the center of mass rather than to the center of gravity. Also, the center of mass has a special significance in calculating the dynamic response of a body to unbalanced forces. This class of problems is discussed at length in Vol. 2 *Dynamics*.

In most problems the calculation of the position of the center of mass may be simplified by an intelligent choice of reference axes. In general the axes should be placed so as to simplify the equations of the boundaries as much as possible. Thus polar coordinates will be useful for bodies having circular boundaries. Another important clue may be taken from considerations of symmetry. Whenever there exists a line or plane of symmetry, a coordinate axis or plane should be chosen to coincide with this line or plane. The center of mass will always lie on such a line or plane, since the moments due to symmetrically located elements will always cancel, and the body can be considered composed of pairs of these elements. Thus the center of mass G of the homogeneous right-circular cone of Fig. 5/5a will lie somewhere on its central axis, which is a line of symmetry. The center of mass of the half right-circular cone lies on its plane of symmetry, Fig. 5/5b. The center of mass of the half ring in Fig. 5/5c lies in both of its planes of symmetry and therefore is situated on line AB. The location of G is always facilitated by the observation of symmetry when it exists.

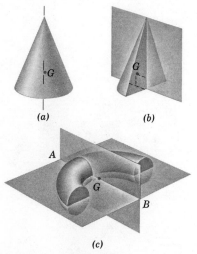

(a)

(b)

(c)

Figure 5/5

5/3 CENTROIDS OF LINES, AREAS, AND VOLUMES. Whenever the density ρ of a body is uniform throughout, it will be a constant factor in both the numerators and denominators of Eqs. 5/3 and will therefore cancel. The expressions that remain define a purely geometrical property of the body, since any reference to its physical properties has disappeared. The term *centroid* is used when the calculation concerns a geometrical shape only. When speaking of an actual physical body, we use the term *center of mass*. If the density is uniform throughout the body, the positions of the centroid and center of mass are identical, whereas if the density varies, these two points will, in general, not coincide.

The calculations of centroids fall within three distinct categories depending on whether the shape of the body involved can be modeled as a line, an area, or a volume.

(a) Lines. For a slender rod or wire of length L, cross-sectional area A, and density ρ, Fig. 5/6, the body approximates a line segment, and $dm = \rho A \, dL$. If ρ and A are constant over the length of the rod, the coordinates of the center of mass also become the coordinates of the centroid C of the line segment which, from Eqs. 5/1, may be written

$$\bar{x} = \frac{\int x \, dL}{L} \qquad \bar{y} = \frac{\int y \, dL}{L} \qquad \bar{z} = \frac{\int z \, dL}{L} \qquad (5/4)$$

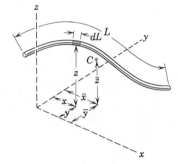

Figure 5/6

It should be noted that, in general, the centroid C will not lie on the line. If the rod lies in a single plane, such as the x-y plane, only two coordinates will require calculation.

(b) Areas. When a body of density ρ has a small but constant thickness t, it can be modeled as a surface area A, Fig. 5/7. The mass of an element becomes $dm = \rho t \, dA$. Again, if ρ and t are constant over the entire area, the coordinates of the center of mass of the body also become the coordinates of the centroid C of the surface area, and from Eqs. 5/1 may be written

$$\bar{x} = \frac{\int x \, dA}{A} \qquad \bar{y} = \frac{\int y \, dA}{A} \qquad \bar{z} = \frac{\int z \, dA}{A} \qquad (5/5)$$

The numerators in Eqs. 5/5 are known as the *first moments of area*.[°] If the surface area is curved, as illustrated in Fig. 5/7 with the shell segment, all three coordinates will be involved. Here again the centroid C for the curved surface will in general not lie on the surface. If the area is a flat surface in, say, the x-y plane, only the coordinates in that plane will be unknown.

[°] Second moments of areas (moments of first moments) appear later in our discussion of area moments of inertia in Appendix A.

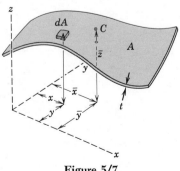

Figure 5/7

(*c*) *Volumes.* For a general body of volume V and density ρ, the element has a mass $dm = \rho\, dV$. The density ρ cancels if it is constant over the entire volume, and the coordinates of the center of mass also become the coordinates of the centroid C of the body. From Eqs. 5/3 or 5/1 they become

$$\bar{x} = \frac{\int x\, dV}{V} \qquad \bar{y} = \frac{\int y\, dV}{V} \qquad \bar{z} = \frac{\int z\, dV}{V} \qquad (5/6)$$

(*d*) *Choice of Element for Integration.* As is often the case the principal difficulty in a theory lies not so much in its concepts but in the procedures for applying it. With mass centers and centroids the concept of the moment principle is simple enough; the difficulties reside primarily with the choice of the differential element and with setting up the integrals. In particular there are five guidelines to be specially observed.

(*1*) *Order of element.* Whenever possible a first-order differential element should be selected in preference to a higher-order element so that only one integration will be required to cover the entire figure. Thus in Fig. 5/8*a* a first-order horizontal strip of area $dA = l\, dy$ will require only one integration with respect to y in order to cover the entire figure. The second-order element $dx\, dy$ will require two integrations, first with respect to x and second with respect to y, to cover the figure. As a further example, for the solid cone in Fig. 5/8*b* we choose a first-order element in the form of a circular slice of volume $dV = \pi r^2\, dy$, which requires only one integration, in preference to choosing a third-order element $dV = dx\, dy\, dz$, which would require three awkward integrations.

(*2*) *Continuity.* Whenever possible we choose an element which can be integrated in one continuous operation to cover the figure.

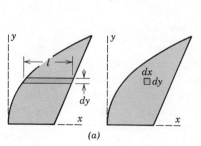

(*a*)

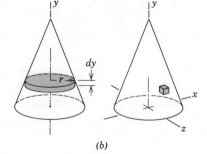

(*b*)

Figure 5/8

Thus the horizontal strip in Fig. 5/8*a* would be preferable to the vertical strip in Fig. 5/9 which, if used, would require two separate integrals because of the discontinuity in the expression for the height of the strip at $x = x_1$.

(3) *Discarding higher-order terms.* Higher-order terms may always be dropped compared with lower-order terms (see Art. 1/7). Thus the vertical strip of area under the curve in Fig. 5/10 is given by the first-order term $dA = y\,dx$, and the second-order triangular area $\frac{1}{2}dx\,dy$ is discarded. In the limit, of course, there is no error.

(4) *Choice of coordinates.* As a general rule we choose the coordinate system which best matches the boundaries of the figure. Thus the boundaries of the area in Fig. 5/11*a* are most easily described in rectangular coordinates, whereas the boundaries of the circular sector of Fig. 5/11*b* are best suited to polar coordinates.

(5) *Centroidal coordinate of element.* When a first- or second-order differential element is adopted, it is essential to use the *coordinate to the centroid of the element* for the moment arm in setting up the moment of the differential element. Thus, for the horizontal strip of area in Fig. 5/12*a* the moment of dA about the *y*-axis is $x_c\,dA$ where x_c is the *x*-coordinate to the centroid C of the element. Note that x_c is *not* the *x* which describes the boundaries of the area. In the *y*-direction for this element the moment arm y_c to the centroid of the element is the same, in the limit, as the *y*-coordinates of the two boundaries.

As a second example, consider the solid half-cone of Fig. 5/12*b* with the semicircular slice of differential thickness as the element of volume. The moment arm for the element in the *x*-direction is the distance x_c to the centroid of the face of the element and not the

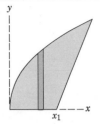

Figure 5/9

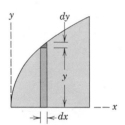

Figure 5/10

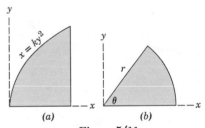

Figure 5/11

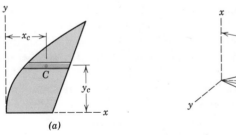

Figure 5/12

x-coordinate to the boundary of the element. On the other hand in the *z*-direction the moment arm z_c to the centroid of the element is the same as the *z*-coordinate of the element.

With these examples in mind we rewrite Eqs. 5/5 and 5/6 in the form

$$\bar{x} = \frac{\int x_c \, dA}{A}$$

$$\bar{y} = \frac{\int y_c \, dA}{A}$$ (5/5a)

$$\bar{z} = \frac{\int z_c \, dA}{A}$$

and

$$\bar{x} = \frac{\int x_c \, dV}{V}$$

$$\bar{y} = \frac{\int y_c \, dV}{V}$$ (5/6a)

$$\bar{z} = \frac{\int z_c \, dV}{V}$$

The subscript c serves as a reminder that the moment arms appearing in the numerators of the integral expressions for moments are *always* the coordinates to the *centroids* of the particular elements chosen.

At this point it would be well for the student to make certain that he understands clearly the principle of moments (Varignon's theorem), which was introduced in Art. 2/4. It is essential that the physical meaning of this principle be recognized as it is applied to the system of parallel weight forces depicted in Fig. 5/4a. With the equivalence clearly in mind between the moment of the resultant weight W and the sum (integral) of the moments of the elemental weights dW, then a mistake in setting up the necessary mathematics is far less likely to occur. Recognition of the principle of moments will provide assurance that the correct expression will be used for the moment arm x_c, y_c, or z_c to the centroid of the particular differential element chosen. Also, with the physical picture of the principle of moments in mind, Eqs. 5/4, 5/5, and 5/6, which are geometric relationships, will be recognized as descriptive also of homogeneous physical bodies where the density ρ has cancelled.

A summary of the centroidal coordinates for some of the commonly used shapes is given in Tables C3 and C4, Appendix C.

Sample Problem 5/1

Centroid of a circular arc. Locate the centroid of a circular arc as shown in the figure.

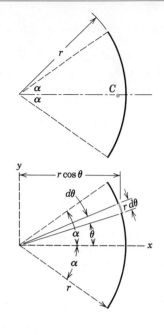

Solution. Choosing the axis of symmetry as the x-axis makes $\bar{y} = 0$. A differential element of arc has the length $dL = r\,d\theta$ expressed in polar coordinates, and the x-coordinate of the element is $r\cos\theta$. Applying the first of Eqs. 5/4 and substituting $L = 2\alpha r$ give

$$[L\bar{x} = \int x\,dL] \qquad (2\alpha r)\bar{x} = \int_{-\alpha}^{\alpha}(r\cos\theta)r\,d\theta$$

$$2\alpha r\bar{x} = 2r^2\sin\alpha$$

$$\bar{x} = \frac{r\sin\alpha}{\alpha} \qquad\qquad Ans.$$

For a semicircular arc $2\alpha = \pi$, which gives $\bar{x} = 2r/\pi$. By symmetry we see immediately that this result also applies to the quarter-circular arc when the measurement is made as shown.

① It should be perfectly evident that polar coordinates are preferable to rectangular coordinates to express the length of a circular arc.

Sample Problem 5/2

Centroid of a triangular area. Determine the distance $\bar{h}$ from the base of a triangle of altitude h to the centroid of its area.

Solution. The x-axis is taken to coincide with the base. A differential strip of area $dA = x\,dy$ is chosen. By similar triangles $x/(h-y) = b/h$. Applying the second of Eqs. 5/5 gives

$$[A\bar{y} = \int y_c\,dA] \qquad \frac{bh}{2}\bar{y} = \int_0^h y\,\frac{b(h-y)}{h}\,dy = \frac{bh^2}{6}$$

and

$$\bar{y} = \frac{h}{3} \qquad\qquad Ans.$$

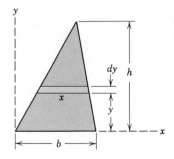

This same result holds with respect to either of the other two sides of the triangle considered a new base with corresponding new altitude. Thus the centroid lies at the intersection of the medians, since the distance of this point from any side is one third the altitude of the triangle with that side considered the base.

① We save one integration here by using the first-order element of area. Recognize that dA must be expressed in terms of the integration variable y; hence, $x = f(y)$ is required.

Sample Problem   5/3

Centroid of the area of a circular sector.   Locate the centroid of the area of a circular sector with respect to its vertex.

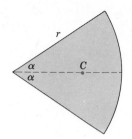

Solution I.   The x-axis is chosen as the axis of symmetry, and $\bar{y}$ is therefore automatically zero. We may cover the area by moving an element in the form of a partial circular ring, as shown in the figure, from the center to the outer periphery. The radius of the ring is r_0 and its thickness is dr_0, so that its area is $dA = 2r_0\alpha\,dr_0$.

The x-coordinate to the centroid of the element from Sample Problem 5/1 is $x_c = r_0 \sin\alpha/\alpha$ where r_0 replaces r in the formula. Thus the first of Eqs. 5/5a gives

$$[A\bar{x} = \int x_c\,dA] \qquad \frac{2\alpha}{2\pi}(\pi r^2)\bar{x} = \int_0^r \left(\frac{r_0 \sin\alpha}{\alpha}\right)(2r_0\alpha\,dr_0)$$

$$r^2\alpha\bar{x} = \tfrac{2}{3}r^3 \sin\alpha$$

$$\bar{x} = \frac{2}{3}\frac{r \sin\alpha}{\alpha} \qquad\qquad \text{Ans.}$$

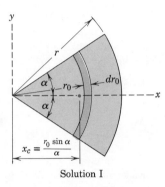

Solution I

① Note carefully that we must distinguish between the variable r_0 and the constant r.

② Be careful not to use r_0 as the centroidal coordinate for the element.

Solution II.   The area may also be covered by swinging a triangle of differential area about the vertex and through the total angle of the sector. This triangle, shown in the illustration, has an area $dA = (r/2)(r\,d\theta)$, where higher-order terms are neglected. From Sample Problem 5/2 the centroid of the triangular element of area is 2/3 of its altitude from its vertex, so that the x-coordinate to the centroid of the element is $x_c = \tfrac{2}{3}r \cos\theta$. Applying the first of Eqs. 5/5a gives

$$[A\bar{x} = \int x_c\,dA] \qquad (r^2\alpha)\bar{x} = \int_{-\alpha}^{\alpha} (\tfrac{2}{3}r \cos\theta)(\tfrac{1}{2}r^2\,d\theta)$$

$$r^2\alpha\bar{x} = \tfrac{2}{3}r^3 \sin\alpha$$

and as before

$$\bar{x} = \frac{2}{3}\frac{r \sin\alpha}{\alpha} \qquad\qquad \text{Ans.}$$

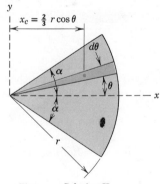

Solution II

For a semicircular area $2\alpha = \pi$, which gives $\bar{x} = 4r/3\pi$. By symmetry we see immediately that this result also applies to the quarter-circular area where the measurement is made as shown.

It should be noted that, if we had chosen a second-order element $r_0\,dr_0\,d\theta$, one integration with respect to θ would yield the ring with which *Solution I* began. On the other hand integration with respect to r_0 initially would give the triangular element with which *Solution II* began.

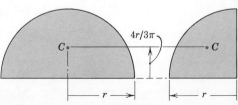

Sample Problem 5/4

Locate the centroid of the area under the curve $x = ky^3$ from $x = 0$ to $x = a$.

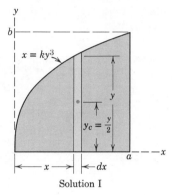

Solution I. A vertical element of area $dA = y\,dx$ is chosen as shown in the figure. The x-coordinate of the centroid is found from the first of Eqs. 5/5a. Thus

① $[A\bar{x} = \int x_c\,dA]$ $\qquad \bar{x}\int_0^a y\,dx = \int_0^a xy\,dx$

Substituting $y = (x/k)^{1/3}$ and $k = a/b^3$ and integrating give

$$\frac{3ab}{4}\bar{x} = \frac{3a^2b}{7}, \qquad \bar{x} = \tfrac{4}{7}a \qquad\qquad Ans.$$

In solving for $\bar{y}$ from the second of Eqs. 5/5a the coordinate to the centroid of the rectangular element is $y_c = y/2$ where y is the height of the strip governed by the equation of the curve $x = ky^3$. Thus the moment principle becomes

$[A\bar{y} = \int y_c\,dA]$ $\qquad \dfrac{3ab}{4}\bar{y} = \int_0^a \left(\dfrac{y}{2}\right)y\,dx$

Substituting $y = b(x/a)^{1/3}$ and integrating give

$$\frac{3ab}{4}\bar{y} = \frac{3ab^2}{10}, \qquad \bar{y} = \tfrac{2}{5}b \qquad\qquad Ans.$$

Solution I

① Note that $x_c = x$ for the vertical element.

Solution II. The horizontal element of area shown in the lower figure may be employed in place of the vertical element. The x-coordinate to the centroid of the rectangular element is the average of the coordinates of the ends or $x_c = (a + x)/2$. Hence

$[A\bar{x} = \int x_c\,dA]$ $\qquad \bar{x}\int_0^b (a - x)\,dy = \int_0^b \left(\dfrac{a + x}{2}\right)(a - x)\,dy$

The value of $\bar{y}$ is found from

$[A\bar{y} = \int y_c\,dA]$ $\qquad \bar{y}\int_0^b (a - x)\,dy = \int_0^b y(a - x)\,dy$

where $y_c = y$ for the horizontal strip. The evaluation of these integrals will check the previous results for $\bar{x}$ and $\bar{y}$.

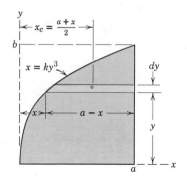

Solution II

Sample Problem 5/5

Hemispherical volume. Locate the centroid of the volume of a hemisphere of radius r with respect to its base.

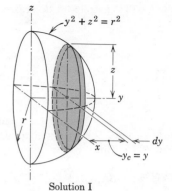

Solution I. With the axes chosen as shown in th figure $\bar{x} = \bar{z} = 0$ by symmetry. The most convenient element is a circular slice of thickness dy parallel to the x-z plane. Since the hemisphere intersects the y-z plane in the circle $y^2 + z^2 = r^2$, the radius of the circular slice is $z = + \sqrt{r^2 - y^2}$. The volume of the elemental slice becomes

$$dV = \pi(r^2 - y^2)\, dy$$

The second of Eqs. 5/6a requires

$$[V\bar{y} = \int y_c\, dV] \qquad \bar{y} \int_0^r \pi(r^2 - y^2)\, dy = \int_0^r y\pi(r^2 - y^2)\, dy$$

where $y_c = y$. Integrating gives

$$\tfrac{2}{3}\pi r^3 \bar{y} = \tfrac{1}{4}\pi r^4, \qquad \bar{y} = \tfrac{3}{8}r \qquad\qquad Ans.$$

Solution I

Solution II. Alternatively we may use for our differential element a cylindrical shell of length y, radius z, and thickness dz, as shown in the lower figure. By expanding the radius of the shell from zero to r, we cover the entire volume. By symmetry the centroid of the elemental shell lies at its center, so that $y_c = y/2$. The volume of the element is $dV = (2\pi z\, dz)(y)$. Expressing y in terms of z from the equation of the circle gives $y = + \sqrt{r^2 - z^2}$. Using the value of $\tfrac{2}{3}\pi r^3$ computed in Solution I for the volume of the hemisphere and substituting in the second of Eqs. 5/6a give us

$$[V\bar{y} = \int y_c\, dV] \qquad (\tfrac{2}{3}\pi r^3)\bar{y} = \int_0^r \frac{\sqrt{r^2 - z^2}}{2}(2\pi z\sqrt{r^2 - z^2})\, dz$$

$$= \int_0^r \pi(r^2 z - z^3)\, dz = \frac{\pi r^4}{4}$$

$$\bar{y} = \tfrac{3}{8}r \qquad\qquad Ans.$$

Solutions I and II are of comparable use since each involves an element of simple shape and requires integration with respect to one variable only.

① Can you identify the higher-order element of volume which is omitted from the expression for dV?

Solution II

Solution III. As an alternative we could use the angle θ as our variable with limits of 0 and $\pi/2$. The radius of either element would become $r\sin\theta$, whereas the thickness of the slice would be $dy = (r\, d\theta)\sin\theta$ and that of the shell would be $dz = (r\, d\theta)\cos\theta$. The length of the shell would be $y = r\cos\theta$.

Solution III

PROBLEMS

5/1 Determine the coordinates of the centroid of the shaded area.   Ans. $\bar{x} = 3b/10$, $\bar{y} = 3b/4$

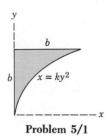

Problem 5/1

5/2 Locate the centroid of the shaded area shown.

Ans. $\bar{x} = 2.09$, $\bar{y} = 1.43$

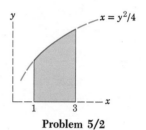

Problem 5/2

5/3 Determine the coordinates of the centroid of the shaded area.   Ans. $\bar{x} = 3a/8$, $\bar{y} = 2b/5$

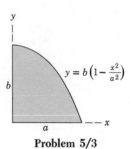

Problem 5/3

5/4 Determine the y-coordinate of the centroid of the area under the sine curve shown.

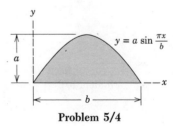

Problem 5/4

5/5 Determine the y-coordinate of the centroid of the area by direct integration.   Ans. $\bar{y} = \dfrac{14R}{9\pi}$

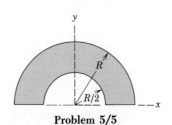

Problem 5/5

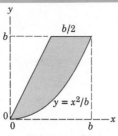

Problem 5/6

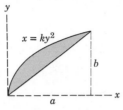

Problem 5/7

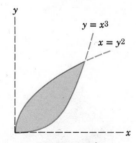

Problem 5/8

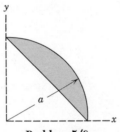

Problem 5/9

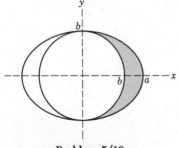

Problem 5/10

5/6 Calculate the x-coordinate of the centroid of the shaded area.

5/7 Calculate the coordinates of the centroid of the area shown.   *Ans.* $\bar{x} = 2a/5,\ \bar{y} = b/2$

5/8 Calculate the coordinates of the centroid of the segment of the circular area.

5/9 Locate the centroid of the shaded area between the two curves.   *Ans.* $\bar{x} = \frac{12}{25},\ \bar{y} = \frac{3}{7}$

5/10 Locate the centroid of the shaded area between the ellipse and the circle.

5/11 Locate the centroid of the area shown in the figure by direct integration. (*Caution:* Observe carefully the proper sign of the radical involved.)

$$Ans. \ \bar{x} = \frac{10 - 3\pi}{4 - \pi} \frac{a}{3}$$

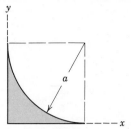

Problem 5/11

5/12 Determine the x-coordinate of the centroid of the shaded area shown. (Observe the caution of Prob. 5/11.) $Ans. \ \bar{x} = 0.339a$

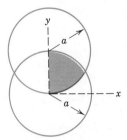

Problem 5/12

5/13 Specify the coordinates of the center of mass of the circular shell section by direct reference to the results of Sample Problem 5/1.

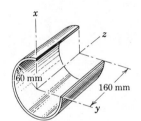

Problem 5/13

5/14 Use the results of Sample Problem 5/3 to compute the coordinates of the mass center of the portion of the solid homogeneous cylinder shown.

$$Ans. \ \bar{x} = \bar{y} = -8/(3\pi) \text{ in.}, \ \bar{z} = 5 \text{ in.}$$

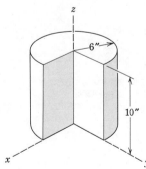

Problem 5/14

5/15 Find the distance $\bar{z}$ from the vertex of the right circular cone to the centroid of its volume.

Problem 5/15

Problem 5/16

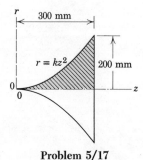

Problem 5/17

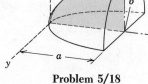

Problem 5/18

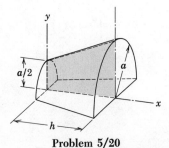

Problem 5/20

5/16 Calculate the distance $\bar{h}$ measured from the base to the centroid of the volume of the frustum of the right-circular cone. *Ans.* $\bar{h} = \frac{11}{56}h$

5/17 Locate the mass center of the homogeneous solid body whose volume is determined by revolving the shaded area through 360° about the z-axis.

5/18 Determine the z-coordinate of the centroid of the volume obtained by revolving the shaded area under the parabola about the z-axis through 180°. *Ans.* $\bar{z} = 2a/3$

5/19 Determine the x-coordinate of the centroid of the volume described in Prob. 5/18.

5/20 Determine the y-coordinate of the centroid of the volume obtained by revolving the shaded area about the x-axis through 180°. *Ans.* $\bar{y} = 15a/(14\pi)$

5/21 Use the results of Sample Problem 5/2 and determine by inspection the distance $\bar{h}$ from the centroid of the lateral area of any cone or pyramid of altitude h to the base of the figure.

5/22 Determine the distance $\bar{z}$ from the base of any cone or pyramid of altitude h to the centroid of its volume.

$$Ans.\ \bar{z} = \frac{h}{4}$$

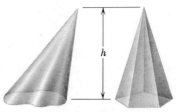

Problem 5/22

5/23 Determine the z-coordinate of the mass center of the homogeneous quarter-spherical shell of radius r.

$$Ans.\ \bar{z} = r/2$$

Problem 5/23

5/24 Determine the z-coordinate of the center of mass of the solid obtained by revolving the quarter-circular area about the z-axis.

$$Ans.\ \bar{z} = \frac{11a}{2(4 + 3\pi)}$$

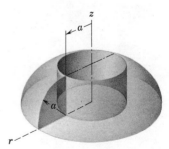

Problem 5/24

5/25 Determine the z-coordinate of the centroid of the volume generated by revolving the quarter-circular area about the z-axis through $90°$.

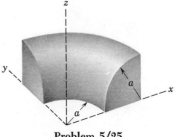

Problem 5/25

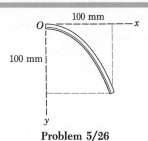

100 mm

O   *x*

100 mm

y

Problem 5/26

5/26 The homogeneous slender rod has a uniform cross section and is bent into the shape of a parabolic arc with the vertex at the origin. Determine the coordinates of the center of mass of the rod. (*Reminder:* A differential arc length is $dL = \sqrt{(dx)^2 + (dy)^2} = \sqrt{1 + (dy/dx)^2}\,dx$.)

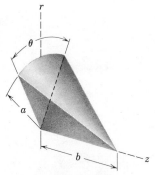

r

θ

a

b

z

Problem 5/27

5/27 Locate the centroid of the conical wedge generated by revolving the right triangle of altitude *a* about its base *b* through an angle *θ*.

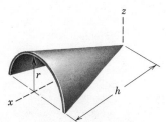

z

r

h

x

Problem 5/28

▶**5/28** Determine the position of the center of mass of the homogeneous thin conical shell shown.

$$\textit{Ans. } \bar{x} = 2h/3, \bar{z} = 4r/3\pi$$

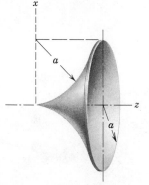

x

a

z

a

Problem 5/29

▶**5/29** Locate the center of mass of the homogeneous bell-shaped shell of uniform but negligible thickness.

$$\textit{Ans. } \bar{z} = \frac{a}{\pi - 2}$$

5/30 Determine the position of the centroid of the volume within the bell-shaped shell of Prob. 5/29.

$$Ans. \; \bar{z} = \frac{a}{2(10 - 3\pi)}$$

5/31 Determine the x-coordinate of the centroid of the top half of a bell-shaped shell.

$$Ans. \; \bar{x} = \frac{a}{\pi}\left(\frac{3\pi - 8}{\pi - 2}\right)$$

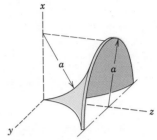

Problem 5/31

5/32 Locate the center of mass G of the steel half ring. (*Hint:* Choose an element of volume in the form of a cylindrical shell whose intersection with the plane of the ends is shown.)

$$Ans. \; \bar{r} = \frac{a^2 + 4R^2}{2\pi R}$$

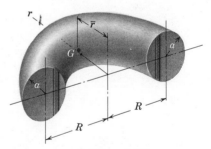

Problem 5/32

5/4 COMPOSITE BODIES AND FIGURES; APPROXIMATIONS.

When a body or figure can be conveniently divided into several parts of simple shape, we may use the principle of Varignon if each part is treated as a finite element of the whole. Thus for such a body, illustrated schematically in Fig. 5/13, whose parts have masses m_1, m_2, m_3 and whose separate coordinates of the centers of mass of these parts in, say, the x-direction are $\bar{x}_1$, $\bar{x}_2$, $\bar{x}_3$, the moment principle gives

$$(m_1 + m_2 + m_3)\bar{X} = m_1\bar{x}_1 + m_2\bar{x}_2 + m_3\bar{x}_3$$

where $\bar{X}$ is the x-coordinate of the center of mass of the whole. Similar relations hold for the other two coordinate directions. We generalize, then, for a body of any number of parts and express the sums in condensed form and obtain the mass-center coordinates

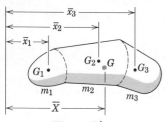

Figure 5/13

$$\bar{X} = \frac{\Sigma m\bar{x}}{\Sigma m} \qquad \bar{Y} = \frac{\Sigma m\bar{y}}{\Sigma m} \qquad \bar{Z} = \frac{\Sigma m\bar{z}}{\Sigma m} \qquad (5/7)$$

Analogous relations hold for composite lines, areas, and volumes, where the m's are replaced by L's, A's, and V's, respectively. It should be pointed out that if a hole or cavity is considered one of the component parts of a composite body or figure, the corresponding mass represented by the cavity or hole is considered a negative quantity.

Frequently in practice the boundaries of an area or volume are not expressible in terms of simple geometrical shapes or in shapes that can be represented mathematically. For such cases we find it necessary to resort to a method of approximation. As an example consider the problem of locating the centroid C of the irregular area shown in Fig. 5/14. The area may be divided into strips of width Δx and variable height h. The area A of each strip, such as the one shown in color, is $h\,\Delta x$ and is multiplied by the coordinates x_c and y_c to its *centroid* to obtain the moments of the element of area. The sum of the moments for all strips divided by the total area of the strips will give

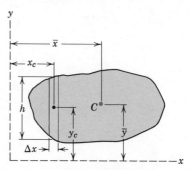

Figure 5/14

the corresponding centroidal component. A systematic tabulation of the results will permit an orderly evaluation of the total area ΣA, the sums $\Sigma A x_c$ and $\Sigma A y_c$, and the centroidal components

$$\bar{x} = \frac{\Sigma A x_c}{\Sigma A} \qquad \bar{y} = \frac{\Sigma A y_c}{\Sigma A}$$

The accuracy of the approximation will be increased by decreasing the widths of the strips used. In all cases the average height of the strip should be estimated in approximating the areas. Although it is usually of advantage to use elements of constant width, it is not necessary to do so. In fact we may use elements of any size and shape which approximate the given area to satisfactory accuracy.

In locating the centroid of an irregular volume the problem may be reduced to one of determining the centroid of an area. Consider the volume shown in Fig. 5/15 where the magnitudes A of the

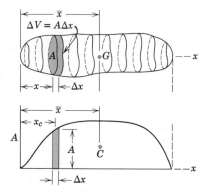

Figure 5/15

cross-sectional areas normal to the x-direction are plotted against x as shown. A vertical strip of area under the curve is $A\,\Delta x$, which equals the corresponding element of volume ΔV. Thus the area under the plotted curve represents the volume of the body, and the x-coordinate to the centroid of the area under the curve is given by

$$\bar{x} = \frac{\Sigma (A\,\Delta x) x_c}{\Sigma A\,\Delta x} \qquad \text{which equals} \qquad \bar{x} = \frac{\Sigma V x_c}{\Sigma V}$$

for the centroid of the actual volume.

Sample Problem 5/6

Locate the center of mass of the bracket-and-shaft combination. The vertical face is made from sheet metal which has a mass of 25 kg/m², the material of the horizontal base has a mass of 40 kg/m², and the steel shaft has a density of 7.83 Mg/m³.

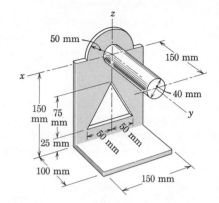

Solution. The composite body may be considered composed of the five elements shown in the lower portion of the illustration. The triangular part will be taken as a negative mass. For the reference axes indicated it is clear by symmetry that the x-coordinate of the center of mass is zero.

The mass m of each part is easily calculated and should need no further explanation. For Part 1 we have from Sample Problem 5/3

$$\bar{z} = \frac{4r}{3\pi} = \frac{4(50)}{3\pi} = 21.2 \text{ mm}$$

For Part 3 from Sample Problem 5/2 we see that the centroid of the triangular mass is $\frac{1}{3}$ of its altitude above its base. Measurement from the coordinate axes becomes

$$\bar{z} = -[150 - 25 - \tfrac{1}{3}(75)] = -100 \text{ mm}$$

The y- and z-coordinates to the mass centers of the remaining parts should be evident by inspection. The terms involved in applying Eqs. 5/7 are best handled in the form of a table as follows:

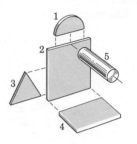

PART	m kg	$\bar{y}$ mm	$\bar{z}$ mm	$m\bar{y}$ kg·mm	$m\bar{z}$ kg·mm
1	0.098	0	21.2	0	2.08
2	0.562	0	−75.0	0	−42.19
3	−0.094	0	−100.0	0	9.38
4	0.600	50.0	−150.0	30.0	−90.00
5	1.476	75.0	0	110.7	0
Totals	2.642			140.7	−120.73

Equations 5/7 are now applied and the results are

$$\left[\bar{Y} = \frac{\Sigma m\bar{y}}{\Sigma m} \right] \qquad \bar{Y} = \frac{140.7}{2.642} = 53.3 \text{ mm} \qquad\qquad Ans.$$

$$\left[\bar{Z} = \frac{\Sigma m\bar{z}}{\Sigma m} \right] \qquad \bar{Z} = \frac{-120.73}{2.642} = -45.7 \text{ mm} \qquad\qquad Ans.$$

PROBLEMS

5/33 Calculate the y-coordinate of the centroid of the triangular area.   *Ans.* $\overline{Y} = 40$ mm

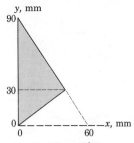

Problem 5/33

5/34 Calculate the y-coordinate of the centroid of the figure shown.

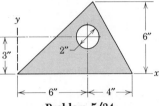

Problem 5/34

5/35 Calculate the x-coordinate of the centroid of the shaded area shown.   *Ans.* $\overline{X} = 6.54$ mm

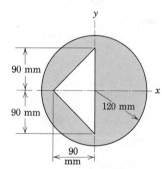

Problem 5/35

5/36 Locate the centroid of the shaded area of Prob. 5/8, repeated here, by the method of this article.

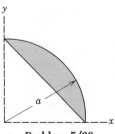

Problem 5/36

5/37 Locate the centroid of the area of Prob. 5/5, repeated here, by the method of this article.

$$Ans. \ \overline{Y} = \frac{14R}{9\pi}$$

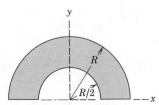

Problem 5/37

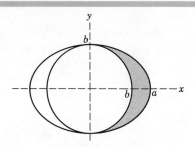

Problem 5/38

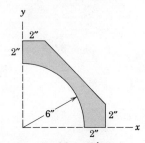

Problem 5/39

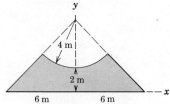

Problem 5/40

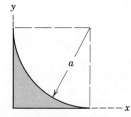

Problem 5/41

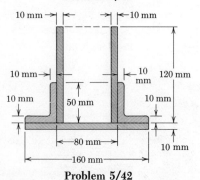

Problem 5/42

5/38 Determine by the method of this article the x-coordinate of the centroid of the shaded area between the ellipse and the circle of Prob. 5/10, shown again here. (Refer to Table C3 in Appendix C for the properties of an elliptical area.)

5/39 Locate the centroid of the shaded area of Prob. 5/11, repeated here, by the method of this article.

$$Ans.\ \overline{X} = \overline{Y} = \frac{10 - 3\pi}{4 - \pi}\frac{a}{3}$$

5/40 Compute the distance $\overline{Y}$ from the x-axis to the centroid of the shaded area.

5/41 Calculate the coordinates of the centroid of the shaded area. $Ans.\ \overline{X} = \overline{Y} = 4.29$ in.

5/42 Determine the distance $\overline{H}$ from the bottom of the base plate to the centroid of the built-up structural section shown. $Ans.\ \overline{H} = 39.3$ mm

/43 A uniform rod is bent into the shape shown and pivoted about O. Find the value of a in terms of the radius r so that the straight section will remain horizontal.

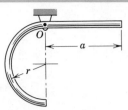

Problem 5/43

/44 Calculate the coordinates of the center of mass of the slender rod bent into the shape shown.

Ans. $\bar{X} = 31.1$ mm, $\bar{Y} = 48.9$ mm, $\bar{Z} = 31.1$ mm

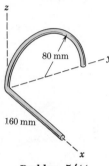

Problem 5/44

/45 By the method of this article determine the height $\bar{H}$ from the base to the mass center of the frustum of the solid cone of Prob. 5/16, repeated here.

Problem 5/45

/46 The hemispherical shell and its semicircular base are formed from the same piece of sheet metal of small thickness. Use the results of Prob. 5/23 and calculate the coordinates of the mass center of the shell and base combined.   *Ans.* $\bar{X} = 0.475r$, $\bar{Y} = r/3$

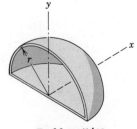

Problem 5/46

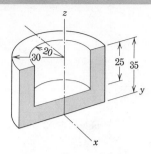

Dimensions in Millimeters
Problem 5/47

5/47 Calculate the coordinates of the mass center of the metal die casting shown.

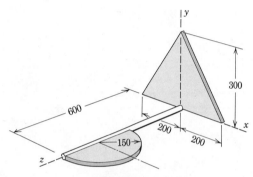

Dimensions in Millimeters
Problem 5/48

5/48 The masses of the three parts of the welded assembly consisting of the triangular plate, uniform rod, and semicircular plate are, respectively, 4.2, 2.2, and 2.5 kg. Calculate the coordinates of the mass center.

Ans. $\bar{X} = 17.9$ mm, $\bar{Y} = 47.2$ mm, $\bar{Z} = 200.6$ mm

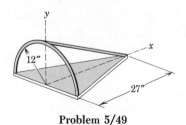

Problem 5/49

5/49 A tubular framework consisting of a semicircular member and a brace is welded to a flat triangular plate as shown. If the tubular material weighs 5 lb per foot of length and the base material weighs 20 lb per square foot of area, calculate the coordinates of the center of gravity of the combined base and frame.

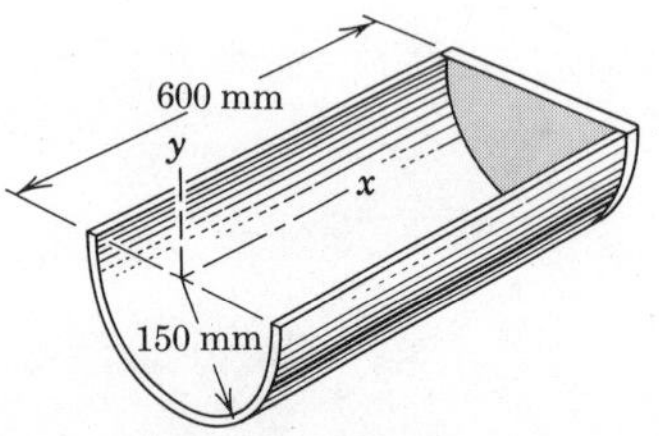

5/50 Determine the position of the center of mass of the cylindrical shell with a closed semicircular end. The shell is made from sheet metal with a mass of 24 kg/m², and the end is made from metal plate with a mass of 36 kg/m².

Ans. $\bar{X} = 348$ mm, $\bar{Y} = -90.5$ mm

Problem 5/50

51 Determine the coordinates of the center of mass of the bracket, which is made from a plate of uniform thickness.   *Ans.* $\bar{X} = -8.3$ mm
$$\bar{Y} = -31.4 \text{ mm}$$
$$\bar{Z} = 10.3 \text{ mm}$$

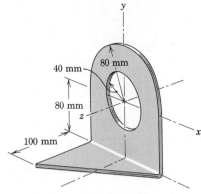

Problem 5/51

52 As an example of the accuracy involved in graphical approximations calculate the percentage error e in determining the x-coordinate of the centroid of the triangular area by using the five approximating rectangles of width $a/5$ in place of the triangle.   *Ans.* $e = 1.00$ percent low

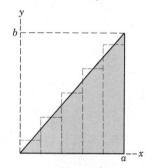

Problem 5/52

53 A sheet-metal pattern has the shape shown. Estimate the location of the centroid of the area visually and note its coordinates. Then check your estimate by a calculation using the superimposed grid.

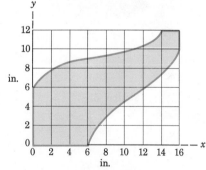

Problem 5/53

54 The circular disk rotates about an axis through its center O and has three holes of diameter d positioned as shown. A fourth hole is to be drilled in the disk at the same radius r so that the disk will be in balance (mass center at O). Determine the required diameter D of the new hole and its angular position.   *Ans.* $D = 1.227d$, $\theta = 84.9°$

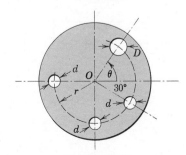

Problem 5/54

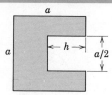

Problem 5/55

5/55 A rectangular piece is removed from the square metal plate of side a. Determine the value of h which will result in the mass center of the remaining plate being as far to the left as possible.

Ans. $h = 0.586a$

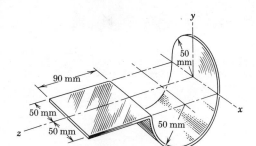

Problem 5/56

5/56 Locate the center of mass of the thin plate formed into the shape shown.

Ans. $\bar{X} = 0,\ \bar{Y} = -14.5$ mm, $\bar{Z} = 73.0$ mm

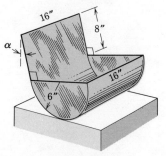

Problem 5/57

▶5/57 A cylindrical container with an extended rectangular back and semicircular ends is all fabricated from the same sheet-metal stock. Calculate the angle α made by the back with the vertical when the container rests in an equilibrium position on a horizontal surface.

Ans. $\alpha = 39.6°$

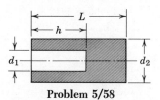

Problem 5/58

▶5/58 A homogeneous charge of explosive is to be formed as a circular cylinder of length L and diameter d_2 with an axial hole of diameter d_1 and depth h as shown in section. Determine the value of h which will place the center of mass of the final charge a maximum distance from the open end.

$$\textit{Ans. } h = \left(\frac{d_2}{d_1}\right)^2 L \left[1 - \sqrt{1 - \left(\frac{d_1}{d_2}\right)^2}\right]$$

5/5 THEOREMS OF PAPPUS. A very simple method exists for calculating the surface area generated by revolving a plane curve about a nonintersecting axis in the plane of the curve. In Fig. 5/16 the line segment of length L in the x-y plane generates a surface when revolved about the x-axis. An element of this surface is the ring generated by dL. The area of this ring is its circumference times its slant height or $dA = 2\pi y\, dL$ and the total area is then

$$A = 2\pi \int y\, dL$$

But since $\bar{y}L = \int y\, dL$, the area becomes

$$\boxed{A = 2\pi \bar{y}L} \tag{5/8}$$

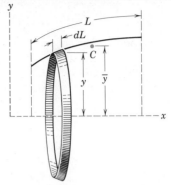

Figure 5/16

where $\bar{y}$ is the y-coordinate of the centroid C for the line of length L. Thus the generated area is the same as the lateral area of a right circular cylinder of length L and radius $\bar{y}$.

In the case of a volume generated by revolving an area about a nonintersecting line in its plane an equally simple relation exists for finding the volume. An element of the volume generated by revolving the area A about the x-axis, Fig. 5/17, is the elemental ring of cross section dA and radius y. The volume of the element is its circumference times dA or $dV = 2\pi y\, dA$, and the total volume is

$$V = 2\pi \int y\, dA$$

But since $\bar{y}A = \int y\, dA$, the volume becomes

$$\boxed{V = 2\pi \bar{y}A} \tag{5/9}$$

where $\bar{y}$ is the y-coordinate of the centroid C of the revolved area A. Thus the generated volume is obtained by multiplying the generating area by the circumference of the circular path described by its centroid.

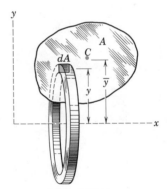

Figure 5/17

The two theorems of Pappus, expressed by Eqs. 5/8 and 5/9, are useful not only in determining areas and volumes of generation, but they are also employed to find the centroids of plane curves and plane areas when the corresponding areas and volumes due to revolution of these figures about a nonintersecting axis are known. Dividing the area or volume by 2π times the corresponding line segment length or plane area will give the distance from the centroid to the axis.

In the event that a line or an area is revolved through an angle θ less than 2π, the generated surface or volume may be found by replacing 2π by θ in Eqs. 5/8 and 5/9. Thus the more general relations are

$$A = \theta \bar{y}L \quad \text{and} \quad V = \theta \bar{y}A$$

where θ is expressed in radians.

---

*Attributed to Pappus of Alexandria, a Greek geometer who lived in the third century A.D. The theorems often bear the name of Guldinus (Paul Guldin, 1577–1643), who claimed original authorship, although the works of Pappus were apparently known to him.

PROBLEMS

5/59 Determine the volume V and lateral area A of a right circular cone of base radius r and altitude h.

5/60 From the known surface area $A = 4\pi r^2$ of a sphere of radius r, determine the radial distance $\bar{r}$ to the centroid of the semicircular arc used to generate the surface.

5/61 From the known volume $V = \frac{4}{3}\pi r^3$ of a sphere of radius r, determine the radial distance $\bar{r}$ to the centroid of the semicircular area used to generate the sphere.

5/62 Use the notation of the half torus of Prob. 5/32, shown again here, and determine the volume V and the surface area A of a complete torus.

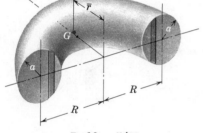

Problem 5/62

5/63 A shell structure has the form of a surface obtained by revolving the circular arc through 360° about the z-axis. Determine the surface area of one side of the complete shell.

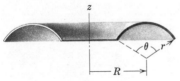

Problem 5/63

5/64 Compute the volume V of the solid generated by revolving the right triangle about the z-axis through 180°.   *Ans.* $V = 3619 \text{ mm}^3$

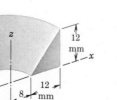

Problem 5/64

5/65 The shaded triangle of base b and altitude h is revolved about its base through an angle θ to generate a portion of a complete solid of revolution. Write the expression for the volume V of the solid generated.

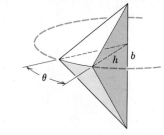

Problem 5/65

5/66 The two circular arcs AB and BC are revolved around the vertical axis to obtain the surface of revolution shown. Compute the area of this surface.

Ans. $A = 157.9$ in.^{2}

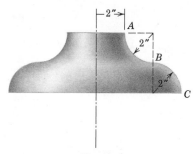

Problem 5/66

5/67 Determine the volume V generated by revolving the quarter-circular area about the z-axis through an angle of $90°$.

Problem 5/67

5/68 Determine the volume V generated by revolving the quarter-circular area about the z-axis through $90°$.

$$\text{Ans. } V = \frac{\pi a^3}{12}(3\pi - 2)$$

Problem 5/68

5/69 Compute the volume V and total surface area A of the complete ring whose square cross section is shown.

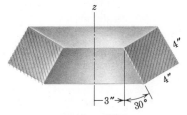

Problem 5/69

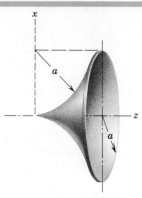

Problem 5/70

5/70 Determine the surface area of one side of the bell-shaped shell of Prob. 5/29, shown again here, using the theorem of Pappus.

5/71 Determine the volume within the bell-shaped shell shown with Prob. 5/70. Use the results cited in Prob. 5/39.

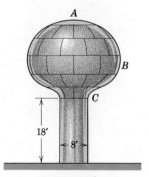

Problem 5/72

5/72 The water storage tank is a shell of revolution and is to be sprayed with two coats of paint which has a coverage of 500 ft² per gallon. The engineer (who remembers mechanics) consults a scale drawing of the tank and determines that the curved line ABC has a length of 34 ft and that its centroid is 8.2 ft from the centerline of the tank. How many gallons of paint will be used for the tank including the vertical cylindrical column? *Ans.* 8.82 gal

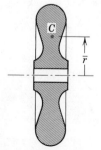

Problem 5/73

5/73 A hand-operated control wheel made of aluminum has the proportions shown in the cross-sectional view. The area of the total section shown is 15 200 mm², and the wheel has a mass of 10.0 kg. Calculate the distance $\bar{r}$ to the centroid of the half-section. The aluminum has a density of 2.69 Mg/m³.

5/74 A surface is generated by revolving the circular arc of 0.8-m radius and subtended angle of 120° completely about the z-axis. The diameter of the neck is 0.6 m. Determine the area A generated.

$Ans.\ A = 4.62\ m^2$

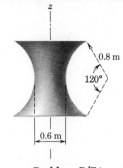

Problem 5/74

5/75 A styrofoam ring, shown in section, is designed for packing a mass-produced item. Calculate the volume of styrofoam used in each ring.

$Ans.\ V = 27.2(10^6)\ mm^3$

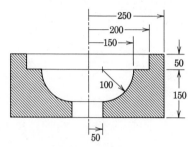

Dimensions in Millimeters

Problem 5/75

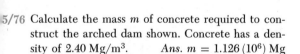

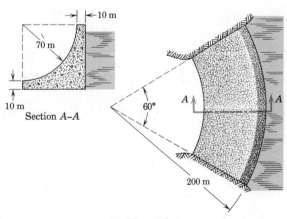

5/76 Calculate the mass m of concrete required to construct the arched dam shown. Concrete has a density of 2.40 Mg/m^3. $Ans.\ m = 1.126\,(10^6)\ Mg$

Problem 5/76

SECTION B. SPECIAL TOPICS

5/6 BEAMS. Structural members that offer resistance to bending caused by applied loads are known as beams. Most beams are long prismatical bars, and the loads are usually applied normal to the axes of the bars. Beams are undoubtedly the most important of all structural members, and the basic theory underlying their design must be thoroughly understood. The analysis of the load-carrying capacities of beams consists, first, in establishing the equilibrium requirements of the beam as a whole and any portion of it considered separately. Second, the relations between the resulting forces and the accompanying internal resistance of the beam to support these forces are established. The first part of this analysis requires the application of the principles of statics, while the second part of the problem involves the strength characteristics of the material and is usually

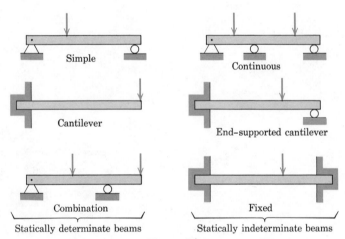

Figure 5/18

treated under the heading of the mechanics of solids or the mechanics of materials. This article concerns the first aspect of the problem only and will enable us to calculate the distribution along the beam of the internal force and moment acting on sections of the beam.

(a) Types of Beams. Beams supported in such a way that their external support reactions can be calculated by the methods of statics alone are called *statically determinate* beams. A beam that has more supports than are necessary to provide equilibrium is said to be *statically indeterminate,* and it is necessary to consider the load-deformation properties of the beam in addition to the equations of statical equilibrium to determine the support reactions. In Fig. 5/18

are shown examples of both types of beams. In this article we will analyze only statically determinate beams.

Beams may also be identified by the type of external loading they support. The beams in Fig. 5/18 are supporting concentrated loads, whereas the beam in Fig. 5/19 is supporting a distributed load. The intensity w of a distributed load may be expressed as force per unit length of beam. The intensity may be constant or variable, continuous or discontinuous. The intensity of the loading in Fig. 5/19 is constant from C to D and variable from A to C and from D to B. The intensity is discontinuous at D where it changes magnitude abruptly. Although the intensity itself is not discontinuous at C, the rate of change of intensity dw/dx is discontinuous.

(b) Shear, Bending, and Torsion. In addition to tension or compression a beam has resistance to shear, bending, and torsion. These three effects are illustrated in Fig. 5/20. The force V is called the *shear* force, the couple M is known as the *bending moment,* and the couple T is called a *torsional moment.* These effects represent the vector components of the resultant of the forces acting on a transverse section of the beam as shown in the lower part of the figure.

We direct our attention now primarily to the shear force V and bending moment M caused by forces applied to the beam in a single plane. The conventions for positive values of shear V and bending moment M shown in Fig. 5/21 are the ones generally used. By the principle of action and reaction we note that the directions of V and M are reversed on the two sections. It is frequently impossible to tell without calculation whether the shear and moment at a particular section are positive or negative. For this reason it will be found advisable to represent V and M in their positive directions on the free-body diagrams and let the algebraic signs of the calculated values indicate the proper directions.

As an aid to the physical interpretation of the bending couple M consider the beam shown in Fig. 5/22 bent by the two equal and opposite positive moments applied at the ends. The cross section of the beam is taken to be that of an H-section with a very narrow center web and heavy top and bottom flanges. For this beam we may

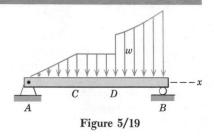

Figure 5/19

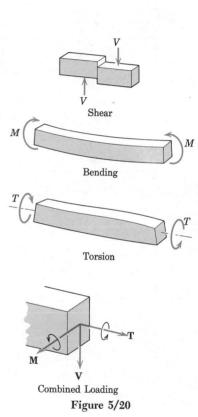

Figure 5/20

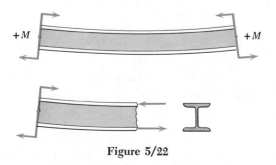

Figure 5/22

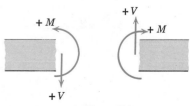

Figure 5/21

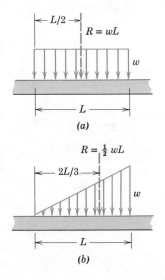

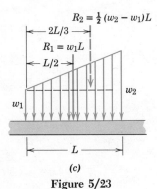

Figure 5/23

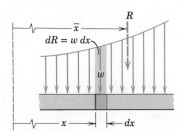

Figure 5/24

neglect the load carried by the small web compared with that carried by the two flanges. It should be perfectly clear that the upper flange of the beam is shortened and is under compression while the lower flange is lengthened and is under tension. The resultant of the two forces, one tensile and the other compressive, acting on any section is a couple and has the value of the bending moment on the section. If a beam of some other cross section were loaded in the same way, the distribution of force over the cross section would be different, but the resultant would be the same couple.

The variation of shear force V and bending moment M over the length of a beam provides information necessary for the design analysis of the beam. In particular, the maximum magnitude of the bending moment is usually the primary consideration in the design or selection of a beam, and its value and position should be determined. The variations in shear and moment are best shown graphically, and the expressions for V and M when plotted against distance along the beam give the *shear-force* and *bending-moment diagrams* for the beam.

The first step in the determination of the shear and moment relations is to establish the values of all external reactions on the beam by applying the equations of equilibrium to a free-body diagram of the beam as a whole. Next, we isolate a portion of the beam, either to the right or to the left of an arbitrary transverse section, with a free-body diagram, and apply the equations of equilibrium to this isolated portion of the beam. These equations will yield expressions for the shear force V and bending moment M acting at the cut section on the part of the beam isolated. The part of the beam which involves the smaller number of forces, either to the right or to the left of the arbitrary section, usually yields the simpler solution. We should avoid using a transverse section that coincides with the location of a concentrated load, as such a position represents a point of discontinuity in the variation of shear or bending moment. Finally, it is important to note that the calculations for V and M on each section chosen should be consistent with the positive convention illustrated in Fig. 5/21.

(*c*) *Distributed Loads.* Loading intensities which are constant or which vary linearly are easily handled. Figure 5/23 illustrates the three most common cases and the resultants of the distributed loads in each case. In each instance we see that the resultant passes through the centroid of the figure formed by the intensity w and the length L over which the force is distributed.

For a more general loading relationship, Fig. 5/24, we must start with a differential increment of force $dR = w\,dx$. The integral $\int w\,dx$ gives the resultant R and the integral $\int xw\,dx$ gives the moment of the distributed force. The principle of moments gives the location of R by $R\bar{x} = \int xw\,dx$.

(d) *General Loading, Shear, and Moment Relationships.* Certain general relationships may be established for any beam with distributed loads which will aid greatly in the construction of the shear and moment distributions. Figure 5/25 represents a portion of a loaded beam, and an element dx of the beam is isolated. The loading w represents the force per unit length of beam. At the location x the shear V and moment M acting on the element are drawn in their positive directions. On the opposite side of the element where the coordinate is $x + dx$ these quantities are also shown in their positive directions but must be labeled $V + dV$ and $M + dM$, since the changes in V and M with x are required. The applied loading w may be considered constant over the length of the element, since this length is a differential quantity and the effect of any change in w disappears in the limit compared with the effect of w itself.

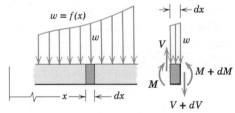

Figure 5/25

Equilibrium of the element requires that the sum of the vertical forces be zero. Thus we have

$$V - w\,dx - (V + dV) = 0$$

or

$$\boxed{w = -\frac{dV}{dx}} \tag{5/10}$$

We see from Eq. 5/10 that the slope of the shear diagram must everywhere be equal to the negative of the value of the applied loading. Equation 5/10 holds on either side of a concentrated load but not at the concentrated load by reason of the discontinuity produced by the abrupt change in shear.

Equilibrium of the element in Fig. 5/25 also requires that the moment sum be zero. Taking moments about the left side of the element gives

$$M + w\,dx\frac{dx}{2} + (V + dV)\,dx - (M + dM) = 0$$

The two M's cancel, and the terms $w(dx)^2/2$ and $dV\,dx$ may be dropped, since they are differentials of higher order than those that remain. This leaves merely

$$V = \frac{dM}{dx} \qquad (5/11)$$

which expresses the fact that the shear everywhere is equal to the slope of the moment curve. We may now express the moment M in terms of the shear V by integrating Eq. 5/11. Thus

$$\int_{M_0}^{M} dM = \int_{x_0}^{x} V\,dx$$

or

$$M = M_0 + (\text{area under shear diagram from } x_0 \text{ to } x)$$

In this expression M_0 is the bending moment at x_0 and M is the bending moment at x. For beams where there is no externally applied moment M_0 at $x_0 = 0$, the total moment at any section equals the area under the shear diagram up to that section. Summing up the area under the shear diagram is usually the simplest way to construct the moment diagram.

When V passes through zero and is a continuous function of x with $dV/dx \neq 0$, the bending moment M will be a maximum or a minimum, since $dM/dx = 0$ at such a point. Critical values of M also occur when V crosses the zero axis discontinuously, which occurs for beams under concentrated loads.

We observe from Eqs. 5/10 and 5/11 that the degree of V in x is one higher than that of w. Also M is of one higher degree in x than is V. Furthermore M is two degrees in x higher than w. Thus for a beam loaded by $w = kx$ which is of the first degree in x, the shear V is of the second degree in x and the bending moment M is of the third degree in x.

Equations 5/10 and 5/11 may be combined to yield

$$\frac{d^2M}{dx^2} = -w \qquad (5/12)$$

Thus, if w is a known function of x, the moment M may be obtained by two integrations, provided that the limits of integration are properly evaluated each time. This method is usable only if w is a continuous function of x.[°]

When bending in a beam occurs in more than a single plane, a separate analysis in each plane may be carried out. The results may then be combined vectorially.

[°] When w is a discontinuous function of x, it is possible to introduce a special set of expressions called *singularity functions* which permit writing analytical expressions for shear V and moment M over a range of discontinuities. These functions are not discussed in this book.

Sample Problem 5/7

Determine the shear and moment distributions produced in the simple beam by the 4-kN concentrated load.

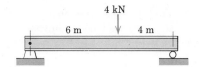

Solution. From the free-body diagram of the entire beam we find the support reactions which are

$$R_1 = 1.6 \text{ kN} \qquad R_2 = 2.4 \text{ kN}$$

A section of the beam of length x is next isolated with its free-body diagram on which we show the shear V and the bending moment M in their positive directions. Equilibrium gives

$$[\Sigma F_y = 0] \qquad 1.6 - V = 0 \qquad V = 1.6 \text{ kN}$$

$$[\Sigma M_{R_1} = 0] \qquad M - 1.6x = 0 \qquad M = 1.6x$$

These values of V and M apply to all sections of the beam to the left of the 4-kN load.

A section of the beam to the right of the 4-kN load is next isolated with its free-body diagram upon which V and M are shown in their positive directions. Equilibrium requires

$$[\Sigma F_y = 0] \qquad V + 2.4 = 0 \qquad V = -2.4 \text{ kN}$$

$$[\Sigma M_{R_2} = 0] \qquad -(2.4)(10 - x) + M = 0 \qquad M = 2.4(10 - x)$$

These results apply only to sections of the beam to the right of the 4-kN load.

The values of V and M are plotted as shown. The maximum bending moment occurs where the shear changes direction. As we move in the positive x-direction starting with $x = 0$, we see that the moment M is merely the accumulated area under the shear diagram.

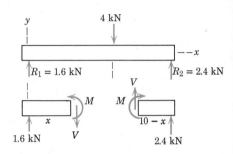

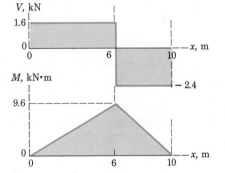

Sample Problem 5/8

Determine the shear and moment distributions induced in the beam by the distributed load whose intensity varies linearly with the length as shown.

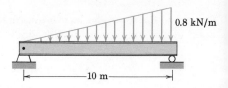

Solution. The resultant R of the linear distribution of load equals the area of the triangular load-distribution diagram and passes through the centroid of this area. Thus $R = \frac{1}{2}(0.8)(10) = 4$ kN. Equilibrium of the beam as a whole, then, gives

$[\Sigma M_{R_1} = 0]$ $0.4(6.67) - R_2(10) = 0,$ $R_2 = 2.67$ kN

$[\Sigma F_y = 0]$ $R_1 + 2.67 - 4 = 0$ $R_1 = 1.33$ kN

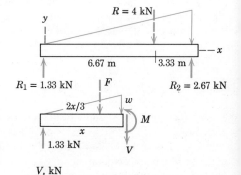

A representative section of the beam is next isolated with its free-body diagram with V and M shown in their positive directions. The linear load intensity is $w = (x/10)(0.8) = 0.08x$ kN/m. The resultant F of the linear loading on the beam section equals the area of the triangular distribution and passes through the centroid of this area. Thus

$$F = \tfrac{1}{2}wx = \tfrac{1}{2}(0.08x)x = 0.04x^2$$

A moment equation about the cut section and a force equation in the y-direction establish the equilibrium of the section and give us

$[\Sigma M = 0]$ $1.33x - 0.04x^2(x/3) - M = 0,$ $M = 1.33x - 0.0133x^3$

$[\Sigma F_y = 0]$ $1.33 - 0.04x^2 - V = 0,$ $V = 1.33 - 0.04x^2$

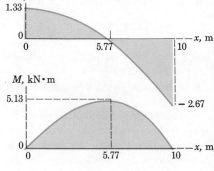

Since the load intensity is a continuous function of x, both expressions for V and M hold throughout the length of the beam.

The shear-force and bending-moment diagrams are now obtained by plotting V and M as shown. We note that M is a maximum when $V = 0$, which occurs when $0 = 1.33 - 0.04x^2$ or $x = \sqrt{1.33/0.04} = 5.77$ m. Substituting this value of x into the expression for M gives

$$M_{max} = 1.33(5.77) - 0.0133(5.77)^3 = 5.13 \text{ kN·m}$$

Again we observe that the value of M at any section is the area under the shear curve up to that section.

If we differentiate the expressions for V and M with respect to x, we have $-dV/dx = 0.08x$ and $dM/dx = 1.33 - 0.04x^2$, which equal w and V, respectively, as required by Eqs. 5/10 and 5/11.

Sample Problem 5/9

Draw the shear-force and bending-moment diagrams for the loaded beam and determine the maximum moment M and its location x from the left end.

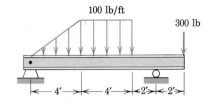

Solution. The support reactions are most easily obtained by considering the resultants of the distributed loads as shown on the free-body diagram of the beam as a whole. The first interval of the beam is analyzed from the free-body diagram of the section for $0 < x < 4$ ft. A vertical summation of forces and a moment summation about the cut section yield

$$[\Sigma F_y = 0] \qquad V = 247 - 12.5x^2$$

$$[\Sigma M = 0] \qquad M + (12.5x^2)\frac{x}{3} - 247x = 0 \qquad M = 247x - 4.167x^3$$

These values of V and M hold for $0 < x < 4$ ft and are plotted for that interval in the shear and moment diagrams shown.

From the free-body diagram of the section for which $4 < x < 8$ ft, equilibrium in the vertical direction and a moment sum about the cut section give

$$[\Sigma F_y = 0] \qquad V + 100(x - 4) + 200 - 247 = 0 \qquad V = 447 - 100x$$

$$[\Sigma M = 0] \qquad M + 100(x - 4)\frac{x - 4}{2} + 200[x - \tfrac{2}{3}(4)] - 247x = 0$$

$$M = -266.7 + 447x - 50x^2$$

These values of V and M are plotted on the shear and moment diagrams for the interval $4 < x < 8$ ft.

The analysis of the remainder of the beam is continued from the free-body diagram of the portion of the beam to the right of a section in the next interval. It should be noted that V and M are represented in their positive directions. A vertical force summation and a moment summation about the section yield

$$V = -353 \text{ lb} \qquad \text{and} \qquad M = 2930 - 353x$$

These values of V and M are plotted on the shear and moment diagrams for the interval $8 < x < 10$ ft.

The last interval may be analyzed by inspection. The shear is constant at $+300$ lb, and the moment follows a straight-line relation beginning with zero at the right end of the beam.

The maximum moment occurs at $x = 4.47$ ft, where the shear curve crosses the zero axis, and the magnitude of M is obtained for this value of x by substitution into the expression for M for the second interval. The maximum moment is

$$M = 731 \text{ lb-ft} \qquad \qquad Ans.$$

As before, note that the moment M at any section equals the area under the shear diagram up to that section. For instance, for $x < 4$ ft,

$$\left[\Delta M = \int V\, dx\right] \qquad M - 0 = \int_0^x (247 - 12.5x^2)\, dx, \quad M = 247x - 4.167x^3$$

which agrees with the results already computed.

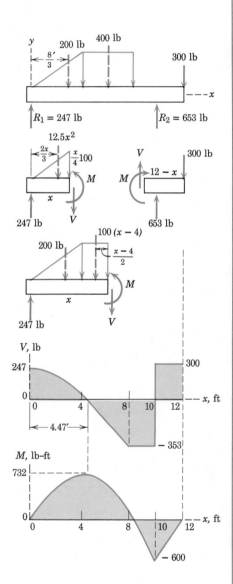

PROBLEMS

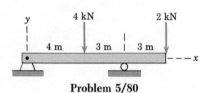

Problem 5/77

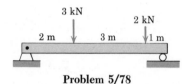

Problem 5/78

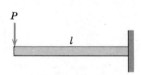

Problem 5/79

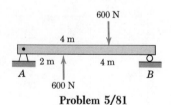

Problem 5/80

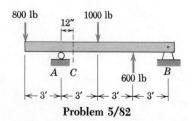

Problem 5/81

Problem 5/82

5/77 Draw the shear and moment diagrams for the diving board which supports the 175-lb man poised for a dive.

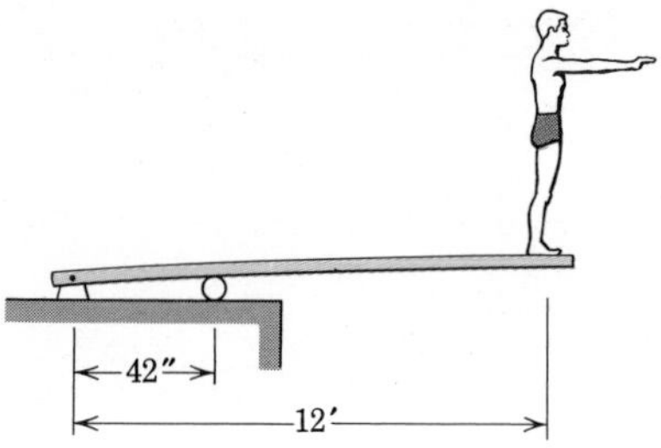

5/78 Draw the shear and moment diagrams for the loaded beam.

5/79 Draw the shear and moment diagrams for the cantilever beam.

5/80 Draw the shear and moment diagrams for the horizontal beam shown and determine the maximum magnitude of the bending moment and its location.
Ans. $|M|_{max} = 6.00$ kN·m at $x = 7$ m

5/81 Draw the shear and moment diagrams for the beam shown.

5/82 Draw the shear and moment diagrams for the beam shown and find the bending moment M at section C.
Ans. $M_c = -1667$ lb-ft

83 Draw the shear and moment diagrams for the beam loaded at its center by the couple C.

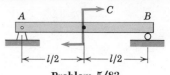

Problem 5/83

84 Draw the shear and moment diagrams for the beam subjected to the end couple. What is the moment M at a section 0.5 m to the right of B?

　　　　　　　　　　　　Ans. $M = -120\,\text{N·m}$

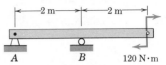

Problem 5/84

85 The angle strut is welded to the beam AB and supports the 1-kN load. Draw the shear and moment diagrams for the beam.

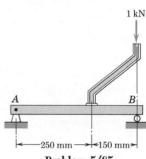

Problem 5/85

86 Construct the moment diagram for the two beams loaded as shown and connected by the hinge joint B.

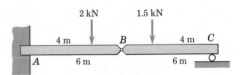

Problem 5/86

87 Draw the shear and moment diagrams for the beam loaded as shown. Determine the bending moment M with the maximum magnitude.

　　　　　　　　　　　　Ans. $M = -2000\,\text{lb-ft}$

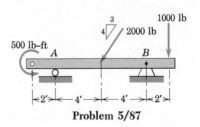

Problem 5/87

88 The resistance of a beam of uniform width to bending is found to be proportional to the square of the beam depth y. For the cantilever beam shown the depth is h at the support. Find the required depth y as a function of the length x in order for all sections to be equally effective in their resistance to bending.　　　　Ans. $y = h\sqrt{x/l}$

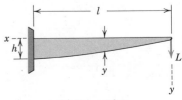

Problem 5/88

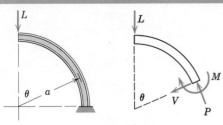

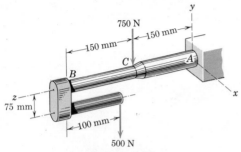

Problem 5/89

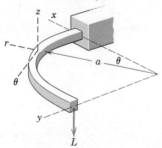

Problem 5/90

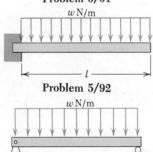

Problem 5/91

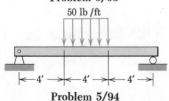

Problem 5/92

Problem 5/93

Problem 5/94

5/89 A curved cantilever beam has the form of a quarter-circular arc. Determine the expressions for the shear V and the bending moment M as functions of θ.

5/90 Construct the bending-moment diagram for the cantilevered shaft AB of the rigid unit shown.

5/91 Write expressions for the torsional moment T and bending moment M in the curved quarter-circular beam under the end load L. Use a notation consistent with the right-handed r-θ-z coordinate system where positive moment vectors are taken in the direction of the positive coordinates.

$Ans.\ M = -La\cos\theta,\ T = -La(1 - \sin\theta)$

5/92 Draw the shear and moment diagrams for the cantilever beam with the uniform load w per unit of length.

5/93 Draw the shear and moment diagrams for the uniformly loaded beam and find the maximum bending moment M.

$Ans.\ M = \dfrac{wl^2}{8}$

5/94 Draw the shear and moment diagrams for the beam, which supports the load of 50 lb per foot of beam length distributed over its midsection.

95 Draw the shear and moment diagrams for the loaded beam and find the maximum magnitude M of the bending moment.   *Ans.* $M = \frac{5}{6}Pl$

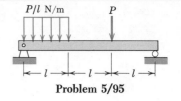

Problem 5/95

96 Draw the shear and moment diagrams for the cantilever beam with the linear loading and find the maximum magnitude M of the bending moment.

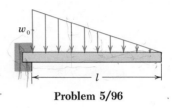

Problem 5/96

97 Draw the shear and moment diagrams for the loaded cantilever beam where the end couple M_1 is adjusted so as to produce zero moment at the fixed end of the beam. Find the bending moment M at $x = 4$ ft.   *Ans.* $M = 4800$ lb-ft

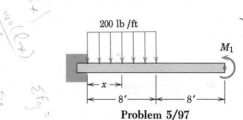

Problem 5/97

98 Draw the shear and moment diagrams for the linearly loaded simple beam shown. Determine the magnitude M of the maximum bending moment.

$$\text{Ans. } M = \frac{w_0 l^2}{12}$$

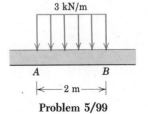

Problem 5/98

99 A certain beam supports a distributed load of constant intensity $w = 3$ kN/m between A and B as shown. If the shear force and bending moment at A are $+2$ kN and -1.5 kN·m, respectively, calculate the shear force and bending moment at B.

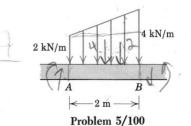

100 The distributed load increases linearly from 2 to 4 kN/m in a distance of 2 m along a certain beam in equilibrium. If the shear force and bending moment at section A are $+3$ kN and $+2$ kN·m, respectively, calculate the shear force and bending moment at section B.

Ans. $V_B = -3$ kN, $M_B = 2.67$ kN·m

Problem 5/100

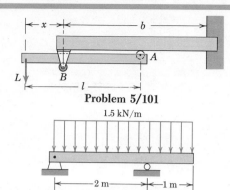

Problem 5/101

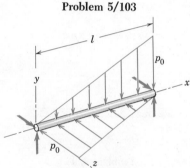

1.5 kN/m

Problem 5/102

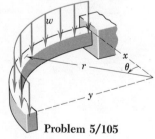

Problem 5/103

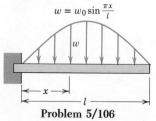

Problem 5/104

Problem 5/105

$$w = w_0 \sin \frac{\pi x}{l}$$

Problem 5/106

5/101 Sketch the bending moment diagrams for the upper support beam for the values of overhang $x = 0$, $x = 0.4b$, and $x = l$.

5/102 Draw the shear and moment diagrams for the beam shown. Determine the distance b, measured from the left end, to the point where the bending moment is zero between the supports.

Ans. $b = 1.5$ m

5/103 Determine the maximum bending moment M and the corresponding value of x in the crane beam and indicate the section where this moment acts.

Ans. $M_A = \dfrac{L}{4l}(l-a)^2$, $x = \dfrac{a+l}{2}$

5/104 The end-supported shaft is subjected to the linearly varying loads in mutually perpendicular planes. Determine the expression for the resultant bending moment M in the shaft.

Ans. $M = \dfrac{p_0}{6l}x(l-x)\sqrt{5l^2 - 2lx + 2x^2}$

▶**5/105** The curved cantilever beam in the form of a quarter-circular arc supports a load of w N/m applied along the curve of the beam on its upper surface. Determine the magnitudes of the torsional moment T and bending moment M in the beam as functions of θ.

Ans. $T = wr^2\left(\dfrac{\pi}{2} - \theta - \cos\theta\right)$

$M = -wr^2(1 - \sin\theta)$

▶**5/106** The cantilever beam supports the half-sine-wave loading $w = w_0 \sin(\pi x/l)$ per unit of length. Determine the bending moment M in terms of x by direct use of Eq. 5/12.

Ans. $M = \dfrac{w_0 l}{\pi}\left(\dfrac{l}{\pi}\sin\dfrac{\pi x}{l} + x - l\right)$

5/7 FLEXIBLE CABLES. One important type of structural member is the flexible cable which is used in suspension bridges, transmission lines, messenger cables for supporting heavy trolley or telephone lines, and many other applications. In the design of these structures it is necessary to know the relations involving the tension, span, sag, and length of the cables. We determine these quantities by examining the cable as a body in equilibrium. In the analysis of flexible cables we assume that any resistance offered to bending is negligible. This assumption means that the force in the cable is always in the direction of the cable.

Flexible cables may support a series of distinct concentrated loads, as shown in Fig. 5/26a, or they may support loads that are continuously distributed over the length of the cable, as indicated by the variable-intensity loading w in Fig. 5/26b. In some instances the weight of the cable is negligible compared with the loads it supports, and in other cases the weight of the cable may be an appreciable load or the sole load, in which case it cannot be neglected. Regardless of which of these conditions is present, the equilibrium requirements of the cable may be formulated in the same manner.

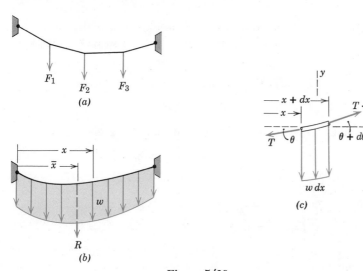

Figure 5/26

(a) General Relationships. If the intensity of the variable and continuous load applied to the cable of Fig. 5/26b is expressed as w units of force per unit of horizontal length x, then the resultant R of the vertical loading is

$$R = \int dR = \int w \, dx$$

where the integration is taken over the desired interval. We find the position of R from the moment principle, so that

$$R\bar{x} = \int x \, dR \qquad \bar{x} = \frac{\int x \, dR}{R}$$

The elemental load $dR = w \, dx$ is represented by an elemental strip of vertical length w and width dx of the shaded area of the loading diagram, and R is represented by the total area. It follows from the foregoing expressions that R passes through the *centroid* of the shaded area.

The equilibrium condition of the cable will be satisfied if each infinitesimal element of the cable is in equilibrium. The free-body diagram of a differential element is shown in Fig. 5/26c. At the general position x the tension in the cable is T, and the cable makes an angle θ with the horizontal x-direction. At the section $x + dx$ the tension is $T + dT$, and the angle is $\theta + d\theta$. Note that the changes in both T and θ are taken positive with positive change in x. The vertical load $w \, dx$ completes the free-body diagram. The equilibrium of vertical and horizontal forces requires, respectively, that

$$(T + dT) \sin (\theta + d\theta) = T \sin \theta + w \, dx$$

$$(T + dT) \cos (\theta + d\theta) = T \cos \theta$$

The trigonometric expansion for the sine and cosine of the sum of two angles and the substitutions $\sin d\theta = d\theta$ and $\cos d\theta = 1$, which hold in the limit as $d\theta$ approaches zero, yield

$$(T + dT)(\sin \theta + \cos \theta \, d\theta) = T \sin \theta + w \, dx$$

$$(T + dT)(\cos \theta - \sin \theta \, d\theta) = T \cos \theta$$

Dropping the second-order terms and simplifying give us

$$T \cos \theta \, d\theta + dT \sin \theta = w \, dx$$

$$-T \sin \theta \, d\theta + dT \cos \theta = 0$$

which we may write as

$$d(T \sin \theta) = w \, dx \qquad \text{and} \qquad d(T \cos \theta) = 0$$

The second relation expresses the fact that the horizontal component of T remains unchanged, which is clear from the free-body diagram. If we introduce the symbol $T_0 = T \cos \theta$ for this constant horizontal force, we may then substitute $T = T_0/\cos \theta$ into the first of the two equations just obtained and get $d(T_0 \tan \theta) = w \, dx$. But $\tan \theta = dy/dx$, so that the equilibrium equation may be written in the form

$$\boxed{\frac{d^2y}{dx^2} = \frac{w}{T_0}} \qquad (5/13)$$

Equation 5/13 is the *differential equation* for the flexible cable. The solution to the equation is that functional relation $y = f(x)$

which satisfies the equation and also satisfies the conditions at the fixed ends of the cable, called *boundary conditions*. This relationship defines the shape of the cable, and we will use it to solve two important and limiting cases of cable loading.

(*b*) *Parabolic Cable.* When the intensity of vertical loading w is constant, the description closely approximates a suspension bridge where the uniform weight of the roadway may be expressed by the constant w. The mass of the cable is not distributed uniformly with the horizontal but is relatively small and its weight is neglected. For this limiting case we will prove that the cable hangs in a parabolic arc. Figure 5/27 shows such a suspension bridge of span L and sag h with origin of coordinates taken at the midpoint of the span. With

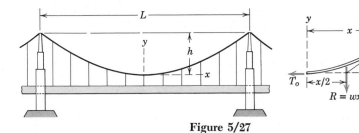

Figure 5/27

both w and T_0 constant, we may integrate Eq. 5/13 once with respect to x to obtain

$$\frac{dy}{dx} = \frac{wx}{T_0} + C$$

where C is a constant of integration. For the coordinate axes chosen, $dy/dx = 0$ when $x = 0$, so that $C = 0$. Hence

$$\frac{dy}{dx} = \frac{wx}{T_0}$$

which defines the slope of the curve as a function of x. One further integration yields

$$\int_0^y dy = \int_0^x \frac{wx}{T_0} dx \qquad \text{or} \qquad \boxed{y = \frac{wx^2}{2T_0}} \qquad (5/14)$$

Readers should make certain that they can obtain the identical results with the indefinite integral together with the evaluation of the constant of integration. Equation 5/14 gives the shape of the cable, which we see is a vertical parabola. The constant horizontal component of cable tension becomes the cable tension at the origin.

Inserting the corresponding values $x = L/2$ and $y = h$ in Eq. 5/14 gives

$$T_0 = \frac{wL^2}{8h} \qquad \text{and} \qquad y = \frac{4hx^2}{L^2}$$

The tension T is found from a free-body diagram of a finite portion of the cable, shown in Fig. 5/27, which requires that

$$T = \sqrt{T_0^2 + w^2 x^2}.$$

Elimination of T_0 gives

$$T = w \sqrt{x^2 + \left(\frac{L^2}{8h}\right)^2} \tag{5/15}$$

The maximum tension occurs when $x = L/2$ and is

$$T_{\max} = \frac{wL}{2} \sqrt{1 + \frac{L^2}{16h^2}} \tag{5/15a}$$

We obtain the length S of the complete cable from the differential relation $ds = \sqrt{(dx)^2 + (dy)^2}$. Thus

$$\int_0^{S/2} ds = \frac{S}{2} = \int_0^{L/2} \sqrt{1 + \left(\frac{dy}{dx}\right)^2}\, dx = \int_0^{L/2} \sqrt{1 + \left(\frac{wx}{T_0}\right)^2}\, dx$$

For convenience in computation we change this expression to a convergent series and then integrate it term by term. From the expansion for a variable x

$$(1 + x)^n = 1 + nx + \frac{n(n-1)}{2!} x^2 + \frac{n(n-1)(n-2)}{3!} x^3 + \cdots$$

the integral may be written as

$$S = 2 \int_0^{L/2} \left(1 + \frac{w^2 x^2}{2 T_0^2} - \frac{w^4 x^4}{8 T_0^4} + \cdots \right) dx$$

$$= L \left(1 + \frac{w^2 L^2}{24 T_0^2} - \frac{w^4 L^4}{640 T_0^4} + \cdots \right)$$

Substitution of $w/T_0 = 8h/L^2$ yields

$$S = L \left[1 + \frac{8}{3} \left(\frac{h}{L}\right)^2 - \frac{32}{5} \left(\frac{h}{L}\right)^4 + \cdots \right] \tag{5/16}$$

When we examine the properties of this series, we find that it converges for all values of $h/L \leq 1/4$. In most cases h is much smaller than $L/4$, so that the three terms of Eq. 5/16 give a sufficiently accurate approximation.

(c) *Catenary Cable.* Consider now a uniform cable, Fig. 5/28, suspended at two points in the same horizontal plane and hanging under the action of its own weight only. The free-body diagram of a finite portion of the cable of length s is shown in the right-hand part of the figure. This free-body diagram differs from that in Fig. 5/27 in that the total vertical force supported is equal to the weight of the section of cable of length s in place of the load distributed uniformly

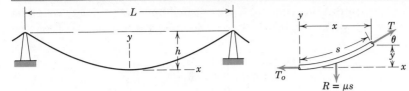

Figure 5/28

with respect to the horizontal. If the cable has a weight μ per unit of its length, the resultant R of the load is $R = \mu s$, and the incremental vertical load $w\,dx$ of Fig. 5/26c is replaced by $\mu\,ds$. Thus we may substitute $w = \mu\,ds/dx$ into the differential relation, Eq. 5/13, for the cable and obtain

$$\frac{d^2y}{dx^2} = \frac{\mu}{T_0}\frac{ds}{dx} \qquad\qquad (5/17)$$

Since $s = f(x, y)$, it is necessary to change this equation to one containing only the two variables.

We may substitute the identity $(ds)^2 = (dx)^2 + (dy)^2$ to obtain

$$\frac{d^2y}{dx^2} = \frac{\mu}{T_0}\sqrt{1 + \left(\frac{dy}{dx}\right)^2} \qquad\qquad (5/18)$$

Equation 5/18 is the differential equation of the curve (catenary) assumed by the cable. Solution of this equation is facilitated by the substitution $p = dy/dx$, which gives

$$\frac{dp}{\sqrt{1 + p^2}} = \frac{\mu}{T_0}dx$$

Integrating this equation gives us

$$\ln\left(p + \sqrt{1 + p^2}\right) = \frac{\mu}{T_0}x + C$$

The constant C is zero since $dy/dx = p = 0$ when $x = 0$. Substituting $p = dy/dx$, changing to exponential form, and clearing the equation of the radical give

$$\frac{dy}{dx} = \frac{e^{\mu x/T_0} - e^{-\mu x/T_0}}{2} = \sinh\frac{\mu x}{T_0}$$

where the hyperbolic function* is introduced for convenience. The slope may be integrated to obtain

$$y = \frac{T_0}{\mu}\cosh\frac{\mu x}{T_0} + K$$

---

* See Tables B8 and B10, Appendix B, for the definition and integral of hyperbolic functions.

The integration constant K is evaluated from the boundary condition $x = 0$ when $y = 0$. This substitution requires that $K = -T_0/\mu$, and hence

$$y = \frac{T_0}{\mu}\left(\cosh\frac{\mu x}{T_0} - 1\right) \qquad (5/19)$$

Equation 5/19 is the equation of the curve (catenary) assumed by the cable hanging under the action of its weight only.

From the free-body diagram in Fig. 5/28 we see that $dy/dx = \tan\theta = \mu s/T_0$. Thus, from the previous expression for the slope,

$$s = \frac{T_0}{\mu}\sinh\frac{\mu x}{T_0} \qquad (5/20)$$

We obtain the tension T in the cable from the equilibrium triangle of the forces in Fig. 5/28. Thus

$$T^2 = \mu^2 s^2 + T_0{}^2$$

which, upon combination with Eq. 5/20, becomes

$$T^2 = T_0{}^2\left(1 + \sinh^2\frac{\mu x}{T_0}\right) = T_0{}^2\cosh^2\frac{\mu x}{T_0}$$

or

$$T = T_0\cosh\frac{\mu x}{T_0} \qquad (5/21)$$

We may also express the tension in terms of y with the aid of Eq. 5/19, which, when substituted into Eq. 5/21, gives

$$T = T_0 + \mu y \qquad (5/22)$$

Equation 5/22 shows us that the increment in cable tension from that at the lowest position depends only on μy.

Most problems dealing with the catenary involve solutions of Eqs. 5/19 through 5/22, which may be handled by a graphical approximation or solved by computer. The graphical procedure is illustrated in the sample problem following this article.

The solution of catenary problems where the sag-to-span ratio is small may be approximated by the relations developed for the parabolic cable. A small sag-to-span ratio means a tight cable, and the uniform distribution of weight along the cable is not much different from the same load intensity distributed uniformly along the horizontal.

Many problems dealing with both the catenary and parabolic cable involve suspension points that are not on the same level. In such cases we may apply the relations just developed to the part of the cable on each side of the lowest point.

Sample Problem 5/10

The light cable supports a mass of 12 kg per meter of horizontal length and is suspended between the two points on the same level 300 m apart. If the sag is 60 m, find the tension at midlength, the maximum tension, and the total length of the cable.

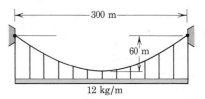

Solution. With a uniform horizontal distribution of load, the solution of part (*b*) of Art. 5/7 applies, and we have a parabolic shape for the cable. For $h = 60$ m, $L = 300$ m, and $w = 12(9.81)(10^{-3})$ kN/m the relation following Eq. 5/14 gives for the midlength tension

$$\left[T_0 = \frac{wL^2}{8h}\right] \qquad T_0 = \frac{0.1177(300)^2}{8(60)} = 22.07 \text{ kN} \qquad \qquad \textit{Ans.}$$

The maximum tension occurs at the supports and is given by Eq. 5/15*a*. Thus

$$\left[T_{\text{max}} = \frac{wL}{2}\sqrt{1 + \left(\frac{L}{4h}\right)^2}\right]$$

$$T_{\text{max}} = \frac{0.1177(300)}{2}\sqrt{1 + \left[\frac{300}{4(60)}\right]^2} = 28.27 \text{ kN} \qquad \textit{Ans.}$$

The sag-to-span ratio is $60/300 = 1/5 < 1/4$. Therefore the series expression developed in Eq. 5/16 is convergent, and we may write for the total length

$$S = 300\left[1 + \frac{8}{3}\left(\frac{1}{5}\right)^2 - \frac{32}{5}\left(\frac{1}{5}\right)^4 + \cdots\right]$$

$$= 300[1 + 0.1067 - 0.01024 + \cdots] = 329 \text{ m} \qquad \textit{Ans.}$$

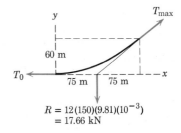

① *Suggestion:* Check the value of T_{max} directly from the free-body diagram of the right-hand half of the cable from which a force polygon may be drawn.

Sample Problem 5/11

Replace the cable of Sample Problem 5/10, which is loaded uniformly along the horizontal, by a cable which has a mass of 12 kg per meter of its own length and supports its own weight only. The cable is suspended between two points on the same level 300 m apart and has a sag of 60 m. Find the tension at midlength, the maximum tension, and the total length of the cable.

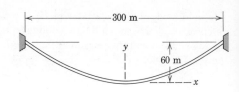

Solution. With a load distributed uniformly along the length of the cable, the solution of part (c) of Art. 5/7 applies, and we have a catenary shape of the cable. Equations 5/20 and 5/21 for the cable length and tension both involve the minimum tension T_0 at midlength which must be found from Eq. 5/19. Thus for $x = 150$ m, $y = 60$ m, and $\mu = 12(9.81)(10^{-3}) = 0.1177$ kN/m, we have

$$60 = \frac{T_0}{0.1177}\left[\cosh\frac{(0.1177)(150)}{T_0} - 1\right]$$

or

$$\frac{7.063}{T_0} = \cosh\frac{17.66}{T_0} - 1$$

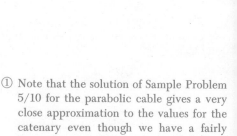

This equation is most easily solved graphically. We compute the expression on each side of the equals sign and plot it as a function of various values of T_0. The intersection of the two curves establishes the equality and determines the correct value of T_0. This plot is shown in the figure accompanying this problem and yields the solution

$$T_0 = 23.2 \text{ kN}$$

The maximum tension occurs for maximum y and from Eq. 5/22 is

$$T_{\max} = 23.2 + (0.1177)(60) = 30.2 \text{ kN} \qquad Ans.$$

① From Eq. 5/20 the total length of the cable becomes

$$2s = 2\frac{23.2}{0.1177}\sinh\frac{(0.1177)(150)}{23.2} = 330 \text{ m} \qquad Ans.$$

① Note that the solution of Sample Problem 5/10 for the parabolic cable gives a very close approximation to the values for the catenary even though we have a fairly large sag. The approximation is even better for smaller sag-to-span ratios.

PROBLEMS

107 The two cables of a suspension bridge with a span of 1000 m and a sag of 150 m support a vertical load of 480 MN distributed uniformly with respect to the horizontal distance between its towers. Compute the tension T_0 in each cable at midspan and the angle θ made by the cables with the horizontal as they approach the support at the top of either tower.   *Ans.* $T_0 = 200$ MN, $\theta = 31.0°$

108 The Golden Gate Bridge in San Francisco has a main span of 4200 ft, a sag of 470 ft, and a total static loading of 21,300 lb per lineal foot of horizontal measurement. The weight of both of the main cables is included in this figure and is assumed uniformly distributed along the horizontal. The angle made by the cable with the horizontal at the top of the tower is the same on each side of each tower. Calculate the midspan tension T_0 in each of the main cables and the compressive force C exerted by each cable on the top of each tower.

109 The cable from A to B carries a load of 80 Mg distributed uniformly along the horizontal. The slope of the cable is zero at A, and its weight is small compared with the load it supports. Calculate the maximum tension T_{max} in the cable.
        Ans. $T_{max} = 1173$ kN

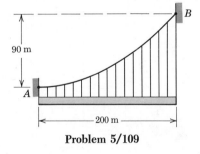

Problem 5/109

110 Expand Eq. 5/19 in a power series for $\cosh (\mu x/T_0)$ and show that the equation for the parabola, Eq. 5/14, is obtained by taking only the first two terms in the series. (See Table B8, Appendix B, for the series expansion of the hyperbolic function.)

111 Strain-gage measurements made on the cables of the suspension bridge at position A indicate an increase of 400,000 lb of tension in *each* of the two main cables because the bridge has been repaved. Determine the total weight w' of added paving material used per foot of roadway.   *Ans.* $w' = 594$ lb/ft

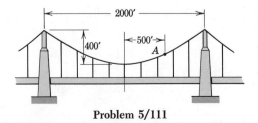

Problem 5/111

5/112 The cable of a suspension bridge with a span of 1000 m is 150 m below the top of the supporting towers at the quarter-span position. Calculate the total length S of each cable between the two identical towers.

5/113 A small suspension bridge for a walkway spans 100 m between supports on the same level. Each of its two cables carries a uniform load of 20 kg per meter of horizontal walkway. If the maximum tension in the cable is found to be 18 kN, compute the tension T_0 at midspan, the sag h, and the angle θ made by the cable with the horizontal at the supports.

Ans. $T_0 = 15.09$ kN, $h = 16.25$ m, $\theta = 33.0°$

5/114 A cable which carries a load uniformly distributed along the horizontal is 112 ft long and is suspended between two points on the same level 100 ft apart. Determine the sag h of the cable.

5/115 A cable supports a load of 50 kg/m uniformly distributed with respect to the horizontal and is suspended from the two fixed points located as shown. Determine the maximum and minimum tensions T and T_0 in the cable.

Ans. $T = 35.61$ kN, $T_0 = 21.04$ kN

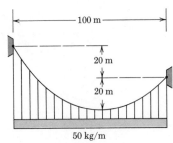

Problem 5/115

5/116 A cable of negligible mass is suspended from the fixed points shown and has a zero slope at its lower end. If the cable supports a unit load w which decreases uniformly with x from w_0 to zero as indicated, determine the equation of the curve assumed by the cable.

Ans. $y = \dfrac{3hx^2}{2l^2}\left(1 - \dfrac{x}{3l}\right)$

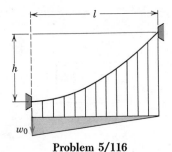

Problem 5/116

5/117 The light cable has zero slope at its lower point of attachment and supports the unit load which varies linearly with x from w_0 to w_1 as shown. Derive the expression for the tension T_0 in the cable at the origin.

Problem 5/117

5/118 The light cable is suspended from two points a distance L apart and on the same horizontal line. If the load per unit of horizontal distance supported by the cable varies from w_0 at the center to w_1 at the ends in accordance with the relation $w = a + bx^2$, derive the equation for the sag h of the cable in terms of the midspan tension T_0.

$$\text{Ans. } h = \frac{L^2}{48\,T_0}(5w_0 + w_1)$$

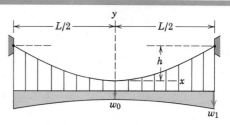
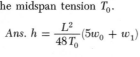

Problem 5/118

5/119 A floating dredge is anchored in position with a single stern cable which has a horizontal direction at the attachment A and extends a horizontal distance of 250 m to an anchorage B on shore. A tension of 300 kN is required in the cable at A. If the cable has a mass of 22 kg per meter of its length, compute the required height H of the anchorage above water level and find the length of cable between A and B.   *Ans.* $H = 24.5$ m, $s = 251$ m

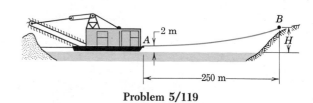

Problem 5/119

5/120 Work Prob. 5/119 using the relations for a parabolic cable as an approximation and compare the results with those cited for Prob. 5/119.

5/121 A flexible cable is secured to point A and passes over a small pulley at B, which is 600 ft higher than A. If it requires a tension $T = 12{,}000$ lb at B to make $\alpha = 0$ at A, determine the weight μ of the cable per foot of its length.   *Ans.* $\mu = 6.35$ lb/ft

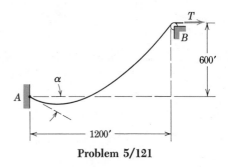

Problem 5/121

5/122 Solve Prob. 121 using the relations for the parabolic cable as an approximation.

5/123 A rope 40 m in length is suspended between two points which are separated by a horizontal distance of 10 m. Compute the distance h to the lowest part of the loop.   *Ans.* $h = 18.5$ m

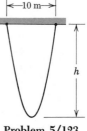

Problem 5/123

5/124 A cable hanging under the action of its own weight is suspended between two points on the same level and 400 m apart. If the sag is 100 m, determine the total length S of the cable. What error is involved if the length is computed from the expression for the parabolic cable using three terms in the series?

5/125 A power line is suspended from two towers 200 m apart on the same horizontal line. The cable has a mass of 18.2 kg per meter of length and has a sag of 32 m at midspan. If the cable can support a maximum tension of 60 kN, determine the mass ρ of ice per meter which can form on the cable without exceeding the maximum cable tension.

Ans. $\rho = 13.4$ kg of ice per meter

5/126 Solve Prob. 5/125 using the relations for the parabolic cable as an approximation.

▶5/127 A cable ship tows a plow A during a survey of the ocean floor for later burial of a telephone cable. The ship maintains a constant low speed with the plow at a depth of 600 ft and with a sufficient length of cable so that it leads horizontally from the plow, which is 1600 ft astern of the ship. The tow cable has an effective weight of 3.10 lb/ft when the buoyancy of the water is accounted for. Also, the forces on the cable due to movement through the water are neglected at the low speed. Compute the horizontal force T_0 applied to the plow and the maximum tension in the cable. Also find the length of the tow cable from point A to point B.

Ans. $T_0 = 6900$ lb, $T_{max} = 8760$ lb, $s = 1740$ ft

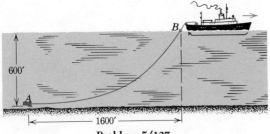

Problem 5/127

▶5/128 The blimp is moored to the ground winch in a gentle wind with 100 m of 12-mm cable which has a mass of 0.51 kg/m. A torque of 400 N·m on the drum is required to start winding in the cable. At this condition the cable makes an angle of 30° with the vertical as it approaches the winch. Calculate the height H of the blimp. The diameter of the drum is 0.5 m.   *Ans.* $H = 90$ m

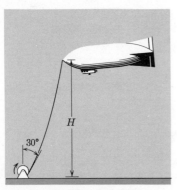

Problem 5/128

5/8 FLUID STATICS. In the work so far, we have directed attention primarily to the action of forces on and between rigid bodies. In this article we shall consider the equilibrium of bodies subjected to forces due to the action of fluid pressures. A fluid is any continuous substance which, when at rest, is unable to support shear force. A shear force is one tangent to the surface upon which it acts and is developed when differential velocities exist between adjacent layers of fluids. Thus a fluid at rest can exert only normal forces on a bounding surface. Fluids may be either gaseous or liquid. The statics of fluids is generally referred to as *hydrostatics* when the fluid is a liquid and as *aerostatics* when the fluid is a gas.

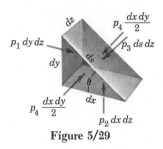

Figure 5/29

(*a*) *Fluid Pressure.* The pressure at any given point in a fluid is the same in all directions (Pascal's law). We may prove this fact by considering the equilibrium of an infinitesimal triangular prism of fluid as shown in Fig. 5/29. The fluid pressures normal to the faces of the element are taken to be p_1, p_2, p_3, and p_4 as shown. With force equal to pressure times area, the equilibrium of forces in the *x*- and *y*-directions gives

$$p_2 \, dx \, dz = p_3 \, ds \, dz \cos \theta \qquad p_1 \, dy \, dz = p_3 \, ds \, dz \sin \theta$$

Since $ds \sin \theta = dy$ and $ds \cos \theta = dx$, these equations require that

$$p_1 = p_2 = p_3 = p$$

By rotating the element through 90° we see that p_4 is also equal to the other pressures. Hence the pressure at any point in a fluid is the same in all directions. In this analysis it is unnecessary to account for the weight of the fluid element since, when the weight per unit volume (density ρ times g) is multiplied by the volume of the element, a differential quantity of third order results which disappears when passing to the limit compared with the second-order pressure force terms.

In all fluids at rest the pressure is a function of the vertical dimension. To determine this function we consider the forces acting on a differential element of a vertical column of fluid of cross-sectional area dA, as shown in Fig. 5/30. The positive direction of vertical measurement h is taken down. The pressure on the upper face is p, and that on the lower face is p plus the change in p, or $p + dp$. The weight of the element equals ρg multiplied by its volume. The normal forces on the lateral surface, which are horizontal and do not affect the balance of forces in the vertical direction, are not shown. Equilibrium of the fluid element in the *h*-direction requires

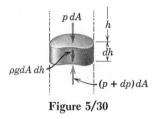

Figure 5/30

$$p \, dA + \rho g \, dA \, dh - (p + dp) \, dA = 0$$

$$dp = \rho g \, dh \qquad (5/23)$$

This differential relation shows us that the pressure in a fluid increases with depth or decreases with increased elevation. Equation 5/23 holds for both liquids and gases and agrees with common observation of air and water pressures.

Fluids that are essentially incompressible are called liquids, and for most practical purposes we may consider their density ρ constant for every part of the liquid.† With ρ a constant Eq. 5/23 may be integrated as it stands, and the result is

$$\boxed{p = p_0 + \rho gh} \tag{5/24}$$

The pressure p_0 is the pressure on the surface of the liquid where $h = 0$. If p_0 is due to atmospheric pressure and the measuring instrument records only the increment above atmospheric pressure,° the measurement gives what is known as "gage pressure" and is $p = \rho gh$.

The common unit for pressure in SI is the kilopascal (kPa), which is the same as a kilonewton per square meter (10^3 N/m^2). In computing pressure, if we use Mg/m^3 for ρ, m/s^2 for g, and m for h, then the product ρgh gives us pressure in kPa directly. For example, the pressure at a depth of 10 m in fresh water is

$$p = \rho gh = \left(1.0 \frac{\text{Mg}}{\text{m}^3}\right)\left(9.81 \frac{\text{m}}{\text{s}^2}\right)(10 \text{ m}) = 98.1 \left(10^3 \frac{\text{kg} \cdot \text{m}}{\text{s}^2} \frac{1}{\text{m}^2}\right)$$

$$= 98.1 \text{ kN/m}^2 = 98.1 \text{ kPa}$$

In the U.S. customary system fluid pressure is generally expressed in pounds per square inch (lb/in.^{2}) or occasionally in pounds per square foot (lb/ft^2). Thus at a depth of 10 ft in fresh water the pressure is

$$p = \rho gh = \left(62.4 \frac{\text{lb}}{\text{ft}^3}\right)\left(\frac{1}{1728} \frac{\text{ft}^3}{\text{in.}^3}\right)(120 \text{ in.}) = 4.33 \text{ lb/in.}^2$$

(b) Hydrostatic Pressure on Submerged Rectangular Surfaces. A surface submerged in a liquid, such as a gate valve in a dam, or the wall of a tank, is subjected to fluid pressure normal to its surface and distributed over its area. In problems where fluid forces are appreciable we must determine the resultant force due to the distribution of pressure on the surface and the position at which this resultant acts. For systems that are open to the earth's atmosphere, the atmospheric pressure p_0 acts over all surfaces and, hence, yields a zero resultant. In such cases, then, we need to consider only the gage pressure $p = \rho gh$ which is the increment above atmospheric pressure.

Consider the special but common case of the action of hydrostatic pressure on the surface of a rectangular plate submerged in a liquid. Figure 5/31a shows such a plate 1-2-3-4 with its top edge horizontal and with the plane of the plate making an arbitrary angle

---

° Atmospheric pressure at sea level may be taken to be 101.3 kPa or 14.7 lb/in.^{2}
† See Table C1, Appendix C, for table of densities.

θ with the vertical plane. The horizontal surface of the liquid is represented by the x-y' plane. The fluid pressure (gage) acting normal to the plate at point 2 is represented by the arrow 6-2 and equals ρg times the vertical depth from the liquid surface to point 2. This same pressure acts at all points along the edge 2-3. At point 1 on the lower edge, the fluid pressure equals ρg times the vertical depth to point 1, and this pressure is the same at all points along edge 1-4. The variation of pressure p over the area of the plate is governed by the linear depth relationship, and we see, therefore, that it is represented by the altitude of the truncated prism 1-2-3-4-5-6-7-8 with the plate as its base. The resultant force produced by this pressure distribution is represented by R, which acts at some point P known as the *center of pressure.*

We see clearly that the conditions that prevail at the vertical section 1-2-6-5 in Fig. 5/31*a* are identical to those at section 4-3-7-8 and at every other vertical section normal to the plate. Thus we may analyze the problem from the two-dimensional view of a vertical section as shown in Fig. 5/31*b* for section 1-2-6-5. For this section the

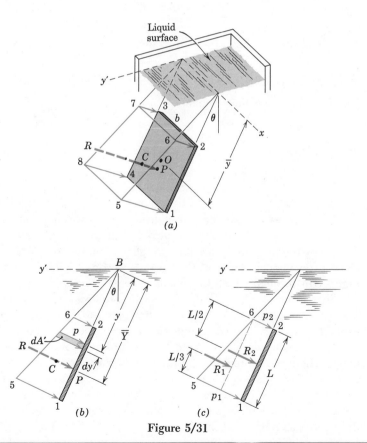

Figure 5/31

pressure distribution is trapezoidal. If b is the horizontal width of the plate, an element of plate area over which the pressure $p = \rho g h$ acts is $dA = b\,dy$, and an increment of the resultant force is $dR = p\,dA = bp\,dy$. But $p\,dy$ is merely the shaded increment of trapezoidal area dA', so that $dR = b\,dA'$. Therefore, we may express the resultant force acting on the entire plate as the trapezoidal area 1-2-6-5 times the width b of the plate,

$$R = b \int dA' = bA'$$

Care must be taken not to confuse the physical area A of the plate with the geometrical area A' defined by the trapezoidal distribution of pressure.

The trapezoidal area representing the pressure distribution is easily expressed by using its average altitude. Therefore the resultant force R may be written in terms of the average pressure $p_{av} = \frac{1}{2}(p_1 + p_2)$ times the plate area A. The average pressure is also the pressure that exists at the average depth to the centroid O of the plate. Therefore an alternative expression for R is

$$R = p_{av}A = \rho g \bar{h} A$$

where $\bar{h} = \bar{y}\cos\theta$.

We obtain the line of action of the resultant force R from the principle of moments. Using the x-axis (point B in Fig. 5/31b) as the moment axis yields $R\bar{Y} = \int y(pb\,dy)$. Substituting $p\,dy = dA'$ and $R = bA'$, and cancelling b give

$$\bar{Y} = \frac{\int y\,dA'}{\int dA'}$$

which is simply the expression for the centroidal coordinate of the trapezoidal area A'. In the two-dimensional view, therefore, the resultant R passes through the centroid C of the trapezoidal area defined by the pressure distribution in the vertical section. Clearly $\bar{Y}$ also locates the centroid C of the truncated prism 1-2-3-4-5-6-7-8 in Fig. 5/31a through which the resultant actually passes.

In treating a trapezoidal distribution of pressure, we may simplify the calculation by considering the resultant composed of two components, Fig. 5/31c. The trapezoid is divided into a rectangle and a triangle, and the force represented by each part is considered separately. The force represented by the rectangular portion acts at the center O of the plate and is $R_2 = p_2 A$ where A is the area 1-2-3-4 of the plate. The force represented by the triangular increment of pressure distribution is $\frac{1}{2}(p_1 - p_2)A$ and acts through the centroid of the triangular portion as shown.

(c) *Hydrostatic Pressure on Cylindrical Surfaces.* For a sub-merged curved surface the resultant R caused by distributed pressure involves more calculation than for a flat surface. As an example consider the submerged cylindrical surface shown in Fig. 5/32a where the elements of the curved surface are parallel to the horizontal surface x-y' of the liquid. Vertical sections perpendicular to the surface all disclose the same curve AB and the same pressure distribution. Hence the two-dimensional representation in Fig. 5/32b may be used. To find R by a direct integration it is necessary for us to

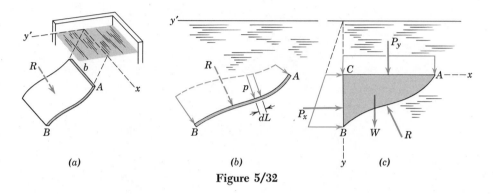

(a)　　　　　(b)　　　　　(c)

Figure 5/32

integrate the x- and y-components of dR along the curve AB, since the pressure continuously changes direction. Thus

$$R_x = b \int (p \, dL)_x = b \int p \, dy \quad \text{and} \quad R_y = b \int (p \, dL)_y = b \int p \, dx$$

A moment equation would now be required to establish the position of R.

A second method for finding R is usually much simpler. The equilibrium of the block of liquid ABC directly above the surface, shown in Fig. 5/32c, is considered. The resultant R is then disclosed as the equal and opposite reaction of the surface upon the block of liquid. The resultants of the pressures along AC and CB are P_y and P_x, respectively, and are easily obtained. The weight W of the liquid block is calculated from the area ABC of its section multiplied by the constant dimension b and ρg. The weight W passes through the centroid of area ABC. The equilibrant R is then determined completely from the equilibrium equations which we apply to the free-body diagram of the fluid block.

(d) *Hydrostatic Pressure on Flat Surfaces of Any Shape.* Figure 5/33a shows a flat plate of any shape submerged in a liquid. The horizontal surface of the liquid is the plane x-y', and the plane of the plate makes an angle θ with the vertical. The force acting on a differential strip of area dA parallel to the surface of the liquid is $dR = p \, dA = \rho g h \, dA$. The pressure p has the same magnitude throughout the length of the strip, since there is no change of depth along the horizontal strip. We obtain the total force acting on the exposed area A by integration which gives

$$R = \int dR = \int p \, dA = \rho g \int h \, dA$$

Substituting the centroidal relation $\bar{h}A = \int h \, dA$ gives us

$$\boxed{R = \rho g \bar{h} A} \tag{5/25}$$

The quantity $\rho g \bar{h}$ represents the pressure that exists at the depth of the centroid O of the area and is the average pressure over the area.

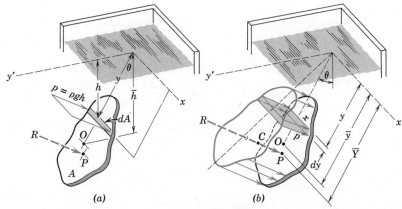

(a)   (b)

Figure 5/33

We may also represent the resultant R geometrically by the volume of the figure shown in Fig. 5/33b. Here the fluid pressure p is represented as an altitude to the plate considered as a base. We see that the resulting volume is a truncated right cylinder. The force dR acting on the differential area $dA = x \, dy$ yields the elemental volume $dV = p \, dA$ shown by the shaded slice, and the total force is, then, represented by the total volume of the cylinder. Hence

$$R = \int dR = \int dV = V$$

We see from Eq. 5/25 that the average altitude of the truncated cylinder is the average pressure $\rho g \bar{h}$ which exists at a depth corresponding to the centroid O of the area exposed to pressure. For problems where the centroid O or the volume V is not readily apparent a direct integration may be performed to obtain R. Thus

$$R = \int dR = \int p \, dA = \int \rho g h x \, dy$$

where the depth h and the length x of the horizontal strip of differential area must be expressed in terms of y to carry out the integration.

The second requirement of the analysis of fluid pressure is the determination of the position of the resultant force in order to account for the moments of the pressure forces. Using the principle of moments with the x-axis of Fig. 5/33b as the moment axis gives

$$R\overline{Y} = \int y\,dR \quad \text{or} \quad \overline{Y} = \frac{\int y\,dV}{V} \tag{5/26}$$

This second relation satisfies the definition of the coordinate $\overline{Y}$ to the centroid of the volume V, and it is concluded, therefore, that the resultant R passes through the centroid C of the volume described by the plate area as base and the linearly varying pressure as altitude. The point P at which R is applied to the plate is the center of pressure. We note carefully that the center of pressure P and the centroid O of the plate area are *not* the same.

(e) *Buoyancy.* The principle of buoyancy, the discovery of which is credited to Archimedes, is easily explained in the following manner for any fluid, gaseous or liquid, in equilibrium. Consider a portion of the fluid defined by an imaginary closed surface, as illustrated by the irregular dotted boundary in Fig. 5/34*a*. If the body of the fluid could be sucked out from within the closed cavity and replaced simultaneously by the forces that it exerted on the boundary of the cavity, Fig. 5/34*b*, there would be no disturbance of the equilibrium of the surrounding fluid. Furthermore, a free-body diagram of the fluid portion before removal, Fig. 5/34*c*, shows us that

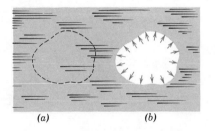
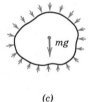

(a)   (b)   (c)

Figure 5/34

the resultant of the pressure forces distributed over its surface must be equal and opposite to its weight mg and must pass through the center of mass of the fluid element. If we replace the fluid element by a body of the same dimensions, the surface forces acting on the body held in this position will be identical with those acting on the fluid element. Thus we see that the resultant force exerted on the surface of an object immersed in a fluid is equal and opposite to the weight of fluid displaced and passes through the center of mass of the displaced fluid. This *resultant force* is called the force of *buoyancy*

$$\boxed{F = \rho g V} \tag{5/27}$$

where ρ is the density of the fluid, g is the acceleration due to gravity, and V is the volume of the fluid displaced. In the case of a liquid whose density is constant the center of mass of the displaced liquid coincides with the centroid of the displaced volume.

From the foregoing discussion we see that when the density of an object is less than the density of the fluid in which it is fully immersed, there will be an unbalance of force in the vertical direction, and the object will rise. When the immersing fluid is a liquid, the object continues to rise until it comes to the surface of the liquid and then comes to rest in an equilibrium position, assuming that the density of the new fluid above the surface is less than the density of the object. In the case of the surface boundary between a liquid and a gas, such as water and air, the effect of the gas pressure on that portion of the floating object above the liquid is balanced by the added pressure in the liquid due to the action of the gas on its surface.

One of the most important problems involving buoyancy is the determination of the stability of a floating object. This analysis may be illustrated by considering a ship's hull shown in cross section in an upright position in Fig. 5/35*a*. Point *B* is the centroid of the displaced volume and is known as the *center of buoyancy*. The resultant of the forces exerted on the hull by the water pressure is the force *F*.

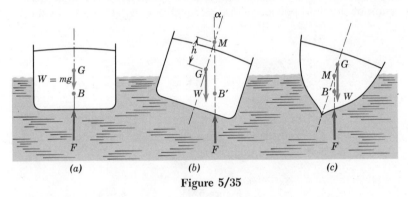

Figure 5/35

Force *F* passes through *B* and is equal and opposite to the weight *W* of the ship. If the ship is caused to list through an angle α, Fig. 5/35*b*, the shape of the displaced volume changes, and the center of buoyancy will shift to some new position such as *B'*. The point of intersection of the vertical line through *B'* with the centerline of the ship is called the *metacenter M*, and the distance *h* of *M* above the center of mass *G* is known as the *metacentric height*. For most hull shapes we find that the metacentric height remains practically constant for angles of list up to about 20°. When *M* is above *G*, as in Fig. 5/35*b*, there is clearly a righting moment which tends to bring the ship back to its original position. The magnitude of this moment for any particular angle of list is a measure of the stability of the ship. If *M* is below *G*, as for the hull of Fig. 5/35*c*, the moment accompanying any list is in the direction to increase the list. This is clearly a condition of instability and must be avoided in the design of any ship.

Sample Problem 5/12

A rectangular plate, shown in vertical section AB, is 4 m high and 6 m wide (normal to the plane of the paper) and blocks the end of a fresh-water channel 3 m deep. The plate is hinged about a horizontal axis along its upper edge through A and is restrained from opening the channel by the fixed ridge at B that bears horizontally against the lower edge of the plate. Calculate the force B exerted against the plate by the ridge.

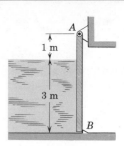

Solution. The free-body diagram of the plate is shown in section and includes the vertical and horizontal components of the force at A, the unspecified weight $W = mg$ of the plate, the unknown horizontal force B, and the resultant R of the triangular distribution of pressure against the vertical face.

The density of fresh water is $\rho = 1.000$ Mg/m^3 so that the average pressure is

① $[p_{av} = \rho g \bar{h}]$ $\qquad p_{av} = 1.000(9.81)(\tfrac{3}{2}) = 14.72$ kPa

The resultant R of the pressure forces against the plate becomes

$[R = p_{av}A]$ $\qquad R = (14.72)(3)(6) = 265$ kN

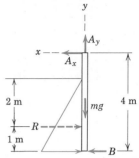

This force acts through the centroid of the triangular distribution of pressure, which is 1 m above the bottom of the plate. A zero moment summation about A establishes the unknown force B. Thus

$[\Sigma M_A = 0]$ $\qquad 3(265) - 4B = 0, \quad B = 198.7$ kN $\qquad$ *Ans.*

① Note that the units of pressure ρgh are

$$\left(10^3 \frac{\text{kg}}{\text{m}^3}\right)\left(\frac{\text{m}}{\text{s}^2}\right)(\text{m}) = \left(10^3 \frac{\text{kg} \cdot \text{m}}{\text{s}^2}\right)\left(\frac{1}{\text{m}^2}\right) = \text{kN/m}^2 = \text{kPa}.$$

Sample Problem 5/13

The air space in the closed fresh-water tank is maintained at a pressure of 0.80 lb/in.^{2} (above atmospheric). Determine the resultant force R exerted by the air and water on the end of the tank.

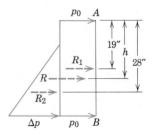

Solution. The pressure distribution on the end surface is shown where $p_0 = 0.80$ lb/in.^{2} The specific weight of fresh water is $\mu = \rho g = 62.4/1728 = 0.0361$ lb/in.^{3} so that the increment of pressure Δp due to the water is

$$\Delta p = \mu \, \Delta h = 0.0361(30) = 1.083 \text{ lb/in.}^2$$

① The resultant forces R_1 and R_2 due to the rectangular and triangular distributions of pressure, respectively, are

$$R_1 = p_0 A_1 = 0.80(38)(25) = 760 \text{ lb}$$

$$R_2 = \Delta p_{av} A_2 = \frac{1.083}{2}(30)(25) = 406 \text{ lb}$$

The resultant is then $R = R_1 + R_2 = 760 + 406 = 1166$ lb $\qquad$ *Ans.*

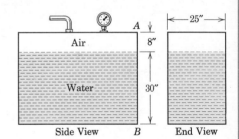

We locate R by applying the moment principle about A noting that R_1 acts through the center of the 38-in. depth and that R_2 acts through the centroid of the triangular pressure distribution which is 20 in. below the surface of the water and $20 + 8 = 28$ in. below A. Thus

$[Rh = \Sigma M_A]$ $\qquad 1166h = 760(19) + 406(28) \qquad h = 22.1$ in. $\qquad$ *Ans.*

① Dividing the pressure distribution into these two parts is decidedly the simplest way in which to make the calculation.

Sample Problem 5/14

Determine completely the resultant force R exerted on the cylindrical dam surface by the water. The density of fresh water is $1.000\ \text{Mg/m}^3$, and the dam has a length b, normal to the paper, of 30 m.

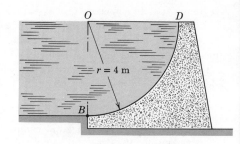

Solution. The circular block of water BDO is isolated and its free-body diagram is drawn. The force P_x is

$$P_x = \rho g \bar{h} A = \frac{\rho g r}{2} br = \frac{(1.000)(9.81)(4)}{2}(30)(4) = 2350\ \text{kN}$$

The weight W of the water passes through the mass center G of the quarter-circular section and is

$$mg = \rho g V = (1.000)(9.81)\frac{\pi (4)^2}{4}(30) = 3700\ \text{kN}$$

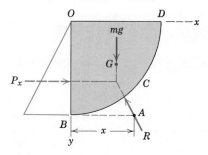

Equilibrium of the section of water requires

$$[\Sigma F_x = 0] \qquad R_x = P_x = 2350\ \text{kN}$$
$$[\Sigma F_y = 0] \qquad R_y = mg = 3700\ \text{kN}$$

The resultant force R exerted by the fluid on the dam is equal and opposite to that shown acting on the fluid and is

$$[R = \sqrt{R_x^2 + R_y^2}] \qquad R = \sqrt{(2350)^2 + (3700)^2} = 4380\ \text{kN} \quad Ans.$$

The x-coordinate of the point A through which R passes may be found from the principle of moments. Using B as a moment center gives

$$P_x \frac{r}{3} + mg\frac{4r}{3\pi} - R_y x = 0, \qquad x = \frac{2350\left(\dfrac{4}{3}\right) + 3700\left(\dfrac{16}{3\pi}\right)}{3700} = 2.55\ \text{m}$$

Alternative solution. The force acting on the dam surface may be obtained by a direct integration of the components

$$dR_x = p\, dA \cos\theta \qquad \text{and} \qquad dR_y = p\, dA \sin\theta$$

where $p = \rho g h = \rho g r \sin\theta$ and $dA = b(r\, d\theta)$. Thus

$$R_x = \int_0^{\pi/2} \rho g r^2 b \sin\theta \cos\theta\, d\theta = -\rho g r^2 b\left[\frac{\cos 2\theta}{4}\right]_0^{\pi/2} = \tfrac{1}{2}\rho g r^2 b$$

$$R_y = \int_0^{\pi/2} \rho g r^2 b \sin^2\theta\, d\theta = \rho g r^2 b\left[\frac{\theta}{2} - \frac{\sin 2\theta}{4}\right]_0^{\pi/2} = \tfrac{1}{4}\pi\rho g r^2 b$$

Thus $R = \sqrt{R_x^2 + R_y^2} = \tfrac{1}{2}\rho g r^2 b \sqrt{1 + \pi^2/4}$. Substituting the numerical values gives

$$R = \tfrac{1}{2}(1.00)(9.81)(4^2)(30)\sqrt{1 + \pi^2/4} = 4380\ \text{kN} \qquad Ans.$$

Since dR always passes through point O, we see that R also passes through O and, therefore, the moments of R_x and R_y about O must cancel. So we write $R_x y_1 = R_y x_1$, which gives us

$$x_1/y_1 = R_x/R_y = (\tfrac{1}{2}\rho g r^2 b)/(\tfrac{1}{4}\pi\rho g r^2 b) = 2/\pi$$

By similar triangles we see that

$$x/r = x_1/y_1 = 2/\pi \qquad \text{and} \qquad x = 2r/\pi = 2(4)/\pi = 2.55\ \text{m} \quad Ans.$$

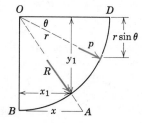

① See note ① in Sample Problem 5/12 if there is any question about the units for $\rho g \bar{h}$.

② This approach by integration is feasible here mainly because of the simple geometry of the circular arc.

Sample Problem 5/15

Determine the resultant force R exerted on the semicircular end of the water tank shown in the figure if the tank is filled to capacity. Express the result in terms of the radius r of the circular end and the density ρ of water.

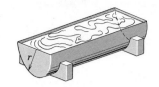

Solution I. We will obtain R first by a direct integration. With a horizontal strip of area $dA = 2x\,dy$ acted upon by the pressure $p = \rho gy$ the increment of the resultant force is $dR = p\,dA$ so that

$$R = \int p\,dA = \int \rho gy(2x\,dy) = 2\rho g \int_0^r y\,\sqrt{r^2 - y^2}\,dy$$

Integrating gives
$$R = \tfrac{2}{3}\rho g r^3 \qquad\qquad Ans.$$

The location of R is determined by using the principle of moments. Taking moments about the x-axis gives

$$[R\overline{Y} = \int y\,dR] \qquad \tfrac{2}{3}\rho g r^3 \overline{Y} = 2\rho g \int_0^r y^2\,\sqrt{r^2 - y^2}\,dy$$

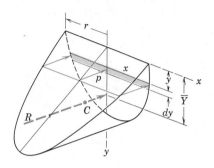

Integrating gives
$$\tfrac{2}{3}\rho g r^3 \overline{Y} = \frac{\rho g r^4}{4}\frac{\pi}{2} \qquad \text{and} \qquad \overline{Y} = \frac{3\pi}{16}r \qquad Ans.$$

Solution II. We may use Eq. 5/25 directly to find R where the average pressure is $\rho g \overline{h}$ and $\overline{h}$ is the coordinate to the centroid of the area over which the pressure acts. For a semicircular area $\overline{h} = 4r/3\pi$. Thus

$$[R = \rho g \overline{h} A] \qquad R = \rho g \frac{4r}{3\pi}\frac{\pi r^2}{2} = \tfrac{2}{3}\rho g r^3 \qquad Ans.$$

This calculation amounts to finding the volume of the pressure-area figure.

The resultant R acts through the centroid C of the volume defined by ① the pressure-area figure. Calculation of the centroidal distance $\overline{Y}$ involves the same integral obtained in *Solution I*.

① Be very careful not to make the mistake of assuming that R passes through the centroid of the area over which the pressure acts.

Sample Problem 5/16

A buoy in the form of a uniform 8-m pole 0.2 m in diameter has a mass of 200 kg and is secured at its lower end to the bottom of a fresh-water lake with 5 m of cable. If the depth of the water is 10 m, calculate the angle θ made by the pole with the horizontal.

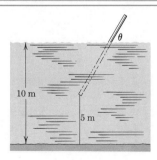

Solution. The free-body diagram of the buoy shows its weight acting ① through G, the vertical tension T in the anchor cable, and the buoyancy force B which passes through centroid C of the submerged portion of the buoy. Let x be the distance from G to the waterline. The density of fresh water is $\rho = 10^3$ kg/m³, so that the buoyancy force is

$$[B = \rho g V] \qquad B = 10^3(9.81)\pi(0.1)^2(4 + x)\ \text{N}$$

Moment equilibrium, $\Sigma M_A = 0$, about A gives

$$200(9.81)(4\cos\theta) - [10^3(9.81)\pi(0.1)^2(4 + x)]\frac{4 + x}{2}\cos\theta = 0$$

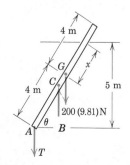

Thus $\quad x = 3.14$ m $\quad$ and $\quad \theta = \sin^{-1}\left(\dfrac{5}{4 + 3.14}\right) = 44.5°$ $\quad$ *Ans.*

① Since the weight and the buoyancy forces are vertical, the only other force, T, must also be vertical for equilibrium.

PROBLEMS

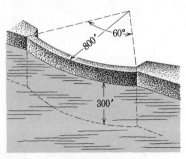

Problem 5/129

5/129 The arched dam has the form of a cylindrical surface of 800-ft radius and subtends an angle of 60°. If the water is 300 ft deep, determine the total force P exerted by the water on the face of the dam.

Ans. $P = 2246(10^6)$ lb

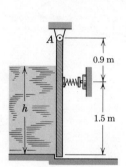

Problem 5/130

5/130 The spring-loaded vertical gate is hinged about a horizontal axis along its upper edge A and closes the end of a rectangular fresh-water channel 1.2 m wide (normal to the plane of the paper). Calculate the preset spring force F that will limit the depth of the water to $h = 1.8$ m.   *Ans.* $F = 38.1$ kN

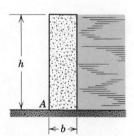

Problem 5/131

5/131 A concrete retaining wall of rectangular section has a height h, a thickness b, and a density ρ_c. The wall holds back mud, which behaves like a liquid of density ρ_m. Determine the minimum thickness b of the wall to prevent it from overturning about the front edge A.

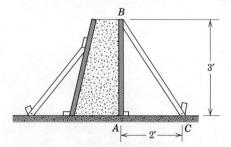

Problem 5/132

5/132 The forms for a small concrete retaining wall are shown in section. There is a brace BC for every 4 ft of wall length. Compute the compression in each brace BC assuming that the joints at A, B, and C act as hinged connections. Wet concrete may be treated as a liquid which weighs 150 lb/ft^3.

Ans. $C = 1622$ lb

133 The cylindrical caisson of radius r and wall thickness t is lowered into place, and the water is then pumped out. Analyze the forces acting on a semi-circular ring of differential height and determine the expression for the compressive stress σ in the cylinder wall for any depth h. The density of water is ρ.

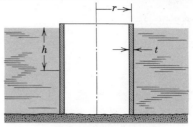

Problem 5/133

134 The hinged gate ABC closes an opening of width b (perpendicular to the paper) in a water channel. The water has free access to the under side as well as the right side of the gate. When the water level rises above a certain value of h, the gate will open. Determine the critical value of h. Neglect the mass of the gate.   *Ans.* $h = a\sqrt{3}$

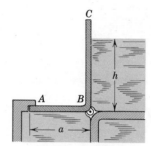

Problem 5/134

135 The figure shows the cross section of a rectangular gate 4 m high and 6 m long (perpendicular to the paper) which blocks a fresh-water channel. The gate has a mass of 8.5 Mg and is hinged about a horizontal axis through C. Compute the vertical force P exerted by the foundation on the lower edge A of the gate. Neglect the mass of the frame to which the gate is attached.

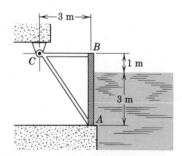

Problem 5/135

136 The solid concrete cylinder 6 ft long and 4 ft in diameter is supported in a half-submerged position in fresh water by a cable which passes over a fixed pulley at A. Compute the tension T in the cable. The cylinder is waterproofed by a plastic coating. (Consult Table C1, Appendix C, as needed.)   *Ans.* $T = 8957$ lb

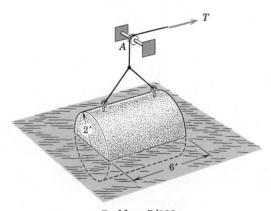

Problem 5/136

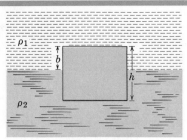

Problem 5/137

5/137 The homogeneous block of density ρ is floating between two liquids of densities $\rho_1 < \rho$ *and* $\rho_2 > \rho$. Determine an expression for the distance b that the block protrudes into the top liquid.

Problem 5/138

5/138 The quonset hut is subjected to a horizontal wind, and the pressure p against the circular roof is approximated by $p_0 \cos \theta$. The pressure is positive on the windward side of the hut and is negative on the leeward side. Determine the total horizontal shear Q on the foundation per unit length of roof measured normal to the paper.   *Ans.* $Q = \frac{1}{2}\pi r p_0$

5/139 The hull of a floating oil-drilling platform consists of two rectangular pontoons and six cylindrical columns that support the working platform. When ballasted, the entire structure has a displacement of 26,000 tons (expressed in long tons of 2240 lb). Calculate the total draft h of the structure when moored in the ocean. The weight density of salt water is 64 lb/ft^3. Neglect the vertical components of the mooring forces.   *Ans.* $h = 74.5$ ft

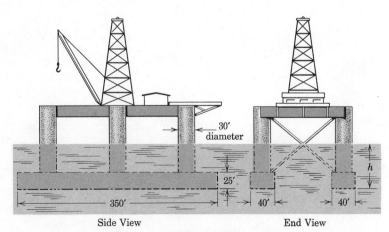

Side View   End View

Problem 5/139

140 The sides of a V-shaped fresh-water trough, shown
in section, are hinged about their common intersection through *O* and are held together by a cable and
turnbuckle placed every 2 m along the length of the
trough. Calculate the tension *T* supported by each
turnbuckle.

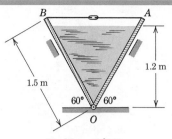

Problem 5/140

141 A fresh-water channel 9 m wide (normal to the
plane of the paper) is blocked by a rectangular
barrier, shown by its section *ACD*. Supporting horizontal struts *BC* are placed every 0.6 m along the
9-m width. Determine the compression *C* in each
strut *BC*. The weight of the barrier is assumed to be
small compared with the other forces acting.

Ans. C = 25.4 kN

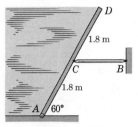

Problem 5/141

142 A rectangular plate, shown in edge view, is 3 m high
and 2.7 m wide (normal to the paper) and separates
reservoirs of sea water and oil. The oil has a specific
gravity (ratio of density to that of fresh water) of
0.85. Determine the depth *h* of the water required
to make the reaction at *B* equal to zero.

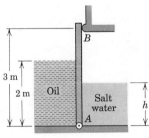

Problem 5/142

143 The rectangular gate shown in section is 10 ft long
(perpendicular to the paper) and is hinged about its
upper edge *B*. The gate divides a channel leading to
a fresh-water lake on the left and a salt-water tidal
basin on the right. Calculate the torque *M* on the
shaft of the gate at *B* required to prevent the gate
from opening when the salt-water level drops to
h = 3 ft.   *Ans. M = 11.22(10⁴) lb-ft*

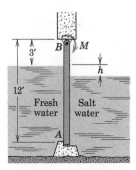

Problem 5/143

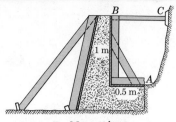

Problem 5/144

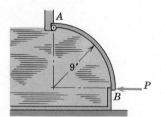

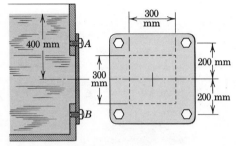

Problem 5/145

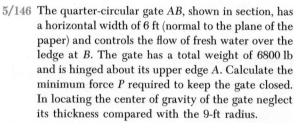

Problem 5/146

Problem 5/147

5/144 The form for a concrete footing is hinged to its lower edge A and held in place every 1.5 m by a horizontal brace BC. Neglect the weight of the form and compute the compression C in each brace resulting from the liquid behavior of wet concrete, which has a density of 2.40 Mg/m³.

5/145 The cover plate for an access opening in a fresh-water tank is bolted in place and the tank is filled to the level shown. Compute the increase in tension in each bolt at A and B due to filling the tank. (Assume no change in the gasket pressure between the cover plate and the tank so that the force due to water pressure is resisted entirely by the increase in the bolt tensions.)   *Ans.* $T_A = 80.0$ N, $T_B = 96.6$ N

5/146 The quarter-circular gate AB, shown in section, has a horizontal width of 6 ft (normal to the plane of the paper) and controls the flow of fresh water over the ledge at B. The gate has a total weight of 6800 lb and is hinged about its upper edge A. Calculate the minimum force P required to keep the gate closed. In locating the center of gravity of the gate neglect its thickness compared with the 9-ft radius.
Ans. $P = 10,830$ lb

5/147 The submersible diving chamber has a total mass of 6.7 Mg including personnel, equipment, and ballast. When the chamber is lowered to a depth of 1.2 km in the ocean, the cable tension is 8 kN. Compute the total volume V displaced by the chamber.

148 Steel chain with a nominal shank diameter of 6 mm has a mass of 93 kg per 100 meters of length and has a breaking strength of 13.5 kN. What length h of chain can be lowered into a deep part of the ocean before the chain breaks under the action of its own weight? *Ans. $h = 1704$ m*

149 A supertanker moves from a freshwater anchorage to a berth in salt water and then takes on an additional 2400 m^3 of fuel oil (specific gravity 0.88). The draft marks on the hull of the tanker read the same in salt water as they did in fresh water before the transfer. Calculate the final displacement m (total mass) of the ship in metric tons (Mg).

150 The deep submersible research vessel has a passenger compartment in the form of a spherical steel shell with a mean radius of 1.000 m and a thickness of 35 mm. Calculate the mass of lead ballast which the vessel must carry so that the combined weight of the steel shell and lead ballast exactly cancels their combined buoyancy. (Consult Table C1, Appendix C, as needed.)
Ans. $m = 1.210$ Mg (metric tons)

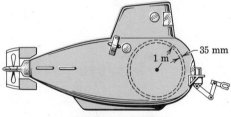

Problem 5/150

151 The solid floating object is composed of a hemisphere and a circular cylinder of equal radii r. If the object floats with the center of the hemisphere above the water surface, determine the maximum height h which the cylinder may have before the object will no longer float in the upright position illustrated. *Ans. $h = r/\sqrt{2}$*

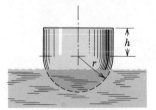

Problem 5/151

152 Determine the total force R exerted on the triangular window by the fresh water in the tank. The water level is even with the top of the window. Also determine the distance H from R to the water level.

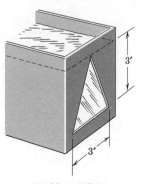

Problem 5/152

Top View

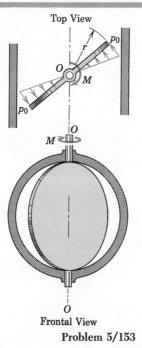

M

O

Frontal View

Problem 5/153

5/153 A large butterfly valve in a horizontal pipe consists of a flat circular disk pivoted about a vertical diametral axis O-O. If the difference in fluid pressure between the two sides of the valve varies linearly with distance from the axis from zero to p_0 as shown, determine the torque M on the shaft at O required to hold the valve in place. (*Hint:* Recognize the applicability of the results of Sample Problem 5/15.)

$$\text{Ans. } M = \frac{\pi}{4}p_0 r^3$$

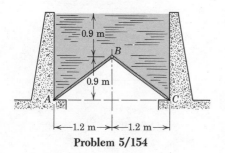

Problem 5/154

5/154 The bottom of a fresh-water channel, shown in section, consists of two uniform rectangular plates AB and BC each having a mass of 1.5 Mg and hinged along their common edge B and hinged also to the base of the fixed channel along their lower edges A and C. The length of the channel is 6 m, measured perpendicular to the plane of the paper. Determine the force P per meter of channel length exerted by each plate on the connecting hinge at B.

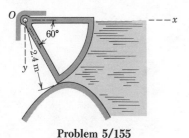

Problem 5/155

5/155 A control gate of cylindrical cross section is used to regulate the flow of water over the spillway of a fresh-water dam as shown. The gate has a mass of 7 Mg and has a length normal to the paper of 6 m. A torque applied to the shaft of the gate at O controls the angular position of the gate. For the position shown determine the horizontal and vertical components F_x and F_y of the total force exerted by the shaft on its bearings at O. Account for the effect of fluid pressure by a direct integration over the cylindrical surface. *Ans.* $F_x = 127.1$ kN, $F_y = 35.4$ kN

156 A flat plate seals a triangular opening in the vertical wall of a tank of liquid of density ρ. The plate is hinged about the upper edge O of the triangle. Determine the force P required to hold the gate in a closed position against the pressure of the liquid.

$$\text{Ans. } P = \frac{\rho g a b}{6}\left(h + \frac{a}{2}\right)$$

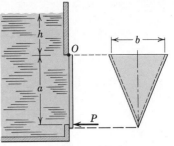

Problem 5/156

157 One end of a uniform pole of length l and density ρ_1 is hinged about a point a distance h above the surface of a liquid of density ρ_2. Find the angle θ assumed by the pole. Assume that $\rho_2 > \rho_1$.

$$\text{Ans. } \theta = \sin^{-1}\left(\frac{h}{l}\sqrt{\frac{\rho_2}{\rho_2 - \rho_1}}\right)$$

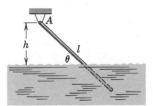

Problem 5/157

158 The accurate determination of the vertical position of the center of mass G of a ship is difficult to achieve by calculation. It is more easily obtained by a simple inclining experiment on the loaded ship. With reference to the figure, a known external mass m_0 is placed a distance d from the centerline, and the angle of list θ is measured by means of the deflection of a plumb bob. The displacement of the ship and the location of the metacenter M are known. Calculate the metacentric height $\overline{GM}$ for a 12000-t ship inclined by a 27-t mass placed 7.8 m from the centerline if a 6-m plumb line is deflected a distance $a = 0.2$ m. The mass m_0 is at a distance $b = 1.8$ m above M. [Note that the metric ton (t) equals 1000 kg and is the same as the megagram (Mg).] $\text{Ans. } \overline{GM} = 0.530$ m

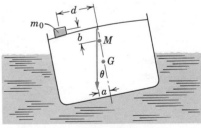

Problem 5/158

5/9 PROBLEM FORMULATION AND REVIEW. In Chapter 5 we have treated various common examples of forces distributed throughout volumes, over areas, and along lines. In all of these problems we are primarily interested in two things—the resultant of the distributed forces and the location of the resultant.

In finding the resultant we begin by multiplying the intensity of the force by the appropriate element of volume, area, or length in terms of which the intensity is expressed. We then sum up (integrate) the incremental forces over the region involved to obtain their resultant.

In finding the location of the line of action of the resultant we depend on the principle of moments (Varignon's theorem). Here we equate the sum of the moments about a convenient axis of all of the increments of force to the moment of the resultant about the same axis, and then solve for the unknown moment arm to the resultant.

When force is distributed throughout a mass, as in the case of gravitational attraction, the intensity is the force of attraction ρg per unit of volume where ρ is the density and g is the gravitational acceleration. For bodies whose density is constant, we saw in Section A that ρg cancels when the moment principle is applied. This leaves us with a strictly geometric problem of finding the centroid of the figure which then coincides with the mass center of the physical body whose boundary defines the figure. For flat plates and shells which are homogeneous and have constant thickness, the problem becomes one of finding the properties of an area. And for slender rods and wires of uniform density and constant cross section, the problem becomes one of finding the properties of a line segment. These mathematical idealizations of the physical bodies are essential elements to recognize in the formulation of actual problems.

For problems that require the integration of differential relationships, we cite four measures which are especially useful. First, select a suitable coordinate system. Generally, the system which provides the simplest description of the boundaries of the region of integration is the best choice. Second, higher-order differential quantities may always be dropped compared with lower-order differential quantities. Third, choose a first-order differential element in preference to a second-order element and a second-order element in preference to a third-order element to reduce the labor of computation. Fourth, wherever possible choose a differential element which avoids discontinuities within the region of integration.

In Section B of Chapter 5 we made use of the foregoing observations along with our principles of equilibrium to solve for the effects of distributed forces in beams, cables, and fluids. In beams and cables we expressed the force intensity as force per unit length. For fluids we expressed the force intensity as force per unit area, or pressure. Although these three problems are physically quite different, in their formulation they embody the common elements cited.

REVIEW PROBLEMS

159 Determine the x-coordinate of the centroid of the shaded area between the quadrant of the ellipse and the straight line. (Use Table C3, Appendix C, as needed.)

$$Ans. \ \bar{X} = \frac{2a/3}{\pi - 2}$$

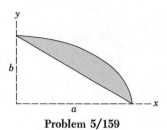

Problem 5/159

160 The uniform semicircular bar AB has a mass ρ per unit of its length and rests with its ends on the smooth horizontal surfaces, one higher than the other. Determine the force F_A under end A.

$$Ans. \ F_A = \frac{\rho g \pi r}{2}\left(1 - \frac{2}{\pi}\right)$$

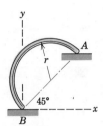

Problem 5/160

161 A small balloon for use in recording wind velocity and its direction has a mass of 0.4 kg. If the balloon is inflated with 0.3 kg of helium and exerts an upward force of 5.3 N on its mooring prior to release, determine the diameter d of the spherical balloon. The density of the air is 1.206 kg/m^3.

$$Ans. \ d = 1.138 \ \text{m}$$

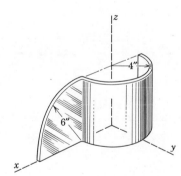

Problem 5/162

162 Determine the coordinates of the centroid of the sheet-metal shape.

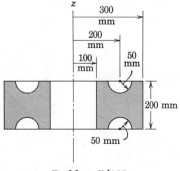

163 Calculate the mass m and total surface area A of the wheel generated by revolving the shaded section about the z-axis. The material of the wheel has a density of 7.21 Mg/m^3.

$$Ans. \ m = 291 \ \text{kg}, \ A = 1.149 \ \text{m}^2$$

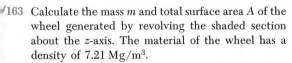

Problem 5/163

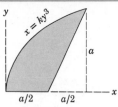

$x = ky^3$

a

$a/2$ $a/2$

Problem 5/164

5/164 Determine the *x*-coordinate of the centroid of the shaded area.

5/165 A cable 300 m long is suspended between two points on the same level 280 m apart. If the cable supports a large load uniformly distributed with respect to the horizontal direction, find the sag *h* in the cable from Eq. 5/16 by successive approximations.   *Ans. h = 47.5 m*

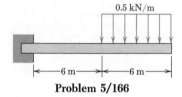

0.5 kN/m

6 m   6 m

Problem 5/166

5/166 Draw the shear and moment diagrams for the cantilever beam loaded as shown.

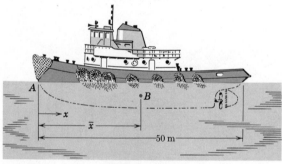

A

x

$\bar{x}$

B

50 m

Problem 5/167

5/167 The center of buoyancy *B* of a ship's hull is the centroid of its submerged volume. The underwater cross-sectional areas *A* of the transverse sections of the tugboat hull shown are tabulated for every five meters of waterline length. Plot these values and determine to the nearest 0.5 m the distance $\bar{x}$ of *B* aft of point *A*.

x, m	A, m²	x, m	A, m²
0	0	25	25.1
5	7.1	30	23.8
10	15.8	35	19.5
15	22.1	40	12.5
20	24.7	45	5.1
		50	0

Ans. $\bar{x}$ = 24 m

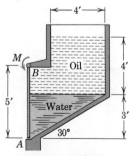

4'

M

B Oil   4'

5'   Water

3'

30°

A

Problem 5/168

5/168 The gate *AB* is a 580-lb rectangular plate 5 ft high and 3.5 ft wide and is used to close the discharge channel at the bottom of an oil reservoir. As a result of condensation in the tank, fresh water collects at the bottom of the channel. Calculate the moment *M* applied about the hinge axis *B* required to close the gate against the hydrostatic forces of the water and oil. The specific gravity of the oil is 0.85.

169 A structure for observation of sea life beneath the ice in polar waters consists of the cylindrical viewing chamber connected to the surface by the cylindrical shaft open at the top for ingress and egress. Ballast is carried in the rack below the chamber. To ensure a stable condition for the structure, it is necessary that its legs bear upon the ice with a force that is at least 15 percent of the total buoyancy force of the submerged structure. If the structure less ballast has a mass of 5.7 Mg, calculate the required mass *m* of lead ballast. The density of lead is 11.37 Mg/m³. *Ans. m* = 4.24 Mg

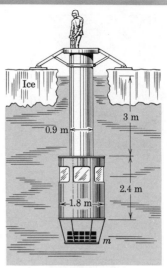

Problem 5/169

170 Draw the shear and moment diagrams for the loaded beam and determine the bending moment *M* that induces the maximum compression in the upper fibers of the beam. *Ans. M* = 52.8 N·m

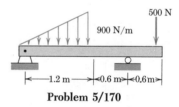

Problem 5/170

171 The beam *AD* is supported and loaded as shown. The cable that holds the 2000-lb load is wrapped around the drum at *E*, and the drum is locked to its bracket and cannot rotate. Neglect the dimensions of the flanges that attach the brackets to the beam at *B* and *D* and construct the shear and moment diagrams for the loaded beam. Determine the bending moment *M* with the greatest magnitude and the distance *x* from the pivot *A* at which *M* occurs.

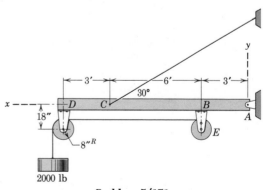

Problem 5/171

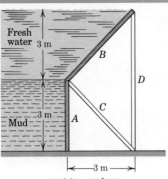

Problem 5/172

5/172 A dam consists of the flat-plate barriers A and B whose masses are small. Supporting struts C and D are placed every 3 m of dam section. A mud sample drawn up to the surface has a density of 1.6 Mg/m³. Determine the compression in C and D. All joints may be assumed to be hinged.

Ans. $C = 474$ kN, $D = 88.3$ kN

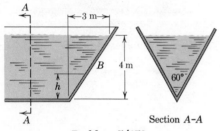

Problem 5/173

5/173 The end of a fresh-water channel with a 60° V-section is closed by the slanted triangular plate B. Calculate the resultant force R exerted on B by the water and the height h of the point on B through which R acts. *Ans.* $R = 151.0$ kN, $h = 2$ m

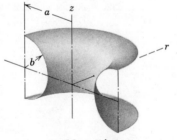

Problem 5/174

▶5/174 The surface is generated by revolving the semicircular arc of radius b through 180° around the z-axis. Determine the distance $\bar{r}$ from the z-axis to the centroid of the surface.

$$Ans. \ \bar{r} = \frac{\pi(2a^2 + b^2) - 8ab}{\pi(\pi a - 2b)}$$

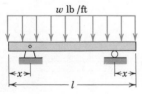

Problem 5/175

▶5/175 The beam supports a uniformly distributed load as shown. Determine the location x of the two supports so as to minimize the maximum bending moment M in the beam. What is M?

Ans. $M = 0.0214wl^2$, $x = 0.207l$

176 While repairs on its cable are being made, the propellers of the cable-laying ship exert a forward thrust of 300 kN to hold a fixed position in calm water. The ocean depth at this location is h, and the cable is observed to make an angle of $\theta = 60°$ with the horizontal where it enters the water at point P. The cable has a mass of 22 kg/m and has a cross-sectional area of 4600 mm^2. The density of salt water is 1.03 Mg/m^3. Calculate the length s of cable from point P to the point where the cable leaves the ocean floor and find the horizontal distance x from this point to a point directly below P. Determine the depth h of the ocean. Also compute the torque M on the 2-m-diameter cable-laying drum to prevent it from turning. (*Note:* The vertical loading is the difference between the cable weight and the weight of the water displaced.)

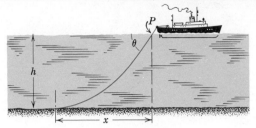

Problem 5/176

$$\text{Ans. } s = 3070 \text{ m}, \qquad x = 2330 \text{ m}$$
$$h = 1770 \text{ m}, \qquad M = 600 \text{ kN·m}$$

FRICTION

6

6/1 INTRODUCTION. In the preceding chapters the forces of action and reaction between contacting surfaces have for the most part been assumed to act normal to the surfaces. This assumption characterizes the interaction between smooth surfaces and was illustrated in example 2 of Fig. 3/1. Although in many instances this ideal assumption involves only a relatively small error, there are a great many problems wherein we must consider the ability of contacting surfaces to support tangential as well as normal forces. Tangential forces generated between contacting surfaces are known as friction forces and are present to some degree with the interaction between all real surfaces. Whenever a tendency exists for one contacting surface to slide along another surface, we find that the friction forces developed are always in a direction to oppose this tendency.

In some types of machines and processes we desire to minimize the retarding effect of friction forces. Examples are bearings of all types, power screws, gears, the flow of fluids in pipes, and the propulsion of aircraft and missiles through the atmosphere. In other situations we wish to maximize the use of friction, as in brakes, clutches, belt drives, and wedges. Wheeled vehicles depend on friction for both starting and stopping, and ordinary walking depends on friction between the shoe and the ground. Friction forces are present throughout nature and exist to a considerable extent in all machines no matter how accurately constructed or carefully lubricated. A machine or process in which friction is small enough to be neglected is often referred to as *ideal*. When friction must be taken into account, the machine or process is termed *real*. In all real cases where sliding motion between parts occurs, the friction forces result in a loss of energy which is dissipated in the form of heat. In addition to the generation of heat and the accompanying loss of energy, friction between mating parts will cause wear.

SECTION A. FRICTIONAL PHENOMENA

6/2 TYPES OF FRICTION. There are a number of separate types of frictional resistance encountered in mechanics, and we will mention each of these types briefly in this article prior to a more detailed account of the most common type of friction in the next article.

(*a*) *Dry Friction.* Dry friction is encountered when the unlubricated surfaces of two solids are in contact under a condition of sliding or tendency to slide. A friction force tangent to the surfaces of contact is developed both during the interval leading up to impending slippage and while slippage takes place. The direction of the force always opposes the motion or impending motion. This type of friction is also called *Coulomb* friction. The principles of dry or Coulomb friction were developed largely from the experiments of Coulomb in 1781 and from the work of Morin from 1831 to 1834. Although a comprehensive theory of dry friction is not available, an analytical model sufficient to handle the vast majority of problems in dry friction is available and is described in Art. 6/3 which follows. This model forms the basis for most of this chapter.

(*b*) *Fluid Friction.* Fluid friction is developed when adjacent layers in a fluid (liquid or gas) are moving at different velocities. This motion gives rise to frictional forces between fluid elements, and these forces depend on the relative velocity between layers. When there is no such relative velocity, there is no fluid friction. Fluid friction depends not only on the velocity gradients within the fluid, but also on the viscosity of the fluid, which is a measure of its resistance to shearing action between fluid layers. Fluid friction is treated in the study of fluid mechanics and will not be developed further in this book.

(*c*) *Internal Friction.* Internal friction is found in all solid materials that are subjected to cyclical loading. For highly elastic materials the recovery from deformation occurs with very little loss of energy caused by internal friction. For materials which have low limits of elasticity and which undergo appreciable plastic deformations during loading, the amount of internal friction that accompanies this deformation may be considerable. The mechanism of internal friction is associated with the action of shear deformation, and the student should consult a reference on materials science for a detailed description of this shear mechanism. Since this book deals primarily with the external effects of forces, we shall not be concerned with internal friction in the work which follows.

6/3 DRY FRICTION.   The remainder of this chapter will be devoted to a description of the effects of dry friction acting on the exterior surfaces of rigid bodies. We shall explain the mechanism of dry friction with the aid of a very simple experiment.

(*a*) *Mechanism of Friction.* Consider a solid block of mass m resting on a horizontal surface, as shown in Fig. 6/1*a*. The contacting surfaces possess a certain amount of roughness. The experiment will involve the application of a horizontal force P which will vary continuously from zero to a value sufficient to move the block and

give it an appreciable velocity. The free-body diagram of the block
for any value of P is shown in Fig. $6/1b$, and the tangential friction
force exerted by the plane on the block is labeled F. This friction
force will *always* be in a direction to oppose motion or the tendency
toward motion of the body on which it acts. There is also a normal
force N which in this case equals mg, and the total force R exerted by
the supporting surface on the block is the resultant of N and F. A
magnified view of the irregularities of the mating surfaces, Fig. $6/1c$,
will aid in visualizing the mechanical action of friction. Support is
necessarily intermittent and exists at the mating humps. The direction

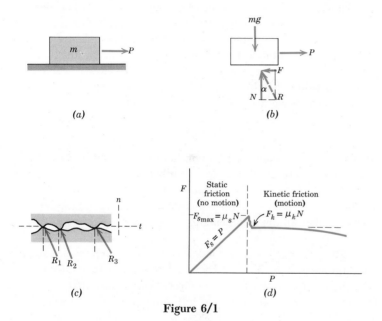

Figure 6/1

of each of the reactions on the block R_1, R_2, R_3, etc., will depend not
only on the geometric profile of the irregularities but also on the
extent of local deformation at each contact point. The total normal
force N is merely the sum of the n-components of the R's, and the
total frictional force F is the sum of the t-components of the R's.
When the surfaces are in relative motion, we can see that the contacts are more nearly along the tops of the humps, and the t-compo-
nents of the R's will be smaller than when the surfaces are at rest
relative to one another. This consideration helps to explain the well-
known fact that the force P necessary to maintain motion is generally
less than that required to start the block when the irregularities are
more nearly in mesh.

Assume now that we perform the experiment indicated and

record the friction force F as a function of P. The resulting experimental relation is indicated in Fig. 6/1d. When P is zero, equilibrium requires that there be no friction force. As P is increased, the friction force must be equal and opposite to P as long as the block does not slip. During this period the block is in equilibrium, and all forces acting on the block must satisfy the equilibrium equations. Finally we reach a value of P which causes the block to slip and to move in the direction of the applied force. At this same time the friction force drops slightly and rather abruptly to a somewhat lower value. Here it remains essentially constant for an interval but then drops off still more with higher velocities.

The region up to the point of slippage or impending motion is known as the range of *static friction*, and the value of the friction force is determined by the *equations of equilibrium*. This force may have any value from zero up to and including, in the limit, the maximum value. For a given pair of mating surfaces we find that this maximum value of static friction $F_{s_{max}}$ is proportional to the normal force N. Hence we may write

$$F_{s_{max}} = \mu_s N$$

where μ_s is the proportionality constant, known as the *coefficient of static friction*. We must observe carefully that this equation describes only the *limiting* or *maximum* value of the static friction force and not any lesser value. Thus the equation applies *only* to cases where it is known that motion is impending with the friction force at its peak value.

After slippage occurs a condition of *kinetic friction* accompanies the ensuing motion. Kinetic friction force is usually somewhat less than the maximum static friction force. We find also that the kinetic friction force F_k is proportional to the normal force. Hence

$$F_k = \mu_k N$$

where μ_k is the *coefficient of kinetic friction*. It follows that μ_k is generally less than μ_s. As the velocity of the block increases, the kinetic friction coefficient decreases somewhat, and when high velocities are reached the effect of lubrication by an intervening fluid film may become appreciable. Coefficients of friction depend greatly on the exact condition of the surfaces as well as on the velocity and are subject to a considerable measure of uncertainty.

It is customary to write the two friction force equations merely as

$$\boxed{F = \mu N} \tag{6/1}$$

We will understand from the problem at hand whether limiting static friction with its corresponding coefficient of static friction or whether kinetic friction with its corresponding kinetic coefficient is involved. We emphasize again that many problems involve a static friction force which is *less* than the maximum value at impending motion, and therefore under these conditions the friction equation *cannot* be used.

From Fig. 6/1*c* we observe that for rough surfaces there is a greater possibility for large angles between the reactions and the *n*-direction than for smoother surfaces. Thus a friction coefficient reflects the roughness of a pair of mating surfaces and incorporates a geometric property of both mating contours. It is meaningless to speak of a coefficient of friction for a single surface.

The direction of the resultant R in Fig. 6/1*b* measured from the direction of N is specified by $\tan \alpha = F/N$. When the friction force reaches its limiting static value, the angle α reaches a maximum value ϕ_s. Thus

$$\tan \phi_s = \mu_s$$

When slippage occurs, the angle α will have a value ϕ_k corresponding to the kinetic friction force. In like manner

$$\tan \phi_k = \mu_k$$

We usually write merely

$$\boxed{\tan \phi = \mu} \qquad (6/2)$$

where application to the limiting static case or to the kinetic case is inferred from the problem at hand. The angle ϕ_s is known as the *angle of static friction*, and the angle ϕ_k is called the *angle of kinetic friction*. This friction angle ϕ for each case clearly defines the limiting position of the total reaction R between two contacting surfaces. If motion is impending, R must be one element of a right circular cone of vertex angle $2\phi_s$, as shown in Fig. 6/2. If motion is not impending, R will be within the cone. This cone of vertex angle $2\phi_s$ is known as the *cone of static friction* and represents the locus of possible positions for the reaction R at impending motion. If motion occurs, the angle of kinetic friction applies, and the reaction must lie on the surface of a slightly different cone of vertex angle $2\phi_k$. This cone is the *cone of kinetic friction*.

Further experiment shows that the friction force is essentially independent of the apparent or projected area of contact. The true contact area is much smaller than the projected value, since only the peaks of the contacting surface irregularities support the load. Relatively small normal loads result in high stresses at these contact points. As the normal force increases, the true contact area also increases as the material undergoes yielding, crushing, or tearing at the points of contact. A comprehensive theory of dry friction must go

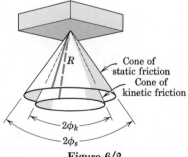

Cone of
static friction
Cone of
kinetic friction

$2\phi_k$
$2\phi_s$

Figure 6/2

beyond the mechanical explanation presented here. For example, there is some evidence to support the theory that molecular attraction may be an important cause of friction under conditions where the mating surfaces are in very intimate contact. Other factors that influence dry friction are the generation of high local temperatures and adhesion at contact points, relative hardness of mating surfaces, and the presence of thin surface films of oxide, oil, dirt, or other substances.

(*b*) *Types of Friction Problems.* From what has been said we may now recognize three distinct types of problems which are encountered in dry friction. The very first step in the solution of a friction problem is to identify which of these categories applies.

(1) In the *first* type the condition of impending motion exists. Here a body which is in equilibrium is on the verge of slipping, and the friction force equals the limiting static friction $F_{s_{\max}} = \mu_s N$. The equations of equilibrium will, of course, also hold.

(2) In the *second* type of problem impending motion need not exist, and therefore the friction force may be smaller than that given by Eq. 6/1 with the static coefficient. In this event the friction force will be determined only by the equations of equilibrium. In such a problem we may be asked whether or not the existing friction is sufficient to maintain the body at rest. To answer the question, we may assume equilibrium and calculate the corresponding friction force F necessary to maintain this condition from the equations of equilibrium. We then compare this friction force with the maximum static friction that the surfaces can support as calculated from Eq. 6/1 with $\mu = \mu_s$. If F is less than that given by Eq. 6/1, we may conclude that the assumed friction force can be supported and therefore the body is at rest. If the calculated value of F is greater than the limiting value, then we conclude that the given surfaces cannot support that much friction force, and therefore motion exists and the friction becomes kinetic.

(3) In the *third* type of problem relative motion exists between the contacting surfaces, and here the kinetic coefficient of friction clearly applies. For this case Eq. 6/1 will always give the kinetic friction force directly.

The foregoing discussion applies to all dry contacting surfaces and, to a limited extent, to moving surfaces which are partially lubricated. Some typical values of the coefficients of friction are given in Table C1, Appendix C. These values are only approximate and are subject to considerable variation, depending on the exact conditions prevailing. They may be used, however, as typical examples of the magnitudes of frictional effects. When a reliable calculation involving friction is required, it is often desirable to determine the appropriate friction coefficient by experiment wherein the surface conditions of the problem are duplicated as closely as possible.

Sample Problem 6/1

Determine the maximum angle θ which the adjustable incline may have with the horizontal before the block of mass m begins to slip. The coefficient of static friction between the block and the inclined surface is μ_s.

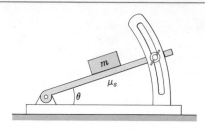

Solution. The free-body diagram of the block shows its weight $W = mg$, the normal force N, and the friction force F exerted by the incline on the block. The friction force acts in the direction to oppose the slipping which would occur if no friction were present.

Equilibrium in the x- and y-directions requires

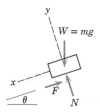

$[\Sigma F_x = 0]$ $\quad mg \sin\theta - F = 0 \quad F = mg \sin\theta$

$[\Sigma F_y = 0]$ $\quad -mg\cos\theta + N = 0 \quad N = mg\cos\theta$

Dividing the first equation by the second gives $F/N = \tan\theta$. Since the maximum angle occurs when $F = F_{s_{max}} = \mu_s N$, for impending motion we have

$$\mu_s = \tan\theta_{max} \quad \text{or} \quad \theta_{max} = \tan^{-1}\mu_s \quad \text{Ans.}$$

① We choose reference axes along and normal to the direction of F to avoid resolving both F and N into components.

② This problem describes a very simple way to determine a static coefficient of friction. The maximum value of θ is known as the *angle of repose*.

Sample Problem 6/2

Determine the range of values which the mass m_0 may have so that the 100-kg block shown in the figure will neither start moving up the plane nor slip down the plane. The coefficient of static friction for the contact surfaces is 0.30.

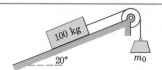

Solution. The maximum value of m_0 will be given by the requirement for motion impending up the plane. The friction force on the block therefore acts down the plane as shown in the free-body diagram of the block for Case I in the figure. With the weight $mg = 100(9.81) = 981$ N, the equations of equilibrium give

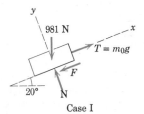

Case I

$[\Sigma F_y = 0]$ $\quad N - 981\cos 20° = 0, \quad N = 922$ N

$[F = \mu N]$ $\quad F = 0.30(922) = 277$ N

$[\Sigma F_x = 0]$ $\quad m_0(9.81) - 277 - 981\sin 20° = 0, \quad m_0 = 62.4$ kg $\quad$ Ans.

The minimum value of m_0 is determined when motion is impending down the plane. The friction force on the block will act up the plane to oppose the tendency to move as shown in the free-body diagram for Case II. Equilibrium in the x-direction requires

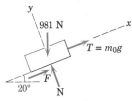

Case II

$[\Sigma F_x = 0]$ $\quad m_0(9.81) + 277 - 981\sin 20° = 0, \quad m_0 = 6.0$ kg $\quad$ Ans.

Thus m_0 may have any value from 6.0 to 62.4 kg, and the block will remain at rest.

In both cases equilibrium requires that the resultant of F and N will be concurrent with the 981-N weight and the applied tension T.

① We see from the results of Sample Problem 6/1 that the block would slide down the incline without the restraint of attachment to m_0 since $\tan 20° > 0.30$. Thus a value of m_0 will be required to maintain equilibrium.

Sample Problem 6/3

Determine the magnitude and direction of the friction force acting on the 100-kg block shown if, first, $P = 500$ N, and, second, $P = 100$ N. The coefficient of static friction is 0.20, and the coefficient of kinetic friction is 0.17. The forces are applied with the block initially at rest.

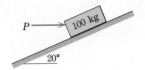

Solution. There is no way of telling from the statement of the problem whether the block will remain in equilibrium or whether it will begin to slip following the application of P. It is therefore necessary that we make an assumption, so we will take the friction force to be up the plane, as shown by the solid arrow. From the free-body diagram a balance of forces in both x- and y-directions gives

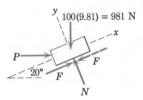

$[\Sigma F_x = 0]$ $P \cos 20° + F - 981 \sin 20° = 0$

$[\Sigma F_y = 0]$ $N - P \sin 20° - 981 \cos 20° = 0$

Case I. $P = 500$ N

Substitution into the first of the two equations gives

$$F = -134 \text{ N}$$

The negative sign tells us that *if* the block is in equilibrium, the friction force acting on it is in the direction opposite to that assumed and therefore is down the plane as represented by the dotted arrow. We cannot reach a conclusion on the magnitude of F, however, until we verify that the surfaces are capable of supporting 134 N of friction force. This may be done by substituting $P = 500$ N into the second equation, which gives

$$N = 1093 \text{ N}$$

The maximum static friction force that the surfaces can support is then

$[F = \mu N]$ $F_{\text{max}} = 0.20(1093) = 219 \text{ N}$

Since this force is greater than that required for equilibrium, we conclude that the assumption of equilibrium was correct. The answer is, then,

$$F = 134 \text{ N down the plane} \qquad Ans.$$

Case II. $P = 100$ N

Substitution into the two equilibrium equations gives

$$F = 231 \text{ N}, \qquad N = 956 \text{ N}$$

But the maximum possible static friction force is

$[F = \mu N]$ $F_{\text{max}} = 0.20(956) = 191 \text{ N}$

It follows that 231 N of friction cannot be supported. Therefore equilibrium cannot exist, and we obtain the correct value of the friction force by using the kinetic coefficient of friction accompanying the motion down the plane. Hence the answer is

$[F = \mu N]$ $F = 0.17(956) = 163 \text{ N up the plane} \qquad Ans.$

① We should note that even though ΣF_x is no longer equal to zero, equilibrium does exist in the y-direction, so that $\Sigma F_y = 0$. Therefore the normal force N is 956 N whether or not the block is in equilibrium.

Sample Problem   6/4

The homogeneous rectangular block of mass m, width b, and height H is placed on the horizontal surface and subjected to a horizontal force P which moves the block along the surface with a constant velocity. The coefficient of kinetic friction between the block and the surface is μ. Determine (a) the greatest value that h may have so that the block will slide without tipping over and (b) the location of a point C on the bottom face of the block through which the resultant of the friction and normal forces acts if $h = H/2$.

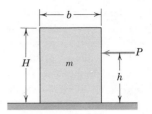

Solution.   (a) With the block on the verge of tipping, we see that the entire reaction between the plane and the block will be at A. The free-body diagram of the block shows this condition. Since slipping occurs, the friction force is the limiting value μN, and the angle θ becomes $\theta = \tan^{-1}\mu$. The resultant of F and N passes through a point B through which P must also pass, since three coplanar forces in equilibrium are concurrent. Hence from the geometry of the block

$$\tan \theta = \mu = \frac{b/2}{h}, \; h = \frac{b}{2\mu} \qquad \textit{Ans.}$$

If h were greater than this value, moment equilibrium about A would not be satisfied, and the block would tip over.

(b) With $h = H/2$ we see from the free-body diagram for case (b) that the resultant of F and N passes through a point C which is a distance x to the left of the vertical centerline through G. The angle θ is still $\theta = \phi = \tan^{-1}\mu$ as long as the block is slipping. Thus, from the geometry of the figure we have

$$\frac{x}{H/2} = \tan \theta = \mu \quad \text{so} \quad x = \mu H/2 \qquad \textit{Ans.}$$

If μ represents the coefficient of static friction, then our solutions describe the condition under which the block is (a) on the verge of tipping and (b) on the verge of slipping, both from a rest position.

① Recall that the equilibrium equations apply to a body moving with a constant velocity (zero acceleration) just as well as to a body at rest.

② Alternatively, we could equate the moments about G to zero which would give us $F(H/2) - Nx = 0$. Thus with $F = \mu N$ we get $x = \mu H/2$.

Sample Problem 6/5

The three flat blocks are positioned on the 30° incline as shown, and a force P parallel to the incline is applied to the middle block. The upper block is prevented from moving by a wire which attaches it to the fixed support. The coefficient of static friction for each of the three pairs of mating surfaces is shown. Determine the maximum value which P may have before any slipping takes place.

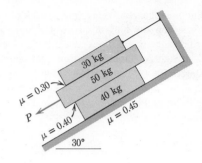

Solution. The free-body diagram of each block is drawn. The friction forces are assigned in the directions to oppose the relative motion which ① would occur if no friction were present. There are two possible conditions for impending motion. Either the 50-kg block slips and the 40-kg block remains in place, or the 50- and 40-kg blocks move together with slipping occurring between the 40-kg block and the incline.

The normal forces, which are in the y-direction, may be determined without reference to the friction forces, which are all in the x-direction. Thus

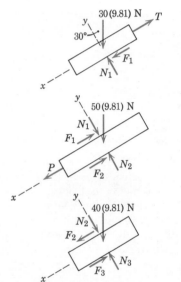

$[\Sigma F_y = 0]$ (30-kg) $N_1 - 30(9.81)\cos 30° = 0$ $N_1 = 255$ N

             (50-kg) $N_2 - 50(9.81)\cos 30° - 255 = 0$ $N_2 = 680$ N

             (40-kg) $N_3 - 40(9.81)\cos 30° - 680 = 0$ $N_3 = 1019$ N

We shall assume arbitrarily that only the 50-kg block slips, so that the 40-kg block remains in place. Thus for impending slippage at both surfaces of the 50-kg block we have

$[F = \mu N]$ $F_1 = 0.30(255) = 76.5$ N $F_2 = 0.40(680) = 272$ N

Equilibrium of forces at impending motion for the 50-kg block gives

$[\Sigma F_x = 0]$ $P - 76.5 - 272 + 50(9.81)\sin 30° = 0, P = 103.1$ N

We now check on the validity of our initial assumption. For the 40-kg block with $F_2 = 272$ N the friction force F_3 would be given by

$[\Sigma F_x = 0]$ $272 + 40(9.81)\sin 30° - F_3 = 0$ $F_3 = 468$ N

But the maximum possible value of F_3 is $F_3 = \mu N_3 = 0.45(1019) = 459$ N. Thus 468 N cannot be supported and our initial assumption was wrong. We conclude, therefore, that slipping occurs between the 40-kg block and the incline. With the corrected value $F_3 = 459$ N equilibrium of the 40-kg block requires

② $[\Sigma F_x = 0]$ $F_2 + 40(9.81)\sin 30° - 459 = 0$ $F_2 = 263$ N

Equilibrium of the 50-kg block gives, finally,

$[\Sigma F_x = 0]$ $P + 50(9.81)\sin 30° - 263 - 76.5 = 0, P = 93.8$ N   *Ans.*

① In the absence of friction the middle block, under the influence of P, would have a greater movement than the 40-kg block, and the friction force F_2 will be in the direction to oppose this motion as shown.

② We see now that F_2 is less than $\mu N_2 = 272$ N.

PROBLEMS

6/1 The 100-lb block rests on the horizontal surface, and a force P whose direction can be varied is applied to the block. If $P = 50$ lb and if the block begins to move when α is increased to $60°$, calculate the static coefficient of friction μ between the block and the surface.   *Ans.* $\mu = 0.577$

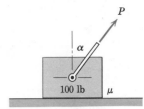

Problem 6/1

6/2 Calculate the force P required to move the 50-kg block up the $30°$ incline.

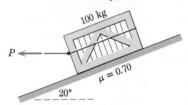

Problem 6/2

6/3 A horizontal force P of 200 N is applied to the 100-kg crate in an effort to slide it down the incline. If the coefficient of friction between the crate and the incline is 0.70, compute the friction force F acting on the crate.   *Ans.* $F = 523$ N

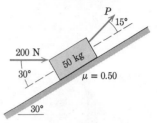

Problem 6/3

6/4 The coefficient of friction between the block and the incline is μ. Determine the value of x for which the force P required to initiate motion up the plane is the least of all possible values. Neglect the radius of the pulley.

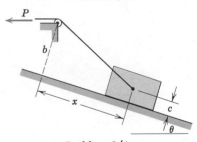

Problem 6/4

6/5 If $P = 50$ lb and $\mu = 0.30$ in Prob. 6/1, determine the value of α, as it is gradually increased, for which the block begins to slide.   *Ans.* $\alpha = 18.4°$

6/6 Determine the maximum distance d in terms of l at which the lower end of the prop of negligible mass may be set without slipping as it supports the heavy hinged panel. The coefficient of friction is 0.30.

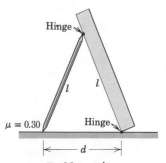

Problem 6/6

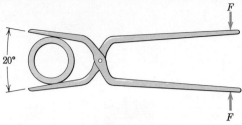

Problem 6/7

6/7 The tongs are used to handle hot steel tubes that are being heat treated in an oil bath. For a 20° jaw opening, what is the minimum coefficient of friction between the jaws and the tube that will enable the tongs to grip the tube without slipping?

Ans. $\mu_{\min} = 0.176$

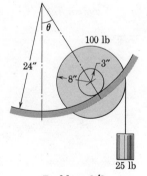

Problem 6/8

6/8 The figure shows a device, called a jam cleat, that secures a rope under tension by reason of large friction forces developed. For the position shown determine the minimum coefficient of friction between the rope and the cam surfaces for which the cleat will be self-locking. Also find the magnitude of the total reaction R on each of the cam bearings.

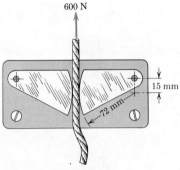

Problem 6/9

6/9 The 100-lb wheel rolls on its hub up the circular incline under the action of the 25-lb weight attached to a cord around the rim. Determine the angle θ at which the wheel comes to rest, assuming that friction is sufficient to prevent slippage. What is the minimum coefficient of friction that will permit this position to be reached with no slipping?

Ans. $\theta = 32.2°$, $\mu_{\min} = 0.630$

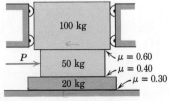

Problem 6/10

6/10 Determine the horizontal force P required to cause slippage to occur. The friction coefficients for the three pairs of mating surfaces are indicated. The top block is free to move vertically.

6/11 The 900-mm-diameter wheel has a mass of 30 kg. Determine the couple M required to roll the wheel over the 90-mm obstruction on the horizontal surface. Specify the minimum value of the coefficient of friction between the wheel and the obstruction so that the wheel does not slip.

Ans. $M = 79.5 \text{ N·m}, \mu_{min} = 0.75$

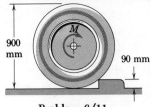

Problem 6/11

6/12 The homogeneous rectangular block of mass m rests on the inclined plane that is hinged about a horizontal axis through O. If the coefficient of static friction between the block and the plane is μ, specify the conditions that determine whether the block tips before it slips or slips before it tips as the angle θ is gradually increased.

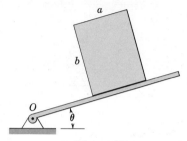

Problem 6/12

6/13 Determine the maximum angle θ at which the V-block may be tilted with the horizontal before the solid cylinder begins to slide. The coefficient of friction is μ.   *Ans.* $\theta = \tan^{-1}(\mu/\cos \alpha)$

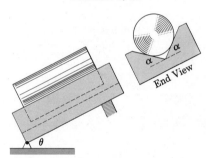

Problem 6/13

6/14 A uniform block of mass m is at rest on an incline θ. Determine the maximum force P that can be applied to the block in the direction shown before slipping begins. The coefficient of friction between the block and the incline is μ. Also determine the angle β between the horizontal direction of P and the direction of initial movement of the block.

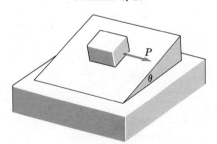

Problem 6/14

6/15 The 300-lb crate with mass center at G is supported on the horizontal surfaces by a skid at A and a roller at B. If a force P of 60 lb is required to initiate motion of the crate, determine the coefficient of friction at A.   *Ans.* $\mu = 0.33$

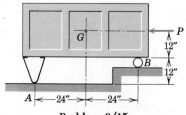

Problem 6/15

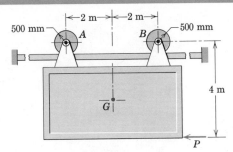

Problem 6/16

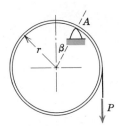

Problem 6/17

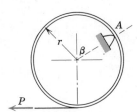

Problem 6/18

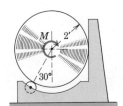

Problem 6/19

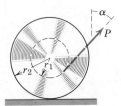

Problem 6/20

6/16 The center of mass of the vertical 800-kg panel is at G. The panel is mounted on wheels which permit ease of horizontal movement along the fixed rail. If the bearing of the wheel at A becomes "frozen" so that the wheel cannot turn, compute the force P required to move the panel. The coefficient of friction between the wheel and the rail is 0.30.

6/17 A metal hoop of negligible thickness has a mass m and mean radius r and hangs from a support at A. If the coefficient of friction between the hoop and the support is μ, determine the vertical force P applied to the periphery of the hoop required to slip the hoop on the support. Also determine the angle β at which slipping occurs.

$$Ans. \ P = mg\frac{\mu}{\sqrt{1 + \mu^2} - \mu}, \ \beta = \tan^{-1}\mu$$

6/18 The hoop of Prob. 6/17 is subjected to a horizontal force P as shown here. If slipping on the support at A occurs when β reaches 60° as P is increased, determine the coefficient of friction μ between the hoop and the support. Also express the value of P in terms of the mass m of the hoop as slipping is initiated.

6/19 The uniform cylinder weighs 400 lb and is supported by the roller, which turns with negligible friction. If the coefficient of friction between the cylinder and the vertical surface is 0.6, calculate the torque M required to turn the cylinder. Also find the reaction R on the bearing of the roller as M is applied. *Ans. M = 206 lb-ft, R = 343 lb*

6/20 The wheel shown will roll to the left when the angle α of the cord is small. When α is large the wheel rolls to the right. Determine by inspection from the geometry of the free-body diagram the angle α for which the wheel will not roll in either direction. If the coefficient of friction is μ and the mass of the wheel is m, determine the value of P for which the wheel will slip for the critical value of α.

6/21 Calculate the force T required to rotate the 200-kg reel of telephone cable that rests on its hubs and bears against a vertical wall. The coefficient of friction for each pair of contacting surfaces is 0.60.

Ans. $T = 727$ N

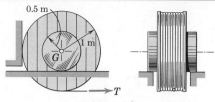

Problem 6/21

6/22 The roll of paper is to be rolled slowly up the incline by a tension T applied to the paper as it is pulled horizontally off the roll. The coefficient of friction between the roll and the incline is 0.20. Prove whether the paper rolls without slipping or whether it slips.

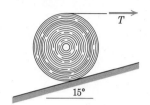

Problem 6/22

6/23 The left-hand jaw of the C-clamp can be slid along the frame to increase the capacity of the clamp. To prevent slipping of the jaw on the frame when the clamp is under load, the dimension x must exceed a certain minimum value. Find this value corresponding to given dimensions a and b and a coefficient of friction μ between the frame and the loose-fitting jaw.

$$Ans. \; x = \frac{a - b\mu}{2\mu}$$

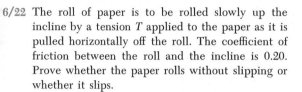

Problem 6/23

6/24 Determine the torque M required to spin the uniform wheel of mass m in its position against the vertical wall. The coefficient of friction for each pair of contacting surfaces is μ.

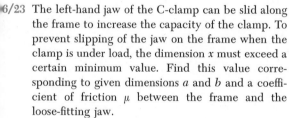

Problem 6/24

6/25 The uniform 100-lb wheel with its hubs resting on inclined rails bears against a fixed support at A. A cord which is wrapped securely around the outer periphery leads horizontally to a hook at B. If the support at A is removed, (*a*) calculate the friction force acting on the wheel. The coefficients of static and kinetic friction between the hub and rail are 0.50 and 0.45, respectively. (*b*) If these friction coefficients were 0.30 and 0.25, what would happen?

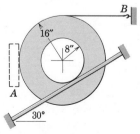

Problem 6/25

Problem 6/26

6/26 The industrial truck is used to move the solid 1200-kg roll of paper up the 30° incline. If the coefficients of friction between the roll and the vertical barrier of the truck and between the roll and the incline are both 0.40, compute the required tractive force P between the tires of the truck and the horizontal surface.   *Ans. P = 22.1 kN*

6/27 Calculate the horizontal force between the tires and the horizontal surface for the industrial truck in Prob. 6/26 if the roll of paper is on the verge of rolling down the incline.

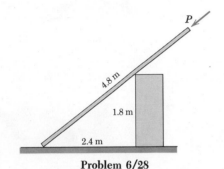

Problem 6/28

6/28 Determine the force P required to move the uniform 50-kg plank from its rest position shown if the coefficient of friction at both contact locations is 0.50.   *Ans. P = 19.62 N*

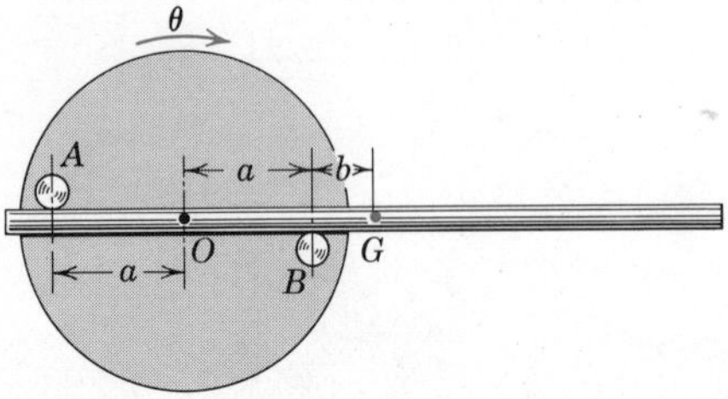

Problem 6/29

6/29 The uniform rod with center of mass at G is supported by the pegs A and B, which are fixed in the wheel. If the coefficient of friction between the rod and the pegs is μ, determine the angle θ through which the wheel may be turned about its horizontal axis through O before the rod begins to slip. Neglect the diameter of the rod compared with the other dimensions.

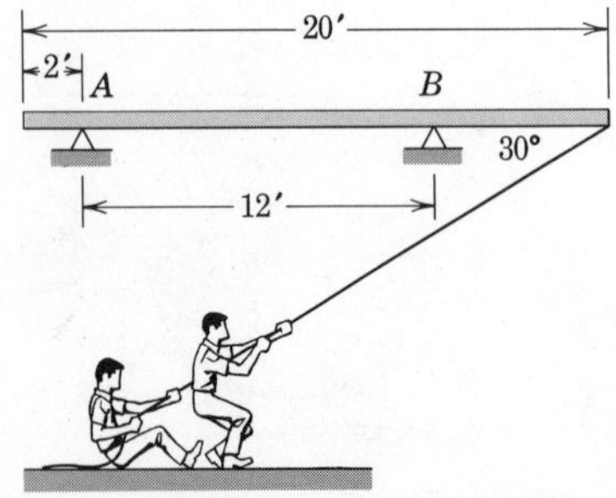

Problem 6/30

6/30 What force P must the two men exert on the rope to slide the uniform 20-ft plank on the overhead rack? The plank weighs 200 lb, and the coefficient of friction between the plank and each support is 0.50.   *Ans. P = 162.3 lb*

6/31 A small roller on end B of the uniform 60-kg bar is constrained to move in the smooth vertical guides. The coefficient of friction between end A and the horizontal supporting surface is 0.80. Determine the horizontal force P required to initiate slipping at A.

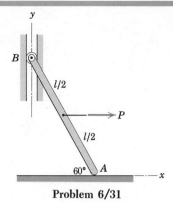

Problem 6/31

6/32 The uniform slender pole of length l and mass m is leaned against the vertical wall at the angle shown. If the coefficient of friction μ is the same at both ends of the pole, calculate the minimum value which μ may have if the pole is not to slip.

Ans. $\mu_{min} = 1/3$

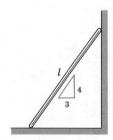

Problem 6/32

6/33 Assuming the worker is strong enough, can he maintain equilibrium of the uniform 7-m plank in the position shown by exerting a horizontal pull on the rope? The coefficient of static friction between the 60-kg plank and the top of the building is 0.30. If the answer is yes, then compute the friction force existing in the equilibrium state.

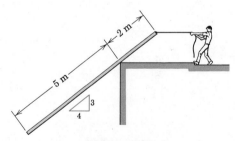

Problem 6/33

Split it up

6/34 Find the greatest height h of the step which the 80-kg man can reach without causing the hinged stepladder to collapse. The coefficient of friction at A is 0.40 and at B is 0.50, and each of the two uniform sections of the hinged ladder has a mass of 10 kg.

Ans. $h = 2.37$ m

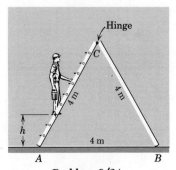

Problem 6/34

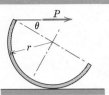

Problem 6/35

6/35 The semicylindrical shell of mass m and radius r is rolled through an angle θ by the horizontal force P applied to its rim. If the coefficient of friction is μ, determine the angle θ at which the shell slips on the horizontal surface as P is gradually increased. What value of μ will permit θ to reach 90°?

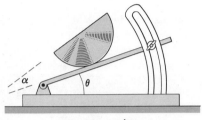

Problem 6/36

6/36 If the coefficient of friction between the solid semicircular cylinder and the incline is 0.30, determine the maximum angle θ with the horizontal at which the incline can be tilted before the cylinder slips. For this condition what is the corresponding angle α through which the cylinder has rolled with respect to the incline?   *Ans.* $\theta = 16.70°$, $\alpha = 25.9°$

6/37 If the worker in Prob. 6/33 relaxes his pull on the rope but maintains its horizontal direction, what angle θ will the plank make with the vertical when it begins to slip? (*Hint:* Approximate the answer with a graphical solution of the equation which governs θ.)   *Ans.* $\theta = 40.6°$

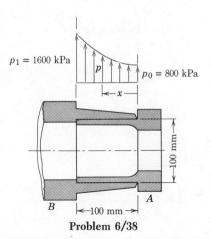

Problem 6/38

6/38 The circular collar A shown in section is mated with part B with a shrink fit which sets up a pressure or compressive stress p between the parts. The pressure has the values shown at the extremities of the overlap and is closely approximated by the relation $p = p_0 + kx^2$ in between. If a torque of 1.3 kN·m is required to turn collar A inside of part B, calculate the effective coefficient of friction μ between the two parts.   *Ans.* $\mu = 0.776$

6/39 A uniform bar of mass m and length l rests on a horizontal surface with its mass evenly distributed along its length. If the coefficient of friction between the bar and the supporting surface is μ, write an expression for the horizontal force P, applied at the end of the bar, required to move the bar and find the distance a to axis O about which the bar is observed to rotate. (*Hint:* The normal force under the bar is uniformly distributed over the length of the bar. The friction force acting on the bar is, therefore, uniformly distributed over the length of each of the two sections on either side of O.)

$\qquad$ *Ans.* $P = 0.414\,\mu mg$, $a = 0.293l$

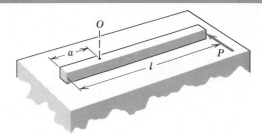

Problem 6/39

6/40 Determine the minimum force P and its corresponding angle α required to tip the uniform 200-lb crate about its front edge. Specify the minimum coefficient of friction which will permit the tipping to take place. $\qquad$ *Ans.* $P_{\min} = 70.7$ lb, $\mu_{\min} = 1/3$

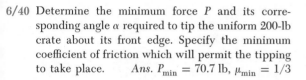

Problem 6/40

6/41 An I-beam of mass m is supported by two fixed horizontal rails as shown. Compute the applied load P that is just sufficient to cause the beam to slip and determine the corresponding friction force at A as slippage begins. The coefficient of friction between the beam and the rails is μ.

$$\textit{Ans.}\; P = \frac{\mu mgb}{2(a+b)}, \; F_A = \frac{\mu mga}{2(a+b)}$$

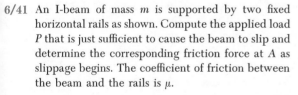

Problem 6/41

►**6/42** The device shown prevents clockwise rotation in the horizontal plane of the central wheel by means of frictional locking of the two small rollers. For given values of R and r and for a common coefficient of friction μ at all contact surfaces, determine the range of values of d for which the device will operate as described.

$$\textit{Ans.}\; \frac{2r + (1 - \mu^2)R}{1 + \mu^2} < d < (R + 2r)$$

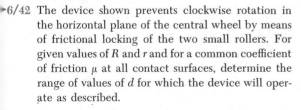
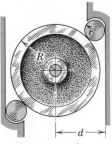
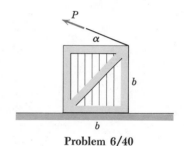

Problem 6/42

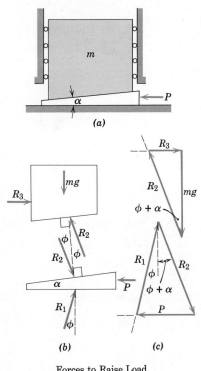

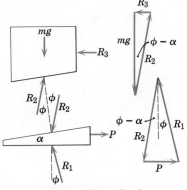

Forces to Raise Load

Figure 6/3

Forces to Lower Load

Figure 6/4

6/4 WEDGES. A wedge is one of the simplest and most useful of machines and is used as a means of producing small adjustments in the position of a body or as a means of applying large forces. Wedges are largely dependent on friction. When sliding of a wedge is impending, the resultant force on each sliding surface of the wedge will be inclined from the normal to the surface by an amount equal to the friction angle. The component of the resultant along the surface is the friction force, which is always in the direction to oppose the motion of the wedge relative to the mating surfaces.

Figure 6/3a shows a wedge that is used to position or lift a large mass m, where the vertical loading is mg. The coefficient of friction for each pair of surfaces is $\mu = \tan\phi$. The force P required to start the wedge is found from the equilibrium triangles of the forces on the load and on the wedge. The free-body diagrams are shown in Fig. 6/3b, where the reactions are inclined at an angle ϕ from their respective normals and are in the direction to oppose the motion. The mass of the wedge is neglected. From these diagrams we may write the force equilibrium equations

$$\mathbf{W} + \mathbf{R}_2 + \mathbf{R}_3 = 0 \quad \text{and} \quad \mathbf{R}_1 + \mathbf{R}_2 + \mathbf{P} = 0$$

where $W = mg$. The solutions of these equations are shown in the c-part of the figure, where R_2 is found first in the upper diagram using the known value of mg. The force P is then found from the lower triangle once the value of R_2 has been established.

If P is removed, the wedge will remain in place as long as α is less than 2ϕ, in which case the wedge is said to be self-locking. If the wedge is self-locking and is to be withdrawn, a pull P on the wedge would be required. In this event the reactions R_1 and R_2 would act on the opposite sides of their normals to oppose the new impending motion, and the solution would proceed along lines similar to those described for the case of raising the load. The free-body diagrams and vector polygons for this condition are shown in Fig. 6/4.

Wedge problems lend themselves to graphical solutions as indicated in the two figures. The accuracy of a graphical solution is easily held to within tolerances consistent with the uncertainty of friction coefficients. Algebraic solutions may also be obtained from the trigonometry of the equilibrium polygons.

6/5 SCREWS. Screws are used for fastenings and for transmitting power or motion. In each case the friction developed in the threads largely determines the action of the screw. For transmitting power or motion the square thread is more efficient than the V-thread, and the analysis illustrated here is confined to the square thread.

Consider the square-threaded jack, Fig. 6/5, under the action of the axial load W and a moment M applied about the axis of the screw. The force R exerted by the thread of the jack frame on a small representative portion of the thread of the screw is shown on the free-body diagram of the screw. Similar reactions exist on all segments of the screw thread where contact occurs with the thread of the base. If M is just sufficient to turn the screw, the thread of the screw will slide around and up on the fixed thread of the frame. The angle ϕ made by R with the normal to the thread will be the angle of friction, so that $\tan \phi = \mu$. The moment of R about the vertical axis of the screw is $Rr \sin (\alpha + \phi)$, and the total moment due to all reactions on the threads is $\Sigma Rr \sin (\alpha + \phi)$. Since $r \sin (\alpha + \phi)$ appears in each term, we may factor it out. The moment equilibrium equation for the screw becomes

$$M = [r \sin (\alpha + \phi)]\Sigma R$$

Equilibrium of forces in the axial direction further requires

$$W = \Sigma R \cos (\alpha + \phi) = [\cos (\alpha + \phi)]\Sigma R$$

Dividing M by W gives us

$$M = Wr \tan (\alpha + \phi) \tag{6/3}$$

We determine the helix angle α by unwrapping the thread of the screw for one turn where we see immediately that $\alpha = \tan^{-1}(L/2\pi r)$.

We may use the unwrapped thread of the screw as an alternative model to simulate the action of the entire screw, as shown in Fig. 6/6a. The equivalent force required to push the movable thread up the fixed incline is $P = M/r$, and the triangle of force vectors gives Eq. 6/3 immediately.

If the moment M is removed, the friction force changes direction so that ϕ is measured to the other side of the normal to the thread. The screw will remain in place and be self-locking provided that $\alpha < \phi$ and will be on the verge of unwinding if $\alpha = \phi$.

In order to lower the load by unwinding the screw, we must reverse the direction of M as long as $\alpha < \phi$. This condition is illustrated in Fig. 6/6b for our simulated thread on the fixed incline, and we see that an equivalent force $P = M/r$ must be applied to the thread to pull it down the incline. Therefore, from the triangle of vectors we get the moment required to lower the screw, which is

$$M = Wr \tan (\phi - \alpha) \tag{6/3a}$$

If $\alpha > \phi$ the screw will unwind by itself, and we see from Fig. 6/6c that the moment required to prevent unwinding would be

$$M = Wr \tan (\alpha - \phi) \tag{6/3b}$$

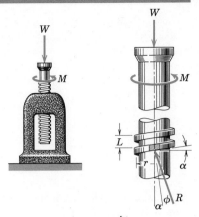

Figure 6/5

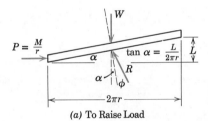

$P = \dfrac{M}{r}$ $\tan \alpha = \dfrac{L}{2\pi r}$

$2\pi r$

(a) To Raise Load

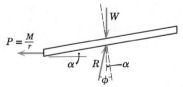

$P = \dfrac{M}{r}$

(b) To Lower Load ($\alpha < \phi$)

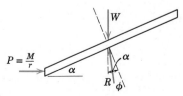

$P = \dfrac{M}{r}$

(c) To Lower Load ($\alpha > \phi$)

Figure 6/6

Sample Problem 6/6

The horizontal position of the 500-kg rectangular block of concrete is adjusted by the 5° wedge under the action of the force **P**. If the coefficient of static friction for both pairs of wedge surfaces is 0.30 and if the coefficient of static friction between the block and the horizontal surface is 0.60, determine the least force P required to move the block.

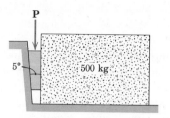

Solution. ① The free-body diagrams of the wedge and the block are drawn with the reactions $\mathbf{R}_1$, $\mathbf{R}_2$, and $\mathbf{R}_3$ inclined with respect to their normals by the amount of the friction angles for impending motion. The friction angle for limiting static friction is given by $\phi = \tan^{-1} \mu$. Each of the two friction angles is computed and shown on the diagram.

We start our vector diagram expressing the equilibrium of the block at a convenient point A and draw the only known vector, the weight **W** of the block. Next we add $\mathbf{R}_3$ whose 31.0° inclination from the vertical is now known. The vector $\mathbf{R}_2$, whose 16.7° inclination from the horizontal is also known, must close the polygon for equilibrium. Thus point B on the lower polygon is determined by the intersection of the known directions of $\mathbf{R}_3$ and $\mathbf{R}_2$ and their magnitudes become known.

For the wedge we draw $\mathbf{R}_2$, which is now known, and add $\mathbf{R}_1$ whose direction is known. The directions of $\mathbf{R}_1$ and **P** intersect at C, thus giving us the solution for the magnitude of **P**.

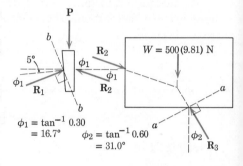

Algebraic solution. The simplest choice of reference axes for calculation purposes is, for the block, in the direction a-a normal to $\mathbf{R}_3$ and, for ② the wedge, in the direction b-b normal to $\mathbf{R}_1$. The angle between $\mathbf{R}_2$ and the a-direction is $16.7° + 31.0° = 47.7°$. Thus for the block

$$[\Sigma F_a = 0] \qquad 500(9.81)\sin 31.0° - R_2 \cos 47.7° = 0$$

$$R_2 = 3747 \text{ N}$$

For the wedge the angle between $\mathbf{R}_2$ and the b-direction is $90° - (2\phi_1 + 5°) = 51.6°$, and the angle between **P** and the b-direction is $\phi_1 + 5° = 21.7°$. Thus

$$[\Sigma F_b = 0] \qquad 3747 \cos 51.6° - P \cos 21.7° = 0$$

$$P = 2505 \text{ N} \qquad\qquad Ans.$$

Graphical solution. The accuracy of a graphical solution is well within the uncertainty of the friction coefficients and provides a simple and direct result. By laying off the vectors to a reasonable scale following the sequence described, the magnitudes of **P** and the **R**'s are easily scaled directly from the diagrams.

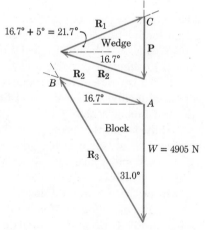

① Be certain to note that the reactions are inclined from their normals in the direction to oppose the motion. Also, we note the equal and opposite reactions $\mathbf{R}_2$.

② It should be evident that we avoid simultaneous equations by eliminating reference to $\mathbf{R}_3$ for the block and $\mathbf{R}_1$ for the wedge.

Sample Problem   6/7

The single-threaded screw of the vise has a mean diameter of 1 in. and has 5 square threads per inch. The coefficient of static friction in the threads is 0.20. A 60-lb pull applied normal to the handle at A produces a clamping force of 1000 lb between the jaws of the vise. (*a*) Determine the frictional moment M_B developed at B due to the thrust of the screw against the body of the jaw. (*b*) Determine the force Q applied normal to the handle at A required to loosen the vise.

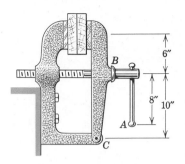

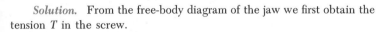

Solution. From the free-body diagram of the jaw we first obtain the tension T in the screw.

$$[\Sigma M_C = 0] \qquad 1000(16) - 10T = 0 \qquad T = 1600 \text{ lb}$$

The helix angle α and the friction angle ϕ for the thread are given by

$$\alpha = \tan^{-1} \frac{L}{2\pi r} = \tan^{-1} \frac{1/5}{2\pi(0.5)} = 3.64°$$

$$\phi = \tan^{-1} \mu = \tan^{-1} 0.20 = 11.31°$$

where the mean radius of the thread is $r = 0.5$ in.

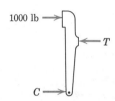

① Be careful to calculate the helix angle correctly. Its tangent is the lead L (advancement per revolution) divided by the mean circumference $2\pi r$ and not by the diameter $2r$.

(*a*) *To tighten.* The isolated screw is simulated by the free-body diagram shown where all of the forces acting on the threads of the screw are represented by a single force R inclined at the friction angle ϕ from the normal to the thread. The moment applied about the screw axis is $60(8) = 480$ lb-in. in the clockwise direction as seen from the front of the vise. The frictional moment M_B due to the friction forces acting on the collar at B is in the counterclockwise direction to oppose the impending motion. From Eq. 6/3 with T substituted for W the net moment acting on the screw is

$$M = Tr \tan (\alpha + \phi)$$

$$480 - M_B = 1600(0.5) \tan (3.64° + 11.31°)$$

$$M_B = 266 \text{ lb-in.} \qquad\qquad\qquad \textit{Ans.}$$

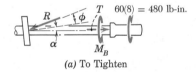

(*a*) To Tighten

(*b*) *To loosen.* The free-body diagram of the screw on the verge of being loosened is shown with R acting at the friction angle from the normal in the direction to counteract the impending motion. Also shown is the frictional moment $M_B = 266$ lb-in. acting in the clockwise direction to oppose the motion. The angle between R and the screw axis is now $\phi - \alpha$, and we use Eq. 6/3a with the net moment equal to the applied moment M' minus M_B. Thus

$$M = Tr \tan (\phi - \alpha)$$

$$M' - 266 = 1600(0.5) \tan (11.31° - 3.64°)$$

$$M' = 374 \text{ lb-in.}$$

Thus the force on the handle required to loosen the vise is

$$Q = M'/d = 374/8 = 46.8 \text{ lb} \qquad\qquad \textit{Ans.}$$

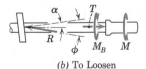

(*b*) To Loosen

② Note that R swings to the opposite side of the normal as the impending motion reverses direction.

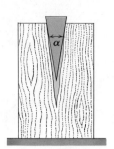

Problem 6/43

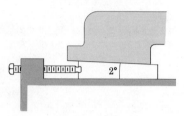

Problem 6/44

Problem 6/46

Problem 6/47

PROBLEMS

6/43 A wedge will be self-locking provided that its angle α is less than a critical value. If the coefficient of friction between the wedge and the material being split is μ, what is the critical value of α?

6/44 The precise alignment of a heavy-duty diesel engine on its foundation is accomplished by screw-adjusted wedges under each of its four mounting flanges. Calculate the horizontal thrust P in the adjusting screw necessary to raise the mounting flange if it supports one-fourth of the total engine mass of 4.8 Mg. The coefficient of friction for both sides of the wedge is 0.25. *Ans. P* = 6.33 kN

6/45 A screw jack with square threads having a mean radius of 25 mm supports a load of 5 kN. If the coefficient of friction is 0.25, what is the greatest lead L (advancement per turn) of the screw for which the screw will not unwind by itself? For this condition what torque M applied to the screw would be required to raise the load?

6/46 The turnbuckle supports a tension T of 15,000 lb. Each of the screws has a mean diameter of 1.354 in. and has a single thread with a lead (advancement per turn) of $\frac{1}{3}$ inch, one being right-handed and the other left-handed. If a torque of 270 lb-ft is required to loosen the turnbuckle by turning the body with both screws prevented from rotating, calculate the effective coefficient of friction μ in the threads. *Ans. μ* = 0.241

6/47 The two 5° wedges shown are used to adjust the position of the column under a vertical load of 5 kN. Determine the magnitude of the forces P required to raise the column if the coefficient of friction for all surfaces is 0.40.

6/48 If the loaded column of Prob. 6/47 is to be lowered, calculate the horizontal forces P' required to withdraw the wedges.   Ans. $P' = 3.51$ kN

6/49 A 5° steel wedge is forced under the end of the 4200 lb machine with a force $P = 1100$ lb. If the coefficient of friction between the wedge and both the machine and the horizontal floor is 0.30, determine the position x of the center of gravity G of the machine. The machine is prevented from sliding horizontally by a rigid ledge at A.

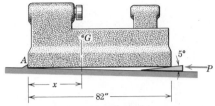

Problem 6/49

6/50 The movable head of a universal testing machine has a mass of 2.2 Mg and is elevated into testing position by two 78-mm-diameter lead screws each with a single thread and a lead of 13 mm. If the coefficient of friction in the threads is 0.25, how much torque M must be supplied to each screw (a) to raise the head and (b) to lower the head? The inner loading columns are not attached to the head during positioning.   Ans. (a) $M = 129.3$ N·m
(b) $M = 81.8$ N·m

Problem 6/50

6/51 The device shown is used as a jack. The screw has a double square thread with a mean diameter of $\frac{7}{8}$ in. and a lead of $\frac{1}{3}$ in. Section A of the screw has right-hand threads, and section B has left-hand threads. For $\theta = 30°$ determine (a) the torque M which must be applied to the screw to raise a load $P = 1500$ lb and (b) the torque M' needed to lower the load. The coefficient of friction in the threads is 0.20.

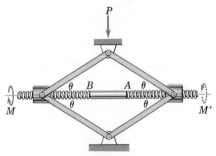

Problem 6/51

6/52 The detent mechanism consists of the spring-loaded plunger with a spherical end which positions the horizontal bar by engaging the spaced notches. If the spring exerts a force of 40 N on the plunger in the position shown and a force $P = 60$ N is required to move the detent bar against the plunger, calculate the coefficient of friction between the plunger and the detent. It is known from earlier tests that the coefficient of friction between the light bar and the horizontal surface is 0.30. Assume that the plunger is well lubricated and accurately fitted so that the friction between it and its guide is negligible.   Ans. $\mu = 0.368$

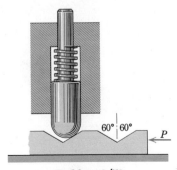

Problem 6/52

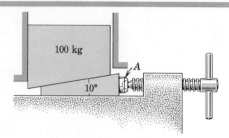

Problem 6/53

6/53 The vertical position of the 100-kg block is adjusted by the screw-activated wedge. Calculate the moment M which must be applied to the handle of the screw to raise the block. The single-threaded screw has square threads with a mean diameter of 30 mm and advances 10 mm for each complete turn. The coefficient of friction for the screw threads is 0.25, and the coefficient of friction for all mating surfaces of the block and wedge is 0.40. Neglect friction at the ball joint A.

6/54 Calculate the moment M' which must be applied to the handle of the screw of Prob. 6/53 to withdraw the wedge and lower the 100-kg load.

Ans. $M' = 3.02$ N·m

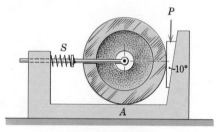

Problem 6/55

6/55 Compute the force P required to move the 20-kg wheel. The coefficient of friction at A is 0.25 and that for both pairs of wedge surfaces is 0.30. Also, the spring S is under a compression of 100 N, and the rod offers negligible support to the wheel.

6/56 Work Prob. 6/55 if the compression in the spring is 200 N. All other conditions remain unchanged.

Ans. $P = 198.8$ N

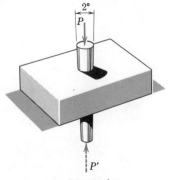

Problem 6/57

▶6/57 The tapered pin is forced into a mating tapered hole in the fixed block with a force $P = 400$ N. If the force required to remove the pin is $P' = 300$ N, calculate the coefficient of friction between the pin and the surface of the hole. (*Hint:* The pressure (stress) normal to the tapered pin surface remains unchanged until the pin actually moves.)

Ans. $\mu = 0.122$

/58 Replace the square thread of the screw jack in Fig. 6/5 by a V-thread as indicated in the figure accompanying this problem and determine the moment M on the screw required to raise the load W. The force R acting on a representative small section of the thread is shown with its relevant projections. The vector R_1 is the projection of R in the plane of the figure containing the axis of the screw. The analysis is begun with an axial force and a moment summation and includes substitutions for the angles γ and β in terms of θ, α, and the friction angle $\phi = \tan^{-1}\mu$. The helix angle of the single thread is exaggerated for clarity.

$$\text{Ans. } M = Wr\,\frac{\tan\alpha + \mu\sqrt{1 + \tan^2\dfrac{\theta}{2}\cos^2\alpha}}{1 - \mu\tan\alpha\sqrt{1 + \tan^2\dfrac{\theta}{2}\cos^2\alpha}}$$

$$\text{where } \alpha = \tan^{-1}\frac{L}{2\pi r}$$

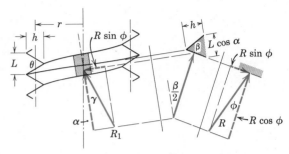

Problem 6/58

6/6 JOURNAL BEARINGS. A journal bearing is a bearing that gives lateral support to a shaft in contrast to axial or thrust support. For dry bearings and for many partially lubricated bearings we may apply the principles of dry friction which give us a satisfactory approximation for design purposes. A dry or partially lubricated journal bearing with contact or near contact between the shaft and the bearing is shown in Fig. 6/7, where the clearance between the shaft and bearing is greatly exaggerated. As the shaft begins to turn in the direction shown, it rolls up the inner surface of the bearing until slippage occurs. Here it remains in a more or less fixed position during rotation. The torque M required to maintain rotation and the radial load L on the shaft will cause a reaction R at the contact point A. For equilibrium in the vertical direction R must equal L but will not be collinear with it. We see that the force R will be tangent to a small circle of radius r_f called the *friction circle*. The angle between R and its normal component N is the friction angle ϕ. Equating the sum of the moments about O to zero gives

$$M = Rr_f = Rr \sin \phi \qquad (6/4)$$

For a small coefficient of friction the angle ϕ is small, and the sine and tangent may be interchanged with only small error. Since $\mu = \tan \phi$, a good approximation to the torque is

$$M = \mu Rr \qquad (6/4a)$$

This relation gives the amount of torque or moment applied to the shaft which is necessary to overcome friction for a dry or partially lubricated journal bearing.

6/7 THRUST BEARINGS; DISK FRICTION. Friction between circular surfaces under normal pressure is encountered in pivot bearings, clutch plates, and disk brakes. To examine this application we consider the two flat circular disks of Fig. 6/8 whose shafts are mounted in bearings (not shown) so that they can be brought into contact under the axial force P. The maximum torque that this clutch can transmit will be equal to the torque M required to slip one disk against the other. If p is the normal pressure at any location between the plates, the frictional force acting on an elemental area is $\mu p\, dA$,

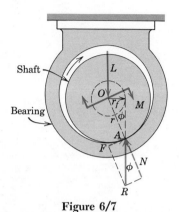

Figure 6/7

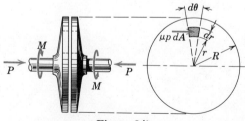

Figure 6/8

where μ is the friction coefficient and dA is the area $r\,dr\,d\theta$ of the element. The moment of this elemental friction force about the shaft axis is $\mu pr\,dA$, and the total moment becomes

$$M = \int \mu pr\,dA$$

where we evaluate the integral over the area of the disk. To carry out this integration we must know the variation of μ and p with r.

In the following examples we shall consider that μ is constant. Furthermore, if the surfaces are new, flat, and well supported, it is reasonable to assume that the pressure p is constant and uniformly distributed so that $\pi R^2 p = P$. Substituting the constant value of p in the expression for M gives us

$$M = \frac{\mu P}{\pi R^2} \int_0^{2\pi} \int_0^R r^2\,dr\,d\theta = \tfrac{2}{3}\mu PR \qquad (6/5)$$

We may interpret this result as being equivalent to the moment due to a friction force μP acting at a distance $\tfrac{2}{3}R$ from the shaft center.

If the friction disks are rings, as in the collar bearing of Fig. 6/9, the limits of integration are the inside and outside radii R_i and R_o, respectively, and the frictional torque becomes

$$M = \tfrac{2}{3}\mu P \frac{R_o{}^3 - R_i{}^3}{R_o{}^2 - R_i{}^2} \qquad (6/5a)$$

After some wear of the surfaces has taken place, we find that the frictional moment decreases somewhat. When the wearing-in period is over, the surfaces retain their new relative shape and further wear is therefore constant over the surface. This wear depends both on the circumferential distance traveled and the pressure p. Since the distance traveled is proportional to r, the expression $rp = K$ may be written, where K is a constant. The value of K is determined from equilibrium of the axial forces or

$$P = \int p\,dA = K \int_0^{2\pi} \int_0^R dr\,d\theta = 2\pi KR$$

With $pr = K = P/(2\pi R)$, we may write the expression for M as

$$M = \int \mu pr\,dA = \frac{\mu P}{2\pi R} \int_0^{2\pi} \int_0^R r\,dr\,d\theta$$

which becomes　　　　　　　　$M = \tfrac{1}{2}\mu PR$ 　　　　　　　(6/6)

The frictional moment for worn-in plates is, therefore, only $(\tfrac{1}{2})/(\tfrac{2}{3})$, or $\tfrac{3}{4}$ as much as for new surfaces.

If the friction disks are rings of inside radius R_i and outside radius R_o, substitution of these limits in the integrations shows that the frictional torque for worn-in surfaces is

$$M = \tfrac{1}{2}\mu P(R_o + R_i) \qquad (6/6a)$$

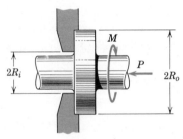

Figure 6/9

Sample Problem 6/8

The bell crank fits over a 100-mm-diameter shaft which is fixed and cannot rotate. The horizontal force T is applied to maintain equilibrium of the crank under the action of the vertical force $P = 100$ N. Determine the maximum and minimum values that T may have without causing the crank to rotate in either direction. The coefficient of static friction μ between the shaft and the bearing surface of the crank is 0.20.

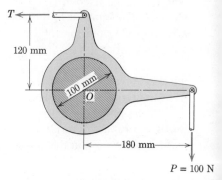

Solution. Impending rotation occurs when the reaction R of the fixed shaft on the bell crank makes an angle $\phi = \tan^{-1}\mu$ with the normal to the bearing surface and is, therefore, tangent to the friction circle. Also, equilibrium requires that the three forces acting on the crank be concurrent at point C. These facts are shown in the free-body diagrams for the two cases of impending motion.

The following calculations are needed:

Friction angle $\phi = \tan^{-1}\mu = \tan^{-1}0.20 = 11.31°$

Radius of friction circle $r_f = r\sin\phi = 50\sin 11.31° = 9.81$ mm

Angle $\theta = \tan^{-1}\dfrac{120}{180} = 33.7°$

Angle $\beta = \sin^{-1}\dfrac{r_f}{OC} = \sin^{-1}\dfrac{9.81}{\sqrt{(120)^2 + (180)^2}} = 2.60°$

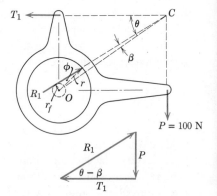

(a) Counterclockwise Motion Impends

(*a*) *Impending counterclockwise motion.* The equilibrium triangle of forces is drawn and gives

$$T_1 = P\,\text{ctn}\,(\theta - \beta) = 100\,\text{ctn}\,(33.7° - 2.60°)$$

$$T_1 = T_{max} = 165.8\ \text{N} \qquad\qquad Ans.$$

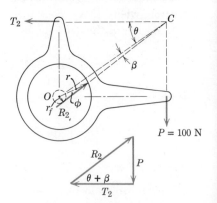

(b) Clockwise Motion Impends

(*b*) *Impending clockwise motion.* The equilibrium triangle of forces for this case gives

$$T_2 = P\,\text{ctn}\,(\theta + \beta) = 100\,\text{ctn}\,(33.7° + 2.60°)$$

$$T_2 = T_{min} = 136.2\ \text{N} \qquad\qquad Ans.$$

PROBLEMS

6/59 The two flywheels are mounted on a common shaft which is supported by a journal bearing between them. Each flywheel weighs 80 lb, and the diameter of the shaft is 1.600 in. If a 25-lb-in. couple M on the shaft is required to maintain rotation of the fly-wheels and shaft at a constant low speed, compute (*a*) the coefficient of friction μ in the bearing and (*b*) the radius r_f of the friction circle.

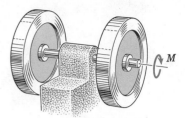

Problem 6/59

6/60 A torque M of 1510 N·m must be applied to the 50-mm-diameter shaft of the hoisting drum to raise the 500-kg load at constant speed. The drum and shaft together have a mass of 100 kg. Calculate the coefficient of friction μ for the bearing.

Ans. $\mu = 0.271$

Problem 6/60

6/61 Calculate the torque M on the shaft of the hoisting drum of Prob. 6/60 that is required to lower the 500-kg load at constant speed. Use the value $\mu = 0.271$ calculated in Prob. 6/60 for the coefficient of friction.

6/62 The front wheels of an experimental rear-drive vehicle have a radius of 300 mm and are equipped with disk-type brakes consisting of a ring A with outside and inside radii of 150 mm and 75 mm, respectively. The ring, which does not turn with the wheel, is forced against the wheel disk with a force P. If the pressure between the ring and the disk wheel is uniform over the mating surfaces, compute the friction force F between each front tire and the horizontal road for an axial force $P = 1$ kN when the vehicle is powered at constant speed with the wheels turning. The coefficient of friction between the disk and ring is 0.35.   *Ans.* $F = 136.1$ N

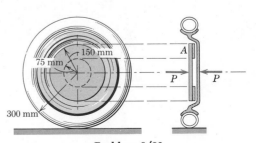

Problem 6/62

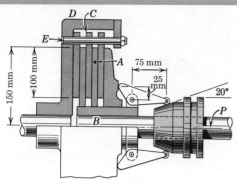

Problem 6/63

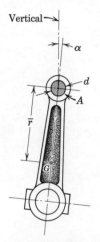

Problem 6/64

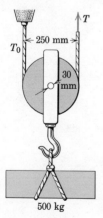

Problem 6/65

6/63 In the figure is shown a multiple-disk clutch for marine use. The driving disks A are splined to the driving shaft B so that they are free to slip along the shaft but must rotate with it. The disks C drive the housing D by means of the bolts E along which they are free to slide. In the clutch shown there are five pairs of friction surfaces. Assume the pressure to be uniformly distributed over the area of the disks and determine the maximum torque M which can be transmitted if the coefficient of friction is 0.15 and $P = 500$ N.   *Ans.* $M = 335$ N·m

6/64 The shaft A fits loosely in the wrist-pin bearing of the connecting rod with center of mass at G as shown. With the rod initially in the vertical position the shaft is rotated slowly until the rod slips at the angle α. Write the exact expression for the coefficient of friction.

6/65 If the coefficient of kinetic friction between the 30-mm-diameter pin and the pulley is 0.25, calculate the tension T required to raise the 500-kg load. Also find the tension T_0 in the stationary part of the cable. Neglect the mass of the pulley.
          Ans. $T = 2.52$ kN, $T_0 = 2.38$ kN

6/66 Calculate the tension T required to lower the load in Prob. 6/65. Also find T_0.

6/67 Each of the four wheels of the vehicle weighs 40 lb and is mounted on a 4-in.-diameter journal (shaft). The total weight of the vehicle is 960 lb, including wheels, and is distributed equally on all four wheels. If a force $P = 16$ lb is required to keep the vehicle rolling at a constant low speed on a horizontal surface, calculate the coefficient of friction which exists in the wheel bearings. (*Hint:* Draw a complete free-body diagram of one wheel.)

Ans. $\mu = 0.204$

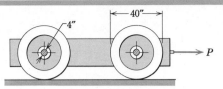

Problem 6/67

6/68 Determine the angle θ with the horizontal made by the steepest slope on which the vehicle of Prob. 6/67 can stand without rolling by itself in the absence of a force P. Take the coefficient of friction in the wheel bearings to be 0.20.

6/69 The linkage is initially at rest under the action of the torques M_1 and M_2. If M_1 is increased gradually until the linkage moves, write the exact expression for the angle α between the resultant compressive force R in link AB and its centerline as motion impends. The coefficient of friction for each bearing is μ.

Ans. $\alpha = \sin^{-1} \dfrac{\mu d}{l\sqrt{1 + \mu^2}}$

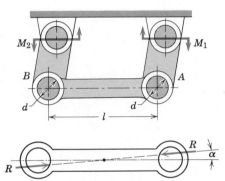

Problem 6/69

6/70 The 20-Mg missile is lowered into its protective silo on a two-screw elevator as shown. Each screw has a mass of 1.07 Mg, is 130 mm in diameter, and has a single square thread with a lead of 13 mm. The screws are turned in synchronism by a motor unit in the base of the silo. The entire mass of the missile, screws, and 4.2-Mg elevator platform is supported equally by flat collar bearings at A, each of which has an outside diameter of 250 mm and an inside diameter of 125 mm. The pressure on the bearings is assumed to be uniformly distributed over the bearing surface. If the coefficient of friction for the collar bearing and the screws at B is 0.15, calculate the torque M which must be applied to each screw (*a*) to raise the elevator and (*b*) to lower the elevator.

Ans. (*a*) $M = 3290$ N·m
(*b*) $M = 2790$ N·m

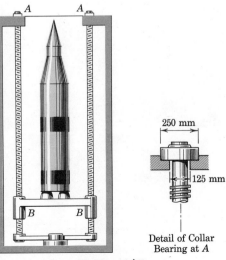

250 mm

125 mm

Detail of Collar
Bearing at A

Problem 6/70

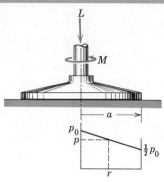

Problem 6/71

6/71 For the flat sanding disk of radius a the pressure p developed between the disk and the sanded surface decreases linearly with r from a value p_0 at the center to $p_0/2$ at $r = a$. If the coefficient of friction is μ, derive the expression for the torque M required to turn the shaft under an axial force L.

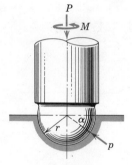

Problem 6/72

6/72 The spherical thrust bearing on the end of the shaft supports an axial load P. Determine the expression for the moment M required to turn the shaft against friction in the bearing. Assume that the pressure p is directly proportional to $\sin \alpha$ and that the coefficient of friction is μ.   *Ans.* $M = \mu Pr$

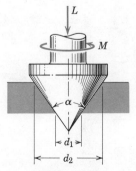

Problem 6/73

6/73 Determine the expression for the torque M required to turn the shaft whose thrust L is supported by a conical pivot bearing. The coefficient of friction is μ, and the bearing pressure is constant.

$$Ans.\ M = \frac{\mu L}{3 \sin \dfrac{\alpha}{2}} \frac{d_2{}^3 - d_1{}^3}{d_2{}^2 - d_1{}^2}$$

6/8 FLEXIBLE BELTS. The impending slippage of flexible members such as belts and ropes over sheaves and drums is of importance in the design of belt drives of all types, band brakes, and hoisting rigs. In Fig. 6/10*a* is shown a drum subjected to the two belt tensions T_1 and T_2, the torque M necessary to prevent rotation, and a bearing reaction R. With M in the direction shown, T_2 is greater than T_1. The free-body diagram of an element of the belt of length $r\,d\theta$ is shown in the *b*-part of the figure. We proceed with the force analysis of this element in a manner similar to that which we have illustrated for other variable-force problems where the equilibrium of a differential element is established. The tension increases from T at the angle θ to $T + dT$ at the angle $\theta + d\theta$. The normal force is a differential dN, since it acts on a differential element of area. Likewise the friction force, which must act on the belt in a direction to oppose slipping, is a differential and is $\mu\,dN$ for impending motion. Equilibrium in the t-direction gives

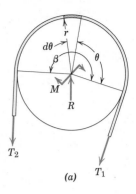

$$T \cos \frac{d\theta}{2} + \mu\,dN = (T + dT) \cos \frac{d\theta}{2}$$

or

$$\mu\,dN = dT$$

since the cosine of a differential quantity is unity. Equilibrium in the n-direction requires that

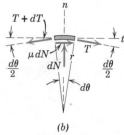

$$dN = (T + dT) \sin \frac{d\theta}{2} + T \sin \frac{d\theta}{2}$$

or

$$dN = T\,d\theta$$

In this reduction we recall that the sine of a differential angle equals the angle and that the product of two differentials must be neglected in the limit compared with the first-order differentials remaining. Combining the two equilibrium relations gives

$$\frac{dT}{T} = \mu\,d\theta$$

Integrating between corresponding limits yields

$$\int_{T_1}^{T_2} \frac{dT}{T} = \int_0^{\beta} \mu\,d\theta$$

or

$$\ln \frac{T_2}{T_1} = \mu\beta$$

where the $\ln (T_2/T_1)$ is a natural logarithm (base e). Solving for T_2 gives

$$T_2 = T_1 e^{\mu\beta} \qquad\qquad (6/7)$$

(a) T_1

(b)

Figure 6/10

We note that β is the total angle of belt contact and must be expressed in radians. If a rope were wrapped around a drum n times, the angle β would be $2\pi n$ radians. Equation 6/7 holds equally well for a noncircular section where the total angle of contact is β. This conclusion is evident from the fact that the radius r of the circular drum of Fig. 6/10 does not enter into the equations for the equilibrium of the differential element of the belt.

The relation expressed by Eq. 6/7 also applies to belt drives where both the belt and the pulley are rotating at constant speed. In this case the equation describes the ratio of belt tensions for slipping or impending slipping. When the speed of rotation becomes large, there is a tendency for the belt to leave the rim, so that Eq. 6/7 will involve some error.

6/9 ROLLING RESISTANCE.

Deformation at the point of contact between a rolling wheel and its supporting surface introduces a resistance to rolling which we will mention only briefly. This resistance is not due to tangential friction forces and therefore is an entirely different phenomenon from that of dry friction.

To describe rolling resistance we consider the wheel of Fig. 6/11 under the action of a load L on the axle and a force P applied at its center to produce rolling. The deformation of the wheel and supporting surfaces as shown is greatly exaggerated. The distribution of pressure p over the area of contact is similar to that indicated, and the resultant R of this distribution will act at some point A and will pass through the center of the wheel for equilibrium. We find the force P necessary to initiate and maintain rolling at constant velocity by equating the moments of all forces about A to zero. This gives us

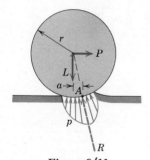

Figure 6/11

$$P = \frac{a}{r}L = \mu_r L$$

where the moment arm of P is taken to be r. The ratio $\mu_r = a/r$ is referred to as the coefficient of rolling resistance. The coefficient as defined is the ratio of resisting force to normal force, and in this respect is analogous to the coefficient of static or kinetic friction. On the other hand there is no slipping or impending slipping in the interpretation of μ_r.

The dimension a depends on many factors which are difficult to quantify, so that a comprehensive theory of rolling resistance is not available. The distance a is a function of the elastic and plastic properties of the mating materials, the radius of the wheel, the speed of travel, and the roughness of the surfaces. Some tests indicate only a small variation with wheel radius, and a is often taken to be independent of the rolling radius. Unfortunately, the quantity a has also been referred to as the coefficient of rolling friction in some references. However, a has the dimension of length and therefore is not a dimensionless coefficient in the usual sense.

Sample Problem   6/9

A flexible cable which supports the 100-kg load is passed over a fixed circular drum and subjected to a force P to maintain equilibrium. The coefficient of static friction μ between the cable and the fixed drum is 0.30. (*a*) For $\alpha = 0$, determine the maximum and minimum values which P may have in order not to raise or lower the load. (*b*) For $P = 500$ N, determine the minimum value which the angle α may have before the load begins to slip.

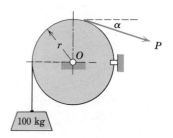

Solution.   Impending slipping of the cable over the fixed drum is given by Eq. 6/7 which is $T_2/T_1 = e^{\mu\beta}$.

(*a*) With $\alpha = 0$ the angle of contact is $\beta = \pi/2$ rad. For impending upward motion of the load, $T_2 = P_{max}$, $T_1 = 981$ N, and we have

$$P_{max}/981 = e^{0.30(\pi/2)}, \qquad P_{max} = 981(1.602) = 1572 \text{ N} \qquad Ans.$$

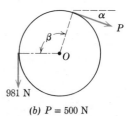

(*a*) $\alpha = 0$

For impending downward motion of the load, $T_2 = 981$ N, $T_1 = P_{min}$. Thus

$$981/P_{min} = e^{0.30(\pi/2)}, \qquad P_{min} = 981/1.602 = 612 \text{ N} \qquad Ans.$$

(*b*) With $T_2 = 981$ N and $T_1 = P = 500$ N, Eq. 6/7 gives us

$$981/500 = e^{0.30\beta}, \quad 0.30\beta = \ln(981/500) = 0.674$$

$$\beta = 2.247 \text{ rad} \quad \text{or} \quad \beta = 2.247\left(\frac{360}{2\pi}\right) = 128.7°$$

$$\alpha = 128.7° - 90° = 38.7° \qquad Ans.$$

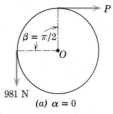

(*b*) $P = 500$ N

① We are careful to note that β must be expressed in radians.

② In our derivation of Eq. 6/7 be certain to note that $T_2 > T_1$.

③ As was noted in the derivation of Eq. 6/7 the radius of the drum does not enter into the calculations. It is only the angle of contact and the coefficient of friction that determine the limiting conditions for impending motion of the flexible cable over the curved surface.

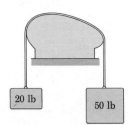

Problem 6/74

Problem 6/76

Problem 6/79

PROBLEMS

6/74 A force $P = 5mg$ is required to raise the load of mass m with the cord making $1\frac{1}{4}$ turns around the fixed shaft. Calculate the coefficient of friction μ between the cord and the shaft.   *Ans.* $\mu = 0.205$

6/75 For a given coefficient of friction and a given number of turns around the shaft for the configuration of Prob. 6/74, a force P of 3 kN is required to raise m and a force P of 0.48 kN is required to lower m. Find m.

6/76 The 20-lb block and the 50-lb block are connected by a rope hung over the fixed curved surface. If the system is on the verge of slipping, calculate the coefficient of friction between the rope and the surface.   *Ans.* $\mu = 0.292$

6/77 A dockworker adjusts a spring line (rope) which keeps a small ship from drifting alongside a wharf. What force T can be supported in the line if he exerts a pull of 240 N on the free end with (*a*) one complete turn around the mooring bit as shown and (*b*) two complete turns? The coefficient of friction between the hemp rope and the cast-steel mooring bit is 0.30.

6/78 If the dockworker of Prob. 6/77 is to support a tension of 16 kN in the rope leading to the ship, how many turns around the mooring bit are necessary? The coefficient of friction between the rope and the bit is 0.30.   *Ans.* $n = 2.23$ turns

6/79 The 180-lb tree surgeon lowers himself with the rope over a horizontal limb of the tree. If the coefficient of friction between the rope and the limb is 0.50, compute the force which the man must exert on the rope to let himself down slowly.

$T \leftarrow$ 240 N $\rightarrow$

Problem 6/77

/80 The tape slides around the two fixed pegs as shown and is under the action of the horizontal tensions $T_1 = 40$ N and $T_2 = 160$ N. Determine the coefficient of friction μ between the tape and the pegs.
Ans. $\mu = 0.313$

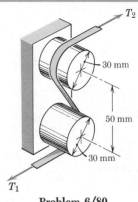

Problem 6/80

/81 The worker hauls a 100-kg log up the 30° ramp by means of the power winch which turns in the direction shown. If the worker exerts a horizontal pull of 160 N on the free end of the rope, compute the coefficient of friction μ_1 between the log and the ramp. The coefficient of friction between the rope and the drum is 0.25.

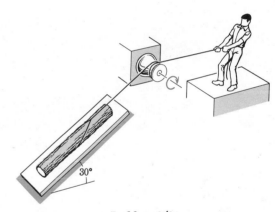

Problem 6/81

/82 A counterclockwise moment $M = 1500$ lb-in. is applied to the flywheel. If the coefficient of friction between the band and the wheel is 0.20, compute the minimum force P necessary to prevent the wheel from rotating.   *Ans.* $P = 142.0$ lb

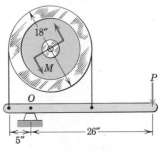

Problem 6/82

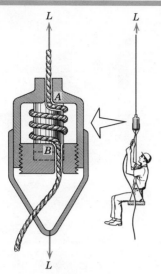

Problem 6/83

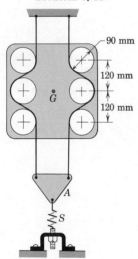

Problem 6/84

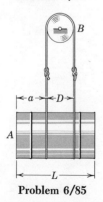

Problem 6/85

6/83 A device for lowering a person in a sling down a rope at a constant controlled rate is shown in the figure. The rope passes around a central shaft fixed to the frame and leads freely out of the lower collar. The number of turns is adjusted by turning the lower collar, which winds or unwinds the rope around the shaft. Entrance of the rope into the upper collar at A is equivalent to $\frac{1}{4}$ of a turn, and passage around the corner at B is also equivalent to $\frac{1}{4}$ of a turn. Friction of the rope through the straight portions of the collars averages 10 N for each collar. If three complete turns around the shaft, in addition to the corner turns, are required for a 75-kg man to lower himself at a constant rate without exerting a pull on the free end of the rope, calculate the coefficient of friction μ between the rope and the contact surfaces of the device. Neglect the small helix angle of the rope around the shaft.

6/84 The assembly shown has a mass of 100 kg with center of mass at G and is supported by the two wires, which pass around the fixed pegs and are kept under equal tensions by the equalizer plate A and the adjustable spring S. Calculate the minimum tension T in the spring which will ensure that the assembly remains suspended as shown. The coefficient of friction between the wires and the pegs is 0.30.   *Ans. T = 555 N*

6/85 The uniform drum A with center of mass at midlength is suspended by a rope which passes over the fixed cylindrical surface B. The coefficient of static friction between the rope and the surface over which it passes is μ. Determine the maximum value that the dimension a may have before the drum tips out of its horizontal position.

/86 Find the couple M required to turn the pipe in the V-block against the action of the flexible band. A force $P = 25$ lb is applied to the lever which is pivoted about O. The coefficient of friction between the band and the pipe is 0.30, and that between the pipe and the block is 0.40. The weights of the parts are negligible.   *Ans.* $M = 1834$ lb-in.

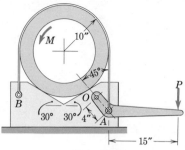
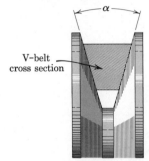

Problem 6/86

/87 Replace the flat belt and pulley of Fig. 6/10 by a V-belt and matching grooved pulley as indicated by the cross-sectional view accompanying this problem. Derive the relation among the belt tensions, the angle of contact, and the coefficient of friction for the V-belt when slipping impends. Use of a V-belt with $\alpha = 35°$ would be equivalent to increasing the coefficient of friction for a flat belt of the same material by what factor n?

$$Ans. \ T_2 = T_1 e^{\mu \beta'} \ \text{where} \ \beta' = \frac{\beta}{\sin(\alpha/2)}$$

$$n = 3.33$$

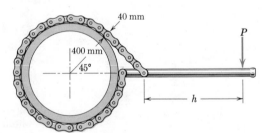

Problem 6/87

6/88 The roller chain is used as a pipe grip. If the coefficient of friction between the chain and the fixed pipe is 0.25, determine the minimum value of h which will ensure that the grip will not slip on the pipe regardless of P. Neglect the masses of the chain and handle, and neglect any friction between the left end of the handle and the pipe.
Ans. $h = 96.9$ mm

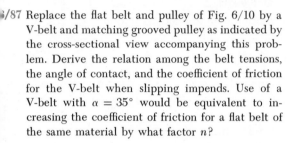

Problem 6/88

6/89 The chain has a mass ρ per unit length. Determine the overhang h below the fixed cylindrical guide for which the chain will be on the verge of slipping. The coefficient of friction is μ. (*Hint:* The resulting differential equation involving the variable chain tension T at the corresponding angle θ is of the form $dT/d\theta + KT = f(\theta)$, a first-order, linear, nonhomogeneous equation with constant coefficient. The solution is

$$T = Ce^{-K\theta} + e^{-K\theta} \int e^{K\theta} f(\theta) \, d\theta$$

where C and K are constants.)

$$Ans. \ h = \frac{2\mu r}{1 + \mu^2}(1 + e^{\mu \pi})$$

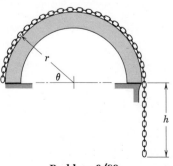

Problem 6/89

6/10 **PROBLEM FORMULATION AND REVIEW.** In our study of friction we have concentrated our attention on dry or Coulomb friction where a simple mechanical model of surface irregularities between the contacting bodies, Fig. 6/1, suffices to explain the phenomenon adequately for most engineering purposes. By having this model clearly in mind we can easily visualize the three types of dry-friction problems which are encountered. These categories are:

1. Static friction less than the maximum possible value and determined by the equations of equilibrium. (Usually requires a check to see that $F < \mu_s N$)
2. Limiting static friction with impending motion ($F_{max} = \mu_s N$)
3. Kinetic friction where sliding motion occurs between contacting surfaces ($F = \mu_k N$)

A coefficient of friction applies to a given pair of mating surfaces. It is meaningless to speak of a coefficient of friction for a single surface. The static coefficient of friction μ_s for a given pair of surfaces is usually greater than the kinetic coefficient μ_k. We generally drop the subscript and understand from the problem at hand which of the two coefficients applies. The friction force which acts on a body is always in the direction to oppose the slipping of the body which takes place or the slipping which would take place in the absence of friction.

When we encounter friction forces distributed in some prescribed manner over a surface or along a line, we select a representative element of the surface or line and evaluate the force and moment effects of the elemental friction force acting on the element. We then integrate these effects over the entire surface or line.

Friction coefficients are subject to considerable variation depending on the exact condition of the mating surfaces. Computation of coefficients of friction to three significant figures represents an accuracy which cannot easily be duplicated by experiment and when cited are included for purposes of computational check only. For design computations in engineering practice the use of a handbook value for a coefficient of static or kinetic friction must be viewed only as an approximation.

In reviewing the foregoing introduction to frictional problems the reader should bear in mind the existence of the other forms of friction mentioned in the introductory article of the chapter. Problems which involve fluid friction, for example, are among the most important of the friction problems encountered in engineering, and a study of this phenomenon is included in the subject of fluid mechanics.

REVIEW PROBLEMS

6/90 A frictional locking device allows bar A to move to the left but prevents movement to the right. If the coefficient of friction between the shoe B and the bar A is 0.40, specify the maximum length b of the link which will permit the device to work as described.   *Ans.* $b = 118.5$ mm

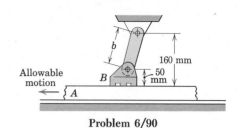

Problem 6/90

6/91 A sliding element of a machine tool consists of a bar with a V-section that is supported by the grooved block. The bar supports a vertical load w per unit of its length. If the coefficient of friction is μ, determine the horizontal force P required to move the bar.

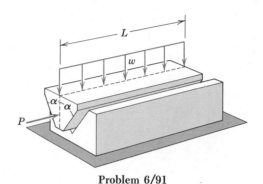

Problem 6/91

6/92 The coefficients of static and kinetic friction between the 100-kg block and the inclined plane are 0.30 and 0.20, respectively. Determine (*a*) the friction force F acting on the block when P is applied with a magnitude of 200 N to the block at rest, (*b*) the force P required to initiate motion up the incline from rest, and (*c*) the friction force F acting on the block if $P = 600$ N.   *Ans.* (*a*) $F = 66.0$ N
(*b*) $P = 516$ N
(*c*) $F = 148$ N

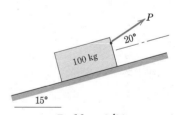

Problem 6/92

6/93 A 3200-lb car with 120-in. wheelbase has a center of gravity 2 ft from the road and midway between the front and rear axles. If the coefficient of friction between the tires and the road is 0.80, find the angle θ with the horizontal made by the steepest grade that the car can climb at constant speed before the rear driving wheels slip. What torque M is applied to each of the 26-in.-diameter rear wheels by the engine under these conditions? Neglect any friction under the front wheels.

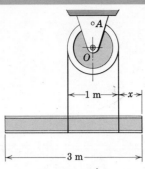

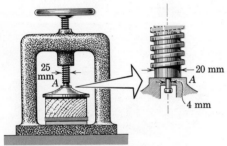

Problem 6/94

Problem 6/95

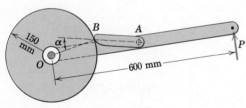

Problem 6/96

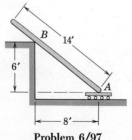

Problem 6/97

6/94 The uniform 3 m beam is suspended by the cable that passes over the large pulley. A locking pin at A prevents rotation of the pulley. If the coefficient of friction between the cable and the pulley is 0.25, determine the minimum value of x for which the cable will not slip on the pulley.

Ans. $x = 0.343$ m

6/95 The screw of the small press has a mean diameter of 25 mm and has a double square thread with a lead of 8 mm. The flat thrust bearing at A is shown in the enlarged view and has surfaces which are well worn. If the coefficient of friction for both the threads and the bearing at A is 0.25, calculate the torque M on the handwheel required (*a*) to produce a compressive force of 4 kN and (*b*) to loosen the press from the 4-kN compression.

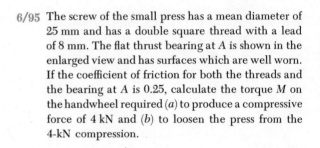

6/96 The figure shows a friction silent ratchet for turning the wheel counterclockwise under the action of a force P applied to the handle. The floating link AB engages the wheel so that $\alpha = 20°$. If a force $P = 150$ N is required to turn the wheel about its bearing at O, determine (*a*) the minimum coefficient of friction between the link and the wheel that will ensure no slipping at B and (*b*) the magnitude R of the force on the pin at A.

Ans. (*a*) $\mu_{\min} = 0.364$, (*b*) $R = 1.754$ kN

6/97 The lower end A of the uniform 100-lb plank rests on rollers that are free to move on the horizontal surface. If the coefficients of static and kinetic friction between the plank and the corner B are 0.80 and 0.70, respectively, compute the friction force F acting at B if the plank is released from rest in the position shown.

6/98 Calculate the torque M which the engine must supply to the rear axle of the car to roll it over the curbing from a rest position if the rear wheels do not slip. Determine the minimum coefficient of friction at the rear wheels to prevent slipping. The car has a mass of 1200 kg.

$Ans.$ $M = 2.02$ kN·m, $\mu_{min} = 1.00$

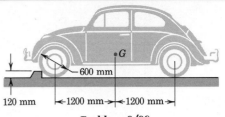

Problem 6/98

6/99 The coefficient of friction between the collar of the drill-press table and the vertical column is 0.30. Will the collar and table slide down the column under the action of the drill thrust if the operator forgets to secure the clamp, or will friction be sufficient to hold it in place? Neglect the weight of the table and collar compared with the drill thrust and assume that contact occurs at the points A and B.

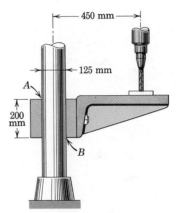

Problem 6/99

6/100 The two 10° wedges are positioned so that a downward force P on the one wedge will result in an elevation of the 1200-lb load. The coefficient of friction for all sliding surfaces is 0.20, and the weights of the wedges are negligible. Determine P.

$Ans.$ $P = 492$ lb

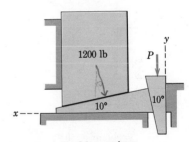

Problem 6/100

6/101 In the figure are shown the elements of a rolling mill. Determine the maximum thickness b which the slab may have and still enter the rolls by means of the friction between the slab and the rolls. The coefficient of friction is μ. (*Hint:* The critical condition for entering occurs when the resultant of the horizontal forces on the slab is zero.)

$$Ans. \ b = a + d\frac{\sqrt{1 + \mu^2} - 1}{\sqrt{1 + \mu^2}}$$

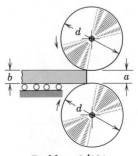

Problem 6/101

Problem 6/102

6/102 Determine the force P that will begin to rotate the cylinder of mass m against the action of friction. The coefficient of friction for both pairs of contacting surfaces is μ.

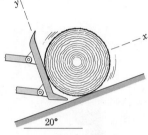

Problem 6/103

6/103 A bulldozer moves the 750-kg log up the 20° incline by pushing with the blade, which is normal to the incline. If the coefficient of friction between the blade and the log is 0.50 and that between the log and the ground is 0.80, calculate the force component P, normal to the blade, that must be exerted against the log. *Ans. $P = 5.03$ kN*

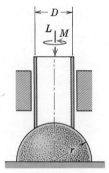

Problem 6/104

6/104 The thin-walled tubular shaft of mean diameter D rotates about the vertical axis under an axial load L and bears against a fixed spherical surface of radius r. If the coefficient of friction between the tube and the spherical surface is μ, write the expression for the moment M on the shaft necessary to overcome friction at the support.

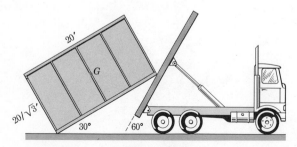

Problem 6/105

6/105 The truck unloads its cargo box by sliding it slowly off the elevated rack. The box has a total weight of 10,000 lb with center of mass at G in the center of the box. The coefficient of friction between the box and the rack is 0.30. Calculate the braking force F between the tires and the level road as the box is on the verge of slipping down the rack from the position shown and the truck is on the verge of rolling forward. No slipping occurs at the lower corner of the box. *Ans. $F = 2034$ lb*

106 A force of 1 kN is developed in the hydraulic cylinder *C* to activate the block brake. If the coefficient of static friction between the blocks and the rim of the wheel is 0.60, compute the maximum torque *M* which can be applied to the wheel without causing rotation. The wheel is mounted in a fixed bearing at its center. Assume that the forces between the blocks and the wheel act at the centers of the contact faces of the blocks. *Ans. M* = 558 N·m

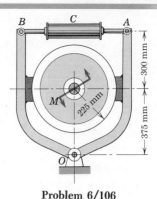

Problem 6/106

107 The two brake shoes and their lining pivot about the points *O* and are expanded against the brake drum through the action of the hydraulic cylinder *C*. The pressure *p* between the drum and the lining may be shown to vary directly as the sine of the angle *θ* measured from the pin *O* for each shoe and has a value p_0 at *θ* = *β*. The width of the lining in contact with the drum is *b*. Write the expression for the braking torque M_f on the wheel if the coefficient of friction between the drum and the lining is *μ*.

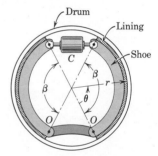

Problem 6/107

108 The friction tongs are designed to lift 1000-lb crates with a nominal width of 48 in. From the configuration of the links determine whether slippage is more likely for crates that are slightly wider (with contact at *A*) or slightly narrower (with contact at *B*) than the nominal size. Determine the minimum coefficient of friction *μ* between the tongs and the crate that will prevent slippage for the case where slippage is more likely, and calculate the corresponding tension *T* in the horizontal chain that connects the jaws of the tongs.

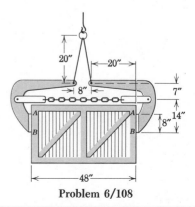

Problem 6/108

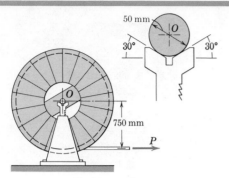

Problem 6/109

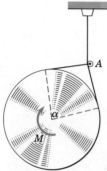

Problem 6/110

6/109 The reel of telephone cable has a mass of 3 Mg and is supported on its shaft in the V-notched blocks on both sides of the reel. The reel is raised off the ground by jacking up the supports so that cable may be pulled off in the horizontal direction as shown. The shaft is fastened to the reel and turns with it. If the coefficient of friction between the shaft and the V-surfaces is 0.30, calculate the pull P in the cable required to turn the reel.   *Ans.* $P = 313$ N

▶6/110 A light flexible cord is passed around the circular disk of mass m and ends in a small pulley that is free to find its equilibrium position on the cord. If the coefficient of friction between the cord and the disk is 0.50, determine the angle α between the normals to the cord at the tangency points for the position where the disk is on the verge of turning under the action of a couple M applied to the disk. (*Suggestion:* Solve the resulting equation for α graphically.)
Ans. $\alpha = 87.3°$

VIRTUAL
WORK

7

7/1 INTRODUCTION.

In the previous chapters we have analyzed the equilibrium of a body by isolating it with a free-body diagram and writing the zero-force and zero-moment summation equations. For the most part this approach has been employed for a body whose equilibrium position was known or specified and where one or more of the external forces was an unknown to be determined.

There is a separate class of problems in which bodies are composed of interconnected members that allow relative motion between the parts, thus permitting various possible equilibrium configurations to be examined. For problems of this type, the force- and moment-equilibrium equations, although valid and adequate, are often not the most direct and convenient approach. Here we find that a method based on the concept of the work done by a force is more useful and direct. Also, the method provides a deeper insight into the behavior of mechanical systems and allows us to examine the question of the stability of systems in equilibrium. We will now develop this approach, called the *method of virtual work*.

7/2 WORK.

We must first define the term work, which is used in a quantitative sense as contrasted to its common nontechnical usage.

(a) *Work of a Force.* Consider the constant force $\mathbf{F}$ acting on the body, Fig. 7/1a, whose movement along the plane from A to A' is represented by the vector $\Delta \mathbf{s}$, called the *displacement* of the body. By definition the work U done by the force $\mathbf{F}$ on the body during this displacement is the component of the force in the direction of the displacement times the displacement or

$$U = (F \cos \alpha)\, \Delta s$$

From Fig. 7/1b we see that the same result is obtained if we multiply the magnitude of the force by the component of the displacement in the direction of the force which is

$$U = F(\Delta s \cos \alpha)$$

Since we obtain the same result regardless of which direction we resolve the vectors, we observe immediately that work U is a scalar quantity.

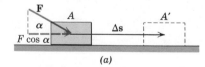

(a)

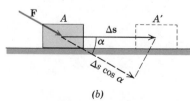

(b)

Figure 7/1

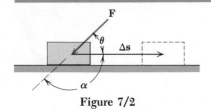

Figure 7/2

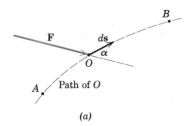

(a)

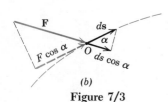

(b)

Figure 7/3

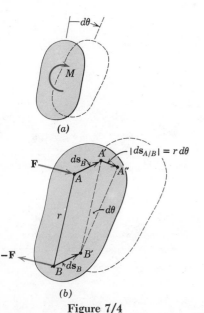

(a)

(b)

Figure 7/4

Work is a positive quantity when the working component of the force is in the same direction as the displacement. When the working component is in the direction opposite to the displacement, Fig. 7/2, the work done is negative. Thus

$$U = (F \cos \alpha) \, \Delta s = -(F \cos \theta) \, \Delta s$$

We now generalize the definition of work to account for conditions under which the direction of the displacement and the magnitude and direction of the force may be variable. Figure 7/3a shows a force **F** acting on a body at a point O which moves along the path shown. During an infinitesimal displacement ds along the path the work done by **F** is

$$dU = F \cos \alpha \, ds \qquad \text{or} \qquad dU = \mathbf{F} \cdot d\mathbf{s} \qquad (7/1)$$

where the dot product of **F** and $d\mathbf{s}$ fits the definition of work. Again, we may interpret this expression as the force component $F \cos \alpha$ in the direction of the displacement times the displacement, or as the displacement component $ds \cos \alpha$ in the direction of the force times the force, as represented in Fig. 7/3b. If we express **F** and $d\mathbf{s}$ in terms of their rectangular components, we have

$$dU = (\mathbf{i}F_x + \mathbf{j}F_y + \mathbf{k}F_z) \cdot (\mathbf{i} \, dx + \mathbf{j} \, dy + \mathbf{k} \, dz)$$

$$= F_x \, dx + F_y \, dy + F_z \, dz$$

To obtain the total work U done by **F** during a finite movement of point O from A to B, Fig. 7/3a, we must integrate dU between these positions. Thus

$$U = \int \mathbf{F} \cdot d\mathbf{s} = \int (F_x \, dx + F_y \, dy + F_z \, dz)$$

or

$$U = \int F \cos \alpha \, ds$$

In order to carry out this integration we must know the relation between the force components and their respective coordinates or the relations between F and s and between $\cos \alpha$ and s.

In the case of concurrent forces acting on a body the work done by their resultant equals the total work done by the several forces. This we conclude from the fact that the component of the resultant in the direction of the displacement equals the sum of the components of the several forces in the same direction.

(b) Work of a Couple. In addition to the work done by forces, couples also may do work. In Fig. 7/4a we have a couple M acting on a body that changes its angular position by an amount $d\theta$. The work done by the couple is easily determined from the combined work of each of the two forces which constitute the couple. In the b-part of the figure we represent the couple by two equal and opposite forces **F** and $-\mathbf{F}$ acting at points A and B such that $M = Fr$. During the

infinitesimal movement in the plane of the figure, line AB moves to $A''B'$. We now take the displacement of A in two steps, first, a displacement ds_B equal to that of B and, second, a displacement $ds_{A/B}$ (read the displacement of A with respect to B) due to the rotation about B. We see that the work done by $\mathbf{F}$ during the displacement from A to A' is equal and opposite to that due to $-\mathbf{F}$ acting through the equal displacement from B to B'. Thus we conclude that no work is done by a couple during a translation (movement without rotation). During the rotation, however, $\mathbf{F}$ does work equal to $\mathbf{F} \cdot ds_{A/B} = Fr\, d\theta$ where $d\theta$ is the infinitesimal angle of rotation in radians. Since $M = Fr$ we have

$$\boxed{dU = M\, d\theta} \tag{7/2}$$

The work of the couple is positive if M has the same sense as $d\theta$ (clockwise in this illustration) and negative if M has a sense opposite to that of the rotation. The total work of a couple during a finite rotation in its plane becomes

$$U = \int M\, d\theta$$

Work has the dimensions of (force) $\times$ (distance). In SI units the unit of work is the joule (J), which is the work done by a force of one newton moving through a distance of one meter in the direction of the force ($\mathrm{J} = \mathrm{N \cdot m}$). In the U.S. customary system the unit of work is the foot-pound (ft-lb), which is the work done by a one-pound force moving through a distance of one foot in the direction of the force. Dimensionally, the work of a force and the moment of a force are the same although they are entirely different physical quantities. We observe carefully that work is a scalar given by the dot product and involves the product of a force and a distance both measured along the same line. Moment, on the other hand, is a vector given by the cross product and involves the product of force and distance measured at right angles to the force. In order to distinguish between these two quantities when we write their units, in SI units we shall use the joule (J) for work and reserve the combined units newton-meter ($\mathrm{N \cdot m}$) for moment. In the U.S. customary system we normally use the sequence foot-pound (ft-lb) for work and pound-foot (lb-ft) for moment.

(c) *Virtual Work.* We consider now a particle whose static equilibrium position is determined by the forces which act upon it. Any assumed and arbitrary small displacement δs away from this natural position is called a *virtual displacement*. The term *virtual* is used to indicate that the displacement does not exist in reality but is only assumed so that we may compare various possible equilibrium positions in the process of selecting the correct one. The work done by any force **F** acting on the particle during the virtual displacement δs is called *virtual work* and is

$$\delta U = \mathbf{F} \cdot \delta \mathbf{s} \qquad \text{or} \qquad \delta U = F \, \delta s \cos \alpha$$

where α is the angle between **F** and δs. The difference between ds and δs is that ds refers to an infinitesimal change in an actual movement and can be integrated, whereas δs refers to an infinitesimal virtual or assumed movement and cannot be integrated. Mathematically both quantities are first-order differentials.

A virtual displacement may also be a rotation $\delta \theta$ of a body. The virtual work done by a couple M during a virtual angular displacement $\delta \theta$ is, then, $\delta U = M \, \delta \theta$.

We may regard the force **F** or couple M as remaining constant during any infinitesimal virtual displacement. If we account for any change in **F** or M during the infinitesimal motion, higher-order terms will result which disappear in the limit. This consideration is the same mathematically as that which permits us to neglect the product $dx \, dy$ when writing $dA = y \, dx$ for the element of area under the curve $y = f(x)$.

7/3 EQUILIBRIUM.

We now express the equilibrium conditions in terms of virtual work, first for a particle, second for a single rigid body, and third for a system of connected rigid bodies.

(*a*) *Particle.* Consider the particle or small body in Fig. 7/5 which finds its equilibrium position as a result of the forces in the springs to which it is attached. If the mass of the particle is significant, then the weight mg would also be included as one of the forces. For an assumed virtual displacement δs of the particle away from its equilibrium position the total virtual work done on the particle will be

$$\delta U = \mathbf{F}_1 \cdot \delta \mathbf{s} + \mathbf{F}_2 \cdot \delta \mathbf{s} + \mathbf{F}_3 \cdot \delta \mathbf{s} + \cdots = \Sigma \mathbf{F} \cdot \delta \mathbf{s}$$

We now express $\Sigma \mathbf{F}$ in terms of its scalar sums and δs in terms of its component virtual displacements in the coordinate directions and write

$$\delta U = \Sigma \mathbf{F} \cdot \delta \mathbf{s} = (\mathbf{i} \Sigma F_x + \mathbf{j} \Sigma F_y + \mathbf{k} \Sigma F_z) \cdot (\mathbf{i} \, \delta x + \mathbf{j} \, \delta y + \mathbf{k} \, \delta z)$$

$$= \Sigma F_x \, \delta x + \Sigma F_y \, \delta y + \Sigma F_z \, \delta z = 0$$

The sum is zero, since $\Sigma \mathbf{F} = \mathbf{0}$ and also $\Sigma F_x = 0$, $\Sigma F_y = 0$, and $\Sigma F_z = 0$. We see therefore that the equation $\delta U = 0$ is an alternative statement of the equilibrium conditions for a particle. This condition of zero virtual work for equilibrium is both necessary and sufficient, since we may apply it to virtual displacements taken one at a time in each of the three mutually perpendicular directions in which case it becomes equivalent to the three known scalar requirements for equilibrium.

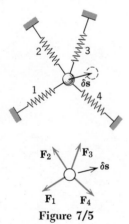

Figure 7/5

   The principle of zero virtual work for the equilibrium of a single particle usually does not simplify this already simple problem since $\delta U = 0$ and $\Sigma \mathbf{F} = 0$ provide the same information. The concept of virtual work for a particle is introduced so that it may be applied to systems of particles in the development that follows.

   (*b*). *Rigid Body.* We easily extend the principle of virtual work for a particle to a system of rigidly connected particles which form our model of a rigid body. Since the virtual work done on each particle of the body in equilibrium is zero, it follows that the virtual work done on the entire rigid body is zero. Only the virtual work done by *external* forces appears in the evaluation of $\delta U = 0$ for the entire body, since the internal forces occur in pairs of equal, opposite, and collinear forces and the net work done by these forces during any movement is zero. This net work is zero because the displacement components of the two particles along the lines of action of the forces are identical for a rigid body.

   Again, as in the case of a particle, we find that the principle of virtual work offers no particular advantage to the solution for a single rigid body in equilibrium. Any assumed virtual displacement defined by a linear or angular movement will appear in each term in $\delta U = 0$ and when canceled will leave us with the same expression as we would have obtained by using one of the force or moment equations of equilibrium directly. This condition is illustrated in Fig. 7/6, where we are asked to determine the reaction R under the roller for the hinged plate of negligible weight under the action of a given force P. A small assumed rotation $\delta\theta$ of the plate about O is consistent with the hinge constraint at O and is taken as the virtual displacement. The work done by P is $-Pa\,\delta\theta$, and the work done by R is $+Rb\,\delta\theta$. Therefore the principle $\delta U = 0$ gives

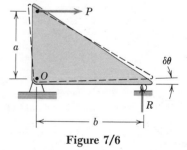

Figure 7/6

$$-Pa\,\delta\theta + Rb\,\delta\theta = 0$$

Canceling out $\delta\theta$ leaves

$$Pa - Rb = 0$$

which is simply the equation of moment equilibrium about O. Therefore nothing is gained by the use of the virtual-work principle for a single rigid body. Use of the principle will, however, provide us with a decided advantage for interconnected bodies as described in the next section.

   (*c*) *Systems of Rigid Bodies.*   We now extend the principle of virtual work to describe the equilibrium of an interconnected system of rigid bodies. Our treatment will be limited to so-called *ideal* systems, which are systems composed of two or more rigid members linked together by mechanical connections which are assumed to be frictionless and incapable of absorbing energy through elongation or compression. Figure 7/7*a* shows a simple example of an ideal system where motion between its two parts is possible and where the equilibrium position is determined by the applied external forces **P** and **F**.

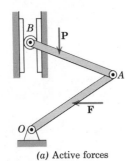

(a) Active forces

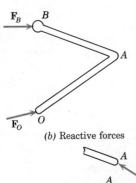

(b) Reactive forces

(c) Internal forces

Figure 7/7

For such an interconnected mechanical system we identify three types of forces which act. They are:

(1) *Active forces* are external forces capable of doing virtual work during possible virtual displacements. In Fig. 7/7*a* forces **P** and **F** are active forces since they would do work as the links move.

(2) *Reactive forces* are forces which act at positions of fixed support where no virtual displacement in the direction of the force takes place. Reactive forces do no work during a virtual displacement. In Fig. 7/7*b* the horizontal force $\mathbf{F}_B$ exerted on the roller end of the member by the vertical guide can do no work since there can be no horizontal displacement of the roller. The force $\mathbf{F}_O$ exerted on the system by the fixed support at O is also a nonworking reactive force since no displacement of O takes place.

(3) *Internal forces* are forces in the connections between members. During any possible movement of the system or its parts we see that the *net work done by the internal forces at the connections is zero*. This is so because the internal forces always exist in pairs of equal and opposite forces, as indicated for the internal forces $\mathbf{F}_A$ and $-\mathbf{F}_A$ at joint A in Fig. 7/7*c*, and the work of one force necessarily cancels the work of the other force during their identical displacements.

With the observation that only the external active forces do work during any possible movement of the system, we may now state the principle of virtual work as follows:

The virtual work done by external active forces on an ideal mechanical system in equilibrium is zero for any and all virtual displacements consistent with the constraints.

By constraint we mean restriction of the motion by the supports. In this form the principle finds its greatest use for ideal systems. We state the principle mathematically by the equation

$$\delta U = 0 \qquad\qquad (7/3)$$

where δU stands for the total virtual work done on the system by all active forces during a virtual displacement.

Only now can we see the real advantages of the method of virtual work. There are essentially two. First, it is not necessary for us to dismember ideal systems in order to establish the relations between the active forces, as is generally the case with the equilibrium method based on force and moment summations. Second, we may determine the relations between the active forces directly without reference to the reactive forces. These advantages make the method of virtual work particularly useful in determining the position of equilibrium of a system under known loads. This type of problem is in contrast with the problem of determining the forces acting on a body whose equilibrium position is fixed or specified.

The method of virtual work is especially useful for the purposes mentioned but requires that the internal friction forces do negligible work during any virtual displacement. Consequently, if internal friction in a mechanical system is appreciable, the method of virtual work will produce error when applied to the system as a whole unless the work done by internal friction is accounted for.

In the method of virtual work a diagram which isolates the system under consideration should be drawn. Unlike the free-body diagram, where all forces are shown, the diagram for the method of work need show only the *active forces*, since the reactive forces do not enter into the application of $\delta U = 0$. Such a drawing will be termed an *active-force diagram*. Figure 7/7*a* is an active-force diagram for the system shown.

(*d*) **Degrees of Freedom.** The number of independent coordinates needed to specify completely the configuration of a mechanical system is referred to as the number of *degrees of freedom* for that system. Figure 7/8*a* shows three examples of one-degree-of-freedom systems where only one coordinate is needed to determine the position of each part of the system. The coordinate can be a distance or an angle. Figure 7/8*b* shows three examples of two-degree-of-freedom systems where two independent coordinates are needed to determine the configuration of the system. By adding more links and removing supporting constraints to the linkage in the right-hand figure there is no limit to the number of degrees of freedom which can be achieved.

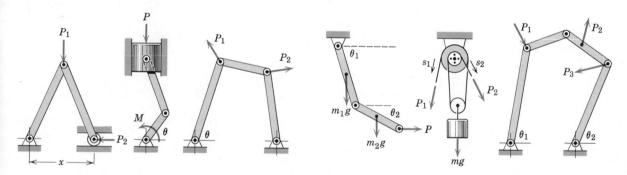

(*a*) Examples of One–Degree–of–Freedom Systems   (*b*) Examples of Two–Degree–of–Freedom Systems

Figure 7/8

The principle of virtual work $\delta U = 0$ may be applied as many times as there are degrees of freedom. With each application we allow only one independent coordinate to change at a time while holding the others constant. In our treatment of virtual work we shall restrict application to one-degree-of-freedom systems.[*]

[*] For examples of solutions to problems of two or more degrees of freedom see Chapter 7 of the author's *Statics, 2nd Edition* 1971 or *SI Version* 1975.

Sample Problem 7/1

Each of the two uniform hinged bars has a mass m and a length l, and is supported and loaded as shown. For a given force P determine the angle θ for equilibrium.

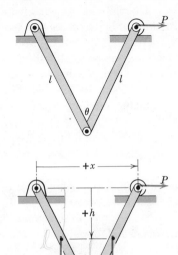

Solution.   The active-force diagram for the system composed of the two members is shown separately and includes the weight mg of each bar in addition to the force P. All other forces acting externally on the system are reactive forces which do no work during a virtual movement δx and are not shown.

The principle of virtual work requires that the total work of all external active forces be zero for any virtual displacement consistent with the constraints. Thus for a movement δx the virtual work becomes

① $[\delta U = 0]$ $\qquad\qquad$ $P\,\delta x + 2mg\,\delta h = 0$

We now express each of these virtual displacements in terms of the variable θ, the required quantity. Hence

$$h = \frac{l}{2}\cos\frac{\theta}{2} \quad\text{and}\quad \delta h = -\frac{l}{4}\sin\frac{\theta}{2}\delta\theta$$

② Similarly,

$$x = 2l\sin\frac{\theta}{2} \quad\text{and}\quad \delta x = l\cos\frac{\theta}{2}\delta\theta$$

Substitution into the equation of virtual work gives us

$$Pl\cos\frac{\theta}{2}\delta\theta - 2mg\frac{l}{4}\sin\frac{\theta}{2}\delta\theta = 0$$

from which we get

$$\tan\frac{\theta}{2} = \frac{2P}{mg} \quad\text{or}\quad \theta = 2\tan^{-1}\frac{2P}{mg} \qquad\text{Ans.}$$

To obtain this result by the principles of force and moment summation, it would be necessary to dismember the frame and take into account all forces acting on each member. Solution by the method of virtual work involves a simpler operation.

① Note carefully that with x positive to the right δx is also positive to the right in the direction of P, so that the virtual work is $P(+\delta x)$. With h positive down δh is also mathematically positive down in the direction of mg, so that the correct mathematical expression for the work is $mg(+\delta h)$. When we express δh in terms of $\delta\theta$ in the next step, δh will have a negative sign, thus bringing our mathematical expression into agreement with the physical observation that the weight mg does negative work as each center of mass moves upward with an increase in x and θ.

② We obtain δh and δx with the same mathematical rules of differentiation with which we may obtain dh and dx.

Sample Problem 7/2

The mass m is brought to an equilibrium position by the application of the couple M to the end of one of the two parallel links that are hinged as shown. The links have negligible mass, and all friction is assumed to be absent. Determine the expression for the equilibrium angle θ assumed by the links with the vertical for a given value of M. Consider the alternative of a solution by force and moment equilibrium.

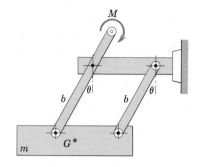

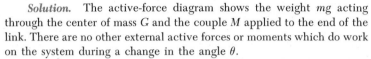

Solution. The active-force diagram shows the weight mg acting through the center of mass G and the couple M applied to the end of the link. There are no other external active forces or moments which do work on the system during a change in the angle θ.

The vertical position of the center of mass G is designated by the distance h below the fixed horizontal reference line and is $h = b \cos \theta + c$. The work done by mg during a movement δh in the direction of mg is

$$+mg\,\delta h = mg\,\delta(b \cos \theta + c)$$

$$= mg(-b \sin \theta\,\delta\theta + 0)$$

$$= -mgb \sin \theta\,\delta\theta$$

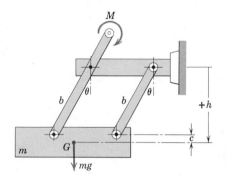

① The minus sign shows that the work is negative for a positive value of $\delta\theta$. The constant c drops out since its variation is zero.

With θ measured positive in the clockwise sense, $\delta\theta$ is also positive clockwise. Thus the work done by the clockwise couple M is $+M\,\delta\theta$. Substitution into the virtual work equation gives us

$[\delta U = 0]$ $\qquad M\,\delta\theta + mg\,\delta h = 0$

which yields

$$M\,\delta\theta = mgb \sin \theta\,\delta\theta$$

$$\theta = \sin^{-1} \frac{M}{mgb} \qquad\qquad \textit{Ans.}$$

① Again, as in Sample Problem 7/1, we are consistent mathematically with our definition of work, and we see that the algebraic sign of the resulting expression agrees with the physical change.

Inasmuch as $\sin \theta$ cannot exceed unity, we see that for equilibrium M is limited to values that do not exceed mgb.

The advantage of the virtual-work solution for this problem is readily seen when we observe what would be involved with a solution by force- and moment-equilibrium. For the latter approach, it would be necessary for us to draw separate free-body diagrams of all of the three moving parts and account for all of the internal reactions at the pin connections. To carry out these steps, it would be necessary for us to include in the analysis the horizontal position of G with respect to the attachment points of the two links, even though reference to this position would finally drop out of the equations when solved. We conclude, then, that the virtual-work method in this problem deals directly with cause and effect and avoids reference to irrelevant quantities.

Sample Problem 7/3

For link *OA* in the horizontal position shown, determine the force *P* on the sliding collar which will prevent *OA* from rotating under the action of the couple *M*. Neglect the mass of the moving parts.

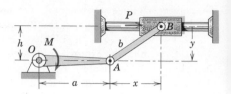

Solution. The given sketch serves as the active-force diagram for the system. All other forces are either internal or nonworking reactive forces due to the constraints.

We will give the crank *OA* a small clockwise angular movement $\delta\theta$ as our virtual displacement and determine the resulting virtual work done by *M* and *P*. From the horizontal position of the crank the angular movement gives a downward displacement of *A* equal to

$$\delta y = a\,\delta\theta$$

where $\delta\theta$ is, of course, expressed in radians.

From the right triangle for which link *AB* is the constant hypotenuse we may write

$$b^2 = x^2 + y^2$$

We now take the differential of the equation and get

$$0 = 2x\,\delta x + 2y\,\delta y \qquad \text{or} \qquad \delta x = -\frac{y}{x}\,\delta y$$

Thus

$$\delta x = -\frac{y}{x}\,a\,\delta\theta$$

and the virtual-work equation becomes

$$[\delta U = 0] \qquad M\,\delta\theta + P\,\delta x = 0 \qquad M\,\delta\theta + P\left(-\frac{y}{x}\,a\,\delta\theta\right) = 0$$

$$P = \frac{Mx}{ya} = \frac{Mx}{ha} \qquad\qquad \textit{Ans.}$$

Again, we observe that the virtual-work method produces a direct relationship between the active force *P* and the couple *M* without involving other forces which are irrelevant to this relationship. Solution by the force and moment equations of equilibrium, although fairly simple in this problem, would require accounting for all forces initially and then eliminating the irrelevant ones.

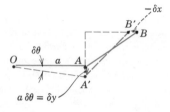

① Note that the displacement $a\,\delta\theta$ of point *A* would no longer equal δy if the crank *OA* were not in a horizontal position.

② The length *b* is constant so that $\delta b = 0$. Notice the negative sign which merely tells us that if one change is positive, the other must be negative.

③ We could just as well use a counterclockwise virtual displacement for the crank which would merely reverse the signs of all terms.

PROBLEMS

(Assume that the negative work of friction is negligible in the following problems unless otherwise indicated.)

7/1 For a given force P determine the angle θ for equilibrium. Neglect the mass of the links.

$$Ans. \ \theta = \cos^{-1} \frac{2P}{mg}$$

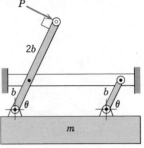

Problem 7/1

7/2 The toggle mechanism is used to position the mass m in the smooth vertical guides. Determine the expression for the horizontal force P required to support m for any value of θ. Would the action be any more effective if P were applied in a direction other than horizontal? $\quad$ *Ans.* $P = mg \tan \theta$

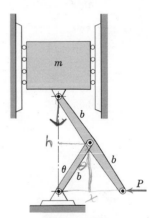

Problem 7/2

7/3 Find the force Q exerted on the paper by the paper punch of Prob. 4/65 repeated here.

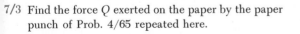

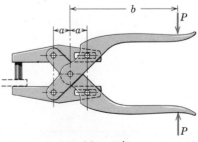

Problem 7/3

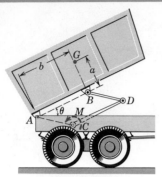

Problem 7/4

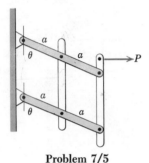

Problem 7/5

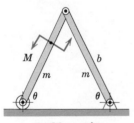

Problem 7/6

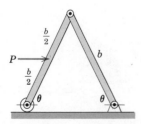

Problem 7/7

7/4 Determine the torque M on the activating lever of the dump truck necessary to balance the load m with center of mass at G when the dump angle is θ. The polygon $ABDC$ is a parallelogram.

7/5 Each of the four uniform links has a mass m. Determine the horizontal force P required to hold them in place in the vertical plane shown.

$$Ans. \ P = \tfrac{5}{2}mg \tan \theta$$

7/6 Each of the uniform links of the frame has a mass m and a length b. The equilibrium position of the frame in the vertical plane is determined by the couple M applied to the left-hand link. Express θ in terms of M.

7/7 Replace the couple M of Prob. 7/6 by the horizontal force P as shown and determine the equilibrium angle θ.

$$Ans. \ \theta = \tan^{-1}\frac{2mg}{3P}$$

7/8 Determine the force F which the patient must apply tangent to the rim of the handwheel of the wheelchair in order to roll up the incline of angle θ. The combined mass of the chair and patient is m. (If s is the displacement of the center of the wheel measured along the incline and β the corresponding angle in radians through which the wheel turns, it is easily shown that $s = R\beta$ if the wheel rolls without slipping.)

Problem 7/8

7/9 The toggle press of Prob. 4/81 is repeated here. Determine the required force F on the handle to produce a compression R on the roller for any given value of θ. *Ans.* $F = 0.8R \cos \theta$

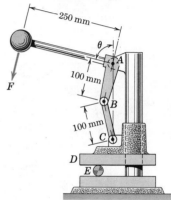

Problem 7/9

7/10 The hydraulic cylinder is used to spread the linkage and elevate the load m. For the position shown determine the compression C in the cylinder. Neglect the mass of all parts other than m.

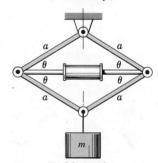

Problem 7/10

7/11 The hydraulic cylinder OA and link OB are arranged to control the tilt of the load which has a mass m and a center of mass at G. The lower corner C is free to roll horizontally as the cylinder linkage elongates. Determine the force P in the cylinder necessary to maintain equilibrium at a given angle θ.

$$Ans.\ P = \frac{mg}{2}\left(1 - \frac{h}{b}\tan\theta\right)$$

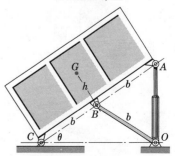

Problem 7/11

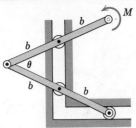

Problem 7/12

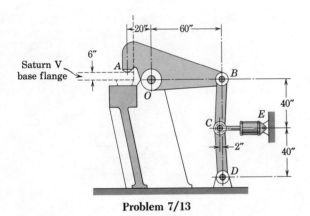

Problem 7/13

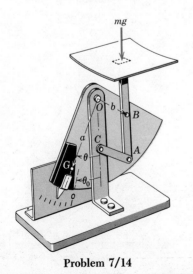

Problem 7/14

7/12 Each of the two uniform bars has a mass m with center rollers confined to move in the smooth vertical guide. With the end roller of the lower bar confined to move in the smooth horizontal guide, determine the angle θ for equilibrium under the action of the couple M.

7/13 The sketch shows the approximate configuration of one of the four toggle-action hold-down assemblies that clamp the base flange of the Saturn V rocket vehicle to the pedestal of its platform prior to launching. Calculate the preset clamping force F at A if the link CE is under tension produced by a fluid pressure of 2000 lb/in.^{2} acting on the left side of the piston in the hydraulic cylinder. The piston has a net area of 16 in.^{2} The weight of the assembly is considerable, but it is small compared with the clamping force produced and is therefore neglected here.   *Ans. $F = 960{,}000$ lb*

7/14 The postal scale consists of a sector of mass m_0 hinged at O and with center of mass at G. The pan and vertical link AB have a mass m_1 and are hinged to the sector at B. End A is hinged to the uniform link AC of mass m_2, which in turn is hinged to the fixed frame. The figure $OBAC$ forms a parallelogram, and the angle GOB is a right angle. Determine the relation between the mass m to be measured and the angle θ assuming that $\theta = \theta_0$ when $m = 0$.

7/15 The claw of the remote-action actuator develops a clamping force C as a result of the tension P in the control rod. Express C in terms of P for the configuration shown where the jaws are parallel.

$$\text{Ans. } C = \frac{P}{2}\frac{e(d+c)}{bc}$$

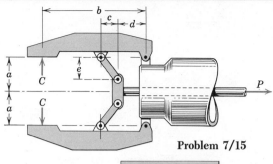

Problem 7/15

7/16 The cargo box of the food-delivery truck for aircraft servicing has a loaded mass m and is elevated by the application of a torque M on the lower end of the link which is hinged to the truck frame. The horizontal slots allow the linkage to unfold as the cargo box is elevated. Express M as a function of h.

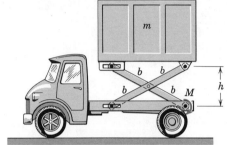

Problem 7/16

7/17 A device for counting the body radiation of a patient is shown. The radiation counter A has a mass m and is positioned by turning the screw of lead L with a torque M which controls the distance BC. Relate the torque M to the load mg for given values of b and θ. Neglect all friction and the mass of the linkage compared with m.

$$\text{Ans. } M = \frac{5mgL}{4\pi}\tan\frac{\theta}{2}$$

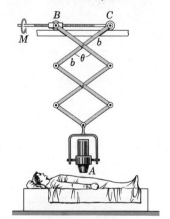

Problem 7/17

7/18 The vertical position of the load of mass m is controlled by the adjusting screw which connects joints A and B. The change in the distance between A and B for one revolution of the screw equals the lead L of the screw (advancement per revolution). If a moment M_f is required to overcome friction in the threads and thrust bearing of the screw, determine the expression for the total moment M, applied to the adjusting screw, necessary to raise the load.

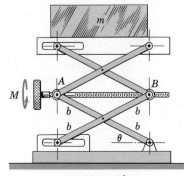

Problem 7/18

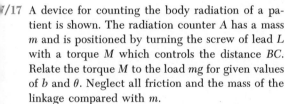

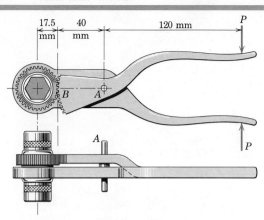

Problem 7/19

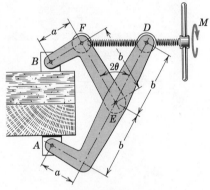

Problem 7/20

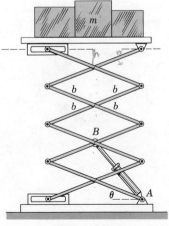

Problem 7/21

7/19 Determine the torque M exerted by the spaceman's antitorque wrench described in Prob. 4/110 and repeated here. The gripping force on the handles is $P = 150$ N.   *Ans.* $M = 7.88$ N·m

7/20 Determine the force F between the jaws of the clamp in terms of a torque M exerted on the handle of the adjusting screw. The screw has a lead (advancement per revolution) L, and friction is to be neglected.

7/21 The elevated platform delivers cargo to a cabin opening in an airliner. If the mass of the framework is neglected compared with the mass m of the cargo, determine the compressive force C in the activating hydraulic cylinder AB for any given angle θ.

$$\text{Ans. } C = \frac{mg\sqrt{1 + 8\sin^2\theta}}{\sin\theta}$$

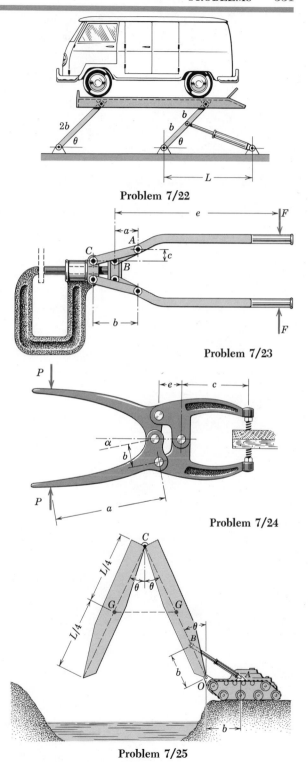

7/22 Express the compression C in the hydraulic cylinder of the car hoist in terms of the angle θ. The mass of the hoist is neglected compared to the mass m of the vehicle.

$$Ans.\ C = 2mg\ \mathrm{ctn}\ \theta\ \sqrt{1 + \left(\frac{b}{L}\right)^2 - 2\frac{b}{L}\cos\theta}$$

Problem 7/22

7/23 Determine the force P developed at the jaws of the rivet squeezer of Prob. 4/85, repeated here.

$$Ans.\ P = \frac{2Feb}{c(b - a)}$$

Problem 7/23

7/24 The toggle pliers of Prob. 4/71 are repeated here with symbolic dimensions. Determine the clamping force C as a function of α for a given handle gripping force P.

$$Ans.\ C = P\frac{e}{c}\left(\frac{a}{b}\ \mathrm{ctn}\ \alpha - 1\right)$$

Problem 7/24

7/25 The scissors-action military bridge of span L is launched from a tank through the action of the hydraulic linkage AB. If each of the two identical bridge sections has a mass m with center of mass at G, relate the force F in the hydraulic cylinder to the angle θ for equilibrium. An internal mechanism at C maintains the angle 2θ between the two sections. Why must the work of the internal moment at C be evaluated? Discuss the choice of the method of virtual work for this problem.

$$Ans.\ F = \frac{mgL\sqrt{2}}{b}\ \tan\theta\ \sqrt{1 + \sin\theta}$$

Problem 7/25

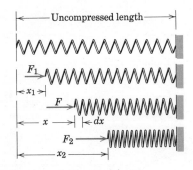

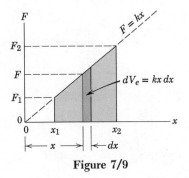

Figure 7/9

7/4 POTENTIAL ENERGY AND STABILITY. In the previous article we have dealt with the equilibrium configuration of mechanical systems composed of individual members that were assumed to be perfectly rigid. We will now extend our method to account for mechanical systems that include elastic elements in the form of springs. For this purpose we find it helpful to introduce the concept of potential energy, which leads us directly to the important problem of determining the stability of equilibrium.

(*a*) *Elastic Potential Energy.* The work done on an elastic member is stored in the member and is called *elastic potential energy* V_e. This energy is potentially available by allowing the member to do work on some other body during the relief of its compression or extension. Consider a spring, Fig. 7/9, which is being compressed by a force F. The spring is assumed to be elastic and linear, that is, the force F is directly proportional to the deflection x. We write this relation as $F = kx$, where k is the *spring constant* or *stiffness* of the spring. The work done on the spring by F during a movement dx is $dU = F\,dx$, so that the elastic potential energy of the spring for a compression x is the total work done on the spring

$$V_e = \int_0^x F\,dx = \int_0^x kx\,dx$$

or

$$\boxed{V_e = \tfrac{1}{2}kx^2} \tag{7/4}$$

Thus we see that the potential energy of the spring equals the triangular area in the diagram of F versus x from 0 to x.

During a change in the compression of the spring from x_1 to x_2 the work done on the spring equals its *change* in elastic potential energy or

$$\Delta V_e = \int_{x_1}^{x_2} kx\,dx = \tfrac{1}{2}k(x_2{}^2 - x_1{}^2)$$

which equals the trapezoidal area from x_1 to x_2.

During a virtual displacement δx of the spring the virtual work done on the spring is the virtual change in elastic potential energy

$$\delta V_e = F\,\delta x = kx\,\delta x$$

If the compressive force is relaxed from $x = x_2$ to $x = x_1$ so that the spring is relieved of its compression, then the *change* (final minus initial) in the potential energy of the spring becomes

$$\Delta V_e = \tfrac{1}{2}k(x_1{}^2 - x_2{}^2) = -\tfrac{1}{2}k(x_2{}^2 - x_1{}^2)$$

Correspondingly if δx is negative, δV_e will also be negative.

When we have a spring in tension rather than compression, the work and energy relations are the same as those for compression where x now represents the elongation of the spring rather than its compression. While the spring is being stretched, we note that the force again acts in the direction of the displacement doing positive work on the spring and increasing its potential energy.

Since the force acting on the movable end of a spring is the negative of the force exerted by the spring on the body to which its movable end is attached, we conclude that the *work done on the body is the negative of the potential energy change of the spring.*

The units of potential energy are the same as those of work and are expressed in joules (J) in SI units and in foot-pounds (ft-lb) in U.S. customary units.

(*b*) *Gravitational Potential Energy.* In the previous article we treated the work of a gravitational force or weight acting on a body in the same way as the work of any other active force. Thus for an upward displacement δh of the body in Fig. 7/10 the weight $W = mg$ does negative work $\delta U = -mg\,\delta h$. Or, if the body has a downward displacement δh, with h measured positive downward, the weight does positive work $\delta U = +mg\,\delta h$.

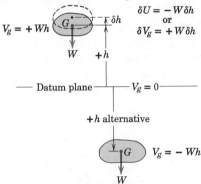

$$\delta U = -W\,\delta h$$
$$\text{or}$$
$$\delta V_g = +W\,\delta h$$

$$V_g = +Wh$$

$$V_g = 0$$

$$V_g = -Wh$$

Figure 7/10

We now adopt an alternative to the foregoing treatment by expressing the work done by gravity in terms of a change in potential energy of the body. This alternative treatment is a useful representation when we describe a mechanical system in terms of its total energy. The *gravitational potential energy* V_g of a body is defined simply as the work done on the body by a force equal and opposite to the weight in bringing the body to the position under consideration from some arbitrary datum plane where the potential energy is defined to be zero. The potential energy, then, is the negative of the work done by the weight. When the body is raised, for example, the work done is converted into energy that is potentially available, since the body is capable of doing work on some other body as it returns to its original lower position. If we take V_g to be zero at $h = 0$, Fig. 7/10, then at a height h above the datum plane the gravitational potential energy of the body is

$$\boxed{V_g = mgh} \tag{7/5}$$

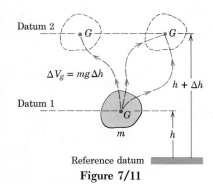

$$\Delta V_g = mg\,\Delta h$$

Datum 2

Datum 1

$h + \Delta h$

m h

Reference datum

Figure 7/11

If the body is a distance h below the datum plane, its gravitational potential energy is $-mgh$.

It is important to observe that the datum plane for zero potential energy is perfectly arbitrary inasmuch as it is only the *change* in energy with which we are concerned, and this change is the same no matter where we take the datum plane. Also, the gravitational potential energy is independent of the path followed in arriving at the particular level h under consideration. Thus the body of mass m in Fig. 7/11 has the same potential-energy change no matter which path it followed in going from datum plane 1 to datum plane 2 since Δh is the same for both. Our measurement of h, of course, is to the center of mass of the body.

The virtual change in gravitational potential energy is simply

$$\delta V_g = mg\,\delta h$$

where δh is the upward virtual displacement of the mass center of the body. If the mass center should have a downward virtual displacement, then δV_g is negative.

The units of gravitational potential energy are the same as those for work and elastic potential energy, joules (J) in SI units and foot-pounds (ft-lb) in U.S. customary units.

(*c*) *Energy Equation.* In the previous two sections we have noted that the work done *by* a spring *on* the body to which its movable end is attached is the negative of the change in the elastic potential energy of the spring. Also, the work done by the gravitational force or weight mg is the negative of the change in gravitational potential energy. Therefore, when we apply the virtual-work equation to systems with springs and with changes in the vertical position of its members, we may replace the work of the spring and the work of the weight by the negative of the respective potential energy changes. Putting these changes on the right-hand side of Eq. 7/3 enables us to write the virtual-work equation as

$$\boxed{\delta U = \delta V_e + \delta V_g} \quad \text{or} \quad \boxed{\delta U = \delta V} \tag{7/6}$$

where $V = V_e + V_g$ stands for the total potential energy of the system. The quantity δU now represents the virtual work done on the system during a virtual displacement by all external active forces *other than* spring forces and gravitational forces.

Thus for a mechanical system with elastic members and members that undergo changes in position, we may restate the principle of virtual work as follows:

The virtual work done by external active forces on a mechanical system in equilibrium equals the corresponding change in the total elastic and gravitational potential energy of the system for any and all virtual displacements consistent with the constraints.

As always, we make the critical decision as to what constitutes our system. With the method of work-energy it is useful to draw the active-force diagram of the system chosen where the boundary of the system clearly distinguishes those members that are a part of the system from other bodies which are not a part of the system. By including an elastic member within the boundary of our system, we see that the forces of interaction between it and the movable member to which it is attached are *internal* to the system and need not be shown. Nor do we show the gravitational forces since their work is accounted for in the V_g-term.

(*d*) *Stability of Equilibrium.* In addition to the replacement of the work of elastic forces and the work of gravitational forces by their corresponding changes in potential energy, we may convert the work of other active forces applied externally to the system to potential energy changes as shown in Fig. 7/12. Here the work done by **F** during a virtual displacement δs of its point of application is equivalent to the change $-mg\,\delta s\cos\alpha$ in potential energy of the mass m for the equivalent system. Hence we may replace the work done by external active forces on a mechanical system by the corresponding potential energy changes with the opposite sign for the equivalent system.

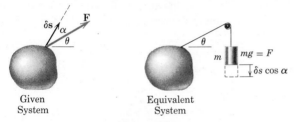

Figure 7/12

With the replacement of the work terms by energy terms, the principle of virtual work for a mechanical system without internal kinetic friction, Eq. 7/6, may now be written as

$$\delta(V_e + V_g) = 0 \qquad \text{or} \qquad \delta V = 0 \qquad (7/7)$$

Equation 7/7 expresses the requirement that the equilibrium configuration of a mechanical system is one for which the total potential energy V of the system has a stationary value. For a system of one degree of freedom where the potential energy and its derivatives are continuous functions of the single variable, say x, that describes the configuration, the equilibrium condition $\delta V = 0$ is equivalent mathematically to the requirement

$$\frac{dV}{dx} = 0 \qquad (7/8)$$

Equation 7/8 states that a mechanical system is in equilibrium when the derivative of its total potential energy is zero. For systems with several degrees of freedom the partial derivative of V with respect to each coordinate in turn must be zero for equilibrium.[*]

There are three conditions under which Eq. 7/8 applies, namely, when the total potential energy is a minimum (stable equilibrium), a maximum (unstable equilibrium), or a constant (neutral equilibrium). We see a simple example of these three conditions in Fig. 7/13 where the potential energy of the roller is clearly a minimum in the stable position, a maximum in the unstable position, and a constant in the neutral position.

We may also characterize the stability of a mechanical system by noting that a small displacement away from the stable position results in an increase in potential energy and a tendency to return to the position of lower energy. On the other hand a small displacement away from the unstable position results in a decrease in potential energy and a tendency to move farther away from the equilibrium position to one of lower energy. For the neutral position a small displacement one way or the other results in no change in potential energy and no tendency to move either way.

When a function and its derivatives are continuous, the second derivative is positive at a point of minimum value of the function and negative at a point of maximum value of the function. Thus we may write the mathematical conditions for the equilibrium and stability for a system with a single degree of freedom x as follows:

Stable   Unstable   Neutral

Figure 7/13

$$
\boxed{
\begin{array}{lll}
\text{Equilibrium} & \dfrac{dV}{dx} = 0 & \\[2mm]
\text{Stable} & \dfrac{d^2V}{dx^2} > 0 & \quad (7/9) \\[2mm]
\text{Unstable} & \dfrac{d^2V}{dx^2} < 0 &
\end{array}
}
$$

Occasionally we may have a situation where the second derivative of V is also zero at the equilibrium position, in which case we must examine the sign of a higher derivative to ascertain the type of equilibrium. When the order of the lowest remaining nonzero derivative is even, the equilibrium will be stable or unstable according to whether the sign of this lowest even-order derivative is positive or negative. If the order of the lowest remaining nonzero derivative is odd, the equilibrium is classified as unstable, and the plot of V versus x for this case appears as an inflection point in the curve with zero slope at the equilibrium value.

Stability criteria for systems with more than one degree of freedom require more advanced treatment. For two degrees of freedom, for example, we use a Taylor-series expansion for two variables.

[*] For examples of two-degree-of-freedom systems see Art. 43, Chapter 7 of the author's *Statics, 2nd Edition* 1971 or *SI Version* 1975.

Sample Problem 7/4

The 10-kg cylinder is suspended by the spring which has a stiffness of 2 kN/m. Plot the potential energy V of the system and show that it is minimum at the equilibrium position.

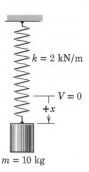

$k = 2$ kN/m

$V = 0$

$+x$

$m = 10$ kg

Solution. (Although the equilibrium position in this simple problem is clearly where the force in the spring equals the weight mg, we will proceed as though this fact were unknown in order to illustrate the energy ① relationships in the simplest way.) We choose the datum plane for zero potential energy at the position for which the spring is unextended.

The elastic potential energy for an arbitrary position x is $V_e = \frac{1}{2}kx^2$ and the gravitational potential energy is $-mgx$, so that the total potential energy is

$$[V = V_e + V_g] \qquad V = \tfrac{1}{2}kx^2 - mgx$$

Equilibrium occurs where

$$\left[\frac{dV}{dx} = 0\right] \qquad \frac{dV}{dx} = kx - mg = 0, \qquad x = mg/k$$

Although we know in this simple case that the equilibrium is stable, we prove it by evaluating the sign of the second derivative of V at the equilibrium position. Thus $d^2V/dx^2 = k$, which is positive, proving that the equilibrium is stable.

Substituting numerical values gives

$$V = \tfrac{1}{2}(2000)x^2 - 10(9.81)x$$

expressed in joules, and the equilibrium value of x is

$$x = 10(9.81)/2000 = 0.049 \text{ m or } 49 \text{ mm} \qquad Ans.$$

We calculate V for various values of x and plot V versus x as shown. ② The minimum value of V occurs at $x = 0.049$ m where $dV/dx = 0$ and d^2V/dx^2 is positive.

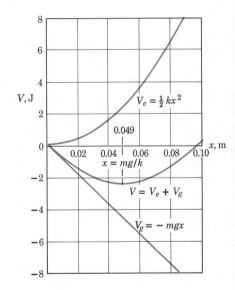

① The choice is arbitrary but simplifies the algebra.

② We could have chosen different datum planes for V_e and V_g without affecting our conclusions. Such a change would merely shift the separate curves for V_e and V_g up or down but would not affect the position of the minimum value of V.

Sample Problem 7/5

The two uniform links, each of mass m, are in the vertical plane and are connected and constrained as shown. As the angle θ between the links increases with the application of the horizontal force P, the light rod, which is connected at A and passes through a pivoted collar at B, compresses the spring of stiffness k. If the spring is uncompressed in the position equivalent to that for which $\theta = 0$, determine the force P which will produce equilibrium at the angle θ.

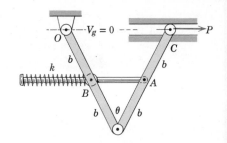

Solution. The given sketch serves as the active-force diagram of the system. The compression x of the spring is the distance that A has moved away from B, which is $x = 2b \sin \dfrac{\theta}{2}$. Thus the elastic potential energy of the spring is

$$[V_e = \tfrac{1}{2}kx^2] \qquad V_e = \tfrac{1}{2}k\left(2b \sin \frac{\theta}{2}\right)^2 = 2kb^2 \sin^2 \frac{\theta}{2}$$

With the datum for zero gravitational potential energy taken through the support at O for convenience, the expression for V_g becomes

$$[V_g = mgh] \qquad V_g = 2mg\left(-b \cos \frac{\theta}{2}\right)$$

The distance between O and C is $4b \sin \dfrac{\theta}{2}$, so that the virtual work done by P is

$$\delta U = P\,\delta\left(4b \sin \frac{\theta}{2}\right) = 2Pb \cos \frac{\theta}{2}\,\delta\theta$$

The virtual-work equation now gives

$$[\delta U = \delta V_e + \delta V_g] \qquad 2Pb \cos \frac{\theta}{2}\,\delta\theta = \delta\left(2kb^2 \sin^2 \frac{\theta}{2}\right) + \delta\left(-2mgb \cos \frac{\theta}{2}\right)$$

$$= 2kb^2 \sin \frac{\theta}{2} \cos \frac{\theta}{2}\,\delta\theta + mgb \sin \frac{\theta}{2}\,\delta\theta$$

Simplifying gives finally

$$P = kb \sin \frac{\theta}{2} + \tfrac{1}{2}mg \tan \frac{\theta}{2} \qquad\qquad Ans.$$

If we had been asked to express the equilibrium value of θ corresponding to a given force P, we would have difficulty solving explicitly for θ in this particular case. For a numerical problem we could resort to an approximation with a graphical plot of numerical values of the sum of the two functions of θ to determine the value of θ for which the sum equals P.

Sample Problem 7/6

The ends of the uniform bar of mass m slide freely in the horizontal and vertical guides as shown. Examine the stability conditions for the positions of equilibrium. The spring of stiffness k is undeformed when $x = 0$.

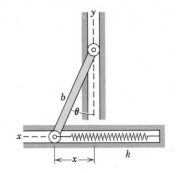

Solution. The system consists of the spring and the bar. Since there are no external active forces, the given sketch serves as the active-force diagram. We shall take the x-axis as the datum for zero gravitational potential energy. In the displaced position the elastic and gravitational potential energies are

$$V_e = \tfrac{1}{2}kx^2 = \tfrac{1}{2}kb^2 \sin^2 \theta \quad \text{and} \quad V_g = mg\frac{b}{2}\cos \theta$$

The total potential energy is, then,

$$V = V_e + V_g = \tfrac{1}{2}kb^2 \sin^2 \theta + \tfrac{1}{2}mgb \cos \theta$$

Equilibrium occurs for $dV/d\theta = 0$ so that

$$\frac{dV}{d\theta} = kb^2 \sin \theta \cos \theta - \tfrac{1}{2}mgb \sin \theta = (kb^2 \cos \theta - \tfrac{1}{2}mgb)\sin \theta = 0$$

The two solutions to this equation are given by

$$\sin \theta = 0 \quad \text{and} \quad \cos \theta = \frac{mg}{2kb}$$

We now determine the stability by examining the sign of the second derivative of V for each of the two equilibrium positions. The second derivative is

$$\frac{d^2V}{d\theta^2} = kb^2(\cos^2 \theta - \sin^2 \theta) - \tfrac{1}{2}mgb \cos \theta$$

$$= kb^2(2\cos^2 \theta - 1) - \tfrac{1}{2}mgb \cos \theta$$

(*Solution I*) $\sin \theta = 0$, $\theta = 0$

$$\frac{d^2V}{d\theta^2} = kb^2(2 - 1) - \tfrac{1}{2}mgb = kb^2\left(1 - \frac{mg}{2kb}\right)$$

$$= \text{positive (stable)} \quad \text{if } k > mg/2b$$

$$= \text{negative (unstable)} \text{ if } k < mg/2b \qquad \textit{Ans.}$$

Thus, if the spring is sufficiently stiff, the bar will return to the vertical position even though there is no force in the spring at that position.

(*Solution II*) $\cos \theta = \dfrac{mg}{2kb}$, $\theta = \cos^{-1}\dfrac{mg}{2kb}$

$$\frac{d^2V}{d\theta^2} = kb^2\left(2\left[\frac{mg}{2kb}\right]^2 - 1\right) - \tfrac{1}{2}mgb\left(\frac{mg}{2kb}\right) = kb^2\left(\left[\frac{mg}{2kb}\right]^2 - 1\right)$$

$$\textit{Ans.}$$

Since the cosine must be less than unity, we see that this solution is limited to the case where $k > mg/2b$, which makes the second derivative of V negative. Hence equilibrium for solution II is never stable. If $k < mg/2b$ we no longer have solution II since the spring will be too weak to maintain equilibrium at a value of θ between 0° and 90°.

① Be careful not to overlook the solution $\theta = 0$ given by $\sin \theta = 0$.

② This result is one that we might not have anticipated without the mathematical analysis of the stability.

③ Again, without the benefit of the mathematical analysis of the stability we might have supposed erroneously that the bar could come to rest in a stable equilibrium position for some value of θ between 0° and 90°.

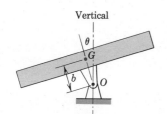

Problem 7/26

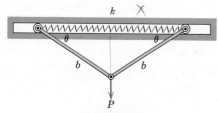

Problem 7/27

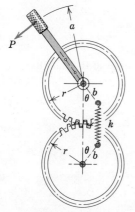

Problem 7/29

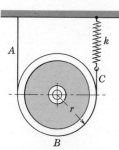

Problem 7/30

PROBLEMS

(Assume that the negative work of friction is negligible in the following problems.)

7/26 The bar of mass m with center of mass at G is pivoted about a horizontal axis through O. Prove the stability conditions for the two positions of equilibrium.

7/27 Determine the force P required to produce equilibrium of the two hinged links at an angle θ. The uncompressed length of the spring is $2b$. Neglect the mass of the links.

$$Ans.\ P = 4kb(1 - \cos\theta)\tan\theta$$

7/28 The potential energies of two mechanical systems are given by $V_1 = C_1 x^4$ and $V_2 = C_2 x^3$, where C_1 and C_2 are positive constants and x is the single coordinate expressing the positions of both systems. Specify the stability of each system at the equilibrium position $x = 0$.

7/29 The uniform wheel of mass m is supported in the vertical plane by the light band ABC and the spring of stiffness k. If the wheel is released initially from the position where the force in the spring is zero, determine the clockwise angle θ through which the wheel rotates from the initial position to the final equilibrium position.

$$Ans.\ \theta = \frac{mg}{4kr}$$

7/30 The handle is fastened to one of the spring-connected gears, which are mounted in fixed bearings. The spring of stiffness k connects two pins mounted in the faces of the gears. When the handle is in the vertical position, $\theta = 0$ and the spring force is zero. Determine the force P required to maintain equilibrium at an angle θ.

7/31 The figure shows the cross section of a uniform 50-kg ventilator door hinged along its upper horizontal edge at O. The door is controlled by the spring-loaded cable which passes over the small pulley at A. The spring has a stiffness of 180 N per meter of stretch and is undeformed when $\theta = 0$. Determine the angle θ for equilibrium.

Ans. $\theta = 48.6°$

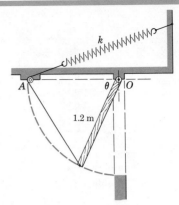

Problem 7/31

7/32 Determine the equilibrium value of x for the spring-supported bar. The spring has a stiffness k and is unstretched when $x = 0$. The force F acts in the direction of the bar, and the mass of the bar is negligible.

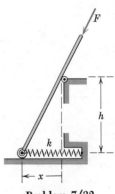

Problem 7/32

7/33 When $u = 0$, the spring of stiffness k is uncompressed. As u increases, the rod slides through the pivoted collar at A and compresses the spring between the collar and the end of the rod. Determine the force P required to produce a given displacement u. Assume the absence of friction and neglect the mass of the rod.

Ans. $P = \left(1 - \dfrac{b}{\sqrt{b^2 + u^2}}\right)ku$

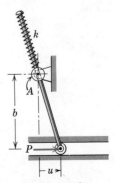

Problem 7/33

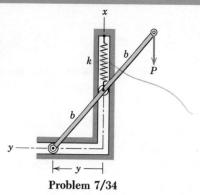

Problem 7/34

7/34 Determine the equilibrium value of the coordinate y for the mechanism under the action of the vertical load P. The spring of stiffness k is unstretched when $y = 0$, and the mass of the link is negligible.

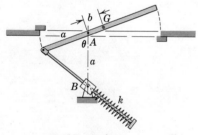

Problem 7/35

7/35 The uniform bar of mass m with center of mass at G is pivoted about O and swings in the vertical plane. Position of the bar is controlled by the light spring-loaded rod that slides through a pivoted collar at A. The spring of stiffness k is uncompressed in the position for which $\theta = 0$. Determine the acute angle θ for equilibrium and prove that the position is stable.

$$\text{Ans. } \theta = \tan^{-1}\frac{mg}{ka}$$

Problem 7/36

7/36 The cross section of a trap door hinged at A and having a mass m and a center of mass at G is shown in the figure. The spring is compressed by the rod which is pinned to the lower end of the door and which passes through the swivel block at B. When $\theta = 0$, the spring is undeformed. Show that with the proper stiffness k of the spring, the door will be in equilibrium for any angle θ.

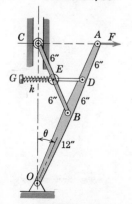

Problem 7/37

7/37 For a horizontal force F of 50 lb, determine the angle θ for equilibrium of the spring-loaded linkage. The rod DG passes through the swivel at E and compresses the spring, which has a stiffness of 25 lb/in. and an uncompressed condition corresponding to $\theta = 0$. Neglect the weights of the members.   *Ans.* $\theta = 19.5°$

7/38 In the mechanism shown, the rod AB slides through the pivoted collar at C and compresses the spring when a couple M is applied to link DE. The spring has a stiffness k and is uncompressed for the position equivalent to $\theta = 0$. Determine the angle θ for equilibrium. The masses of the parts are negligible.

Ans. $\theta = \sin^{-1} \dfrac{M}{kb^2}$

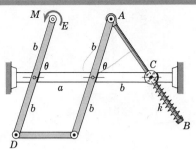

Problem 7/38

7/39 Each of the two gears carries an eccentric mass m and is free to rotate in the vertical plane about its bearing. Determine the values of θ for equilibrium and identify the type of equilibrium for each position.

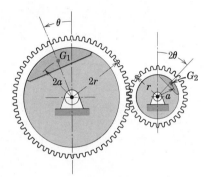

Problem 7/39

7/40 The center of mass G of the uniform link AB is constrained to move in the smooth vertical guide. Link OA has negligible mass, and the spring of stiffness k is uncompressed for $\theta = 0$. Determine the equilibrium positions and their stability.

Ans.

$\theta = \pi$, stable if $k < \dfrac{mg}{2b}$ and unstable if $k > \dfrac{mg}{2b}$

$\theta = 2 \sin^{-1} \dfrac{mg}{2kb}$ requires $k > \dfrac{mg}{2b}$, stable

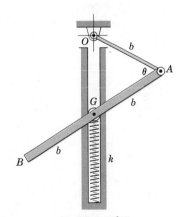

Problem 7/40

7/41 The figure shows the cross section of a container composed of a hemispherical shell of radius r and a cylindrical shell of height h, both made from the same material. Specify the limitation of h for stability in the upright position when the container is placed on the horizontal surface.

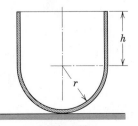

Problem 7/41

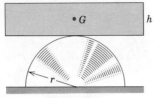

Problem 7/42

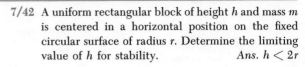

7/42 A uniform rectangular block of height h and mass m is centered in a horizontal position on the fixed circular surface of radius r. Determine the limiting value of h for stability.   *Ans.* $h < 2r$

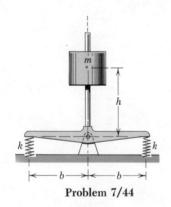

Problem 7/43

7/43 Predict through calculation whether the homogeneous semicylinder and the half-cylindrical shell will remain in the positions shown or whether they will roll off the lower cylinder.

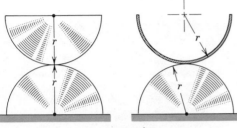

Problem 7/44

7/44 Determine the maximum height h of the mass m for which the inverted pendulum will be stable in the vertical position shown. Each of the springs has a stiffness k, and they have equal precompressions in this position. Neglect the mass of the remainder of the mechanism.

$$Ans.\ h_{\max} = \frac{2kb^2}{mg}$$

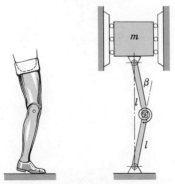

Problem 7/45

7/45 One of the critical requirements in the design of an artificial leg for an amputee is to prevent the knee joint from buckling under load when the leg is straight. As a first approximation, simulate the artificial leg by the two light links with a torsion spring at their common joint. The spring develops a torque $M = K\beta$, which is proportional to the angle of bend β at the joint. Determine the minimum value of K that will ensure stability of the knee joint for $\beta = 0$.

$$Ans.\ K_{\min} = \tfrac{1}{2}mgl$$

7/46 In the figure is shown a small industrial lift with a foot release. There are four identical springs, two on each side of the central shaft. The stiffness of each pair of springs is $2k$. Specify the value of k that will ensure stable equilibrium when the lift supports a load L in the position shown with no force on the pedal. The springs have an equal initial precompression and may be assumed to act in the horizontal direction at all times.

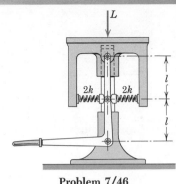

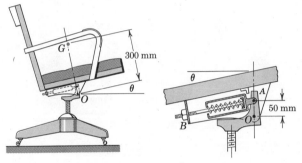

Problem 7/46

7/47 The figure shows a tilting desk chair together with a detail of the spring-loaded tilting mechanism. The frame of the seat is pivoted about the fixed point O on the base. The increase in distance between A and B as the chair tilts back about O is the increase in compression of the spring. The spring, which has a stiffness of 96 kN/m, is uncompressed when $\theta = 0$. For small angles of tilt it may be assumed with negligible error that the axis of the spring remains parallel to the seat. The center of mass of an 80-kg person who sits in the chair is at G on a line through O perpendicular to the seat. Determine the angle of tilt θ for equilibrium. (*Hint:* The deformation of the spring may be visualized by allowing the base to tilt through the required angle θ about O while the seat is held in a fixed position.)   Ans. $\theta = 11.19°$

Problem 7/47

7/48 The slender bar of length l and mass m is pivoted freely about a horizontal axis through O. The spring has an unstretched length of $l/2$. Determine the equilibrium positions, excluding $\theta = \pi$, and determine the maximum value of the spring stiffness k for stability in the position $\theta = 0$.

$$Ans. \ \theta = 0, \ \theta = 2\cos^{-1}\frac{1/2}{1 - \dfrac{2mg}{kl}}, \ k_{max} = \frac{4mg}{l}$$

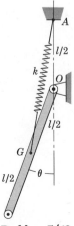

Problem 7/48

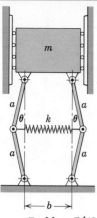

Problem 7/49

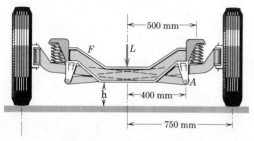

Problem 7/50

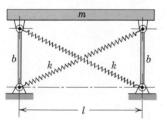

Problem 7/51

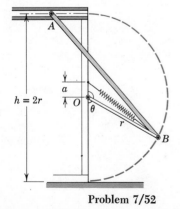

Problem 7/52

7/49 The mass m moves in smooth vertical guides and is supported by the four spring-loaded links of negligible mass. The spring of stiffness k is unstretched in the position for which $\theta = 0$. Specify the stability of the system for its equilibrium positions.

7/50 The front-end suspension of Prob. 4/82 is repeated here. The frame F must be jacked up so that $h = 350$ mm in order to relieve the compression in the coil springs. Determine the value of h when the jack is removed. Each spring has a stiffness of 120 kN/m. The load L is 12 kN, and the central frame F has a mass of 40 kg. Each wheel and attached link has a mass of 35 kg with a center of mass 680 mm from the vertical centerline.

Ans. $h = 265$ mm

▶**7/51** The platform of mass m is supported by equal legs and braced by the two springs as shown. If the masses of the legs and springs are negligible, determine the minimum stiffness k of each spring that will ensure stability of the platform in the position shown. Each spring has a tensile preset deflection Δ.

$$Ans.\ k_{\min} = \frac{mg}{2b}\left(1 + \frac{b^2}{l^2}\right)$$

▶**7/52** The uniform garage door AB shown in section has a mass m and is equipped with two of the spring-loaded mechanisms shown, one on each side of the door. The arm OB has negligible mass, and the upper corner A of the door is free to move horizontally on a roller. The unstretched length of the spring is $r - a$, so that in the top position with $\theta = \pi$ the spring force is zero. In order to ensure smooth action of the door as it reaches the vertical closed position $\theta = 0$, it is desirable that the door be insensitive to movement in this position. Determine the required spring stiffness k.

$$Ans.\ k = \frac{mg(r + a)}{8a^2}$$

7/5 PROBLEM FORMULATION AND REVIEW.

When various configurations are possible for a body or a system of interconnected bodies as a result of applied forces, the equilibrium position attained is usually best found by applying the principle of virtual work developed in this chapter. We have seen repeatedly that the only forces which need to be considered when determining the equilibrium position by this method are those which do work (active forces) during the assumed differential movement of the body or system away from its equilibrium position. Those external forces which do no work (reactive forces) need not be involved. For this reason we have constructed the active-force diagram of the body or system (rather than the free-body diagram) to focus attention on only those external forces which do work during the virtual displacements.

Relating the corresponding virtual displacements, linear and angular, of the parts of a mechanical system during a virtual movement consistent with the constraints is often the most difficult part of the analysis. First, the geometric relationships which describe the configuration of the system must be written. Next, the differential changes in the positions of parts of the system are established by the process of differentiation of the geometrical relationship to obtain expressions for the differential virtual movements.

In the method of virtual work we take special note of the fact that a virtual displacement is a first-order differential change in a length or an angle. This change is ficticious in that it is an assumed movement which need not take place in reality. Mathematically, a virtual displacement is treated the same as a differential change in an actual movement. We use the symbol δ for the differential virtual change and the usual symbol d for the differential change in a real movement.

In Chapter 7 we have restricted our attention to mechanical systems for which the positions of the members can be specified by a single variable (single-degree-of-freedom systems). For two or more degrees of freedom, we would apply the virtual-work equation as many times as there are degrees of freedom, allowing one variable to change at a time while holding the remaining ones constant.

We have found that the concept of potential energy, both gravitational (V_g) and elastic (V_e), is useful in solving equilibrium problems where changes in the vertical position of the mass centers of the bodies occur and where corresponding changes in the elongation or compression of elastic members (springs) also result during the virtual displacement. Here we obtain an expression for the total potential energy V of the system in terms of the variable that specifies the possible position of the system. The first and second derivatives of V are used to establish, respectively, the position of equilibrium and the type of stability which exists.

REVIEW PROBLEMS

Plane of each figure is vertical. Size and mass of each member and applied force are known.

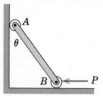

Find θ for equilibrium

(a)

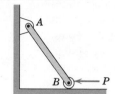

Find reactions at A and B

(b)

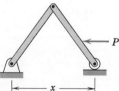

Find x for equilibrium

(c)

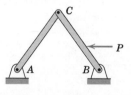

Find forces at $A, B,$ and C

(d)

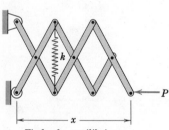

Find x for equilibrium

(e)

Determine maximum k for stable equilibrium at $\theta = 0$

(f)

Problem 7/53

7/53 Identify which of the problems (a) through (f) are best solved (A) by the force and moment equilibrium equations and (B) by virtual work. Outline briefly the procedure for each solution.

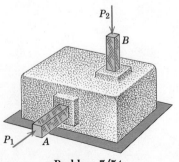

Problem 7/54

7/54 A "black box" contains a series of interconnected racks and pinions, gears, and other internal mechanical elements which transfer the linear motion of the pushrod A to produce the linear motion of the pushrod B. For every unit of inward movement of A under the action of force P_1, rod B moves outward from the box one-third of a unit against the action of force P_2. If $P_1 = 100$ N, calculate P_2 for equilibrium. Neglect all friction and assume all mechanical components are ideally connected light rigid bodies.

7/55 If the internal mechanism in the box described in Prob. 7/54 is rearranged so that an inward movement of one unit of pushrod A results in a movement of one-third of a unit of rod B also inward, compute the tension T supported by rod B for equilibrium if $P_1 = 100$ N.

7/56 A control mechanism consists of an input shaft at A which is turned by applying a couple M and an output slider B which moves in the x-direction against the action of force P. The mechanism is arranged so that the linear movement of B is proportional to the angular movement of A with x increasing 80 mm for every complete turn of A. If $M = 10$ N·m, determine P for equilibrium. Neglect internal friction and assume all mechanical components are ideally connected rigid bodies.

$$\text{Ans. } P = 785 \text{ N}$$

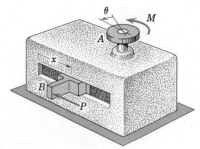

Problem 7/56

7/57 The bar is free to rotate through a complete vertical circle. Verify mathematically the stability conditions that are evident at the two equilibrium positions.

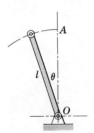

Problem 7/57

7/58 The potential energy of a mechanical system with negligible friction is given by $V = b \sin^2 \theta + c \cos \theta$ where θ is the angular coordinate that defines the position of the system and where b and c are positive constants. Establish the equilibrium position or positions and determine the type of equilibrium for each one.

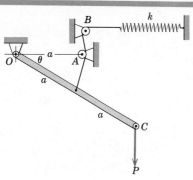

Problem 7/59

7/59 The light bar OC is pivoted at O and swings in the vertical plane. When $\theta = 0$, the spring of stiffness k is unstretched. Determine the equilibrium angle corresponding to a given vertical force P applied to the end of the bar. Neglect the mass of the bar and the diameter of the small pulleys.

$$Ans. \ \theta = \tan^{-1} \frac{2P}{ka}$$

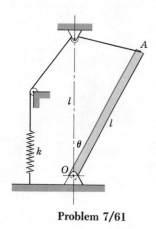

Problem 7/60

7/60 A power-operated loading platform for the back of a truck is shown in the figure. The position of the platform is controlled by the hydraulic cylinder, which applies force at C. The links are pivoted to the truck frame at A, B, and F. Determine the force P supplied by the cylinder in order to support the platform in the position shown. The mass of the platform and links may be neglected compared with that of the 250-kg crate with center of mass at G.

$$Ans. \ P = 3.5 \text{ kN}$$

7/61 The figure shows the edge view of a uniform sky-light door of mass m with center of mass midway between A and O. The door is counterbalanced by the action of the spring, which has a stiffness k and which is unstretched when the door is in the vertical plane with $\theta = 0$. Determine the necessary stiffness k to balance the door in an equilibrium position for $\theta > 0$.

$$Ans. \ k = \frac{mg}{2l}, \text{ independent of } \theta$$

Problem 7/61

VOLUME 2
DYNAMICS

I
DYNAMICS OF PARTICLES

INTRODUCTION TO DYNAMICS

1

/1 HISTORY AND MODERN APPLICATIONS. Dynamics is that branch of mechanics which deals with the motion of bodies under the action of forces. The study of dynamics in engineering usually follows the study of statics, which deals with the action of forces on bodies at rest. Dynamics has two distinct parts—*kinematics*, which is the study of motion without reference to the forces which cause motion, and *kinetics*, which relates the action of forces on bodies to their resulting motions. The student of engineering will find that a thorough comprehension of dynamics will provide him with one of his most useful and powerful tools for analysis in engineering.

Historically, dynamics is a relatively recent subject compared with statics. The beginning of a rational understanding of dynamics is credited to Galileo (1564–1642) who made careful observations concerning bodies in free fall, motion on an inclined plane, and motion of the pendulum. He was largely responsible for bringing a scientific approach to the investigation of physical problems. Galileo was continually under severe criticism for refusing to accept the established beliefs of his day, such as the philosophies of Aristotle which held, for example, that heavy bodies fall more rapidly than light bodies. The lack of accurate means for the measurement of time was a severe handicap to Galileo, and further significant development in dynamics awaited the invention of the pendulum clock by Huygens in 1657. Newton (1642–1727), guided by Galileo's work, was able to make an accurate formulation of the laws of motion and, hence, to place dynamics on a sound basis. Newton's famous work was published in the first edition of his *Principia*,° which is generally recognized as one of the greatest of all recorded contributions to knowledge. In addition to stating the laws governing the motion of a particle, Newton was the first to formulate correctly the law of universal gravitation. Although his mathematical description was accurate, he felt that the concept of remote transmission of gravitational force without a supporting medium was an absurd notion. Following Newton's time, important contributions to mechanics

---

° The original formulations of Sir Isaac Newton may be found in the translation of his *Principia* (1687), revised by F. Cajori, University of California Press, 1934.

were made by Euler, D'Alembert, Lagrange, Laplace, Poinsot, Coriolis, Einstein, and others.

In terms of engineering application, dynamics is an even more recent science. Only since machines and structures have operated with high speeds and appreciable accelerations has it been necessary to make calculations based on the principles of dynamics rather than on the principles of statics. The rapid technological developments of the present day require increasing application of the principles of mechanics, particularly dynamics. These principles are basic to the analysis and design of moving structures, to fixed structures subject to shock loads, to high-speed computer mechanisms, to automatic control systems, to rockets, missiles, and spacecraft, to ground and air transportation vehicles, to electron ballistics of electrical devices, and to machinery of all types such as turbines, pumps, reciprocating engines, hoists, machine tools, etc. Students whose interests lead them into one or more of these and many other activities will find a constant need for applying the basic knowledge of dynamics.

1/2 **BASIC CONCEPTS.**   The concepts basic to mechanics were set forth in Art. 1/2 of Vol. 1 *Statics*. They are summarized here along with additional comments of special relevance to the study of dynamics.

Space is the geometric region occupied by bodies. Position in space is determined relative to some geometric reference system by means of linear and angular measurements. The basic frame of reference for the laws of Newtonian mechanics is the *primary inertial system* or *astronomical frame of reference*, which is an imaginary set of rectangular axes assumed to have no translation or rotation in space. Measurements show that the laws of Newtonian mechanics are valid for this reference system as long as any velocities involved are negligible compared with the speed of light.° Measurements made with respect to this reference are said to be *absolute*, and this reference system may be considered "fixed" in space. A reference frame attached to the surface of the earth has a somewhat complicated motion in the primary system, and a correction to the basic equations of mechanics must be applied for measurements made relative to the earth's reference frame. In the calculation of rocket and space flight trajectories, for example, the absolute motion of the earth becomes an important parameter. For most engineering problems of machines and structures which remain on the earth's surface, the corrections are extremely small and may be neglected. For these problems the laws of mechanics may be applied directly for measurements made relative to the earth, and in a practical sense such measurements will

---

° For velocities of the same order as the speed of light, 300 000 km/s or 186,000 mi/sec, the theory of relativity must be applied. See Art. 3/2 for a brief discussion of this theory and a numerical example of its effect.

be referred to as *absolute*.

Time is a measure of the succession of events and is considered an absolute quantity in Newtonian mechanics.

Mass is the quantitative measure of the inertia or resistance to change in motion of a body. Mass is also the property which gives rise to gravitational attraction.

Force is the vector action of one body on another. The properties of forces have been thoroughly treated in Vol. 1 *Statics*.

A *particle* is a body of negligible dimensions. Also, when the dimensions of a body are irrelevant to the description of its motion or the action of forces acting on it, the body may be treated as a particle. An airplane, for example, may be treated as a particle for the description of its flight path.

A *rigid body* is a body whose changes in shape are negligible compared with the overall dimensions of the body or with the changes in position of the body as a whole. As an example of the assumption of rigidity, the small flexural movement of the wing tip of an airplane flying through turbulent air is clearly of no consequence to the description of the motion of the airplane as a whole along its flight path. For this purpose, then, the treatment of the airplane as a rigid body offers no complication. On the other hand, if the problem is one of examining the internal stresses in the wing structure due to changing dynamic loads, then the deformable characteristics of the structure would have to be examined, and for this purpose the airplane could no longer be considered a rigid body.

Vector and *scalar* quantities have been treated extensively in Vol. 1 *Statics*, and their distinction should be perfectly clear by now. Scalar quantities are printed in lightface italic type, and vectors are shown in boldface type. Thus V is the scalar magnitude of the vector **V**. It is important that we use an identifying mark, such as an underline $\underline{V}$, for all handwritten vectors to take the place of the boldface designation in print. For two nonparallel vectors recall, for example, that $\mathbf{V}_1 + \mathbf{V}_2$ and $V_1 + V_2$ have two entirely different meanings.

It is assumed that the reader is familiar with the geometry and algebra of vectors through previous study of statics and mathematics. Students who need to review these topics will find a brief summary of them in Appendix B along with other mathematical relations which find frequent use in mechanics. Experience has shown that the geometry of mechanics is often a source of difficulty. Mechanics by its very nature is geometrical, and students should bear this in mind as they review their mathematics. In addition to vector algebra dynamics requires the use of vector calculus, and the essentials of this topic will be developed in the text as they are needed.

Dynamics involves the frequent use of time derivatives of both vectors and scalars. As a notational shorthand a dot over a quantity will frequently be used to indicate a derivative with respect to time. Thus $\dot{x}$ means dx/dt and $\ddot{x}$ stands for d^2x/dt^2.

1/3 NEWTON'S LAWS. Newton's three laws of motion, stated in Art. 1/4 of Vol. 1 *Statics*, are restated here because of their special significance to dynamics. In modern terminology they are

Law I. A particle remains at rest or continues to move in a straight line with a uniform velocity if there is no unbalanced force acting on it.

Law II. The acceleration of a particle is proportional to the resultant force acting on it and is in the direction of this force.°

Law III. The forces of action and reaction between interacting bodies are equal in magnitude, opposite in direction, and collinear.

These laws have been verified by countless physical measurements. The first two laws hold for measurements made in an absolute frame of reference but are subject to some correction when the motion is measured relative to a reference system having acceleration, such as one attached to the earth's surface.

Newton's second law forms the basis for most of the analysis in dynamics. For a particle of mass m subjected to a resultant force $\mathbf{F}$ the law may be stated as

$$\boxed{\mathbf{F} = m\mathbf{a}} \tag{1/1}$$

where $\mathbf{a}$ is the resulting acceleration measured in a nonaccelerating frame of reference. Newton's first law is a consequence of the second law since there is no acceleration when the force is zero, and the particle is either at rest or is moving with constant velocity. The third law constitutes the principle of action and reaction with which we should be thoroughly familiar from our work in statics.

1/4 UNITS. Both the International System of metric units (SI) and the U.S. customary system of units are defined and used in Vol. 2 *Dynamics*, although a stronger emphasis is placed on the metric system since it is replacing the U.S. customary system. To become familiar with each system it is necessary to think directly in each system. Although in engineering practice numerical conversion from one system to the other must be accomplished for some years, it would be a mistake to assume that familiarity with the new system can be achieved simply by the conversion of numerical results from the old system. Tables defining the SI units and giving numerical conversions between U.S. customary and SI units are included inside the front cover of the book. Charts comparing selected quantities in SI and U.S. customary units are included inside the back cover of the book to facilitate conversion and to help establish a feel for the relative size of units in both systems.

° To some it is preferable to interpret Newton's second law as meaning that the resultant force acting on a particle is proportional to the time rate of change of momentum of the particle and that this change is in the direction of the force. Both formulations are equally correct when applied to a particle of constant mass.

The four fundamental quantities of mechanics and their units and symbols for the two systems are summarized in the following table.

QUANTITY	DIMENSIONAL SYMBOL	SI UNITS		U.S. CUSTOMARY UNITS	
		UNIT	SYMBOL	UNIT	SYMBOL
Mass	M	Base units { kilogram	kg	slug	—
Length	L	meter°	m	Base units { foot	ft
Time	T	second	s	second	sec
Force	F	newton	N	pound	lb

° Also spelled *metre*.

In SI the units for mass, length, and time are taken as base units and the units for force were derived from Newton's second law of motion, Eq. 1/1. In the U.S. customary system the units for force, length, and time are base units and the units for mass are derived from the second law. The SI system is termed an *absolute* system since mass is taken to be an absolute or base quantity. The U.S. customary system is termed a *gravitational* system since force (as measured from gravitational pull) is taken as a base quantity. This distinction is a fundamental difference between the two systems.

In SI units by definition one newton is that force which will give a one-kilogram mass an acceleration of one meter per second squared. In the U.S. customary system a 32.1740-pound mass (1 slug) will have an acceleration of one foot per second squared when acted upon by a force of one pound. Thus for each system we have from Eq. 1/1

SI UNITS	U.S. CUSTOMARY UNITS
$(1 \text{ N}) = (1 \text{ kg})(1 \text{ m/s}^2)$	$(1 \text{ lb}) = (1 \text{ slug})(1 \text{ ft/sec}^2)$
$\text{N} = \text{kg} \cdot \text{m/s}^2$	$\text{slug} = \text{lb} \cdot \text{sec}^2/\text{ft}$

In SI units each unit and symbol stands for only one quantity. The kilogram is to be used *only* as a unit of mass and *never* as a unit of force. On the other hand in U.S. customary units the *pound* is used both as a unit of force (lbf) and a unit of mass (lbm). The distinction between the meaning as force or mass is usually clear from the situation at hand. The symbol for pound force is most frequently written merely lb.

Additional quantities used in mechanics and their equivalent base units will be defined as they are introduced in the chapters which follow. However, for convenient reference these quantities along with their SI base units are listed in one place in the first table inside the front cover of the book.

Detailed guidelines have been established for the consistent use of SI units, and these guidelines have been followed throughout this book. The most essential ones are summarized inside the front cover, and the student should observe these rules carefully.

1/5 GRAVITATION.   Newton's law of gravitation which governs the mutual attraction between bodies is

$$F = K\,\frac{m_1 m_2}{r^2}$$ (1/2)

where F = the mutual force of attraction between two particles
$\quad\quad K$ = a universal constant called the constant of gravitation
$\quad m_1, m_2$ = the masses of the two particles
$\quad\quad\quad r$ = the distance between the centers of the particles

The value of the gravitational constant from experimental data is $K = 6.673(10^{-11})\,\text{m}^3/(\text{kg}\cdot\text{s}^2)$. The only gravitational force of appreciable magnitude is the force due to the attraction of the earth. It was shown in Vol. 1 *Statics*, for example, that each of two iron spheres 100 mm in diameter is attracted to the earth with a gravitational force of 37.9 N, which is called its weight, but the force of mutual attraction between them if they are just touching is only 0.000 000 099 4 N.

Since the gravitational attraction or weight of a body is a force, it should always be expressed in force units, newtons (N) in SI units and pounds force (lb) in U.S. customary units. Unfortunately the mass unit kilogram (kg) has been used widely in common practice as a measure of weight. When expressed in kilograms the word "weight" technically means mass. To avoid confusion the word "weight" in this book shall be restricted to mean the force of gravitational attraction, and it will always be expressed in newtons or pounds force.

The force of gravitational attraction of the earth on a body depends on the position of the body relative to the earth. If the earth were a perfect sphere of the same volume, a body with a mass of exactly 1 kg would be attracted to the earth by a force of 9.824 N on the earth's surface, 9.821 N at an altitude of 1 km, 9.523 N at an altitude of 100 km, 7.340 N at an altitude of 1000 km, and 2.456 N at an altitude equal to the mean radius 6371 km of the earth. It is at once apparent that the variation in gravitational attraction of high-altitude rockets and spacecraft becomes a major consideration.

Every object which is allowed to fall in a vacuum at a given position on the earth's surface will have the same acceleration g as can be seen by combining Eqs. 1/1 and 1/2 and canceling the term representing the mass of the falling object. This combination gives

$$g = \frac{Km_0}{r^2}$$

where m_0 is the mass of the earth and r is the radius of the earth.° The

---

° It can be proved that the earth, when taken as a sphere with a symmetrical distribution of mass about its center, may be considered a particle with its entire mass concentrated at its center.

mass m_0 and the mean radius r of the earth have been found through experimental measurements to be $5.976(10^{24})$ kg and $6.371(10^6)$ m, respectively. These values, together with the value of K already cited, when substituted into the expression for g, give

$$g = 9.824 \text{ m/s}^2$$

The acceleration due to gravity as determined from the gravitational law is the acceleration which would be measured from a set of axes whose origin is at the center of the earth but which do not rotate with the earth. With respect to these "fixed" axes, then, this value may be termed the absolute value of g. Because of the fact that the earth rotates, the acceleration of a freely falling body as measured from a position attached to the earth's surface is slightly less than the absolute value. Accurate values of the gravitational acceleration as measured relative to the earth's surface account for the fact that the earth is a rotating oblate spheroid with flattening at the poles. These values are given by the International Gravity Formula which is

$$g = 9.780\ 49(1 + 0.005\ 288\ 4 \sin^2 \gamma - 0.000\ 005\ 9 \sin^2 2\gamma)$$

where γ is the latitude and g is expressed in meters per second squared. The constants account for the deviation of the earth's shape from that of a sphere and also for the effect of the earth's rotation. The absolute acceleration due to gravity as determined for a nonrotating earth may be computed from the relative values to a close approximation by adding $3.382(10^{-2}) \cos^2 \gamma$ m/s^2, which removes the effect of the earth's rotation. The variation of both the absolute and the relative values of g with latitude is shown in Fig. 1/1 for sea-level conditions.° The standard value adopted internationally for the gravitational accelera-

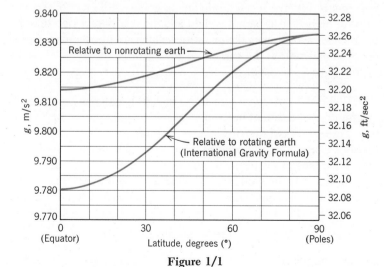

Figure 1/1

° Students will be able to derive these relations for a spherical earth following their study of relative motion in Chapter 3.

tion relative to the rotating earth at sea level and at a latitude of 45° is 9.806 65 m/s² or 32.1740 ft/sec².

The proximity of large land masses and the variations in the density of the earth's crust also influence the local value of g by a small but detectable amount. In almost all engineering problems where measurements are made on the surface of the earth the difference between the absolute and relative values of the gravitational acceleration and the effect of local variations are neglected, and 9.81 m/s² in SI units and 32.2 ft/sec² in U.S. customary units are used for the sea-level value of g.

The variation of g with altitude is easily determined by the gravitational law. If g_0 represents the absolute acceleration due to gravity at sea level, the absolute value at an altitude h is

$$g = g_0 \frac{r^2}{(r + h)^2}$$

where r is the radius of the earth.

The earth's gravitational attraction on a body may be calculated from the results of the simple gravitational experiment. If the gravitational force of attraction or true weight of a body is **W**, then, since the body will fall with an absolute acceleration **g** in a vacuum, Eq. 1/1 gives

$$\boxed{\mathbf{W} = m\mathbf{g}} \tag{1/3}$$

The apparent weight of a body as determined by a spring balance, calibrated to read the correct force and attached to the surface of the earth, will be slightly less than its true weight. The difference is due to the rotation of the earth. The ratio of the apparent weight to the apparent or relative acceleration due to gravity still gives the correct value of mass. The apparent weight and the relative acceleration due to gravity are, of course, the quantities which are measured in experiments conducted on the surface of the earth.

1/6 DIMENSIONS. A given dimension such as length can be expressed in a number of different units such as meters, millimeters, or kilometers. Thus the word *dimension* is distinguished from the word *unit*. Physical relations must always be dimensionally homogeneous, that is, the dimensions of all terms in an equation must be the same. It is customary to use the symbols *L*, *M*, *T*, and *F*, to stand for length, mass, time, and force, respectively. In SI units force is a derived quantity and from Eq. 1/1 has the dimensions of mass times acceleration or

$$F = ML/T^2$$

One important use of the theory of dimensions is found in checking the dimensional correctness of some derived physical relation. The following expression for the velocity v of a body of mass m which is moved from rest a horizontal distance x by a force F may be derived:

$$Fx = \tfrac{1}{2}mv^2$$

where the $\frac{1}{2}$ is a dimensionless coefficient resulting from integration. This equation is dimensionally correct since substitution of L, M, and T gives

$$[MLT^{-2}][L] = [M][LT^{-1}]^2$$

Dimensional homogeneity is a necessary condition for correctness, but it is not sufficient since the correctness of dimensionless coefficients cannot be checked in this way.

1/7 FORMULATION AND SOLUTION OF DYNAMICS PROBLEMS.

The study of dynamics is directed toward the understanding and description of the various quantities involved in the motions of bodies. This description, which is largely mathematical, enables predictions to be made of dynamical behavior. A dual thought process is necessary in formulating this description. It is necessary to think both in terms of the physical situation and in terms of the corresponding mathematical description. Analysis of every problem will require this repeated transition of thought between the physical and the mathematical. Without question, one of the greatest difficulties encountered by students is the inability to make this transition of thought freely. They should recognize that the mathematical formulation of a physical problem represents an ideal and limiting description, or model, which approximates but never quite matches the actual physical situation.

In the course of constructing the idealized mathematical model for any given engineering problem, certain approximations will always be involved. Some of these approximations may be mathematical, whereas others will be physical. For instance, it is often necessary to neglect small distances, angles, or forces compared with large distances, angles, or forces. If the change in velocity of a body with time is nearly uniform, then an assumption of constant acceleration may be justified. An interval of motion which cannot be easily described in its entirety is often divided into small increments each of which can be approximated. The retarding effect of bearing friction on the motion acquired by a machine as the result of applied forces or moments may often be neglected if the friction forces are small. However, these same friction forces cannot be neglected if the purpose of the inquiry is a determination of the drop in efficiency of the machine due to the friction process. Thus the degree of assumption involved depends on what information is desired and on the accuracy required. The student should be constantly alert to the various assumptions called for in the formulation of real problems. The ability to understand and make use of the appropriate assumptions in the course of the formulation and solution of engineering problems is, certainly, one of the most important characteristics of a successful engineer. Along with the development of the principles and analytical tools needed for modern dynamics, one of the major aims of this book is to provide a maximum of opportunity to develop ability in formulating good mathematical models. Strong emphasis is placed on a wide range of practical problems which not only require the full exercise of theory but also force consideration of the decisions which must be made concerning relevant assumptions.

An effective method of attack on dynamics problems, as in all engineering problems, is essential. The development of good habits in formulating problems and in representing their solutions will prove to be an invaluable asset. Each solution should proceed with a logical sequence of steps from hypothesis to conclusion, and its representation should include a clear statement of the following parts, each clearly identified:

1. Given data
2. Results desired
3. Necessary diagrams
4. Calculations
5. Answers and conclusions

In addition it is well to incorporate a series of checks on the calculations at intermediate points in the solution. The reasonableness of numerical magnitudes should be observed, and the accuracy and dimensional homogeneity of terms should be checked frequently. It is also important that the arrangement of work be neat and orderly. Careless solutions which cannot be read easily by others are of little or no value. It will be found that the discipline involved in adherence to good form will in itself be an invaluable aid to the development of the abilities for formulation and analysis. Many problems which at first may seem difficult and complicated become clear and straightforward once they are begun with a logical and disciplined method of attack.

The subject of dynamics is based upon a surprisingly few fundamental concepts and principles which, however, are extended and applied over an exceedingly wide range of conditions. One of the most valuable aspects of the study of dynamics is the experience afforded in reasoning from fundamentals. This experience cannot be obtained merely by memorizing the kinematical and dynamical equations which describe various motions. It must be obtained through exposure to a wide variety of problem situations which force the choice, use, and extension of basic principles to meet the given conditions.

In describing the relations between forces and the motions they produce, it is essential that the system to which a principle is applied be clearly defined. At times a single particle or a rigid body is the system to be isolated, whereas at other times two or more bodies taken together constitute the system. The definition of the system to be analyzed is made clear by constructing its *free-body diagram*. This diagram consists of a closed outline of the external boundary of the system defined. All bodies which contact and exert forces on the system but are not a part of it are removed and replaced by vectors representing the forces they exert *on* the system isolated. In this way we make a clear distinction between the action and reaction of each force, and account is taken of *all* forces on and external to the system. It is assumed that all students are familiar with the technique of drawing free-body diagrams from their prior work in statics.

In applying the laws of dynamics, numerical values of the quantities may be used directly in proceeding toward the solution, or algebraic symbols may be used to represent the quantities involved and the answer left as a formula. With numerical substitution the magnitudes of all quantities expressed in their particular units are evident at each stage of the calculation. This approach offers advantage when the practical significance of the magnitude of each term is important. The symbolic solution, however, has several advantages over the numerical solution. First, the abbreviation achieved by the use of symbols aids in focusing attention on the interconnection between the physical situation and its related mathematical description. Second, a symbolic solution permits a dimensional check to be made at every step, whereas dimensional homogeneity may not be checked when numerical values are used. Third, a symbolic solution may be used repeatedly for obtaining answers to the same problem when different sets and sizes of units are used. Facility with both forms of solution is essential, and ample practice with each should be sought in the problem work.

Students will find that solutions to the various equations of dynamics may be obtained in one of three ways. First, a direct mathematical solution by hand calculation may be carried out with answers appearing either as algebraic symbols or as numerical results. The large majority of the problems come under this category. Second, certain problems are readily handled by graphical solutions, such as with the determination of velocities and accelerations in two-dimensional relative motion of rigid bodies. Third, there are a number of problems in Vol. 2 *Dynamics* which are appropriate for computer solution, and students who have access to digital computation facilities may wish to use them for some of the problems. The choice of the most expedient method of solution is an important aspect of the experience to be gained from the problem work.

KINEMATICS OF PARTICLES

2

2/1 INTRODUCTION. Kinematics is that branch of dynamics which describes the motion of bodies without reference to the forces which either cause the motion or are generated as a result of the motion. Kinematics is often referred to as the "geometry of motion." The design of cams, gears, linkages, and other machine elements to control or produce certain desired motions and the calculation of flight trajectories for aircraft, rockets, and spacecraft are a few examples of kinematic problems which engage the attention of engineers. A thorough working knowledge of kinematics is an absolute prerequisite to kinetics, which is the study of the relationships between motion and the corresponding forces which cause or accompany the motion.

We start our study of kinematics by first discussing in this chapter the motions of points or particles. A particle is a body whose physical dimensions are so small compared with the radius of curvature of its path that we can treat the motion of the particle as that of a point. For example, the wingspan of a jet transport flying between Los Angeles and New York is of no consequence compared with the curvature of its flight path, and the treatment of the airplane as a particle or point should raise no question.

There are a number of ways in which the motion of a particle can be described, and the choice of the most convenient or appropriate method depends a great deal on experience and on how the data are given. Let us get an overview of the several methods developed in this chapter by referring, first, to Fig. 2/1 which shows a particle P moving along some general path in space. If the particle is confined to a specified path, as with a bead sliding along a fixed wire, its motion is said to be *constrained*. If there are no physical guides, the motion is said to be *unconstrained*. A small rock tied to the end of a string and whirled in a circle undergoes constrained motion until the string breaks, after which instant its motion is unconstrained.

The position of particle P at any time t can be described by specifying its rectangular coordinates° x, y, z, its cylindrical coordi-

---

°Often called *Cartesian* coordinates, named after René Descartes (1596–1650), a French mathematician who was one of the inventors of analytic geometry.

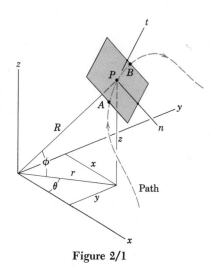

Figure 2/1

nates r, θ, z, or its spherical coordinates R, θ, ϕ. The motion of P may also be described by measurements along the tangent t and normal n to the curve. These measurements are known as *path variables*. The direction of n lies in the local plane of the curve.[°]

The motion of particles (or rigid bodies) may be described by using coordinates measured from fixed reference axes (*absolute-motion* analysis) or by using coordinates measured from moving reference axes (*relative-motion* analysis). Both descriptions will be developed and applied in the articles which follow.

With this conceptual picture of the description of particle motion in mind, we will now restrict our attention in the remainder of this chapter to the case of *plane motion* where all movement occurs in or can be represented as occurring in a single plane. A large proportion of the motions of machines and structures in engineering can be represented as plane motion. Later, in Chapter 7 an introduction to three-dimensional motion is presented. We will begin our discussion of plane motion with *rectilinear motion,* which is motion along a straight line, and follow it with a description of motion along a plane curve.

2/2 RECTILINEAR MOTION. Consider a particle P moving along a straight line, Fig. 2/2. The position of P at any instant of time t may be specified by its distance s measured from some convenient reference point O fixed on the line. At time $t + \Delta t$ the particle has moved to P' and its coordinate becomes $s + \Delta s$. The change in the position coordinate during the interval Δt is called the *displacement* Δs of the particle. The displacement would be negative if the particle moved in the negative s-direction.

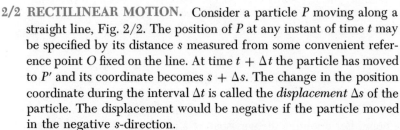

Figure 2/2

The average velocity of the particle during the interval Δt is the displacement divided by the time interval or $v_{av} = \Delta s / \Delta t$. As Δt becomes smaller and approaches zero in the limit, the average velocity approaches the *instantaneous velocity* of the particle, which is

$$v = \lim_{\Delta t \to 0} \frac{\Delta s}{\Delta t} \text{ or}$$

$$v = \frac{ds}{dt} = \dot{s} \tag{2/1}$$

Thus the velocity is the time rate of change of the position coordinate s. The velocity is positive or negative depending on whether the corresponding displacement is positive or negative.

[°]This plane is known as the *osculating* plane, which comes from the Latin word *osculari* meaning "to kiss." The plane which contains P and the two points A and B, one on either side of P, becomes the osculating plane as the distances between the points approach zero.

The average acceleration of the particle during the interval Δt is the change in its velocity divided by the time interval or $a_{\text{av}} = \Delta v / \Delta t$. As Δt becomes smaller and approaches zero in the limit, the average acceleration approaches the *instantaneous acceleration* of the particle, which is $a = \lim\limits_{\Delta t \to 0} \dfrac{\Delta v}{\Delta t}$ or

$$\boxed{a = \frac{dv}{dt} = \dot{v}} \quad \text{or} \quad \boxed{a = \frac{d^2s}{dt^2} = \ddot{s}} \qquad (2/2)$$

The acceleration is positive or negative depending on whether the velocity is increasing or decreasing. Note that the acceleration would be positive if the particle had a negative velocity which was becoming less negative. If the particle is slowing down but its velocity is still positive, the particle is said to be *decelerating*. In this case the acceleration would be negative and would be in the direction opposite to the velocity.

By eliminating the time dt between Eq. 2/1 and the first of Eqs. 2/2 a differential equation relating displacement, velocity, and acceleration results which is

$$\boxed{v \, dv = a \, ds} \quad \text{or} \quad \boxed{\dot{s} \, d\dot{s} = \ddot{s} \, ds} \qquad (2/3)$$

Equations 2/1, 2/2, and 2/3 are the differential equations for the rectilinear motion of a particle. Problems in rectilinear motion involving finite changes in the motion variables are solved by integration of these basic differential relations. The position coordinate s, the velocity v, and the acceleration a are algebraic quantities, so that their signs, positive or negative, must be carefully observed. Note that the positive directions for v and a are the same as the positive direction for s.

Interpretation of the differential equations governing rectilinear motion is considerably clarified by representing the relationships among s, v, a, and t graphically. Figure 2/3a represents a schematic plot of the variation of s with t from time t_1 to time t_2 for some given rectilinear motion. By constructing the tangent to the curve at any time t we obtain the slope, which is the velocity $v = ds/dt$. Thus the velocity may be determined at all points on the curve and plotted against the corresponding time as shown in Fig. 2/3b. Similarly the slope dv/dt of the v-t curve at any instant gives the acceleration at that instant, and the a-t curve can therefore be plotted as in Fig. 2/3c.

We now see from Fig. 2/3b that the area under the v-t curve during time dt is $v \, dt$ which from Eq. 2/1 is the displacement ds. Consequently the net displacement of the particle during the interval

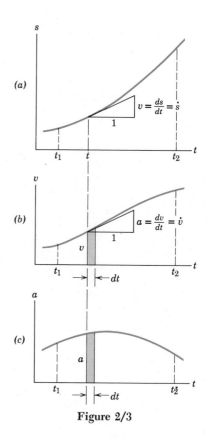

Figure 2/3

from t_1 to t_2 is the corresponding area under the curve which is

$$\int_{s_1}^{s_2} ds = \int_{t_1}^{t_2} v\, dt \qquad \text{or} \qquad s_2 - s_1 = (\text{area under } v\text{-}t \text{ curve})$$

Similarly, from Fig. 2/3c we see that the area under the a-t curve during time dt is $a\, dt$ which, from the first of Eqs. 2/2, is dv. Thus the net change in velocity between t_1 and t_2 is the corresponding area under the curve, which is

$$\int_{v_1}^{v_2} dv = \int_{t_1}^{t_2} a\, dt \qquad \text{or} \qquad v_2 - v_1 = (\text{area under } a\text{-}t \text{ curve})$$

Two additional graphical relations are noted. When the acceleration a is plotted as a function of the position coordinate s, Fig. 2/4a, the area under the curve during a displacement ds is $a\, ds$ which, from Eq. 2/3, is $v\, dv = d(v^2/2)$. Thus the net area under the curve between position coordinates s_1 and s_2 is

$$\int_{v_1}^{v_2} v\, dv = \int_{s_1}^{s_2} a\, ds \qquad \text{or} \qquad \tfrac{1}{2}(v_2{}^2 - v_1{}^2) = (\text{area under } a\text{-}s \text{ curve})$$

When the velocity v is plotted as a function of the position coordinate s, Fig. 2/4b, the slope of the curve at any point A is dv/ds. By constructing the normal AB to the curve at this point, we see from the similar triangles that $\overline{CB}/v = dv/ds$. Thus, from Eq. 2/3, $\overline{CB} = v(dv/ds) = a$, the acceleration. It is necessary that the velocity and position coordinate axes have the same numerical scales so that the acceleration read on the position coordinate scale in meters (or feet), say, will represent the actual acceleration in the units meters (or feet) per second squared.

The graphical representations described are useful not only in visualizing the relationships among the several motion quantities but also in approximating results by graphical integration or differentiation when a lack of knowledge of the mathematical relationship prevents its expression as an explicit mathematical function. Experimental data and motions which involve discontinuous relationships between the variables are frequently analyzed graphically.

If the position coordinate s is known for all values of the time t, then successive mathematical or graphical differentiation with respect to t gives the velocity v and acceleration a. In many problems, however, the functional relationship between position coordinate and time is unknown, and we must determine it by successive integration from the acceleration. Acceleration is determined by the forces which act on moving bodies and is computed from the equations of kinetics discussed in subsequent chapters. Depending on the nature of

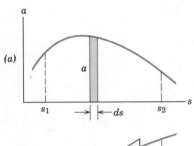

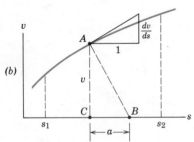

Figure 2/4

the forces, the acceleration may be specified as a function of time, velocity, position coordinate, or as a combined function of these quantities. The procedure for integrating the differential equation in each case is indicated as follows:

(*a*) *Constant Acceleration.* When *a* is constant, the first of Eqs. 2/2 and 2/3 may be integrated directly. For simplicity with $s = s_0$, $v = v_0$, and $t = 0$ designated at the beginning of the interval, then for a lapse of time *t* the integrated equations become

$$\int_{v_0}^{v} dv = a \int_{0}^{t} dt \quad \text{or} \quad v = v_0 + at$$

$$\int_{v_0}^{v} v \, dv = a \int_{s_0}^{s} ds \quad \text{or} \quad v^2 = v_0^2 + 2a(s - s_0)$$

Substitution of the integrated expression for *v* into Eq. 2/1 and integration with respect to *t* give

$$\int_{s_0}^{s} ds = \int_{0}^{t} (v_0 + at) \, dt \quad \text{or} \quad s = s_0 + v_0 t + \tfrac{1}{2}at^2$$

These relations are necessarily restricted to the special case where the acceleration is constant. The integration limits depend on the initial and final conditions and for a given problem may be different from those used here. It may be more convenient, for instance, to begin the integration at some specified time t_1 rather than at time $t = 0$.

Caution: One of the most frequent mistakes made by students is the attempt to use the foregoing equations for problems of variable acceleration where they do not apply since they have been integrated for constant acceleration only.

(*b*) *Acceleration Given as a Function of Time, $a = f(t)$.* Substitution of the function into the first of Eqs. 2/2 gives $f(t) = dv/dt$ which permits separating the variables and integrating. Thus

$$\int_{v_0}^{v} dv = \int_{0}^{t} f(t) \, dt \quad \text{or} \quad v = v_0 + \int_{0}^{t} f(t) \, dt$$

From this integrated expression for *v* as a function of *t*, the position coordinate *s* is obtained by integrating Eq. 2/1 which, in form, would be

$$\int_{s_0}^{s} ds = \int_{0}^{t} v \, dt \quad \text{or} \quad s = s_0 + \int_{0}^{t} v \, dt$$

If the indefinite integral is employed, the end conditions are used to establish the constants of integration with results which are identical with those obtained by using the definite integral.

If desired, the displacement s may be obtained by a direct solution of the second-order differential equation $\ddot{s} = f(t)$ obtained by substitution of $f(t)$ into the second of Eqs. 2/2.

(*c*) *Acceleration Given as a Function of Velocity,* $a = f(v)$. Substitution of the function into the first of Eqs. 2/2 gives $f(v) = dv/dt$, which permits separating the variables and integrating. Thus

$$t = \int_0^t dt = \int_{v_0}^v \frac{dv}{f(v)}$$

This result gives t as a function of v. Then it would be necessary to solve for v as a function of t so that Eq. 2/1 can be integrated to obtain the position coordinate s as a function of t.

Alternatively the function $a = f(v)$ can be substituted into the first of Eqs. 2/3 giving $v \, dv = f(v) \, ds$. The variables are now separated and the equation integrated in the form

$$\int_{v_0}^v \frac{v \, dv}{f(v)} = \int_{s_0}^s ds \qquad \text{or} \qquad s = s_0 + \int_{v_0}^v \frac{v \, dv}{f(v)}$$

Note that this equation gives s in terms of v without explicit reference to t.

(*d*) *Acceleration Given as a Function of Displacement,* $a = f(s)$. Substitution of the function into Eq. 2/3 and integrating gives the form

$$\int_{v_0}^v v \, dv = \int_{s_0}^s f(s) \, ds \qquad \text{or} \qquad v^2 = v_0{}^2 + 2 \int_{s_0}^s f(s) \, ds$$

Next we solve for v to give $v = g(s)$, a function of s. Now we can substitute ds/dt for v, separate variables, and integrate in the form

$$\int_{s_0}^s \frac{ds}{g(s)} = \int_0^t dt \qquad \text{or} \qquad t = \int_{s_0}^s \frac{ds}{g(s)}$$

which gives t as a function of s. Lastly we can rearrange to obtain s as a function of t.

In each of the foregoing cases when the acceleration varies according to some functional relationship, the ability to solve the equations by direct mathematical integration will depend on the form of the function. In cases where the integration is excessively awkward or difficult, integration by graphical, numerical, or computer methods may be utilized.

Sample Problem 2/1

The position coordinate of a particle which is confined to move along a straight line is given by $s = 2t^3 - 24t + 6$ where s is measured in meters from a convenient origin and where t is in seconds. Determine (a) the time required for the particle to reach a velocity of 72 m/s from its initial condition at $t = 0$, (b) the acceleration of the particle when $v = 30$ m/s, and (c) the net displacement of the particle during the interval from $t = 1$ s to $t = 4$ s.

Solution. The velocity and acceleration are obtained by successive differentiation of s with respect to the time. Thus

$[v = \dot{s}]$ $\qquad\qquad$ $v = 6t^2 - 24$ m/s

$[a = \dot{v}]$ $\qquad\qquad$ $a = 12t$ m/s^2

(a) Substituting $v = 72$ m/s into the expression for v gives us $72 = 6t^2 - 24$ from which $t = \pm 4$ s. The negative root describes a mathematical solution for t before the initiation of motion, so is of no physical interest. Thus the desired result is

$$t = 4 \text{ s} \qquad\qquad Ans.$$

(b) Substituting $v = 30$ m/s into the expression for v gives $30 = 6t^2 - 24$ from which the positive root is $t = 3$ s, and the corresponding acceleration is

$$a = 12(3) = 36 \text{ m/s}^2 \qquad\qquad Ans.$$

(c) The net displacement during the specified interval is

$$\Delta s = s_4 - s_1 \qquad \text{or}$$

$$\Delta s = [2(4^3) - 24(4) + 6] - [2(1^3) - 24(1) + 6]$$

$$= 54 \text{ m} \qquad\qquad Ans.$$

which represents the net advancement of the particle along the s-axis from the position it occupied at $t = 1$ s to its position at $t = 4$ s.

To help visualize the motion the values of s, v, and a are plotted against the time t as shown. Because the area under the v-t curve represents displacement, we see that the net displacement from $t = 1$ s to $t = 4$ s is the positive area Δs_{2-4} less the negative area Δs_{1-2}.

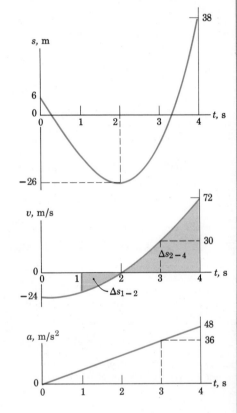

① Be alert to the proper choice of sign when taking a square root. When the situation calls for only one answer, the positive root is not always the one you may need.

② Note carefully the distinction between italic s for the position coordinate and the vertical s for seconds.

③ Note from the graphs that the values for v are the slopes ($\dot{s}$) of the s-t curve and that the values for a are the slopes ($\dot{v}$) of the v-t curve. *Suggestion:* Integrate $v\,dt$ for each of the two intervals and check the answer for Δs. Show that the total distance traveled during the interval $t = 1$ s to $t = 4$ s is 74 m.

Sample Problem   2/2

A particle moves along the x-axis with an initial velocity $v_x = 50$ ft/sec at the origin when $t = 0$. For the first 4 sec it has no acceleration, and thereafter it is acted upon by a retarding force which gives it a constant acceleration $a_x = -10$ ft/sec². Calculate the velocity and the x-coordinate of the particle for the conditions of $t = 8$ sec and $t = 12$ sec and find the maximum positive x-coordinate reached by the particle.

① Learn to be flexible with symbols. The position coordinate x is just as good as s.

Solution.   The velocity of the particle after $t = 4$ sec is computed from

$$\left[\int dv = \int a\, dt \right] \qquad \int_{50}^{v_x} dv_x = -10 \int_{4}^{t} dt, \qquad v_x = 90 - 10t \text{ ft/sec}$$

② Note that we integrate to a general time t and then substitute specific values.

and is plotted as shown. At the specified times the velocities are

$$t = 8 \text{ sec}, \qquad v_x = 90 - 10(8) = 10 \text{ ft/sec}$$

$$t = 12 \text{ sec}, \qquad v_x = 90 - 10(12) = -30 \text{ ft/sec} \qquad \text{Ans.}$$

The x-coordinate of the particle at any time greater than 4 sec is the distance traveled during the first 4 sec plus the distance traveled after the discontinuity in acceleration occurred. Thus

$$\left[\int ds = \int v\, dt \right] \qquad x = 50(4) + \int_{4}^{t} (90 - 10t)\, dt = -5t^2 + 90t - 80 \text{ ft}$$

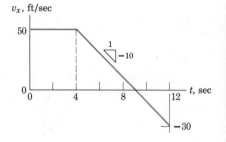

For the two specified times

$$t = 8 \text{ sec}, \qquad x = -5(8^2) + 90(8) - 80 = 320 \text{ ft}$$

$$t = 12 \text{ sec}, \qquad x = -5(12^2) + 90(12) - 80 = 280 \text{ ft} \qquad \text{Ans.}$$

The x-coordinate for $t = 12$ sec is less than that for $t = 8$ sec since the motion is in the negative x-direction after $t = 9$ sec. The maximum positive x-coordinate is, then, the value of x for $t = 9$ sec which is

$$x_{\text{max}} = -5(9^2) + 90(9) - 80 = 325 \text{ ft} \qquad \text{Ans.}$$

③ These displacements are seen to be the net positive areas under the v-t graph up to the value of t in question.

③ Show that the total distance traveled by the particle in the 12 seconds is 370 ft.

Sample Problem 2/3

The spring-mounted slider moves in the horizontal guide with negligible friction and has a velocity v_0 in the s-direction as it crosses the mid-position where $s = 0$ and $t = 0$. The two springs together exert a retarding force to the motion of the slider which gives it an acceleration proportional to the displacement but oppositely directed and equal to $a = -k^2 s$ where k is constant. (The constant is arbitrarily squared for later convenience in the form of the expressions.) Determine the expressions for the displacement s and velocity v as functions of the time t.

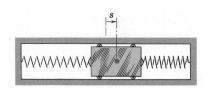

Solution I. Since the acceleration is specified in terms of the displacement, the differential relation $v \, dv = a \, ds$ may be integrated. Thus

$$\int v \, dv = \int -k^2 s \, ds + C_1 \text{ a constant,} \quad \text{or} \quad \frac{v^2}{2} = -\frac{k^2 s^2}{2} + C_1$$

When $s = 0$, $v = v_0$, so that $C_1 = v_0^2/2$, and the velocity becomes

$$v = +\sqrt{v_0^2 - k^2 s^2}$$

① We have used an indefinite integral here and evaluated the constant of integration. For practice obtain the same results by using the definite integral with the appropriate limits.

The plus sign of the radical is taken when v is positive (in the plus s-direction). This last expression may be integrated by substituting $v = ds/dt$. Thus,

$$\int \frac{ds}{\sqrt{v_0^2 - k^2 s^2}} = \int dt + C_2 \text{ a constant,} \quad \text{or} \quad \frac{1}{k} \sin^{-1} \frac{ks}{v_0} = t + C_2$$

② Again try the definite integral here as above.

With the requirement of $t = 0$ when $s = 0$, the constant of integration becomes $C_2 = 0$, and we may solve the equation for s so that

$$s = \frac{v_0}{k} \sin kt \qquad \qquad Ans.$$

The velocity is $v = \dot{s}$ which gives

$$v = v_0 \cos kt \qquad \qquad Ans.$$

Solution II. Since $a = \ddot{s}$, the given relation may be written at once as

$$\ddot{s} + k^2 s = 0$$

This is an ordinary linear differential equation of second order for which the solution is well known and is

$$s = A \sin Kt + B \cos Kt$$

where A, B, and K are constants. Substitution of this expression into the differential equation shows that it satisfies the equation provided that $K = k$. The velocity is $v = \dot{s}$ which becomes

$$v = Ak \cos kt - Bk \sin kt$$

The boundary condition $v = v_0$ when $t = 0$ requires that $A = v_0/k$, and the condition $s = 0$ when $t = 0$ gives $B = 0$. Thus the solution is

③ This motion is called *simple harmonic motion* and is characteristic of all oscillations where the restoring force, and hence the acceleration, is proportional to the displacement but opposite in sign.

$$s = \frac{v_0}{k} \sin kt \qquad \text{and} \qquad v = v_0 \cos kt \qquad \qquad Ans.$$

Sample Problem 2/4

A freighter is moving at a speed of 8 knots when its engines are suddenly stopped. If it takes 10 minutes for the freighter to reduce its speed to 4 knots, determine and plot the distance s in nautical miles moved by the ship and its speed v in knots as functions of the time t during this interval. The deceleration of the ship is proportional to the square of its speed, so that $a = -kv^2$.

① Recall that one knot is the speed of one nautical mile (6080 ft) per hour. Work directly in the units of nautical miles and hours.

Solution. The speeds and the time are given, so we may substitute the expression for acceleration directly into the basic definition $a = dv/dt$ and integrate. Thus

$$-kv^2 = \frac{dv}{dt}, \qquad \frac{dv}{v^2} = -k\,dt, \qquad \int_8^v \frac{dv}{v^2} = -k\int_0^t dt$$

$$-\frac{1}{v} + \frac{1}{8} = -kt, \qquad v = \frac{8}{1 + 8kt}$$

Now we substitute the end limits of $v = 4$ knots and $t = \frac{10}{60} = \frac{1}{6}$ hour and get

$$4 = \frac{8}{1 + 8k(1/6)}, \qquad k = \frac{3}{4}\text{ mi}^{-1}, \qquad v = \frac{8}{1 + 6t} \qquad Ans.$$

② We choose to integrate to a general value of v and its corresponding time t so that we may obtain the variation of v with t.

The speed is plotted against the time as shown.

The distance is obtained by substituting the expression for v into the definition $v = ds/dt$ and integrating. Thus

$$\frac{8}{1 + 6t} = \frac{ds}{dt}, \qquad \int_0^t \frac{8\,dt}{1 + 6t} = \int_0^s ds, \qquad s = \frac{4}{3}\ln(1 + 6t) \qquad Ans.$$

The distance s is also plotted against the time as shown, and we see that the ship has moved through a distance $s = \frac{4}{3}\ln(1 + \frac{6}{6}) = \frac{4}{3}\ln 2 = 0.924$ mi (nautical) during the 10 minutes.

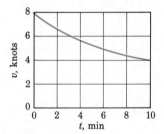

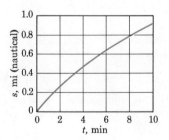

PROBLEMS

2/1 The motion of a particle is governed by the equation $s = \frac{1}{3}t^3 - 2t^2 - 6$ where s is the position coordinate in feet and t is the time in seconds. Plot the velocity v of the particle against the time t and find the acceleration for $t = 0$, 2, and 4 seconds.

　　　　Ans. $a_0 = -4$ ft/sec^2, $a_2 = 0$, $a_4 = 4$ ft/sec^2

2/2 The position of a certain particle is described by $s = 4t^3 + 15t^2 - 18t + 6$ where s is in meters and t is in seconds. Determine the time t and the acceleration a when $v = 0$. Consider positive values of t only.

2/3 The position of a particle in millimeters is given by $s = -9 + 4t - t^2/3$ where t is in seconds. Plot the s-t and v-t curves and determine the velocity v and coordinate s when $t = 9$ s. Also find the displacement Δs and the total distance D traveled during the interval from $t = 3$ s to $t = 9$ s. What can you tell about the acceleration by inspection of the degree of the s-t relation?

　　　　　Ans. $v = -2$ mm/s, $s = 0$, $D = 6$ mm

2/4 In the final stages of a moon landing the lunar module descends under retro-thrust of its descent engine to within $h = 5$ m of the lunar surface where it has a downward velocity of 4 m/s. If the descent engine is cut off abruptly at this point, compute the impact velocity of the landing gear with the moon. Lunar gravity is $\frac{1}{6}$ of the earth's gravity.

2/5 A projectile is fired vertically with an initial velocity of 800 ft/sec. Calculate the maximum altitude h reached by the projectile and the time t after firing for it to return to the ground. Neglect air resistance and take the gravitational acceleration to be constant at 32.2 ft/sec^2.

　　　　　　Ans. $h = 9938$ ft, $t = 49.7$ sec

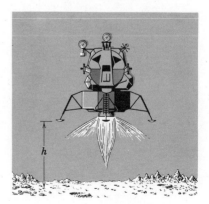

Problem 2/4

2/6 The pilot of a jet transport brings his engines to full takeoff power before releasing the brakes as the aircraft is standing on the runway. The jet thrust remains constant, and the aircraft has a near-constant acceleration of 0.6g. If the takeoff speed is 120 mi/hr, calculate the distance s and time t from rest to takeoff.

2/7 A retarding force of 4-s duration acts on a particle moving initially with a velocity of 100 m/s. The oscilloscope record of the deceleration is shown. Approximate the velocity of the particle at $t = 12$ s.
Ans. $v_{12} = -96.2$ m/s

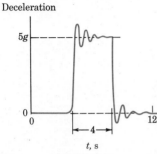

Problem 2/7

2/8 A jet aircraft with a landing speed of 200 km/h has a maximum of 600 m of available runway after touchdown in which to reduce its ground speed to 30 km/h. Compute the average acceleration a required of the aircraft during braking.

2/9 A single-stage rocket is launched vertically from rest, and its thrust is programmed to give the rocket a constant upward acceleration of 6 m/s². If the fuel is exhausted 20 s after launch, calculate the maximum velocity v_m and the subsequent maximum altitude h reached by the rocket.
Ans. $v_m = 120$ m/s, $h = 1.934$ km

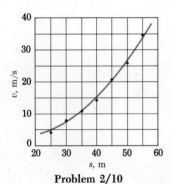

Problem 2/10

2/10 Experimental data for the motion of a particle along a straight line yield measured values of the velocity v for various position coordinates s. A smooth curve is drawn through the points as shown in the graph. Determine the acceleration of the particle when $s = 40$ m.   *Ans.* $a = 15$ m/s²

2/11 The velocity v of a particle varies with the position coordinate s as shown in the graph for an interval of its motion. Find the velocity v of the particle for the instant when $s = 40$ m if its velocity is decreasing 30 m/s per second at this position.

Problem 2/11

2/12 In a test of air resistance under conditions of deceleration a small test vehicle is given an initial velocity against the air flow in a wind tunnel, and its position x from the start is measured as a function of time and tabulated. Approximate the velocity v and acceleration a in the x-direction at $t = 2$ sec. Draw a smooth curve through the plotted points and solve graphically.

t, sec	x, in.	t, sec	x, in.
0	0	3.5	80
0.5	19	4.0	84
1.0	32	4.5	83
1.5	49	5.0	82
2.0	57	5.5	79
2.5	68	6.0	69
3.0	75		

Ans. $v = 21$ in./sec, $a = -7.5$ in./sec^2

2/13 A particle which moves along the x-axis has a coordinate in meters given by $x = 3t^3 - 18t^2 + 24t$ where t is measured in seconds. The variation of x with t is shown in the figure. First, by strictly graphical procedures, plot the v-t and then the a-t curve for the motion. Then check your results by plotting your mathematical expressions for v and a on the same figures.

2/14 We are given the s-t curve for the motion of a particle. We are also given that the slope of the curve at $t = 2$ min and $t = 6$ min is 0.33 km/min. Construct the corresponding v-t curve to within the accuracy of the given information using km/h for v.

2/15 The spring-loaded body oscillates in the vertical direction with a y-coordinate specified by $y = y_0 \sin 2\pi nt$ where y is measured from the equilibrium position. The maximum value of y is y_0, and n is the constant number of complete cycles per unit of time t. Determine the velocity $v_{1/4}$ and acceleration $a_{1/4}$ of the body when $t = 1/(4n)$ which is the time for $\frac{1}{4}$ of a complete oscillation.

Ans. $v_{1/4} = 0$, $a_{1/4} = -4\pi^2 n^2 y_0$

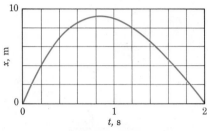

Problem 2/13

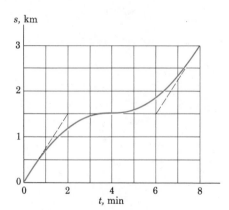

Problem 2/14

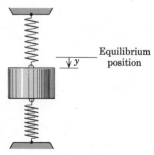

Problem 2/15

2/16 The velocity v in meters per second of a sliding part of a machine varies with the time t in seconds according to $v = 0.4(t - 2)$. If the position coordinate s of the part is 0.6 m when $t = 0$, determine the displacement Δs and the total distance D traveled during the first 3 seconds of motion starting at $t = 0$. Plot s versus t for the first 4 seconds.

2/17 A particle which moves along a straight line has a velocity in millimeters per second given by $v = 300 - 75t^2$ where t is in seconds. Calculate the total distance D covered during the interval from $t = 0$ to $t = 3$ s and find the net displacement Δs of the particle during this same interval.
     Ans. $D = 575$ mm, $\Delta s \doteq 225$ mm

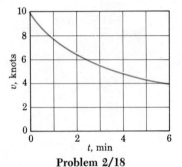

Problem 2/18

2/18 A ship reduces speed as it approaches the entrance to a harbor in still water. The speed is reduced from 10 knots as the ship passes the first entrance buoy to 4 knots 6 minutes later as it passes the second buoy. Approximate the distance Δs between the buoys in nautical miles by using the graph directly. (Recall that 1 knot equals one nautical mile per hour.)
     Ans. $\Delta s = 0.60$ mi (nautical)

2/19 An object moves along the x-axis with a constant acceleration. When $t = 0$, $x = -6$ m and $\dot{x} = 4$ m/s. Also when $t = 10$ s, a maximum value of x is recorded. Determine x_{max} and the value of x when $t = 15$ s.

2/20 A particle moves along the y-axis with a constant acceleration for a period of time. When $t = 0$, $y = 16$ ft, and when $t = 8$ sec, $y = 0$. Also the particle comes to rest momentarily when $t = 5$ sec. Determine the initial velocity v_0 when $t = 0$. *Suggestion:* Make use of the y-t and $\dot{y}$-t graphs as a guide to the mathematics.   *Ans.* $v_0 = -10$ ft/sec

2/21 A particle which moves along the x-axis with constant acceleration has a velocity of 60 in./sec in the negative x-direction at time $t = 0$ when its x-coordinate is 45 in. Three seconds later the particle passes the origin going in the positive x-direction. How far to the negative side of the origin does the particle travel?

22 A motorcycle starts from rest at point A and travels 300 m along a straight horizontal track to point B where it comes to a stop. If the acceleration of the motorcycle is limited to $0.7g$ and its deceleration is limited to $0.6g$, calculate the least possible time t to cover the distance. What maximum velocity v is reached?   *Ans.* $t = 13.76$ s, $v = 157.0$ km/h

23 The magnitude of the acceleration and deceleration of an express elevator is limited to $0.4g$, and maximum vertical speed is 1200 ft/min. Calculate the minimum time t required for the elevator to go from rest at the 10th floor to a stop at the 30th floor, a distance of 300 ft.

24 A vacuum-propelled capsule for a high-speed tube transportation system of the future is being designed for operation between two stations 8 km apart. If maximum acceleration and deceleration are to have a limiting magnitude of $0.7g$ and if velocities are to be limited to 400 km/h, determine the minimum time t for the capsule to make the 8-km trip.
Ans. $t = 1.47$ min

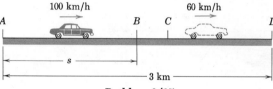

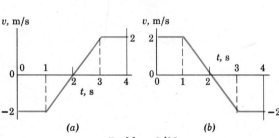

Problem 2/25

25 In traveling a distance of 3 km between points A and D a car is driven at 100 km/h from A to B for t seconds and at 60 km/h from C to D also for t seconds. If the brakes are applied for 4 s between B and C to give the car a uniform deceleration, calculate t and the distance s between A and B.
Ans. $t = 65.5$ s, $s = 1.819$ km

26 The v-t curves for two different particle motions are shown. Plot the corresponding s-t curves, taking $s = 0$ when $t = 0$ for both instances.

Problem 2/26

27 A particle with an initial velocity of -0.3 km/min at $t = 0$ is given the discontinuous acceleration shown. Plot the variation of v in km/min with t and determine the values of t when $v = 0$.
Ans. $t = 2$ min and $t = 7$ min

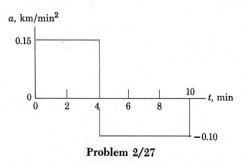

Problem 2/27

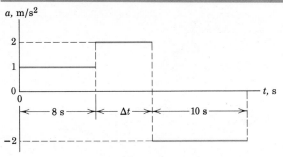

Problem 2/28

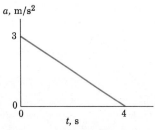

Problem 2/29

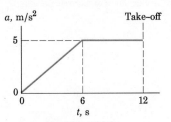

Problem 2/30

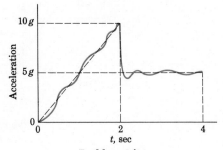

Problem 2/31

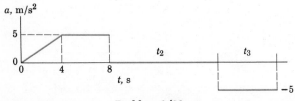

Problem 2/32

2/28 A subway train travels between two of its scheduled station stops with the acceleration schedule shown. Determine the time interval Δt during which the acceleration is 2 m/s^2. What is the distance s between stations?   *Ans.* $\Delta t = 6 \text{ s}$, $s = 216 \text{ m}$

2/29 A particle has an acceleration which varies linearly with the time as shown. If the position coordinate s is 2 m and the velocity v is -4 m/s when $t = 0$, determine and plot the variations of s and v with t for the 4-second interval.

2/30 In an experiment on takeoff performance the engine thrust of a jet aircraft is programmed so that the acceleration varies, as shown in the graph, for the 12 seconds required to take off from a rest position. Calculate the required takeoff velocity v and the corresponding distance s along the runway for takeoff.   *Ans.* $v = 162 \text{ km/h}$, $s = 210 \text{ m}$

2/31 The record of acceleration measurements made on an experimental vehicle during its rectilinear motion is shown in the full line. The vehicle starts from rest at $t = 0$. Use the dotted approximation and draw the v-t curve for the 4-sec interval and determine the total distance s traveled.

2/32 The acceleration of a subway train between two stops a distance of 2 km apart varies as shown. Determine (*a*) the maximum speed v_m of the train, (*b*) the time t_3 during deceleration, and (*c*) the time t_2 during which the speed is constant.
          Ans. (*a*) $v_m = 30 \text{ m/s}$, (*b*) $t_3 = 6 \text{ s}$, (*c*) $t_2 = 60.6 \text{ s}$

2/33 Show that the acceleration is constant when the relationship between the position coordinate s and the square of the velocity v is linear. Find the acceleration a for the case illustrated.

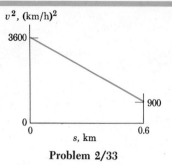

Problem 2/33

2/34 A self-propelled vehicle of mass m whose engine delivers constant power P has an acceleration $a = P/(mv)$ where all frictional resistance is neglected. Determine expressions for the distance s traveled and the corresponding time t required by the vehicle to increase its speed from v_1 to v_2.

$$Ans.\ s = \frac{m}{3P}(v_2{}^3 - v_1{}^3),\ t = \frac{m}{2P}(v_2{}^2 - v_1{}^2)$$

2/35 Approximate the v-t data for the decelerating ship of Prob. 2/18 by the form of the expression derived in Sample Problem 2/4, namely, $v = v_0/(1 + v_0 kt)$ and calculate the distance s between the two buoys.

$$Ans.\ s = 0.61\ \text{mi (nautical)}$$

2/36 Approximate the v-t data for the decelerating ship of Prob. 2/18 by using the exponential relation $v = v_0 e^{-kt}$ where v_0 and k are constants determined by the initial and final conditions. Then calculate the distance s between the buoys.

$$Ans.\ s = 0.65\ \text{mi (nautical)}$$

2/37 A projectile is fired horizontally into a resisting medium with a velocity v_1, and the resulting deceleration is equal to cv^n, where c and n are constants and v is the velocity within the medium. Find the expression for the velocity v of the projectile in terms of the time t of penetration.

2/38 A retarding force is applied to a body moving in a straight line so that, during an interval of its motion, its speed v decreases with increased position coordinate s according to the relation $v^2 = k/s$, where k is a constant. If the body has a forward speed of 2 in./sec and its position coordinate is 9 in. at time $t = 0$, determine the speed v at $t = 3$ sec.

$$Ans.\ v = 1.59\ \text{in./sec}$$

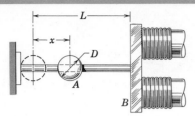

Problem 2/39

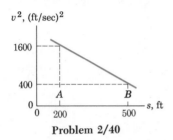

Problem 2/40

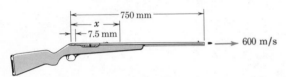

Problem 2/41

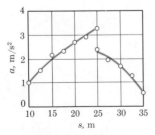

Problem 2/43

2/39 The steel ball A of diameter D slides freely on the horizontal rod which leads to the pole face of the electromagnet. The force of attraction obeys an inverse-square law, and the resulting acceleration of the ball is $a = K/(L - x)^2$ where K is a measure of the strength of the magnetic field. If the ball is released from rest at $x = 0$, determine the velocity v with which it strikes the pole face.

$$\text{Ans. } v = 2\sqrt{\frac{K(L - D/2)}{LD}}$$

2/40 A body moves in a straight line with a velocity whose square decreases linearly with the displacement between two points A and B which are 300 ft apart as shown. Determine the displacement Δs of the body during the last 2 sec before arrival at B.

2/41 To a close approximation the pressure behind a rifle bullet varies inversely with the distance traveled by the bullet along the barrel. Thus the acceleration of the bullet may be written as $a = k/x$ where k is a constant. If the bullet starts from rest at $x = 7.5$ mm and if the muzzle velocity of the bullet is 600 m/s at the end of the 750-mm barrel, compute the acceleration of the bullet as it passes the midpoint of the barrel at $x = 375$ mm.   *Ans.* $a = 104.2$ km/s^2

2/42 A retarding force is applied to a body moving in a straight line so that, during an interval of its motion, its speed v decreases with increased position coordinate s according to the relation $v^2 = k/s$, where k is a constant. If the body has a forward speed of 50 mm/s and a position coordinate of 225 mm at time $t = 0$, determine the speed v at $t = 3$ s.

2/43 The forward acceleration a of a certain test vehicle is measured experimentally during an interval of its motion and plotted against its position coordinate as shown. At the 25-m position the driving mechanism shifts abruptly, and a discontinuity in the acceleration occurs. If the vehicle has a velocity of 7.2 km/h at $s = 10$ m, plot the velocity during the measured motion. Find the velocity at $s = 35$ m.

Ans. $v_{35} = 10.2$ m/s

44 A certain lake is proposed as a landing area for large
jet aircraft. The touchdown speed of 160 km/h upon
contact with the water is to be reduced to 30 km/h
in a distance of 400 m. If the deceleration is proportional to the square of the velocity of the aircraft
through the water, $a = -Kv^2$, find the value of the
design parameter K, which would be a measure of
the size and shape of the landing gear vanes which
plow through the water. Also find the time t elapsed
during the specified interval.

45 If the velocity v of a particle moving along a
straight line decreases linearly with its position
coordinate s from 40 ft/sec to a value approaching
zero at $s = 80$ ft, show that the particle never
reaches the 80-ft position. What is the acceleration
a of the particle when $s = 60$ ft?

$$Ans. \ a = -5 \text{ ft/sec}^2$$

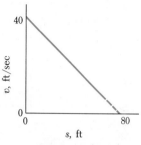

Problem 2/45

46 In a test drop in still air, a small steel ball is released
from rest at a considerable height. Its initial acceleration g is reduced by the increment kv^2, where k is
a constant and v is the downward velocity. Determine the maximum velocity attainable by the ball
and find the vertical drop y in terms of the time t
from release.

47 The horizontal motion of the plunger and shaft is
arrested by the resistance of the attached disk
which moves through the oil bath. If the velocity of
the plunger is v_0 in the position A where $x = 0$ and
$t = 0$, and if the deceleration is proportional to v so
that $a = -kv$, derive expressions for the velocity v
and position coordinate x in terms of the time t. Also
express v in terms of x.

$$Ans. \ v = v_0e^{-kt}, \ x = \frac{v_0}{k}(1 - e^{-kt}), \ v = v_0 - kx$$

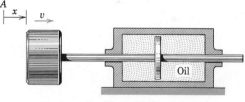

Problem 2/47

48 A small object is released from rest in a tank of oil.
The downward acceleration of the object is $g - kv$
where g is the constant acceleration due to gravity,
k is a constant which depends on the viscosity of the
oil and shape of the object, and v is the downward
velocity of the object. Derive expressions for the
velocity v and vertical drop y as functions of the
time t after release.

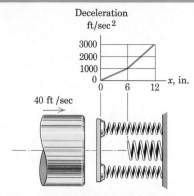

Problem 2/49

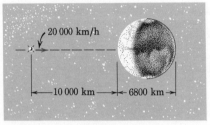

Problem 2/50

▶ 2/49 A bumper, consisting of a nest of three springs, is used to arrest the horizontal motion of a large mass which is traveling at 40 ft/sec as it contacts the bumper. The two outer springs cause a deceleration proportional to the spring deformation. The center spring increases the deceleration rate when the compression exceeds 6 in. as shown on the graph. Determine the maximum compression x of the outer springs.   *Ans.* $x = 0.831$ ft

▶ 2/50 A spacecraft is designed for Martian impact on a course heading directly toward the center of the planet. If the craft is 10 000 km from the planet and is traveling at a relative speed of 20 000 km/h, compute the impact velocity v. The gravitational acceleration at the surface of Mars is 3.73 m/s², and the planet has a diameter of approximately 6800 km. Neglect any atmospheric resistance.
                                   Ans. $v = 25\,400$ km/h

▶ 2/51 A particle which is constrained to move in a straight line is subjected to an accelerating force which increases with time and a retarding force which increases directly with the position coordinate x. The resulting acceleration is $a = Kt - k^2x$ where K and k are positive constants and where both x and $\dot{x}$ are zero when $t = 0$. Determine x as a function of t.

$$Ans.\ x = \frac{K}{k^3}(kt - \sin kt)$$

▶ 2/52 The vertical acceleration of a certain solid-fuel rocket is given by $a = ke^{-bt} - cv - g$, where k, b, and c are constants, v is the vertical velocity acquired, and g is the gravitational acceleration, essentially constant for atmospheric flight. The exponential term represents the effect of a decaying thrust as fuel is burned, and the term $-cv$ approximates the retardation due to atmospheric resistance. Determine the expression for the vertical velocity of the rocket t seconds after firing.

$$Ans.\ v = \frac{g}{c}(e^{-ct} - 1) + \frac{k}{c - b}(e^{-bt} - e^{-ct})$$

√3 PLANE CURVILINEAR MOTION. We now begin our description of the motion of a particle along a curved path which lies in a single plane. This motion is seen to be a special case of the more general three-dimensional motion introduced in Art. 2/1 and illustrated in Fig. 2/1. If we let the plane of motion be the x-y plane, for instance, then the coordinates z and ϕ of Fig. 2/1 are both zero and R becomes the same as r. As mentioned previously the vast majority of the motions of points or particles encountered in engineering practice may be represented as plane motion.

Before pursuing the description of plane curvilinear motion in any specific set of coordinates, we will first describe the motion with the aid of vector analysis, the results of which are independent of any particular coordinate system. What follows in this article will constitute one of the most basic of all treatments in dynamics, namely, development of the *time derivative of a vector*. A large fraction of the entire subject of dynamics hinges directly on the time rates of change of vector quantities, and all students of dynamics are well advised to master this topic at the outset because they will have frequent occasion to use it.

Consider now the continuous motion of a particle along a plane curve as represented in Fig. 2/5. At time t the particle is at position A, which is located by the *position vector* $\mathbf{r}$ measured from some convenient fixed origin O. If both the magnitude and direction of $\mathbf{r}$ are known at time t, then the position of the particle is completely specified. At time $t + \Delta t$ the particle is at A', located by the position vector $\mathbf{r} + \Delta\mathbf{r}$. We note, of course, that this combination is vector addition and not scalar addition. The *displacement* of the particle during time Δt is the vector $\Delta\mathbf{r}$ which represents the vector change of position and is clearly independent of the choice of origin. If an origin were chosen at some different location, the position vector $\mathbf{r}$ would be changed, but $\Delta\mathbf{r}$ would be unchanged. The *distance* actually traveled by the particle as it moves along the path from A to A' is the scalar length Δs measured along the path. Thus we distinguish between the vector displacement $\Delta\mathbf{r}$ and the scalar distance Δs.

The *average velocity* of the particle between A and A' is defined as $\mathbf{v}_{av} = \Delta\mathbf{r}/\Delta t$, which is a vector whose direction is that of $\Delta\mathbf{r}$ and whose magnitude is the magnitude of $\Delta\mathbf{r}$ divided by Δt. The average *speed* of the particle between A and A' is the scalar quotient $\Delta s/\Delta t$. Clearly the magnitude of the average velocity and the speed approach one another as the interval Δt decreases and A and A' become closer together.

The *instantaneous velocity* $\mathbf{v}$ of the particle is defined as the limiting value of the average velocity as the time interval approaches zero. Thus

$$\mathbf{v} = \lim_{\Delta t \to 0} \frac{\Delta\mathbf{r}}{\Delta t}$$

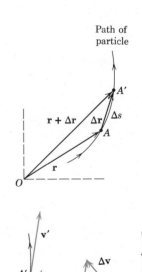

Figure 2/5

We observe that the direction of $\Delta\mathbf{r}$ approaches that of the tangent to the path as Δt approaches zero and, hence, the velocity $\mathbf{v}$ is always a vector tangent to the path. We now extend the basic definition of the derivative of a scalar quantity to include a vector quantity and write

$$\mathbf{v} = \frac{d\mathbf{r}}{dt} = \dot{\mathbf{r}} \qquad (2/4)$$

The derivative of a vector is itself a vector having both a magnitude and a direction. The magnitude of $\mathbf{v}$ is called the *speed* and is the scalar

$$v = |\mathbf{v}| = \frac{ds}{dt} = \dot{s}$$

At this point we make a careful distinction between the magnitude of the derivative and the derivative of the magnitude. The magnitude of the derivative may be written in any one of the several ways $|d\mathbf{r}/dt| = |\dot{\mathbf{r}}| = \dot{s} = |\mathbf{v}| = v$ and represents the magnitude of the velocity or the speed of the particle. On the other hand the derivative of the magnitude is written $d|\mathbf{r}|/dt = dr/dt = \dot{r}$, and represents the rate at which the length of the position vector $\mathbf{r}$ is changing. Thus these two derivatives have two entirely different meanings, and we must be extremely careful to distinguish between them in our thinking and in our notation. For this and other reasons students are urged to adopt a consistent notation for their handwritten work for all vector quantities to distinguish them from scalar quantities. For simplicity the underline $\underline{v}$ is recommended. Other handwritten symbols such as $\vec{v}$, $\underset{\sim}{v}$, and $\hat{v}$ are sometimes used.

With the concept of velocity as a vector established, we return to Fig. 2/5 and denote the velocity of the particle at A by the tangent vector $\mathbf{v}$ and the velocity at A' by the tangent vector $\mathbf{v}'$. Clearly there is a vector change in the velocity during the time Δt. The velocity $\mathbf{v}$ at A plus (vectorially) the change $\Delta\mathbf{v}$ must equal the velocity at A', so we may write $\mathbf{v}' - \mathbf{v} = \Delta\mathbf{v}$. Inspection of the vector diagram shows that $\Delta\mathbf{v}$ depends both on the change in magnitude (length) of $\mathbf{v}$ and on the change in direction of $\mathbf{v}$. These two changes are fundamental characteristics of the derivative of a vector.

The *average acceleration* of the particle between A and A' is defined as $\Delta\mathbf{v}/\Delta t$, which is a vector whose direction is that of $\Delta\mathbf{v}$. The magnitude of this average acceleration is the magnitude of $\Delta\mathbf{v}$ divided by Δt.

The *instantaneous acceleration* $\mathbf{a}$ of the particle is defined as the limiting value of the average acceleration as the time interval approaches zero. Thus

$$\mathbf{a} = \lim_{\Delta t \to 0} \frac{\Delta\mathbf{v}}{\Delta t}$$

By definition of the derivative, then, we write

$$\mathbf{a} = \frac{d\mathbf{v}}{dt} = \dot{\mathbf{v}} \qquad (2/5)$$

As the interval Δt becomes smaller and approaches zero, the direction of the change $\Delta \mathbf{v}$ approaches that of the differential change $d\mathbf{v}$ and, hence, $\mathbf{a}$. The acceleration $\mathbf{a}$, then, includes the effect of change in magnitude of $\mathbf{v}$ and the change of direction of $\mathbf{v}$. It is apparent, in general, that the direction of the acceleration of a particle in curvilinear motion is neither tangent to the path nor normal to the path. We do observe, however, that the acceleration component which is normal to the path points toward the center of curvature of the path.

A further approach to the visualization of acceleration is shown in Fig. 2/6, where the position vectors to three arbitrary positions on the path of the particle are shown for illustrative purpose. There is a velocity vector tangent to the path corresponding to each position vector, and the relation is $\mathbf{v} = \dot{\mathbf{r}}$. Now if these velocity vectors are plotted from some arbitrary point C, then a curve, known as the *hodograph*, is formed. The derivatives of these velocity vectors will be the acceleration vectors $\mathbf{a} = \dot{\mathbf{v}}$ which are tangent to the hodograph. We see that the acceleration bears the same relation to the velocity as the velocity bears to the position vector.

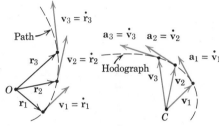

Figure 2/6

The geometric portrayal of the derivatives of the position vector $\mathbf{r}$ and velocity vector $\mathbf{v}$ in Fig. 2/5 can be used to describe the derivative of any vector quantity with respect to t or with respect to any other scalar variable. With the introduction of the derivative of a vector in the definitions of velocity and acceleration it is important at this point to establish the rules under which the differentiation of vector quantities may be carried out. These rules are the same as for the differentiation of scalar quantities except for the case of the cross product where the order of the terms must be preserved. These rules are covered in section B7 of Appendix B and should be reviewed at this point.

For curvilinear motion of a particle in a plane there are three different coordinate systems which are in common use to describe the vector relationships developed in this article. An important lesson to be learned from the study of these coordinate systems is the proper choice of reference system for a given problem. This choice is usually revealed by the manner in which the motion is generated or by the form in which the data are specified. Each of the three coordinate systems will now be developed and illustrated.

2/4　RECTANGULAR COORDINATES (x-y).　This system of coordinates is particularly useful for describing motions where the x- and y-components of acceleration are independently generated or determined. The resulting curvilinear motion is, then, obtained by a vector

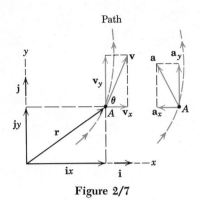

Figure 2/7

combination of the x- and y-components of the position vector, the velocity, and the acceleration.

The particle path of Fig. 2/5 is shown again in Fig. 2/7 along with x- and y-axes. The position vector $\mathbf{r}$, the velocity $\mathbf{v}$, and the acceleration $\mathbf{a}$ of the particle as developed in Art. 2/3 are represented in Fig. 2/7 together with their x- and y-components. With the aid of the unit vectors $\mathbf{i}$ and $\mathbf{j}$ we may write the vectors $\mathbf{r}$, $\mathbf{v}$, and $\mathbf{a}$ in terms of their x- and y-components. Thus

$$\begin{aligned} \mathbf{r} &= \mathbf{i}x + \mathbf{j}y \\ \mathbf{v} &= \dot{\mathbf{r}} = \mathbf{i}\dot{x} + \mathbf{j}\dot{y} \\ \mathbf{a} &= \dot{\mathbf{v}} = \ddot{\mathbf{r}} = \mathbf{i}\ddot{x} + \mathbf{j}\ddot{y} \end{aligned} \qquad (2/6)$$

As we differentiate with respect to time we observe that the unit vectors have no time derivatives since their magnitudes and directions remain constant. The scalar magnitudes of the components of $\mathbf{v}$ and $\mathbf{a}$ are merely $v_x = \dot{x}$, $v_y = \dot{y}$ and $a_x = \dot{v}_x = \ddot{x}$, $a_y = \dot{v}_y = \ddot{y}$. (As drawn in Fig. 2/7, a_x is in the negative x-direction, so that $\ddot{x}$ would be a negative number.)

As observed previously the direction of the velocity is always tangent to the path, and from the figure it is clear that

$$v^2 = v_x{}^2 + v_y{}^2 \qquad v = \sqrt{v_x{}^2 + v_y{}^2} \qquad \tan\theta = \frac{v_y}{v_x}$$

$$a^2 = a_x{}^2 + a_y{}^2 \qquad a = \sqrt{a_x{}^2 + a_y{}^2}$$

If the angle θ is measured counterclockwise from the x-axis to $\mathbf{v}$ for the configuration of axes shown, then we may also observe that $dy/dx = \tan\theta = v_y/v_x$.

If the coordinates x and y are each known independently as a function of time, $x = f_1(t)$ and $y = f_2(t)$, then for any value of the time we may combine them to obtain $\mathbf{r}$. Similarly we combine their first derivatives $\dot{x}$ and $\dot{y}$ to obtain $\mathbf{v}$ and their second derivatives $\ddot{x}$ and $\ddot{y}$ to obtain $\mathbf{a}$. Or, if the acceleration components a_x and a_y are given as functions of the time, we may integrate each one separately with respect to time, once to obtain v_x and v_y and again to obtain $x = f_1(t)$ and $y = f_2(t)$. Elimination of the time t between these last two parametric equations gives the equation of the curved path $y = f(x)$.

From the foregoing discussion it should be recognized that the rectangular-coordinate representation of curvilinear motion is merely the superposition of the components of two simultaneous rectilinear motions in the x- and y-directions. Therefore, everything covered in Art. 2/2 on rectilinear motion may be applied separately to the x-motion and to the y-motion.

Sample Problem 2/5

The curvilinear motion of a particle is defined by $v_x = 50 - 16t$ and $y = 100 - 4t^2$ where v_x is in meters per second, y is in meters, and t is in seconds. It is also known that $x = 0$ when $t = 0$. Plot the path of the particle and determine its velocity and acceleration when the position $y = 0$ is reached.

Solution. The x-coordinate is obtained by integrating the expression for v_x, and the x-component of the acceleration is obtained by differentiating v_x. Thus

$$\left[\int dx = \int v_x \, dt \right] \qquad \int_0^x dx = \int_0^t (50 - 16t)dt \qquad x = 50t - 8t^2 \text{ m}$$

$$[a_x = \dot{v}_x] \qquad a_x = \frac{d}{dt}(50 - 16t) \qquad a_x = -16 \text{ m/s}^2$$

The y-components of velocity and acceleration are

$$[v_y = \dot{y}] \qquad v_y = \frac{d}{dt}(100 - 4t^2) \qquad v_y = -8t \text{ m/s}$$

$$[a_y = \dot{v}_y] \qquad a_y = \frac{d}{dt}(-8t) \qquad a_y = -8 \text{ m/s}^2$$

We now calculate corresponding values of x and y for various values of t and plot x against y to obtain the path as shown.

When $y = 0, 0 = 100 - 4t^2$, so $t = 5$ s. For this value of the time we have

$$v_x = 50 - 16(5) = -30 \text{ m/s}$$

$$v_y = -8(5) = -40 \text{ m/s}$$

$$v = \sqrt{(-30)^2 + (-40)^2} = 50 \text{ m/s}$$

$$a = \sqrt{(-16)^2 + (-8)^2} = 17.9 \text{ m/s}^2$$

The velocity and acceleration components and their resultants are shown on the separate diagrams for point A where $y = 0$.

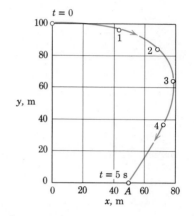

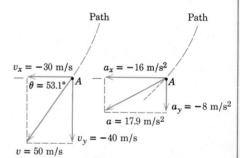

We observe that the velocity vector lies along the tangent to the path as it should but that the acceleration vector is not tangent to the path. Note especially that the acceleration vector has a component which points toward the inside of the curved path. We concluded from our diagram in Fig. 2/5 that it is impossible for the acceleration to have a component which points toward the outside of the curve.

Sample Problem 2/6

A rocket has expended all its fuel when it reaches position A where it has a velocity u at an angle θ with respect to the horizontal. It then begins unpowered flight and attains a maximum added height h at position B after traveling a horizontal distance s from A. Determine the expressions for h and s, the time t of flight from A to B, and the equation of the path. For the interval concerned, assume a flat earth with a constant gravitational acceleration g and neglect any atmospheric resistance.

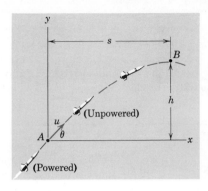

Solution. Since all motion components are directly expressible in terms of horizontal and vertical coordinates, a rectangular set of axes x-y will be employed. With the neglect of atmospheric resistance, $a_x = 0$ and $a_y = -g$, and the resulting motion is a direct superposition of two rectilinear motions with constant acceleration. Thus

$$[dx = v_x\, dt] \qquad x = \int_0^t u \cos \theta\, dt \qquad x = ut \cos \theta$$

$$[dv_y = a_y\, dt] \qquad \int_{u \sin \theta}^{v_y} dv_y = \int_0^t (-g)\, dt \qquad v_y = u \sin \theta - gt$$

$$[dy = v_y\, dt] \qquad y = \int_0^t (u \sin \theta - gt)\, dt \qquad y = ut \sin \theta - \tfrac{1}{2}gt^2$$

Position B is reached when $v_y = 0$, which occurs for $0 = u \sin \theta - gt$ or

$$t = (u \sin \theta)/g \qquad \qquad \textit{Ans.}$$

Substitution of this value for the time into the expression for y gives the maximum added altitude

$$h = u\left(\frac{u \sin \theta}{g}\right) \sin \theta - \frac{1}{2}g\left(\frac{u \sin \theta}{g}\right)^2 \qquad h = \frac{u^2 \sin^2 \theta}{2g} \qquad \textit{Ans.}$$

The horizontal distance is seen to be

$$s = u\left(\frac{u \sin \theta}{g}\right) \cos \theta \qquad s = \frac{u^2 \sin 2\theta}{2g} \qquad \textit{Ans.}$$

which is clearly a maximum when $\theta = 45°$. The equation of the path is obtained by eliminating t from the expressions for x and y, which gives

$$y = x \tan \theta - \frac{gx^2}{2u^2} \sec^2 \theta \qquad \qquad \textit{Ans.}$$

This equation describes a vertical parabola as indicated in the figure.

① Note that this problem is simply the description of projectile motion neglecting atmospheric resistance.

② We see that the total range and time of flight for a projectile fired above a horizontal plane would be twice the respective values of s and t given here.

③ If atmospheric resistance were to be accounted for, the dependency of the acceleration components on the velocity would have to be established before an integration of the equations could be carried out. This becomes a much more difficult problem.

PROBLEMS

53 The x- and y-coordinates of a particle moving with plane curvilinear motion are given by $x = 2t^2 + 3t$ and $y = t^3/3 - 8$ where x and y are in meters and t is in seconds. Determine the magnitudes of the velocity **v** and acceleration **a** and the angles which the vectors make with the x-axis when $t = 3$ s.

$$Ans.\ v = 17.49 \text{ m/s with } \theta_x = 31.0°$$
$$a = 7.21 \text{ m/s}^2 \text{ with } \theta_x = 56.3°$$

54 The x-coordinate to a particle in curvilinear motion is given by $x = 3t^3 - 2t$ where x is in meters and t is in seconds. If the particle has an acceleration in the y-direction given by $a_y = 8t$ and if it has a velocity component of 6 m/s in the y-direction when $t = 0$, determine the magnitudes of the velocity **v** and the acceleration **a** of the particle when $t = 1$ s.

55 The position vector of a point which moves in the x-y plane is given by

$$\mathbf{r} = \left(\frac{2}{3}t^3 - \frac{3}{2}t^2\right)\mathbf{i} + \frac{t^4}{12}\mathbf{j}$$

where **r** is in meters and t is in seconds. Determine the velocity **v** and the acceleration **a** when $t = 3$ s. What do the results tell about the curvature of the path at this instant?

$$Ans.\ \mathbf{v} = 9(\mathbf{i} + \mathbf{j}) \text{ m/s}, \ \mathbf{a} = 9(\mathbf{i} + \mathbf{j}) \text{ m/s}^2$$

56 A particle P moves on the circular path of radius r as shown. If $\dot{\theta} = \omega$, a constant, prove that the acceleration of P is always directed toward the center.

57 A rock is thrown horizontally from a tower at A and hits the ground 3.5 sec later at B. The line of sight from A to B makes an angle of 50° with the horizontal. Compute the initial velocity u of the rock.

$$Ans.\ u = 47.3 \text{ ft/sec}$$

58 With what minimum horizontal velocity u can a boy throw a rock at A and have it just clear the obstruction at B?

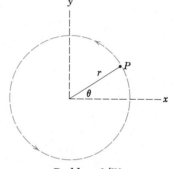

Problem 2/56

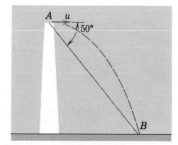

Problem 2/57

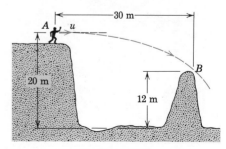

Problem 2/58

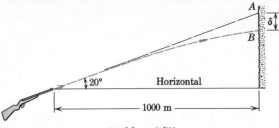

Problem 2/59

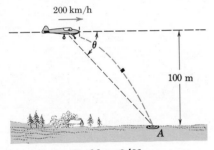

Problem 2/60

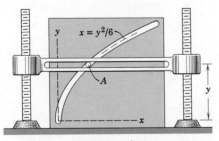

Problem 2/61

Problem 2/64

2/59 If the barrel of the rifle shown is aimed at point A, compute the distance δ below A to the point B where the bullet strikes. The muzzle velocity is 600 m/s. *Ans.* $\delta = 15.43$ m

2/60 Calculate the minimum possible muzzle velocity u which a projectile must have when fired from point A to reach a target B on the same horizontal plane 10 km away.

2/61 The pilot of an airplane carrying a package of mail to a remote outpost wishes to release the package at the right moment to hit the recovery location A. What angle θ with the horizontal should the pilot's line of sight to the target make at the instant of release? The airplane is flying horizontally at an altitude of 100 m with a velocity of 200 km/h. *Ans.* $\theta = 21.7°$

2/62 A particle moves along the positive branch of the curve $y = x^2/2$ with x governed by $x = t^2/2$, where x and y are measured in inches and t is in seconds. Determine the velocity and acceleration of the particle when $t = 2$ sec. Express the results in vector notation. *Ans.* $\mathbf{v} = 2\mathbf{i} + 4\mathbf{j}$ in./sec, $\mathbf{a} = \mathbf{i} + 6\mathbf{j}$ in./sec^2

2/63 A particle starts from rest at $y = 0$ and at time $t = 0$ and moves along the positive-y branch of the curve $x = y^2/12 - 3$ with a y-component of velocity governed by $v_y = 2t$ where x and y are in meters and t is in seconds. Determine the velocity and acceleration of the particle when $y = 9$ m and show them as vectors on a sketch of the particle path.

2/64 For a certain interval of motion the pin A is forced to move in the fixed parabolic slot by the horizontal slotted arm which is elevated in the y-direction at the constant rate of 3 in./sec. All measurements are in inches and seconds. Calculate the velocity v and acceleration a of pin A when $x = 6$ in. *Ans.* $v = 3\sqrt{5}$ in./sec, $a = 3$ in./sec^2

65 A particle moves in the x-y plane with a y-component of velocity in feet per second given by $v_y = 8t$ with t in seconds. The acceleration of the particle in the x-direction in feet per second squared is given by $a_x = 4t$ with t in seconds. When $t = 0$, $y = 2$ ft, $x = 0$, and $v_x = 0$. Find the equation of the path of the particle and calculate the magnitude of the velocity **v** of the particle for the instant when its x-coordinate reaches 18 ft.

66 In the cathode-ray tube electrons traveling horizontally from their source with the velocity v_0 are deflected by an electric field E due to the voltage gradient across the plates P. The deflecting force causes an acceleration in the vertical direction on the sketch equal to eE/m where e is the electron charge and m is its mass. When clear of the plates the electrons travel in straight lines. Determine the expression for the deflection δ for the tube and plate dimensions shown.

$$Ans. \ \delta = \frac{eEl}{mv_0{}^2}\left(\frac{l}{2} + b\right)$$

67 For a given muzzle velocity u find the maximum range R on a horizontal plane which can be achieved by the projectile without exceeding an altitude h.

68 To pass inspection small ball bearings must bounce through an opening of limited size at the top of their trajectory when rebounding from a heavy plate as shown. Calculate the angle θ made by the rebound velocity with the horizontal and the velocity v of the balls as they pass through the opening.

$$Ans. \ \theta = 68.2°, \ v = 1.25 \ m/s$$

69 Calculate the firing angle θ of the anti-aircraft gun with a muzzle velocity of 1800 ft/sec if a direct hit is to be scored on an aircraft flying horizontally at 600 mi/hr at an altitude of 20,000 ft. The gun is fired at the instant when the aircraft is directly overhead. Find the time t required for the shell to reach the aircraft.

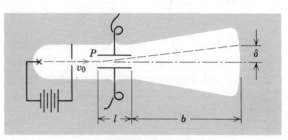

Problem 2/66

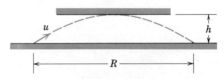

Problem 2/67

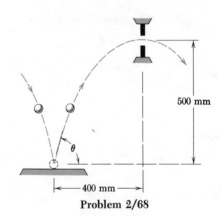

Problem 2/68

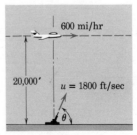

Problem 2/69

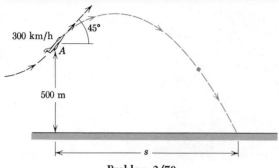

Problem 2/70

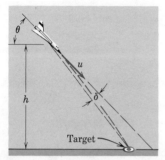

Problem 2/71

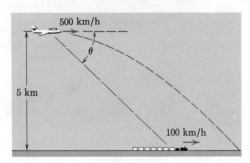

Problem 2/72

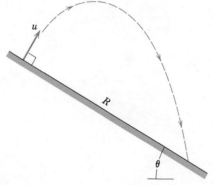

Problem 2/73

2/70 The pilot of an airplane pulls into a steep 45° climb at 300 km/h and releases a package at position A. Calculate the horizontal distance s and the time t from the point of release to the point at which the package strikes the ground.

Ans. $s = 1.046$ km, $t = 17.75$ s

2/71 By what angle δ should the pilot of a dive bomber lead his target if he is to release his bomb at an altitude $h = 1.8$ km at a speed $u = 1000$ km/h while diving at the angle $\theta = 45°$?

2/72 A bomber flying with a horizontal velocity of 500 km/h at an altitude of 5 km is to score a direct hit on a train moving with a constant velocity of 100 km/h in the same direction and in the same vertical plane. Calculate the angle θ between the line of sight to the target and the horizontal at the instant the bomb should be released.

Ans. $\theta = 54.6°$

2/73 A projectile is fired with a velocity u at right angles to the slope, which is inclined at an angle θ with the horizontal. Derive an expression for the distance R to the point of impact.

2/74 An object is projected up the incline at the angle shown with an initial velocity of 100 ft/sec. Calculate the distance s up the incline at which the object lands.   *Ans. $s = 207$ ft*

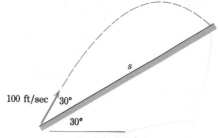

Problem 2/74

2/75 The muzzle velocity of a long-range rifle at A is $u = 400$ m/s. Determine the two angles of elevation θ which will permit the projectile to hit the mountain target B.

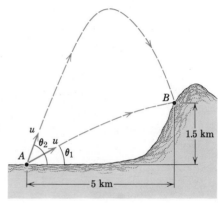

Problem 2/75

2/76 A rocket is released at point A from a jet aircraft flying horizontally at 600 mi/hr at a 5000-ft altitude. If the rocket thrust gives it a constant horizontal acceleration of 0.5g, determine the angle θ between the horizontal and the line of sight to the target for a direct hit.   *Ans. $\theta = 15.5°$*

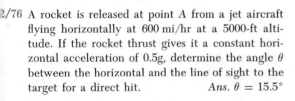

Problem 2/76

2/77 An object which is released from rest relative to the earth at A at a distance h above a horizontal surface will appear not to fall straight down by virtue of the effect of the earth's rotation. It may be shown that the object has an eastward acceleration relative to the horizontal surface on the earth equal to $2v_y \omega \cos \gamma$, where v_y is the freefall downward velocity, ω is the angular velocity of the earth, and γ is the north latitude. Determine the expression for the distance δ to the east of the vertical line through A at which the object strikes the ground.

$$\text{Ans. } \delta = \frac{2\sqrt{2}}{3} \omega h \sqrt{\frac{h}{g}} \cos \gamma$$

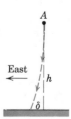

Problem 2/77

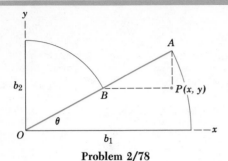

Problem 2/78

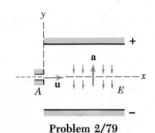

Problem 2/79

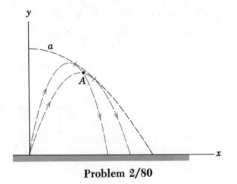

Problem 2/80

▶2/78   Two collinear line segments, OA of length b_1 and OB of length b_2, make an angle θ with the x-axis and revolve about the origin O at the constant rate $\dot{\theta} = \omega$. A point P is defined by the coordinates $x = b_1 \cos \theta$ and $y = b_2 \sin \theta$. Show that P travels on an elliptical path and prove that its acceleration is always directed toward the origin O. Take $\theta = 0$ when $t = 0$.

▶2/79   At time $t = 0$ an electron is emitted at A with a velocity $\mathbf{u}$ in the x-direction into an electric field $E = E_0 \sin pt$ at right angles to $\mathbf{u}$. The electron has an acceleration in the y-direction equal to eE/m, where e is the electron charge and m is its mass. Find the x- and y-coordinates of the electron at the end of the first complete cycle of E. The field frequency $f = p/2\pi$ is constant.

$$Ans.\ x = 2\pi u/p,\ y = 2\pi \frac{eE_0}{mp^2}$$

▶2/80   Determine the equation for the envelope a of the parabolic trajectories of a projectile fired at any angle but with the same muzzle velocity u. (*Hint*. Substitute $m = \tan \theta$, where θ is the firing angle with the horizontal, into the equation of the trajectory. The two roots m_1 and m_2 of the equation written as a quadratic in m give the two firing angles for the two trajectories shown such that the shells pass through the same point A. Point A will approach the envelope a as the two roots approach equality.) Neglect air resistance and assume g is constant.

$$Ans.\ y = \frac{u^2}{2g} - \frac{gx^2}{2u^2}$$

2/5 NORMAL AND TANGENTIAL COORDINATES (*n-t*). In the introduction to this chapter, Art. 2/1, it was mentioned that one of the common descriptions of curvilinear motion involves the use of path variables, which are measurements made along the tangent *t* and normal *n* to the path of the particle. These coordinates provide a very natural description for curvilinear motion and are frequently the most direct and convenient coordinates to use. The *n*- and *t*-coordinates are considered to move along the path with the particle as seen in Fig. 2/8 where the particle advances from *A* to *B* to *C*. The positive direction for *n* at any position is always taken toward the center of curvature of the path. As seen from Fig. 2/8 the positive *n*-direction will shift from one side of the curve to the other side if the curvature changes direction.

The coordinates *n* and *t* will now be used to describe the velocity **v** and acceleration **a** which were introduced in Art. 2/3 for the curvilinear motion of a particle. For this purpose we introduce unit vectors $\mathbf{n}_1$ in the *n*-direction and $\mathbf{t}_1$ in the *t*-direction, as shown in Fig. 2/9*a* for the position of the particle at point *A* on its path. During a differential increment of time *dt* the particle moves a differential distance *ds* along the curve from *A* to *A′*. With the radius of curvature of the path at this position designated by ρ, we see that $ds = \rho \, d\theta$, where θ is in radians, so that the magnitude of the velocity may be written $v = ds/dt = \rho \, d\theta/dt$. Thus we may write the velocity as the vector

$$\mathbf{v} = v\mathbf{t}_1 = \rho\dot{\theta}\mathbf{t}_1 \qquad (2/7)$$

The acceleration **a** of the particle was defined in Art. 2/3 as $\mathbf{a} = d\mathbf{v}/dt$, and we observed from Fig. 2/5 that the acceleration is a vector which reflects both the change in magnitude and the change in direction of **v**. We now differentiate **v** in Eq. 2/7 by applying the ordinary rule for the differentiation of the product of a scalar and a vector[*] and get

$$\mathbf{a} = \frac{d\mathbf{v}}{dt} = \frac{d(v\mathbf{t}_1)}{dt} = v\dot{\mathbf{t}}_1 + \dot{v}\mathbf{t}_1 \qquad (2/8)$$

where the unit vector $\mathbf{t}_1$ now has a derivative since its direction changes. To find $\dot{\mathbf{t}}_1$ we analyze the change in $\mathbf{t}_1$ during a differential increment of motion as the particle moves from *A* to *A′* in Fig. 2/9*a*. The unit vector $\mathbf{t}_1$ correspondingly changes to $\mathbf{t}_1′$, and the vector difference $d\mathbf{t}_1$ is shown in the *b*-part of the figure. The vector $d\mathbf{t}_1$ in the limit has a magnitude equal to the length of the arc $|\mathbf{t}_1|d\theta = d\theta$ obtained by swinging the unit vector $\mathbf{t}_1$ through the angle $d\theta$ ex-

[*] See Section B7 of Appendix B.

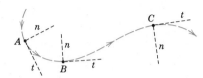

Figure 2/8

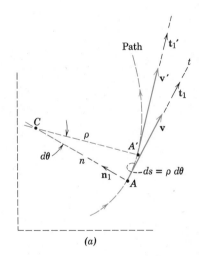

(a)

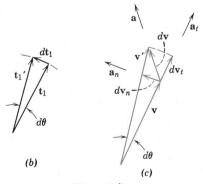

(b)

(c)

Figure 2/9

pressed in radians. The direction of $d\mathbf{t}_1$ is given by $\mathbf{n}_1$. Thus we may write $d\mathbf{t}_1 = \mathbf{n}_1\,d\theta$. Dividing by $d\theta$ gives

$$\frac{d\mathbf{t}_1}{d\theta} = \mathbf{n}_1$$

or dividing by dt gives $d\mathbf{t}_1/dt = (d\theta/dt)\mathbf{n}_1$ which may be written

$$\dot{\mathbf{t}}_1 = \dot{\theta}\mathbf{n}_1 \tag{2/9}$$

With the substitution of Eq. 2/9 and $\dot{\theta}$ from the relation $v = \rho\dot{\theta}$ Eq. 2/8 for the acceleration becomes

$$\mathbf{a} = \frac{v^2}{\rho}\mathbf{n}_1 + \dot{v}\mathbf{t}_1 \tag{2/10}$$

where

$$a_n = \frac{v^2}{\rho} = \rho\dot{\theta}^2 = v\dot{\theta}$$

$$a_t = \dot{v} = \ddot{s}$$

$$a = \sqrt{a_n{}^2 + a_t{}^2}$$

We may also note that $a_t = \dot{v} = d(\rho\dot{\theta})/dt = \rho\ddot{\theta} + \dot{\rho}\dot{\theta}$. However this relation finds little use since we seldom have reason to compute $\dot{\rho}$.

A clear understanding of Eq. 2/10 comes only when the geometry of the physical changes which it describes can be clearly seen. Figure 2/9c shows the velocity vector $\mathbf{v}$ when the particle is at A and $\mathbf{v}'$ when it is at A'. The vector change in the velocity is $d\mathbf{v}$, which establishes the direction of the acceleration $\mathbf{a}$. The n-component of $d\mathbf{v}$ is labeled $d\mathbf{v}_n$, and in the limit its magnitude equals the length of the arc generated by swinging the vector $\mathbf{v}$ as a radius through the angle $d\theta$. Thus $|d\mathbf{v}_n| = v\,d\theta$ and the n-component of acceleration is $a_n = |d\mathbf{v}_n|/dt = v(d\theta/dt) = v\dot{\theta}$ as before. The t-component of $d\mathbf{v}$ is labeled $d\mathbf{v}_t$, and its magnitude is simply the change dv in the magnitude or length of the velocity vector. Therefore the t-component of acceleration is $a_t = dv/dt = \dot{v} = \ddot{s}$ again as before. The acceleration vectors which result from their corresponding vector changes in velocity are also shown in Fig. 2/9c.

It is especially important to observe that the normal component of acceleration a_n is *always directed toward the center of curvature C.* The tangential component of acceleration, on the other hand, will be in the positive t-direction of motion if the speed v is increasing and in the negative t-direction if the speed is decreasing. In Fig. 2/10 are shown schematic representations of the variation in the acceleration vector for a particle moving from A to B with (a) increasing speed and (b) decreasing speed. At an inflection point on the curve the normal acceleration v^2/ρ goes to zero since ρ becomes infinite.

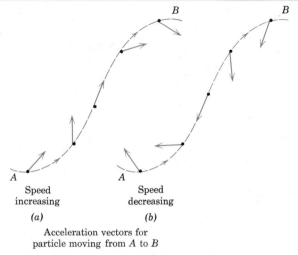

(a) Speed increasing

(b) Speed decreasing

Acceleration vectors for
particle moving from A to B

Figure 2/10

Circular motion is an important special case of plane curvilinear motion where the radius of curvature ρ becomes the constant radius r of the circle, Fig. 2/11. The velocity and the acceleration components for the circular motion of the particle P become

$$v = r\dot{\theta}$$
$$a_n = v^2/r = r\dot{\theta}^2 = v\dot{\theta} \qquad (2/11)$$
$$a_t = \dot{v} = r\ddot{\theta}$$

We find repeated use for Eqs. 2/10 and 2/11 in dynamics, and these relations and the principles behind them should be *mastered thoroughly*.

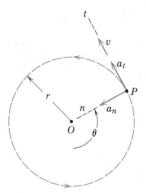

Figure 2/11

Sample Problem 2/7

To anticipate the dip and hump in the road the driver of a car applies her brakes to produce a uniform deceleration. Her speed is 100 km/h at the bottom A of the dip and 50 km/h at the top C of the hump, which is 120 m along the road from A. If the passengers experience a total acceleration of 3 m/s² at A and if the radius of curvature of the hump at C is 150 m, calculate (a) the radius of curvature ρ at A, (b) the acceleration at the inflection point B, and (c) the total acceleration at C.

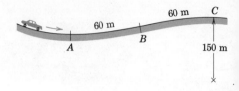

Solution. The dimensions of the car are small compared with those of the path, so we will treat the car as a particle. The velocities are

$$v_A = 100/3.6 = 27.78 \text{ m/s} \qquad v_C = 50/3.6 = 13.89 \text{ m/s}$$

We find the constant deceleration along the path from

$$\left[\int v \, dv = \int a_t \, ds \right] \qquad \int_{v_A}^{v_C} v \, dv = a_t \int_0^s ds$$

$$a_t = \frac{1}{2s}(v_C{}^2 - v_A{}^2) = \frac{(13.89)^2 - (27.78)^2}{2(120)} = -2.41 \text{ m/s}^2$$

(a) *Condition at A.* With the total acceleration given and a_t determined we can easily compute a_n and hence ρ from

$$[a^2 = a_n{}^2 + a_t{}^2] \qquad a_n{}^2 = 3^2 - (2.41)^2 = 3.19 \qquad a_n = 1.78 \text{ m/s}^2$$

$$[a_n = v^2/\rho] \qquad \rho = v^2/a_n = (27.78)^2/1.78 = 432 \text{ m} \qquad \qquad Ans.$$

(b) *Condition at B.* Since the radius of curvature is infinite at the inflection point, $a_n = 0$ and

$$a = a_t = -2.41 \text{ m/s}^2 \qquad \qquad Ans.$$

(c) *Condition at C.* The normal and total acceleration become

$$[a_n = v^2/\rho] \qquad a_n = (13.89)^2/150 = 1.29 \text{ m/s}^2$$

$$[a = \sqrt{a_n{}^2 + a_t{}^2}] \qquad a = \sqrt{(1.29)^2 + (-2.41)^2} = 2.73 \text{ m/s}^2 \qquad Ans.$$

The acceleration vectors representing the conditions at each of the three points are shown for clarification.

① Actually the radius of curvature to the road differs by about one meter from that to the path followed by the center of mass of the passengers, but we have neglected this relatively small difference.

② To convert from km/h to m/s divide by

$$\frac{3600 \text{ s/h}}{1000 \text{ m/km}} = 3.6 \text{ ks/h}. \text{ To convert from}$$

m/s to km/h multiply by 3.6 ks/h.

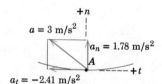

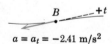

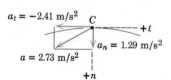

Sample Problem 2/8

A certain rocket maintains a horizontal attitude of its axis during the powered phase of its flight at high altitude. The thrust imparts a horizontal component of acceleration of 20 ft/sec², and the downward acceleration component is the acceleration due to gravity at that altitude, which is $g = 30$ ft/sec². At the instant represented the velocity of the mass center G of the rocket along the 15-deg direction of its trajectory is 12,000 mi/hr. For this position determine (a) the radius of curvature of the flight trajectory, (b) the rate at which the speed v is increasing, and (c) the angular rate $\dot{\theta}$ of the radial line from G to the center of curvature C.

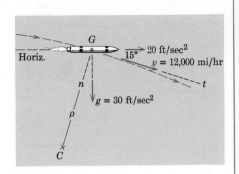

Solution. We observe that the radius of curvature appears in the expression for the normal component of acceleration, so we use n- and t-coordinates to describe the motion of G. The n- and t-components of the total acceleration are obtained by resolving the given horizontal and vertical accelerations into their n- and t-components and then combining. From the figure we get

① Alternatively we could find the resultant acceleration and then resolve it into n- and t-components.

$$a_n = 30 \cos 15° - 20 \sin 15° = 23.8 \text{ ft/sec}^2$$

$$a_t = 30 \sin 15° + 20 \cos 15° = 27.1 \text{ ft/sec}^2$$

(a) We may now compute the radius of curvature from

$$[a_n = v^2/\rho] \qquad \rho = \frac{v^2}{a_n} = \frac{[(12,000)(44/30)]^2}{23.8} = 13.01(10^6) \text{ ft} \qquad Ans.$$

② To convert from mi/hr to ft/sec multiply by $\dfrac{5280 \text{ ft/mi}}{3600 \text{ sec/hr}} = \dfrac{44 \text{ ft/sec}}{30 \text{ mi/hr}}$ which is easily remembered as 30 mi/hr equals 44 ft/sec.

(b) The rate at which v is increasing is simply the t-component of acceleration.

$$[\dot{v} = a_t] \qquad \dot{v} = 27.1 \text{ ft/sec}^2 \qquad Ans.$$

(c) The angular rate $\dot{\theta}$ of line GC depends on v and ρ and is given by

$$[v = \rho \dot{\theta}] \qquad \dot{\theta} = v/\rho = \frac{12,000(44/30)}{13.01(10^6)} = 13.53(10^{-4}) \text{ rad/sec} \qquad Ans.$$

PROBLEMS

2/81   A racing car traveling at a constant speed negotiates a horizontal turn where the radius of curvature is 300 m. What is the maximum speed v of the car if its normal acceleration cannot exceed $0.8g$ without the car sliding?   *Ans.* $v = 174.7$ km/h

Problem 2/82

2/82   The driver of the truck has an acceleration of $0.4g$ as the truck passes over the top A of the hump in the road at constant speed. The radius of curvature of the road at the top of the hump is 98 m, and the center of mass G of the driver (considered as a particle) is 2 m above the road. Calculate the speed v of the truck.

2/83   A car rounds a curve at a constant speed. An accelerometer mounted in the car records an acceleration of 2 ft/sec^2 at a point A on the curve and an acceleration of 4 ft/sec^2 at a point B on the curve. Determine the ratio n of the radius of curvature at A to that at B.

2/84   The muzzle velocity for a certain rifle is 600 m/s. If the rifle is pointed vertically upward and fired from an automobile moving horizontally at a speed of 72 km/h, determine the radius of curvature ρ of the path of the bullet at its maximum altitude. Neglect air resistance.   *Ans.* $\rho = 40.8$ m

Problem 2/84

2/85   To simulate a condition of "weightlessness" in its cabin, a jet transport traveling at 720 km/h moves on a sustained vertical curve as shown. At what rate $\dot{\theta}$ in degrees per second should the pilot drop his longitudinal line of sight to effect the desired condition? The maneuver takes place at a mean altitude of 8 km, and the gravitational acceleration may be taken as 9.81 m/s^2.

Problem 2/85

2/86   A ship which moves at a steady 20-knot speed (1 knot = 1.852 km/h) executes a turn to port by changing its compass heading at a constant counterclockwise rate. If it requires 60 s to alter course 90°, calculate the acceleration a of the ship during the turn.   *Ans.* $a = 0.269$ m/s^2

2/87 A car rounds a turn of constant curvature between *A* and *B* with a steady speed of 45 mi/hr. If an accelerometer were mounted in the car, what magnitude of acceleration would it record between *A* and *B*?

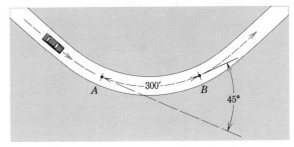

Problem 2/87

2/88 Consider the polar axis of the earth to be fixed in space and compute the acceleration *a* of a point *P* on the earth's surface at latitude 40° north. The mean diameter of the earth is 12 742 km and its angular velocity is $0.729(10^{-4})$ rad/s.

Ans. $a = 0.0259$ m/s^2

2/89 The car *C* increases its speed at the constant rate of 1.5 m/s^2 as it rounds the curve shown. If the magnitude of the total acceleration of the car is 2.5 m/s^2 at point *A* where the radius of curvature is 200 m, compute the speed *v* of the car at this point.

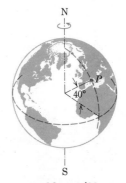

Problem 2/88

2/90 A car is traveling around a circular track of 600-ft radius and is increasing its speed at the constant rate of 5 mi/hr each second. What is the speed of the car when the magnitude of its total acceleration reaches 10 ft/sec^2?   *Ans. $v = 43.5$ mi/hr*

2/91 At the bottom *A* of the vertical inside loop the magnitude of the total acceleration of the airplane is 3*g*. If the airspeed is 800 km/h and is increasing at the rate of 20 km/h per second, calculate the radius of curvature *ρ* of the path at *A*.

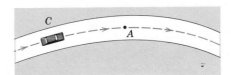

Problem 2/89

2/92 A point on the rim of a flywheel has a peripheral speed of 10 m/s at an instant when this speed is decreasing at the rate of 60 m/s^2. If the magnitude of the total acceleration of the point at this instant is 100 m/s^2, find the radius *r* of the flywheel.

Ans. $r = 1.25$ m

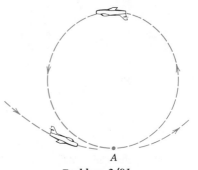

Problem 2/91

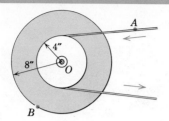

Problem 2/93

2/93 The wheel and attached pulley rotate about the fixed shaft at O and are driven by the belt shown. At a certain instant the velocity and acceleration of a point A on the belt are 2 ft/sec and 6 ft/sec², respectively, both in the direction shown. Calculate the magnitude of the total acceleration of point B on the wheel for this instant. Observe that the linear motion of point A on the belt and the tangential motion of a point on the 4-in.-radius circle are identical and that $\dot{\theta}$ and $\ddot{\theta}$ for the radial lines to all points on the wheel are the same.

2/94 The position of a particle moving along a curved path is determined by its distance s in meters measured along the curve and given by $s = 4 + t^3/3$ where t is in seconds. When $t = 3$ s the particle is at a position where the radius of curvature to the path is 9 m. Calculate the magnitude of the total acceleration of the particle as it passes this position.

Ans. $a = 10.82$ m/s²

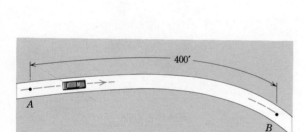

Problem 2/95

2/95 A car travels along the level curved road with a speed which is decreasing at the constant rate of 1.5 ft/sec each second. The speed of the car as it passes point A is 40 ft/sec. Calculate the magnitude of the total acceleration of the car as it passes point B which is 400 ft along the road from A. The radius of curvature of the road at B is 200 ft.

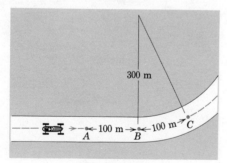

Problem 2/96

2/96 A race driver traveling at a speed of 200 km/h on the straightaway applies his brakes at point A and reduces his speed at a uniform rate to 150 km/h at C in a distance of $100 + 100 = 200$ m. Calculate the magnitude of the total acceleration of the race car an instant after it passes point B.

Ans. $a = 8.72$ m/s²

Problem 2/97

2/97 A particle which moves along the curved path shown passes point O with a speed of 12 m/s and slows down to 6 m/s at point A in a distance of 18 m, measured along the curve from O. The deceleration measured along the curve is proportional to the distance from O. If the magnitude of the total acceleration of the particle is 10 m/s² as it passes A, determine the radius of curvature ρ of the path at A.

Ans. $\rho = 4.5$ m

2/98 A particle moves on a circular path of radius $r = 2$ ft with a constant speed of 10 ft/sec. During the interval from A to B the velocity undergoes a vector change $\Delta \mathbf{v}$. Divide this change by the time interval between the two points to obtain the average normal acceleration for (a) $\Delta \theta = 30°$, (b) $\Delta \theta = 15°$, and (c) $\Delta \theta = 5°$. Compare the results with the instantaneous normal acceleration.

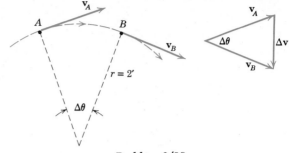

Problem 2/98

2/99 A scientific satellite is injected into an elliptical orbit about the earth and has a velocity of 32 000 km/h at its point of nearest approach (perigee) 400 km above the earth. Calculate the radius of curvature ρ of the orbit at this point. The absolute gravitational acceleration at the surface of the earth is 9.821 m/s², and the mean diameter of the earth is 12 742 km.   *Ans.* $\rho = 9090$ km

2/100 A satellite which moves in a circular orbit around the earth at a height $h = 200$ mi above the surface of the earth must have a velocity of 17,260 mi/hr. Calculate the gravitational acceleration g for that altitude. The mean radius R of the earth is 3958 mi. (Check your answer by computing g from the gravitational law, which gives $g = g_0(R/[R + h])^2$ where $g_0 = 32.22$ ft/sec² from Table C2 in Appendix C.)

2/101 The command module of a lunar mission is orbiting the moon in a circular path at an altitude of 160 km above the moon's surface. Consult Table C2, Appendix C, for information about the moon as is needed and compute the orbital velocity v of the module with respect to the moon.   *Ans.* $v = 5781$ km/h

2/102 Magnetic tape is being transferred from reel A to reel B and passes around idler pulleys C and D. At a certain instant point P_1 on the tape is in contact with pulley C and point P_2 is in contact with pulley D. If the normal component of acceleration of P_1 is 40 m/s² and the tangential component of acceleration of P_2 is 30 m/s² at this instant, compute the corresponding speed v of the tape, the magnitude of the total acceleration a_1 of P_1, and the magnitude of the total acceleration a_2 of P_2.

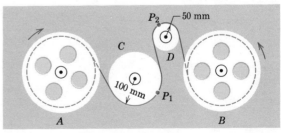

Problem 2/102

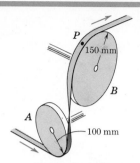

Problem 2/103

Problem 2/104   **Problem 2/105**

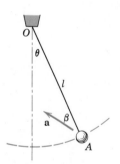

Problem 2/106

2/103 The direction of motion of a flat tape in a numerical-control device is changed by the two pulleys *A* and *B* shown. If the speed of the tape increases uniformly from 2 m/s to 18 m/s while 8 meters of tape pass over the pulleys, calculate the magnitude of the acceleration of point *P* on the tape in contact with pulley *B* at the instant when the tape speed is 3 m/s.   *Ans.* $a = 63.2$ m/s^2

2/104 The rocket is coasting in unpowered flight above the atmosphere at a height where the attraction of the earth gives it an acceleration of 9.6 m/s^2 vertically down. If the velocity of the rocket at this instant is $v = 4000$ m/s in the direction shown, compute the radius of curvature ρ of its trajectory at this position and the rate at which v is changing with time.

2/105 A rocket traveling above the atmosphere at an altitude of 1000 km would have a free-fall acceleration $g = 7.32$ m/s^2 in the absence of forces other than gravitational attraction. Because of thrust, however, the rocket has an additional acceleration component a_1 of 7.62 m/s^2 tangent to its trajectory, which makes an angle of 30° with the vertical at the instant considered. If the velocity v of the rocket is 40 000 km/h at this position, compute the radius of curvature ρ of the trajectory and the rate at which v is changing with time.   *Ans.* $\rho = 33\ 700$ km
$\dot{v} = 1.28$ m/s^2

2/106 The tangential acceleration of the simple pendulum equals $g \sin \theta$. For a given angle θ if the total acceleration makes an angle β with respect to OA, derive expressions for the corresponding values of $\dot{\theta}$ and $\ddot{\theta}$.

2/107 A projectile is fired at an angle of 30 deg above the horizontal with a muzzle velocity of 1500 ft/sec. Determine the radius of curvature ρ of its path 10 seconds after firing. Neglect atmospheric resistance so that the gravitational attraction is the only force to be considered. The acceleration of the projectile, consequently, is g down.
Ans. $\rho = 61,200$ ft

108 The position vector of a particle which moves in the
x-y plane is given by $\mathbf{r} = 2t^2\mathbf{i} + \frac{2}{3}t^3\mathbf{j}$ where $\mathbf{r}$ is in
inches and t is in seconds. Calculate the radius of
curvature ρ of the path for the position when
$t = 2$ sec. Sketch the velocity $\mathbf{v}$ and the curvature of
the path for the instant considered.

Ans. $\rho = 45.3$ in.

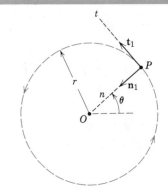

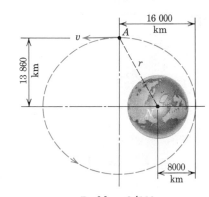

Problem 2/109

109 Derive the vector expressions for the velocity $\mathbf{v}$ and
acceleration $\mathbf{a}$ of point P moving in a circular path
of radius r with varying speed. Introduce unit vec-
tors $\mathbf{t}_1$ in the tangential direction and $\mathbf{n}_1$ in the
normal direction and start by writing the position
vector of P as $\mathbf{r} = r(-\mathbf{n}_1)$ which gives the proper
magnitude and direction for the vector $\overrightarrow{OP}$. Then
obtain $\mathbf{v} = \dot{\mathbf{r}}$ and $\mathbf{a} = \dot{\mathbf{v}}$. Work out the proper expressions for $\dot{\mathbf{n}}_1$ and $\dot{\mathbf{t}}_1$ as needed.

110 An earth satellite which moves in the elliptical
equatorial orbit shown has a velocity v in space of
17 970 km/h when it passes the end of the semiminor axis at A. The earth has an absolute surface
value of g of 9.821 m/s² and has a radius of 6370 km.
Determine the radius of curvature ρ of the orbit
at A.

Ans. $\rho = 18\ 480$ km

Problem 2/110

111 The mine skip is being hauled to the surface over
the curved track by the cable around the 30-in.
drum which turns at the constant clockwise speed of
120 rev/min. The shape of the track is defined by
$y = x^2/40$ where x and y are in feet. Calculate the
magnitude of the total acceleration of the skip as it
reaches a level of 2 ft below the top. Neglect the
dimensions of the skip compared with those of the
path. (Recall that the radius of curvature is given
by $\rho = [1 + (dy/dx)^2]^{3/2}/(d^2y/dx^2)$.)

Problem 2/111

112 The motion of the pin A in the fixed circular slot is
controlled by the guide B which is being elevated
by its lead screw with a constant upward velocity
$v_0 = 2$ m/s for an interval of its motion. Calculate
both the normal and tangential components of acceleration of pin A as it passes the position for
which $\theta = 30°$.

Ans. $a_n = 21.3$ m/s², $a_t = 12.32$ m/s²

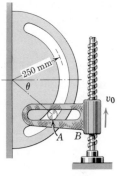

Problem 2/112

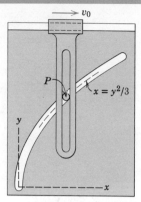

Problem 2/113

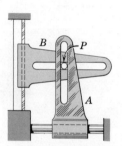

Problem 2/114

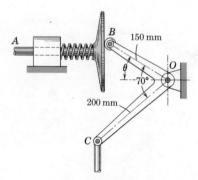

Problem 2/116

▶2/113 The vertical slot moves to the right with a constant speed v_0 in meters per second for an interval of motion and causes the pin P to move along the parabolic slot $x = y^2/3$ where x and y are in meters. Calculate the radius of curvature ρ to the path for the position where $y = 2$ m and calculate the tangential acceleration a_t of the pin when it passes this position.   *Ans.* $\rho = 6.94$ m, $a_t = -0.169v_0^2$ m/s^2

▶2/114 The pin P is constrained to move in the slotted guides which move at right angles to one another. At the instant represented, A has a velocity to the right of 0.2 m/s which is decreasing at the rate of 0.75 m/s each second. At the same time B is moving down with a velocity of 0.15 m/s which is decreasing at the rate of 0.5 m/s each second. For this instant determine the radius of curvature ρ of the path followed by P. Is it possible to determine also the time rate of change of ρ?   *Ans.* $\rho = 1.25$ m

▶2/115 If the guide B in Prob. 2/112 has an upward acceleration $\dot{v}_0 = 14$ m/s^2 and an upward velocity $v_0 = 1.5$ m/s at the instant when $\theta = 30°$, determine the magnitude of the total acceleration of pin A and the value of $\ddot{\theta}$ for this instant.
       Ans. $a = 15.14$ m/s^2, $\ddot{\theta} = -37.0$ rad/s^2

▶2/116 The horizontal plunger A, which operates the 70° bell crank BOC, has a velocity to the right of 75 mm/s and is speeding up at the rate of 100 mm/s per second at the position for which $\theta = 30°$. Compute the angular acceleration $\ddot{\theta}$ of the bell crank at this instant.   *Ans.* $\ddot{\theta} = -0.399$ rad/s^2

2/6 POLAR COORDINATES $(r\text{-}\theta)$. We now consider the third description of plane curvilinear motion, namely, polar coordinates where the particle is located by the radial distance r from a fixed pole and by an angular measurement θ to the radial line. Polar coordinates are particularly useful when a motion is constrained through the control of a radial distance and an angular position or when an unconstrained motion is observed by measurements of a radial distance and an angular position.

Figure 2/12*a* shows the polar coordinates r and θ which locate a particle traveling on a curved path. An arbitrary fixed line, such as the *x*-axis, is used as a reference for the measurement of θ. Unit vectors $\mathbf{r}_1$ and $\boldsymbol{\theta}_1$ are established in the positive *r*- and θ-directions, respectively. The position vector $\mathbf{r}$ to the particle at A has a magnitude equal to the radial distance r and a direction specified by the unit vector $\mathbf{r}_1$. Thus we express the location of the particle at A by the vector

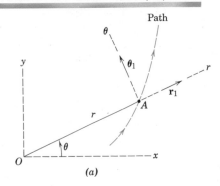

(a)

$$\mathbf{r} = r\mathbf{r}_1$$

As we differentiate this relation with respect to time to obtain $\mathbf{v} = \dot{\mathbf{r}}$ and $\mathbf{a} = \dot{\mathbf{v}}$, we will need expressions for the time derivatives of both unit vectors $\mathbf{r}_1$ and $\boldsymbol{\theta}_1$. We obtain $\dot{\mathbf{r}}_1$ and $\dot{\boldsymbol{\theta}}_1$ in exactly the same way that we derived $\dot{\mathbf{t}}_1$ in the preceding article. During time dt the coordinate directions rotate through the angle $d\theta$, and the unit vectors also rotate through the same angle from $\mathbf{r}_1$ and $\boldsymbol{\theta}_1$ to $\mathbf{r}_1{}'$ and $\boldsymbol{\theta}_1{}'$ as shown in Fig. 2/12*b*. We note that the vector change $d\mathbf{r}_1$ is in the plus θ-direction and that $d\boldsymbol{\theta}_1$ is in the minus *r*-direction. Since their magnitudes in the limit are equal to the unit vector as radius times the angle $d\theta$ in radians, we may write them as $d\mathbf{r}_1 = \boldsymbol{\theta}_1\, d\theta$ and $d\boldsymbol{\theta}_1 = -\mathbf{r}_1\, d\theta$. If we divide these equations by $d\theta$ we have

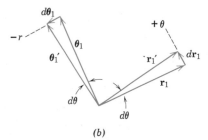

(b)

Figure 2/12

$$\frac{d\mathbf{r}_1}{d\theta} = \boldsymbol{\theta}_1 \quad \text{and} \quad \frac{d\boldsymbol{\theta}_1}{d\theta} = -\mathbf{r}_1$$

Or, if we divide them by dt, we have $d\mathbf{r}_1/dt = (d\theta/dt)\boldsymbol{\theta}_1$ and $d\boldsymbol{\theta}_1/dt = -(d\theta/dt)\mathbf{r}_1$ or simply

$$\boxed{\dot{\mathbf{r}}_1 = \dot{\theta}\boldsymbol{\theta}_1 \quad \text{and} \quad \dot{\boldsymbol{\theta}}_1 = -\dot{\theta}\mathbf{r}_1} \qquad (2/12)$$

We are now ready to differentiate $\mathbf{r} = r\mathbf{r}_1$ with respect to time. Using the rule for differentiating the product of a scalar and a vector gives

$$\mathbf{v} = \dot{\mathbf{r}} = \dot{r}\mathbf{r}_1 + r\dot{\mathbf{r}}_1$$

With the substitution of $\dot{\mathbf{r}}_1$ from Eq. 2/12 the vector expression for the velocity becomes

$$\boxed{\mathbf{v} = \dot{r}\mathbf{r}_1 + r\dot{\theta}\boldsymbol{\theta}_1} \tag{2/13}$$

where

$$v_r = \dot{r}$$
$$v_\theta = r\dot{\theta}$$
$$v = \sqrt{v_r^2 + v_\theta^2}$$

The r-component of $\mathbf{v}$ is merely the rate at which the vector $\mathbf{r}$ stretches. The θ-component of $\mathbf{v}$ is due to the rotation of $\mathbf{r}$.

We now differentiate the expression for $\mathbf{v}$ to obtain the acceleration $\mathbf{a} = \dot{\mathbf{v}}$. Note that the derivative of $r\dot{\theta}\boldsymbol{\theta}_1$ will produce three terms since all three factors are variable. Thus

$$\mathbf{a} = \dot{\mathbf{v}} = (\ddot{r}\mathbf{r}_1 + \dot{r}\dot{\mathbf{r}}_1) + (\dot{r}\dot{\theta}\boldsymbol{\theta}_1 + r\ddot{\theta}\boldsymbol{\theta}_1 + r\dot{\theta}\dot{\boldsymbol{\theta}}_1)$$

Substitution of $\dot{\mathbf{r}}_1$ and $\dot{\boldsymbol{\theta}}_1$ from Eq. 2/12 and collecting terms give

$$\boxed{\mathbf{a} = (\ddot{r} - r\dot{\theta}^2)\mathbf{r}_1 + (r\ddot{\theta} + 2\dot{r}\dot{\theta})\boldsymbol{\theta}_1} \tag{2/14}$$

where

$$a_r = \ddot{r} - r\dot{\theta}^2$$
$$a_\theta = r\ddot{\theta} + 2\dot{r}\dot{\theta}$$
$$a = \sqrt{a_r^2 + a_\theta^2}$$

We may write the θ-component alternatively as

$$a_\theta = \frac{1}{r}\frac{d}{dt}(r^2\dot{\theta})$$

which can be verified easily by carrying out the differentiation. This form for a_θ will be found useful when we treat the angular momentum of particles in the next chapter.

An adequate appreciation of the terms in Eq. 2/14 comes only when the geometry of the physical changes can be clearly seen. For this purpose Fig. 2/13a is developed to show the velocity vectors and their r- and θ-components at position A and at position A' after an infinitesimal movement. Each of these components undergoes a change in magnitude and direction as shown in Fig. 2/13b. In this figure we see the following changes:

Magnitude change of $\mathbf{v}_r$. This change is simply the increase in length of v_r or $dv_r = d\dot{r}$, and the corresponding acceleration term is $d\dot{r}/dt = \ddot{r}$ in the positive r-direction.

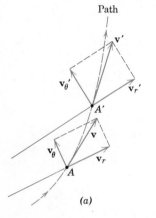

(a)

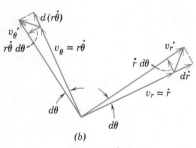

(b)

Figure 2/13

Direction change of v_r. The magnitude of this change is seen from the figure to be $v_r \, d\theta = \dot{r} \, d\theta$, and its contribution to the acceleration becomes $\dot{r} \, d\theta/dt = \dot{r}\dot{\theta}$ which is in the positive θ-direction.

Magnitude change of v_θ. This term is the change in length of v_θ or $d(r\dot{\theta})$, and its contribution to the acceleration is $d(r\dot{\theta})/dt = r\ddot{\theta} + \dot{r}\dot{\theta}$ and is in the positive θ-direction.

Direction change of v_θ. The magnitude of this change is $v_\theta \, d\theta = r\dot{\theta} \, d\theta$, and the corresponding acceleration term is seen to be $r\dot{\theta}(d\theta/dt) = r\dot{\theta}^2$ in the negative r-direction.

Collecting terms gives $a_r = \ddot{r} - r\dot{\theta}^2$ and $a_\theta = r\ddot{\theta} + 2\dot{r}\dot{\theta}$ as obtained previously. We see that the term $\ddot{r}$ is the acceleration which the particle would have along the radius in the absence of a change in θ. The term $-r\dot{\theta}^2$ is the normal component of acceleration if r were constant as in circular motion. The term $r\ddot{\theta}$ is the tangential acceleration which the particle would have if r were constant but is only a part of the acceleration due to the change in magnitude of v_θ when r is variable. Lastly, the term $2\dot{r}\dot{\theta}$ is composed of two effects. The first effect comes from that portion of the change in magnitude $d(r\dot{\theta})$ of v_θ due to the change in r, and the second effect comes from the change in direction of v_r. The term $2\dot{r}\dot{\theta}$ represents, therefore, a combination of changes and is not so easily perceived as are the other acceleration terms. We also note carefully that a_r is not $\dot{v}_r$ and that a_θ is not $\dot{v}_\theta$.

The total acceleration **a** and its components are represented in Fig. 2/14. If **a** has a component normal to the path, we know from our analysis of n- and t-components in the previous article that the sense of the n-component *must* be toward the center of curvature.

For motion in a circular path with r constant the components of Eqs. 2/13 and 2/14 become simply

$$v_r = 0 \qquad\qquad v_\theta = r\dot{\theta}$$

$$a_r = -r\dot{\theta}^2 \qquad a_\theta = r\ddot{\theta}$$

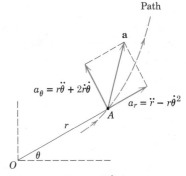

Path

$$a_\theta = r\ddot{\theta} + 2\dot{r}\dot{\theta}$$

$$a_r = \ddot{r} - r\dot{\theta}^2$$

Figure 2/14

This description is the same as that obtained with n- and t-components where the θ- and t-directions coincide but where the positive r-direction is in the negative n-direction. Hence $a_r = -a_n$ for circular motion.

The expressions for a_r and a_θ in scalar form may also be obtained by direct differentiation of the coordinate relations $x = r\cos\theta$ and $y = r\sin\theta$ to get $a_x = \ddot{x}$ and $a_y = \ddot{y}$. Each of these rectangular components of acceleration may then be resolved into r- and θ-components which, when combined, will yield the expressions of Eq. 2/14. This approach is straightforward but has the disadvantage of somewhat more algebraic manipulation than the foregoing proofs.

Sample Problem 2/9

Rotation of the radially slotted arm is governed by $\theta = 0.2t + 0.02t^3$ where θ is in radians and t is in seconds. Simultaneously the power screw in the arm engages the slider B and controls its distance from O according to $r = 0.2 + 0.04t^2$ where r is in meters and t is in seconds. Calculate the velocity and acceleration of the slider for the instant when $t = 3$ s.

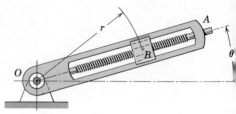

Solution. The coordinates and their time derivatives which appear in the expressions for velocity and acceleration in polar coordinates are obtained first and evaluated for $t = 3$ s.

$$r = 0.2 + 0.04t^2 \qquad r_3 = 0.2 + 0.04(3^2) = 0.56 \text{ m}$$

$$\dot{r} = 0.08t \qquad \dot{r}_3 = 0.08(3) = 0.24 \text{ m/s}$$

$$\ddot{r} = 0.08 \qquad \ddot{r}_3 = 0.08 \text{ m/s}^2$$

$$\theta = 0.2t + 0.02t^3 \qquad \theta_3 = 0.2(3) + 0.02(3^3) = 1.14 \text{ rad}$$

$$\text{or } \theta = 1.14(180/\pi) = 65.3°$$

$$\dot{\theta} = 0.2 + 0.06t^2 \qquad \dot{\theta}_3 = 0.2 + 0.06(3^2) = 0.74 \text{ rad/s}$$

$$\ddot{\theta} = 0.12t \qquad \ddot{\theta}_3 = 0.12(3) = 0.36 \text{ rad/s}^2$$

The velocity components are obtained from Eq. 2/13 and for $t = 3$ s are

$$[v_r = \dot{r}] \qquad v_r = 0.24 \text{ m/s}$$

$$[v_\theta = r\dot{\theta}] \qquad v_\theta = 0.56(0.74) = 0.414 \text{ m/s}$$

$$[v = \sqrt{v_r^2 + v_\theta^2}] \quad v = \sqrt{(0.24)^2 + (0.414)^2} = 0.479 \text{ m/s} \qquad \textit{Ans.}$$

The velocity and its components are shown for the specified position of the arm.

The acceleration components are obtained from Eq. 2/14 and for $t = 3$ s are

$$[a_r = \ddot{r} - r\dot{\theta}^2] \qquad a_r = 0.08 - 0.56(0.74)^2 = -0.227 \text{ m/s}^2$$

$$[a_\theta = r\ddot{\theta} + 2\dot{r}\dot{\theta}] \qquad a_\theta = 0.56(0.36) + 2(0.24)(0.74) = 0.557 \text{ m/s}^2$$

$$[a = \sqrt{a_r^2 + a_\theta^2}] \qquad a = \sqrt{(-0.227)^2 + (0.557)^2} = 0.601 \text{ m/s}^2 \qquad \textit{Ans.}$$

The acceleration and its components are also shown for the 65.3° position of the arm.

① We see that this problem is an example of constrained motion where the center B of the slider is mechanically constrained by the rotation of the slotted arm and by engagement with the turning screw.

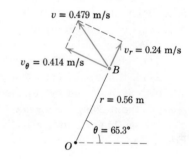

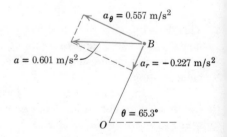

Sample Problem 2/10

A tracking radar lies in the vertical plane of the path of a rocket which is coasting in unpowered flight above the atmosphere. For the instant when $\theta = 30°$ the tracking data give $r = 25(10^4)$ ft, $\dot{r} = 4000$ ft/sec, and $\dot{\theta} = 0.80$ deg/sec. The acceleration of the rocket is due only to gravitational attraction and for its particular altitude is 31.4 ft/sec² vertically down. For these conditions determine the velocity v of the rocket and the values of $\ddot{r}$ and $\ddot{\theta}$.

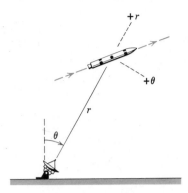

Solution. The components of velocity from Eq. 2/13 are

$[v_r = \dot{r}]$ $\qquad v_r = 4000$ ft/sec

$[v_\theta = r\dot{\theta}]$ $\qquad v_\theta = 25(10^4)(0.80)\left(\dfrac{\pi}{180}\right) = 3490$ ft/sec

$[v = \sqrt{v_r^2 + v_\theta^2}]$ $\quad v = \sqrt{(4000)^2 + (3490)^2} = 5310$ ft/sec $\qquad$ *Ans.*

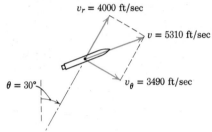

Since the total acceleration of the rocket is $g = 31.4$ ft/sec² down, we can easily find its r- and θ-components for the given position. As shown in the figure they are

$a_r = -31.4 \cos 30° = -27.2$ ft/sec²

$a_\theta = 31.4 \sin 30° = 15.7$ ft/sec²

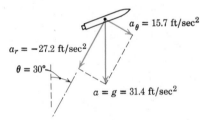

We now equate these values to the polar-coordinate expressions for a_r and a_θ which contain the unknowns $\ddot{r}$ and $\ddot{\theta}$. Thus from Eq. 2/14

$[a_r = \ddot{r} - r\dot{\theta}^2]$ $\quad -27.2 = \ddot{r} - 25(10^4)\left(0.80 \dfrac{\pi}{180}\right)^2$

$\ddot{r} = 21.5$ ft/sec²

$[a_\theta = r\ddot{\theta} + 2\dot{r}\dot{\theta}]$ $\quad 15.7 = 25(10^4)\ddot{\theta} + 2(4000)\left(0.80 \dfrac{\pi}{180}\right)$

$\ddot{\theta} = -3.84(10^{-4})$ rad/sec² $\qquad$ *Ans.*

① We observe that the angle θ in polar coordinates need not always be taken positive in a counterclockwise sense.

② Note that the r-component of acceleration is in the negative r-direction, so it carries a minus sign.

③ We must be careful to convert $\dot{\theta}$ from deg/sec to rad/sec.

PROBLEMS

2/117 The polar coordinates of a particle are governed by $r = 4 + t^2$ and $\theta = t^2/2$ where r is in meters, θ is in radians, and t is in seconds. Determine the velocity $\mathbf{v}$ and acceleration $\mathbf{a}$ of the particle when $t = 2$ s.
Ans. $\mathbf{v} = 4\mathbf{r}_1 + 16\boldsymbol{\theta}_1$ m/s, $\mathbf{a} = -30\mathbf{r}_1 + 24\boldsymbol{\theta}_1$ m/s^2

2/118 A particle P moves on a curved path defined by polar coordinates r and θ. For a certain position of the particle $r = 20$ mm, $\dot{r} = 30$ mm/s, $\ddot{r} = 300$ mm/s^2, $\theta = 45°$, $\dot{\theta} = -2$ rad/s, and $\ddot{\theta} = 16$ rad/s^2. Determine the magnitude of the velocity of the particle and the r- and θ-components of acceleration. Sketch the position of P and draw vectors representing the velocity and acceleration.

2/119 A particle P moves on a curved path defined by polar coordinates. At a certain instant when $\theta = 5\pi/4$, the θ-component of acceleration is 16 in./sec^2. Also $r = 5$ in, $\ddot{r} = 6$ in./sec^2, and $\dot{\theta} = 0.5$ rad/sec and is increasing at the rate of 4 rad/sec each second. Use whatever information you need and compute the magnitude of the velocity of the particle.   *Ans.* $v = 4.72$ in./sec

2/120 A pin which is constrained by two machine parts which it joins has plane curvilinear motion governed by its polar coordinates $r = r_0 e^{kt}$ and $\theta = Ct$ where r_0, k, and C are constants and θ is in radians. Determine the expressions for the magnitudes of the velocity $\mathbf{v}$ and acceleration $\mathbf{a}$ of the pin in terms of the time t.

2/121 The motion of a point in the vertical plane is given by its polar coordinates $r = 2t^2$ and $\theta = 0.3 \sin \pi t/3$, where r is in meters, θ is in radians, and t is in seconds. Determine the velocity $\mathbf{v}$ and acceleration $\mathbf{a}$ of the particle when $t = 2$ s.
Ans. $\mathbf{v} = 8\mathbf{r}_1 - 1.257\boldsymbol{\theta}_1$ m/s
$\mathbf{a} = 3.80\mathbf{r}_1 - 4.79\boldsymbol{\theta}_1$ m/s^2

122 At the instant represented the slotted arm is rotating at a speed $N = 80$ rev/min and is slowing down
at the rate of 280 rev/min per second. Also at the
same time the radial distance r to the slider P is
250 mm and is decreasing at the constant rate of
300 mm/s. For this instant determine the acceleration $\mathbf{a}$ of P.

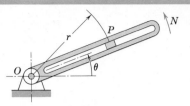

Problem 2/122

123 For the mechanism of Prob. 2/122, if $\dot{\theta} = 4$ rad/s,
$\ddot{\theta} = 16$ rad/s², and $r = 250$ mm at a certain instant
of time, what would be the corresponding time
derivatives of r if the slider P had zero acceleration
at this instant?   *Ans.* $\dot{r} = -0.5$ m/s, $\ddot{r} = 4$ m/s²

124 As the hydraulic cylinder rotates around O the
length l of the piston rod P is controlled by the
action of oil pressure in the cylinder. If the cylinder
rotates at the constant rate $\dot{\theta} = 60$ deg/sec and l is
decreasing at the constant rate of 6 in./sec, calculate the magnitudes of the velocity $\mathbf{v}$ and acceleration $\mathbf{a}$ of end B when $l = 5$ in.

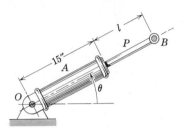

Problem 2/124

125 For the hydraulic cylinder of Prob. 2/124 calculate
the magnitudes of the velocity $\mathbf{v}$ and acceleration $\mathbf{a}$
of pin B for the following conditions: $l = 5$ in.,
$\theta = 45$ deg, $\dot{l} = 6$ in./sec, $\dot{\theta} = 45$ deg/sec,
$\ddot{l} = -4$ in./sec², and $\ddot{\theta} = -10$ deg/sec².
                Ans. $v = 16.82$ in./sec, $a = 17.38$ in./sec²

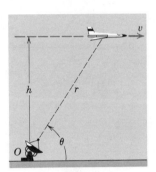

Problem 2/126

126 A jet plane flying at a constant velocity v at an
altitude $h = 8$ km is being tracked by radar located
at O directly below the line of flight. If the angle θ
is decreasing at the rate of 0.025 rad/s when
$\theta = 60°$, determine the value of $\ddot{r}$ at this instant and
the magnitude of the velocity $\mathbf{v}$ of the plane.

127 The rocket is fired vertically and tracked by the
radar shown. When θ reaches 60°, other corresponding measurements give the values $r =$
30,000 ft, $\ddot{r} = 70$ ft/sec², and $\dot{\theta} = 0.02$ rad/sec.
Calculate the velocity and acceleration of the
rocket at this position.
                Ans. $v = 1200$ ft/sec,
                $a = 67.0$ ft/sec²

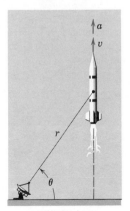

Problem 2/127

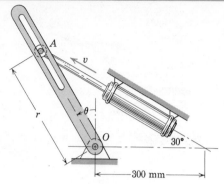

Problem 2/128

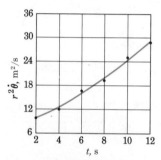

Problem 2/129

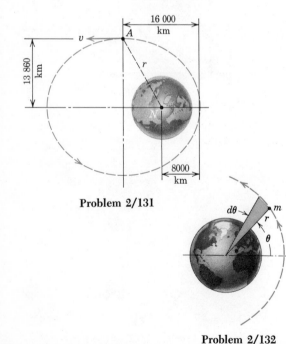

Problem 2/131

Problem 2/132

2/128 The hydraulic cylinder gives pin A a constant velocity $v = 2$ m/s along its axis for an interval of motion and in turn causes the slotted arm to rotate about O. Determine the values of $\dot{r}$, $\ddot{r}$, and $\ddot{\theta}$ for the instant when $\theta = 30°$. (*Hint:* Recognize that all acceleration components are zero when the velocity is constant.)

2/129 For a certain curvilinear motion of a particle expressed in polar coordinates the product $r^2\dot{\theta}$ in m²/s varies with the time t in seconds measured over a short period as shown. Approximate the θ-component of the acceleration of the particle at the instant when $t = 7$ s at which time $r = 0.5$ m.

Ans. $a_\theta = 4$ m/s²

2/130 A particle moving along a plane curve has a position vector $\mathbf{r}$, a velocity $\mathbf{v}$, and an acceleration $\mathbf{a}$. Unit vectors in the r- and θ-directions are $\mathbf{r}_1$ and $\boldsymbol{\theta}_1$, and both r and θ are changing with time. Explain why each of the following statements has been marked wrong.

$$\dot{\mathbf{r}} \neq v \qquad \ddot{\mathbf{r}} \neq a \qquad \dot{\mathbf{r}} \neq \dot{r}\mathbf{r}_1$$
$$\dot{r} \neq v \qquad \ddot{r} \neq a \qquad \ddot{\mathbf{r}} \neq \ddot{r}\mathbf{r}_1$$
$$\dot{r} \neq \mathbf{v} \qquad \ddot{r} \neq \mathbf{a} \qquad \dot{\mathbf{r}} \neq r\dot{\theta}\boldsymbol{\theta}_1$$

2/131 The earth satellite of Prob. 2/110, shown again here, has a velocity $v = 17\,970$ km/h as it passes the end of the semiminor axis at A. Gravitational attraction produces an acceleration $a = a_r = -1.556$ m/s² as calculated from the gravitational law. For this position calculate the rate $\dot{v}$ at which the speed of the satellite is changing and the quantity $\ddot{r}$. *Ans.* $\dot{v} = -0.778$ m/s², $\ddot{r} = -0.388$ m/s²

2/132 A satellite m moves in an elliptical orbit around the earth. There is no force on the satellite in the θ-direction, so that $a_\theta = 0$. Prove Kepler's second law of planetary motion which says that the radial line r sweeps through equal areas in equal times. The area dA swept by the radial line during time dt is shaded in the figure.

133 A particle *P* moves along a path, given by $r = f(\theta)$, which is symmetrical about the line $\theta = 0$. As the particle passes the position $\theta = 0$ where the radius of curvature of the path is ρ, the velocity of *P* is *v*. Derive an expression for $\ddot{r}$ in terms of *v*, *r*, and ρ for the motion of the particle at this point.

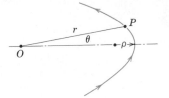

Problem 2/133

134 The cam has a shape such that the center of the roller *A* which follows the contour moves on a limaçon defined by $r = b - c \cos \theta$, where $b > c$. If the cam does not rotate, determine the magnitude of the total acceleration of *A* in terms of θ if the slotted arm revolves with a constant counterclockwise angular rate $\dot{\theta} = \omega$.

$$Ans. \ a = \omega^2 \sqrt{4c^2 - 4bc \cos \theta + b^2}$$

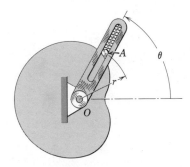

Problem 2/134

135 The path of a fluid particle *P* in a certain centrifugal pump with radial vanes is to be approximated by the spiral $r = r_0 e^{n\theta}$, where *n* is a dimensionless constant. If the pump turns at a constant rate $\dot{\theta} = K$, determine the expression for the total acceleration of the particle just prior to leaving the vane in terms of *R*, *K*, and *n*.

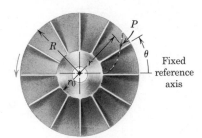

Problem 2/135

136 The slotted arm *OA* forces the small pin to move in the fixed spiral guide defined by $r = K\theta$. Arm *OA* starts from rest at $\theta = \pi/4$ and has a constant counterclockwise angular acceleration $\ddot{\theta} = \alpha$. Determine the magnitude of the acceleration of the pin when $\theta = 3\pi/4$.   *Ans. a = 10.76K\alpha*

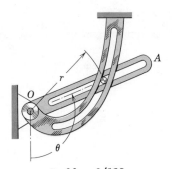

Problem 2/136

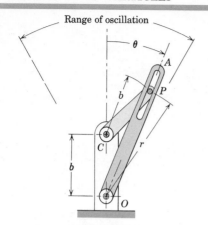

Problem 2/137

2/137 The slotted arm OA oscillates about O within the limits shown and drives the crank CP through the pin P. For an interval of the motion $\dot{\theta} = K$, a constant. Determine the magnitude of the corresponding total acceleration of P for any value of θ within the range for which $\dot{\theta} = K$. Use polar coordinates r and θ. Show that the magnitude of the velocity of P in its circular path is constant.

2/138 The slotted arm is pivoted at O and carries the slider C. The position of C in the slot is governed by the cord which is fastened at D and remains taut. The arm turns counterclockwise with a constant angular rate $\dot{\theta} = 4$ rad/sec during an interval of its motion. The length DBC of the cord equals R, which makes $r = 0$ when $\theta = 0$. Determine the magnitude of the acceleration of the slider at the position for which $\theta = 30°$. The distance R is 15 in.
Ans. $a = 489$ in./sec^2

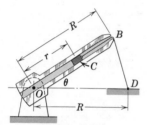

Problem 2/138

2/139 A cable attached to the vehicle at A passes over the small fixed pulley at B and around the drum at C. If the vehicle moves with a constant speed $v_0 = \dot{x}$, determine an expression for the acceleration of a point P on the cable between B and C in terms of θ. Also express $\ddot{\theta}$ in terms of θ. (*Hint:* Observe that the r- and θ-components of the acceleration of A are both zero.)

2/140 For the mechanism of Prob. 2/137 the slotted arm OA is given a periodic oscillation governed by $\theta = \theta_0 \sin(2\pi t/\tau)$ where τ is the period (time for one complete oscillation) and t is the time. The total range of oscillation is $2\theta_0$. Determine the expression for the magnitude of the acceleration of the pin P when (*a*) $t = \tau/4$ and (*b*) when $t = \tau/2$.
Ans. (*a*) $a = 8\pi^2 \theta_0 b/\tau^2$,
(*b*) $a = -a_r = 16\pi^2 \theta_0^2 b/\tau^2$

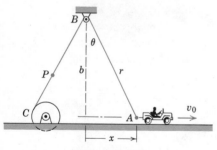

Problem 2/139

141 The circular disk rotates about its center O with a constant angular velocity $\omega = \dot{\theta}$ and carries the two spring-loaded plungers shown. The distance b which each plunger protrudes from the rim of the disk varies according to $b = b_0 \sin 2\pi nt$ where b_0 is the maximum protrusion, n is the constant frequency of oscillation of the plungers in the radial slots, and t is the time. Determine the maximum magnitudes of the r- and θ-components of the acceleration of the ends A of the plungers during their motion.

$Ans.$ $|a_r|_{\text{max}} = (4\pi^2 n^2 + \omega^2)b_0 + r_0\omega^2$
$|a_\theta|_{\text{max}} = 4\pi b_0 n\omega$

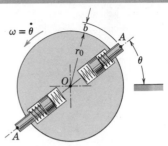

Problem 2/141

142 During reentry a space capsule A is tracked by radar station B located in the vertical plane of the trajectory. The values of r and θ are read against time and are recorded in the following table. Determine the velocity v of the capsule when $t = 40$ s. Explain how the acceleration of the capsule could be found from the given data.

t, s	r, km	θ, deg	t, s	r, km	θ, deg
0	36.4	110.5	60	19.0	45.0
5	29.9	100.0	70	18.8	38.5
10	26.2	91.0	80	18.7	32.8
15	24.1	83.7	90	18.7	27.0
20	22.7	77.7	100	18.7	21.6
30	20.9	67.7	110	18.8	16.8
40	20.1	58.6	120	19.0	12.0
50	19.3	52.0			

$Ans.$ $v = 1020$ km/h

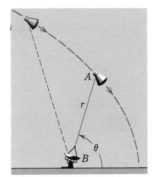

Problem 2/142

143 A centrifugal pump with radial vanes, similar to that illustrated with Prob. 2/135, rotates at the constant rate $\dot{\theta} = K$. An element of the fluid being pumped will be considered here as a smooth particle P which is introduced at the radius r_0 without radial velocity and which moves outward along the vane without friction. With no force on the particle in the radial direction, its acceleration in this direction is zero. Under these conditions, determine the equation of the path of the particle if the time t is zero for $r = r_0$. $Ans.$ $r = r_0 \cosh Kt$

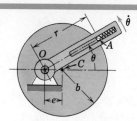

Problem 2/144

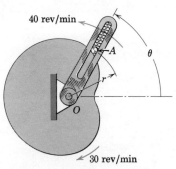

40 rev/min

30 rev/min

Problem 2/145

▶2/144 The slotted arm is pivoted at O and revolves counterclockwise with a constant angular velocity $\dot{\theta} = \omega$ about the eccentrically mounted circular cam which is fixed and does not rotate. Determine the expressions for the magnitudes of the velocity and acceleration of the pin A for the position $\theta = \pi/2$. The pin has negligible diameter and always contacts the cam.

$$Ans. \ v = b\omega, \ a = \frac{b^2\omega^2}{\sqrt{b^2 - e^2}}$$

▶2/145 If the slotted arm (Prob. 2/134) is revolving counterclockwise at the constant rate of 40 rev/min and the cam is revolving clockwise at the constant rate of 30 rev/min, determine the magnitude of the acceleration of the center of the roller A when the cam and arm are in the relative position for which $\theta = 30°$. The limaçon has the dimensions $b = 100$ mm and $c = 75$ mm. (*Caution:* Redefine the coordinates as necessary after noting that the θ in the expression $r = b - c \cos \theta$ is not the absolute angle appearing in Eq. 2/14.)

$$Ans. \ a = 3.68 \ \text{m/s}^2$$

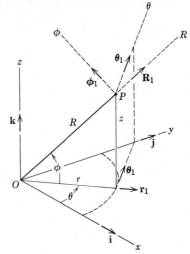

2/7 SPACE CURVILINEAR MOTION. The general case of three-dimensional motion of a particle along a space curve was introduced in Art. 2/1 and illustrated in Fig. 2/1. Mention was made of the three coordinate systems, rectangular (*x-y-z*), cylindrical (*r-θ-z*), and spherical (*R-θ-φ*) which are commonly used to describe this motion. These systems are indicated in Fig. 2/15, which also shows the unit vectors for the three coordinate systems.

Before describing the use of these coordinate systems we may observe that a path-variable description, using *n*- and *t*-coordinates, which we developed in Art. 2/5, may be applied in the osculating plane shown in Fig. 2/1. We defined this plane earlier as the plane which contains the curve at the location in question. We see that the velocity **v**, which is along the tangent *t* to the curve, lies in the osculating plane. The acceleration **a** also lies in the osculating plane and, as in the case of plane motion, has a component $a_t = \dot{v}$ tangent to the path due to the change in magnitude of the velocity and a component $a_n = v^2/\rho$ normal to the curve due to the change in direction of the velocity. As before ρ is the radius of curvature to the path at the point in question and would be measured in the osculating plane. This description of motion, which we found to be natural and direct for many plane-motion problems, finds little use for space motion because the osculating plane continually shifts its orientation, thus making its use as a reference awkward. We shall confine our attention, therefore, to the three fixed coordinate systems shown in Fig. 2/15.

(a) Rectangular Coordinates (x-y-z). The extension from two to three dimensions offers no particular difficulty. We merely add the *z*-coordinate and its two time derivatives to the two-dimensional expressions of Eqs. 2/6 so that the position vector **R**, the velocity **v**, and the acceleration **a** become

$$\boxed{\begin{aligned} \mathbf{R} &= \mathbf{i}x + \mathbf{j}y + \mathbf{k}z \\ \mathbf{v} = \dot{\mathbf{R}} &= \mathbf{i}\dot{x} + \mathbf{j}\dot{y} + \mathbf{k}\dot{z} \\ \mathbf{a} = \dot{\mathbf{v}} = \ddot{\mathbf{R}} &= \mathbf{i}\ddot{x} + \mathbf{j}\ddot{y} + \mathbf{k}\ddot{z} \end{aligned}} \qquad (2/15)$$

(b) Cylindrical Coordinates (r-θ-z). If we understand the polar-coordinate description of plane motion, then there should be no difficulty with cylindrical coordinates because all that is required is the addition of the *z*-coordinate and its two time derivatives. The position vector **R** to the particle for cylindrical coordinates is simply

$$\mathbf{R} = r\mathbf{r}_1 + z\mathbf{k}$$

Figure 2/15

In place of Eq. 2/13 for plane motion we may write the velocity as

$$\mathbf{v} = \dot{r}\mathbf{r}_1 + r\dot{\theta}\boldsymbol{\theta}_1 + \dot{z}\mathbf{k} \qquad (2/16)$$

where
$$v_r = \dot{r}$$
$$v_\theta = r\dot{\theta}$$
$$v_z = \dot{z}$$
$$v = \sqrt{v_r^2 + v_\theta^2 + v_z^2}$$

Similarly the acceleration is written by adding the z-component to Eq. 2/14, which gives us

$$\mathbf{a} = (\ddot{r} - r\dot{\theta}^2)\mathbf{r}_1 + (r\ddot{\theta} + 2\dot{r}\dot{\theta})\boldsymbol{\theta}_1 + \ddot{z}\mathbf{k} \qquad (2/17)$$

where
$$a_r = \ddot{r} - r\dot{\theta}^2$$
$$a_\theta = r\ddot{\theta} + 2\dot{r}\dot{\theta} = \frac{1}{r}\frac{d}{dt}(r^2\dot{\theta})$$
$$a_z = \ddot{z}$$
$$a = \sqrt{a_r^2 + a_\theta^2 + a_z^2}$$

Whereas the unit vectors $\mathbf{r}_1$ and $\boldsymbol{\theta}_1$ have time derivatives due to the change in their directions, we note that the unit vector $\mathbf{k}$ in the z-direction remains fixed in direction and therefore has no time derivative.

(c) *Spherical Coordinates* (R-θ-φ). When a radial distance and two angles are used to locate the position of a particle, as in the case of radar measurements, for example, spherical coordinates R, θ, ϕ are used. Derivation of the expression for the velocity $\mathbf{v}$ is easily obtained, but the expression for the acceleration $\mathbf{a}$ is more complex because of the added geometry. Consequently only the results will be cited here.[*] First we designate unit vectors $\mathbf{R}_1$, $\boldsymbol{\theta}_1$, $\boldsymbol{\phi}_1$ as shown in Fig. 2/15. Note that $\mathbf{R}_1$ is in the direction which the particle P would move if R increases but θ and ϕ are held constant. Also $\boldsymbol{\theta}_1$ is in the direction which P would move if θ increases while R and ϕ are held constant. Lastly $\boldsymbol{\phi}_1$ is in the direction which P would move if ϕ increases while R and θ are held constant. The resulting expressions for $\mathbf{v}$ and $\mathbf{a}$ are

$$\mathbf{v} = v_R\mathbf{R}_1 + v_\theta\boldsymbol{\theta}_1 + v_\phi\boldsymbol{\phi}_1 \qquad (2/18)$$

[*] For a complete derivation of $\mathbf{v}$ and $\mathbf{a}$ in spherical coordinates see the author's book *Dynamics*, 2nd Edition 1971 or SI Version 1975 (John Wiley & Sons, Inc.)

where
$$v_R = \dot{R}$$
$$v_\theta = R\dot{\theta}\cos\phi$$
$$v_\phi = R\dot{\phi}$$

and
$$\boxed{\mathbf{a} = a_R\mathbf{R}_1 + a_\theta\boldsymbol{\theta}_1 + a_\phi\boldsymbol{\phi}_1}$$
(2/19)

where

$$a_R = \ddot{R} - R\dot{\phi}^2 - R\dot{\theta}^2\cos^2\phi$$

$$a_\theta = \frac{\cos\phi}{R}\frac{d}{dt}(R^2\dot{\theta}) - 2R\dot{\theta}\dot{\phi}\sin\phi$$

$$a_\phi = \frac{1}{R}\frac{d}{dt}(R^2\dot{\phi}) + R\dot{\theta}^2\sin\phi\cos\phi$$

We should mention that linear algebraic transformations between any two of the three coordinate-system expressions for velocity or acceleration may be developed. These transformations make it possible to express the motion components in rectangular coordinates, for example, if the components are known in spherical coordinates, or vice versa.[*] These transformations are easily handled with the aid of matrix algebra and a simple computer program.

---

[*] These coordinate transformations are developed and illustrated in the author's book *Dynamics*, 2nd Edition, 1971 or SI Version 1975 (John Wiley & Sons, Inc.)

Sample Problem 2/11

The power screw starts from rest and is given a rotational speed $\dot{\theta}$ which increases uniformly with time t according to $\dot{\theta} = kt$ where k is a constant. Determine the expressions for the velocity v and acceleration a of the center of ball A when the screw has turned through one complete revolution from rest. The lead of the screw (advancement per revolution) is L.

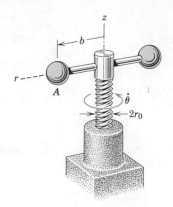

Solution. The center of ball A moves in a helix on the cylindrical surface of radius b, and the cylindrical coordinates r, θ, z are clearly indicated.

Integrating the given relation for $\dot{\theta}$ gives $\theta = \Delta\theta = \int \dot{\theta}\, dt = \frac{1}{2}kt^2$. For one revolution from rest we have

$$2\pi = \tfrac{1}{2}kt^2 \quad \text{giving} \quad t = 2\sqrt{\pi/k}$$

Thus the angular rate at one revolution is

$$\dot{\theta} = kt = k(2\sqrt{\pi/k}) = 2\sqrt{\pi k}$$

① The helix angle γ of the path followed by the center of the ball governs the relation between the θ- and the z-components of velocity and is given by $\tan\gamma = L/(2\pi b)$. Now from the figure we see that $v_\theta = v\cos\gamma$. Substituting $v_\theta = r\dot{\theta} = b\dot{\theta}$ from Eq. 2/16 gives ② $v = v_\theta/\cos\gamma = b\dot{\theta}/\cos\gamma$. With $\cos\gamma$ obtained from $\tan\gamma$ and with $\dot{\theta} = 2\sqrt{\pi k}$ we have for the one-revolution position

$$v = 2b\sqrt{\pi k}\,\frac{\sqrt{L^2 + 4\pi^2 b^2}}{2\pi b} = \sqrt{\frac{k}{\pi}}\sqrt{L^2 + 4\pi^2 b^2} \quad Ans.$$

The acceleration components from Eq. 2/17 become

③ $[a_r = \ddot{r} - r\dot{\theta}^2]$ $a_r = 0 - b(2\sqrt{\pi k})^2 = -4b\pi k$

$[a_\theta = r\ddot{\theta} + 2\dot{r}\dot{\theta}]$ $a_\theta = bk + 2(0)(2\sqrt{\pi k}) = bk$

$[a_z = \ddot{z} = \dot{v}_z]$ $a_z = \dfrac{d}{dt}(v_z) = \dfrac{d}{dt}(v_\theta \tan\gamma) = \dfrac{d}{dt}(b\dot{\theta}\tan\gamma)$

$$= (b\tan\gamma)\ddot{\theta} = b\frac{L}{2\pi b}k = \frac{kL}{2\pi}$$

Now we combine the components to give the total acceleration, which becomes

$$a = \sqrt{(-4b\pi k)^2 + (bk)^2 + \left(\frac{kL}{2\pi}\right)^2}$$

$$= bk\sqrt{(1 + 16\pi^2) + L^2/(4\pi^2 b^2)} \quad Ans.$$

① Unless we recognize that the helix angle of the path governs the z-motion in relation to the rotation of the screw, we cannot complete the problem. We must also be careful to divide the lead L by the circumference $2\pi b$ and not the diameter $2b$ to obtain $\tan\gamma$. If in doubt unwrap one turn of the helix traced by the center of the ball.

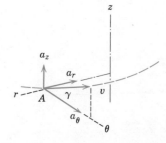

② Sketch a right triangle and recall that for $\tan\beta = a/b$ the cosine of β becomes $b/\sqrt{a^2 + b^2}$.

③ The negative sign for a_r is consistent with our previous knowledge of the normal component of acceleration.

PROBLEMS

146 Consider the power screw of Sample Problem 2/11 with a lead $L = 1\frac{1}{4}$ in. If $b = 6$ in. and if the screw turns at a constant rate of 4 rev/sec, calculate the magnitudes of the velocity and acceleration of the center of ball A.   *Ans.* $v = 150.9$ in./sec, $a = |a_r| = 3790$ in./sec^2

147 The velocity and acceleration of a particle moving on a space curve at a certain instant are given by $\mathbf{v} = 2\mathbf{i} - 3\mathbf{j} - 2\mathbf{k}$ ft/sec and $\mathbf{a} = 2\mathbf{i} + 2\mathbf{j} - \mathbf{k}$ ft/sec^2. Show that $\mathbf{v}$ and $\mathbf{a}$ are perpendicular and use this fact to determine $\dot{v}$. Also determine the corresponding radius of curvature ρ of the path in the osculating plane of the particle.

Ans. $\dot{v} = 0$, $\rho = 5.67$ ft

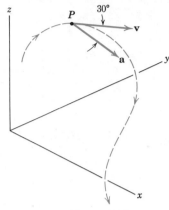

Problem 2/148

148 The particle P moves along a space curve and has a velocity $\mathbf{v} = 4\mathbf{i} + 3\mathbf{j}$ m/s at the position represented. At this same instant the total acceleration $\mathbf{a}$ of P has a magnitude of 10 m/s^2 and makes an angle of 30° with $\mathbf{v}$. Calculate the radius of curvature ρ measured in the osculating plane for this instant and determine the acceleration of P tangent to its path.

149 An amusement ride called the "corkscrew" takes the passengers through the upside-down curve of a horizontal cylindrical helix. The velocity of the cars as they pass position A is 15 m/s, and the component of their acceleration measured along the tangent to the path is $g \cos \gamma$ at this point. The effective radius of the cylindrical helix is 5 m, and the helix angle is $\gamma = 40°$. Compute the magnitude of the acceleration of the passengers as they pass position A.   *Ans.* $a = 27.5$ m/s^2

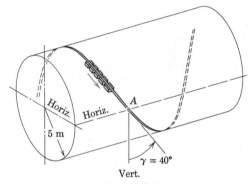

Problem 2/149

150 A particle moves along the cylindrical helix shown. As it passes point A the magnitude of its total acceleration is 10 m/s^2, and the particle is increasing its speed along the path at the rate of 8 m/s^2. For this position compute the velocity v, $\dot{\theta}$, $\ddot{\theta}$, and a_z.

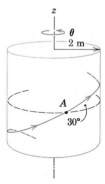

Problem 2/150

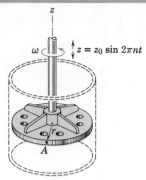

Problem 2/151

2/151 The rotating element in a mixing chamber is given a periodic axial movement $z = z_0 \sin 2\pi nt$ while it is rotating at the constant angular velocity $\dot{\theta} = \omega$. Determine the expression for the maximum magnitude of the acceleration of a point A on the rim of radius r. The frequency n of vertical oscillation is constant.   *Ans.* $a_{max} = \sqrt{r^2\omega^4 + 16n^4\pi^4z_0{}^2}$

2/152 The rotating nozzle sprays a large circular area and turns with the constant angular rate $\dot{\theta} = K$. Particles of water move along the tube at the constant rate $\dot{l} = c$ relative to the tube. Write expressions for the magnitudes of the velocity and acceleration of a water particle P for a given position l in the rotating tube.

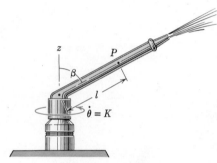

Problem 2/152

▶**2/153** The power screw shown in Sample Problem 2/11 is given a rotational oscillation $\theta = \pi \sin 2\pi nt$ where the angle θ and the time t are measured from the mid position and the ball travels between C and D as shown in the separate figure here. The constant frequency (number of complete oscillations per unit time) is n. Determine the magnitude of the acceleration of the center of ball A when it reaches D at $\theta = \pi$. The lead of the screw is L.
Ans. $a = 2\pi^2n^2\sqrt{4\pi^2b^2 + L^2}$

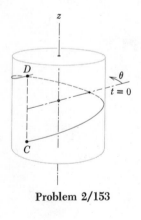

Problem 2/153

▶**2/154** The particle P moves down the spiral path which is wrapped around the surface of a right circular cone of base radius b and altitude h. The angle γ between the tangent to the curve at any point and a horizontal tangent to the cone at this point is constant. Also the motion of the particle is controlled so that $\dot{\theta}$ is constant. Determine the expression for the radial acceleration a_r of the particle for any value of θ.
Ans. $a_r = b\dot{\theta}^2(\tan^2\gamma \sin^2\beta - 1)e^{-\theta\tan\gamma\sin\beta}$
where $\beta = \tan^{-1}(b/h)$

▶**2/155** Assign unit vectors $\mathbf{R}_1$, $\boldsymbol{\theta}_1$, and $\boldsymbol{\phi}_1$ along the spherical coordinate directions, Fig. 2/15, and determine $\dot{\mathbf{R}}_1$, $\dot{\boldsymbol{\theta}}_1$, and $\dot{\boldsymbol{\phi}}_1$.
Ans. $\dot{\mathbf{R}}_1 = \boldsymbol{\phi}_1\dot{\phi} + \boldsymbol{\theta}_1\dot{\theta}\cos\phi$
$\dot{\boldsymbol{\theta}}_1 = -\mathbf{R}_1\dot{\theta}\cos\phi + \boldsymbol{\phi}_1\dot{\theta}\sin\phi$
$\dot{\boldsymbol{\phi}}_1 = -\mathbf{R}_1\dot{\phi} - \boldsymbol{\theta}_1\dot{\theta}\sin\phi$

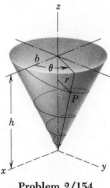

Problem 2/154

▶**2/156** Use the results of Prob. 2/155 to obtain the acceleration components in spherical coordinates, Eqs. 2/19, by a direct differentiation of the position vector $\mathbf{R} = R\mathbf{R}_1$.

57 The revolving crane has a boom of length $OP = 24$ m and is turning about its vertical axis at a constant rate of 2 rev/min. At the same time, the boom is being lowered at the constant rate $\dot{\beta} = 0.10$ rad/s. Calculate the magnitude of the velocity and acceleration of the end P of the boom at the instant when the position $\beta = 30°$ is passed.
Ans. $v = 3.48$ m/s, $a = 1.104$ m/s²

58 The rod OA is held at the constant angle $\beta = 30°$ while it rotates about the vertical with a constant angular rate of 120 rev/min. Simultaneously the slider P oscillates along the rod with a variable distance from the fixed pivot O given in inches by $R = 8 + 2 \sin 2\pi nt$ where the frequency n of oscillation along the rod is a constant 2 cycles per second and where the time t is in seconds. Calculate the magnitude of the acceleration of the slider for an instant when its velocity $\dot{R}$ along the rod is a maximum.
Ans. $a = 706$ in./sec²

59 Solve Prob. 2/158 for the magnitude of the acceleration of the slider P in the position where its acceleration component $\ddot{R}$ along the rod has a maximum magnitude and is directed toward O.
Ans. $a = 986$ in./sec²

60 In a test of the actuating mechanism for a telescoping antenna on a spacecraft the supporting shaft rotates about the fixed z-axis with an angular rate $\dot{\theta}$. Determine the R-, θ-, and ϕ-components of the acceleration **a** of the end of the antenna at the instant when $L = 1.2$ m and $\beta = 45°$ if the rates $\dot{\theta} = 2$ rad/s, $\dot{\beta} = \frac{3}{2}$ rad/s, and $\dot{L} = 0.9$ m/s are constant during the motion.
Ans. $a_R = -5.10$ m/s²
$a_\theta = 7.64$ m/s²
$a_\phi = -0.3$ m/s²

61 As a further test of the operation of the control mechanism for the telescoping antenna of Prob. 2/160, the vertical shaft is made to oscillate about the fixed z-axis according to $\theta = \pi/4 + 0.12 \sin 4\pi t$, where θ is in radians and t is in seconds. During the oscillation β is increasing at the rate of $\frac{3}{2}$ rad/s, and L is increasing at the rate of 0.9 m/s, both constant. Determine the magnitude of the acceleration **a** of the tip of the antenna when $\beta = 60°$ if this occurs at the position $\theta = \pi/4$ and when $L = 1.2$ m.
Ans. $a = 7.11$ m/s²

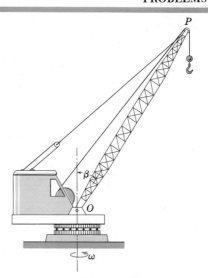

Problem 2/157

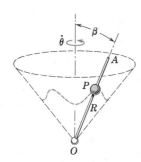

Problem 2/158

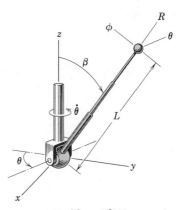

Problem 2/160

2/8 RELATIVE MOTION (TRANSLATING AXES). In the previous articles of this chapter we have described particle motion using coordinates which were referred to fixed reference axes. The displacements, velocities, and accelerations so determined are termed *absolute*. But it is not always possible or convenient to use a fixed set of axes for the observation of motion, and there are many engineering problems for which the analysis of motion is simplified by using measurements made with respect to a moving reference system. These measurements, when combined with the absolute motion of the moving coordinate system, permit us to determine the absolute motion in question. This approach is known as a *relative-motion* analysis.

The motion of the moving coordinate system is specified with respect to a fixed coordinate system. Strictly speaking, this fixed system in Newtonian mechanics is the primary inertial system which is a set of reference axes that are assumed to have no motion in space. From an engineering point of view the fixed system may be taken as any system whose absolute motion is negligible for the problem at hand. For most earthbound engineering problems it is sufficiently precise for us to take for the fixed reference system a set of axes attached to the earth, in which case we neglect the motion of the earth. For problems involving the motion of satellites around the earth, a nonrotating coordinate system with origin on the earth's axis of rotation is a convenient reference. For a description of interplanetary travel, a nonrotating coordinate system fixed to the sun would be appropriate. Hence the choice of the fixed system depends on the type of problem involved.

We will confine our attention in this article to moving reference systems which translate but do not rotate. Motion measured in rotating systems will be discussed in Art. 6/7 of Chapter 6 on rigid-body kinematics where this approach finds special but important application. We will also confine our attention here to the relative motion analysis of plane curvilinear motion.

Consider now two particles A and B which may have separate curvilinear motions in a given plane or in parallel planes, Fig. 2/16. We will arbitrarily attach the origin of a set of translating (nonrotating) axes x-y to particle B and observe the motion of A from our moving position on B. The position vector of A as measured relative to the frame x-y is $\mathbf{r}_{A/B} = \mathbf{i}x + \mathbf{j}y$ where the subscript notation "A/B" means "A relative to B" or "A with respect to B." The unit vectors along the x- and y-axes are $\mathbf{i}$ and $\mathbf{j}$, and x and y are the coordinates of A measured in x-y. The absolute position of B is defined by the vector $\mathbf{r}_B$ measured from the origin of the fixed axes X-Y. The absolute position of A is seen, therefore, to be determined by the vector

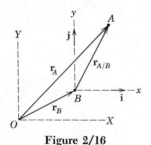

Figure 2/16

$$\mathbf{r}_A = \mathbf{r}_B + \mathbf{r}_{A/B}$$

We now differentiate this vector equation once with respect to time

to obtain velocities and twice to obtain accelerations. Thus

$$\dot{\mathbf{r}}_A = \dot{\mathbf{r}}_B + \dot{\mathbf{r}}_{A/B} \qquad \text{or} \qquad \boxed{\mathbf{v}_A = \mathbf{v}_B + \mathbf{v}_{A/B}} \qquad (2/20)$$

$$\ddot{\mathbf{r}}_A = \ddot{\mathbf{r}}_B + \ddot{\mathbf{r}}_{A/B} \qquad \text{or} \qquad \boxed{\mathbf{a}_A = \mathbf{a}_B + \mathbf{a}_{A/B}} \qquad (2/21)$$

In Eq. 2/20 the velocity which we observe A to have from our position at B attached to the moving axes x-y is $\dot{\mathbf{r}}_{A/B} = \mathbf{v}_{A/B} = \mathbf{i}\dot{x} + \mathbf{j}\dot{y}$. This term is the velocity of A with respect to B. Similarly in Eq. 2/21 the acceleration which we observe A to have from our nonrotating position on B is $\ddot{\mathbf{r}}_{A/B} = \dot{\mathbf{v}}_{A/B} = \mathbf{i}\ddot{x} + \mathbf{j}\ddot{y}$. This term is the acceleration of A with respect to B. We note carefully that the unit vectors $\mathbf{i}$ and $\mathbf{j}$ have no derivatives because their directions as well as their magnitudes remain unchanged. (Later when we discuss rotating reference axes we will be obliged to account for the derivatives of the unit vectors as we have done previously in this chapter.)

In words, Eq. 2/20 (or 2/21) states that the absolute velocity (or acceleration) of A equals the absolute velocity (or acceleration) of B plus, vectorially, the velocity (or acceleration) of A relative to B. The relative term is the velocity (or acceleration) measurement which an observer attached to the moving coordinate system x-y would make. We may express the relative motion terms in whatever coordinate system is convenient—rectangular, normal and tangential, or polar—and the formulations in the preceding articles may be used for this purpose. The appropriate fixed system of the previous articles becomes the moving system in the present article.

The selection of the moving point B for attachment of the reference coordinate system is arbitrary. As shown in Fig. 2/17 point A could be used just as well for the attachment of the moving system, in which case the three corresponding relative motion equations for position, velocity, and acceleration are

$$\mathbf{r}_B = \mathbf{r}_A + \mathbf{r}_{B/A} \qquad \mathbf{v}_B = \mathbf{v}_A + \mathbf{v}_{B/A} \qquad \mathbf{a}_B = \mathbf{a}_A + \mathbf{a}_{B/A}$$

It is seen, therefore, that $\mathbf{r}_{B/A} = -\mathbf{r}_{A/B}$, $\mathbf{v}_{B/A} = -\mathbf{v}_{A/B}$, and $\mathbf{a}_{B/A} = -\mathbf{a}_{A/B}$.

An important observation to be made in relative motion analysis is that the acceleration of a particle as observed in a translating system x-y is the same as that observed in a fixed system X-Y if the moving system has a constant velocity. This conclusion broadens the application of Newton's second law of motion, treated in Chapter 3. We conclude, consequently, that a set of axes which has a constant absolute velocity may be used in place of a "fixed" system for the determination of accelerations. A translating reference system which has no acceleration is known as an *inertial system*.

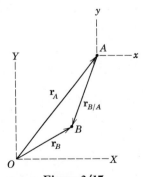

Figure 2/17

Sample Problem   2/12

Passengers in the jet transport A flying east at a speed of 800 km/h observe a second jet plane B which passes under the transport in horizontal flight. Although the nose of B is pointed in the 45° northeast direction, plane B appears to the passengers in A to be moving away from the transport at the 60° angle as shown. Determine the true velocity of B.

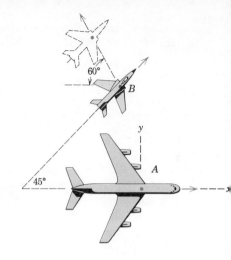

Solution.   The moving reference axes x-y are attached to A from which the relative observations are made. We write, therefore,

$$\mathbf{v}_B = \mathbf{v}_A + \mathbf{v}_{B/A}$$

Next we identify the knowns and unknowns. The velocity $\mathbf{v}_A$ is given both in magnitude and direction. The 60° direction of $\mathbf{v}_{B/A}$, the velocity which B appears to have to the moving observers in A, is known, and the true velocity of B is in the 45° direction in which it is heading. The two remaining unknowns are the magnitudes of $\mathbf{v}_B$ and $\mathbf{v}_{B/A}$. We may solve the vector equation in any one of three ways.

(I) *Graphical.*   We start the vector sum at some point P by drawing $\mathbf{v}_A$ to a convenient scale and then construct a line through the tip of $\mathbf{v}_A$ with the known direction of $\mathbf{v}_{B/A}$. The known direction of $\mathbf{v}_B$ is then drawn through P, and the intersection C yields the unique solution enabling us to complete the vector triangle and scale off the unknown magnitudes, which are found to be

$$v_{B/A} = 586 \text{ km/h} \quad \text{and} \quad v_B = 717 \text{ km/h} \qquad Ans.$$

(II) *Trigonometric.*   A sketch of the vector triangle is made to reveal the trigonometry, which gives

$$\frac{v_B}{\sin 60°} = \frac{v_A}{\sin 75°} \qquad v_B = 800 \frac{\sin 60°}{\sin 75°} = 717 \text{ km/h} \qquad Ans.$$

(III) *Vector algebra.*   Using unit vectors $\mathbf{i}$ and $\mathbf{j}$ we express each of the velocities in vector form as

$$\mathbf{v}_A = 800\mathbf{i} \text{ km/h}, \qquad \mathbf{v}_B = (v_B \cos 45°)\mathbf{i} + (v_B \sin 45°)\mathbf{j}$$

$$\mathbf{v}_{B/A} = (v_{B/A} \cos 60°)(-\mathbf{i}) + (v_{B/A} \sin 60°)\mathbf{j}$$

Substituting these relations into the relative-velocity equation and solving separately for the $\mathbf{i}$ and $\mathbf{j}$ terms give

$$(\mathbf{i}\text{-terms}) \qquad v_B \cos 45° = 800 - v_{B/A} \cos 60°$$

$$(\mathbf{j}\text{-terms}) \qquad v_B \sin 45° = v_{B/A} \sin 60°$$

Solving simultaneously yields the unknown velocity magnitudes

$$v_{B/A} = 586 \text{ km/h} \quad \text{and} \quad v_B = 717 \text{ km/h} \qquad Ans.$$

It is worth noting the solution of this problem from the viewpoint of an observer in B. With reference axes attached to B we would write $\mathbf{v}_A = \mathbf{v}_B + \mathbf{v}_{A/B}$. The apparent velocity of A as observed by B is then $\mathbf{v}_{A/B}$, which is the negative of $\mathbf{v}_{B/A}$.

① We treat each airplane as a particle.

② We assume no side slip due to cross wind.

③ Students should become familiar with all three solutions.

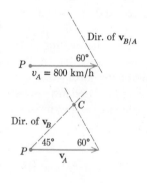

④ We must be prepared to recognize the appropriate trigonometric relation, which here is the law of sines.

⑤ We can see that the graphical or trigonometric solution is shorter than the vector algebra solution in this particular problem.

Sample Problem 2/13

Car A is accelerating in the direction of its motion at the rate of 3 ft/sec². Car B is rounding a curve of 440-ft radius at a constant speed of 30 mi/hr. Determine the velocity and acceleration which car B appears to have to an observer in car A if car A has reached a speed of 45 mi/hr for the positions represented.

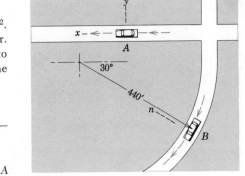

Solution. We choose nonrotating reference axes attached to car A since the motion of B with respect to A is desired.

Velocity. The relative-velocity equation is

$$\mathbf{v}_B = \mathbf{v}_A + \mathbf{v}_{B/A}$$

and the velocities of A and B for the position considered have the magnitudes

$$v_A = 45 \frac{5280}{60^2} = 45 \frac{44}{30} = 66 \text{ ft/sec} \qquad v_B = 30 \frac{44}{30} = 44 \text{ ft/sec}$$

The triangle of velocity vectors is drawn in the sequence required by the equation, and application of the law of cosines and the law of sines gives

$$v_{B/A} = 58.2 \text{ ft/sec} \qquad \theta = 40.9° \qquad \textit{Ans.}$$

Acceleration. The relative-acceleration equation is

$$\mathbf{a}_B = \mathbf{a}_A + \mathbf{a}_{B/A}$$

The acceleration of A is given, and the acceleration of B is normal to the curve in the n-direction and has the magnitude

$$[a_n = v^2/\rho] \qquad a_B = (44)^2/440 = 4.40 \text{ ft/sec}^2$$

The triangle of acceleration vectors is drawn in the sequence required by the equation as illustrated. Solving for the x- and y-components of $\mathbf{a}_{B/A}$ gives us

$$(a_{B/A})_x = 4.4 \cos 30° - 3 = 0.810 \text{ ft/sec}^2$$

$$(a_{B/A})_y = 4.4 \sin 30° = 2.2 \text{ ft/sec}^2$$

from which $a_{B/A} = \sqrt{(0.810)^2 + (2.2)^2} = 2.344 \text{ ft/sec}^2$ *Ans.*

The direction of $\mathbf{a}_{B/A}$ may be specified by the angle β which, by the law of sines, becomes

$$\frac{4.4}{\sin \beta} = \frac{2.344}{\sin 30°} \qquad \beta = \sin^{-1}\left(\frac{4.4}{2.344}0.5\right) = 110.2° \qquad \textit{Ans.}$$

① Alternatively we could use either a graphical or a vector algebraic solution.

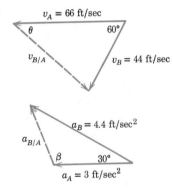

② Be careful to choose between the two values 69.8 deg and $180 - 69.8 = 110.2$ deg.

Suggestion: To gain familiarity with the manipulation of vector equations, it is suggested that the student rewrite the relative-motion equations in the form $\mathbf{v}_{B/A} = \mathbf{v}_B - \mathbf{v}_A$ and $\mathbf{a}_{B/A} = \mathbf{a}_B - \mathbf{a}_A$ and redraw the vector polygons to conform with these alternative relations.

Caution: So far we are only prepared to handle motion relative to *nonrotating* axes. If we had attached the reference axes rigidly to car B, they would rotate with the car, and we would find that the velocity and acceleration terms relative to the rotating axes are *not* the negative of those measured from the nonrotating axes moving with A. Rotating axes are treated in Art. 6/7

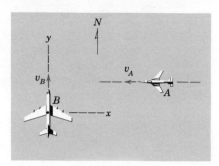

Problem 2/162

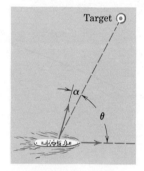

Problem 2/164

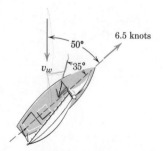

Problem 2/165

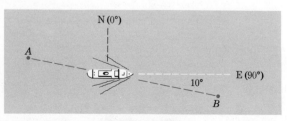

Problem 2/166

PROBLEMS

2/162 The aircraft A is flying west with a velocity $v_A = 900$ km/h, while aircraft B is flying north with a velocity $v_B = 600$ km/h at approximately the same altitude. Determine the magnitude and direction of the velocity which A appears to have to a passenger riding in B.

Ans. $v_{A/B} = 1082$ km/h directed 33.7° south of west

2/163 A ship capable of making a speed of 13 knots through calm water is to maintain a true easterly course while encountering a 2-knot current running from north to south. What should be the heading of the ship (measured clockwise from the north to the nearest degree)? How long does it take the ship to proceed 20 nautical miles due east?

2/164 The destroyer moves at 30 knots (1 knot = 1.852 km/h) and fires a rocket at an angle which trails the line of sight to the fixed target by the angle α. The launching velocity is 75 m/s relative to the ship and has an angle of elevation of 30° above the horizontal. If the missile continues to move in the same vertical plane as that determined by its absolute velocity at launching, determine α for $\theta = 60°$.

Ans. $\alpha = 11.88°$

2/165 A sailboat moving in the direction shown is tacking to windward against a north wind. The log registers a hull speed of 6.5 knots. A "telltale" (light string tied to the rigging) indicates that the direction of the apparent wind is 35° from the center line of the boat. What is the true wind velocity v_w?

2/166 A small boat capable of making a speed of 6 knots through calm water maintains an easterly heading while being set to the south by an ocean current. The actual course of the boat is from A to B, a distance of 10 nautical miles which requires exactly 2 hours. Determine the speed v_w of the current and its direction measured clockwise from the north.

Ans. $v_w = 1.38$ knots; 231°

167 Ship *A* is headed west at a speed of 15 knots, and ship *B* is headed southeast. The relative bearing θ of *B* with respect to *A* is 20° and is unchanging. If the distance between *A* and *B* is 10 nautical miles at 2:00 P.M., when would collision occur if neither ship altered course?

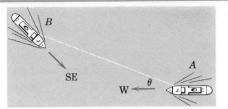

Problem 2/167

168 Airplane *A* is flying north with a constant horizontal velocity of 500 km/h. Airplane *B* is flying southwest at the same altitude with a velocity of 500 km/h. From the frame of reference of *A* determine the magnitude v_r of the apparent or relative velocity of *B*. Also find the magnitude of the apparent velocity v_n with which *B* appears to be moving sideways or normal to its center line. Would the results be different if the two airplanes were flying at different but constant altitudes?

$$\text{Ans. } v_r = 924 \text{ km/h}, \quad v_n = 354 \text{ km/h}$$

169 In Prob. 2/162 if aircraft *B* is accelerating in its northward direction at the rate of 4.5 km/h each second while aircraft *A* is slowing down at the rate of 3 km/h each second in its westward direction of flight, determine the acceleration in m/s² which *B* appears to have to an observer in *A*.

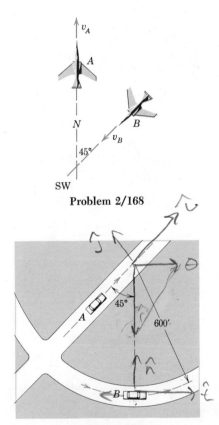

Problem 2/168

170 For the instant represented car *A* has an acceleration in the direction of its motion and car *B* has a speed of 45 mi/hr which is increasing. If the acceleration of *B* as observed from *A* is zero for this instant, determine the acceleration of *A* and the rate at which the speed of *B* is changing.

$$\text{Ans. } a_A = 10.27 \text{ ft/sec}^2, \quad \dot{v}_B = 7.26 \text{ ft/sec}^2$$

171 For the airplanes shown with Prob. 2/168, if *A* has a forward acceleration of 45 km/h per second and if *B* has a forward acceleration of 30 km/h per second, determine the expression in vector notation for the acceleration which A appears to have from a reference frame attached to *B*. Each airplane is in horizontal flight. Take the *x*-direction north and the *y*-direction west.

Problem 2/170

172 Car *A* rounds a curve of 150-m radius at a constant speed of 54 km/h. At the instant represented, car *B* is moving at 81 km/h but is slowing down at the rate of 3 m/s². Determine the velocity and acceleration of car *A* as observed from car *B*.

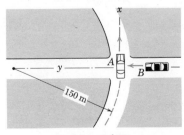

Problem 2/172

2/173 Car A travels at a constant speed of 60 mi/hr along a straight road, and car B travels at the same constant speed on a road which makes a circular arc of 1000-ft radius. For the instant when the cars are in the relative positions shown, determine the x-component of the acceleration which car A appears to have as seen by a nonrotating observer in car B.

Ans. $(a_{A/B})_x = 3.87$ ft/sec²

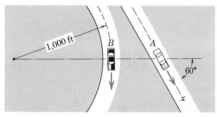

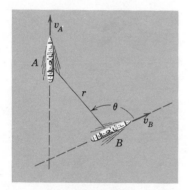

Problem 2/173

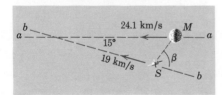

Problem 2/175

Problem 2/178

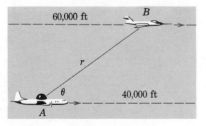

Problem 2/179

2/174 Work Prob. 2/173 for the conditions where car B is speeding up at the rate of 7.5 mi/hr per second at the instant represented. All other conditions remain the same as in Prob. 2/173.

2/175 Two ships A and B are moving with constant speeds v_A and v_B, respectively, along straight intersecting courses. The navigator of ship B notes the time rates of change of the separation distance r between the ships and the bearing angle θ. Show that $\ddot{\theta} = -2\dot{r}\dot{\theta}/r$ and $\ddot{r} = r\dot{\theta}^2$.

2/176 For the coplanar or parallel-plane motion of three particles A, B, and C prove that the relative velocities and relative accelerations measured from nonrotating axes obey the equations $\mathbf{v}_{A/B} = \mathbf{v}_{A/C} + \mathbf{v}_{C/B}$ and $\mathbf{a}_{A/B} = \mathbf{a}_{A/C} + \mathbf{a}_{C/B}$.

2/177 An earth satellite is put into a circular polar orbit at an altitude of 240 km, which requires an orbital velocity of 27 940 km/h with respect to the center of the earth considered fixed in space. In going from south to north, when the satellite passes over an observer on the equator, in which direction does the satellite appear to be moving? The equatorial radius of the earth is 6378 km, and the angular velocity of the earth is 0.729(10⁻⁴) rad/s.

Ans. Apparent direction 3.43° west of north

2/178 The spacecraft S approaches the planet Mars along a trajectory b-b in the orbital plane of Mars with an absolute velocity of 19 km/s. Mars has a velocity of 24.1 km/s along its trajectory a-a. Determine the angle β between the line of sight S-M and the trajectory b-b when Mars appears from the spacecraft to be approaching it head on.

Ans. $\beta = 55.6°$

▶2/179 The aircraft A with radar detection equipment is flying horizontally at an altitude of 40,000 ft and is increasing its speed at the rate of 4 ft/sec each second. Its radar locks onto an aircraft flying in the same direction and in the same vertical plane at an altitude of 60,000 ft. If A has a speed of 600 mi/hr at the instant that $\theta = 30°$, determine the values of $\ddot{r}$ and $\ddot{\theta}$ at this same instant if B has a constant speed of 900 mi/hr.

Ans. $\ddot{r} = -2.25$ ft/sec²
$\ddot{\theta} = 1.548(10^{-4})$ rad/sec²

2/9 PROBLEM FORMULATION AND REVIEW. In Chapter 2 we have developed and illustrated the basic elements of theory and their use for the description of particle motion. It is important for the student to review this material before proceeding to consolidate and enlarge his perspective of particle kinematics. The concepts and procedures developed and illustrated in this chapter form the basis for building much of the entire subject of dynamics, and satisfactory progress in the topics ahead will be contingent upon a firm grasp of Chapter 2.

By far the most important concept in Chapter 2 is the time derivative of a vector. The independent contributions of magnitude change and direction change to the time derivative are, indeed, basic. As we proceed in our study of dynamics we will have further occasion to examine the time derivatives of vectors other than position and velocity vectors, and the principles and procedures developed in Chapter 2 will be utilized directly.

When one concentrates primarily on the individual parts of a subject, the relationships among its parts are not always clearly grasped. When guided by a specific topical assignment the student automatically recognizes the problem category and the indicated method of solution. But without this guide he must choose his own method of solution. The ability to choose the most appropriate method is certainly a key to the successful formulation and solution of engineering problems. One of the best ways to acquire this ability is to compare alternative methods after each is studied.

To facilitate this comparison it will be helpful if we recognize the following breakdowns:

(a) *Type of Motion.* The three conspicuous categories are
Rectilinear motion (one coordinate)
Plane curvilinear motion (two coordinates)
Space curvilinear motion (three coordinates)
Generally the geometry of a given problem permits this identification to be made readily. One exception to this categorization is encountered when only the magnitudes of the motion quantities measured along the path are of interest. In this event we may use the single distance coordinate measured along the curved path together with its scalar time derivatives giving the speed $\dot{s}$ and the tangential acceleration $\ddot{s}$.

Plane motion is easier to generate and control, particularly in machinery, than is space motion, so that a large fraction of our motion problems come under the plane-curvilinear or rectilinear categories.

(b) *Reference Fixity.* We commonly make motion measurements with respect to fixed reference axes (absolute motion) and moving axes (relative motion). The acceptable choice of the fixed axes depends on the problem. Axes attached to the earth are sufficiently "fixed" for most engineering problems although important exceptions include earth-satellite and interplanetary motion, accurate projectile trajectories, navigation, and other problems. The equations of relative motion discussed in Chapter 2 are restricted to translating reference axes.

(*c*) *Coordinates.*   Of prime importance is the choice of coordinates. We have developed the description of motion using

Rectangular (Cartesian) coordinates (x-y) and (x-y-z)
Normal and tangential coordinates (n-t)
Polar coordinates (r-θ)
Cylindrical coordinates (r-θ-z)
Spherical coordinates (R-θ-ϕ)

When the coordinates are not specified, the choice usually depends on how the motion is generated or measured. Thus for a particle which slides radially along a rotating rod polar coordinates are the natural ones to use. Radar tracking calls for polar or spherical coordinates. When measurements are made along a curved path, normal and tangential coordinates are indicated. An x-y plotter clearly involves rectangular coordinates.

(*d*) *Approximations.*   One of the most important abilities to acquire is that of making appropriate approximations. The assumption of constant acceleration is valid when the forces which cause the acceleration do not vary appreciably. When motion data are acquired experimentally, we must utilize the nonexact data to acquire the best possible description, often with the aid of graphical or numerical approximations.

(*e*) *Mathematical Method.*   We frequently have a choice of solution using scalar algebra, vector algebra, trigonometric geometry, or graphical geometry. All of these methods have been illustrated and all are important to learn. The choice of method will depend on the geometry of the problem, how the motion data are given, and the accuracy desired. Inasmuch as mechanics by its very nature is geometrical, the student is encouraged to develop his facility to sketch vector relationships both as an aid to the disclosure of appropriate geometrical and trigonometrical relations and as a means of solving vector equations graphically. Geometric portrayal is the most direct representation of the vast majority of mechanics problems.

The problems which follow are included to facilitate a review of Chapter 2 and have a random arrangement of motion category.

REVIEW PROBLEMS

180 A particle P moves along a plane curve with a speed v, measured in feet per second, given by $v = 2 + 0.3t^2$ where t is the time in seconds after P passes a certain fixed point on the curve. If the total acceleration of P is 2.4 ft/sec^2 when $t = 2$ sec, compute the radius of curvature ρ of the curve for the position of the particle at this instant.

Ans. $\rho = 4.93$ ft

181 The tangential acceleration for a simple pendulum is $a_t = g \sin \theta$. For the pendulum shown $l = 200$ mm, and when $\theta = 30°$, $\dot{\theta} = -9$ rad/s. Calculate the magnitude of the total acceleration of the pendulum for this position.

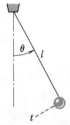

Problem 2/181

182 The third stage of a rocket is injected by its booster with a velocity u of 15 000 km/h at A into an unpowered coasting flight to B. At B its rocket motor is ignited when the trajectory makes an angle of 20° with the horizontal. Operation is effectively above the atmosphere, and the gravitational acceleration during this interval may be taken as 9 m/s^2, constant in magnitude and direction. Determine the time t to go from A to B. (This quantity is needed in the design of the ignition control system.) Also determine the corresponding increase h in altitude.

Ans. $t = 3$ min 28 s, $h = 418$ km

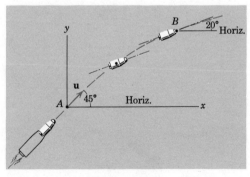

Problem 2/182

183 At a certain instant the velocity and acceleration of a particle are given by $\mathbf{v} = 3\mathbf{i} + 2\mathbf{j} - 9\mathbf{k}$ m/s and $\mathbf{a} = 4\mathbf{i} + 3\mathbf{j} + 2\mathbf{k}$ m/s^2 respectively. Show that $\mathbf{v}$ and $\mathbf{a}$ are perpendicular at this instant and describe the characteristic of this motion as viewed from the osculating plane. (Recall that two vectors are perpendicular if their scalar product is zero.)

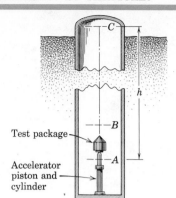

Test package

Accelerator piston and cylinder

"Zero–g" test facility

Problem 2/184

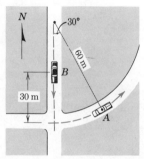

Problem 2/185

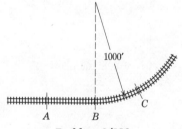

Problem 2/186

2/184 To test the effects of "weightlessness" for short periods of time a test facility is designed which accelerates a test package vertically up from A to B by means of a gas-activated piston and allows it to ascend and descend from B to C to B under free-fall conditions. The test chamber consists of a deep well and is evacuated to eliminate any appreciable air resistance. If a constant acceleration of $40g$ from A to B is provided by the piston and if the total test time for the "weightless" condition from B to C to B is 10 s, calculate the required working height h of the chamber. Upon returning to B the test package is recovered in a basket filled with polystyrene pellets and inserted in the line of fall.

Ans. $h = 125.7$ m

2/185 Car A negotiates a curve of 60-m radius at a constant speed of 50 km/h. When A passes the position shown, car B is 30 m from the intersection and is accelerating south toward the intersection at the rate of 1.5 m/s². Determine the acceleration which A appears to have when observed by an occupant of B at this instant.

Ans. $a_{A/B} = 4.58$ m/s², 20.6° west of north

2/186 An inexperienced designer of a roadbed for a new high-speed train proposes to join a straight section of track to a circular section of 1000-ft radius as shown. For a train which would travel at a constant speed of 90 mi/hr plot the magnitude of its acceleration as a function of distance along the track between points A and C and explain why this design is unacceptable.

2/187 A ship with a total displacement of 16 kt starts from rest in still water under a constant propeller thrust $T = 250$ kN. The ship develops a total resistance to motion through the water given by $R = 4.50v^2$ where R is in kilonewtons and v is in meters per second. The acceleration of the ship is $a = (T - R)/m$ where m equals the mass of the ship. Compute the distance s in nautical miles which the ship must go to reach a speed of 14 knots. (1 nautical mile = 1.852 km; 1 knot = 1 nautical mile per hour.)   *Ans.* $s = 2.60$ mi (nautical)

188 Three ships, A, B, and C, are cruising on straight courses in the vicinity of one another. With the x-direction as east and the y-direction as north the relative velocities in knots of A with respect to B and of C with respect to B are given by $\mathbf{v}_{A/B} = 6.8\mathbf{i} - 3.6\mathbf{j}$ and $\mathbf{v}_{C/B} = -8.2\mathbf{i} - 10.6\mathbf{j}$. Find both the magnitude and the direction of the velocity that A appears to have to an observer on C. Express the direction in terms of the clockwise angle β measured from the north.

189 The coordinates of a particle which moves in the X-Y plane are recorded every second, and the signals are fed through a computer to the automatic plotter which draws simultaneous graphs of both coordinates against the time. Each millimeter of x-motion plotted on the graph represents 0.25 m of actual X-movement, and each millimeter of y-motion on the graph represents 1.25 m of actual Y-movement. On the horizontal time scale t each millimeter on the graph represents 0.04 s of real time. The resulting graphs with all measurements in millimeters are closely approximated by the relations $x = 50 + 0.1t^2$ and $y = 50 + 0.0015t^3/3 - 0.04t^2/2$. Determine the magnitude of the actual acceleration of the particle for the condition corresponding to $t = 2$ s and find the angle θ_X made by the velocity vector of the particle with the X-direction at this same instant.

$$\text{Ans. } a = 91.4 \text{ m/s}^2, \, \theta_X = 41.2°$$

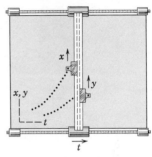

Problem 2/189

2/190 A small object is thrown down the slope as shown. Determine the magnitude u of the initial velocity.

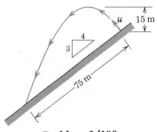

Problem 2/190

2/191 A rocket propels a space probe for measuring micrometeorite density vertically upward from the north pole and separates from the probe at burn-out conditions. The probe continues to move upward subject to the influence of the earth's gravity and has an upward velocity of 16 000 km/h at an altitude of 160 km above the earth's surface at the pole. Calculate the maximum distance h from the earth's surface reached by the probe. The polar radius of the earth is 6357 km, and the value of g at the pole is 9.833 m/s². 　　　　$\text{Ans. } h = 1420$ km

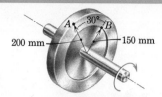

Problem 2/192

2/192 The flywheel is revolving with a changing angular velocity. At a certain instant point A has a component of acceleration tangent to its path of 1 m/s^2 and point B has a component of acceleration normal to its path of 0.6 m/s^2. For this instant compute the rim speed of point A and the total acceleration of point B. Note that $\dot{\theta}$ and $\ddot{\theta}$ are common to the two points.

Problem 2/193

2/193 Small objects are released from rest at A and slide with negligible friction down the cylindrical spiral chute of constant helix angle $\gamma = \tan^{-1}(h/2\pi r)$. The component of acceleration measured tangent to the path is $g \sin \gamma$. Determine the radial component of acceleration a_r for each object as it passes B after one complete turn.

Ans. $a_r = -2\pi g \sin 2\gamma$

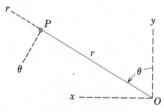

Problem 2/194

2/194 A particle P moving in plane curvilinear motion is located by the polar coordinates shown. At a particular instant $r = 2 \text{ m}$, $\theta = 60°$, $v_r = 3 \text{ m/s}$, $v_\theta = 4 \text{ m/s}$, $a_r = -10 \text{ m/s}^2$, and $a_\theta = -5 \text{ m/s}^2$. For this instant calculate the radius of curvature ρ of the path of the particle. Also locate the center of curvature C graphically.

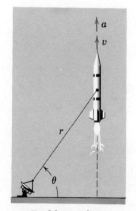

Problem 2/195

2/195 A rocket is fired vertically and tracked by the radar shown with Prob. 2/127 and repeated here. At the instant when $\theta = 60°$, measurements give $\dot{\theta} = 0.03 \text{ rad/sec}$ and $r = 25,000 \text{ ft}$, and the vertical acceleration of the rocket is found to be $a = 64 \text{ ft/sec}^2$. For this instant determine the values of $\ddot{r}$ and $\ddot{\theta}$.

Ans. $\ddot{r} = 77.9 \text{ ft/sec}^2$, $\ddot{\theta} = -1.838(10^{-3}) \text{ rad/sec}^2$

/196 The guide with the horizontal slot is made to move up the vertical edge of the fixed plate at the constant rate $\dot{y} = 2$ m/s before reversing the direction of its motion at $y = 175$ mm. Pin P is constrained to move in both the horizontal and circular slots. Calculate the angular acceleration $\ddot{\theta}$ of line OP for the instant when $y = 100$ mm.

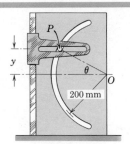

Problem 2/196

/197 For the instant represented the particle P has a velocity $v = 6$ ft/sec in the direction shown and has acceleration components $a_x = 15$ ft/sec² and $a_\theta = -15$ ft/sec². Determine a_r, a_y, a_t, and a_n and the radius of curvature ρ of the path for this position. (*Hint:* Draw the related acceleration components of the total acceleration of the particle and take advantage of the simplified geometry for your calculations.)

$$\text{Ans. } a_r = 5\sqrt{3} \text{ ft/sec}^2, \ a_y = -5\sqrt{3} \text{ ft/sec}^2,$$

$$a_t = 0, \ a_n = 10\sqrt{3} \text{ ft/sec}^2,$$

$$\rho = 6\sqrt{3}/5 \text{ ft}$$

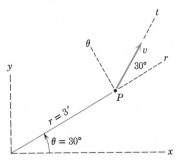

Problem 2/197

/198 The radar tracking antenna oscillates about its vertical axis according to $\theta = \theta_0 \cos pt$ where p is the constant circular frequency and $2\theta_0$ is the double amplitude of oscillation. Simultaneously, the angle of elevation ϕ is increasing at the constant rate $\dot{\phi} = K$. Determine the expression for the magnitude a of the acceleration of the signal horn (a) as it passes position A and (b) as it passes the top position B, assuming that $\theta = 0$ at this instant.

$$\text{Ans. } (a) \ a = b\sqrt{K^4 + p^4\theta_0^2 \cos^2 \phi}$$

$$(b) \ a = bK\sqrt{K^2 + 4p^2\theta_0^2}$$

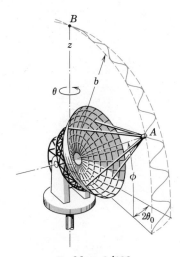

Problem 2/198

KINETICS OF PARTICLES

3

3/1 **INTRODUCTION.** According to Newton's second law when a particle is subjected to unbalanced forces it will accelerate. Kinetics is the study of the relations between unbalanced forces and the changes in motion which they produce. In Chapter 3 we shall study the kinetics of particles. This topic requires that we combine our knowledge of two previously learned parts of mechanics, namely, the properties of forces which we developed in our earlier study of statics, and the kinematics of particle motion which we have just covered in Chapter 2. With the aid of Newton's second law of motion we are now ready to combine these two topics and prepare for the solution of engineering problems involving force, mass, and motion.

There are three general approaches to the solution of kinetics problems: (A) direct application of Newton's second law (called the force-mass-acceleration method), (B) use of work and energy principles, and (C) solution by impulse and momentum methods. Each approach has its special characteristics and advantages, and Chapter 3 is subdivided into three sections according to these three methods of solution. Before proceeding it is strongly recommended that the student review carefully the definitions and concepts of Chapter 1, as they are fundamental to the developments which follow.

SECTION A. FORCE, MASS, AND ACCELERATION

3/2 **NEWTON'S SECOND LAW.** The basic relation between force and acceleration is found in Newton's second law of motion, Eq. 1/1, the verification of which is entirely experimental. We will describe the fundamental meaning of this law by an ideal experiment in which force and acceleration are assumed to be measured without error. A mass particle is isolated in the primary inertial system° and is subjected to the action of a single force $\mathbf{F}_1$. The acceleration $\mathbf{a}_1$ of the particle is measured, and the ratio F_1/a_1 of the magnitudes of the

---

° The primary inertial system or astronomical frame of reference is an imaginary set of reference axes which are assumed to have no translation or rotation in space. See Art. 1/2, Chapter 1.

force and the acceleration will be some number C_1 whose value depends on the units used for measurement of force and acceleration. The experiment is now repeated by subjecting the same particle to a different force $\mathbf{F}_2$ and measuring the corresponding acceleration $\mathbf{a}_2$. Again the ratio of the magnitudes F_2/a_2 will produce a number C_2. The experiment is repeated as many times as desired. We draw two important conclusions from the results. First, the ratios of applied force to corresponding acceleration all equal the *same* number, provided the units used for measurement are not changed in the experiments. Thus,

$$\frac{F_1}{a_1} = \frac{F_2}{a_2} = \cdots = \frac{F}{a} = C, \qquad \text{a constant}$$

We conclude that the constant C is a measure of some property of the particle which does not change. This property is the *inertia* of the particle which is its *resistance to rate of change of velocity*. For a particle of high inertia (large C) the acceleration will be small for a given force F. On the other hand, if the inertia is small, the acceleration will be large. The mass m is used as a quantitative measure of inertia, and therefore the expression $C = km$ may be written, where k is a constant to account for the units used. Thus the experimental relation becomes

$$F = kma \tag{3/1}$$

where F is the magnitude of the resultant force acting on the particle of mass m, and a is the magnitude of the resulting acceleration of the particle.

The second conclusion we draw from the ideal experiment is that the acceleration is always in the direction of the applied force. Thus Eq. 3/1 becomes a *vector* relation and may be written

$$\mathbf{F} = km\mathbf{a} \tag{3/2}$$

Although an actual experiment cannot be performed in the ideal manner described, the conclusions are inferred from the measurements of countless accurately performed experiments where the results are correctly predicted from the hypothesis of the ideal experiment. One of the most accurate checks lies in the precise prediction of the motions of planets based on Eq. 3/2.

(*a*) *Inertial System.* Whereas the results of the ideal experiment are obtained for measurements made relative to the "fixed" primary inertial system, they are equally valid for measurements made with respect to any nonrotating reference system which translates with a constant velocity with respect to the primary system. From our study of relative motion in Art. 2/8 we saw that the acceleration measured in a system translating with no acceleration is the same as that measured in the primary system. Thus Newton's second law holds equally well in a nonaccelerating system, so that we may define an *inertial system* as any system in which Eq. 3/2 is valid.

If the ideal experiment described were performed on the surface of the earth and all measurements were made relative to a reference system attached to the earth, the measured results would show a slight discrepancy upon substitution into Eq. 3/2. This discrepancy would be due to the fact that the measured acceleration would not be the correct absolute acceleration. The discrepancy would disappear when we introduced the corrections due to the acceleration components of the earth. These corrections are negligible for most engineering problems which involve the motions of structures and machines on the surface of the earth. In such cases the accelerations measured with respect to reference axes attached to the surface of the earth may be treated as "absolute," and Eq. 3/2 may be applied with negligible error to experimental measurements made on the surface of the earth.°

There are an increasing number of problems, particularly in the fields of rocket and spacecraft design, where the acceleration components of the earth are of primary concern. For this work it is essential that the fundamental basis of Newton's law be thoroughly understood and that the appropriate absolute acceleration components be employed.

Before 1905 the laws of Newtonian mechanics had been verified by innumerable physical experiments and were considered the final description of the motion of bodies. The concept of *time*, considered an absolute quantity in the Newtonian theory, received a basically different interpretation in the theory of relativity announced by Einstein in 1905. The new concept called for a complete reformulation of the accepted laws of mechanics. The theory of relativity was subjected to early ridicule but has had experimental check and is now universally accepted by scientists the world over. Although the difference between the mechanics of Newton and that of Einstein is basic, there is a practical difference in the results given by the two theories only when velocities of the order of the speed of light

---

° An example of the magnitude of the error introduced by neglect of the motion of the earth may be cited for the case of a particle which is allowed to fall from rest (relative to the earth) at a height h above the ground. We can show that the rotation of the earth gives rise to an eastward acceleration (Coriolis acceleration) relative to the earth and, neglecting air resistance, that the particle falls to the ground a distance

$$x = \frac{2}{3} \omega \sqrt{\frac{2h^3}{g}} \cos \gamma$$

east of the point on the ground directly under that from which it was dropped. The angular velocity of the earth is $\omega = 0.729(10^{-4})$ rad/s, and the latitude, north or south, is γ. At a latitude of 45° and from a height of 200 m, this eastward deflection would be $x = 43.9$ mm.

$(300 \times 10^6$ m/s) are encountered.° Important problems dealing with atomic and nuclear particles, for example, involve calculations based on the theory of relativity and are of basic concern to both scientists and engineers.

(*b*) *Units.*   It is customary to take *k* equal to unity in Eq. 3/2, thus putting the relation in the usual form of Newton's second law

$$\boxed{\mathbf{F} = m\mathbf{a}} \qquad\qquad [1/1]$$

A system of units for which *k* is unity is known as a *kinetic* system. Thus for a kinetic system the units of force, mass, and acceleration are not independent. In SI units, as explained in Art. 1/4, the units of force (newtons, N) are derived by Newton's second law from the base units of mass (kilograms, kg) times acceleration (meters per second squared, m/s²). Thus N = kg·m/s². This system is known as an *absolute* system since the unit for force is dependent on the absolute value of mass.

In U.S. customary units, on the other hand, the units of mass (slugs) are derived from the units of force (pounds force, lb) divided by acceleration (feet per second squared, ft/sec²). Thus the mass units are slugs = lb-sec²/ft. This system is known as a gravitational system since mass is derived from force as determined from gravitational attraction.

For measurements made relative to the rotating earth the relative value of *g* should be used. The internationally accepted value of *g* relative to the earth at sea level and at a latitude of 45° is 9.806 65 m/s². Except where greater precision is required the value of 9.81 m/s² will be used for *g*. For measurements relative to a nonrotating earth the absolute value of *g* should be used. At a latitude of 45° and at sea level the absolute value is 9.8236 m/s². The sea-level variation in both the absolute and relative values of *g* with latitude is shown in Fig. 1/1 of Art. 1/5.

In the U.S. customary system the standard value of *g* relative to the rotating earth at sea level and at a latitude of 45° is 32.1740 ft/sec². The corresponding value relative to a nonrotating earth is 32.2295 ft/sec².

By virtue of our need to use both SI units and U.S. customary units, we must be absolutely sure that we have a clear understanding of the correct force and mass units in each system. These units were explained in Art. 1/4, but it will be helpful to illustrate them here

---

° The theory of relativity demonstrates that there is no such thing as a preferred primary inertial system and that measurements of time which are made in two coordinate systems which have a velocity relative to one another are different. On this basis, for example, the principles of relativity show that a clock carried by the pilot of a spacecraft traveling around the earth in a circular polar orbit of 644 km altitude at a velocity of 27 080 km/h would be slow compared with a clock at the pole by 0.000 001 85 s for each orbit.

using simple numbers before moving ahead to the applications of Newton's second law. Consider, first, the free-fall experiment as depicted in Fig. 3/1a where we release an object from rest at the earth's surface and allow it to fall freely under the influence of the force of gravitational attraction W on the body, which we call its weight. In SI units for a mass $m = 1$ kg the weight is $W = 9.81$ N, and the corresponding downward acceleration a is $g = 9.81$ m/s^2. In U.S. customary units for a mass $m = 1$ lbm ($1/32.2$ slug) the weight is $W = 1$ lbf and the resulting gravitational acceleration is $g = 32.2$ ft/sec^2. For a mass $m = 1$ slug (32.2 lbm) the weight is $W = 32.2$ lbf and the acceleration, of course, is also $g = 32.2$ ft/sec^2.

In Fig. 3/1b we illustrate the proper units with the simplest example where we accelerate an object of mass m along the horizontal with a force F. In the coherent system of SI units a force $F = 1$ N causes a mass $m = 1$ kg to accelerate at the rate $a = 1$ m/s^2. Thus 1 N $= 1$ kg $\cdot$ m/s^2. In the U.S. customary system (which is a gravitational, noncoherent system) a force $F = 1$ lbf causes a mass $m = 1$ lbm ($1/32.2$ slugs) to accelerate at the rate $a = 32.2$ ft/sec^2, whereas a force $F = 1$ lbf causes a mass $m = 1$ slug (32.2 lbm) to accelerate at the rate $a = 1$ ft/sec^2.

We note that in SI units where the mass is expressed in kilograms (kg) the weight W of the body in newtons (N) is given by $W = mg$

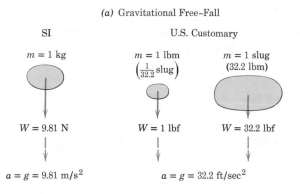

(a) Gravitational Free–Fall

(b) Newton's Second Law

Figure 3/1

where $g = 9.81$ m/s^2. In U.S. customary units the weight W of a body is expressed in pounds force (lbf), and the mass in slugs (lbf-sec^2/ft) is given by $m = W/g$ where $g = 32.2$ ft/sec^2. In U.S. customary units we frequently speak of the weight of a body when we really mean mass. It is entirely proper to specify the mass of a body in pounds (lbm) which must be converted to mass in slugs before substituting into Newton's second law. Unless otherwise stated, the pound (lb) is normally used as the unit of force (lbf).

3/3 EQUATION OF MOTION AND SOLUTION OF PROBLEMS.

When a particle of mass m is subjected to the action of concurrent forces $\mathbf{F}_1$, $\mathbf{F}_2$, $\mathbf{F}_3$, . . . whose vector sum is $\Sigma\mathbf{F}$, Eq. 1/1 becomes

$$\boxed{\Sigma\mathbf{F} = m\mathbf{a}} \tag{3/3}$$

In the solution of problems Eq. 3/3 is usually expressed in scalar component form using one of the coordinate systems developed in Chapter 2. The choice of the appropriate coordinate system is dictated by the type of motion involved and is a vital step in the formulation of any problem. Equation 3/3, or any one of the component forms of the force-mass-acceleration equation, is usually referred to as the *equation of motion*. The equation of motion gives the instantaneous value of the acceleration corresponding to the instantaneous values of the forces which are acting.

(*a*) *The Two Problems of Dynamics.* We encounter two types of problems when applying Eq. 3/3. In the first type the acceleration is either specified or can be determined directly from known kinematic conditions. The corresponding forces which act on the particle whose motion is specified are then determined by direct substitution into Eq. 3/3. This problem is generally quite straightforward.

In the second type of problem one or more of the forces are specified and the resulting motion is to be determined. If the forces are constant, the acceleration is constant and is easily found from Eq. 3/3. When the forces are functions of time, position, velocity, or acceleration, Eq. 3/3 becomes a differential equation which must be integrated to determine the velocity and displacement. Problems of this second type are often more formidable, as the integration may be difficult to carry out, particularly when the force is a mixed function of two or more motion variables. In practice it is frequently necessary to resort to approximate integration techniques, either numerical or graphical, particularly when experimental data are involved. The procedures for a mathematical integration of the acceleration when it is a function of the motion variables were developed in Art. 2/2, and these same procedures apply when the force is a specified function of these same parameters, since force and acceleration differ only by the constant factor of the mass.

(*b*) *Constrained and Unconstrained Motion.* There are two physically distinct types of motion, both described by Eq. 3/3. The first type is *unconstrained* motion where the particle is free of mechanical guides and follows a path determined by its initial motion and by the forces which are applied to it from external sources. An airplane or rocket in flight and an electron moving in a charged field are examples of unconstrained motion. The second type is *constrained* motion where the path of the particle is partially or totally determined by restraining guides. An ice-hockey puck moves with the partial constraint of the ice. A train moving along its track and a collar sliding along a fixed shaft are examples of fully constrained motion. Some of the forces acting on a particle during constrained motion may be applied from outside sources and others may be the reactions on the particle from the constraining guides. *All forces,* both applied and reactive, which act *on* the particle must be accounted for in applying Eq. 3/3.

The choice of a coordinate system is frequently indicated by the number and geometry of the constraints. Thus, if a particle is free to move in space, as with the center of mass of the airplane or rocket in free flight, the particle is said to have *three degrees of freedom* since three independent coordinates are required to specify the position of the particle at any instant. All three of the scalar components of the equation of motion would have to be applied and integrated to obtain the space coordinates as a function of time. If a particle is constrained to move along a surface, as with the hockey puck or a marble sliding on the curved surface of a bowl, only two coordinates are needed to specify its position, and in this case it is said to have *two degrees of freedom*. If a particle is constrained to move along a fixed linear path, as with the collar sliding along a fixed shaft, its position may be specified by the coordinate measured along the shaft. In this case the particle would have only *one degree of freedom*.

(c) *Free-Body Diagram.* In applying any of the force-mass-acceleration equations of motion it is absolutely necessary to account correctly for *all* forces acting on the particle. The only forces which we may neglect are those whose magnitudes are negligible compared with other forces acting, such as the forces of mutual attraction between two particles compared with their attraction to a celestial body such as the earth. The vector sum $\Sigma\mathbf{F}$ of Eq. 3/3 means the vector sum of *all* forces acting *on* the particle in question. Likewise, the corresponding scalar force summation in any one of the component directions means the sum of the components of *all* forces acting *on* the particle in that particular direction. The only reliable way to account accurately and consistently for every force is to *isolate* the particle under consideration from *all* contacting and influencing bodies and replace the bodies removed by the forces they exert on the particle isolated. The resulting *free-body diagram* is the means by which every force, known and unknown, which acts on the particle is represented and hence accounted for. Only after this vital step has been completed should the appropriate equation or equations of motion be written. The free-body diagram serves the same key purpose in dynamics as it does in statics. This purpose is simply to establish a *thoroughly reliable method* for the correct evaluation of the resultant of all actual forces acting on the particle or body in question. In statics this resultant equals zero, whereas in dynamics it is equated to the product of mass and acceleration. If we recognize that the equations of motion must be interpreted literally and exactly, and if in so doing we respect the full scalar and vector meaning of the equals sign in the motion equation, then a minimum of difficulty will be experienced. Every experienced student of engineering mechanics recognizes that careful and consistent observance of the *free-body method* is the *most important single lesson* to be learned in the study of engineering mechanics. As a part of the drawing of a free-body diagram, the coordinate axes and their positive directions should be clearly indicated. When the equations of motion are written, all force summations should be consistent with the choice of these positive directions. Also, as an aid to the identification of external forces which act on the body in question, these forces are shown as heavy red vectors in the illustrations throughout the remainder of the book. Sample Problems 3/1 through 3/5 in the next article contain five examples of free-body diagrams which may be easily reviewed as a reminder of their construction.

In solving problems the student often wonders how to get started and what sequence of steps to follow in arriving at the solution. This difficulty may be minimized by forming the habit of first recognizing some relationship between the desired unknown quantity in the problem and other quantities, known and unknown. Additional relationships between these unknowns and other quantities, known and unknown, are then perceived. Finally the dependence upon the original data is established, and the procedure for the analysis and computation is indicated. A few minutes spent organizing the plan of attack through recognition of the dependence of one quantity upon another will be time well spent and will usually prevent groping for the answer with irrelevant calculations.

3/4 RECTILINEAR MOTION.   We now apply the concepts discussed in Arts. 3/2 and 3/3 to problems in particle motion starting with rectilinear motion in this article and treating curvilinear motion in Art. 3/5. In both articles we shall analyze the motions of bodies which can be treated as particles. This simplification is permitted as long as we are interested in the linear motion of the mass center of the body, in which case the forces may be treated as concurrent through the mass center. When we discuss the kinetics of rigid bodies in Chapter 7, we shall account for the action of nonconcurrent forces on the motions of bodies.

If we choose the x-direction, for example, as the direction of the rectilinear motion of a particle of mass m, the acceleration in the y- and z-directions will be zero and the scalar components of Eq. 3/3 become

$$\Sigma F_x = ma_x$$
$$\Sigma F_y = 0 \qquad\qquad (3/4)$$
$$\Sigma F_z = 0$$

For cases where we are not free to choose a coordinate direction along the motion, we would have in the general case all three component equations

$$\Sigma F_x = ma_x$$
$$\Sigma F_y = ma_y \qquad\qquad (3/5)$$
$$\Sigma F_z = ma_z$$

where the acceleration and resultant force are given by

$$\mathbf{a} = \mathbf{i}a_x + \mathbf{j}a_y + \mathbf{k}a_z$$
$$a = \sqrt{a_x{}^2 + a_y{}^2 + a_z{}^2}$$
$$\Sigma\mathbf{F} = \mathbf{i}\Sigma F_x + \mathbf{j}\Sigma F_y + \mathbf{k}\Sigma F_z$$
$$|\Sigma\mathbf{F}| = \sqrt{(\Sigma F_x)^2 + (\Sigma F_y)^2 + (\Sigma F_z)^2}$$

When we apply Eqs. 3/4 or 3/5 to solve problems in the rectilinear motion of a particle, we must be careful to use Newton's second law exactly as described in Art. 3/3. The left-hand or force side of the equation is obtained by summing in the particular direction chosen the components of *all* forces acting on the particle as disclosed by the free-body diagram. On the right-hand or motion side of the equation we multiply the mass m by the corresponding component of the acceleration of the particle in the direction of our force summation. It is essential that we be consistent algebraically so that the positive direction chosen for the force summation coincides with the positive direction of the corresponding acceleration. With these precautions most of the difficulties commonly experienced may be avoided.

Sample Problem 3/1

A 75-kg man stands on a spring scale in an elevator. During the first 3 s of motion from rest the tension T in the hoisting cable is 8300 N. Find the reading R of the scale in newtons during this interval and the upward velocity v of the elevator at the end of the 3 seconds. The total mass of the elevator, man, and scale is 750 kg.

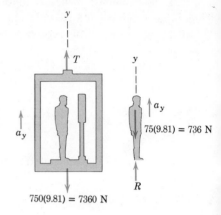

Solution. The force registered by the scale and the velocity both depend on the acceleration of the elevator, which is constant during the interval for which the forces are constant. From the free-body diagram of the elevator, scale, and man taken together, the acceleration is found to be

$[\Sigma F_y = ma_y]$ $8300 - 7360 = 750\, a_y$ $a_y = 1.257 \text{ m/s}^2$

The scale reads the downward force exerted on it by the man's feet. The equal and opposite reaction R to this action is shown on the free-body diagram of the man alone together with his weight, and the equation of motion for him gives

① $[\Sigma F_y = ma_y]$ $R - 736 = 75(1.257)$ $R = 830 \text{ N}$ *Ans.*

The velocity reached at the end of the 3 seconds is

$\left[\Delta v = \int a\, dt\right]$ $v - 0 = \int_0^3 1.257\, dt$ $v = 3.77 \text{ m/s}$ *Ans.*

① If the scale is calibrated in kilograms it would read $830/9.81 = 84.6$ kg which, of course, is not his true mass since the measurement was made in a noninertial (accelerating) system. *Suggestion:* Rework this problem in U.S. customary units.

Sample Problem 3/2

A small inspection car with a mass of 200 kg runs along the fixed overhead cable and is controlled by the attached cable at A. Determine the acceleration of the car when the control cable is horizontal and under a tension $T = 2.4$ kN. Also find the total force P exerted by the supporting cable on the wheels.

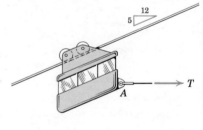

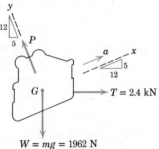

Solution. The free-body diagram of the car and wheels taken together and treated as a particle discloses the 2.4 kN tension T, the weight $W = mg = 200(9.81) = 1962$ N, and the force P exerted on the wheel assembly by the cable.

The car is in equilibrium in the y-direction since there is no acceleration in this direction. Thus

$[\Sigma F_y = 0]$ $P - 2.4(\tfrac{5}{13}) - 1.962(\tfrac{12}{13}) = 0$ $P = 2.73 \text{ kN}$ *Ans.*

In the x-direction the equation of motion gives

① $[\Sigma F_x = ma_x]$ $2400(\tfrac{12}{13}) - 1962(\tfrac{5}{13}) = 200a$ $a = 7.30 \text{ m/s}^2$ *Ans.*

① By choosing our coordinate axes along and normal to the direction of the acceleration we are able to solve the two equations independently. Would this be so if x and y were chosen as horizontal and vertical?

Sample Problem 3/3

The 250-lb concrete block A is released from rest in the position shown and pulls the 400-lb log up the 30-deg ramp. If the coefficient of kinetic friction between the log and the ramp is 0.5, determine the velocity of the log as the block hits the ground at B.

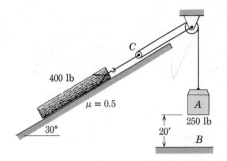

Solution. The velocity of the log will be half the velocity of the block A since the center of the small pulley at C to which the log is attached moves only half as far as the upper of the two cables which is attached to A. From this consideration we see also that the acceleration of the log will be half that of the block. We assume here that the masses of the pulleys are negligible and that they turn with negligible friction. With these assumptions the free-body diagram of the pulley C discloses force and moment equilibrium. Hence the tension in the cable attached to the log is twice that applied to the block.

Equilibrium of the log in the y-direction gives

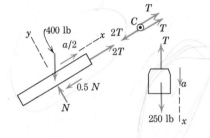

$$[\Sigma F_y = 0] \qquad N - 400 \cos 30° = 0 \qquad N = 346 \text{ lb}$$

and its equation of motion in the x-direction gives

$$[\Sigma F_x = ma_x] \qquad 2T - 0.5(346) - 400 \sin 30° = \frac{400}{32.2}\frac{a}{2}$$

For the block in its x-direction of motion we have

$$[\Sigma F_x = ma_x] \qquad 250 - T = \frac{250}{32.2}a$$

Solving these two equations simultaneously gives us

$$a = 5.83 \text{ ft/sec}^2 \qquad T = 205 \text{ lb}$$

For the 20-ft drop with constant acceleration the block acquires a velocity

$$[v^2 = 2ax] \qquad v_A = \sqrt{2(5.83)(20)} = 15.3 \text{ ft/sec}$$

Hence the velocity of the log is

$$v = \tfrac{1}{2}v_A = 7.65 \text{ ft/sec} \qquad \qquad Ans.$$

① We can verify that the log will indeed move up the ramp by calculating the force in the cable necessary to initiate motion from the equilibrium condition. This force is $2T = 0.5N + 400 \sin 30° = 373$ lb or $T = 186.5$ lb, which is less than the 250-lb weight of block A. Hence the log will move up.

② Note the serious error in assuming that $T = 250$ lb, in which case the block would not accelerate.

③ As long as the forces on a mechanical system remain constant, any resulting acceleration will also remain constant.

④ *Reminder:* The free-body diagrams are essential steps and should always be carefully and completely drawn prior to application of Newton's second law. Also the kinematical relationships between the motions of the two parts must be correctly incorporated.

Sample Problem 3/4

The design model for a new ship has a mass of 10 kg and is tested in an experimental towing tank to determine its resistance to motion through the water at various speeds. The test results are plotted on the accompanying graph and the resistance R may be closely approximated by the dotted parabolic curve shown. If the model is released when it has a speed of 2 m/s, determine the time t required for it to reduce its speed to 1 m/s and the corresponding distance x it has gone after release.

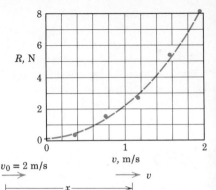

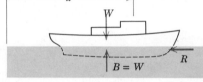

Solution. We approximate the resistance-velocity relation by $R = kv^2$ and find k by substituting $R = 8$ N and $v = 2$ m/s into the equation which gives $k = 8/2^2 = 2$ N·s²/m². Thus $R = 2v^2$.

The only horizontal force on the model is R, so that

① $[\Sigma F_x = ma_x]$ $-R = ma_x$ or $-2v^2 = 10\dfrac{dv}{dt}$

We separate the variables and integrate to obtain

$$\int_0^t dt = -5 \int_2^v \frac{dv}{v^2} \qquad t = 5\left(\frac{1}{v} - \frac{1}{2}\right) \qquad \text{seconds}$$

Thus when $v = v_0/2 = 1$ m/s, the time is $t = 5(\frac{1}{1} - \frac{1}{2}) = 2.5$ s   *Ans.*

① Be careful to observe the minus sign for R.

The distance traveled during the 2.5 seconds is obtained by integrating $v = dx/dt$. Thus $v = 10/(5 + 2t)$ so that

② $\displaystyle\int_0^x dx = \int_0^{2.5} \frac{10}{5 + 2t}\,dt \qquad x = \frac{10}{2}\ln(5 + 2t)\Big]_0^{2.5} = 3.47$ m   *Ans.*

② *Suggestion:* Express the distance x after release in terms of the velocity v and see if you agree with the resulting relation $x = 5\ln(v_0/v)$.

Sample Problem 3/5

The collar of mass m slides up the vertical shaft under the action of a force F of constant magnitude but variable direction. If $\theta = kt$ where k is a constant and if the collar starts from rest with $\theta = 0$, determine the magnitude F of the force which will result in the collar coming to rest as θ reaches $\pi/2$.

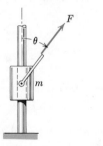

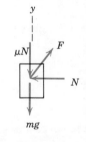

Solution. After drawing the free-body diagram, we apply the equation of motion in the y-direction to get

① $[\Sigma F_y = ma_y]$ $F\cos\theta - \mu N - mg = m\dfrac{dv}{dt}$

where equilibrium in the horizontal direction requires that $N = F\sin\theta$. Substituting $\theta = kt$ and integrating first between general limits give

$$\int_0^t (F\cos kt - \mu F\sin kt - mg)\,dt = m\int_0^v dv$$

which becomes $\dfrac{F}{k}[\sin kt + \mu(\cos kt - 1)] - mgt = mv$

For $\theta = \pi/2$ the time becomes $t = \pi/2k$ and $v = 0$ so that

② $\dfrac{F}{k}[1 + \mu(0 - 1)] - \dfrac{mg\pi}{2k} = 0$ and $F = \dfrac{mg\pi}{2(1 - \mu)}$ *Ans.*

① If θ were expressed as a function of the vertical displacement y instead of the time t, the acceleration would become a function of the displacement and we would use $v\,dv = a\,dy$.

② We see that the results do not depend on k, the rate at which the force changes direction.

PROBLEMS

3/1 Calculate the vertical acceleration a of the 100-lb cylinder for each of the two cases illustrated. Neglect friction and the mass of the pulleys.

Ans. (a) $a = 6.44 \text{ ft/sec}^2$

(b) $a = 16.1 \text{ ft/sec}^2$

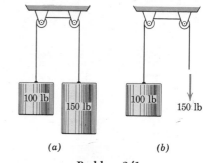

(a)　　　　　　　　(b)

Problem 3/1

3/2 The 10-Mg truck hauls the 20-Mg trailer. If the unit starts from rest on a level road with a tractive force of 20 kN between the driving wheels and the road, compute the tension T in the horizontal drawbar and the acceleration a of the rig.

Problem 3/2

3/3 A man pulls himself up the 15° incline by the method shown. If the combined weight of the man and cart is 250 lb, determine the acceleration of the cart while the man exerts a pull of 60 lb on the rope. Neglect all friction and the mass of the rope, pulleys, and wheels.　　*Ans.* $a = 14.85 \text{ ft/sec}^2$

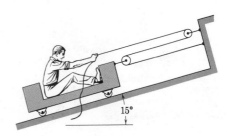

Problem 3/3

3/4 A 3-kg particle is moving in the horizontal y-direction with a velocity of 4 m/s when acted upon by a resultant force $F_y = 5 + 2t^2$ newtons where t is the time of application in seconds. Calculate the velocity v of the particle when $t = 3$ s.

3/5 A cesium-ion engine for deep-space propulsion is designed to produce a constant thrust of 2.5 N for long periods of time. If the engine is to propel a 70-Mg spacecraft on an interplanetary mission, compute the time t for it to increase its speed from 40 000 km/h to 65 000 km/h. Also find the distance s traveled during this interval. Assume that the spacecraft is moving in a remote region of space where the thrust from its ion engine is the only force acting on the spacecraft in the direction of its motion.　　*Ans.* $t = 6.16$ years, $s = 2.84(10^9)$ km

3/6 Beginning at time $t = 0$ a force F_y is applied for 3 s in the positive y-direction to a 3.5-kg particle moving initially in the negative y-direction with a velocity of 3 m/s. If the force in newtons increases with the time in seconds according to $F_y = 16t$ and if F_y is the only force acting on the particle in the y-direction, determine the net displacement Δy of the particle during the 3 seconds.

3/7 The device shown is used as an accelerometer and consists of a 0.30-lb plunger A which deflects the spring as the housing of the unit is given an upward acceleration a. Specify the necessary spring stiffness k which will permit the plunger to deflect 0.20 in. beyond the equilibrium position and touch the electrical contact when the steadily but slowly increasing upward acceleration reaches 5g. Friction may be neglected.   *Ans.* $k = 7.5$ lb/in.

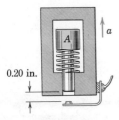

0.20 in.

Problem 3/7

3/8 The coefficient of friction between the flat bed of the truck and the crate it carries is 0.30. Determine the minimum stopping distance s which the truck can have from a speed of 70 km/h with constant deceleration if the crate is not to slip forward.

Problem 3/8

3/9 If the coefficient of friction between the bed of the truck in Prob. 3/8 and its load is 0.30, determine the maximum speed v which the truck can acquire from rest in a distance of 50 m up a 10-percent grade if the load is not to slip.   *Ans.* $v = 50.3$ km/h

3/10 A car is climbing the hill of slope θ_1 at a constant speed v. If the slope decreases abruptly to θ_2 at point A, determine the acceleration a of the car just after passing point A if the driver does not change the throttle setting or shift into a different gear.

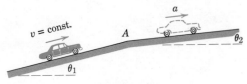

$v = $ const.

Problem 3/10

3/11 The 3000-lb car enters a 10-percent grade at 60 mi/hr and slows down to 30 mi/hr in a distance of $\frac{1}{4}$ mi along the road. Neglect wind resistance and compute the net tractive force F between the tires and the road, assuming F to remain constant during the interval.   *Ans.* $F = 93.5$ lb

$\frac{1}{4}$ mi

60 mi/hr   30 mi/hr

Problem 3/11

12 Small objects are delivered to the 2-m inclined chute by a conveyer belt A which moves at a speed $v_1 = 0.4$ m/s. If the conveyer belt B has a speed $v_2 = 1$ m/s and the objects are delivered to this belt with no slipping, calculate the coefficient of friction μ between the objects and the chute.

Ans. $\mu = 0.553$

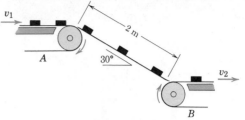

Problem 3/12

13 The block shown is observed to have a velocity $v_1 = 20$ ft/sec as it passes point A and a velocity $v_2 = 10$ ft/sec as it passes point B on the incline. Calculate the coefficient of friction μ between the block and the incline if $x = 30$ ft and $\theta = 15$ deg.

Ans. $\mu = 0.429$

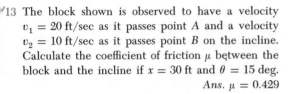

Problem 3/13

14 Derive an expression for the velocity v_2 of the sliding block shown with Prob. 3/13 in terms of its initial velocity v_1, the distance traveled x, the coefficient of friction μ between the block and the incline, and the angle θ.

15 For the sliding block shown with Prob. 3/13 calculate the angle θ of the incline which will result in a final velocity $v_2 = 10$ m/s at $x = 12$ m if $v_1 = 6$ m/s and the coefficient of friction between the block and the incline is 0.3. (*Suggestion:* Solve the resulting equation by approximate graphical means.)

Ans. $\theta = 31.8°$

16 The acceleration of the 50-kg carriage A in its smooth vertical guides is controlled by the tension T exerted on the control cable which passes around the two circular pegs fixed to the carriage. Determine the value of T required to limit the downward acceleration of the carriage to 1.2 m/s² if the coefficient of friction between the cable and the pegs is 0.20. (Recall the relation between the tensions in a flexible cable which is on the verge of slipping on a fixed peg: $T_2 = T_1 e^{\mu\beta}$.)

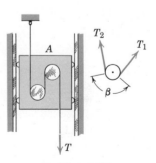

Problem 3/16

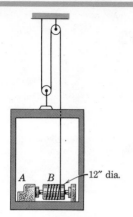

Problem 3/17

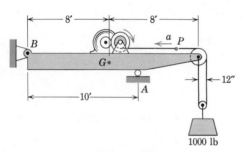

Problem 3/18

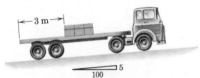

Problem 3/19

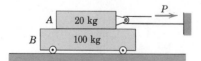

Problem 3/20

3/17 The workman's cage weighs 1200 lb fully loaded. If the motor unit A supplies a constant torque of 215 lb-ft to the shaft of the hoisting drum B, calculate the vertical acceleration a of the cage. The mass of the drum is small so that it may be treated as though it were in rotational equilibrium.

Ans. $a = 2.41$ ft/sec^2

3/18 A 3.6-Mg flat-bed truck carries a 750-kg box. As the truck starts from rest with constant acceleration, the box slides 3 m to the edge of the bed in the time that it takes the truck to acquire a velocity of 40 km/h in a distance of 15 m up the incline. Determine the coefficient of friction μ between the box and the truck bed.   *Ans.* $\mu = 0.386$

3/19 The beam and attached hoisting mechanism together weigh 2400 lb with center of gravity at G. If the initial acceleration a of a point P on the hoisting cable is 20 ft/sec^2, calculate the corresponding reaction at the support A.

3/20 If the coefficient of friction between the 20-kg block A and the 100-kg cart B is 0.50, determine the acceleration of each part for (a) $P = 60$ N and (b) $P = 40$ N.

Ans. (a) $a_A = 1.10$ m/s^2, $a_B = 0.98$ m/s^2
(b) $a_A = a_B = a = 0.67$ m/s^2

21 A force $P = 80$ lb is applied to the cable to slide the two crates up the incline. Compute the acceleration a of the 150-lb crate. The friction and mass of the pulleys may be neglected.

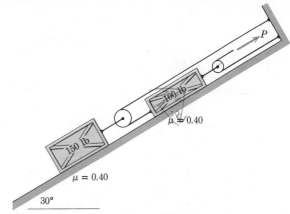

22 The steel ball is suspended from the accelerating frame by the two cords A and B. Determine the acceleration a of the frame which will cause the tension in A to be twice that in B.

$$\text{Ans. } a = \frac{g}{3\sqrt{3}}$$

Problem 3/21

23 Calculate the acceleration a of the 100-lb block up the incline if $P = 50$ lb.

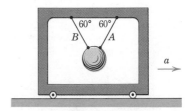

Problem 3/22

24 Determine the tension P in the cable which will give the block shown with Prob. 3/23 an acceleration of 5 ft/sec² up the incline.   *Ans. P = 43.8 lb*

25 If the cylinder of mass m and supporting carriage are released from rest and move without friction down the incline, prove that the contact forces at A and B are equal for all values of θ and β less than $\pi/2$ and determine an expression for their value.

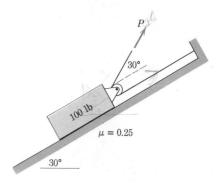

Problem 3/23

26 For the cylinder and carriage of Prob. 3/25 with $\beta = 45°$ and $\theta = 30°$ calculate the maximum acceleration a in meters per second squared which the carriage may be given up the incline so that the cylinder does not lose contact at B.

$$\text{Ans. } a = 3.59 \text{ m/s}^2$$

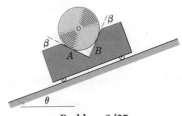

Problem 3/25

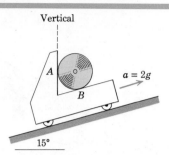

Vertical

$a = 2g$

15°

Problem 3/27

T R

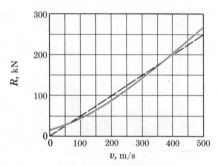

Problem 3/28

x

Problem 3/29

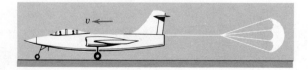

v

Problem 3/30

3/27 The smooth 25-kg metal cylinder is supported in a carriage which is given an acceleration of $2g$ up the 15° incline. Calculate the forces of contact at A and B independently of one another (i.e., use no simultaneous equations).

3/28 The total resistance R to horizontal motion of a rocket test sled is shown by the solid line in the graph. Approximate R by the dotted line and determine the distance x which the sled must travel along the track from rest to reach a velocity of 400 m/s. The forward thrust T of the rocket motors is essentially constant at 300 kN, and the total mass of the sled is also nearly constant at 2 Mg.

Ans. $x = 1037$ m

3/29 The 2-kg collar is released from rest against the light elastic spring, which has a stiffness of 1.75 kN/m and which has been compressed a distance of 0.15 m. Determine the acceleration a of the collar as a function of the vertical displacement x of the collar measured in meters from the point of release. Find the velocity v of the collar when $x = 0.15$ m. Friction is negligible.

3/30 A jet fighter plane with a mass of 5 Mg has a touchdown speed of 300 km/h at which instant the braking parachute is deployed and the power shut off. If the total drag on the aircraft varies with velocity as shown in the accompanying graph, calculate the distance x along the runway required to reduce the speed to 150 km/h. Approximate the variation of the drag by an equation of the form $D = kv^2$ where k is a constant.

Ans. $x = 201$ m

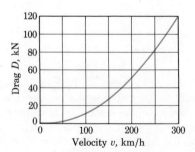

/31 A heavy chain with a mass ρ per unit length is pulled along a horizontal surface consisting of a smooth section and a rough section by the constant force P. If the chain is initially at rest on the smooth surface with $x = 0$ and if the coefficient of friction between the chain and the rough surface is μ, determine the velocity v of the chain when $x = L$. Assume that the chain remains taut and thus moves as a unit throughout the motion. What is the minimum value of P which will permit the chain to remain taut? (*Hint:* The acceleration must not become negative.)

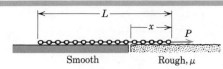

Problem 3/31

/32 If the rough and smooth surfaces of Prob. 3/31 are reversed, as shown here, determine the velocity v of the chain when all of it is transferred to the smooth surface. What is the minimum velocity v corresponding to the least value of P to initiate motion?

$$Ans.\ v = \sqrt{\frac{2P}{\rho} - \mu g L},\ v_{min} = \sqrt{\mu g L}$$

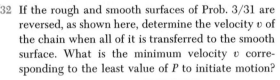

Problem 3/32

/33 The chain is released from rest with an overhanging length b. Neglect all friction and determine the velocity v of the chain when its last link leaves the edge.

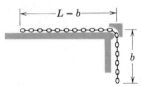

Problem 3/33

/34 The chain of Prob. 3/33 is released from rest in the position shown with barely enough overhanging links to initiate motion. The coefficient of friction between the links and the horizontal surface is μ. Determine the velocity v of the chain when the last link leaves the edge. Neglect friction at the corner.

$$Ans.\ v = \sqrt{\frac{gL}{1 + \mu}}$$

/35 In a test of resistance to motion in an oil bath a small steel ball of mass m is released from rest at the surface ($y = 0$). If the resistance to motion is given by $R = kv$ where k is a constant, derive an expression for the depth h required for the ball to reach a velocity v.

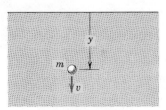

Problem 3/35

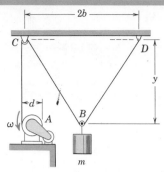

Problem 3/37

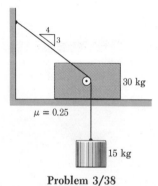

Problem 3/38

3/36 If the steel ball of Prob. 3/35 is released from rest at the surface of a liquid in which the resistance to motion is $R = cv^2$ where c is a constant and v is the downward velocity of the ball, determine the depth h required to reach a velocity v.

$$\text{Ans. } h = \frac{m}{2c} \ln \frac{mg}{mg - cv^2}$$

▶**3/37** The hoisting drum A has a diameter d and turns counterclockwise with a constant angular velocity ω. Determine the tension T in the cable which attaches the mass m to the small pulley B. Express the result in terms of the variable y. The size, mass, and friction of the pulleys at C and B are negligible.

$$\text{Ans. } T = mg \left[1 + \frac{\omega^2 b^2 d^2}{16 \, gy^3} \right]$$

▶**3/38** The system is released from rest in the position shown. Calculate the tension T in the cord and the acceleration a of the 30-kg block. The small pulley attached to the block has negligible mass and friction. (*Suggestion:* First establish the kinematic relationship between the accelerations of the two bodies.)   Ans. $T = 138.0$ N, $a = 0.77$ m/s²

▶**3/39** The lunar landing module has a mass of 17.5 Mg. As it descends onto the moon's surface, its retro-engine is shut off when it reaches a hovering condition a few meters above the moon's surface. The module then falls under the action of lunar gravity, and the coiled springs in the landing pads cushion the impact. If the stiffness of each of the three springs is 15 kN/m, determine the maximum magnitude of the "jerk" J (time rate of change of the acceleration) during compression of the springs. The acceleration due to lunar gravity is 1.62 m/s², and touchdown velocity is to be 1.5 m/s.   Ans. $J = 4.65$ m/s³

Problem 3/39

/5 CURVILINEAR MOTION.   We turn our attention now to the kinetics of particles which move along plane curvilinear paths. In applying Newton's second law, Eq. 3/3, we shall make use of the three coordinate descriptions of acceleration in curvilinear motion which we developed and applied in Arts. 2/4, 2/5, and 2/6.

The choice of coordinate system depends on the conditions of the problem and is one of the basic decisions to be made in solving curvilinear-motion problems. We now rewrite Eq. 3/3 in three ways the choice of which depends on which coordinate system is most appropriate.

Rectangular coordinates (Art. 2/4)

$$\left.\begin{array}{l} \Sigma F_x = ma_x \\ \Sigma F_y = ma_y \end{array}\right\} \tag{3/6}$$

where $a_x = \ddot{x}$ and $a_y = \ddot{y}$

Normal and tangential coordinates (Art. 2/5)

$$\left.\begin{array}{l} \Sigma F_n = ma_n \\ \Sigma F_t = ma_t \end{array}\right\} \tag{3/7}$$

where $a_n = \rho\dot{\theta}^2 = v^2/\rho = v\dot{\theta},$ $a_t = \dot{v},$ and $v = \rho\dot{\theta}$

Polar coordinates (Art. 2/6)

$$\left.\begin{array}{l} \Sigma F_r = ma_r \\ \Sigma F_\theta = ma_\theta \end{array}\right\} \tag{3/8}$$

where $a_r = \ddot{r} - r\dot{\theta}^2,$ $a_\theta = r\ddot{\theta} + 2\dot{r}\dot{\theta}$

In making application of these motion equations the general procedure established in the previous article on rectilinear motion should be followed here. After the motion is identified and the coordinate system is designated, the free-body diagram of the body to be treated as a particle should be drawn. The appropriate force summations are then obtained from this diagram in the usual way. The free-body diagram should be complete so that no mistakes are made in figuring the force summations. We should note especially that the sense assigned to a given force summation must be consistent with the corresponding sense of the acceleration term. Thus for the first of Eqs. 3/7, for example, the positive sense for the force sum ΣF_n is that of the n-component of acceleration $a_n = \rho\dot{\theta}^2$, which is *toward* the center of curvature.

Successful application of the equations of motion depends as much on a thorough knowledge of the kinematics of the motion as it does on the correct analysis of all forces which act on the particle as obtained from the free-body diagram. Therefore a review of Arts. 2/4, 2/5, and 2/6 should be undertaken at this point if needed.

Sample Problem 3/6

A 1500-kg car enters a section of curved road in the horizontal plane and slows down at a uniform rate from a speed of 100 km/h at A to a speed of 50 km/h as it passes C. The radius of curvature ρ of the road at A is 400 m and at C is 80 m. Determine the total horizontal force exerted by the road on the tires at positions A, B, and C. Point B is the inflection point where the curvature changes direction.

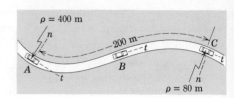

Solution. The car will be treated as a particle so that the effect of all forces exerted by the road on the tires will be treated as a single force. Since the motion is described along the direction of the road, normal and tangential coordinates will be used to specify the acceleration of the car. We will then determine the forces from the accelerations.

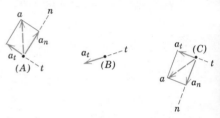

The constant tangential acceleration is in the negative t-direction and its magnitude is given by

① $[v_C{}^2 = v_A{}^2 + 2a_t\Delta s]$ $\quad a_t = \left|\dfrac{(50/3.6)^2 - (100/3.6)^2}{2(200)}\right| = 1.45 \text{ m/s}^2$

① Recognize the numerical value of the conversion factor from km/h to m/s as 1000/3600 or 1/3.6.

The normal components of acceleration at A, B, and C are

② $[a_n = v^2/\rho]$ $\quad$ At A, $\quad a_n = \dfrac{(100/3.6)^2}{400} = 1.93 \text{ m/s}^2$

② Note that a_n is always directed toward the center of curvature.

At B, $\quad a_n = 0$

At C, $\quad a_n = \dfrac{(50/3.6)^2}{80} = 2.41 \text{ m/s}^2$

Application of Newton's second law in both the n- and t-directions to the free-body diagrams of the car gives

$[\Sigma F_t = ma_t]$ $\quad\quad\quad F_t = 1500(1.45) = 2170 \text{ N}$

③ $[\Sigma F_n = ma_n]$ $\quad$ At A, $\quad F_n = 1500(1.93) = 2894 \text{ N}$

At B, $\quad F_n = 0$

At C, $\quad F_n = 1500(2.41) = 3617 \text{ N}$

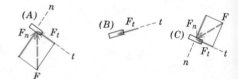

③ Note that the direction of F_n must agree with that of a_n.

Thus the total horizontal force acting on the tires becomes

At A, $\quad F = \sqrt{F_n{}^2 + F_t{}^2} = \sqrt{(2894)^2 + (2170)^2} = 3617 \text{ N}$ $\quad$ *Ans.*

At B, $\quad F = F_t = 2170 \text{ N}$ $\quad$ *Ans.*

④ At C, $\quad F = \sqrt{F_n{}^2 + F_t{}^2} = \sqrt{(3617)^2 + (2170)^2} = 4218 \text{ N}$ $\quad$ *Ans.*

④ The angle made by **a** and **F** with the direction of the path can be computed if desired.

Sample Problem 3/7

Determine the height h and tension T in the cord for the conical pendulum of mass m and length l which rotates about the vertical axis at the constant angular rate $\dot\theta = \omega$.

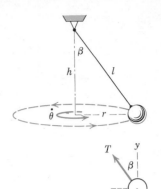

Solution. We will neglect the dimensions of the ball compared with the radius r of its circular path and, therefore, treat the ball as a particle. The ball executes circular motion in the horizontal plane and has an acceleration $a_n = r\dot\theta^2 = r\omega^2$. From the free-body diagram we have

$[\Sigma F_n = ma_n]$ $\qquad T \sin\beta = mr\omega^2$

$[\Sigma F_y = 0]$ $\qquad T \cos\beta = mg$

Substitution of $r = l \sin\beta$ into the first equation gives

$$T = ml\omega^2 \qquad\qquad Ans.$$

Division of the first equation by the second to eliminate T and substitution of $h = l \cos\beta$ give

$$\sin\beta/\cos\beta = (l\omega^2 \sin\beta)/g \quad\text{or}\quad h = g/\omega^2 \qquad Ans.$$

① This last result shows that all conical pendulums which rotate at the same rate ω will have the same h regardless of l. Also, for a given length l the minimum angular rate for which the mass will act as a conical pendulum is seen to be $\sqrt{g/l}$ as β approaches zero.

Sample Problem 3/8

Small objects are released from rest at A and slide down the smooth circular surface of radius R to a conveyor B. Determine the expression for the normal contact force N between the guide and each object in terms of θ and specify the correct angular velocity ω of the conveyor pulley of radius r to prevent any sliding on the belt as the objects transfer to the conveyor.

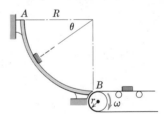

Solution. The free-body diagram of the object is shown together with the coordinate directions n and t. The normal force N depends on the n-component of the acceleration which, in turn, depends on the velocity. The velocity will be cumulative according to the tangential acceleration a_t. Hence we will find a_t first for any general position.

$[\Sigma F_t = ma_t]$ $\qquad mg \cos\theta = ma_t \qquad a_t = g \cos\theta$

Now we can find the velocity by integrating

$$[v\,dv = a_t\,ds] \qquad \int_0^v v\,dv = \int_0^\theta g \cos\theta \, d(R\theta) \qquad v^2 = 2gR \sin\theta$$

We obtain the normal force by summing forces in the positive n-direction, which is the direction of the n-component of acceleration.

$[\Sigma F_n = ma_n] \quad N - mg\sin\theta = m\dfrac{v^2}{R} \qquad N = 3\,mg\sin\theta \qquad Ans.$

The conveyor pulley must turn at the rate $v = r\omega$ for $\theta = \pi/2$, so that

$$\omega = \sqrt{2gR}/r \qquad\qquad Ans.$$

① It is essential here that we recognize the need to express the tangential acceleration as a function of position so that v may be found by integrating the kinematical relation $v\,dv = a_t\,ds$ in which all quantities are measured along the path.

Sample Problem 3/9

Tube A rotates about the vertical O-axis with a constant angular rate $\dot{\theta} = \omega$ and contains a small cylindrical plug B of mass m whose radial position is controlled by the cord which passes freely through the tube and shaft and is wound around the drum of radius b. Determine the tension T in the cord and the horizontal component F_θ of force exerted by the tube on the plug if the constant angular rate of rotation of the drum is ω_0 first in the direction for case (a) and second in the direction for case (b). Neglect friction.

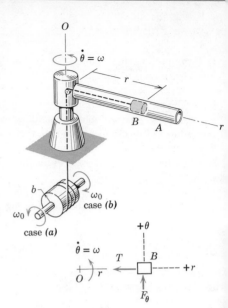

case (b)

case (a)

Solution. With r a variable we use the polar-coordinate form of the equations of motion, Eqs. 3/8. The free-body diagram of B is shown in the horizontal plane and discloses only T and F_θ. The equations of motion are

① ② $[\Sigma F_r = ma_r]$ $\qquad -T = m(\ddot{r} - r\dot{\theta}^2)$

$[\Sigma F_\theta = ma_\theta]$ $\qquad F_\theta = m(r\ddot{\theta} + 2\dot{r}\dot{\theta})$

Case (a) With $\dot{r} = +b\omega_0$, $\ddot{r} = 0$, and $\ddot{\theta} = 0$ the forces become

$$T = mr\omega^2 \qquad F_\theta = 2mb\omega_0\omega \qquad \textit{Ans.}$$

③ **Case (b)** With $\dot{r} = -b\omega_0$, $\ddot{r} = 0$, and $\ddot{\theta} = 0$ the forces become

$$T = mr\omega^2 \qquad F_\theta = -2mb\omega_0\omega \qquad \textit{Ans.}$$

① The weight and vertical reaction of the tube on B are not seen in the horizontal plane. The force F_θ is arbitrarily shown in the plus θ-direction.

② Note that we are summing forces in the plus r-direction so we use $-T$.

③ The minus sign shows that F_θ is in the direction opposite to that shown on the free-body diagram.

PROBLEMS

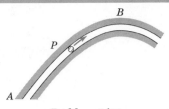

Problem 3/40

3/40 A particle P having a mass of 2 kg is given an initial velocity at A in the smooth guide, which is curved in the vertical plane. If the particle has a horizontal velocity of 4.8 m/s at the top B of the guide, where its radius of curvature is 0.5 m, calculate the force R exerted on the particle by the guide at this point.

Ans. $R = 72.5$ N down

3/41 The small 2-lb block A slides down the curved path and crosses the lowest point B with a velocity of 4 ft/sec. If the force normal to the surface between the block and the guide is 2.5 lb as this position is passed, compute the radius of curvature ρ of the path at B.

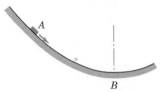

Problem 3/41

3/42 Determine the maximum speed which the sliding block may have as it passes over the top C of the circular portion of the path without losing contact with the path.

Ans. $v = 4.20$ m/s

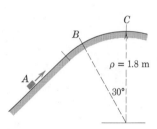

Problem 3/42

3/43 In the design of a space station to operate outside the earth's gravitational field, it is desired to give the structure a rotational speed N which will simulate the effect of the earth's gravity for members of the crew. If the centers of the crew's quarters are to be located 8 m from the axis of rotation, calculate the necessary rotational speed N of the space station in revolutions per minute.

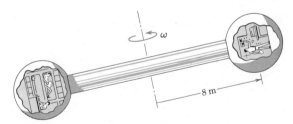

Problem 3/43

3/44 An 80-kg stunt pilot executes a vertical inside loop. If his horizontal velocity is 300 km/h in the upside-down position at the top of the loop and if the force between him and his seat has dropped to one-fourth of his weight at this point, calculate the radius of curvature ρ of the top of the loop.

Ans. $\rho = 566$ m

3/45 The simple pendulum is released from the dotted position with an initial velocity. As it swings past the vertical, the tension T in the supporting wire is k times the weight of the pendulum. Determine the expression for the velocity v of the pendulum at the bottom position in terms of k.

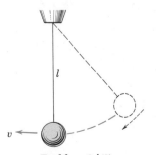

Problem 3/45

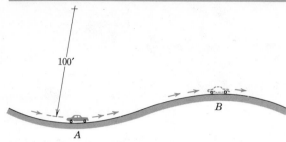

Problem 3/46

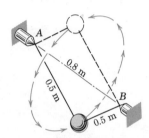

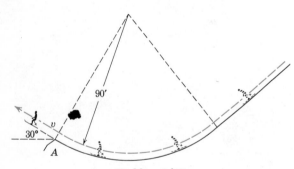

Problem 3/47

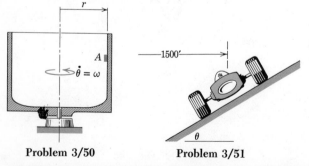

Problem 3/49

Problem 3/50   **Problem 3/51**

3/46 A car weighing 3220 lb goes through the bottom A of a dip in the road at 30 mi/hr. If the radius of curvature of the path at the dip is 100 ft, calculate the total vertical force N exerted on the wheels by the road. If N is reduced to half of the weight of the car as it passes over the top B of the next hump—also of 100-ft radius of curvature—what is the speed v of the car?   *Ans.* $N = 5160$ lb, $v = 27.4$ mi/hr

3/47 A small 0.8-kg ball is attached to two 0.5-m cords fastened to fixed points A and B, which are 0.8 m apart and on the same horizontal line. The ball is set into circular motion in the vertical plane. If the velocity of the ball is 4 m/s at the bottom position, calculate the corresponding tension T in each cord.

3/48 If the velocity of the ball of Prob. 3/47 is 3.18 m/s at the top of its path (dotted position), what is the corresponding tension T in each cord?
    Ans. $T = 15.9$ N

3/49 If the 180-lb ski-jumper attains a speed of 80 ft/sec as he approaches the takeoff position, calculate the normal force N exerted by the snow on his skis just before he reaches A.   *Ans.* $N = 553$ lb

3/50 A small object A is held against the vertical side of the rotating cylindrical container of radius r due to centrifugal action. If the coefficient of friction between the object and the container is μ, determine the expression for the minimum rotational rate $\dot{\theta} = \omega$ of the container which will keep the object from slipping down the vertical side.

3/51 At what angle θ should the racetrack turn of 1500-ft radius be banked in order that a racecar traveling at 120 mi/hr will have no tendency to slip sideways when rounding the turn?   *Ans.* $\theta = 32.7°$

52 The barrel of a rifle is rotating in a horizontal plane about the vertical z-axis at the constant angular rate $\dot{\theta} = 0.5$ rad/s when a 60-g bullet is fired. If the velocity of the bullet relative to the barrel is 600 m/s just before it reaches the muzzle A, determine the resultant horizontal side thrust P exerted by the barrel on the bullet just before it emerges from A. On which side of the barrel does P act?

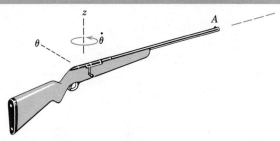

Problem 3/52

53 Calculate the tension T in a metal hoop rotating in its plane about its geometric axis with a large rim speed v. Analyze the differential element of the hoop as a particle. The mass per unit length of rim is ρ.

Problem 3/53

54 A 150-lb woman is standing on a scale placed on the cabin floor of a jet transport. At what rate $\dot{\theta}$ in degrees per second is the pilot dropping his line of sight if the scales read 100 lb when the aircraft is at the top of its trajectory and moving at 600 mi/hr? Also calculate the radius of curvature ρ of the flight path at this same condition.
> *Ans.* $\dot{\theta} = 0.70$ deg/sec, $\rho = 13.66$ mi

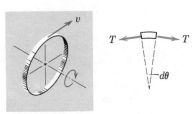

Problem 3/54

55 The slider A of Problem 3/42 has a mass of 2 kg and is given an initial velocity up the incline. If it crosses position B with a velocity of 3 m/s and if the magnitude of the velocity is decreasing at the rate of 6 m/s² at this position, determine the coefficient of friction μ between the slider and its supporting surface. The radius of curvature of the path at B is 1.8 m.   *Ans.* $\mu = 0.313$

56 The small mass m and its supporting wire become a simple pendulum when the horizontal cord is severed. Determine the ratio k of the tension T in the supporting wire immediately after the cord is cut to that in the wire before the cord is cut.

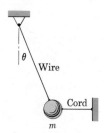

Problem 3/56

57 The small sphere of mass m is suspended initially at rest by the two wires. If one wire is suddenly cut, determine the ratio k of the tension in the remaining wire just after the other wire is cut to the initial equilibrium tension.   *Ans.* $k = 3/2$

Problem 3/57

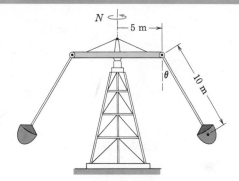

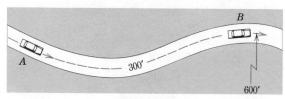

Problem 3/58

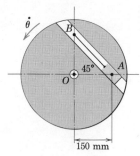

Problem 3/59

Problem 3/60

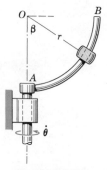

Problem 3/62

3/58 Calculate the necessary rotational speed N for the aerial ride in the amusement park in order that the arms of the gondolas will assume an angle $\theta = 60°$ with the vertical. Neglect the mass of the arms to which the gondolas are attached and treat each gondola as a particle.

3/59 A 3220-lb car enters an S-curve at A with a speed of 60 mi/hr with brakes applied to reduce the speed to 45 mi/hr at a uniform rate in a distance of 300 ft measured along the curve from A to B. The radius of curvature of the road at B is 600 ft. Calculate the total friction force exerted by the road on the tires at B. The road at B lies in a horizontal plane.

Ans. F = 920 lb

3/60 The 3-kg slider A fits loosely in the smooth 45° slot in the disk, which rotates in a horizontal plane about its center O. If A is held in position by a cord secured to point B, determine the tension T in the cord for a constant rotational velocity $\dot{\theta} = 300$ rev/min. Would the direction of the velocity make any difference? *Ans. T = 314 N; No*

3/61 Determine the tension T in the cord in Prob. 3/60 as the disk starts from rest with $\ddot{\theta} = 120$ rad/s². Also calculate the initial horizontal force N exerted by the slot on the slider.

3/62 A small slider is free to move along the circular rod AB with negligible friction. If the rod is rotating about the vertical axis OA with a constant angular velocity $\dot{\theta}$, determine the angle β which locates the stable position assumed by the slider.

$$\text{Ans. } \beta = \cos^{-1}\frac{g}{r\dot{\theta}^2}, \dot{\theta} \geq \sqrt{g/r}$$

3/63 A 2-lb particle moves in a horizontal plane with its polar coordinates given by $r = 2t^3 - 3t + 4$ and $\theta = 2\sin\frac{1}{2}\pi t$ where r is in inches, θ is in radians, and t is in seconds. Compute the resultant horizontal force on the particle at the instant $t = 2$ sec.

64 An object of mass m is given an initial velocity u up the incline at point A. An instant after it passes point B, the normal force of contact between it and the supporting surface drops to one-half of the value it had when the object was approaching B. The coefficient of friction between the object and the incline is 0.30. Calculate u.   *Ans.* $u = 7.75$ m/s

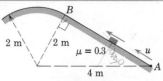

Problem 3/64

65 The rocket moves in a vertical plane and is being propelled by a thrust T of 32 kN. It is also subjected to an atmospheric resistance R of 9.6 kN. If the rocket has a velocity of 3 km/s and if the gravitational acceleration is 6 m/s² at the altitude of the rocket, calculate the radius of curvature ρ of its path for the position described and the time rate of change of the magnitude v of the velocity of the rocket. The mass of the rocket at the instant considered is 2000 kg.

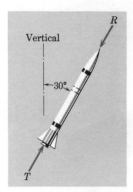

Problem 3/65

66 The slotted arm rotates about its center in a horizontal plane at the constant angular rate $\dot{\theta} = 10$ rad/sec and carries a 3.22-lb spring-mounted slider which oscillates freely in the slot. If the slider has a speed of 24 in./sec relative to the slot as it crosses the center, calculate the horizontal side thrust P exerted by the slotted arm on the slider at this instant. Determine which side A or B of the slot is in contact with the slider.

    Ans. $P = 4$ lb, Side A

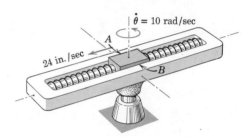

Problem 3/66

67 The spring-mounted 0.8-kg collar A oscillates along the horizontal rod which is rotating at the constant angular rate $\dot{\theta} = 6$ rad/s. At a certain instant r is increasing at the rate of 800 mm/s. If the coefficient of friction between the collar and the rod is 0.40, calculate the friction force F exerted by the rod on the collar at this instant.

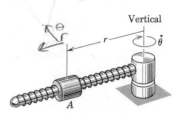

Problem 3/67

68 A small coin is placed on the horizontal surface of the rotating disk. If the disk starts from rest and is given a constant angular acceleration $\ddot{\theta} = \alpha$, determine an expression for the number of revolutions N through which the disk turns before the coin slips. The coefficient of friction between the coin and the disk is μ.

    Ans. $N = \dfrac{1}{4\pi}\sqrt{\left(\dfrac{\mu g}{r\alpha}\right)^2 - 1}$

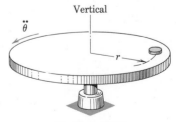

Problem 3/68

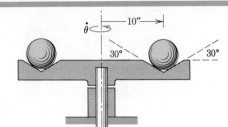

Problem 3/69

Problem 3/72

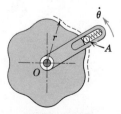

Problem 3/73

3/69 The disk with the circular groove rotates about the vertical axis with a constant speed of 30 rev/min and carries the two 8-lb spheres. Calculate the larger of the two forces of contact between the disk and each sphere. (Can this result be reached by using only one force equation?)

3/70 Calculate the maximum rotational speed which the disk of Prob. 3/69 may have without causing the spheres to leave their groove.

Ans. N = 45.1 rev/min

3/71 A body at rest relative to the surface of the earth rotates with the earth and therefore moves in a circular path about the polar axis of the earth considered as fixed. Derive an expression for the ratio k of the apparent weight of such a body as measured by a spring scale at the equator (calibrated to read the actual force applied) to the true weight of the body, which is the absolute gravitational attraction to the earth. The absolute acceleration due to gravity at the equator is $g = 9.815$ m/s². The radius of the earth at the equator is $R = 6378$ km, and the angular velocity of the earth is $\omega = 0.729(10^{-4})$ rad/s. If the true weight is 100 N, what is the apparent measured weight W'?

$$Ans. \ k = 1 - \frac{R\omega^2}{g}, \ W' = 99.655 \text{ N}$$

3/72 The member OA rotates about a horizontal axis through O with a constant counterclockwise velocity $\dot{\theta} = 3$ rad/sec. As it passes the position $\theta = 0$, a small mass m is placed upon it at a radial distance $r = 18$ in. If the mass is observed to slip at $\theta = 45°$, determine the coefficient of friction μ between the mass and the member.

3/73 The slotted arm revolves with a constant angular velocity $\dot{\theta} = \omega$ about a vertical axis through the center O of the fixed cam. The cam is cut so that the radius to the path of the center of the pin A varies according to $r = r_0 + b \sin N\omega t$ where N equals the number of lobes—six in this case. If $\omega = 12$ rad/s, $r_0 = 100$ mm, and $b = 10$ mm, and if the spring compression varies from 11.5 N to 19.1 N from valley to crest, calculate the force R between the cam and the 100-g pin A as it passes over the top of the lobe in the position shown. Neglect friction.

Ans. R = 12.33 N

74 The small vehicle has a mass of 30 kg and is given an initial velocity $v_0 = 8$ m/s at the bottom of the circular track. Calculate the velocity v of the vehicle and the normal reaction R on its wheels as it passes the position at which $\theta = 30°$. Neglect friction.

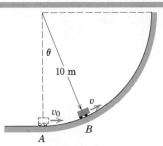

Problem 3/74

75 A small vehicle enters the top A of the circular path with a horizontal velocity v_0 and gathers speed as it moves down the path. Determine the expression for the angle β to the position where the vehicle leaves the path and becomes a projectile. Neglect friction and treat the vehicle as a particle.

$$\text{Ans. } \beta = \cos^{-1}\left(\frac{2}{3} + \frac{v_0{}^2}{3gR}\right)$$

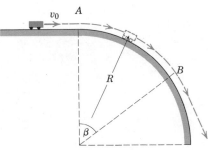

Problem 3/75

76 In theory an object projected vertically up from the surface of the earth with a sufficiently high velocity can escape from the earth's gravity field. Calculate the minimum value v of this escape velocity on the basis of the absence of an atmosphere to offer resistance due to air friction and considering the earth to be nonrotating. Take the radius of the earth to be 6370 km and the gravitational acceleration at the earth's surface to be 9.821 m/s².

77 A stretch of highway includes a succession of evenly spaced dips and humps, the contour of which may be represented by the relation $y = b \sin \dfrac{2\pi x}{L}$. What is the maximum speed at which the car A can go over a hump and still maintain contact with the road? If the car maintains this critical speed, what is the total reaction N under its wheels at the bottom of a dip? The mass of the car is m.

$$\text{Ans. } v = \frac{L}{2\pi}\sqrt{g/b}, \; N = 2mg$$

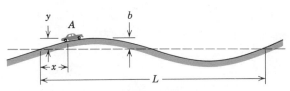

Problem 3/77

78 A small rocket-propelled vehicle of mass m travels down the circular path of effective radius r under the action of its weight and a constant thrust T from its rocket motor. If the vehicle starts from rest at A, determine its velocity v when it reaches B and the force N exerted by the guide on the wheels just prior to reaching B. Neglect any friction and any loss of mass of the rocket.

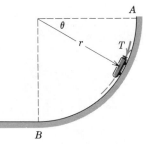

Problem 3/78

Problem 3/79

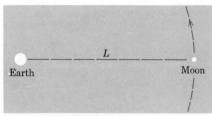

Problem 3/80

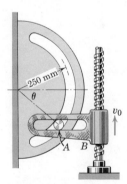

Problem 3/81

Problem 3/82

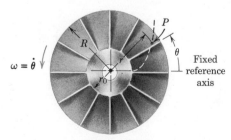

Problem 3/83

3/79 A satellite S, which weighs 322 lb when launched, is put into an elliptical orbit around the earth. At the instant represented the satellite is a distance $r = 6667$ mi from the center of the earth and has a velocity $v = 16{,}610$ mi/hr making an angle $\beta = 60°$ with the radial line through S. The only force acting on the satellite is the gravitational attraction of the earth, which amounts to 114 lb for these conditions. Calculate the value of $\ddot{r}$ for the condition described.   *Ans.* $\ddot{r} = 1.24$ ft/sec^2

3/80 The mean radius R of the earth is 6370 km, and the mean distance L between the centers of the earth and the moon is 384 398 km. Determine the period τ of the orbit of the moon about the earth considered as fixed. Assume a circular orbit.

$$Ans. \ \tau = \frac{2\pi L}{R}\sqrt{\frac{L}{g}}, \ \tau = 27.5 \text{ days}$$

3/81 Derive an expression for the magnitude v of the velocity of the artificial satellite S which is to operate in a circular orbit at an altitude h above the surface of the earth. The altitude is sufficiently great to permit neglect of the effect of atmospheric resistance. The mean radius R of the earth is 6370 km. What is the period τ of the satellite's motion? Compute v and τ for $h = 640$ km.

$$Ans. \ v = R\sqrt{\frac{g}{R+h}} = 27\ 143 \text{ km/h}$$

$$\tau = 2\pi \frac{(R+h)^{3/2}}{R\sqrt{g}} = 1 \text{ h } 37 \text{ min } 22 \text{ s}$$

3/82 The motion of the 0.5-kg pin A in the circular slot is controlled by the guide B which is being elevated by its lead screw with a constant upward velocity v_0 of 2 m/s for an interval of its motion. Calculate the force P exerted on the pin by the circular guide as the pin passes the position for which $\theta = 30°$. Friction is negligible.   *Ans.* $P = 14.22$ N

3/83 The centrifugal pump with smooth radial vanes rotates about its vertical axis with an angular velocity $\dot{\theta} = \omega$. Find the force N exerted by a vane on a particle of mass m as it moves out along the vane. The particle is introduced at $r = r_0$ without radial velocity. Assume that the particle contacts the side of the vane only.

84 A flexible bicycle-type chain of length $\pi r/2$ has a mass ρ per unit length and is released after being held by its upper end in an initial rest condition in the smooth circular channel. Determine the acceleration a_t which all links experience just after release. Also find the expression for the tension T in the chain as a function of θ for the condition immediately following release. Isolate a differential element of the chain as a free body and apply the appropriate motion equation.

$$Ans.\ a_t = \frac{2g}{\pi},\ T = \rho g r\left[\frac{2\theta}{\pi} - \sin\theta\right]$$

Problem 3/84

85 An experimental rocket which operates above the earth's atmosphere is programmed for a constant pitching rate so that $\theta = kt$, where k is a constant and t is the time. Take $x = y = 0$ at the position for which $t = 0$ and determine x and y as functions of the time. Assume that the gravitational attraction mg and thrust T remain constant in magnitude. The vertical y-velocity of the rocket was u when $t = 0$.

$$Ans.\ x = \frac{T}{mk}\left(t - \frac{1}{k}\sin kt\right)$$

$$y = \frac{T}{mk^2}(1 - \cos kt) - \tfrac{1}{2}gt^2 + tu$$

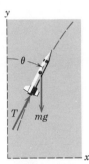

Problem 3/85

86 The slotted arm of Prob. 2/134, repeated here, rotates in a horizontal plane around the fixed cam with a constant counterclockwise velocity $\omega = 20$ rad/s. The spring has a stiffness of 5.4 kN/m and is uncompressed with $\theta = 0$. The cam has the shape $r = b - c\cos\theta$. If $b = 100$ mm, $c = 75$ mm, and the smooth roller A has a mass of 0.5 kg, find the force P exerted on A by the smooth sides of the slot for the position in which $\theta = 60°$.

Ans. $P = 231$ N

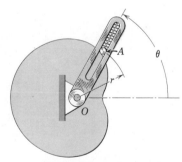

Problem 3/86

87 A small object is released from rest at A and slides with friction down the circular path. If the coefficient of friction is $\frac{1}{5}$, determine the velocity of the object as it passes B. [*Hint:* Write the equations of motion in the n- and t-directions, eliminate N, and substitute $v\,dv = a_t r\,d\theta$, first changing variables to $u = v^2$ so that $du = 2a_t r\,d\theta$. The resulting equation is a linear nonhomogeneous differential equation of the form $dy/dx + f(x)y = g(x)$, the solution of which is well known.]

Ans. $v = 5.52$ m/s

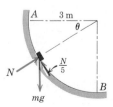

Problem 3/87

SECTION B. WORK AND ENERGY

3/6 WORK AND KINETIC ENERGY.   In the previous two articles we applied Newton's second law $\mathbf{F} = m\mathbf{a}$ to various problems of particle motion to establish the instantaneous relationship between the net force acting on a particle and the resulting acceleration of the particle. When intervals of motion were involved where the change in velocity or the corresponding displacement of the particle was required, we integrated the computed acceleration over the interval by using the appropriate kinematical equations.

There are two general classes of problems where the cumulative effects of unbalanced forces acting on a particle over an interval of motion are of interest to us. These cases involve, respectively, integration of the forces with respect to the displacement of the particle and integration of the forces with respect to the time they are applied. We may incorporate the results of these integrations directly into the governing equations of motion so that it becomes unnecessary to solve directly for the acceleration. Integration with respect to displacement leads to the equations of work and energy, which are the subject of this article. Integration with respect to time leads to the equations of impulse and momentum, discussed in Art. 3/8.

(a) Work. The quantitative meaning of the term work will now be developed.° The work done by a force $\mathbf{F}$ during a differential displacement $d\mathbf{s}$ of its point of application O, Fig. 3/2a, is by definition the scalar quantity

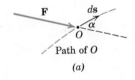

(a)

$$dU = \mathbf{F} \cdot d\mathbf{s}$$

The magnitude of this dot product of the vector force and the vector displacement is $dU = F\,ds\cos\alpha$, where α is the angle between $\mathbf{F}$ and $d\mathbf{s}$. This expression may be interpreted as the displacement multiplied by the force component $F_t = F\cos\alpha$ in the direction of the displacement, as represented by the dotted lines in Fig. 3/2b. Alternatively the work dU may be interpreted as the force multiplied by the displacement component $ds\cos\alpha$ in the direction of the force, as represented by the full lines in Fig. 3/2b. With this definition of work, it should be noted that the component $F_n = F\sin\alpha$ normal to the displacement does no work. Work is positive if the working component F_t is in the direction of the displacement and negative if it is in the opposite direction.

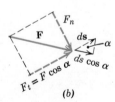

(b)

Figure 3/2

In SI units work has the units of force (N) times displacement (m) or N·m. This unit is given the special name *joule* (J) which is defined

° The concept of work was also developed in the study of virtual work in Chapter 7 of Vol. 1 *Statics.*

as the work done by a force of 1 N moving through a distance of 1 m
in the direction of the force. Consistent use of the joule for work (and
energy) rather than the units $N \cdot m$ will avoid possible ambiguity with
the units of moment of a force or torque which are also written $N \cdot m$.

In the U.S. customary system work has the units of ft-lb. Dimensionally, work and moment are the same. In order to distinguish
between the two quantities it is recommended that work be expressed
as foot pounds (ft-lb) and moment as pound feet (lb-ft). It should be
noted that work is a scalar as given by the dot product and involves
the product of a force and a distance, both measured along the same
line. Moment, on the other hand, is a vector as given by the cross
product and involves the product of force and distance measured at
right angles to the force.

During a finite movement s of the point of application of a force
the force does an amount of work equal to

$$U = \int \mathbf{F} \cdot d\mathbf{s} = \int (F_x \, dx + F_y \, dy + F_z \, dz)$$

or

$$U = \int F_t \, ds$$

In order to carry out this integration it is necessary to know the
relation between the force components and their respective coordinates or the relations between F_t and s. If the functional relationship
is not known as a mathematical expression which can be integrated
but is specified in the form of approximate or experimental data, then
the work may be evaluated by carrying out a numerical or graphical
integration which would be represented by the area under the curve
of F_t versus s, as indicated in Fig. 3/3.

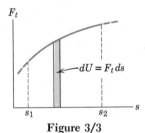

Figure 3/3

A common example of the work done by a variable force is
found with the extension or compression of an elastic member. In the
case of an elastic spring of stiffness k and negligible mass, Fig. 3/4,
the force F supported by the spring at any deformation x, either
compression or extension, is $F = kx$. Thus the work U done *on* a
spring during an increase of its compression from x_1 to x_2 or during an
increase of its extension from x_1 to x_2 is

$$U = \int F \, dx = \int_{x_1}^{x_2} kx \, dx = \tfrac{1}{2}k(x_2^2 - x_1^2)$$

which is represented by the trapezoidal area on the F-x diagram. We
note that *positive* work is done on the spring when it is being compressed or extended since in each case the force and the displacement
are in the same direction. In the reverse process when the spring is
being relieved from its compression or extension, *negative* work is
done on the spring since the force is in the direction opposite to the
displacement.

The spring, in turn, does work on the body to which its movable
end is attached. Since the force exerted by the spring on this body is

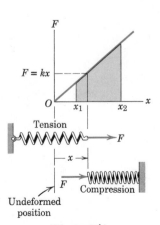

Figure 3/4

in the direction opposite to that exerted by the body on the spring, it follows that the spring does *negative* work

$$U = -\tfrac{1}{2}k(x_2{}^2 - x_1{}^2)$$

on the body which compresses or stretches it. Conversely when the spring is relieved from its compression or extension, it does *positive* work on the movable body to which it is attached.

The expression $F = kx$ is actually a static relationship which is true only when elements of the spring have no acceleration. Dynamic behavior of a spring when its mass is accounted for is a fairly complex problem which will not be treated here. We shall assume that the mass of the spring is small compared with the masses of other accelerating parts of the system, in which case the linear static relationship will not involve appreciable error.

We now consider the work done on a particle of mass m, Fig. 3/5, moving along a curved path under the action of the force $\mathbf{F}$ which stands for the resultant $\Sigma\mathbf{F}$ of all forces acting on the particle. The position of m is established by the position vector $\mathbf{r}$, and its displacement along its path during time dt is represented by the change $d\mathbf{r}$ in its position vector. The work dU done by $\mathbf{F}$ during this displacement is defined by the dot product

$$dU = \mathbf{F} \cdot d\mathbf{r}$$

This expression represents the scalar magnitude $F_t|d\mathbf{r}| = F_t \, ds = (F\cos\alpha)\,ds$ where s is the scalar distance measured along the curve.

If F_t is in the direction of $d\mathbf{r}$ as pictured, the work is positive. If F_t is in the direction opposite to $d\mathbf{r}$, the work is negative. The component F_n of the force normal to the direction of the path can do no work since its dot product with the displacement $d\mathbf{r}$ is zero.

The work done by $\mathbf{F}$ during a finite movement of the particle from point 1 to point 2 is

$$U = \int \mathbf{F} \cdot d\mathbf{r} = \int_{s_1}^{s_2} F_t \, ds$$

where the limits specify the initial and final end points of the interval of motion involved. When we substitute Newton's second law $\mathbf{F} = m\mathbf{a}$, the expression for the work of all forces becomes

$$U = \int \mathbf{F} \cdot d\mathbf{r} = \int m\mathbf{a} \cdot d\mathbf{r}$$

But $\mathbf{a} \cdot d\mathbf{r} = a_t \, ds$ where a_t is the tangential component of the acceleration of m and ds is the magnitude of $d\mathbf{r}$. In terms of the velocity v of the particle, Eq. 2/3 gives $a_t \, ds = v \, dv$. Thus the expression for the work of $\mathbf{F}$ becomes

$$U = \int \mathbf{F} \cdot d\mathbf{r} = \int mv \, dv$$

or

$$U = \int \mathbf{F} \cdot d\mathbf{r} = \tfrac{1}{2}m(v_2{}^2 - v_1{}^2) \tag{3/9}$$

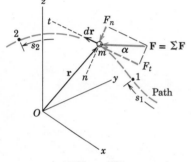

Figure 3/5

where the integration is carried out between points 1 and 2 along the curve at which points the velocities have the magnitudes v_1 and v_2, respectively.

(b) Kinetic Energy. The *kinetic energy* T of the particle is defined as

$$T = \tfrac{1}{2}mv^2 \tag{3/10}$$

and is the total work which must be done on the particle to bring it from a state of rest to a velocity v. Kinetic energy T is a scalar quantity with the units of N·m or *joules* (J) in SI units and ft-lb in U.S. customary units. Kinetic energy is *always* positive regardless of the direction of the velocity. Equation 3/9 may be restated simply as

$$U = \Delta T \tag{3/11}$$

which is the *work-energy equation* for a particle.

The equation states that the *total work done* by all forces acting on a particle during an interval of its motion equals the corresponding *change in kinetic energy* of the particle. Although T is always positive, the change ΔT may be positive, negative, or zero.

We now see from Eq. 3/11 that a major advantage of the method of work and energy is that it avoids the necessity of computing the acceleration and leads directly to the velocity changes as functions of the forces which do work. Further, the work-energy equation involves only those forces which do work and thus give rise to changes in the magnitude of the velocities.

We consider now two particles joined together by a connection which is frictionless and incapable of any deformation. The forces in the connection constitute a pair of equal and opposite forces, and the points of application of these forces necessarily have identical displacement components in the direction of the forces. Hence the net work done by these internal forces is zero during any movement of the system of the two connected particles. Thus Eq. 3/11 is applicable to the entire system, where U is the total or net work done on the system by forces external to it and ΔT is the change in the total kinetic energy of the system. The total kinetic energy is the sum of the kinetic energies of both elements of the system. It may now be observed that a further advantage of the work-energy method is that it permits the analysis of a system of particles joined in the manner described without dismembering the system.

Application of the work-energy method calls for an isolation of the particle or system under consideration. For a single particle a *free-body diagram* showing all externally applied forces should be drawn. For a system of particles rigidly connected without springs an *active-force diagram* which shows only those external forces which do work (active forces) on the system may be drawn.°

° The active-force diagram was introduced in the method of virtual work in statics. See Chapter 7 of Vol. 1 *Statics*.

(c) *Power.* The capacity of a machine is measured by the time rate at which it can do work or deliver energy. The total work or energy output is not a measure of this capacity since a motor, no matter how small, can deliver a large amount of energy if given sufficient time. On the other hand a large and powerful machine is required to deliver a large amount of energy in a short period of time. Thus the capacity of a machine is rated by its *power*, which is defined as the *time rate of doing work.*

Accordingly, the power P developed by a force $\mathbf{F}$ which does an amount of work U is $P = dU/dt = \mathbf{F} \cdot d\mathbf{r}/dt$. Since $d\mathbf{r}/dt$ is the velocity $\mathbf{v}$ of the point of application of the force, we have

$$\boxed{P = \mathbf{F} \cdot \mathbf{v}} \qquad (3/12)$$

Power is clearly a scalar quantity, and in SI units it has the units of $N \cdot m/s = J/s$. The special unit for power is the *watt* (W), which equals one joule per second (J/s). In U.S. customary units the unit for mechanical power is the *horsepower* (hp). These units and their numerical equivalences are

$$1\,W = 1\,J/s$$

$$1\,hp = 550\,ft\text{-}lb/sec = 33{,}000\,ft\text{-}lb/min$$

$$1\,hp = 746\,W = 0.746\,kW$$

(d) *Efficiency.* The ratio of the work done *by* a machine to the work done *on* the machine during the same time interval is called the *mechanical efficiency* e_m of the machine. This definition assumes that the machine operates uniformly so that there is no accumulation or depletion of energy within it. Efficiency is always less than unity since every device operates with some loss of energy and since energy cannot be created within the machine. In mechanical devices which involve moving parts there will always be some loss of energy due to the negative work of kinetic friction forces. This work is converted to heat energy which in turn is dissipated to the surroundings. The mechanical efficiency at any instant of time may be expressed in terms of mechanical power P by

$$e_m = \frac{P_{\text{output}}}{P_{\text{input}}} \qquad (3/13)$$

In addition to energy loss by mechanical friction there may also be electrical and thermal energy loss in which case the *electrical efficiency* e_e and *thermal efficiency* e_t are also involved. The *overall efficiency* e in such instances would be

$$e = e_m e_e e_t$$

Sample Problem 3/10

Calculate the velocity v of the 50-kg crate when it reaches the bottom of the chute at B if it is given an initial velocity of 4 m/s down the chute at A. The coefficient of friction is 0.30.

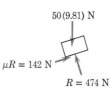

Solution. The free-body diagram of the crate is drawn and includes the normal force R and the kinetic friction force F calculated in the usual manner. The work done by the component of the weight down the plane is positive, whereas that done by the friction force is negative. The total work done on the crate during the prescribed motion is

$[U = Fs]$ $U = [50(9.81)\sin 15° - 142]10 = -152$ J

The change in kinetic energy is

$[T = \frac{1}{2}mv^2]$ $\Delta T = \frac{1}{2}(50)(v^2 - 4^2)$

The work-energy equation gives

$[U = \Delta T]$ $-152 = 25(v^2 - 16)$

$v^2 = 9.93$ (m/s)^2 $v = 3.15$ m/s   *Ans.*

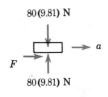

50(9.81) N

$\mu R = 142$ N

$R = 474$ N

① The coefficient of friction is, of course, a kinetic coefficient because motion occurs.

② Since the net work done is negative, we obtain a decrease in the kinetic energy.

Sample Problem 3/11

The flatbed truck, which carries an 80-kg crate, starts from rest and attains a speed of 72 km/h in a distance of 75 m on a level road with constant acceleration. Calculate the work done by the friction force acting on the crate during this interval if the static and kinetic coefficients of friction between the crate and the truck bed are (a) 0.30 and 0.28, respectively, or (b) 0.25 and 0.20, respectively.

Solution. If the crate does not slip on the bed, its acceleration will be that of the truck, which is

$[v^2 = 2as]$ $a = \dfrac{v^2}{2s} = \dfrac{(72/3.6)^2}{2(75)} = 2.67$ m/s^2

Case (a). This acceleration requires a friction force on the block of

$[F = ma]$ $F = 80(2.67) = 213$ N

which is less than the maximum possible value of $\mu_s N = 0.30(80)(9.81) = 235$ N. Therefore the crate does not slip and the work done by the actual static friction force of 213 N is

$[U = Fs]$ $U = 213(75) = 16\,000$ J or 16.0 kJ

Case (b). For $\mu_s = 0.25$ the maximum possible friction force is $0.25(80)(9.81) = 196$ N, which is slightly less than the value of 213 N required for no slipping. Therefore we conclude that the crate slips, and the friction force is governed by the kinetic coefficient and is $F = 0.20(80)(9.81) = 157$ N. The acceleration becomes

$[F = ma]$ $a = F/m = 157/80 = 1.96$ m/s^2

The distances traveled by the crate and the truck are in proportion to their accelerations. Thus the crate has a displacement of $(1.96/2.67)75 = 55.2$ m, and the work done by kinetic friction is

$[U = Fs]$ $U = 157(55.2) = 8660$ J   or   8.66 kJ   *Ans.*

80(9.81) N

F a

80(9.81) N

① We note that static friction forces do no work when the contacting surfaces are both at rest. When they are in motion, however, as in this problem, the static friction force acting on the crate does positive work and that acting on the truck bed does negative work.

② This problem shows that a kinetic friction force can do positive work when the surface which supports the object and generates the friction force is in motion. If the supporting surface is at rest, then the kinetic friction force acting on the moving part always does negative work.

Sample Problem 3/12

The 50-kg block at A is mounted on rollers so that it moves along the fixed horizontal rail with negligible friction under the action of the constant 300-N force in the cable. The block is released from rest at A with the spring to which it is attached extended an initial amount $x_1 = 0.233$ m. The spring has a stiffness $k = 80$ N/m. Calculate the velocity v of the block as it reaches position B.

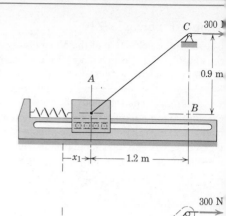

Solution. It will be assumed initially that the stiffness of the spring is small enough to allow the block to reach position B. The active-force diagram for the system composed of the block and the cable is shown for a general position. The spring force $80x$ and the 300-N tension are the only forces external to this system which do work on the system. The force exerted on the block by the rail, the weight of the block, and the reaction of the small pulley on the cable do no work on the system and are not included on the active-force diagram.

As the block moves from $x = 0.233$ m to $x = 0.233 + 1.2 = 1.433$ m, the work done by the spring force acting on the block is negative and equals

① $[U = \int F\,dx]$ $U = -\int_{0.233}^{1.433} 80x\,dx = -40x^2 \Big]_{0.233}^{1.433} = -80.0$ J

The work done on the system by the constant 300-N force in the cable is the force times the net horizontal movement of the cable over pulley C, which is $\sqrt{(1.2)^2 + (0.9)^2} - 0.9 = 0.6$ m. Thus the work done is $300(0.6) = 180$ J. We now apply the work-energy equation to the system and get

$[U = \Delta T]$ $-80.0 + 180 = \tfrac{1}{2}(50)(v^2 - 0)$, $v = 2.0$ m/s   *Ans.*

We take special note of the advantage to our choice of system. If the block alone had constituted the system, the horizontal component of the 300-N cable tension acting on the block would have to be integrated over the 1.2-m displacement. This step would require considerably more effort than was needed in the solution as presented. If there had been appreciable friction between the block and its guiding rail, we would have found it necessary to isolate the block alone in order to compute the variable normal force and, hence, the variable friction force. Integration of the friction force over the displacement would then be required to evaluate the negative work which it would do.

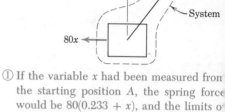

① If the variable x had been measured from the starting position A, the spring force would be $80(0.233 + x)$, and the limits of integration would be 0 and 1.2 m.

Sample Problem 3/13

The power winch A hoists the 800-lb log up the 30° incline at a constant speed of 4 ft/sec. If the power output of the winch is 6 hp, compute the coefficient of friction μ between the log and the incline. If the power is suddenly increased to 8 hp, what is the corresponding instantaneous acceleration a of the log?

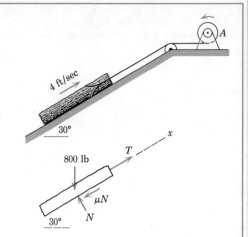

Solution. From the free-body diagram of the log we get $N = 800 \cos 30° = 693$ lb, and the friction force becomes 693μ. For constant speed the forces are in equilibrium so that

$$[\Sigma F_x = 0] \quad T - 693\mu - 800 \sin 30° = 0, \quad T = 693\mu + 400$$

The power output of the winch gives the tension in the cable

$$[P = Tv] \qquad T = P/v = 6(550)/4 = 825 \text{ lb}$$

Substituting T gives

$$825 = 693\mu + 400 \qquad \mu = 0.613 \qquad \textit{Ans.}$$

When the power is increased the tension momentarily becomes

$$[P = Tv] \qquad T = P/v = 8(550)/4 = 1100 \text{ lb}$$

and the corresponding acceleration is

$$[\Sigma F_x = ma_x] \quad 1100 - 693(0.613) - 800 \sin 30° = \frac{800}{32.2} a$$

$$a = 11.07 \text{ ft/sec}^2 \qquad \textit{Ans.}$$

① Note the conversion from horsepower to ft-lb/sec.

② As the speed increases the acceleration will drop until the speed stabilizes at a value higher than 4 ft/sec.

Sample Problem 3/14

A satellite of mass m is put into an elliptical orbit around the earth. At point A its distance from the earth is $h_1 = 500$ km and it has a velocity $v_1 = 30\,000$ km/h. Determine the velocity v_2 of the satellite as it reaches point B, a distance $h_2 = 1200$ km from the earth.

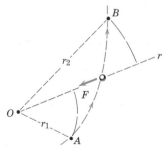

Solution. The satellite is moving outside of the earth's atmosphere so that the only force acting on it is the gravitational attraction of the earth. With the mass and radius of the earth expressed by m_0 and R, respectively, the gravitational law of Eq. 1/2 gives $F = Kmm_0/r^2 = gR^2m/r^2$ when the substitution $Km_0 = gR^2$ is made for the surface values $F = mg$ and $r = R$. The work done by F is due only to the radial component of motion along the line of action of F and is negative for increasing r.

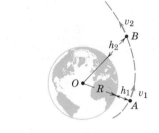

$$U = -\int_{r_1}^{r_2} F\,dr = -mgR^2\int_{r_1}^{r_2} \frac{dr}{r^2} = mgR^2\left(\frac{1}{r_2} - \frac{1}{r_1}\right)$$

The work-energy equation $U = \Delta T$ gives

$$mgR^2\left(\frac{1}{r_2} - \frac{1}{r_1}\right) = \tfrac{1}{2}m(v_2{}^2 - v_1{}^2), \qquad v_2{}^2 = v_1{}^2 + 2gR^2\left(\frac{1}{r_2} - \frac{1}{r_1}\right)$$

Substituting the numerical values gives

$$v_2{}^2 = (30\,000/3.6)^2 + 2(9.81)[(6371)(10^3)]^2\left(\frac{10^{-3}}{6371 + 1200} - \frac{10^{-3}}{6371 + 500}\right)$$

$$= 69.44(10^6) - 10.72(10^6) = 58.73(10^6) \text{ (m/s)}^2$$

$$v_2 = 7663 \text{ m/s or } v_2 = 7663(3.6) = 27\,590 \text{ km/h} \qquad \textit{Ans.}$$

① Note that the result is independent of the mass of the satellite.

② Consult Table C2, Appendix C, to find the radius R of the earth.

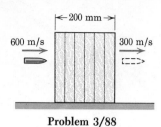

Problem 3/88

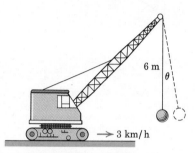

Problem 3/89

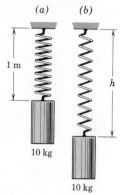

Problem 3/90

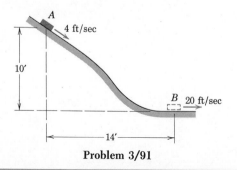

Problem 3/91

PROBLEMS

3/88 A 60-g bullet is fired through a stack of fiberboard sheets 200 mm thick. If the bullet approaches the stack with a velocity of 600 m/s and emerges with a velocity of 300 m/s, calculate the average resistance R to penetration of the bullet over the 200 mm.

Ans. $R = 40.5$ kN

3/89 The crawler wrecking crane is moving with a constant speed of 3 km/h when it is suddenly brought to a stop. Compute the maximum angle through which the cable of the wrecking ball swings.

3/90 The 10-kg cylinder is released from rest in position (a) with the spring unstretched. Determine the distance h in position (b) where the cylinder has reached its lowest position. The stiffness of the spring is $k = 450$ N/m.   *Ans. $h = 1.436$ m*

3/91 The 64.4-lb crate slides down the curved path in the vertical plane. If the crate has a velocity of 4 ft/sec down the incline at A and a velocity of 20 ft/sec at B, compute the work U_f done on the block by friction during the motion from A to B.

3/92 The ball is released from position *A* with a velocity of 2 m/s and swings in a vertical plane. At the bottom position the cord strikes the fixed bar at *B*, and the ball continues to swing in the dotted arc. Calculate the velocity v_C of the ball as it passes position *C*.   *Ans.* $v_C = 2.63$ m/s

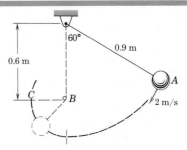

Problem 3/92

3/93 The 15-lb cylindrical collar is released from rest in the position shown and drops onto the spring. Calculate the velocity *v* of the cylinder when the spring has been compressed 2 in.

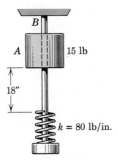

Problem 3/93

3/94 The pressure *p* in a certain rifle barrel varies with the position of the bullet as shown in the graph. If the diameter of the bore is 7.5 mm and the mass of the bullet is 14g, determine the muzzle velocity *v* of the bullet. Neglect the effect of friction in the barrel compared with the force of the gases on the bullet. Observe that one megapascal (MPa) equals 10^6 newtons per square meter (N/m^2).
            Ans. $v = 940$ m/s

3/95 Skid marks left by all four wheels of a car on level pavement are measured by a police officer to be 20 m in length. The driver of the car claims that she was not exceeding the speed limit of 80 km/h before she applied the brakes. Furthermore the driver released the brakes before coming to a stop and was observed to continue at a speed of 40 km/h after her skid. It is known from prevailing conditions that the coefficient of kinetic friction between the tires and the road was at least 0.80. Can it be shown that the driver was guilty of speeding?

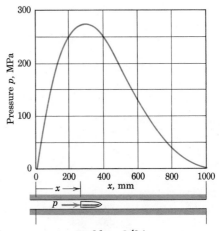

Problem 3/94

3/96 A car is traveling at 50 km/h down a 10 percent grade when the brakes on all four wheels lock. If the kinetic coefficient of friction between the tires and the road is 0.70, find the distance *s* measured along the road which the car skids before coming to a stop.   *Ans.* $s = 16.47$ m

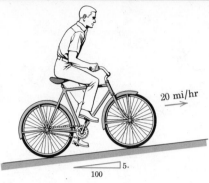

Problem 3/97

3/97 The man and his bicycle together weigh 200 lb. What power P is the man developing in riding up a 5 percent grade at a constant speed of 20 mi/hr?

3/98 The motor unit A is used to hoist the two 100-kg bodies. If the lower one is given a constant upward speed of 0.5 m/s and the wattmeter B shows an electrical input of 1.6 kW, compute the combined electrical and mechanical efficiency of the system.
Ans. $e = 0.92$

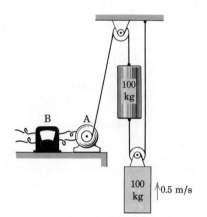

Problem 3/98

3/99 A department store escalator handles a steady load of 36 people per minute in elevating them from the first to the second floor through a vertical rise of 30 ft. The average person weighs 140 lb. If the motor which drives the unit delivers 5 hp, calculate the mechanical efficiency of the system.

3/100 The position vector of a particle is given by $\mathbf{r} = 1.2t\mathbf{i} + 0.9t^2\mathbf{j} - 0.6(t^3 - 1)\mathbf{k}$ where t is the time in seconds from the start of the motion and where $\mathbf{r}$ is expressed in meters. For the condition when $t = 4$ s, determine the power P developed by the force $\mathbf{F} = 60\mathbf{i} - 25\mathbf{j} - 40\mathbf{k}$ N which acts on the particle.
Ans. $P = 1.044$ kW

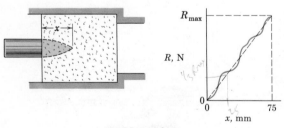

Problem 3/101

3/101 The resistance R to penetration x of a 0.25-kg projectile fired with a velocity of 600 m/s into a certain block of fibrous material is shown in the graph. Represent this resistance by the dotted line and compute the velocity v of the projectile for the instant when $x = 25$ mm if the projectile is brought to rest after a total penetration of 75 mm.

102 Sliders *A* and *B* are of equal mass and are confined to move, respectively, in the vertical and horizontal guides. Their connecting link has negligible mass. If they are released from rest in the position for which $x = y$ and if they slide with negligible friction, calculate the velocity v_A of *A* as it passes the horizontal line through *B*.   *Ans.* $v_A = 3.53$ m/s

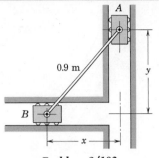

Problem 3/102

103 The 0.5-lb slider moves freely along the fixed curved rod from *A* to *B* in the vertical plane under the action of the constant 1.2-lb tension in the cord. If the slider is released from rest at *A*, calculate its velocity *v* as it reaches *B*.

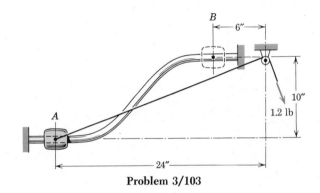

Problem 3/103

104 The 10-kg collar slides freely on the fixed circular guide. Determine the velocity *v* of the collar as it reaches *B* if it is elevated from rest at *A* by the action of a constant 250-N force in the cable.   *Ans.* $v = 8.54$ m/s

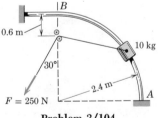

Problem 3/104

105 Small metal blocks are discharged with a velocity of 0.45 m/s to a ramp by the upper conveyor shown. If the coefficient of friction between the blocks and the ramp is 0.30, calculate the angle θ which the ramp must make with the horizontal so that the blocks will transfer without slipping to the lower conveyor moving at the speed of 0.15 m/s.

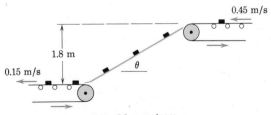

Problem 3/105

80 N

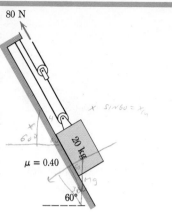

$\mu = 0.40$

60°

Problem 3/106

3/106 The pulleys around which the cables pass have negligible mass and friction. Determine the velocity v of the 20-kg block after it has moved 4 m from rest under the action of the 80-N force. The coefficient of kinetic friction is 0.40.   *Ans. $v = 6.66$ m/s*

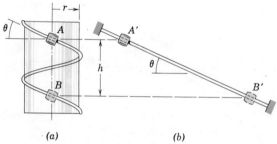

(a)   *(b)*

Problem 3/107

3/107 A bead is released from rest at point A and slides down the wire bent into the form of a helix, as shown in the *a*-part of the figure. The experiment is repeated in the *b*-part of the figure where the wire has been unwrapped to form a straight line with a slope equal to that of the helix. What is the velocity reached by the bead as it passes point B or B' in each case in the absence of friction? In the presence of some friction does the velocity depend on the path, and if so, for which case would the velocity at the lower position be greater?

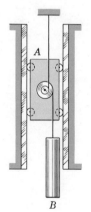

A

B

Problem 3/108

3/108 The slider A has a mass of 10 kg and the cylinder B has a mass of 14 kg. The connecting cables are securely wrapped around the two pulleys of negligible mass which are fastened together and turn as a unit on A. The diameters of the pulleys are in the ratio of 2:1. If the system is released from rest, determine the velocity v of B after it has dropped 4 m. Neglect friction. Note that the top cable has no motion and that A moves up as far as B moves down.   *Ans. $v = 3.62$ m/s*

109 The 300-lb carriage has an initial velocity of 9 ft/sec down the incline at A when a constant force of 110 lb is applied to the hoisting cable as shown. Calculate the velocity of the carriage when it reaches B. Show that in the absence of friction this velocity is independent of whether the initial velocity of the carriage at A was up or down the incline.

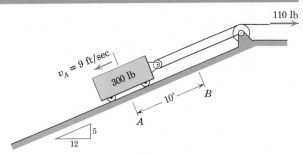

Problem 3/109

110 Determine the velocity v of the 50-kg sliding block after it has moved 1.2 m along the incline from rest under the action of the constant force $P = 480$ N. The light cables are wrapped securely around the integral pulleys of negligible mass. Pulley diameters are in the ratio of 2:1, friction in the bearing is negligible, and the coefficient of friction μ between the block and the incline is 0.40. Note that the center of the pulley moves up the incline the same distance that the point of application of P on the cable moves down the incline.

Ans. $v = 1.764$ m/s

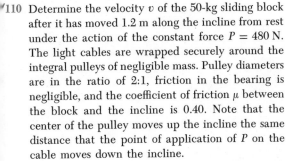

Problem 3/110

111 A small rocket-propelled test vehicle with a total mass of 100 kg starts from rest at A and moves with negligible friction along the track in the vertical plane as shown. If the propelling rocket exerts a constant thrust T of 2 kN from A to position B where it is shut off, determine the distance s which the vehicle rolls up the incline before stopping. The loss of mass due to the expulsion of gases by the rocket is small and may be neglected.

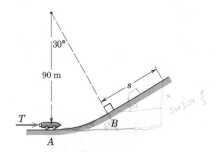

Problem 3/111

112 The 2-kg slider moves along the smooth fixed rod under the action of its weight and a constant externally applied force $\mathbf{F} = -15\mathbf{i} + 10\mathbf{j} + 15\mathbf{k}$ N. If the slider starts from A with a velocity $\mathbf{v}_A = -2\mathbf{i}$ m/s, determine the magnitude of its velocity as it reaches B. *Ans.* $v_B = 4.83$ m/s

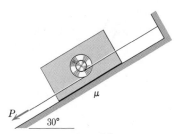

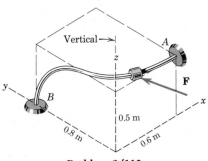

Problem 3/112

113 For the conditions of Problem 3/93 calculate the maximum tension T in the rod at B as the cylindrical collar falls on the spring. The rod weighs 6 lb.

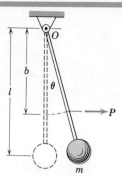

Problem 3/114

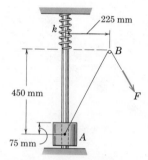

Problem 3/115

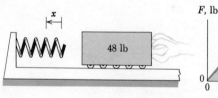

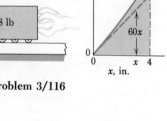

Problem 3/116

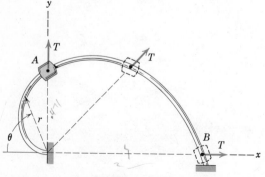

Problem 3/118

3/114 The small sphere of mass m is fastened to the end of the rigid rod of negligible mass freely pivoted at O. A horizontal force P, constant in magnitude and direction, is applied to the rod initially at rest in the vertical position $\theta = 0$. Calculate the velocity of the sphere when $\theta = 30°$ if $P = 20$ N, $m = 2$ kg, $b = 0.6$ m, and $l = 0.8$ m.   *Ans. $v = 1.97$ m/s*

3/115 The 7-kg collar A slides with negligible friction on the fixed vertical shaft. When the collar is released from rest at the bottom position shown, it moves up the shaft under the action of the constant force $F = 200$ N applied to the cable. Calculate the stiffness k which the spring must have if its maximum compression is to be limited to 75 mm. The position of the small pulley at B is fixed.

3/116 Calculate the horizontal velocity v with which the 48-lb carriage must strike the spring in order to compress it a maximum of 4 in. The spring is known as a "hardening" spring, since its stiffness increases with deflection as shown in the accompanying graph.   *Ans. $v = 7.80$ ft/sec*

3/117 A truck with a mass of 6 metric tons accelerates uniformly from 30 km/h to 90 km/h in a distance of 140 m on a level road. The horizontal force required to tow the truck with its engine disengaged is a measure of the frictional resistance and equals 4 percent of the weight of the truck. Determine the power output P of the engine as the truck reaches a speed of 60 km/h.

3/118 The 0.5-kg collar slides with negligible friction along the fixed spiral rod which lies in the vertical plane. The rod has the shape of the spiral $r = 0.3\theta$ where r is in meters and θ is in radians. The collar is released from rest at A and slides to B under the action of a constant radial force $T = 10$ N. Calculate the velocity v of the slider as it reaches B.   *Ans. $v = 5.30$ m/s*

119 The 50-lb slider in the position shown has an initial velocity $v_0 = 2$ ft/sec on the inclined rail and slides under the influence of gravity and friction. The coefficient of friction between the slider and the rail is 0.5. Calculate the velocity of the slider as it passes the position for which the spring is compressed a distance $x = 4$ in. The spring offers a compressive resistance C and is known as a "hardening" spring, since its stiffness increases with deflection as shown in the accompanying graph.   *Ans.* $v = 5.44$ ft/sec

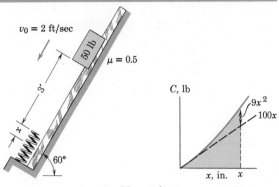

Problem 3/119

120 A bullet of mass m is fired with a velocity v from a rifle at rest relative to the ground. With the rifle attached to a vehicle moving with a speed u an identical bullet is fired in the direction of u. The kinetic energy of the bullet fired from the ground is $\frac{1}{2}mv^2$ and that of the bullet fired from the vehicle is $\frac{1}{2}m(u + v)^2 = \frac{1}{2}mu^2 + \frac{1}{2}mv^2 + muv$. Explain the significance of the muv term. Assume the velocity of the vehicle remains constant during the firing.

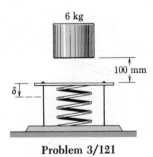

121 The 6-kg cylinder is released from rest in the position shown and falls on the spring which has been initially precompressed 50 mm by the light strap and restraining wires. If the stiffness of the spring is 4 kN/m, compute the additional deflection δ of the spring produced by the falling cylinder before it rebounds.   *Ans.* $\delta = 29.4$ mm

Problem 3/121

122 The 80-lb cylinder is released from rest in the position for which the spring is under a tension of 9 lb. Compute the velocity of the cylinder after it has fallen 6 in. and determine the maximum downward displacement δ of the cylinder. Neglect the mass of the pulleys.   *Ans.* $v = 3.23$ ft/sec, $\delta = 10.33$ in.

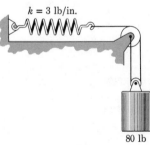

Problem 3/122

123 The car of mass m accelerates on a level road under the action of the driving force F from a speed v_1 to a higher speed v_2 in a distance s. If the engine develops a constant power output P, determine v_2. Treat the car as a particle under the action of the single horizontal force F.

$$Ans.\ v_2 = \left(\frac{3Ps}{m} + v_1{}^3\right)^{1/3}$$

Problem 3/123

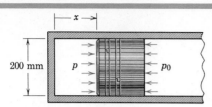

Problem 3/124

3/124 As a test of piston-ring friction the 5-kg piston is released from rest with $x = 50$ mm under an initial absolute pressure $p = 350$ kPa and an atmospheric pressure $p_0 = 100$ kPa. The cylinder pressure p varies according to $pV = $ constant where V is the cylinder volume to the left of the piston. If the piston moves a distance of 350 mm before coming to rest, compute the constant frictional resistance F to the motion of the piston.   *Ans.* $F = 124.8$ N

3/125 Solve Prob. 3/109 if the wheels of the carriage are removed and it slides on the incline with a coefficient of friction of 0.30. Note that the answer would be different if the initial velocity of the sliding carriage were up rather than down the incline.
Ans. $v_B = 7.45$ ft/sec

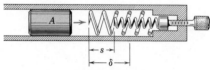

Problem 3/126

3/126 The nest of two springs is used to bring the 0.5-kg plunger A to a stop from a speed of 5 m/s and reverse its direction of motion. The inner spring increases the deceleration, and the adjustment of its position is used to control the exact point at which the reversal takes place. If this point is to correspond to a maximum deflection $\delta = 200$ mm for the outer spring, specify the adjustment of the inner spring by determining the distance s. The outer spring has a stiffness of 300 N/m and the inner one a stiffness of 150 N/m.   *Ans.* $s = 142.3$ mm

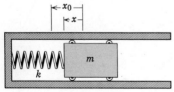

Problem 3/127

3/127 The carriage of mass m is released from rest in its horizontal guide against the spring of stiffness k which has been compressed an amount x_0. Derive an expression for the power P developed by the spring in terms of the deflection x of the spring. Also determine the maximum power P_{max} and the corresponding value of x.

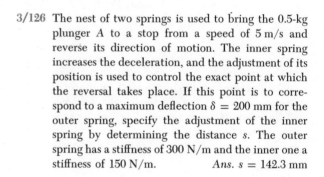

$$\text{Ans. } P = kx\sqrt{\frac{k}{m}(x_0{}^2 - x^2)}$$

$$P_{max} = \frac{k}{2}\sqrt{\frac{k}{m}}x_0{}^2 \text{ at } x = \frac{x_0}{\sqrt{2}}$$

3/7 POTENTIAL ENERGY. In the previous article on work and kinetic energy a particle or a combination of joined particles was isolated and the work done by gravity forces and by spring forces acting on the particle or system was determined along with the work done by other externally applied forces when evaluating U in the work-energy equation. In the present article we shall treat the work done by gravity forces and by spring forces by introducing the concept of *potential energy*. This concept will simplify the analysis of many problems.

(*a*) *Gravitational Potential Energy.* We consider, first, the motion of a particle of mass m in close proximity to the earth's surface where the gravitational attraction (weight) mg is essentially constant, Fig. 3/6*a*. The *gravitational potential energy* V_g of the particle is defined as the work mgh done *against* the gravitational field to elevate the particle a distance h above some arbitrary reference plane where V_g is taken to be zero. Thus we write the potential energy as

$$V_g = mgh \tag{3/14}$$

This work is called potential energy because it may be converted into energy if the particle is allowed to do work on a supporting body while it returns to its lower original datum plane. In going from one level at $h = h_1$ to a higher level at $h = h_2$ the *change* in potential energy becomes

$$\Delta V_g = mg(h_2 - h_1) = mg\Delta h$$

The corresponding work done *by* the gravitational force on the particle is $-mg\Delta h$. Thus the work done by the gravitational force is the negative of the change in potential energy.

When large changes in altitude in the earth's field are encountered, Fig. 3/6*b*, the gravitational force $Kmm_0/r^2 = mgR^2/r^2$ is no longer constant. The work done *against* this force to change the radial position of the particle from r to r' is the change $V_g' - V_g$ in gravitational potential energy

$$\int_r^{r'} mgR^2 \frac{dr}{r^2} = mgR^2 \left(\frac{1}{r} - \frac{1}{r'} \right) = V_g' - V_g$$

It is customary to take $V_g' = 0$ when $r' = \infty$, so that with this datum we have

$$V_g = -\frac{mgR^2}{r} \tag{3/15}$$

In going from r_1 to r_2 the corresponding change in potential energy is

$$\Delta V_g = mgR^2 \left(\frac{1}{r_1} - \frac{1}{r_2} \right)$$

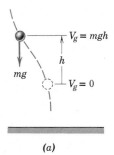

(*a*)

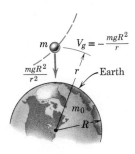

(*b*)

Figure 3/6

which, again, is the *negative* of the work done *by* the gravitational force. We note that the potential energy of a given particle depends only on its position, h or r, and not on the particular path it followed in reaching that position.

(b) *Elastic Potential Energy.* The second example of potential energy is found in the deformation of an elastic body. For the one-dimensional elastic spring of stiffness k, which we discussed in Art. 3/6 and illustrated in Fig. 3/4, the force supported by the spring at any deformation x, compression or extension, from its undeformed position is $F = kx$. Thus we found that the work done on an elastic spring during an interval of compression or extension is

$$\int F \, dx = \int_{x_1}^{x_2} kx \, dx = \tfrac{1}{2}kx_2{}^2 - \tfrac{1}{2}kx_1{}^2$$

which represents the trapezoidal area in the *F-x* diagram of Fig. 3/4. We noted that positive work is done on a spring by the force which deforms it whether it is extended or compressed. It follows that the work done by the equal and opposite force on the body to which the spring is attached is negative. Conversely, during a release from its tension or compression, negative work is done on the spring which means that the spring does positive work on the body to which it is attached. The work done on a spring to deform it an amount x from its undeformed state equals the recoverable energy stored in the spring. This energy is known as the *elastic potential energy* of the spring and is

$$\boxed{V_e = \tfrac{1}{2}kx^2} \tag{3/16}$$

The change in potential energy of the spring is

$$\Delta V_e = \tfrac{1}{2}k(x_2{}^2 - x_1{}^2)$$

and will be positive, negative, or zero, respectively, when x_2 is greater than, less than, or equal to x_1.

The work done *by* the spring *on* the body to which its movable end is attached is the negative of the work done *on* the spring. Therefore the work done by the spring on the body may be replaced by the negative of the potential energy term provided that the spring is now included within the system.

(c) *Work-Energy Equation.* We now modify the work-energy equation to account for the potential energy terms. If U stands for the work of all external forces *other* than gravitational forces and spring forces, we may write Eq. 3/11 as $U + (-\Delta V_g) + (-\Delta V_e) = \Delta T$ or

$$\boxed{U = \Delta T + \Delta V_g + \Delta V_e} \tag{3/17}$$

This alternative form of the work-energy equation is often far more convenient to use than Eq. 3/11, since the work of both gravity and spring forces is accounted for by focusing attention on the end-point

positions of the particle and on the length of the elastic spring. The
path followed between these end-point positions is of no consequence
in the evaluation of ΔV_g and ΔV_e.

To help clarify the difference between the use of Eqs. 3/11 and
3/17, Fig. 3/7 shows schematically a particle of mass m constrained
to move along a fixed path under the action of forces F_1 and F_2, the
gravitational force $W = mg$, the spring force F, and the normal
reaction N. In the b-part of the figure the particle is isolated with its
free-body diagram, and the work done by each of the forces F_1, F_2, W,
and the spring force $F = kx$ is evaluated for the interval of motion in
question, say from A to B, and equated to the change ΔT in kinetic
energy using Eq. 3/11. The constraint reaction N, if normal to the
path, will do no work. In the c-part of the figure for the alternative
approach the spring is included as a part of the isolated system. The
work done during the interval by F_1 and F_2 constitutes the U-term of
Eq. 3/17 with the changes in elastic and gravitational potential
energies included on the energy side of the equation. We note with
the first approach that the work done by $F = kx$ could require a
somewhat awkward integration to account for the changes in magnitude and direction of F as the particle moved from A to B. With the
second approach, however, only the initial and final lengths of the
spring would be required to evaluate ΔV_e.

We may rewrite the alternative work-energy relation, Eq. 3/17,
for a particle-and-spring system as

$$U = \Delta(T + V_g + V_e) = \Delta E \qquad (3/17a)$$

where $E = T + V_g + V_e$ is the total mechanical energy of the particle and its attached linear spring. Equation 3/17a states that the net
work done on the system by all forces other than gravitational forces
and elastic forces equals the change in the total mechanical energy of
the system. For problems where the only forces are gravitational,
elastic, and nonworking constraint forces, the U-term will be zero,
and the energy equation becomes merely

$$\Delta E = 0 \qquad \text{or} \qquad E = \text{const.} \qquad (3/18)$$

When E is constant, it is seen that transfers of energy between kinetic
and potential may take place as long as the total mechanical energy
$T + V_g + V_e$ does not change. Equation 3/18 expresses the *law of
conservation of dynamical energy*.

(d) *Conservative Force Fields.* We have observed that the work
done against a gravitational or an elastic force depends only on the net
change of position and not on the particular path followed in reaching
the new position. Forces with this characteristic are associated with
conservative force fields which possess an important mathematical
property. Consider a force field where the force $\mathbf{F}$ is a function of the

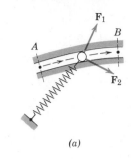

(a)

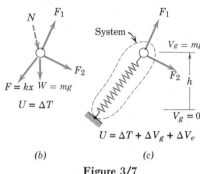

(b)   (c)

Figure 3/7

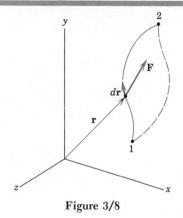

Figure 3/8

coordinates, Fig. 3/8. The work done by $\mathbf{F}$ during a displacement $d\mathbf{r}$ of its point of application is $dU = \mathbf{F} \cdot d\mathbf{r}$, and the total work done along its path from 1 to 2 is

$$U = \int \mathbf{F} \cdot d\mathbf{r} = \int (F_x \, dx + F_y \, dy + F_z \, dz)$$

The integral $\int \mathbf{F} \cdot d\mathbf{r}$ is a line integral dependent, in general, upon the particular path followed between any two points 1 and 2 in space. If, however, $\mathbf{F} \cdot d\mathbf{r}$ is an *exact differential*° $-dV$ of some scalar function V of the coordinates, then

$$U = \int_{V_1}^{V_2} -dV = -(V_2 - V_1) \tag{3/19}$$

which depends only on the end points of the motion and which is thus *independent* of the path followed. The minus sign before dV is arbitrary but is chosen to agree with the customary designation of the sign of potential energy change in the earth's gravity field. If V exists, the differential change in V may be written as

$$dV = \frac{\partial V}{\partial x} dx + \frac{\partial V}{\partial y} dy + \frac{\partial V}{\partial z} dz$$

Comparison with $-dV = \mathbf{F} \cdot d\mathbf{r} = F_x \, dx + F_y \, dy + F_z \, dz$ gives us

$$F_x = -\frac{\partial V}{\partial x} \qquad F_y = -\frac{\partial V}{\partial y} \qquad F_z = -\frac{\partial V}{\partial z}$$

The force may also be written as the vector

$$\mathbf{F} = -\nabla V \tag{3/20}$$

where the symbol ∇ stands for the vector operator "del" which is

$$\nabla = \mathbf{i} \frac{\partial}{\partial x} + \mathbf{j} \frac{\partial}{\partial y} + \mathbf{k} \frac{\partial}{\partial z}$$

The quantity V is known as the *potential function*, and the expression ∇V is known as the *gradient of the potential function*.

When the components of a force are derivable from a potential in the manner described, the force is said to be *conservative*, and it follows that the work done by $\mathbf{F}$ between any two points is independent of the path followed.

° Recall that a function $d\phi = P \, dx + Q \, dy + R \, dz$ is an exact differential in the coordinates x-y-z if

$$\frac{\partial P}{\partial y} = \frac{\partial Q}{\partial x}, \qquad \frac{\partial P}{\partial z} = \frac{\partial R}{\partial x}, \qquad \frac{\partial Q}{\partial z} = \frac{\partial R}{\partial y}.$$

Sample Problem 3/15

The 10-kg slider A moves with negligible friction up the inclined guide. The attached spring has a stiffness of 60 N/m and is stretched 0.6 m in position A where the slider is released from rest. The 250-N force is constant and the pulley offers negligible resistance to the motion of the cord. Calculate the velocity v of the slider as it passes point C.

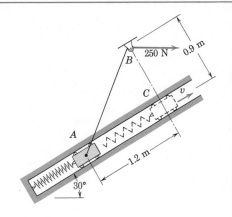

Solution. The slider and inextensible cord together with the attached spring will be analyzed as a system, which permits the use of Eq. 3/17. The only force acting on this system which does work is the 250-N tension applied to the cord. While the slider moves from A to C the point of application of the 250-N force moves a distance of $\overline{AB} - \overline{BC}$ or $1.5 - 0.9 = 0.6$ m. Thus

$$U = 250(0.6) = 150 \text{ J}$$

The change in kinetic energy of the slider is

$$\Delta T = \tfrac{1}{2}m(v^2 - v_0^2) = \tfrac{1}{2}(10)v^2$$

where the initial velocity v_0 is zero. The change in gravitational potential energy is

$$\Delta V_g = mg(\Delta h) = 10(9.81)(1.2 \sin 30°) = 58.9 \text{ J}$$

The change in elastic potential energy is

$$\Delta V_e = \tfrac{1}{2}k(x_2^2 - x_1^2) = \tfrac{1}{2}(60)([1.2 + 0.6]^2 - [0.6]^2) = 86.4 \text{ J}$$

Substitution into the alternative work-energy equation gives

$$[U = \Delta T + \Delta V_g + \Delta V_e] \qquad 150 = \tfrac{1}{2}(10)v^2 + 58.9 + 86.4$$
$$v = 0.974 \text{ m/s} \qquad\qquad \textit{Ans.}$$

① The reactions of the guides on the slider are normal to the direction of motion and do no work.

② Since the center of mass of the slider has an upward component of displacement, ΔV_g is positive.

③ Be very careful not to make the mistake of using $\tfrac{1}{2}k(x_2 - x_1)^2$ for ΔV_e. We want the difference of the squares and not the square of the difference.

Sample Problem 3/16

The 6-lb slider is released from rest at point A and slides with negligible friction in a vertical plane along the circular rod. The attached spring has a stiffness of 2 lb/in. and has an unstretched length of 24 in. Determine the velocity of the slider as it passes position B.

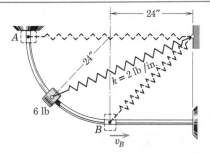

Solution. The work done by the weight and the spring force on the slider will be treated as changes in the potential energies, and the reaction of the rod on the slider is normal to motion and does no work. Hence $U = 0$. The changes in the potential and kinetic energies for the system of slider and spring are

$$\Delta V_e = \tfrac{1}{2}k(x_B^2 - x_A^2) = \tfrac{1}{2}(2)\{(24[\sqrt{2} - 1])^2 - (24)^2\} = -477 \text{ in.-lb}$$

$$\Delta V_g = W\Delta h = 6(-24) = -144 \text{ in.-lb}$$

$$\Delta T = \frac{1}{2}m(v_B^2 - v_A^2) = \frac{1}{2}\frac{6}{(32.2)(12)}(v_B^2 - 0) = \frac{v_B^2}{128.8}$$

$$[\Delta T + \Delta V_g + \Delta V_e = 0] \quad \frac{v_B^2}{128.8} - 144 - 477 = 0, \ v_B = 283 \text{ in./sec} \quad \textit{Ans.}$$

① Note that if we evaluated the work done by the spring force acting on the slider by means of the integral $\int \mathbf{F} \cdot d\mathbf{r}$ it would necessitate a lengthy computation to account for the change in the magnitude of the force along with the change in the angle between the force and the tangent to the path. Note further that v_B depends only on the end conditions of the motion and does not require knowledge of the shape of the path since the slider moves in a conservative field of force.

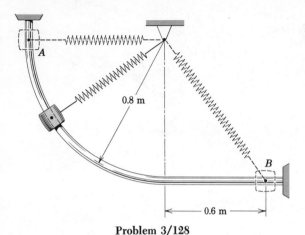

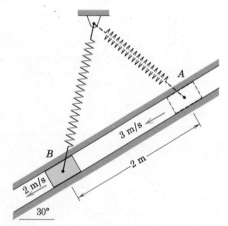

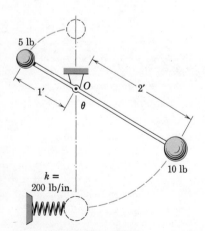

Problem 3/128

Problem 3/129

Problem 3/130

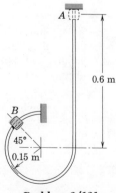

Problem 3/131

PROBLEMS

3/128 The spring has an unstretched length of 0.4 m and a stiffness of 200 N/m. The 3-kg slider and attached spring are released from rest at A. Calculate the velocity v of the slider as it reaches B in the absence of friction.   *Ans. $v = 1.54$ m/s*

3/129 The 4-kg block has a velocity of 3 m/s at A and slides down the inclined guide where it has a velocity of 2 m/s at B. The attached spring has a stiffness of 120 N/m and is stretched 0.6 m at A and 1.0 m at B. Calculate the work done by friction between A and B and, hence, the loss of energy.

3/130 The light rod is pivoted at O and carries the 5- and 10-lb particles. If the rod is released from rest at $\theta = 60°$ and swings in the vertical plane, calculate (a) the velocity v of the 10-lb particle just before it hits the spring in the dotted position and (b) the maximum compression x of the spring. Assume that x is small so that the position of the rod when the spring is compressed is essentially vertical.
Ans. (a) $v = 6.55$ ft/sec, (b) $x = 0.949$ in.

3/131 A bead with a mass of 0.25 kg is released from rest at A and slides down and around the fixed smooth wire. Determine the force N between the wire and the bead as it passes point B.

132 The light rod and its fixed 5-lb ball are released from rest in the 45° position and rotate in the vertical plane about O under the action of the constant 3-lb horizontal force. Compute the spring stiffness k necessary to stop the rod in the vertical position with the spring compressed 2 in. Why is the result independent of b? Treat the ball as a particle.

Ans. $k = 13.89$ lb/in.

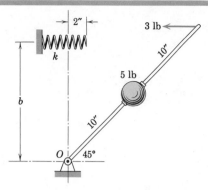

Problem 3/132

133 The collar has a mass of 2 kg and is attached to the light spring which has a stiffness of 24 N/m and an unstretched length of 1.5 m. The collar is released from rest at A and slides up the smooth rod under the action of the constant 40-N force. Calculate the velocity v of the collar as it passes position B.

134 The 2.5-kg sliding collar C with attached spring moves with friction from A to B along the fixed rod. If C has a velocity of 1.8 m/s at A and a velocity of 2.4 m/s as it reaches B, determine the frictional energy loss U_f. The spring has a stiffness of 30 N/m and an unstretched length of 0.9 m. Also calculate the average friction force F_{av} between the collar and the rod over the distance of travel. The x-y plane is horizontal. Ans. $U_f = -8.69$ J, $F_{av} = 4.97$ N

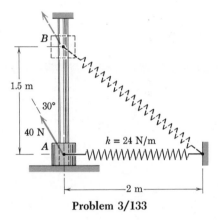

Problem 3/133

135 The two wheels consisting of hoops and spokes of negligible mass rotate about their respective centers and are pressed together sufficiently to prevent any slipping. The 3-lb and 2-lb eccentric masses are mounted on the rims of the wheels. If the wheels are given a slight nudge from rest in the equilibrium positions shown, compute the angular velocity $\dot{\theta}$ of the larger of the two wheels when it has revolved through a quarter of a revolution and put the eccentric masses in the dotted positions shown. Note that the angular velocity of the small wheel is twice that of the large wheel. Neglect any friction in the wheel bearings.

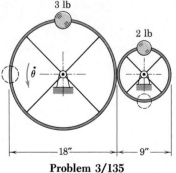

Problem 3/135

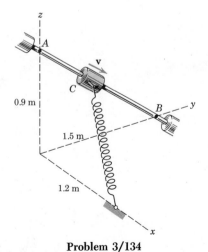

Problem 3/134

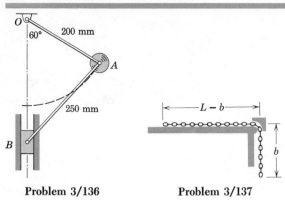

Problem 3/136

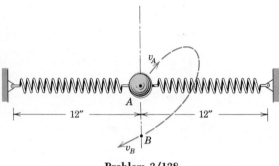

Problem 3/137

3/136 The small bodies A and B each of mass m are connected and supported by the pivoted links of negligible mass. If A is released from rest in the position shown, calculate its velocity v_A as it crosses the vertical center line. Neglect any friction.

Ans. $v_A = 2.30$ m/s

3/137 The figure for Prob. 3/33 is shown again here. Solve for the velocity v of the chain as the last link leaves the edge if the chain is released from rest in the position shown. Friction is negligible.

Problem 3/138

3/138 The 3-lb ball is given an initial velocity $v_A = 8$ ft/sec in the vertical plane at position A where the two horizontal attached springs are unstretched. The ball follows the dotted path shown and crosses point B which is 5 in. directly below A. Calculate the velocity v_B of the ball at B. Each spring has a stiffness of 10 lb/in.   *Ans.* $v_B = 8.54$ ft/sec

3/139 The chain of length L is released from rest on the smooth incline with $x = 0$. Determine the velocity v of the links in terms of x.

Problem 3/139

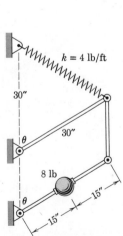

Problem 3/140

3/140 The 8-lb sphere is rigidly mounted on the lower of the two hinged rods of negligible mass which lie in the vertical plane. The spring has an unstretched length of 30 in. and a stiffness of 4 lb/ft. If the assembly is released from rest in the position for which $\theta = 60°$, determine the velocity v of the sphere when the position $\theta = 90°$ is reached.

Ans. $v = 4.79$ ft/sec

141 The rope shown in the *a*-part of the figure is re-leased from rest in the smooth semicircular guide and acquires a velocity *v* when all of it is clear of the guide as seen in the *b*-part of the figure. Find *v*. (*Hint:* In finding the potential energy change imagine that the semicircular portion of the rope in the initial position becomes the bottom part of the rope of equivalent length in the final position.)

Ans. $v = 4.92$ m/s

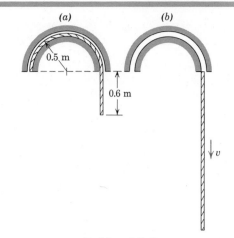

Problem 3/141

142 The ball of mass *m* is supported by the hinged bars of negligible mass and by the two springs each of stiffness *k*. Each spring has an unstretched length *b*. If the assembly is released from rest in the position for which $\theta = 0$, determine the rate $\dot{\theta}$ at which the bars are swinging when $\theta = 90°$. What limitation is placed on the stiffness *k* to ensure that the bottom position is reached?

143 The shank of the 5-lb vertical plunger occupies the dotted position when resting in equilibrium against the spring of stiffness $k = 10$ lb/in. The upper end of the spring is welded to the plunger, and the lower end is welded to the base plate. If the plunger is lifted $1\frac{1}{2}$ in. above its equilibrium position and released from rest, calculate its velocity *v* as it strikes the button *A*. Friction is negligible.

Ans. $v = 3.43$ ft/sec

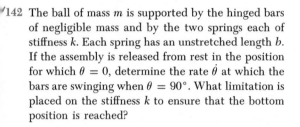

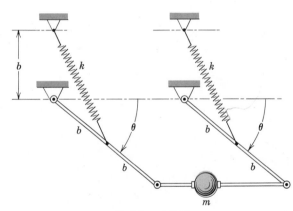

Problem 3/142

144 The sphere of mass *m* is attached to the light rod which pivots about a horizontal axis through *O*. Link *BC* slides through a collar pivoted at *A* and compresses the spring as the sphere assumes a lower position. If the spring is uncompressed when the mechanism is released from rest in the position $\theta = 60°$, determine the angular rate $\dot{\theta}$ of the rod *BO* for the position $\theta = 90°$.

Ans. $\dot{\theta} = \sqrt{\dfrac{g}{b} - \dfrac{k}{m}(3 - 2\sqrt{2})}$

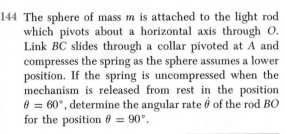

Problem 3/143

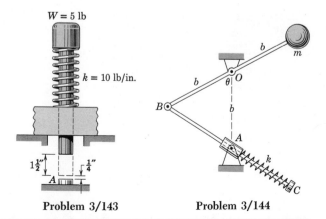

Problem 3/144

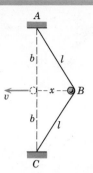

Problem 3/145

3/145 A particle of mass m is released from rest in the position shown and is accelerated to the left by the action of the elastic cord ABC which lies in the horizontal plane and which is unstretched at $x = 0$. The cord has a unit stiffness K, which is the ratio of its tension force to its unit elongation or strain (change in length divided by original length). Determine the expression for the velocity v reached by the particle at the instant x returns to zero.

3/146 An instrument package of mass m is attached at A to two springs each of stiffness k and unstretched length b. The package is released from rest at this point and falls a short distance. The deflection y at any instant of time is very small compared with b, so that the stretch of the spring is very nearly given by $x = y \sin \theta$ where $\theta = \cos^{-1}(c/b)$. Determine the velocity $\dot{y}$ of the package as a function of y and find the maximum deflection y_{max} of the package.

$$\text{Ans. } \dot{y} = \sqrt{2y\left(g - \frac{k}{m}y\,\frac{b^2 - c^2}{b^2}\right)}, \ y_{max} = \frac{mg}{k}\,\frac{b^2}{b^2 - c^2}$$

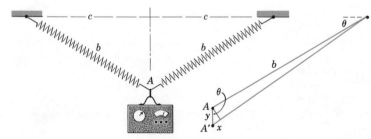

Problem 3/146

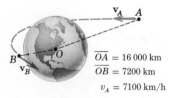

$\overline{OA} = 16\,000$ km
$\overline{OB} = 7200$ km
$v_A = 7100$ km/h

Problem 3/147

3/147 Upon its return voyage to earth, a space capsule has an absolute velocity of 7100 km/h at point A, which is 16 000 km from the center of the earth. Determine the absolute velocity of the capsule when it reaches point B, which is 7200 km from the earth's center. The trajectory between these two points is outside the effect of the earth's atmosphere.

3/148 A projectile is fired vertically up from the north pole with a velocity v_0. Calculate the least value of v_0 which will allow the projectile to escape from the earth's gravitational pull assuming no atmospheric resistance. The earth's radius is 6370 km. Use the absolute value $g = 9.824$ m/s^2.

Ans. $v_0 = 11.19$ km/s

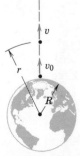

Problem 3/148

149 A satellite is put into an elliptical orbit around the earth and has a velocity v_P at the perigee position P. Determine the expression for the velocity v_A at the apogee position A. The radii to A and P are, respectively, r_A and r_P. Note that the total energy remains constant.

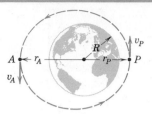

Problem 3/149

150 A rocket launches an unpowered space capsule at point A with an absolute velocity $v_A = 8000$ mi/hr at an altitude of 25 mi. After the capsule has traveled a distance of 250 mi measured along its absolute space trajectory, its velocity at B is 7600 mi/hr and its altitude is 50 mi. Determine the average resistance P to motion in the rarified atmosphere. The earth weight of the capsule is 48 lb, and the earth's mean radius is 3958 mi. Consider the center of the earth as fixed in space.   *Ans. $P = 2.87$ lb*

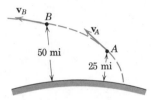

Problem 3/150

151 Derive an expression for the net amount of energy E which would be required to move an earth satellite of mass m from a rest condition on the earth's surface at the equator to a circular orbit of altitude h. Assume that the transfer is made without atmospheric friction loss. The radius of the earth is R and its angular velocity is ω. (See Prob. 3/81 for the satellite velocity.)

$$\text{Ans. } E = \frac{1}{2}mR\left(g\frac{R + 2h}{R + h} - R\omega^2\right)$$

152 A spacecraft of mass m travels in a circular orbit of radius r_1 around the earth and enters an elliptical transfer orbit at A with a burst of speed. It is then inserted into the circular orbit of radius r_2 at B with a second burst of speed. Determine the total energy ΔE needed to effect the transfer.

$$\text{Ans. } \Delta E = \tfrac{1}{2}mgR^2\left(\frac{1}{r_1} - \frac{1}{r_2}\right)$$

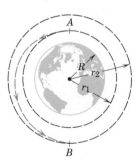

Problem 3/152

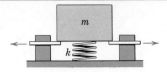

Problem 3/153

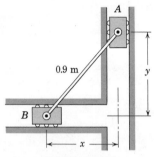

Problem 3/154

3/153 The block of mass m is supported by the two fingers and is barely in contact with the uncompressed elastic spring of stiffness k and negligible mass. If the fingers are suddenly removed, determine the maximum velocity v reached by the block, the maximum force R transmitted to the floor through the spring, and the maximum deformation δ of the spring.   *Ans.* $v = g\sqrt{m/k}$, $R = 2mg$, $\delta = 2mg/k$

3/154 Calculate the maximum velocity $v_{B_{max}}$ of slider B during the motion described in Prob. 3/102, for which the figure is repeated here. Note that the velocities of A and B are related through the time derivatives $\dot{x}$ and $\dot{y}$ determined from the equation $x^2 + y^2 = (0.9)^2$.   *Ans.* $v_{B_{max}} = 0.962$ m/s

▶3/155 A body rotates with a constant angular velocity ω about an axis fixed in space. At any point within the body a radial force $F_r = -r\omega^2$ may be associated with a unit-mass particle in order to hold it in place. Show that this force field is conservative and derive the potential function V from which the force components may be obtained by differentiation. Take $V = 0$ when $r = 0$.

▶3/156 If the surface upon which the chain in Prob. 3/137 slides is not smooth and if the chain starts from rest with a sufficient number of links hanging over the edge barely to initiate motion, determine the velocity v of the chain as the last link leaves the edge. The coefficient of friction is μ. Neglect friction at the edge.

$$Ans.\ v = \sqrt{\frac{gL}{1 + \mu}}$$

SECTION C. IMPULSE AND MOMENTUM

3/8 IMPULSE AND MOMENTUM. In the previous two articles we focused attention on the equations of work and energy which are obtained by integrating the equation of motion $\mathbf{F} = m\mathbf{a}$ with respect to the displacement of the particle. In consequence we found that the velocity changes could be expressed directly in terms of the work done or in terms of the overall changes in energy. In the present article we will direct our attention to the integration of the equation of motion with respect to time rather than displacement. This approach leads to the equations of impulse and momentum. We will find that these equations greatly facilitate the solution of many problems where the applied forces act during specified intervals of time.

(*a*) *Linear Impulse and Linear Momentum.* Consider again the general curvilinear motion in space of a particle of mass m, Fig. 3/9, where the particle is located by its position vector $\mathbf{r}$ measured from a fixed origin O. The velocity of the particle is $\mathbf{v} = \dot{\mathbf{r}}$ and is tangent to its path (shown as a dashed line). The resultant $\Sigma\mathbf{F}$ of all forces on m is in the direction of its acceleration $\dot{\mathbf{v}}$. We may now write the basic equation of motion for the particle, Eq. 3/3, as

$$\Sigma\mathbf{F} = m\dot{\mathbf{v}} = \frac{d}{dt}(m\mathbf{v}) \qquad \text{or} \qquad \boxed{\Sigma\mathbf{F} = \dot{\mathbf{G}}} \qquad (3/21)$$

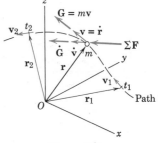

Figure 3/9

where the product of the mass and velocity is defined as the *linear momentum* $\mathbf{G} = m\mathbf{v}$ of the particle. Equation 3/21 states that *the resultant of all forces acting on a particle equals its time rate of change of linear momentum.* In SI the units of linear momentum $m\mathbf{v}$ are seen to be kg·m/s, which also equals N·s. In U.S. customary units the units of linear momentum $m\mathbf{v}$ are [lb/(ft/sec²)][ft/sec] = lb-sec.

Since Eq. 3/21 is a vector equation, we recognize that, in addition to the equality of the magnitudes of $\Sigma\mathbf{F}$ and $\dot{\mathbf{G}}$, the direction of the resultant force coincides with the direction of the *change* in linear momentum, which is the direction of the change in velocity. Equation 3/21 is one of the most useful and important relationships in dynamics, and it is valid as long as the mass m of the particle is not changing with time. The case where m changes with time is discussed in Art. 5/4 of Chapter 5. We now write the three scalar components of Eq. 3/21 as

$$\Sigma F_x = \dot{G}_x \qquad \Sigma F_y = \dot{G}_y \qquad \Sigma F_z = \dot{G}_z \qquad (3/22)$$

These equations may be applied independently of one another.

All that we have done so far in this article is to rewrite Newton's second law in an alternative form in terms of momentum. But we are now able to describe the effect of the resultant force $\Sigma\mathbf{F}$ on the linear momentum of the particle over a finite period of time simply by integrating Eq. 3/21 with respect to the time t. Multiplying the equation by dt gives $\Sigma \mathbf{F}\, dt = d\mathbf{G}$ which we integrate from time t_1 to time t_2 to obtain

$$\int_{t_1}^{t_2} \Sigma \mathbf{F}\, dt = \mathbf{G}_2 - \mathbf{G}_1 \qquad (3/23)$$

Here the linear momentum at time t_2 is $\mathbf{G}_2 = m\mathbf{v}_2$ and the linear momentum at time t_1 is $\mathbf{G}_1 = m\mathbf{v}_1$. The product of force and time is defined as *linear impulse* of the force, and Eq. 3/23 states that *the total linear impulse on m equals the corresponding change in linear momentum of m.*

The impulse integral is a vector which, in general, may involve changes in both magnitude and direction during the time interval. Under these conditions it will be necessary to express $\Sigma \mathbf{F}$ and $\mathbf{G}$ in component form and then combine the integrated components. The components of Eq. 3/23 become the scalar equations

$$\int_{t_1}^{t_2} \Sigma F_x\, dt = (mv_x)_2 - (mv_x)_1$$

$$\int_{t_1}^{t_2} \Sigma F_y\, dt = (mv_y)_2 - (mv_y)_1 \qquad (3/24)$$

$$\int_{t_1}^{t_2} \Sigma F_z\, dt = (mv_z)_2 - (mv_z)_1$$

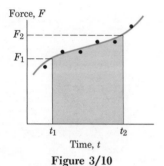

Force, F

Time, t

Figure 3/10

These three scalar impulse-momentum equations are completely independent.

There are times when a force acting on a particle varies with the time in a manner determined by experimental measurements or by other approximate means. In this case a graphical or numerical integration must be accomplished. If, for example, a force F acting on a particle in a given direction varies with the time t as indicated in Fig. 3/10, then the impulse of this force from t_1 to t_2, which equals $\int_{t_1}^{t_2} F\, dt$, becomes the shaded area under the curve.

In evaluating the resultant impulse it is necessary to include the effect of *all* forces acting on m except those whose magnitudes are negligible. The student should be fully aware at this point that the only reliable method of accounting for the effects of *all* forces is to isolate the particle in question by drawing its *free-body diagram.*

(*b*) *Angular Impulse and Angular Momentum.* In addition to the equations of linear impulse and linear momentum, there exists a parallel set of equations for angular impulse and angular momentum. First, we define the term angular momentum. In Fig. 3/11*a* is shown a particle P of mass m moving along a curve in space. The particle is

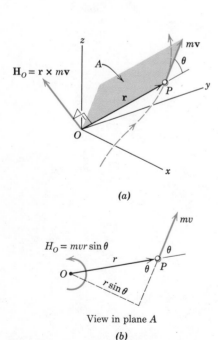

$\mathbf{H}_O = \mathbf{r} \times m\mathbf{v}$

(*a*)

$H_O = mvr \sin \theta$

View in plane A

(*b*)

Figure 3/11

located by its position vector **r** with respect to a convenient origin O of fixed coordinates x-y-z. The velocity of the particle is $\mathbf{v} = \dot{\mathbf{r}}$, and its linear momentum is $\mathbf{G} = m\mathbf{v}$. The *moment* of the *linear momentum vector* $m\mathbf{v}$ about the origin O is defined as the *angular momentum* $\mathbf{H}_O$ of P about O and is given by the cross-product relation for the moment of a vector

$$\boxed{\mathbf{H}_O = \mathbf{r} \times m\mathbf{v}} \qquad (3/25)$$

The angular momentum, then, is a vector perpendicular to plane A defined by **r** and **v**. The sense of $\mathbf{H}_O$ is clearly defined by the right-hand rule for cross products.

The scalar components of angular momentum may be obtained from the expansion

$$\mathbf{H}_O = \mathbf{r} \times m\mathbf{v} = \mathbf{i}m(v_z y - v_y z) + \mathbf{j}m(v_x z - v_z x) + \mathbf{k}m(v_y x - v_x y)$$

or

$$\mathbf{H}_O = m\begin{vmatrix} \mathbf{i} & \mathbf{j} & \mathbf{k} \\ x & y & z \\ v_x & v_y & v_z \end{vmatrix} \qquad (3/26)$$

so that

$$H_x = m(v_z y - v_y z), \qquad H_y = m(v_x z - v_z x), \qquad H_z = m(v_y x - v_x y)$$

Each of these expressions for angular momentum may be checked easily from Fig. 3/12, which shows the three linear-momentum components, by taking the moments of these components about the respective axes.

To help visualize angular momentum we show in Fig. 3/11*b* a two-dimensional representation in plane A of the vectors shown in the *a*-part of the figure. The motion is viewed in plane A defined by **r** and **v**. The magnitude of the moment of $m v$ about O is simply the linear momentum $m v$ times the moment arm $r \sin \theta$ or $m v r \sin \theta$, which is the magnitude of the cross product $\mathbf{H}_O = \mathbf{r} \times m\mathbf{v}$.

Angular momentum is the moment of linear momentum and must not be confused with linear momentum. In SI units angular momentum has the units $\text{kg} \cdot (\text{m/s}) \cdot \text{m} = \text{kg} \cdot \text{m}^2/\text{s} = \text{N} \cdot \text{m} \cdot \text{s}$. In the U.S. customary system angular momentum has the units $[\text{lb}/(\text{ft/sec}^2)][\text{ft/sec}][\text{ft}] = \text{ft-lb-sec}$.

We are now ready to relate the moment of the forces acting on the particle P to its angular momentum. If $\Sigma\mathbf{F}$ represents the resultant of *all* forces acting on the particle P of Fig. 3/11, the moment $\mathbf{M}_O$ about the origin O is the vector cross product.

$$\Sigma\mathbf{M}_O = \mathbf{r} \times \Sigma\mathbf{F} = \mathbf{r} \times m\dot{\mathbf{v}}$$

where Newton's second law $\mathbf{F} = m\dot{\mathbf{v}}$ has been substituted. We now

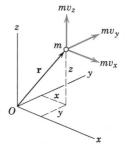

Figure 3/12

differentiate Eq. 3/25 with time, using the rule for the differentiation of a cross product (see item 9 Art. B7, Appendix B) and get

$$\dot{\mathbf{H}}_O = \dot{\mathbf{r}} \times m\mathbf{v} + \mathbf{r} \times m\dot{\mathbf{v}} = \mathbf{v} \times m\mathbf{v} + \mathbf{r} \times m\dot{\mathbf{v}}$$

The term $\mathbf{v} \times m\mathbf{v}$ is zero since the cross product of equal and, hence, parallel vectors is identically zero. Substituting into the expression for $\Sigma \mathbf{M}_O$ gives

$$\boxed{\Sigma \mathbf{M}_O = \dot{\mathbf{H}}_O} \qquad (3/27)$$

Equation 3/27 states that the *moment about the fixed point O of all forces acting on m equals the time rate of change of angular momentum of m about O*. This relation, particularly when extended to a system of particles, rigid or nonrigid, constitutes one of the most powerful tools of analysis in all of dynamics. Equation 3/27 is a vector equation with scalar components

$$\Sigma M_{O_x} = \dot{H}_{O_x} \qquad \Sigma M_{O_y} = \dot{H}_{O_y} \qquad \Sigma M_{O_z} = \dot{H}_{O_z} \qquad (3/28)$$

Equation 3/27 gives the instantaneous relation between the moment and the time rate of change of angular momentum. To obtain the effect of the moment $\Sigma \mathbf{M}_O$ on the angular momentum of the particle over a finite period of time, Eq. 3/27 may be integrated from time t_1 to time t_2. Multiplying the equation by dt gives $\Sigma \mathbf{M}_O \, dt = d\mathbf{H}_O$ which we integrate to obtain

$$\boxed{\int_{t_1}^{t_2} \Sigma \mathbf{M}_O \, dt = \mathbf{H}_{O_2} - \mathbf{H}_{O_1}} \qquad (3/29)$$

where $\mathbf{H}_{O_2} = \mathbf{r}_2 \times m\mathbf{v}_2$ and $\mathbf{H}_{O_1} = \mathbf{r}_1 \times m\mathbf{v}_1$. The product of moment and time is defined as *angular impulse*, and Eq. 3/29 states that the *total angular impulse on m about the fixed point O equals the corresponding change in angular momentum of m about O*. The units of angular impulse are clearly those of angular momentum, which are $\text{N} \cdot \text{m} \cdot \text{s}$ or $\text{kg} \cdot \text{m}^2/\text{s}$ in SI units and ft-lb-sec in U.S. customary units.

As in the case of linear impulse and linear momentum, the equation of angular impulse and angular momentum is a vector equation where changes in direction as well as magnitude may occur during the interval of integration. Under these conditions it is necessary to express $\Sigma \mathbf{M}_O$ and $\mathbf{H}_O$ in component form and then combine the integrated components. Thus the x-component of Eq. 3/29 becomes

$$\int_{t_1}^{t_2} \Sigma M_{O_x} \, dt = (H_{O_x})_2 - (H_{O_x})_1$$

$$= m[(v_z y - v_y z)_2 - (v_z y - v_y z)_1]$$

where the subscripts 1 and 2 refer to the values of the respective quantities at times t_1 and t_2. Similar expressions exist for the y- and z-components of the angular momentum integral.

The foregoing angular-impulse and angular-momentum relations have been developed in their general three-dimensional sense. Most of the applications with which we shall be concerned, on the other hand, can be analyzed as plane-motion problems where moments are taken about a single axis normal to the plane of motion. In this event the angular momentum may change magnitude and sense, but the direction of the vector remains unaltered. Thus for a particle of mass m moving along a curved path in the x-y plane, Fig. 3/13, the angular momenta about O at points 1 and 2 have the magnitudes $H_{O_1} = |\mathbf{r}_1 \times m\mathbf{v}_1| = mv_1d_1$ and $H_{O_2} = |\mathbf{r}_2 \times m\mathbf{v}_2| = mv_2d_2$, respectively. In the illustration both H_{O_1} and H_{O_2} are represented in the counterclockwise sense according to the direction of the moment of the linear momentum. The scalar form of Eq. 3/29 applied to the motion between points 1 and 2 during the time interval t_1 to t_2 becomes

$$\int_{t_1}^{t_2} \Sigma M_O \, dt = H_{O_2} - H_{O_1} \quad \text{or} \quad \int_{t_1}^{t_2} \Sigma Fr \sin \theta \, dt = mv_2d_2 - mv_1d_1$$

This illustration should help to clarify the relation between the scalar and vector forms of the angular impulse-momentum relations.

Equations 3/21 and 3/27 add no new basic information since they are merely alternative forms of Newton's second law. However, we will discover in subsequent chapters that the motion equations expressed in terms of time rate of change of momentum are applicable to the motion of rigid and nonrigid bodies and constitute a very general and powerful approach to many problems. The full generality of Eq. 3/27 is usually not required to describe the motion of a single particle or the plane motion of rigid bodies, but it does find important use in the analysis of the space motion of rigid bodies, an introduction to which is presented in Chapter 8.

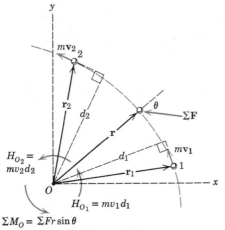

Figure 3/13

Sample Problem 3/17

A 2-lb particle moves in the vertical y-z plane (z up, y horizontal) under the action of its weight and a force **F** which varies with the time. The linear momentum of the particle in lb-sec is given by the expression $\mathbf{G} = \frac{3}{2}(t^2 + 3)\mathbf{j} - \frac{2}{3}(t^3 - 4)\mathbf{k}$ where t is the time in seconds. Determine **F** for the instant when $t = 2$ sec.

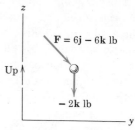

Solution. The weight expressed as a vector is $-2\mathbf{k}$ lb. Thus the force-momentum equation becomes

① $[\Sigma\mathbf{F} = \dot{\mathbf{G}}]$ $\qquad \mathbf{F} - 2\mathbf{k} = \dfrac{d}{dt}[\frac{3}{2}(t^2 + 3)\mathbf{j} - \frac{2}{3}(t^3 - 4)\mathbf{k}]$

① Don't forget that $\Sigma\mathbf{F}$ includes *all* forces on the particle of which the weight is one.

$$= 3t\mathbf{j} - 2t^2\mathbf{k}$$

For $t = 2$ sec, $\qquad \mathbf{F} = 2\mathbf{k} + 3(2)\mathbf{j} - 2(2^2)\mathbf{k} = 6\mathbf{j} - 6\mathbf{k}$ lb

Thus $\qquad\qquad F = \sqrt{6^2 + 6^2} = 6\sqrt{2}$ lb $\qquad\qquad$ *Ans.*

Sample Problem 3/18

A particle with a mass of 0.5 kg has a velocity $u = 10$ m/s in the x-direction at time $t = 0$. Forces $\mathbf{F}_1$ and $\mathbf{F}_2$ act on the particle, and their magnitudes change with time according to the graphical schedule shown. Determine the velocity $\mathbf{v}$ of the particle at the end of the 3 seconds.

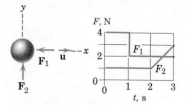

Solution. The impulse-momentum equation is applied in component form and gives for the x- and y-directions, respectively,

① $\left[\int \Sigma F_x\, dt = m\,\Delta v_x\right]$ $\qquad -[4(1) + 2(3 - 1)] = 0.5(v_x - 10)$

$$v_x = -6 \text{ m/s}$$

$\left[\int \Sigma F_y\, dt = m\,\Delta v_y\right]$ $\qquad [1(2) + 2(3 - 2)] = 0.5(v_y - 0)$

$$v_y = 8 \text{ m/s}$$

Thus

$$\mathbf{v} = -6\mathbf{i} + 8\mathbf{j} \text{ m/s} \qquad \text{and} \qquad v = \sqrt{6^2 + 8^2} = 10 \text{ m/s}$$

$$\theta_x = \tan^{-1}\frac{8}{-6} = 126.9° \qquad\qquad \textit{Ans.}$$

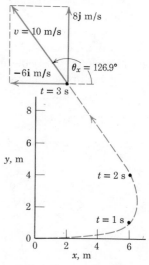

② Although not called for, the path of the particle for the first 3 seconds is plotted in the figure. The velocity at $t = 3$ s is shown together with its components.

① The impulse in each direction is the corresponding area under the force-time graph. Note that F_1 is in the negative x-direction so its impulse is negative.

② It is important to note that the algebraic signs must be carefully respected in applying the momentum equations. Also we must recognize that impulse and momentum are vector quantities in contrast to work and energy, which are scalars.

Sample Problem 3/19

The loaded 150-kg skip is rolling down the incline at 4 m/s when a force P is applied to the cable as shown at time $t = 0$. The force P is increased uniformly with the time until it reaches 600 N at $t = 4$ s after which time it remains constant at this value. Calculate (a) the time t_1 at which the skip reverses its direction and (b) the velocity v of the skip at $t = 8$ s. Treat the skip as a particle.

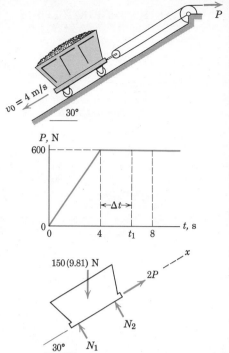

Solution. The stated variation of P with the time is plotted, and the free-body diagram of the skip is drawn.

Part (a). The skip reverses direction when its velocity becomes zero. We will assume that this condition occurs at $t = 4 + \Delta t$ seconds. The impulse-momentum equation applied in the positive x-direction gives

$$\left[\int \Sigma F_x \, dt = m \, \Delta v_x \right] \quad \tfrac{1}{2}(4)(2)(600) + 2(600)\Delta t$$

$$- 150(9.81)\sin 30°(4 + \Delta t) = 150(0 - [-4])$$

$$464\Delta t = 1143, \quad \Delta t = 2.46 \text{ s}$$

$$t = 4 + 2.46 = 6.46 \text{ s} \quad\quad Ans.$$

Part (b). Applying the impulse-momentum equation to the entire interval gives

$$\left[\int \Sigma F_x \, dt = m \, \Delta v_x \right] \quad \tfrac{1}{2}(4)(2)(600) + 4(2)(600) - 150(9.81)\sin 30°(8)$$

$$= 150(v - [-4])$$

$$150v = 714 \quad v = 4.76 \text{ m/s} \quad\quad Ans.$$

The same result would be obtained by analyzing the interval from t_1 to 8 s.

① The free-body diagram keeps us from making the error of using the impulse of one P rather than $2P$ or of forgetting the impulse of the component of the weight. The first term in the equation is the triangular area of the P-t relation for the first 4 s, doubled for the force of $2P$. Assignment of the coordinate direction keeps the algebraic signs straight.

Sample Problem 3/20

The small 3-lb block slides on a smooth horizontal surface under the action of the force in the spring and a force F. The angular momentum of the block about O varies with time as shown in the graph. When $t = 6.5$ sec, it is known that $r = 6$ in. and $\beta = 60°$. Determine F for this instant.

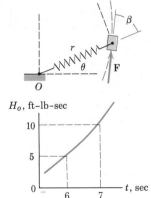

Solution. The only moment of the forces about O is due to F since the spring force passes through O. Thus $\Sigma M_O = Fr \sin \beta$. From the graph the time rate of change of H_O for $t = 6.5$ sec is very nearly $(8 - 4)/(7 - 6)$ or $\dot{H}_O = 4$ ft-lb. The moment-angular momentum relation gives

$$[\Sigma M_O = \dot{H}_O] \quad F\left(\frac{6}{12}\right)\sin 60° = 4, \quad F = 9.24 \text{ lb} \quad\quad Ans.$$

① We do not need vector notation here since we have plane motion where the direction of the vector $\mathbf{H}_O$ does not change.

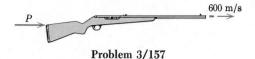

Problem 3/157

Problem 3/158

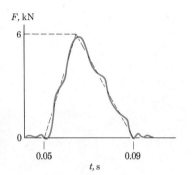

Problem 3/159

Problem 3/160

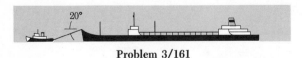

Problem 3/161

PROBLEMS

3/157 A rifle fires a 60-g bullet with a horizontal muzzle velocity of 600 m/s. If it takes 3×10^{-3} s for the bullet to travel the length of the barrel, determine the average horizontal force P applied to the gun to hold it in place during the firing.

Ans. $P = 12$ kN

3/158 The jet fighter weighs 15,000 lb and requires 12 sec from rest to reach its takeoff speed of 150 mi/hr under a constant jet thrust of 9000 lb. Compute the time average of the combined air and ground resistance R during takeoff. Neglect the relatively small loss of mass during takeoff.

3/159 A 5-kg object, which is moving on a smooth horizontal plane with a velocity of 20 m/s to the right, is struck with an impulsive force F that acts to the left on the body. The magnitude of F is represented in the graph. Approximate the loading by the dotted line shown and determine the final velocity v of the object.   *Ans. $v = 4.00$ m/s to the left*

3/160 A jet-propelled airplane with a mass of 10 Mg is flying horizontally at a constant speed of 1000 km/h under the action of the engine thrust T and the equal and opposite air resistance R. The pilot ignites two rocket-assist units, each of which develops a forward thrust of 8 kN for 9 s. If the velocity of the airplane in its horizontal flight is 1050 km/h at the end of the 9 seconds, calculate the time-average increase ΔR in air resistance. The mass of the rocket fuel used is negligible compared with that of the airplane.

3/161 The supertanker has a total displacement (mass) of 150,000 long tons (one long ton equals 2240 lb) and is lying still in the water when the tug commences a tow. If a constant tension of 50,000 lb is developed in the tow cable, compute the time required to bring the tanker to a speed of 1 knot from rest. At this low speed hull resistance to motion through the water is very small and may be neglected. (1 knot = 1.152 mi/hr)   *Ans. $t = 6.25$ min*

62 The lunar landing craft has a mass of 200 kg and is equipped with a retro-rocket that provides a constant upward thrust T capable of yielding a total impulse of 18 kN·s during a 10-s period. If the lunar craft was 150 m above the surface of the moon and was descending at the rate of 45 m/s, determine the time t required to reduce the descent rate to 1.5 m/s. The absolute acceleration due to gravity near the surface of the moon is 1.62 m/s².

63 A ship with a displacement (mass) of 15 000 metric tons is cruising at a speed of 10 knots when the propeller-shaft power is increased to produce a propeller thrust of 250 kN. If the resistance R to motion of the ship through the water varies with speed as shown, pick out an appropriate average value of R and calculate the time t required for the ship to increase its speed from 10 to 12 knots. (1 knot = 1.852 km/h)   *Ans. t = 3.7 min*

64 An automobile with a mass of 1800 kg has a velocity of 80 km/h down a 10 percent grade when the brakes are applied. If the car is slowed to a velocity of 20 km/h in 8 s, compute the average of the total braking force F exerted by the road on all tires. Treat the car as a particle and neglect air resistance.

65 The inspection gondola for a cableway is being drawn up the sloping cable at the speed of 12 ft/sec. If the control cable at A suddenly breaks, calculate the time t after the break required for the gondola to reach a speed of 24 ft/sec down the incline. Neglect friction and treat the gondola as a particle.   *Ans. t = 2.91 sec*

66 The linear momentum of a certain particle is given by the equation $G = 6t^3i + 15t^2j$, where t is the time in seconds after the particle starts from rest and where the units of G are kg·m/s. Calculate the magnitude of the resultant force F that acts on the particle at time $t = 3$ s.

67 At a certain instant the linear momentum of a particle is given by $G = -2i - 2j + 3k$ lb-sec and its position vector is $r = 2i + 4j - 3k$ ft. Determine the magnitude H_O of the angular momentum of the particle about the origin of coordinates.   *Ans. H_O = 7.21 ft-lb-sec*

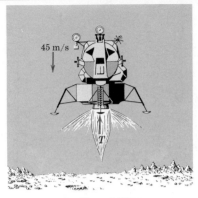

Problem 3/162

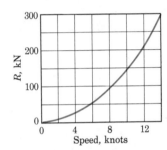

Problem 3/163

Problem 3/164

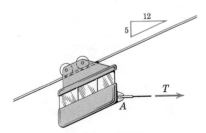

Problem 3/165

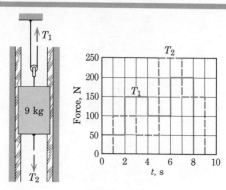

Problem 3/168

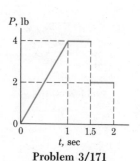

Problem 3/169

Problem 3/171

Problem 3/172

3/168 The motion of the 9-kg carriage in the smooth vertical guides is controlled by the tensions T_1 and T_2 in the cables. If the carriage has a downward velocity of 1.2 m/s at time $t = 0$, determine its velocity v when $t = 10$ s.

3/169 A 3-Mg rocket sled is propelled by six rocket motors each with an impulse rating of 100 kN·s. The rockets are fired at $\frac{1}{4}$-s intervals starting with the sled at rest, and the duration of each rocket firing is 1.5 s. If the velocity of the sled is 150 m/s in 3 s from the start, compute the time average R of the total resistance to motion. Neglect the loss of mass due to exhaust gases compared with the total mass of the sled.

3/170 Assume that the gondola of Prob. 3/165 has a mass of 200 kg and is moving down the fixed inclined cable with a speed of 2 m/s. If a horizontal tension $T = 900$ N is applied to the attached cable at A, calculate the time t required for the gondola to acquire a speed of 3 m/s up the incline. Neglect friction and treat the gondola as a particle.

3/171 A 16.1-lb body is traveling in a horizontal straight line with a velocity of 8 ft/sec when a horizontal force P is applied to it at right angles to the initial direction of motion. If P varies according to the accompanying graph, remains constant in direction, and is the only force acting on the body in its plane of motion, find the magnitude of the velocity of the body when $t = 2$ sec and the angle θ it makes with the direction of P.

Ans. $v = 12.81$ ft/sec, $\theta = 38.7°$

3/172 The third and fourth stages of a rocket are coasting in space with a velocity of 15 000 km/h when the fourth-stage engine ignites causing separation of the stages under a thrust T and its reaction. If the velocity v of the fourth stage is 10 m/s greater than the velocity v_1 of the third stage at the end of the $\frac{1}{2}$-s separation interval, calculate the average thrust T during this period. The masses of the third and fourth stages at separation are 30 kg and 50 kg, respectively. Assume that the entire blast of the fourth-stage engine impinges against the third stage with a force equal and opposite to T.

173 The force P, which varies linearly with time as shown, is applied to the 10-kg block initially at rest. If the coefficients of static and kinetic friction are both 0.40, determine the velocity of the block when $t = 4$ s.   *Ans.* $v = 7.38$ m/s

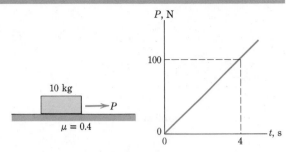

Problem 3/173

174 The hydraulic braking system for the truck and trailer is set to produce equal braking force for each of the two units. If the brakes are applied uniformly for 5 sec to bring the rig to a stop from a speed of 20 mi/hr down the 10 percent grade, determine the force P in the coupling between the trailer and the truck. The truck weighs 20,000 lb and the trailer weighs 15,000 lb.

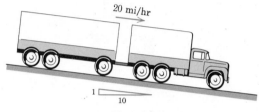

Problem 3/174

175 The 1.6-kg particle has an initial velocity of 10 m/s in the horizontal x-y plane in the direction shown. A force in newtons given by $F = t^2/2 - 8$ where t is in seconds is applied to the particle as indicated. Determine the magnitude of the velocity **v** of the particle for $t = 8$ s, and find the angle θ made by **v** with respect to the x-axis.   *Ans.* $v = 10.85$ m/s, $\theta = 47.5°$

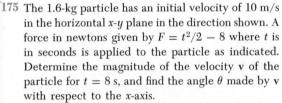

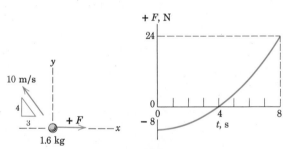

Problem 3/175

176 The ice-hockey puck with a mass of 0.20 kg has a velocity of 12 m/s before being struck by the hockey stick. After the impact the puck moves in the new direction shown with a velocity of 18 m/s. If the stick is in contact with the puck for 0.04 s, compute the magnitude of the average force **F** exerted by the stick on the puck during contact, and find the angle β made by **F** with the x-direction.

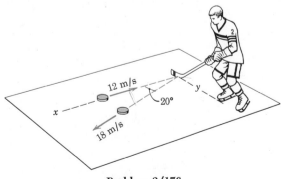

Problem 3/176

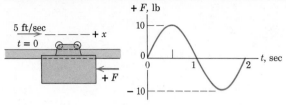

Problem 3/177

3/177 The carriage weighs 64.4 lb and moves with negligible friction along the horizontal rail. A sinusoidal force $F = 10 \sin \pi t$ in pounds measured positive to the left acts on the carriage as shown. If the carriage has a velocity of 5 ft/sec to the right at $t = 0$, determine the velocity v of the carriage when $t = 1$ sec. By inspection what is the velocity of the carriage when $t = 2$ sec?

Ans. $v = 1.817$ ft/sec for $t = 1$ sec

3/178 A particle of mass m starts from rest and moves in a horizontal straight line under the action of a constant force P. Resistance to motion is proportional to the square of the velocity and is $R = kv^2$. Determine the total impulse I on the particle from the time it starts until it reaches its maximum velocity. (*Hint:* First prove that v_{max} is reached when $P = R$.)

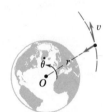

Problem 3/179

3/179 The only force acting on an earth satellite traveling outside of the earth's atmosphere is the radial gravitational attraction. The moment of this force is zero about the earth's center taken as a fixed point. Prove that $r^2\dot{\theta}$ remains constant for the motion of the satellite.

3/180 A spacecraft with a mass of 260 kg is moving with a velocity $u = 30\,000$ km/h in the fixed x-direction remote from any attracting celestial body. The spacecraft is spin-stabilized and rotates about the z-axis at the constant rate $\dot{\theta} = \pi/10$ rad/s. During a quarter of a revolution from $\theta = 0$ to $\theta = \pi/2$, a jet is activated which produces a 600-N thrust of constant magnitude. Determine the y-component of the velocity of the spacecraft when $\theta = \pi/2$. Neglect the small change in mass due to the loss of exhaust gas through the control nozzle and treat the spacecraft as a particle. *Ans.* $v_y = 7.35$ m/s

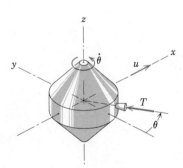

Problem 3/180

3/181 The 60-lb crate is sliding down the incline with a velocity of 20 ft/sec when the horizontal force $P = 40$ lb is applied to the crate. How long does it take to bring the crate to a stop after P is applied?

Ans. $t = 1.96$ sec

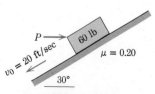

Problem 3/181

182 If the resistance R to the motion of a freight train of total mass m increases with velocity according to $R = R_0 + Kv$, where R_0 is the initial resistance to be overcome in starting the train and K is a constant, find the time t required for the train to reach a velocity v from rest on a level track under the action of a constant tractive force F.

183 A particle with a mass of 5 kg has a position vector in meters given by $\mathbf{r} = 4t^2\mathbf{i} - 2t\mathbf{j} - 3t\mathbf{k}$ where t is the time in seconds. For $t = 3$ s determine the magnitude of the angular momentum of the particle and the magnitude of the moment of all forces on the particle, both about the origin of coordinates.

        Ans. $H = 649$ N·m·s, $M = 433$ N·m

184 The 10-kg block is resting on the horizontal surface when the force T is applied to it for 7 seconds. The variation of T with time is shown. Calculate the maximum velocity reached by the block and the total time Δt during which the block is in motion. The coefficients of static and kinetic friction are both 0.50.   *Ans.* $v_{\text{max}} = 5.19$ m/s, $\Delta t = 5.54$ s

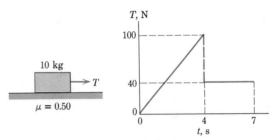

Problem 3/184

185 The loaded mine skip has a gross mass of 3 Mg. The hoisting drum produces a tension T in the cable according to the time schedule shown. If the skip is at rest against A when the drum is activated, determine the speed v of the skip when $t = 6$ s. Friction loss may be neglected.   *Ans.* $v = 9.13$ m/s

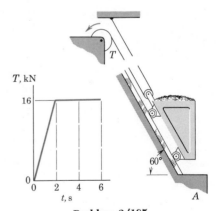

Problem 3/185

186 A charged particle of mass m starts from rest at time $t = 0$. It is subjected to a radial force F due to a rotating field of force whose intensity increases linearly with time, $F = Kt$, and whose direction with the x-axis is given by the angle $\theta = \omega t$ where $\omega = \dot{\theta} = $ constant. Determine the magnitude of the velocity $\mathbf{v}$ acquired by the particle during one half of a revolution of the force field starting from $t = 0$. Also specify the corresponding angle β made by $\mathbf{v}$ with the positive x-axis at this instant.

$$\text{Ans. } v = \frac{K}{m\omega^2}\sqrt{4 + \pi^2}, \ \beta = 122.5°$$

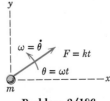

Problem 3/186

3/9 CONSERVATION OF MOMENTUM. If the resultant force on a particle is zero during an interval of time, we see that Eq. 3/21 requires that its linear momentum **G** remain constant. Similarly if the resultant moment about a fixed point O of all forces acting on a particle is zero during an interval of time, Eq. 3/27 requires that its angular momentum $\mathbf{H}_O$ about that point remain constant. In the first case the linear momentum of the particle is said to be conserved, and in the second case the angular momentum of the particle is said to be conserved. Linear momentum may be conserved in one coordinate direction, such as x, but not necessarily in the y- or z-directions. Also angular momentum may be conserved about one axis but not about another axis.

A careful examination of the free-body diagram of the particle will disclose whether the total linear impulse on the particle in a particular direction is zero. If it is, the corresponding linear momentum is unchanged (conserved). Similarly, the free-body diagram will tell us whether the moment of the resultant force on the particle about a fixed point is zero, in which case the angular momentum about that point is unchanged (conserved).

Consider now the motion of two particles of mass m_a and m_b moving with velocities $\mathbf{v}_a$ and $\mathbf{v}_b$. If the particles interact during an interval of time t and if the forces of interaction $\mathbf{F}$ and $-\mathbf{F}$ are the only unbalanced forces acting on the particles during the interval, then we may write Eq. 3/23 for each particle as

$$\int_0^t \mathbf{F}\, dt = m_a\, \Delta\mathbf{v}_a \qquad \text{and} \qquad \int_0^t -\mathbf{F}\, dt = m_b\, \Delta\mathbf{v}_b$$

where $\Delta\mathbf{v}_a$ and $\Delta\mathbf{v}_b$ are the vector changes in the respective velocities of m_a and m_b during the period of interaction. The impulse integrals differ only in sign, so that $m_a\, \Delta\mathbf{v}_a = -m_b\, \Delta\mathbf{v}_b$ or $\Delta(m_a\mathbf{v}_a) = -\Delta(m_b\mathbf{v}_b)$ where m_a and m_b are constant. Thus $\Delta(m_a\mathbf{v}_a) + \Delta(m_b\mathbf{v}_b) = \mathbf{0}$ or

$$\boxed{\Delta\mathbf{G} = \mathbf{0}} \tag{3/30}$$

where the total linear momentum of the system of two particles is $\mathbf{G} = \mathbf{G}_a + \mathbf{G}_b = m_a\mathbf{v}_a + m_b\mathbf{v}_b$. Equation 3/30 states that the linear momentum of the system remains unchanged during an interval if no forces external to the system act upon it. This statement constitutes the *principle of conservation of linear momentum*. The principle can apply to one direction only, such as the x-direction, if no external forces act on the system in that direction. In Chapter 5 we will extend the principle to cover the motion of a system containing any number of interacting particles where all forces acting on the particles are actions and reactions internal to the system.

Sample Problem 3/21

The 50-g bullet traveling at 600 m/s strikes the 4-kg block centrally and is embedded within it. If the block is sliding on a smooth horizontal plane with a velocity of 12 m/s in the direction shown just before the impact, determine the velocity **v** of the block and bullet combined and its direction θ immediately after impact.

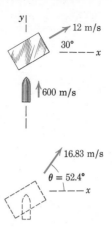

Solution. Since the force of impact is internal to the system composed of the block and bullet and since there are no other external forces acting on the system in the plane of motion, it follows that the linear momentum of the system is conserved in both the x- and y-directions. Thus

$[\Delta G_x = 0]$ $\qquad 4(12 \cos 30°) + 0 = (4 + 0.050)v_x$

$\qquad\qquad v_x = 10.26 \text{ m/s}$

$[\Delta G_y = 0]$ $\quad 4(12 \sin 30°) + 0.050(600) = (4 + 0.050)v_y$

$\qquad\qquad v_y = 13.33 \text{ m/s}$

The final velocity and its direction are given by

$[v = \sqrt{v_x{}^2 + v_y{}^2}]$ $\quad v = \sqrt{(10.26)^2 + (13.33)^2} = 16.83 \text{ m/s}$ *Ans.*

$\left[\tan \theta = \dfrac{v_y}{v_x}\right]$ $\quad \tan \theta = \dfrac{13.33}{10.26} = 1.30, \quad \theta = 52.4°$ *Ans.*

① Writing the change in momentum equals zero, $\Delta G = 0$, is the same as saying that the initial and the final momenta are equal.

Sample Problem 3/22

A small mass particle is given an initial velocity v_0 tangent to the horizontal rim of a smooth hemispherical bowl at a radius r_0 from the vertical center line, as shown at point A. As the particle slides past point B, a distance h below A and a distance r from the vertical center line, its velocity **v** makes an angle θ with the horizontal tangent to the bowl through B. Determine θ.

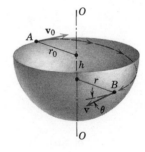

Solution. The forces on the particle are its weight and the normal reaction exerted by the smooth surface of the bowl. Neither force exerts a moment about the axis O-O, so that angular momentum is conserved about that axis. Thus

$[\Delta H_O = 0]$ $\qquad mv_0 r_0 = mvr \cos \theta$

Also, energy is conserved so that

$[\Delta T + \Delta V = 0]$ $\quad \tfrac{1}{2}m(v^2 - v_0{}^2) - mgh = 0, \quad v = \sqrt{v_0{}^2 + 2gh}$

Eliminating v and substituting $r^2 = r_0{}^2 - h^2$ give

$$v_0 r_0 = \sqrt{v_0{}^2 + 2gh}\ \sqrt{r_0{}^2 - h^2} \cos \theta$$

$$\theta = \cos^{-1} \frac{1}{\sqrt{1 + \dfrac{2gh}{v_0{}^2}}\ \sqrt{1 - \dfrac{h^2}{r_0{}^2}}} \qquad Ans.$$

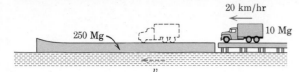

Problem 3/187

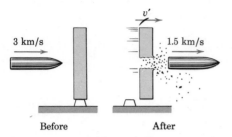

Before   After

Problem 3/188

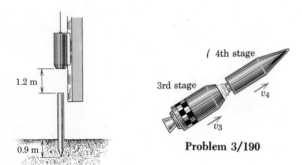

Problem 3/189

Problem 3/190

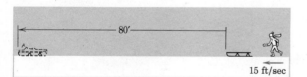

Problem 3/191

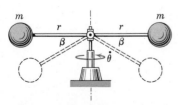

Problem 3/192

PROBLEMS

3/187 The 10-Mg truck drives onto the 250-Mg barge from the dock at 20 km/h and brakes to a stop on the deck. The barge is free to move in the water, which offers negligible resistance to motion at low speeds. Calculate the speed of the barge after the truck has come to rest on it. *Ans.* $v = 0.77$ km/h

3/188 The 180-g projectile is fired with a velocity of 3 km/s at the center of the 0.96-kg disk, which rests on a smooth support. If the projectile passes through the disk and emerges with a velocity of 1.5 km/s, determine the velocity v' of the disk immediately after the projectile clears the disk.

3/189 The 400-kg ram of a pile driver falls 1.2 m from rest and strikes the top of a 250-kg pile embedded 0.9 m in the ground. Upon impact the ram is seen to move with the pile with no noticeable rebound. Determine the velocity v of the pile and ram immediately after impact. *Ans.* $v = 2.99$ m/s

3/190 The third and fourth stages of a rocket are coasting in space with a velocity of 15 000 km/h when a small explosive charge between the stages separates them. Immediately after separation the fourth stage has increased its velocity to $v_4 = 15\,050$ km/h. What is the corresponding velocity v_3 of the third stage? At separation the third and fourth stages have masses of 500 and 80 kg, respectively.

3/191 A boy weighing 100 lb runs and jumps on his 20-lb sled with a horizontal velocity of 15 ft/sec. If the sled and boy coast 80 ft on the level snow before coming to rest, compute the coefficient of friction μ between the snow and the runners of the sled.
Ans. $\mu = 0.030$

3/192 The two balls each of mass m are mounted on the light rods which are fixed to rotate in a horizontal plane at the angular rate $\dot{\theta}$ about the vertical axis. If an internal mechanism lowers the rods to the dotted positions shown without interfering with the free rotation about the vertical axis, determine the new rate of rotation $\dot{\theta}'$ in the lowered position.

193 For a perigee altitude of 300 km and an apogee altitude of 600 km an earth satellite must have a velocity at perigee $v_P = 28\,137$ km/h. Calculate the apogee velocity v_A.

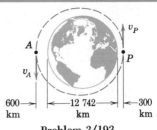

Problem 3/193

194 The ballistic pendulum is a simple device for determining ballistic velocities. The 60-g bullet is fired into the 30-kg box of sand suspended as a simple pendulum. If the pendulum is observed to swing through a maximum angle $\theta = 20°$ after the impact, compute the muzzle velocity of the bullet and the percent loss of energy of the system during impact with the sand.

    Ans. $v = 771$ m/s, 99.8% loss of energy

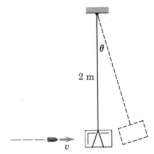

Problem 3/194

195 Freight car A with a gross weight of 150,000 lb is moving along the horizontal track in a switching yard at 2 mi/hr. Freight car B with a gross weight of 120,000 lb and moving at 3 mi/hr overtakes car A and is coupled to it. Determine (a) the common velocity v of the two cars as they move together after being coupled and (b) the loss of energy ΔE due to the impact.

    Ans. $v = 2.44$ mi/hr, $\Delta E = 2229$ ft-lb loss

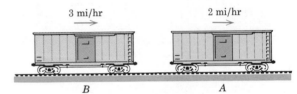

Problem 3/195

196 The central attractive force F on an earth satellite can exert no moment about the center O of the earth. For the particular elliptical orbit with major and minor axes as shown, a satellite will have a velocity of 33 880 km/h at the perigee altitude of 390 km. Determine the velocity of the satellite at point B and at apogee A. The radius of the earth is 6370 km.

    Ans. $v_B = 19\,540$ km/h
      $v_A = 11\,290$ km/h

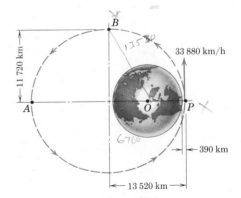

Problem 3/196

197 The 500-ton tug is towing the 900-ton coal barge at a steady speed of 6 knots. For a short period of time the stern winch takes in the towing cable at the rate of 2 ft/sec. Calculate the reduced speed v of the tug during this interval. Assume the tow cable to be horizontal. (Recall 1 knot = 1.690 ft/sec)

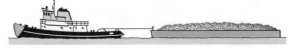

Problem 3/197

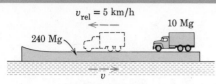

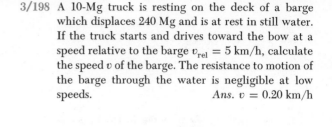

Problem 3/198

3/198 A 10-Mg truck is resting on the deck of a barge which displaces 240 Mg and is at rest in still water. If the truck starts and drives toward the bow at a speed relative to the barge $v_{rel} = 5$ km/h, calculate the speed v of the barge. The resistance to motion of the barge through the water is negligible at low speeds. *Ans.* $v = 0.20$ km/h

Problem 3/199

3/199 The block of mass m is resting on the rough inclined surface $(\mu > \tan \theta)$ when struck by a bullet of mass m_0 fired along the incline with a velocity v_0. Derive an expression for the distance s which the block and embedded bullet move up the incline.

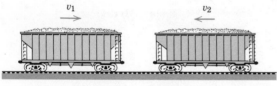

Problem 3/200

3/200 The two freight cars of equal mass approach each other with velocities v_1 and v_2 as shown and are coupled together. Derive an expression for the fraction $\Delta T/T$ of the original kinetic energy which is lost during the coupling.

$$Ans. \quad \frac{\Delta T}{T} = \frac{1}{2} + \frac{v_1 v_2}{v_1{}^2 + v_2{}^2}$$

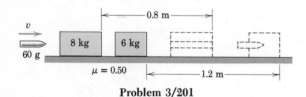

Problem 3/201

3/201 The 60-g bullet is fired at the two blocks resting on a surface where the coefficient of friction is 0.50. The bullet passes through the 8-kg block and lodges in the 6-kg block. The blocks slide the distances shown. Compute the initial velocity v of the bullet.

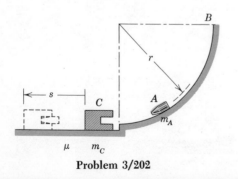

Problem 3/202

3/202 The cylindrical plug A of mass m_A is released from rest at B and slides down the smooth circular guide. The plug strikes the block C and becomes embedded in it. Write the expression for the distance s which the block and plug slide before coming to rest. The coefficient of friction between the block and the horizontal surface is μ.

$$Ans. \quad s = \frac{r}{\mu} \left(\frac{m_A}{m_A + m_C} \right)^2$$

203 The two mine cars of equal mass are connected by a rope which is initially slack. Car A is given a shove which imparts to it a velocity of 4 ft/sec with car B initially at rest. When the slack is taken up the rope suffers a tension impact which imparts a velocity to car B and reduces the velocity of car A. (a) If 40 percent of the kinetic energy of car A is lost during the rope impact, calculate the velocity v_B imparted to car B. (b) Following the initial impact car B overtakes car A and the two are coupled together. Calculate their final common velocity v_C.
       Ans. $v_B = 2.89$ ft/sec, $v_C = 2$ ft/sec

Problem 3/203

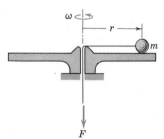

Problem 3/204

204 The small particle of mass m and its restraining cord are spinning with an angular velocity ω on the horizontal surface of a smooth disk, shown in section. As the force F is slightly relaxed, r increases and ω changes. Determine the rate of change of ω with respect to r and show that the work done by F during a movement dr equals the change in kinetic energy of the particle.

205 Each of the 4-kg balls is mounted on the frame of negligible mass and is rotating freely at a speed of 90 rev/min about the vertical with $\theta = 60°$. If the force F on the vertical control rod is increased so that the frame rotates with $\theta = 30°$, determine the new rotational speed N and the work U done by F. Point O on the rotating collar remains fixed.
       Ans. $N = 270$ rev/min, $U = 56.6$ J

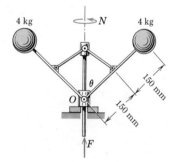

Problem 3/205

206 Two barges each with a displacement (mass) of 500 metric tons are loosely moored in calm water. A stunt driver starts his 1500-kg car from rest at A, drives along the deck, and leaves the end of the 15° ramp at a speed of 50 km/h relative to the barge and ramp. The driver successfully jumps the gap and brings his car to rest relative to barge 2 at B. Calculate the velocity v_2 imparted to barge 2 just after the car has come to rest on the barge. Neglect the resistance of the water to motion at the low velocities involved.   *Ans.* $v_2 = 40.0$ mm/s

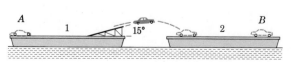

Problem 3/206

3/10 PROBLEM FORMULATION AND REVIEW. In Chapter 3 we have developed the three basic methods of solution to problems in particle kinetics. This experience is central to the study of dynamics and lays the foundation for the subsequent study of rigid-body and nonrigid-body dynamics. First, we applied Newton's second law $\Sigma \mathbf{F} = m\mathbf{a}$ to determine the instantaneous relation between forces and the acceleration they produce. With the background of Chapter 2 for identifying the kind of motion and with the aid of our familiar free-body diagram to be certain that all forces are accounted for, we were able to solve a large variety of problems using *x-y*, *n-t*, and *r-θ* coordinates for plane-motion problems and *x-y-z*, *r-θ-z*, and *R-θ-φ* coordinates for space problems.

Second, we integrated the basic equation of motion $\Sigma \mathbf{F} = m\mathbf{a}$ with respect to displacement and derived the scalar equations for work and energy. These equations enabled us to relate the initial and final velocities to the work done during the interval by forces external to our defined system. We expanded this approach to include potential energy, both elastic and gravitational. With these tools we discovered that the energy approach is especially valuable for conservative systems, that is, systems wherein the loss of energy due to friction or other forms of dissipation is negligible.

Third, we rewrote Newton's second law in the form of force equals time rate of change of linear momentum and moment equals time rate of change of angular momentum. Then we integrated these relations with respect to time and derived the impulse and momentum equations. These equations were then applied to motion intervals where the forces were functions of the time. We also investigated the interactions between particles under conditions where the linear momentum is conserved and where the angular momentum is conserved.

Difficulties encountered in the solution of problems in particle kinetics are frequently traceable to difficulty with the corresponding kinematics. It was for this reason that we strongly urged all students to review Chapter 2 as they study Chapter 3. Another common source of difficulty resides with problems which involve intermediate steps in going from the given information to the desired result. Students often wonder how to get started on a problem—what to do first. To ascertain the relation between the given and required quantities, it is frequently helpful to start by identifying the intermediate quantity or quantities upon which the desired result depends. Tracing the dependency of one quantity on another will lead back to the given information and establish in reverse the sequence of steps to be followed in carrying out the solution. This sequence of reasoning is an essential part of the vital process of problem formulation.

REVIEW PROBLEMS

207 A small vessel with a displacement (mass) of 20 t (metric tons) is moving through still water at a speed of 15 knots when its engine is put into reverse. If the propeller develops a constant reverse thrust of 10 kN and the vessel slows down to a speed of 3 knots in 8 s, compute the time average of the resistance R of the water on the hull during the 8-s interval. (1 knot = 1.852 km/h)

$$\text{Ans. } R = 5.43 \text{ kN}$$

208 The small 2-kg carriage is moving freely along the horizontal with a speed of 4 m/s at time $t = 0$. A force applied to the carriage in the direction opposite to motion produces two impulse "peaks," one after the other, as shown by the graphical plot of the readings of the instrument which measured the force. Approximate the loading by the dotted lines and determine the velocity v of the carriage for $t = 1.5$ s.

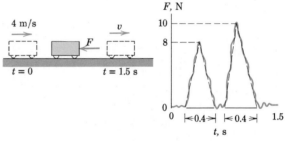

Problem 3/208

209 Assume that the lunar lander of Prob. 3/162 has an earth weight of 3220 lb and is descending onto the moon's surface with a velocity of 20 ft/sec when its retro-engine is fired. If the engine produces a thrust T for 4 sec which varies with the time as shown and then cuts off, calculate the velocity of the lander when $t = 5$ sec, assuming that it has not yet landed. Gravitational acceleration at the moon's surface is 5.32 ft/sec². $\qquad$ *Ans. $v = 10.6$ ft/sec*

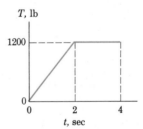

Problem 3/209

210 The figure shows a centrifugal clutch consisting in part of a rotating spider A which carries four plungers B. As the spider is made to rotate about its center with a speed ω, the plungers move outward and bear against the interior surface of the rim of wheel C, causing it to rotate. The wheel and spider are independent except for frictional contact. If each plunger has a mass of 2 kg with center of mass at G, and if the coefficient of friction between the plungers and the wheel is 0.40, calculate the maximum moment M which can be transmitted to wheel C for a spider speed of 3000 rev/min.

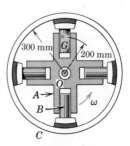

Problem 3/210

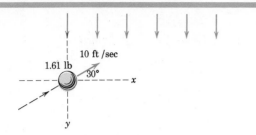

Problem 3/211

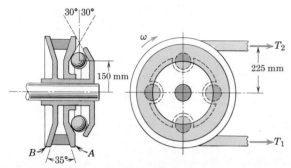

Problem 3/212

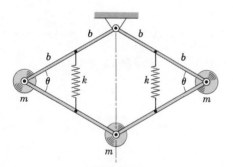

Problem 3/213

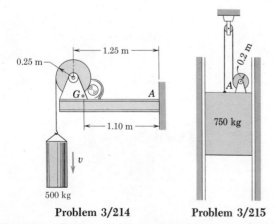

Problem 3/214   **Problem 3/215**

3/211 A 1.61-lb particle has a velocity of 10 ft/sec in the direction shown in the horizontal *x-y* plane and encounters a steady flow of air in the *y*-direction at time $t = 0$. If the *y*-component of the force exerted on the particle by the air is essentially constant and equal to 0.1 lb, determine the time *t* required for the particle to cross the fixed *x*-axis again.

Ans. $t = 5$ sec

3/212 When a V-belt drives a pulley at high speed, the centrifugal action tends to lessen its contact with the pulley and hence to reduce the capacity to transmit torque. A device to compensate for this effect is shown and consists of a cage and four balls which rotate with the pulley. The balls press against the two 30° conical surfaces and force the inner side *A* to slide to the left toward the opposite side *B*, thus tightening the belt. Part *A* is splined to *B* along the hub, so rotates with the remainder of the pulley but is free to slide on *B*. Compute the axial force *F* on *A* caused by the action of the balls for a speed of 600 rev/min, if the mass of each of the four balls is 2.5 kg. *Ans.* $F = 5130$ N

3/213 The mechanism consists of three small cylinders each of mass *m* connected by the light hinged bars. Each of the two springs has a stiffness *k* and is unstretched in the initial position shown. If the mechanism is released from rest at this position, determine the expression for the maximum downward movement *y* of the lower cylinder.

Ans. $y = 8mg/k$

3/214 The mass of the hoisting drum, motor drive, and beam together is 240 kg with a combined center of mass 1.10 m to the left of the welded attachment at *A*. If the drum is lowering the 500-kg load with a velocity *v* which increases uniformly from 0.6 m/s to 6 m/s in 4 s, calculate the bending moment *M* supported by the beam at *A* during this interval.

3/215 The 750-kg elevator is operated through the action of the rotating drum *A* around which is wrapped the hoisting cable. Determine the constant torque *M* which must be supplied to the shaft of the drum by its motor (not shown) in order to give the elevator an upward velocity of 3 m/s in a vertical rise of 4 m from rest. The mass of the drum is small, and it may be analyzed as though it were in rotational equilibrium. Neglect friction. *Ans.* $M = 820$ N·m

216 The 2-kg cylinder is released from rest against a coiled spring which has been compressed 0.5 m from its uncompressed position. If the stiffness of the spring is 120 N/m, determine (a) the maximum height h reached by the cylinder above its release position, and (b) the maximum velocity v reached by the cylinder. Neglect the mass of the spring.

$$\text{Ans. } (a) \ h = 0.764 \text{ m}$$
$$(b) \ v = 2.61 \text{ m/s}$$

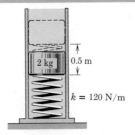

Problem 3/216

217 A spacecraft m is heading toward the center of the moon with a velocity of 3000 km/h at a distance from the moon's surface equal to the radius R of the moon. Compute the impact velocity v with the surface of the moon if the spacecraft is unable to fire its retro-rockets. Consider the moon fixed in space and neglect the gravitational attraction of the earth. The radius R of the moon is 1738 km, and the acceleration due to lunar gravity at its surface is 1.62 m/s².

$$\text{Ans. } v = 6740 \text{ km/h}$$

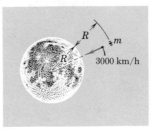

Problem 3/217

218 A body rotates with an angular acceleration α about an axis fixed in space and is starting from rest with $\omega = 0$. At any point within the body a tangential force $F_t = r\alpha$ may be associated with a unit-mass particle in order to hold it in place relative to the body. Show that this force field is nonconservative and, hence, that a force potential does not exist.

219 The small object is placed on the surface of the conical dish at the radius shown. If the coefficient of friction between the object and the conical surface is 0.30, for what range of angular velocities ω about the vertical axis will the block remain on the dish without slipping? Assume that speed changes are made slowly so that any angular acceleration may be neglected.

$$\text{Ans. } 3.41 < \omega < 7.21 \text{ rad/s}$$

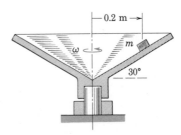

Problem 3/219

220 The rotating slotted arm and the fixed circular cam of Prob. 2/144 are repeated here. If the arm revolves around a vertical axis through O with the constant speed $\omega = \dot{\theta}$ and if the pin A has a mass m, write the expression for the contact force R exerted by the slotted arm on the pin when $\theta = \pi/2$ at which position the spring force is P. Neglect friction and pin diameter.

$$\text{Ans. } R = me\omega^2 \frac{b^2}{b^2 - e^2} + \frac{Pe}{\sqrt{b^2 - e^2}}$$

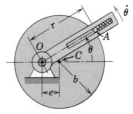

Problem 3/220

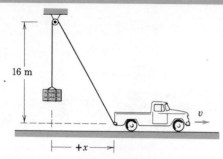

16 m

+x

Problem 3/221

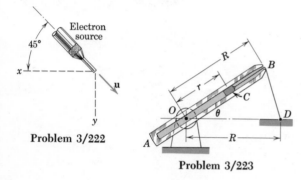

Electron source

45°

x

u

y

Problem 3/222

R

B

r

C

O

θ

D

R

A

Problem 3/223

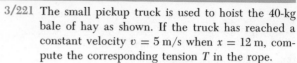

3/221 The small pickup truck is used to hoist the 40-kg bale of hay as shown. If the truck has reached a constant velocity $v = 5$ m/s when $x = 12$ m, compute the corresponding tension T in the rope.

Ans. $T = 424$ N

3/222 Electrons of mass m are introduced with a velocity $\mathbf{u} = (u/\sqrt{2})(-\mathbf{i} + \mathbf{j})$ into an electric field. The electric field or voltage gradient is the vector $\mathbf{E} = \mathbf{i}E_0 \sin pt + \mathbf{j}E_0 \cos pt$. If an electron is admitted into the field at time $t = 0$, determine its velocity $\mathbf{v}$ after $\frac{1}{4}$ cycle has elapsed $(t = \pi/2p)$. The force on the electron in the direction of the field equals $\mathbf{E}$ multiplied by the electron charge e.

3/223 The mechanism of Prob. 2/138 is repeated here. If the arm rotates about a horizontal axis through O and if the slider C has a mass m, determine the expression in terms of θ for the tension T in the cord at its attachment to C. The angular velocity of OB is $\omega = \dot{\theta}$ and is constant during the motion interval concerned. Also $r = 0$ when $\theta = 0$. All surfaces are assumed to be smooth. What is the maximum speed $\dot{\theta}$, for a given θ, of the arm before T goes to zero?

Ans. $T = m \sin \dfrac{\theta}{2} \left(2g \cos \dfrac{\theta}{2} - \dfrac{5}{2}R\dot{\theta}^2 \right)$

$$\dot{\theta}_{max} = 2\sqrt{\dfrac{g \cos (\theta/2)}{5R}}$$

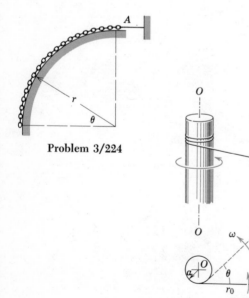

A

r

θ

Problem 3/224

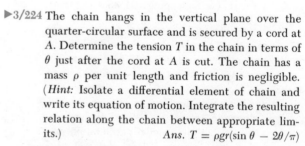

O

m

O

ω

a

O

θ

r_0

m

ω_0

Problem 3/225

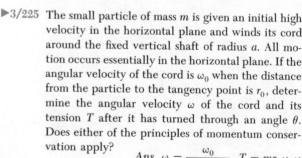

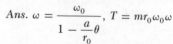

▶**3/224** The chain hangs in the vertical plane over the quarter-circular surface and is secured by a cord at A. Determine the tension T in the chain in terms of θ just after the cord at A is cut. The chain has a mass ρ per unit length and friction is negligible. (*Hint:* Isolate a differential element of chain and write its equation of motion. Integrate the resulting relation along the chain between appropriate limits.)   *Ans.* $T = \rho gr(\sin \theta - 2\theta/\pi)$

▶**3/225** The small particle of mass m is given an initial high velocity in the horizontal plane and winds its cord around the fixed vertical shaft of radius a. All motion occurs essentially in the horizontal plane. If the angular velocity of the cord is ω_0 when the distance from the particle to the tangency point is r_0, determine the angular velocity ω of the cord and its tension T after it has turned through an angle θ. Does either of the principles of momentum conservation apply?

Ans. $\omega = \dfrac{\omega_0}{1 - \dfrac{a}{r_0}\theta}$, $T = mr_0\omega_0\omega$

SPECIAL APPLICATIONS

4

4/1 INTRODUCTION. The basic principles and methods of particle kinetics were developed and illustrated in Chapter 3. This treatment included the direct use of Newton's second law, the equations of work and energy, and the equations of impulse and momentum. Special attention was paid to the kind of problem for which each of the approaches was most appropriate.

There are several topics of specialized interest in particle kinetics which will be treated briefly in this separate chapter. These topics involve further extension and application of the principles of Chapter 3, and their study will help to broaden one's background in dynamics.

4/2 CENTRAL FORCE MOTION. When a particle moves under the influence of a force directed toward a fixed center of attraction, the motion is called central-force motion. The most common example of central-force motion is found with the orbits of planets and satellites. The laws which govern this motion were deduced from observation of the motions of the planets by J. Kepler (1571–1630). The dynamics of central-force motion is basic to the design of high-altitude rockets, earth satellites, and space vehicles of all types.

Consider a particle of mass m, Fig. 4/1, moving under the action of the central gravitational attraction

$$F = K\frac{mm_0}{r^2}$$

where m_0 is the mass of the attracting body assumed to be fixed, K is the universal gravitational constant, and r is the distance between the centers of the masses. The particle of mass m could represent the earth moving about the sun, the moon moving about the earth, or a satellite in its orbital motion about the earth above the atmosphere. The most convenient coordinate system to use is polar coordinates in the plane of motion since $\mathbf{F}$ will always be in the negative r-direction and there is no force in the θ-direction.

Equations 3/8 may be applied directly for the r- and θ-directions to give

$$-K\frac{mm_0}{r^2} = m(\ddot{r} - r\dot{\theta}^2)$$

$$0 = m(r\ddot{\theta} + 2\dot{r}\dot{\theta})$$

$$(4/1)$$

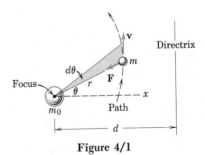

Figure 4/1

The second of the two equations when multiplied by r/m is seen to be the same as $d(r^2\dot{\theta})/dt = 0$, which is integrated to give

$$r^2\dot{\theta} = h, \text{ a constant} \tag{4/2}$$

The physical significance of Eq. 4/2 is made clear when we note that the angular momentum $\mathbf{r} \times m\mathbf{v}$ of m about m_0 has the magnitude $mr^2\dot{\theta}$. Thus Eq. 4/2 merely states that the angular momentum of m about m_0 remains constant, or is conserved. This statement is easily deduced from Eq. 3/27 where it is observed that the angular momentum $\mathbf{H}_O$ remains constant (is conserved) if there is no moment acting on the particle about a fixed point O.

We observe that during time dt the radius vector sweeps out an area, shaded in Fig. 4/1, equal to $dA = (\frac{1}{2}r)(r\,d\theta)$. Therefore the rate at which area is swept by the radius vector is $\dot{A} = \frac{1}{2}r^2\dot{\theta}$, which is constant according to Eq. 4/2. This conclusion is expressed in Kepler's *second law* of planetary motion which states that the areas swept through in equal times are equal.

The shape of the path followed by m may be obtained by solving the first of Eqs. 4/1 with the time t eliminated through combination with Eq. 4/2. To this end the mathematical substitution $r = 1/u$ is useful. Thus $\dot{r} = -(1/u^2)\dot{u}$, which from Eq. 4/2 becomes $\dot{r} = -h(\dot{u}/\dot{\theta})$ or $\dot{r} = -h(du/d\theta)$. The second time derivative is $\ddot{r} = -h(d^2u/d\theta^2)\dot{\theta}$, which, by combining with Eq. 4/2, becomes $\ddot{r} = -h^2u^2(d^2u/d\theta^2)$. Substitution into the first of Eqs. 4/1 now gives

$$-Km_0u^2 = -h^2u^2\frac{d^2u}{d\theta^2} - \frac{1}{u}h^2u^4$$

or

$$\frac{d^2u}{d\theta^2} + u = \frac{Km_0}{h^2} \tag{4/3}$$

which is a nonhomogeneous linear differential equation. The solution of this familiar second order equation may be verified by direct substitution and is

$$u = \frac{1}{r} = C\cos(\theta + \delta) + \frac{Km_0}{h^2}$$

where C and δ are the two integration constants. The phase angle δ may be eliminated by choosing the x-axis so that r is a minimum when $\theta = 0$. Thus

$$\frac{1}{r} = C\cos\theta + \frac{Km_0}{h^2} \tag{4/4}$$

The interpretation of Eq. 4/4 requires a knowledge of the equations for conic sections. We recall that a conic section is formed by the locus of a point which moves so that the ratio e of its distance from a point (focus) to a line (directrix) is constant. Thus from Fig. 4/1, $e = r/(d - r \cos \theta)$, which may be rewritten as

$$\frac{1}{r} = \frac{1}{d} \cos \theta + \frac{1}{ed} \qquad (4/5)$$

which is the same form as Eq. 4/4. Hence it is seen that the motion of m is along a conic section with $d = 1/C$ and $ed = h^2/(Km_0)$ or

$$e = \frac{h^2 C}{Km_0} \qquad (4/6)$$

There are three cases to be investigated corresponding to $e < 1$ (ellipse), $e = 1$ (parabola), and $e > 1$ (hyperbola). The trajectory for each of these cases is shown in Fig. 4/2.

Case I: ellipse $(e < 1)$. From Eq. 4/5 it is seen that r is a minimum when $\theta = 0$ and is a maximum when $\theta = \pi$. Thus,

$$2a = r_{\min} + r_{\max} = \frac{ed}{1 + e} + \frac{ed}{1 - e} \qquad \text{or} \qquad a = \frac{ed}{1 - e^2}$$

With the distance d expressed in terms of a, Eq. 4/5 and the maximum and minimum values of r may be written as

$$\frac{1}{r} = \frac{1 + e \cos \theta}{a(1 - e^2)} \qquad (4/7)$$

$$r_{\min} = a(1 - e) \qquad r_{\max} = a(1 + e)$$

In addition, the relation $b = a\sqrt{1 - e^2}$, which comes from the geometry of the ellipse, gives the expression for the semiminor axis. It is seen that the ellipse becomes a circle with $r = a$ when $e = 0$. Equation 4/7 is an expression of Kepler's *first law* which says that the planets move in elliptical orbits around the sun as a fucus.

The period τ for the elliptical orbit is the total area A of the ellipse divided by the constant rate $\dot{A}$ at which area is swept through. Thus

$$\tau = A/\dot{A} = \frac{\pi ab}{\frac{1}{2} r^2 \dot{\theta}} \qquad \text{or} \qquad \tau = \frac{2\pi ab}{h}$$

by Eq. 4/2. Substitution of Eq. 4/6, the identity $d = 1/C$, the geometric relationships $a = ed/(1 - e^2)$ and $b = a\sqrt{1 - e^2}$ for the ellipse, and the equivalence $Km_0 = gR^2$ yields upon simplification

$$\boxed{\tau = 2\pi \frac{a^{3/2}}{R\sqrt{g}}} \qquad (4/8)$$

In this equation it is noted that R is the mean radius of the central attracting body and g is the absolute value of the acceleration due to gravity at the surface of the attracting body.

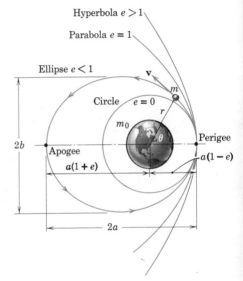

Figure 4/2

Equation 4/8 expresses Kepler's *third law* of planetary motion which states that the square of the period of motion is proportional to the cube of the semimajor axis of the orbit.

Case II: parabola $(e = 1)$. Equations 4/5 and 4/6 become

$$\frac{1}{r} = \frac{1}{d}(1 + \cos\theta) \qquad \text{and} \qquad h^2 C = K m_0$$

The radius vector and the dimension a become infinite as θ approaches π.

Case III; hyperbola $(e > 1)$. From Eq. 4/5 it is seen that the radial distance r becomes infinite for the two values of the polar angle θ_1 and $-\theta_1$ defined by $\cos\theta_1 = -1/e$. Only branch I corresponding to $-\theta_1 < \theta < \theta_1$, Fig. 4/3, represents a physically possible motion. Branch II corresponds to angles in the remaining sector (with r negative). For this branch positive r's may be used if θ is replaced by $\theta - \pi$ and $-r$ by r. Thus Eq. 4/5 becomes

$$\frac{1}{-r} = \frac{1}{d}\cos(\theta - \pi) + \frac{1}{ed} \qquad \text{or} \qquad \frac{1}{r} = -\frac{1}{ed} + \frac{\cos\theta}{d}$$

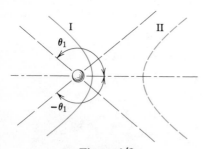

Figure 4/3

But this expression contradicts the form of Eq. 4/4 where Km_0/h^2 is necessarily positive. Hence branch II does not exist (except for repulsive forces).

Now consider the energies of particle m. The system is conservative, and the constant energy E of m is the sum of its kinetic energy T and potential energy V. The kinetic energy is $T = \frac{1}{2}mv^2 = \frac{1}{2}m(\dot{r}^2 + r^2\dot{\theta}^2)$ and the potential energy from Eq. 3/15 is $V = -mgR^2/r$. Recall that g is the absolute acceleration due to gravity measured at the surface of the attracting body, R is the radius of the attracting body, and $Km_0 = gR^2$. Thus

$$E = \frac{1}{2}m(\dot{r}^2 + r^2\dot{\theta}^2) - \frac{mgR^2}{r}$$

This constant value of E can be determined from its value at $\theta = 0$ where $\dot{r} = 0$, $1/r = C + gR^2/h^2$ from Eq. 4/4, and $r\dot{\theta} = h/r$ from Eq. 4/2. Substitution into the expression for E and simplification yield

$$\frac{2E}{m} = h^2 C^2 - \frac{g^2 R^4}{h^2}$$

Now eliminate C by substitution of Eq. 4/6, which may be written as $h^2 C = egR^2$, and obtain

$$e = +\sqrt{1 + \frac{2Eh^2}{mg^2R^4}} \tag{4/9}$$

The plus value of the radical is mandatory since by definition e is positive. It is now seen that for the elliptical orbit $e < 1$, E is negative

parabolic orbit $e = 1$, E is zero

hyperbolic orbit $e > 1$, E is positive

These conclusions, of course, depend on the arbitrary selection of the datum condition for zero potential energy ($V = 0$ when $r = \infty$).

The expression for the velocity v of m may be found from the energy equation, which is

$$\tfrac{1}{2}mv^2 - \frac{mgR^2}{r} = E$$

The total energy E is obtained from Eq. 4/9 by combining Eq. 4/6 and $1/C = d = a(1 - e^2)/e$ to give for the elliptical orbit

$$E = -\frac{gR^2m}{2a} \qquad (4/10)$$

Substitution into the energy equation yields

$$v^2 = 2gR^2\left(\frac{1}{r} - \frac{1}{2a}\right) \qquad (4/11)$$

from which the magnitude of the velocity may be computed for a particular orbit in terms of the radial distance r. Combination of the expressions for $r_{\min}$ and $r_{\max}$ corresponding to perigee and apogee, Eq. 4/7, with Eq. 4/11 gives for the respective velocities at these two positions for the elliptical orbit

$$v_P = R\sqrt{\frac{g}{a}}\sqrt{\frac{1 + e}{1 - e}} = R\sqrt{\frac{g}{a}}\sqrt{\frac{r_{\max}}{r_{\min}}}$$

$$\qquad (4/12)$$

$$v_A = R\sqrt{\frac{g}{a}}\sqrt{\frac{1 - e}{1 + e}} = R\sqrt{\frac{g}{a}}\sqrt{\frac{r_{\min}}{r_{\max}}}$$

Selected numerical data pertaining to the solar system are included in Appendix C and will be found useful in applying the foregoing relationships to problems in planetary motion.

The foregoing analysis is based on the assumption that m_0 is fixed in space. For an artificial satellite in earth orbit the error of this assumption is clearly negligible since the ratio of the mass of the earth to that of the satellite is so great. However, for the earth-moon system a small but significant error is involved in the assumption of a fixed earth. Accounting for the motion of both attracting bodies is known as the "two-body problem," and the student is referred to more advanced treatments[*] for this analysis.

[*] See, for example, the author's *Dynamics*, 2nd Edition 1971 (also *SI Version* 1975) John Wiley & Sons.

Sample Problem 4/1

An artificial satellite is launched from point B on the equator by its carrier rocket and inserted into an elliptical orbit with a perigee altitude of 2000 km. If the apogee altitude is to be 4000 km, compute (*a*) the necessary perigee velocity v_P and the corresponding apogee velocity v_A, (*b*) the velocity at point C where the altitude of the satellite is 2500 km, and (*c*) the period τ for a complete orbit.

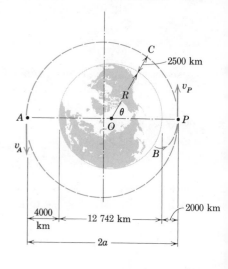

Solution. (*a*) The perigee and apogee velocities for specified altitudes are given by Eqs. 4/12 where

$$r_{max} = 6371 + 4000 = 10\ 371 \text{ km}$$

①

$$r_{min} = 6371 + 2000 = 8371 \text{ km}$$

$$a = (r_{min} + r_{max})/2 = 9371 \text{ km}$$

Thus

$$v_P = R\sqrt{\frac{g}{a}}\sqrt{\frac{r_{max}}{r_{min}}} = 6371(10^3)\sqrt{\frac{9.824}{9371(10^3)}}\sqrt{\frac{10\ 371}{8371}}$$

$$= 7261 \text{ m/s} \quad \text{or} \quad 26\ 140 \text{ km/h} \qquad Ans.$$

$$v_A = R\sqrt{\frac{g}{a}}\sqrt{\frac{r_{min}}{r_{max}}} = 6371(10^3)\sqrt{\frac{9.824}{9371(10^3)}}\sqrt{\frac{8371}{10\ 371}}$$

$$= 5861 \text{ m/s} \quad \text{or} \quad 21\ 098 \text{ km/h} \qquad Ans.$$

(*b*) For an altitude of 2500 km the radial distance from the earth's center is $r = 6371 + 2500 = 8871$ km. From Eq. 4/11 the velocity at point C becomes

②

$$v_C{}^2 = 2gR^2\left(\frac{1}{r} - \frac{1}{2a}\right) = 2(9.824)[(6371)(10^3)]^2\left(\frac{1}{8871} - \frac{1}{18\ 742}\right)\frac{1}{10^3}$$

$$= 47.348(10^6) \text{ (m/s)}^2$$

$$v_C = 6881 \text{ m/s} \quad \text{or} \quad 24\ 772 \text{ km/h} \qquad Ans.$$

(*c*) The period of the orbit is given by Eq. 4/8, which becomes

③

$$\tau = 2\pi\frac{a^{3/2}}{R\sqrt{g}} = 2\pi\frac{[(9371)(10^3)]^{3/2}}{(6371)(10^3)\sqrt{9.824}} = 9026 \text{ s}$$

$$\text{or} \quad \tau = 2.507 \text{ h} \qquad Ans.$$

① The mean radius of $12\ 742/2 = 6371$ km from Table C2 in Appendix C is used. Also the absolute acceleration due to gravity $g = 9.824$ m/s^2 from Art. 1/5 will be used.

② We must be careful with units. It is often safer to work in base units, meters in this case, and convert later.

③ We should observe here that the time interval between successive overhead transits of the satellite as recorded by an observer on the equator is longer than the period calculated here since the observer will have moved in space due to the counterclockwise rotation of the earth as seen looking down on the north pole.

PROBLEMS

(Unless otherwise indicated the velocities mentioned in the problems which follow are measured from a nonrotating reference frame moving with the center of the attracting body. Use $g = 9.824 \text{ m/s}^2$ (32.22 ft/sec^2) for the absolute gravitational acceleration at the surface of the earth.)

4/1 Identify the location on the earth and the direction of flight trajectory which require the least expenditure of fuel for the launching of a satellite into an earth orbit.

4/2 Calculate the velocity of a spacecraft which orbits the moon in a circular path of 80-km altitude.
$$Ans. \ v = 5910 \text{ km/h}$$

4/3 Calculate the time τ required for the spacecraft of Prob. 4/2 to make one complete circular orbit around the moon at the altitude of 80 km.

4/4 Show that Eq. 4/11 reduces to the equation for the velocity of a satellite in a circular orbit of altitude H when $e = 0$ and $a = R + H$ (see Prob. 3/81).

4/5 For a certain satellite with an apogee altitude of 1242 mi the ratio of its maximum to its minimum orbital velocity is 1.2. Compute the perigee altitude H_p.
$$Ans. \ H_p = 375 \text{ mi}$$

4/6 For an earth satellite moving in an elliptical orbit, show that the distance r from the center of the earth to the satellite is given by the expression $r = r_{max}r_{min}/a$ for the position corresponding to an angular displacement $\theta = \pi/2$ of the r-vector beyond the perigee position. The distance between the perigee and apogee (major axis) is $2a$.

4/7 A rocket is orbiting the earth in a circular path of altitude H. If the jet engine is activated to give the rocket a sudden burst of speed, write the expression for the velocity v which must be attained in order to escape from the influence of the earth.
$$Ans. \ v = R\sqrt{\frac{2g}{R + H}}$$

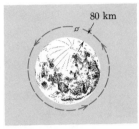

80 km

Problem 4/2

4/8 Compute the minimum velocity in mi/hr for an earth satellite in its orbit if its perigee and apogee altitudes above the earth are equal to the radius and two times the radius of the earth, respectively.

Ans. $v = 9136$ mi/hr

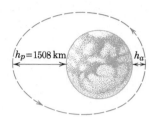

Problem 4/9

4/9 The Mars orbiter for the Viking mission was designed to make one complete trip around the planet in exactly the same time that it takes Mars to revolve once about its own axis. This time is 24 h, 37 min, 23 s. Therefore in this way it becomes possible for the orbiter to pass over the landing site of the lander capsule at the same time in each Martian day at the orbiter's minimum (periapsis) altitude. For the Viking I mission the periapsis altitude of the orbiter was 1508 km. Make use of the data in Table C2 in Appendix C and compute the maximum (apoapsis) altitude h_a for the orbiter in its elliptical path.

4/10 If the perigee altitude of an earth satellite is 240 km and the apogee altitude is 400 km, compute the eccentricity e of the orbit and the period τ of one complete orbit in space.

Ans. $e = 0.011\,96$, $\tau = 1$ h 30 min 46 s

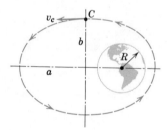

Problem 4/11

4/11 Prove that the velocity of a satellite traveling in an elliptical orbit equals $R\sqrt{g/a}$ when it reaches point C on the end of the semiminor axis. Also show that this velocity is the same as for a circular orbit of radius a.

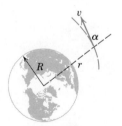

Problem 4/12

4/12 If an earth satellite has a velocity v whose direction makes an angle α with the radius r from the center of the earth to the satellite, derive an expression for the radius of curvature ρ of the path at this particular position. The radius of the earth is R.

$$Ans. \ \rho = \frac{r^2 v^2}{gR^2 \sin \alpha}$$

Problem 4/13

4/13 An artificial satellite is injected into orbit at an altitude H with an absolute velocity v_0 as shown. Determine the expression for the velocity v of the satellite during its orbit in terms of its radial distance r from the center of the earth.

/14 A satellite passes over the north pole at the perigee altitude of 500 km in an elliptical orbit of eccentricity $e = 0.7$. Calculate the absolute velocity v of the satellite as it crosses the equator.

   Ans. $v = 25\ 670$ km/h

4/15 A space capsule which executes a polar orbit comes within 150 km of the north pole at its point of closest approach to the earth. If the capsule passes over the pole once every 90 min, calculate its velocity v over the north pole.

4/16 In one of the orbits of the Apollo spacecraft about the moon its distance from the lunar surface varied from 60 to 180 mi. Compute the maximum velocity of the spacecraft in this orbit.

   Ans. $v_P = 3745$ mi/hr

4/17 It is proposed to place an observation satellite in a circular orbit around the earth so that it will remain above the same spot on the earth's surface at all times. Compute the required altitude H and absolute velocity v at which injection into orbit should occur. What limitations on latitude exist for the spot beneath the satellite?

4/18 Determine the angle β made by the velocity vector **v** with respect to the θ-direction for an earth satellite traveling in an elliptical orbit of eccentricity e. Express β in terms of the angle θ measured from perigee.

   Ans. $\tan \beta = \dfrac{e \sin \theta}{1 + e \cos \theta}$

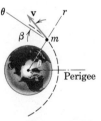

Problem 4/18

4/19 If the earth were suddenly deprived of its orbital velocity around the sun, how long would it take for the earth to "fall" into the sun? (*Hint:* The time would be one half the period of a degenerate elliptical orbit around the sun with the semiminor diameter approaching zero.) Assume a fixed sun.

4/20 If the satellite of Sample Problem 4/1 has a mass of 500 kg, compute the net amount of energy E imparted to the satellite from earth sources to put it into its orbit.   *Ans. $E = 20.6$ GJ*

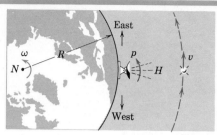

Problem 4/21

4/21 A satellite moving in a west-to-east equatorial orbit is observed by a tracking station located on the equator. If the satellite has a perigee altitude $H = 150$ km with velocity v directly over the station and an apogee altitude of 1500 km, determine an expression for the angular rate p (relative to the earth) at which the antenna dish must be rotated when the satellite is directly overhead. Compute p. The angular velocity of the earth is ω.

$$\text{Ans. } p = \frac{v - R\omega}{H} - \omega = 0.0513 \text{ rad/s}$$

4/22 Assume that a tower could be erected to a height of 800 km at the equator. Compute the muzzle velocity u (relative to the gun) of a bullet to be fired horizontally from the tower in order that it orbit the earth in a circular path if the bullet is aimed (a) south and (b) west.   *Ans.* (a) $u = 26\,770$ km/h
(b) $u = 28\,680$ km/h

4/23 Derive an expression for the radius of curvature ρ of the path of an earth satellite at its nearest point of approach to the earth. The perigee altitude above the earth is H, the radius of the earth is R, and the eccentricity of the orbit is e.

$$\text{Ans. } \rho = (R + H)(1 + e)$$

4/24 A satellite is moving in a circular orbit at an altitude of 200 mi above the surface of the earth. If a rocket motor on the satellite is activated to produce a velocity increase of 1000 ft/sec in the direction of its motion during a very short interval of time, calculate the altitude H of the satellite at its new apogee position.   *Ans.* $H = 928$ mi

4/25 A satellite is placed in a circular polar orbit a distance H above the earth. As the satellite goes over the north pole at A its retrojet is activated to produce a burst of negative thrust which reduces its velocity to a value that will ensure an equatorial landing. Derive the expression for the required reduction Δv_A of velocity at A. Note that A is the apogee of the elliptical path.

$$\text{Ans. } \Delta v_A = R\sqrt{\frac{g}{R + H}}\left(1 - \sqrt{\frac{R}{R + H}}\right)$$

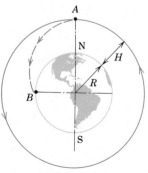

Problem 4/25

4/26 A spacecraft with a mass of 800 kg is traveling in a circular orbit 6000 km above the earth. It is desired to change the orbit to an elliptical one with a perigee altitude of 3000 km as shown. The transition is made by firing its retro-engine at A with a reverse thrust of 2000 N. Calculate the required time t for the engine to be activated.   *Ans. $t = 162$ s*

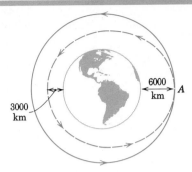

Problem 4/26

4/27 A space vehicle moving in a circular orbit of radius r_1 transfers to a larger circular orbit of radius r_2 by means of an elliptical path between A and B. (This transfer path is known as the Hohmann transfer ellipse.) The transfer is accomplished by a burst of speed Δv_A at A and a second burst of speed Δv_B at B. Write expressions for Δv_A and Δv_B in terms of the radii shown and the value g of the acceleration due to gravity at the earth's surface. If each Δv is positive, how can the velocity for path 2 be less than the velocity for path 1? Compute each Δv if $r_1 = (6370 + 500)$ km and $r_2 = (6370 + 1000)$ km.

$$Ans. \ \Delta v_A = R\sqrt{\frac{g}{r_1}}\left(\sqrt{\frac{2r_2}{r_1 + r_2}} - 1\right)$$

$$= 132.6 \ \text{m/s}$$

$$\Delta v_B = R\sqrt{\frac{g}{r_2}}\left(1 - \sqrt{\frac{2r_1}{r_1 + r_2}}\right)$$

$$= 130.2 \ \text{m/s}$$

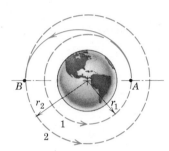

Problem 4/27

4/28 A space station S is being assembled in its circular orbit 1200 km above the earth. The next payload P of parts along with its carrier rocket has a mass of 800 kg and is injected into a coasting orbit at P, which is 450 km above the earth. Determine the angle θ which specifies the relative position between S and P at injection in order that rendezvous will occur at A with parallel paths. If the carrier rocket can develop a thrust of 900 N, determine the time t during which the engines should be fired as P approaches S at A in order that the relative velocity between P and S will be zero at rendezvous.   *Ans. $\theta = 13.2°$, $t = 170$ s*

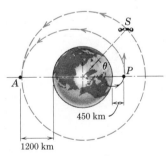

Problem 4/28

4/3 IMPACT. The principles of impulse and momentum find important use in describing the behavior of colliding bodies. *Impact* refers to the collision between two bodies and is characterized by the generation of relatively large contact forces which act over a very short interval of time. In times past impact theory has been exceedingly difficult to verify experimentally by virtue of the short time intervals available for measurements. However, with the advent of modern electronic instrumentation reliable data on many impact problems have become available in more recent years.

(a) Direct Central Impact. As an introduction to impact we consider the collinear motion of two spheres of masses m_1 and m_2, Fig. 4/4a, traveling with velocities v_1 and v_2. If v_1 is greater than v_2, collision occurs with the contact forces directed along the line of centers. This condition is called *direct central impact.* Following contact a short period of deformation takes place until the contact area between the spheres ceases to increase. At this instant both spheres, Fig. 4/4b, are moving with the same velocity v_0. During the remainder of contact a period of restoration occurs during which the contact area reduces to zero. The final condition is shown in the *c*-part of the figure where the spheres now have new velocities v_1' and v_2' where of necessity v_1' is less than v_2'. All velocities are arbitrarily assumed positive to the right, so that with this scalar notation a velocity to the left would carry a negative sign. If the impact is not overly severe and if the spheres are highly elastic, they will regain their original shape following the restoration. With a more severe impact and with less elastic bodies a permanent deformation may result.

Inasmuch as the contact forces are equal and opposite during impact, the linear momentum of the system remains unchanged, as discussed in Art. 3/9. Thus we apply the law of conservation of momentum $\Delta G = 0$ and write

$$m_1 v_1 + m_2 v_2 = m_1 v_1' + m_2 v_2' \qquad (4/13)$$

It is assumed that any forces other than the large internal forces of contact which may act on the spheres during impact are relatively small and produce negligible impulses compared with the impulse associated with each internal impact force.

For given initial conditions the momentum equation contains two unknowns, v_1' and v_2'. Clearly an additional relationship is required before the final velocities can be found. This relationship must reflect the capacity of the contacting bodies to recover from the impact and can be expressed by the ratio e of the magnitude of the restoration impulse to the magnitude of the deformation impulse. This ratio is called the *coefficient of restitution.* If F_r and F_d represent the magnitudes of the contact forces during the restoration and deformation periods, respectively, as shown in Fig. 4/5, for particle 1

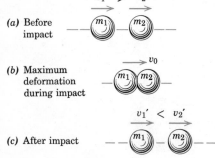

(a) Before impact

$$v_1 > v_2$$

(b) Maximum deformation during impact

$$v_0$$

(c) After impact

$$v_1' < v_2'$$

Figure 4/4

Deformation period

$$v_1 \qquad v_2$$
$$\leftarrow F_d \rightarrow$$

Restoration period

$$v_0 \qquad v_0$$
$$\leftarrow F_r \rightarrow$$
$$v_1' \qquad v_2'$$

Figure 4/5

the definition of e together with the impulse-momentum equation give us

$$e = \frac{\int_{t_0}^{t} F_r \, dt}{\int_{0}^{t_0} F_d \, dt} = \frac{m_1(-v_1' - [-v_0])}{m_1(-v_0 - [-v_1])} = \frac{v_0 - v_1'}{v_1 - v_0}$$

Similarly for particle 2 we have

$$e = \frac{\int_{t_0}^{t} F_r \, dt}{\int_{0}^{t_0} F_d \, dt} = \frac{m_2(v_2' - v_0)}{m_2(v_0 - v_2)} = \frac{v_2' - v_0}{v_0 - v_2}$$

We are careful in these equations to express the change of momentum (and hence Δv) in the same direction as the impulse (and hence the force). The time for the deformation is taken as t_0 and the total time of contact is t. Eliminating v_0 between the two expressions for e gives us

$$e = \frac{v_2' - v_1'}{v_1 - v_2} = \frac{|\text{relative velocity of separation}|}{|\text{relative velocity of approach}|} \qquad (4/14)$$

In addition to the initial conditions, if e is known for the impact condition at hand, then Eqs. 4/13 and 4/14 give us two equations in the two unknown final velocities.

Impact phenomena are almost always accompanied by energy loss, which may be calculated by subtracting the kinetic energy of the system just after impact from that just before impact. Energy is lost through the generation of heat during the localized inelastic deformation of the material, through the generation and dissipation of elastic stress waves within the bodies, and through the generation of sound energy.

According to this classical theory of impact the value $e = 1$ means that the capacity of the two particles to recover equals their tendency to deform. This condition is one of *elastic impact* with no energy loss. The value $e = 0$, on the other hand, describes *inelastic* or *plastic impact* where the particles cling together after collision and the loss of energy is a maximum. All impact conditions lie somewhere between these two extremes. Also it should be noted that a coefficient of restitution must be associated with a *pair* of contacting bodies.

The coefficient of restitution is frequently considered a constant for given geometries and a given combination of contacting materials. Actually it depends upon the impact velocity and approaches unity as the impact velocity approaches zero as shown schematically in Fig. 4/6. A handbook value for e is generally unreliable.

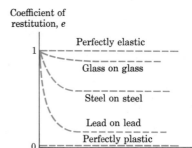

Figure 4/6

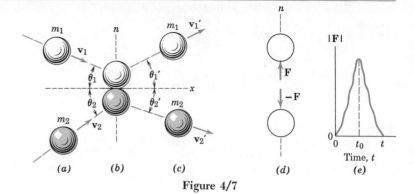

Figure 4/7

(*b*) *Oblique Central Impact.*   We now extend the relationships developed for direct central impact to the case where the initial and final velocities are not parallel, Fig. 4/7. Here spherical particles of mass m_1 and m_2 have initial velocities v_1 and v_2 in the same plane and approach each other on a collision course, as shown in the *a*-part of the figure. The directions of the velocity vectors are measured arbitrarily from the direction tangent to the contacting surfaces, Fig. 4/7*b*. The final rebound conditions are shown in the *c*-part of the figure. The impact forces are **F** and $-$**F** as seen in the *d*-part of the figure. They vary from zero to their peak value during the deformation portion of the impact and again back to zero during the restoration period as indicated in the *e*-part of the figure where t is the duration of the impact interval.

For given initial conditions of m_1, m_2, v_1, v_2, θ_1, and θ_2, there will be four unknowns, namely, v_1', v_2', θ_1', and θ_2'. The four needed equations are:

(1) Momentum of the system is conserved in the *n*-direction giving

$$-m_1 v_1 \sin \theta_1 + m_2 v_2 \sin \theta_2 = m_1 v_1' \sin \theta_1' - m_2 v_2' \sin \theta_2'$$

(2) and (3) The momentum for each particle is conserved in the *x*-direction since there is no impulse on either particle in the *x*-direction. Thus

$$m_1 v_1 \cos \theta_1 = m_1 v_1' \cos \theta_1'$$
$$m_2 v_2 \cos \theta_2 = m_2 v_2' \cos \theta_2'$$

(4) The coefficient of restitution, as in the case of direct central impact, is the ratio of the magnitudes of the recovery impulse to the deformation impulse. Equation 4/14 applies, then, to the velocity components in the *n*-direction, and for the notation adopted with Fig. 4/7 we have

$$e = \frac{v_1' \sin \theta_1' + v_2' \sin \theta_2'}{v_1 \sin \theta_1 + v_2 \sin \theta_2}$$

Sample Problem 4/2

The ram of a pile driver has a mass of 800 kg and is released from rest 2 m above the top of the 2400-kg pile. If the ram is observed to rebound to a height of 0.1 m immediately after impact with the pile, calculate (a) the velocity v_p' of the pile immediately after impact, (b) the coefficient of restitution e which applies, and (c) the percentage loss of energy due to the impact.

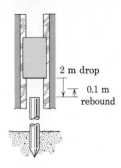

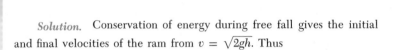

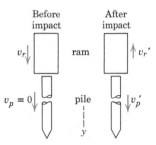

Solution. Conservation of energy during free fall gives the initial and final velocities of the ram from $v = \sqrt{2gh}$. Thus

$$v_r = \sqrt{2(9.81)(2)} = 6.26 \text{ m/s}, \qquad v_r' = \sqrt{2(9.81)(0.1)} = 1.40 \text{ m/s}$$

(a) Conservation of momentum for the ram and pile gives

$$[\Delta G_y = 0] \qquad 800(6.26) + 0 = 800(-1.40) + 2400v_p'$$

$$v_p' = 2.55 \text{ m/s} \qquad\qquad Ans.$$

① Note carefully the negative sign for the momentum of the ram immediately after impact since its velocity is in the minus y-direction. The impulses of the weights of the ram and pile are very small compared with the impulses of the impact forces so are neglected.

(b) The coefficient of restitution gives

$$e = \frac{|\text{rel. vel. separation}|}{|\text{rel. vel. approach}|} \qquad e = \frac{2.55 + 1.40}{6.26 + 0} = 0.63 \qquad Ans.$$

(c) The kinetic energy of the system just before impact is the same as the potential energy of the ram above the pile and is

$$T = V_g = mgh = 800(9.81)(2) = 15\ 700 \text{ J}$$

The kinetic energy T' just after impact is

$$T' = \tfrac{1}{2}(800)(1.40)^2 + \tfrac{1}{2}(2400)(2.55)^2 = 8590 \text{ J}$$

The percentage loss of energy is, therefore,

$$\frac{15\ 700 - 8590}{15\ 700}(100) = 45.3\% \qquad\qquad Ans.$$

Sample Problem 4/3

A steel ball is projected onto the heavy metal plate with a velocity of 50 ft/sec at the 30° angle shown. If the coefficient of restitution between the ball and the plate is 0.5, compute the rebound velocity v' and its angle θ'.

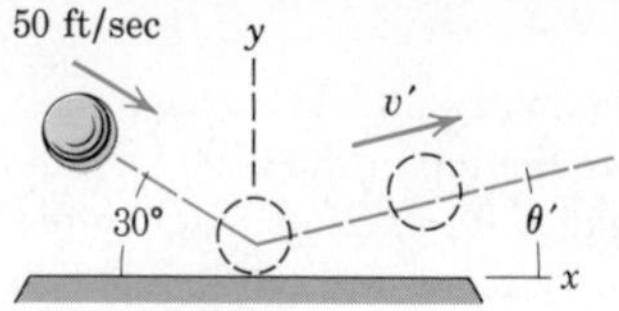

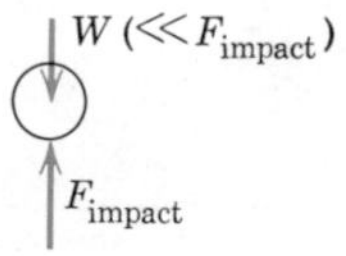

Solution. The mass of the heavy plate may be considered infinite and its corresponding velocity zero after impact. The coefficient of restitution is applied to the velocity components normal to the plate in the direction of the impact force and gives

① $e = \dfrac{|\text{rel. vel. separation}|}{|\text{rel. vel. approach}|}$, $\quad 0.5 = \dfrac{v' \sin \theta' - 0}{50 \sin 30° + 0}$, $\quad v' \sin \theta' = 12.5 \text{ ft/sec}$

Momentum of the ball in the x-direction is unchanged since, with assumed smooth surfaces, there is no force acting on the ball in that direction. Thus

$$m(50 \cos 30°) = m(v' \cos \theta') \qquad \text{or} \qquad v' \cos \theta' = 43.3 \text{ ft/sec}$$

Solution of the two equations gives $\tan \theta' = 12.5/43.3$, $\theta' = 16.1°$

and $\qquad\qquad v' = \dfrac{12.5}{\sin 16.1°} = 45.1 \text{ ft/sec.}$ *Ans.*

① We observe here that with infinite mass there is no way of applying the principle of conservation of momentum for the system in the y-direction. From the free-body diagram of the ball during impact we note that the impulse of the weight W is neglected since W is very small compared with the impact force.

Sample Problem 4/4

Spherical particle 1 has a velocity $v_1 = 6$ m/s in the direction shown in the a-part of the figure and collides with spherical particle 2 of equal mass and diameter and initially at rest. If the coefficient of restitution for these conditions is 0.6, determine the resulting motion of each particle following impact. Also calculate the percentage loss of energy due to the impact.

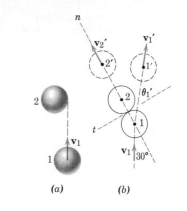

(a)   (b)

Solution. The geometry of the spheres indicates that the normal n to the contacting surfaces makes an angle of $30°$ with the direction of $\mathbf{v}_1$ as indicated in the figure.

Conservation of momentum of the system of two particles in the n-direction gives

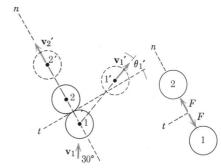

$$[\Delta G_n = 0] \quad (m_2 v_2' + m_1 v_1' \sin \theta_1') - m_1(6 \cos 30°) + 0 = 0$$

or, since $m_1 = m_2$,

$$v_2' + v_1' \sin \theta_1' - 3\sqrt{3} = 0 \qquad (a)$$

Conservation of momentum for each particle in the t-direction occurs since, with assumed smooth surfaces, the contact force F on it has no t-component. Thus for particle 1

$$[\Delta G_t = 0] \qquad m_1(v_1' \cos \theta_1') - m_1(6 \sin 30°) = 0 \qquad (b)$$

The given coefficient of restitution requires

$$0.6 = \frac{v_2' - v_1' \sin \theta_1'}{6 \cos 30° - 0} \qquad (c)$$

Solution of Eqs. a, b, and c for the unknown quantities gives

$$v_1' = 3.18 \text{ m/s} \qquad v_2' = 4.16 \text{ m/s} \qquad \theta_1' = 19.11° \qquad Ans.$$

The kinetic energy just after impact is

$$T' = \tfrac{1}{2}m_1(3.18)^2 + \tfrac{1}{2}m_2(4.16)^2$$

and that just before impact is

$$T = \tfrac{1}{2}m_1(6)^2 + 0$$

With the masses being equal they cancel, and the percentage energy loss becomes

$$\frac{\Delta E}{E}(100) = \frac{T - T'}{T}(100) = \frac{6^2 - [(3.18)^2 + (4.16)^2]}{6^2}(100)$$

$$= 23.8 \text{ percent} \qquad Ans.$$

① Be sure to verify the geometry of the impact positions.

② We note that particle 2 has no initial or final velocity in the t-direction, so we get no additional information from this observation.

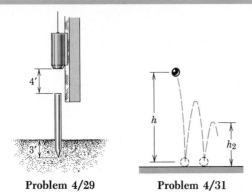

Problem 4/29   **Problem 4/31**

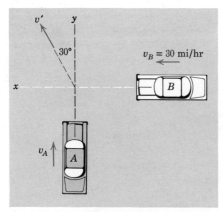

Problem 4/32

Problem 4/33

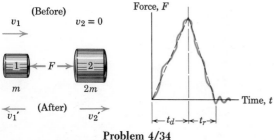

Problem 4/34

PROBLEMS

4/29 The 500-lb ram of a pile driver falls 4 ft from rest and strikes the top of an 800-lb pile embedded 3 ft in the ground. Immediately after impact the ram is seen to have no velocity. Determine the coefficient of restitution e and the velocity v' of the pile immediately after impact.

4/30 A ball is released from rest and drops a distance h onto the horizontal surface of a heavy steel plate. If the ball rebounds to a height h', determine the coefficient of restitution e. What is the fraction n of the original energy which it lost?

Ans. $n = 1 - e^2$

4/31 Determine the coefficient of restitution e for a steel ball dropped from rest at a height h above a heavy horizontal steel plate if the height of the second rebound is h_2.

4/32 Cars A and B of equal mass collide at right angles in the intersection of two icy roads. The cars become entangled and move off together with a common velocity v' in the direction indicated. If car B was traveling 30 mi/hr at the instant of impact, compute the corresponding velocity of car A just prior to the collision.   *Ans.* $v_A = 52.0$ mi/hr

4/33 Three identical steel cylinders are free to slide horizontally on the fixed horizontal shaft. Cylinders 2 and 3 are at rest and are approached by cylinder 1 at a speed u. Express the final speed v of cylinder 3 in terms of u and the coefficient of restitution e.

4/34 Cylinder 1 of mass m moving with a velocity v_1 strikes cylinder 2 of mass $2m$ initially at rest. The impact force F varies with time as shown, where t_d is the duration of the deformation period and t_r is the duration of the restoration period. Determine the velocity v_2' of cylinder 2 immediately after impact in terms of the initial velocity v_1 of cylinder 1 for (a) $t_r = t_d$, (b) $t_r = 0.5t_d$, and (c) $t_r = 0$.

/35 Show that the common velocity v_0 of cylinders 1 and 2 of Prob. 4/34 at the instant of maximum deformation is $v_0 = v_1/3$ and that v_0 is independent of the coefficient of restitution.

/36 Two steel balls of the same diameter are connected by a rigid bar of negligible mass as shown and are dropped in the horizontal position from a height of 150 mm above the heavy steel and brass base plates. If the coefficient of restitution between the ball and the steel base is 0.6 and that between the other ball and the brass base is 0.4, determine the angular velocity ω of the bar immediately after impact. Assume the two impacts are simultaneous.

Ans. $\omega = 0.57$ rad/s counterclockwise

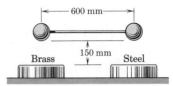

Problem 4/36

/37 In the selection of the ram of a pile driver, similar to that shown with Prob. 4/29, it is desired that the ram lose all of its kinetic energy at each blow. Hence the velocity of the ram is zero immediately after impact. The mass of each pile to be driven is 300 kg, and experience has shown that a coefficient of restitution of 0.3 can be expected. What should be the mass m of the ram? Compute the velocity v of the pile immediately after impact if the ram is dropped from a height of 4 m onto the pile. Also compute the energy loss ΔE due to impact at each blow.

/38 The steel ball strikes the heavy steel plate with a velocity $v_0 = 24$ m/s at an angle of $60°$ with the horizontal. If the coefficient of restitution is $e = 0.8$, compute the velocity v and its direction θ with which the ball rebounds from the plate.

Ans. $v = 20.5$ m/s, $\theta = 54.2°$

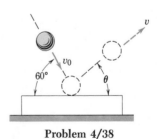

Problem 4/38

/39 The sphere of mass m travels with an initial velocity v_1 and collides centrally with a sphere of mass $2m$ initially at rest. For a coefficient of restitution which makes the loss of kinetic energy of the system a maximum, determine the velocities v_1' and v_2' of m and $2m$ after impact.

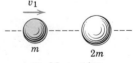

Problem 4/39

/40 Freight car A of mass m_A is rolling to the right when it collides with freight car B of mass m_B initially at rest. If the two cars are coupled together at impact, show that the fractional loss of energy equals $m_B/(m_A + m_B)$.

Problem 4/40

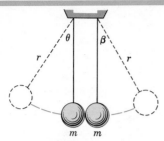

Problem 4/42

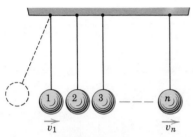

Problem 4/43

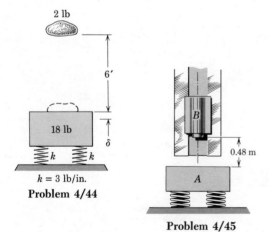

$k = 3$ lb/in.
Problem 4/44

Problem 4/45

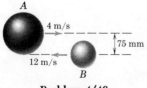

Problem 4/46

4/41 With a coefficient of restitution which would result in no loss of energy for the ram and pile of Prob. 4/29, what would be the pile velocity v following impact? Would this value of e be reasonable?

Ans. $v = 12.35$ ft/sec, No.

4/42 Two spheres of equal mass m are suspended by wires of equal length r measured to their centers. The left-hand sphere is released from rest in the dotted position. The angle through which the wire of the right-hand sphere swings after impact is β. Show that $(1 - \cos \beta)/(1 - \cos \theta) = (1 + e)^2/4$ where e is the coefficient of restitution.

4/43 The figure shows n spheres of equal mass m suspended in a line by wires of equal length so that the spheres are almost touching each other. If sphere 1 is released from the dotted position and strikes sphere 2 with a velocity v_1, write an expression for the velocity v_n of the nth sphere immediately after being struck by the one adjacent to it. The common coefficient of restitution is e.

$$Ans.\ v_n = \left(\frac{1 + e}{2}\right)^{n-1} v_1$$

4/44 The 2-lb piece of putty is dropped 6 ft onto the 18-lb block initially at rest on the two springs, each with a stiffness $k = 3$ lb/in. Calculate the additional deflection δ of the springs due to the impact of the putty which adheres to the block upon contact.

4/45 The 3-Mg anvil A for a drop forge is mounted on a nest of four springs with a combined stiffness of 2.88 MN/m. The 500-kg hammer B falls 0.48 m from rest and strikes the anvil which is then observed to deflect 20 mm from its equilibrium position. Determine the height h of rebound of the hammer and the coefficient of restitution e which applies.

Ans. $h = 21.5$ mm, $e = 0.414$

4/46 Sphere A has a mass of 23 kg and a radius of 75 mm, while sphere B has a mass of 4 kg and a radius of 50 mm. If the spheres are traveling initially along the parallel paths with the speeds shown, determine the velocities of the spheres immediately after impact. The coefficient of restitution is 0.4 and friction is neglected.

Ans. $v_A' = 2.46$ m/s
$v_B' = 9.16$ m/s

47 The two identical steel balls moving with initial velocities v_1 and v_2 as shown collide in such a way that the line joining their centers is in the direction of v_2. From previous experiment the coefficient of restitution is known to be 0.60. Determine the velocity of each ball immediately after impact and find the percentage loss of kinetic energy of the system as a result of the impact.

Ans. $v_1' = 3.70$ ft/sec, $v_2' = 3.36$ ft/sec
Kinetic energy loss is 52%

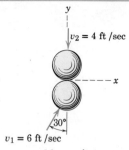

Problem 4/47

48 To pass inspection, steel balls for ball bearings must clear the fixed bar A at the top of their rebound when dropped from rest through the vertical distance $H = 900$ mm onto the heavy inclined steel plate. If balls which have a coefficient of restitution of less than 0.7 with the rebound plate are to be rejected, determine the position of the bar by specifying h and s. Neglect any friction during impact.

Ans. $h = 379$ mm, $s = 339$ mm

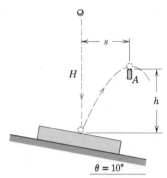

Problem 4/48

49 Show that the loss of energy due to direct central impact of two masses m_1 and m_2 having velocities v_1 and v_2 directed toward each other is given by

$$\Delta E = \frac{1 - e^2}{2} \frac{m_1 m_2}{m_1 + m_2} (v_1 + v_2)^2$$

where e is the coefficient of restitution for these particular impact conditions and the internal vibrational energy is neglected. (*Hint:* The energy loss depends on the relative impact velocity $v_1 + v_2$. Thus the center of mass of the system may be taken at rest to simplify the algebra so that $m_1 v_1 = m_2 v_2$.)

50 The 2-kg sphere is projected horizontally with a velocity of 10 m/s against the 10-kg carriage which is backed up by the spring with a stiffness of 1600 N/m. The carriage is initially at rest with the spring uncompressed. If the coefficient of restitution is 0.60, calculate the rebound velocity v', the rebound angle θ, and the maximum travel δ of the carriage after impact.

Ans. $v' = 6.04$ m/s, $\theta = 85.9°$, $\delta = 165$ mm

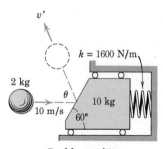

Problem 4/50

4/4 RELATIVE MOTION. Up to this point in our development of the kinetics of particle motion we have applied Newton's second law and the equations of work-energy and impulse-momentum to problems where all measurements of motion were made with respect to a reference system which was considered fixed. The nearest we can come to a "fixed" reference system is the primary inertial system or astronomical frame of reference, which is an imaginary set of axes attached to the fixed stars. All other reference systems, then, are considered to have motion in space, including any reference system attached to the moving earth.

The accelerations of points attached to the earth as measured in the primary system are quite small, however, and we normally neglect them for most earth-surface measurements. For example, the acceleration of the center of the earth in its near-circular orbit around the sun considered fixed is 0.00593 m/s² (or 0.01946 ft/sec²), and the acceleration of a point on the equator at sea level with respect to the center of the earth considered as fixed is 0.0399 m/s² (or 0.1111 ft/sec²). Clearly these accelerations are small compared with g and with most other significant accelerations in engineering work. Thus we make only a small error when we assume that our earth-attached reference axes are equivalent to a fixed reference system.

(*a*) *Relative-Motion Equation.* We now consider a particle A of mass m, Fig. 4/8, whose motion is observed from a set of axes x-y-z which have translatory motion with respect to a fixed reference frame X-Y-Z. Thus the x-y-z directions always remain parallel to the X-Y-Z directions. A discussion of motion relative to a rotating reference system is reserved for a later treatment in Arts. 6/7 and 8/7. The acceleration of the origin B of x-y-z is $\mathbf{a}_B$. The acceleration of A as observed from or relative to x-y-z is $\mathbf{a}_{\text{rel}} = \mathbf{a}_{A/B} = \ddot{\mathbf{r}}_{A/B}$, and by the relative-motion principle of Art. 2/8 the absolute acceleration of A is

$$\mathbf{a}_A = \mathbf{a}_B + \mathbf{a}_{\text{rel}}$$

Thus Newton's second law $\Sigma\mathbf{F} = m\mathbf{a}_A$ becomes

$$\Sigma\mathbf{F} = m(\mathbf{a}_B + \mathbf{a}_{\text{rel}}) \tag{4/15}$$

The force sum $\Sigma\mathbf{F}$ is disclosed, as always, by a complete free-body diagram which will appear the same to an observer in x-y-z or to one in X-Y-Z as long as only the real forces acting on the particle are represented. We can conclude immediately that Newton's second law does not hold with respect to an accelerating system since $\Sigma\mathbf{F} \neq m\mathbf{a}_{\text{rel}}$.

D'Alembert's Principle. When a particle is observed from a fixed set of axes X-Y-Z, Fig. 4/9a, its absolute acceleration $\mathbf{a}$ is measured and the familiar relation $\Sigma\mathbf{F} = m\mathbf{a}$ is applied. When the particle is observed from a moving system x-y-z to which it is attached at the

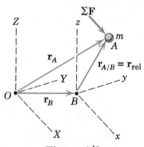

Figure 4/8

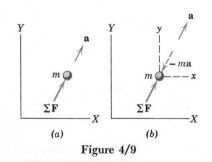

(*a*)   (*b*)

Figure 4/9

origin, Fig. 4/9*b*, the particle of necessity appears to be at rest or in equilibrium in *x*-*y*-*z*. Thus the observer who is accelerating with *x*-*y*-*z* concludes that a force −*m***a** acts on the particle to balance Σ**F**. This point of view which permits the treatment of a dynamics problem by the methods of statics was an outgrowth of the work of D'Alembert contained in his *Traité de Dynamique* published in 1743. This approach merely amounts to rewriting the equation of motion as Σ**F** − *m***a** = **0**, which assumes the form of a zero force summation if −*m***a** is treated as a force. This fictitious force is known as the *inertia force,* and the artificial state of equilibrium created is known as *dynamic equilibrium.* The apparent transformation of a problem in dynamics to one in statics has become known as *D'Alembert's principle.*

Opinion differs concerning the original interpretation of D'Alembert's principle, but the principle in the form in which it is generally known is regarded in this book as being mainly of historical interest. It was evolved during a time when understanding and experience with dynamics were extremely limited and was a means of explaining dynamics in terms of the principles of statics which during earlier times were more fully understood. This excuse for using an artificial situation to describe a real one is open to question, as there is today a wealth of knowledge and experience with the phenomena of dynamics to support strongly the direct approach of thinking in terms of dynamics rather than in terms of statics. It is somewhat difficult to justify the long persistence in the acceptance of statics as a way of understanding dynamics particularly in view of the continued search for the understanding and description of physical phenomena in their most direct form.

Only one simple example of the method known as D'Alembert's principle will be cited. The conical pendulum of mass *m*, Fig. 4/10*a*, is swinging in a horizontal circle with its radial line *r* having an angular velocity *ω*. In the straightforward application of the equation of motion Σ**F** = *m***a**$_n$ in the direction *n* of the acceleration, the

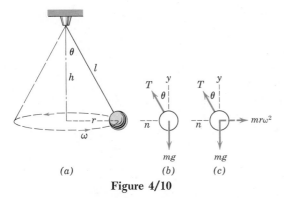

Figure 4/10

free-body diagram in the *b*-part of the figure shows that $T \sin \theta = mr\omega^2$. When the equilibrium requirement in the *y*-direction, $T \cos \theta - mg = 0$, is introduced, the unknowns T and θ can be found. But if the reference axes are attached to the particle, the particle will appear to be in equilibrium relative to these axes. Accordingly, the inertia force $-m\mathbf{a}$ must be added, which amounts to visualizing the application of $mr\omega^2$ in the direction opposite to the acceleration, as shown in the *c*-part of the figure. With this pseudo free-body diagram a zero force summation in the *n*-direction gives $T \sin \theta - mr\omega^2 = 0$ which, of course, gives us the same result as before. We may conclude that no advantage results from this alternative formulation. The author recommends against using it since it introduces no simplification and adds a nonexistent force to the diagram. In the case of a particle moving in a circular path this hypothetical inertia force is known as the *centrifugal force* since it is directed away from the center and is opposite to the direction of the acceleration. The student is urged to recognize that there is no actual centrifugal force acting on the particle. The only actual force which may properly be called centrifugal is the horizontal component of the tension T exerted *on* the cord *by* the particle.

(*b*) *Constant-Velocity Translating Systems.* In discussing particle motion relative to moving reference systems, it is of importance for us to note the special case where the reference system has a constant translational velocity. If the *x-y-z* axes of Fig. 4/8 have a constant velocity, then $\mathbf{a}_B = \mathbf{0}$ and the acceleration of the particle is $\mathbf{a}_A = \mathbf{a}_{\text{rel}}$. Therefore we may write Eq. 4/15 as

$$\boxed{\Sigma \mathbf{F} = m\mathbf{a}_{\text{rel}}} \tag{4/16}$$

which tells us that Newton's second law holds for measurements made in a system moving with a constant velocity. Such a system is known as a Newtonian frame of reference. Observers in the moving system and in the fixed system will also agree on the designation of the resultant force acting on the particle from their identical free-body diagrams, provided they avoid the use of any so-called "inertia forces."

We will now examine the parallel question concerning the validity of the work-energy equation and the impulse-momentum equation relative to a constant velocity, nonrotating system. Again we take the *x-y-z* axes of Fig. 4/8 to be moving with a constant velocity $\mathbf{v}_B = \dot{\mathbf{r}}_B$ relative to the fixed axes *X-Y-Z*. The path of the particle *A* relative to *x-y-z* is governed by $\mathbf{r}_{\text{rel}}$ and is represented schematically in Fig. 4/11. The work done by $\Sigma \mathbf{F}$ relative to *x-y-z* is $dU_{\text{rel}} = \Sigma \mathbf{F} \cdot d\mathbf{r}_{\text{rel}}$. But $\Sigma \mathbf{F} = m\mathbf{a}_A = m\mathbf{a}_{\text{rel}}$ since $\mathbf{a}_B = \mathbf{0}$. Also $\mathbf{a}_{\text{rel}} \cdot d\mathbf{r}_{\text{rel}} = \mathbf{v}_{\text{rel}} \cdot d\mathbf{v}_{\text{rel}}$ for the same reason that $a_t \, ds = v \, dv$ in Art. 2/5 on curvilinear motion. Thus we have

$$dU_{\text{rel}} = m\mathbf{a}_{\text{rel}} \cdot d\mathbf{r}_{\text{rel}} = mv_{\text{rel}} \, dv_{\text{rel}} = d(\tfrac{1}{2}mv_{\text{rel}}^2)$$

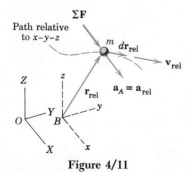

Figure 4/11

We define the kinetic energy relative to x-y-z as $T_{rel} = \frac{1}{2}mv_{rel}^2$ so that we now have

$$\boxed{dU_{rel} = dT_{rel}} \quad \text{or} \quad \boxed{U_{rel} = \Delta T_{rel}} \quad (4/17)$$

which shows that the work-energy equation holds for measurements made relative to a constant-velocity system.

Relative to x-y-z the impulse on the particle during time dt is $\Sigma \mathbf{F}\, dt = m\mathbf{a}_A\, dt = m\mathbf{a}_{rel}\, dt$. But $m\mathbf{a}_{rel}\, dt = m\, d\mathbf{v}_{rel} = d(m\mathbf{v}_{rel})$ so

$$\Sigma \mathbf{F} \cdot dt = d(m\mathbf{v}_{rel})$$

We define the linear momentum of the particle relative to x-y-z as $\mathbf{G}_{rel} = m\mathbf{v}_{rel}$ which gives us $\Sigma \mathbf{F}\, dt = d\mathbf{G}_{rel}$. Dividing by dt and integrating give

$$\boxed{\Sigma \mathbf{F} = \dot{\mathbf{G}}_{rel}} \quad \text{and} \quad \boxed{\int \Sigma \mathbf{F}\, dt = \Delta \mathbf{G}_{rel}} \quad (4/18)$$

Thus the impulse-momentum equations for a fixed reference system also hold for measurements made relative to a constant-velocity system.

Finally, we define the relative angular momentum of the particle about a point in x-y-z, such as the origin B, as the moment of the relative linear momentum. Thus $\mathbf{H}_{B_{rel}} = \mathbf{r}_{rel} \times \mathbf{G}_{rel}$. The time derivative is $\dot{\mathbf{H}}_{B_{rel}} = \dot{\mathbf{r}}_{rel} \times \mathbf{G}_{rel} + \mathbf{r}_{rel} \times \dot{\mathbf{G}}_{rel}$. The first term is nothing more than $\mathbf{v}_{rel} \times m\mathbf{v}_{rel} = \mathbf{0}$ and the second term is $\mathbf{r}_{rel} \times \Sigma \mathbf{F} = \Sigma \mathbf{M}_B$, the sum of the moments about B of all forces on m. Thus we have

$$\boxed{\Sigma \mathbf{M}_B = \dot{\mathbf{H}}_{B_{rel}}} \quad (4/19)$$

which shows that the moment-angular momentum relation holds with respect to a constant-velocity system.

Although the work-energy and impulse-momentum equations hold relative to a system translating with a constant velocity, the individual expressions for work, kinetic energy, and momentum differ between the fixed and the moving systems. Thus

$$(dU = \Sigma \mathbf{F} \cdot d\mathbf{r}_A) \neq (dU_{rel} = \Sigma \mathbf{F} \cdot d\mathbf{r}_{rel})$$
$$(T = \tfrac{1}{2}mv_A^2) \neq (T_{rel} = \tfrac{1}{2}mv_{rel}^2)$$
$$(\mathbf{G} = m\mathbf{v}_A) \neq (\mathbf{G}_{rel} = m\mathbf{v}_{rel})$$

Equations 4/16 through 4/19 are formal proof of the validity of the Newtonian equations of kinetics in any constant-velocity, nonrotating system. We might have surmised these conclusions from the fact that $\Sigma \mathbf{F} = m\mathbf{a}$ depends on acceleration and not velocity. We are also ready to conclude that there is no experiment which can be conducted in and relative to a constant-velocity system (Newtonian frame of reference) which discloses its absolute velocity. Any mechanical experiment will achieve the same results in any Newtonian system.

Sample Problem 4/5

A simple pendulum of mass m and length r is mounted on the flatcar which has a constant horizontal acceleration a_0 as shown. If the pendulum is released from rest relative to the flatcar at the position $\theta = 0$, determine the expression for the tension T in the supporting light rod for any value of θ. Also find T for $\theta = \pi/2$ and $\theta = \pi$.

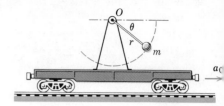

Solution. We attach our moving x-y coordinate system to the translating car with origin at O for convenience. Relative to this system n- and t-coordinates are the natural ones to use since the motion is circular within x-y. The acceleration of m is given by the relative-acceleration equation

$$\mathbf{a} = \mathbf{a}_0 + \mathbf{a}_{\mathrm{rel}}$$

where $\mathbf{a}_{\mathrm{rel}}$ is the acceleration which would be measured by an observer riding with the car. He would measure an n-component equal to $r\dot\theta^2$ and a t-component equal to $r\ddot\theta$. The three components of the absolute acceleration of m are shown in the separate view.

First, we apply Newton's second law to the t-direction and get

① $[\Sigma F_t = ma_t]$ $\quad mg\cos\theta = m(r\ddot\theta - a_0\sin\theta)$
$$r\ddot\theta = g\cos\theta + a_0\sin\theta$$

The angular rate $\dot\theta$ depends on $\ddot\theta$ and its change with θ. Thus

② $[\dot\theta\,d\dot\theta = \ddot\theta\,d\theta]$ $\quad \displaystyle\int_0^{\dot\theta}\dot\theta\,d\dot\theta = \int_0^{\theta}\frac{1}{r}(g\cos\theta + a_0\sin\theta)\,d\theta$

$$\frac{\dot\theta^2}{2} = \frac{1}{r}[g\sin\theta + a_0(1 - \cos\theta)]$$

We now apply Newton's second law to the n-direction noting that the n-component of the absolute acceleration is $r\dot\theta^2 - a_0\cos\theta$.

$[\Sigma F_n = ma_n]$ $\quad T - mg\sin\theta = m(r\dot\theta^2 - a_0\cos\theta)$
$$= m[2g\sin\theta + 2a_0(1 - \cos\theta) - a_0\cos\theta]$$
$$T = m[3g\sin\theta + a_0(2 - 3\cos\theta)] \qquad Ans.$$

For $\theta = \pi/2$ and $\theta = \pi$ we have

$$T_{\pi/2} = m[3g(1) + a_0(2 - 0)] = m(3g + 2a_0) \qquad Ans.$$
$$T_{\pi} = m[3g(0) + a_0(2 - 3[-1])] = 5a_0 m \qquad Ans.$$

Free–body diagram

Acceleration components

① We choose the t-direction first since the n-direction equation, which contains the unknown T, will involve $\dot\theta^2$, which, in turn, is obtained from an integration of $\ddot\theta$.

② Be sure to recognize that $\dot\theta\,d\dot\theta = \ddot\theta\,d\theta$ may be obtained from $v\,dv = a_t\,ds$ by dividing by r.

Sample Problem 4/6

The flatcar moves with a constant speed v_0 and carries a winch which produces a constant tension P in the cable attached to the small carriage. The carriage has a mass m and rolls freely on the horizontal surface starting from rest relative to the flatcar at $x = 0$ at which instant $X = x_0 = b$. Apply the work-energy equation to the carriage, first, as an observer moving with the frame of reference of the car and, second, as an observer on the ground. Show the compatibility of the two expressions.

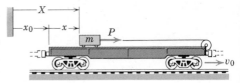

Solution. To the observer on the flatcar the work done by P is

$$U_{rel} = \int_0^x P \, dx = Px \qquad \text{for constant } P$$

The change in kinetic energy relative to the car is

$$\Delta T_{rel} = \tfrac{1}{2}m(\dot{x}^2 - 0)$$

The work-energy equation for the moving observer becomes

$$[U_{rel} = \Delta T_{rel}] \qquad Px = \tfrac{1}{2}m\dot{x}^2$$

To the observer on the ground the work done by P is

$$U = \int_b^X P \, dX = P(X - b)$$

The change in kinetic energy for the ground measurement is

$$\Delta T = \tfrac{1}{2}m(\dot{X}^2 - v_0^2)$$

The work-energy equation for the fixed observer gives

$$[U = \Delta T] \qquad P(X - b) = \tfrac{1}{2}m(\dot{X}^2 - v_0^2)$$

To reconcile this equation with that for the moving observer we can make the following substitutions.

$$X = x_0 + x, \quad \dot{X} = v_0 + \dot{x}, \quad \ddot{X} = \ddot{x}$$

$$P(X - b) = Px + P(x_0 - b) = Px + m\ddot{x}(x_0 - b)$$

$$= Px + m\ddot{x}v_0 t = Px + mv_0\dot{x}$$

$$\dot{X}^2 - v_0^2 = (v_0^2 + \dot{x}^2 + 2v_0\dot{x} - v_0^2) = \dot{x}^2 + 2v_0\dot{x}$$

The work-energy equation for the fixed observer now becomes

$$Px + mv_0\dot{x} = \tfrac{1}{2}m\dot{x}^2 + mv_0\dot{x}$$

which is merely $Px = \tfrac{1}{2}m\dot{x}^2$ as concluded by the moving observer. We see, therefore, that the difference between the two work-energy expressions is

$$U - U_{rel} = T - T_{rel} = mv_0\dot{x}$$

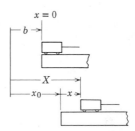

① The only coordinate which the moving observer can measure is x.

② To the ground observer the initial velocity of the carriage is v_0 so its initial kinetic energy is $\tfrac{1}{2}mv_0^2$.

③ The symbol t stands for the time of motion from $x = 0$ to $x = x$. The displacement $x_0 - b$ of the carriage is its velocity v_0 times the time t or $x_0 - b = v_0 t$. Also, since the constant acceleration times the time equals the velocity change, $\ddot{x}t = \dot{x}$.

PROBLEMS

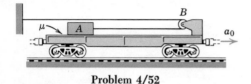

Problem 4/52

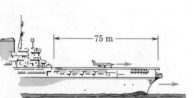

Problem 4/53

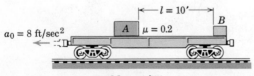

Problem 4/55

4/51 Apply the equation of impulse and momentum to the carriage of Sample Problem 4/6 for the first t seconds of the motion starting when $x = 0$. Write the equation both as an observer riding with the car and as a fixed observer on the ground. Reconcile the two relations.

4/52 If the flatcar is given an acceleration $a_0 = 6$ ft/sec^2 starting from rest, compute the corresponding tension T in the cable attached to the 100-lb crate A. Neglect the mass of the pulley at B and its friction. The coefficient of friction between the crate and the horizontal surface of the flatcar is $\mu = 0.30$.

Ans. $T = 67.3$ lb

4/53 The aircraft carrier is moving at a constant speed and launches a jet plane with a mass of 3 Mg in a distance of 75 m along the deck by means of a steam-driven catapult. If the plane leaves the deck with a velocity of 240 km/h relative to the carrier and if the jet thrust is constant at 22 kN during takeoff, compute the constant force P exerted by the catapult on the airplane during the 75-m travel of the launch carriage.

4/54 Two jet planes A and B are flying alongside one another at a constant speed of 1000 km/h when the pilot of A ignites a constant-thrust jet-assist unit which burns for 5 s. The navigator of B observes that A is traveling 16 km/h faster than B at the end of the 5 s. Calculate the constant forward thrust F of the jet-assist unit if plane A has an average total mass of 5 Mg during the acceleration period. Neglect the small increment of air resistance due to the higher speed. What is the momentum G_{rel} of A with respect to B's system at the end of the 5-s period?

4/55 The block A and the flatcar are initially at rest when the car is given a constant horizontal acceleration of 8 ft/sec^2. Determine the velocity of the car when the block strikes the stop at B. The coefficient of friction between the block and the car is 0.20.

√56 The slider A has a mass of 2 kg and moves with negligible friction in the 30° slot in the vertical sliding plate. What horizontal acceleration a_0 should be given to the plate so that the absolute acceleration of the slider will be vertically down? What is the value of the corresponding force R exerted on the slider by the slot?

$$\text{Ans. } a_0 = 16.99 \text{ m/s}^2, R = 0$$

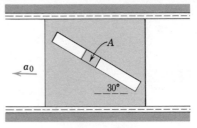

Problem 4/56

√57 The slider A of Prob. 4/56 with a mass of 2 kg moves with negligible friction in the slot of the vertical plate. If the plate has a constant acceleration $a_0 = 8$ m/s² to the left and the slider is released from rest relative to the plate, calculate the force R exerted by the side of the slot on the slider during the motion. Also find the acceleration a_{rel} of the slider relative to the slot.

√58 If the aircraft carrier of Prob. 4/53 is moving with a constant speed u and its catapult launches a plane of mass m with a velocity v relative to the carrier and in the direction of u, the kinetic energy of the plane with respect to the land is $\frac{1}{2}m(v + u)^2 = \frac{1}{2}mv^2 + \frac{1}{2}mu^2 + muv$ when launching is complete. If the plane is launched by the same catapult with the same accelerating force when the carrier is not moving, the kinetic energy of the plane is $\frac{1}{2}mv^2$. Explain the difference between the two kinetic-energy expressions.

√59 The ball A of mass $m = \frac{1}{4}$ lb-sec²/ft and attached light rod of length $l = 2$ ft are free to rotate in a horizontal plane about the vertical shaft attached to the vehicle. The vehicle, rod, and ball are initially at rest with $\theta = 0$ when the vehicle is given a constant acceleration $a_0 = 4$ ft/sec². Write an expression for the tension T in the rod as a function of θ and calculate T for the position $\theta = \pi/2$.

$$\text{Ans. } T = 3ma_0 \sin \theta, T_{\pi/2} = 3 \text{ lb}$$

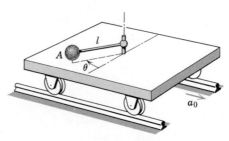

Problem 4/59

√60 The point of support B for a simple pendulum of mass m and length l has a constant horizontal acceleration a as shown. If the pendulum is released from rest relative to the moving system with $\theta = 0$, determine the expression for the tension T in the cord of the pendulum as a function of θ.

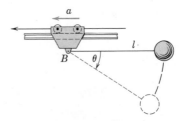

Problem 4/60

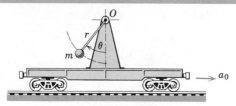

Problem 4/61

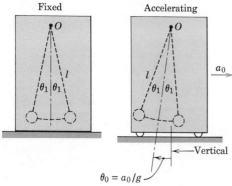

Problem 4/62

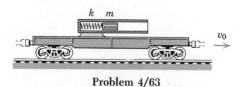

Problem 4/63

4/61 If the pendulum of Sample Problem 4/5 is released from rest in the vertical position relative to the accelerating flatcar, write the expression for the tension T in the light rod in terms of θ where θ is now measured clockwise from the vertical as indicated here (rather than from the horizontal as previously).

Ans. $T = m\,[g(3\cos\theta - 2) + 3a_0\sin\theta\,]$

4/62 For small amplitudes θ_1 the simple pendulum with fixed support O will oscillate with simple harmonic motion about the vertical with a period $\tau = 2\pi\sqrt{l/g}$. If the support O is given a small but steady horizontal acceleration a_0, show that the pendulum will oscillate about the inclined axis $\theta_0 = a_0/g$ with the same period as with the fixed support provided that the amplitude of motion θ_1 is small.

4/63 The capsule of mass m is released from rest relative to the horizontal tube with the spring of stiffness k initially compressed a distance δ. The flatcar has a constant velocity v_0 to the right. Apply the work-energy equation to the motion of the capsule both as an observer riding with the car and as an observer fixed to the ground. Reconcile the two equations. (*Hint:* For the second case include the spring with the capsule as your system.)

4/64 The sliding block of mass m is attached to the frame by a spring of stiffness k and slides horizontally along the frame with negligible friction. The frame and block are both initially at rest with $x = x_0$, the unstretched length of the spring. If the frame is given a constant acceleration a_0, determine the maximum velocity of the block $(v_{\text{rel}})_{\text{max}}$ relative to the frame.   *Ans.* $(v_{\text{rel}})_{\text{max}} = (kx_0 + ma_0)/\sqrt{km}$

Problem 4/64

65 When a particle is dropped from rest relative to the earth's surface at a latitude γ, the initial apparent acceleration is the relative acceleration due to gravity g_{rel}. The absolute acceleration due to gravity g is directed toward the center of the earth. Derive an expression for g_{rel} in terms of g, R, ω, and γ where R is the radius of the earth treated as a sphere and ω is the constant angular velocity of the earth about the polar axis considered as fixed. (Although axes x-y-z are attached to the earth and, hence, rotate, we may use Eq. 4/15 as long as the particle has no velocity relative to x-y-z.) (*Hint.* Use the first two terms of the binomial expansion for the approximation.)

$$Ans. \; g_{rel} = g - R\omega^2 \cos^2 \gamma \left(1 - \frac{R\omega^2}{2g}\right) + \cdots$$

$$= 9.824 - 0.03381 \cos^2 \gamma \; \text{m/s}^2$$

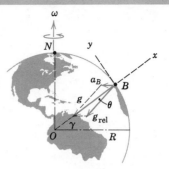

Problem 4/65

66 A plumb bob is suspended from a fixed point on the surface of the earth at north latitude γ. Refer to the figure for Prob. 4/65 and determine the expression for the small angle θ made by the cord which supports the bob with the true vertical due to the angular velocity ω of the earth. The radius of the earth is R. Calculate the value of θ for $\gamma = 30°$. Let g stand for the absolute acceleration due to gravity. (Draw the free-body diagram of the bob and relate the vectors to the acceleration vectors shown with Prob. 4/65.)

$$Ans. \; \tan \theta = \frac{R\omega^2}{2g} \frac{\sin 2\gamma}{1 - (R\omega^2/g)\cos^2 \gamma}$$

$$\theta = 5'$$

4/5 VIBRATION AND TIME RESPONSE. An important class of problems in dynamics consists of the motions of bodies which vibrate or otherwise respond to applied disturbances in the presence of restoring forces. The response of a structure to earthquake or blast loading, the periodic vibration of the mounting of a motor which is out of balance, and the aeolian vibrations of power lines are but a few examples of this class of problems.

A useful engineering description of the time response of masses is accomplished by the solution of a mathematical model of an equivalent system. For example, the transmission of force to a foundation from an unbalanced spring-mounted machine may usually be described by considering the machine as though it were a concentrated mass mounted on a single equivalent spring.

The detailed study of vibrations and time response is a large subject for specialized study, and only a bare introduction to the topic will be given in this article. For a more complete treatment the student is directed to references on mechanical vibrations,[*] linear systems, electric circuits, nonlinear oscillations, and pulse techniques.

By far the most common equivalent system for modeling vibration problems consists of a concentrated spring-mounted mass subjected to a disturbing force and a retarding force. Such a system is represented in Fig. 4/12a with the mass m in a general position which is displaced a distance x from the neutral or equilibrium position where the force in the spring of stiffness k would be zero. The mass is acted upon by an applied force $F = f(t)$ which is expressed as a function of the time t. Also, the mass is retarded by a force with a magnitude proportional to the velocity $\dot{x}$. This type of frictional retardation is termed *viscous damping* and is represented by the action of a dashpot or fluid damper under laminar flow conditions. Other types of damping forces may be encountered such as dry friction or Coulomb damping which is essentially independent of velocity, internal damping due to material hysteresis losses, turbulent-flow damping where the retarding force is more nearly proportional to the square of the velocity, and magnetic damping.

The free-body diagram discloses the applied force F, the restoring force $-kx$, and the retarding or damping force $-c\dot{x}$. The constant c is called the *viscous damping coefficient*. We now apply Newton's second law for motion in the x-direction to give

$$F - kx - c\dot{x} = m\ddot{x}$$

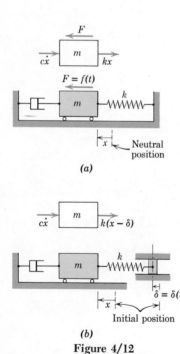

(a)

(b)

Figure 4/12

which may be written as

$$m\ddot{x} + c\dot{x} + kx = F \qquad (4/20)$$

[*] See, for example, Chapter 9 of the author's book *Dynamics,* 2nd Edition 1971 (also *SI Version* 1975), John Wiley & Sons.

A system similar to that of Fig. 4/12a is shown in Fig. 4/12b where the applied force is transmitted through the spring attached to a foundation which has a displacement $\delta = \delta(t)$ from the initial position. If x is the absolute displacement of the mass measured from the equilibrium position when $\delta = 0$, then the spring has a tension $k(x - \delta)$, and the free-body diagram requires that

$$-c\dot{x} - k(x - \delta) = m\ddot{x}$$

or

$$m\ddot{x} + c\dot{x} + kx = k\delta \qquad (4/21)$$

We see, therefore, that Eq. 4/20 is equivalent to Eq. 4/21 when F is replaced by $k\delta$. The solutions of either of Eqs. 4/20 or 4/21 for various values of c, k, and F or δ cover a wide variety of oscillations and responses which can be used to describe the behavior of many engineering systems. Equations 4/20 and 4/21 are seen to be linear, second-order differential equations, and solution can be obtained by several standard procedures. We now summarize very briefly the most common solutions to Eqs. 4/20 and 4/21.

(a) Free Response, No Damping. With $c = 0$ and $F = 0$ the mass vibrates freely without energy loss, and Eqs. 4/20 and 4/21 become

$$\boxed{m\ddot{x} + kx = 0} \qquad (4/22)$$

This equation is the characteristic description for a *simple harmonic* oscillation in which the acceleration is proportional to the negative of the displacement. The general solution to the second-order equation has two constants of integration and may be written as

$$x = C_1 \sin pt + C_2 \cos pt \qquad \text{or} \qquad x = C \sin(pt + \phi)$$

where C_1 and C_2 or C and ϕ are the integration constants which depend on the manner in which the motion was begun. These solutions are easily verified by direct substitution into Eq. 4/22 whereupon it is seen that

$$\boxed{p = \sqrt{k/m}}$$

If we let $t = 0$ when $x = 0$ and let $C_1 = x_0$, the amplitude of the motion, then $C_2 = 0$ and the solution is simply

$$\boxed{x = x_0 \sin pt} \qquad (4/23)$$

This motion is represented graphically in Fig. 4/13 by the projection x of the rotating vector of length x_0. The angular velocity p of this rotating reference vector is known as the *circular frequency,* and the time required for one complete revolution or oscillation is the period $\tau = 2\pi/p = 2\pi\sqrt{m/k}$. The number of cycles per unit time is the *frequency* $f = 1/\tau = \sqrt{k/m}/(2\pi)$ and is expressed in hertz (1 Hz =

1 cycle per second). If we had let $t = 0$ in the position where $x = x_0$ and $\dot{x} = 0$, then we would have $x = x_0 \cos pt$. Or, if we had measured time from a position where the reference vector made an angle ϕ with the horizontal axis in Fig. 4/13, the solution would be $x = x_0 \sin (pt + \phi)$.

Energy considerations may be used to determine the period or frequency of oscillation for a linear conservative system without having to derive the differential equation of motion. Since energy is conserved, the maximum kinetic energy occurs at the position $x = 0$ and must equal the maximum potential energy at $x = x_0$ or

$$T_{max} = V_{max}$$

where V is taken to be zero when $T = T_{max}$. From the solution for simple harmonic motion, Eq. 4/23, the maximum velocity is $\dot{x}_{max} = x_0 p$, so that

$$\tfrac{1}{2}m(x_0 p)^2 = \tfrac{1}{2}kx_0^2 \quad \text{from which} \quad p = \sqrt{k/m}$$

This direct determination of circular frequency may be used for any linear undamped oscillator.

(b) *Damped Free Response.* When the damping force is not negligible, the free response from Eq. 4/20 or 4/21 is described by

$$m\ddot{x} + c\dot{x} + kx = 0 \qquad (4/24)$$

The solution of Eq. 4/24 is obtained by standard procedures for treating linear differential equations and, again, if we measure time from the instant when $x = 0$, the solution may be written as

$$x = x_0 e^{-bt} \sin qt \qquad (4/25)$$

where

$$b = \frac{c}{2m} \qquad q = \sqrt{\frac{k}{m} - \left(\frac{c}{2m}\right)^2}, \qquad c^2 < 4km$$

The period of the motion is clearly $\tau = 2\pi/q$. If $c^2 > 4km$ a non-oscillatory motion ensues. In this event the motion is said to be

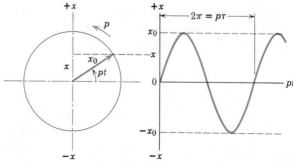

Figure 4/13

overdamped, and upon being released from a displaced position, the mass will creep back toward its neutral position without oscillation. Equation 4/25 is plotted in Fig. 4/14 which shows the decay in amplitude due to the action of the viscous damping force.

(c) *Forced Response.* When either the applied force $F = f(t)$ or the foundation movement $\delta = \delta(t)$ of the linear spring-mass system of Fig. 4/12 is not zero, a forced response results which may be described by a solution to the complete Eq. 4/20 or 4/21.

We will consider only the case where damping is negligible $(c = 0)$ and where $F = F_0 \sin \omega t$ or $\delta = \delta_0 \sin \omega t$ is a steady periodic function of time. The resulting *steady-state* vibration is, then, described by Eqs. 4/20 and 4/21, which become

$$\ddot{x} + \frac{k}{m}x = \frac{F_0}{m}\sin \omega t \quad \text{and} \quad \ddot{x} + \frac{k}{m}x = \frac{k\delta_0}{m}\sin \omega t \quad (4/26)$$

The particular solution which describes the steady-state motion for the first of the two equations may be written as $x = A \sin \omega t$. Substitution of $\ddot{x}$ and x into the differential equation requires that

$$A = \frac{F_0/m}{k/m - \omega^2} = \frac{F_0/k}{1 - (\omega/p)^2} \quad \text{where} \quad p = \sqrt{k/m}$$

so that the steady-state response is

$$x = \frac{F_0/k}{1 - (\omega/p)^2}\sin \omega t \quad\quad (4/27)$$

For the second of Eqs. 4/26 which describes the forced vibration of the foundation we merely replace F_0/k in Eq. 4/27 by δ_0.

If we let the symbol δ_0 also stand for the static deflection of the mass against the spring under the action of the force F_0, then the amplitude x_0 of Eq. 4/27 applied to either of the two cases is given by

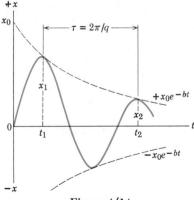

Figure 4/14

$$\frac{x_0}{\delta_0} = \frac{1}{1 - (\omega/p)^2}$$

(4/28)

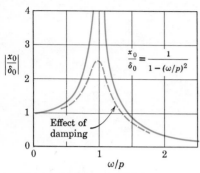

Figure 4/15

This ratio is known as the *magnification factor*, which compares the actual amplitude of the deflection with the static deflection. For $\omega < p$ the magnification factor is positive and the vibration is in phase with the forcing function or foundation vibration. For $\omega > p$ the magnification factor is negative and the vibration is 180 degrees out of phase with the forcing function or foundation vibration. The magnitude of this factor is plotted in Fig. 4/15 against the *frequency ratio* ω/p and discloses the important phenomenon of *resonance*. The value $\omega = p$ is known as the *resonant frequency* or *critical frequency* and represents an operating condition of the forcing function to be avoided since the resulting vibration tends to become excessively large. Thus, if rotating machinery is operating at or near the natural frequency for free vibration of the system, abnormally severe vibrations are likely to occur.

It is usually desirable to reduce as much as possible the forced vibrations which are generated in engineering structures and machines. Vibration reduction is normally accomplished in any of four ways: (1) reduction or elimination of the exciting force by balancing or other removal, (2) introduction of sufficient damping to limit the amplitude, (3) isolation of the body from the vibration source by

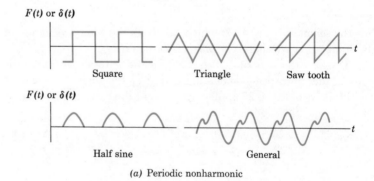

(a) Periodic nonharmonic

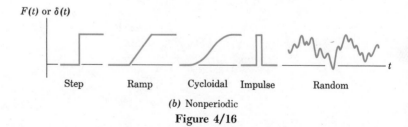

(b) Nonperiodic

Figure 4/16

providing elastic mountings of the proper stiffness, and (4) operation at a forced frequency sufficiently different from the natural frequency so as to avoid resonance.

When the forcing function $F(t)$ or $\delta(t)$ of Eq. 4/20 or 4/21 is periodic and regular but nonharmonic, the response will be a steady-state motion, but it will no longer be represented by a single sine or cosine expression. In Fig. 4/16a are shown five examples of forcing functions or excitations which are periodic but nonharmonic. Figure 4/16b shows five examples of forcing functions which result in nonperiodic response. Methods for the analysis of nonharmonic forcing functions will not be developed here but are described in detail in a large number of references on vibrations, linear systems, and circuit analysis.

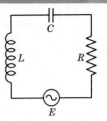

Figure 4/17

An important analogy exists between electric circuits and mechanical spring-mass systems. Figure 4/17 shows a series circuit consisting of a voltage E which is a function of time, an inductance L, a capacitance C, and a resistance R. If we denote the charge by the symbol q, the equation which governs the charge is

$$L\ddot{q} + R\dot{q} + \frac{1}{C}q = E \qquad (4/29)$$

This equation has the same form as the equation for the mechanical system. Thus by a simple interchange of symbols, the behavior of the electrical circuit may be used to predict the behavior of the mechanical system, or vice versa. The mechanical and electrical equivalents in the following table are worth noting.

MECHANICAL-ELECTRICAL EQUIVALENTS

MECHANICAL			ELECTRICAL		
QUANTITY	SYMBOL	SI UNIT	QUANTITY	SYMBOL	SI UNIT
Mass	m	kg	Inductance	L	H henry
Spring stiffness	k	N/m($= kg/s^2$)	1/Capacitance	$1/C$	1/F 1/farad
Force	F	N($= kg \cdot m/s^2$)	Voltage	E	V volt
Velocity	$\dot{x}$	m/s	Current	I	A ampere
Displacement	x	m	Charge	q	C coulomb
Viscous damping constant	c	N/(m/s)($= kg/s$)	Resistance	R	Ω ohm

Sample Problem 4/7

The small sphere of mass m is mounted on the light rod pivoted at O and supported at end A by the vertical spring of stiffness k. End A is displaced downward a small distance y_0 below the horizontal equilibrium position and released. Derive the differential equation for small oscillations of the rod and determine the expression for the frequency f.

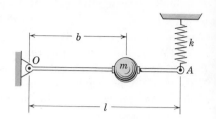

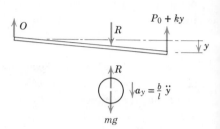

Solution. The equation of motion of the sphere involves the force R exerted on it by the light rod. Since the mass of the rod is negligible, its acceleration is due to a negligible unbalance of forces and moments so that we may treat the rod as though it were in equilibrium. A zero moment sum about O for the rod gives $Rb = (P_0 + ky)l$ where the equilibrium value P_0 is simply mgb/l. Thus $R = mg + kly/b$. Therefore Newton's second law for the sphere gives

① $[\Sigma F_y = ma_y]$ $\qquad mg - R = m\left(\dfrac{b}{l}\ddot{y}\right), \qquad mg - \left(mg + k\dfrac{l}{b}y\right) = m\dfrac{b}{l}\ddot{y}$

or $\qquad\qquad\qquad \ddot{y} + \dfrac{k}{m}\left(\dfrac{l}{b}\right)^2 y = 0 \qquad\qquad$ *Ans.*

Comparing this equation with Eq. 4/22 enables us to write the circular frequency p immediately as the square root of the coefficient of y. Thus p and the frequency become

$$p = \frac{l}{b}\sqrt{\frac{k}{m}} \quad \text{and} \quad f = \frac{p}{2\pi} = \frac{l}{2\pi b}\sqrt{\frac{k}{m}} \qquad \textit{Ans.}$$

We will now solve for the frequency using the energy principle which tells us that the change in potential energy from the neutral position to the position of maximum displacement equals the corresponding change in kinetic energy. Since the spring was stretched a distance P_0/k in the neutral position, its total stretch in the position of maximum displacement is $P_0/k + y_0$. Therefore the change in potential energy is

② $\Delta V = \Delta V_e + \Delta V_g = \left[\frac{1}{2}k\left(\dfrac{P_0}{k} + y_0\right)^2 - \frac{1}{2}k\left(\dfrac{P_0}{k}\right)^2\right] - mg\left(\dfrac{b}{l}y_0\right)$

Substituting the equilibrium value $P_0 = mgb/l$ and simplifying give

$$\Delta V = \tfrac{1}{2}ky_0{}^2$$

The corresponding change in kinetic energy is

③ $\Delta T = T_{\text{max}} = \frac{1}{2}m\left(\dfrac{b}{l}\dot{y}\right)^2_{\text{max}} = \frac{1}{2}m\left(\dfrac{b}{l}y_0\right)^2 p^2$

Equating energy changes gives

$|\Delta V| = |\Delta T| \qquad \frac{1}{2}ky_0{}^2 = \frac{1}{2}m\left(\dfrac{b}{l}y_0\right)^2 p^2$

and $\qquad\qquad p = \dfrac{l}{b}\sqrt{\dfrac{k}{m}}, \qquad f = \dfrac{p}{2\pi} = \dfrac{l}{2\pi b}\sqrt{\dfrac{k}{m}} \qquad$ *Ans.*

① The downward displacement of the sphere from the neutral position is $(b/l)y$ so that the downward acceleration is $(b/l)\ddot{y}$. Remember that we are dealing with small displacements so that the arc length through which A moves is very nearly equal to the vertical movement of A and the additional stretch y of the spring.

② The change ΔV_g is negative since the mass drops a distance $(b/l)y_0$.

③ Recall from the theory where $x = x_0 \sin pt$ that $\dot{x}_{\text{max}} = x_0 p$. Hence we may write $(b\dot{y}/l)_{\text{max}} = (by_0/l)p$.

Sample Problem 4/8

A 2-lb mass is vibrating freely with viscous damping so that the amplitudes of vibration are successively smaller as shown. If two successive amplitudes are $x_1 = 0.042$ in. and $x_2 = 0.038$ in. and if the frequency of vibration is 2 cycles per second, compute the viscous damping constant c.

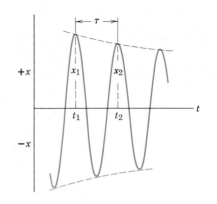

Solution. From Eq. 4/25 we have

$$x_1 = x_0 e^{-bt_1} \quad \text{and} \quad x_2 = x_0 e^{-b(t_1 + \tau)}$$

where $b = c/(2m)$, τ is the period, and $\sin qt$ has the same value for each peak displacement. The ratio of the peaks is

$$\frac{x_1}{x_2} = \frac{x_0 e^{-bt_1}}{x_0 e^{-b(t_1 + \tau)}} = e^{b\tau} \quad \text{or} \quad \frac{c\tau}{2m} = \ln(x_1/x_2)$$

Substituting numerical values with $\tau = 1/f = 1/2$ sec gives

$$c = \frac{2(2/32.2)}{1/2} \ln \frac{0.042}{0.038} = 0.0249 \text{ lb/(ft/sec)} \qquad \textit{Ans.}$$

Sample Problem 4/9

The delicate instrument shown has a mass of 20 kg and is spring-mounted to the horizontal base. If the amplitude of vertical vibration of the base is 0.20 mm, calculate the range of frequencies of the base vibration which must be prohibited if the amplitude of vertical vibration of the instrument is not to exceed 0.30 mm. Each of the four spring mounts has a stiffness of 3.0 kN/m.

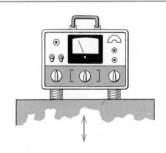

Solution. The magnification ratio, Eq. 4/28, has the magnitude

$$\frac{x_0}{\delta_0} = \left| \frac{1}{1 - (\omega/p)^2} \right|$$

Thus the specified value $(x_0/\delta_0)_{\max} = 0.30/0.20 = 1.5$ will govern the permissible values of ω/p.

For the instrument the square of the natural circular frequency of free vibration is

$$p^2 = k/m = 4(3)(10^3)/20 = 600 \text{ (rad/s)}^2$$

The square of the frequency ratio becomes

$$(\omega/p)^2 = (2\pi f/p)^2 = 4\pi^2 f^2/600 = 0.0658 f^2$$

where f is the frequency of the foundation in hertz (cycles per second).

For $\omega < p$, $\quad +1.5 = \dfrac{1}{1 - 0.0658 f^2}$, $\quad f = 2.25$ Hz

For $\omega > p$, $\quad -1.5 = \dfrac{1}{1 - 0.0658 f^2}$, $\quad f = 5.03$ Hz

① Note that when $\omega > p$ the quantity $1/[1 - (\omega/p)^2]$ is negative. *Suggestion:* Calculate the corresponding frequency ratios and check the amplification ratio from Fig. 4/15.

Thus the prohibited range of operating frequencies is

$$2.25 < f < 5.03 \qquad \textit{Ans.}$$

Problem 4/67

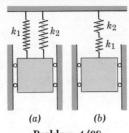

(a)   (b)

Problem 4/68

PROBLEMS

(*Undamped free vibrations*)

4/67 In the equilibrium position the 4-lb cylinder causes the spring to deform 0.80 in. Calculate the frequency f of vertical vibration of the cylinder when it is set into motion.   *Ans.* $f = 3.50$ cycles/sec

4/68 Replace the springs in each of the two cases shown by a single spring of stiffness k (equivalent spring stiffness) which will cause each mass to vibrate with its original frequency.

$$\text{*Ans.* } (a) \ \ k = k_1 + k_2$$
$$(b) \ \ \frac{1}{k} = \frac{1}{k_1} + \frac{1}{k_2}$$

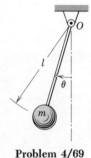

Problem 4/69

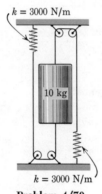

Problem 4/70

4/69 Derive the differential equation for the motion of the simple pendulum for large amplitudes. Simplify the equation for small amplitudes and determine the corresponding period τ.

4/70 Calculate the natural frequency f of vertical oscillation of the spring-loaded cylinder when set into motion. Both springs are in tension at all times.   *Ans.* $f = 3.90$ Hz

4/71 Derive the differential equation for the vertical vibration of the cylinder and determine its frequency f. Neglect the friction and mass of the pulley.

4/72 The cylinder of mass m is given a vertical displacement y_0 from its equilibrium position and released. Write the differential equation for the vertical vibration of the cylinder and find the period τ of its motion. Neglect the friction and mass of the pulley.

$$\text{*Ans.* } \tau = \pi \sqrt{\frac{m}{k}}$$

Problem 4/71

Problem 4/72

√73 The cylindrical buoy floats in salt water (density 1030 kg/m³) and has a mass of 800 kg with a low center of mass to keep it stable in the upright position. Determine the frequency f of vertical oscillation of the buoy. Assume the water level remains undisturbed adjacent to the buoy.

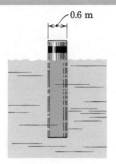

0.6 m

Problem 4/73

√74 The time-rate-of-change of acceleration is called "jerk" J. Specify the maximum value of the amplitude x_0, measured from the neutral position, for the undamped linear harmonic oscillator of mass m which will limit the jerk to a value J_0. The stiffness of the elastic supports is k.

$$\text{Ans. } |x_0| = \left(\frac{m}{k}\right)^{3/2} J_0$$

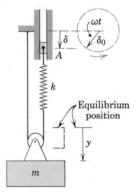

Problem 4/75

√75 The equilibrium position of the mass m occurs where $y = 0$ and $\delta = 0$. When the attachment A is given a steady vertical motion $\delta = \delta_0 \sin \omega t$, the mass m will acquire a steady vertical oscillation. Derive the differential equation of motion for m and specify the circular frequency ω_{cr} for which the oscillations of m tend to become excessively large. The stiffness of the spring is k, and the mass and friction of the pulley are negligible.

$$\text{Ans. } \ddot{y} + 4\frac{k}{m}y = \frac{2k}{m}\delta_0 \sin \omega t, \quad \omega_{cr} = 2\sqrt{\frac{k}{m}}$$

√76 Derive the differential equation for small oscillations of the spring-loaded pendulum about the pivot O and find the period τ. Let θ stand for the angular displacement. Each spring of stiffness k is under a compressive force C in the equilibrium position shown. Neglect the mass of the rod.

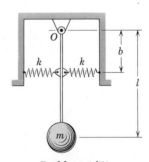

Problem 4/76

√77 The weighing platform has a mass m and is connected to the spring of stiffness k by the system of levers shown. Derive the differential equation for small vertical oscillation of the platform and find its period τ. Designate y as the platform displacement from the equilibrium position and neglect the mass of the levers.

$$\text{Ans. } \tau = 2\pi \left(\frac{b}{c}\right)^2 \sqrt{\frac{m}{k}}$$

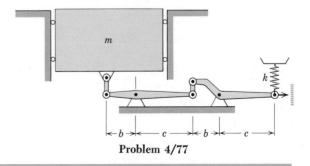

Problem 4/77

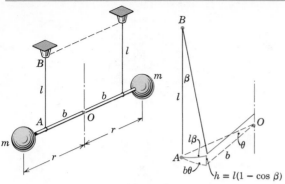

Problem 4/78

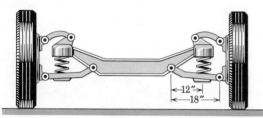

Problem 4/79

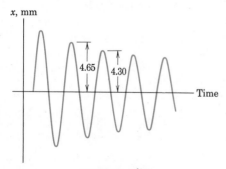

Problem 4/83

4/78 Two small balls each of mass m are fastened to a rod of negligible mass, which is supported in the horizontal plane by a bifilar suspension. The rod and balls are set into small oscillation about the vertical axis through O. Derive the expression for the period τ. Proceed by equating the maximum potential energy to the maximum kinetic energy using the angle θ as the variable. (*Hint:* From the auxiliary sketch note that the balls rise a distance h corresponding to an angle of twist θ. Also $l\beta \approx b\theta$ for small angles, and $\cos \beta$ may be replaced by the first two terms in its series expansion. A simple harmonic solution of the form $\theta = \theta_0 \sin pt$ may be used for small angles.)

4/79 The front-end suspension of an automobile is shown. Each of the coil springs has a stiffness of 270 lb/in. If the weight of the front-end frame and equivalent portion of the body attached to the front end is 1800 lb, determine the natural frequency of vertical oscillation of the frame and body in the absence of shock absorbers. (*Hint:* To relate the spring deflection to the deflection of the frame and body, consider the frame as fixed and let the ground and wheels move vertically.)

Ans. $f = 1.14$ cycles/sec

4/80 Solve for the period τ of the platform of Prob. 4/77 by using the energy method.

4/81 Solve for the frequency f of the vibrating cylinder of Prob. 4/71 by using the energy method.

(Damped free vibrations)

4/82 The period τ of damped linear oscillation for a certain 1-kg mass is 0.32 s. If the stiffness of the supporting linear spring is 850 N/m, calculate the damping coefficient c.   *Ans.* $c = 43.1$ N·s/m

4/83 A linear harmonic oscillator having a mass of 1.10 kg is set into motion with viscous damping. If the frequency is 10 Hz and if two successive amplitudes a full cycle apart are measured to be 4.65 mm and 4.30 mm as shown, compute the viscous damping coefficient c.

84 The damping constant for a spring-supported 1-lb mass is 0.020 lb-sec/in. The mass is displaced 0.75 in. from its equilibrium position and then released. If the frequency of the vibration is observed to be 3 cycles/sec, compute the amplitude of the 3rd cycle following release.   *Ans.* $x_3 = 0.0157$ in.

85 A linear oscillator with mass m, spring stiffness k, and damping coefficient c is set into motion when released from a displaced position. Derive an expression for the energy loss E in one cycle in terms of the amplitude x_1 at the start of the cycle. (See Fig. 4/14.)

$$Ans.\ E = \tfrac{1}{2}kx_1{}^2(1 - e^{-2\pi c/(mq)})$$

$$\text{where } q = \sqrt{\frac{k}{m} - \left(\frac{c}{2m}\right)^2}$$

(Forced vibrations)

86 Each 0.5-kg ball is attached to the end of the light elastic rod and deflects 4 mm when a 2-N force is statically applied to the ball. If the central collar is given a vertical harmonic movement with a frequency of 4 Hz and an amplitude of 3 mm, find the amplitude y_0 of vertical vibration of each ball.

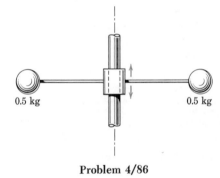

0.5 kg   0.5 kg

Problem 4/86

87 A static horizontal force of 2 lb gives the 3-lb weight A a deflection of 0.50 in. against the elasticity of the light cantilever spring to which it is attached. If the base B of the spring is given a horizontal harmonic oscillation with a frequency of 2 cycles/sec and an amplitude of $\frac{1}{4}$ in., calculate the amplitude x_0 of the resulting vibration of A. Assume negligible damping.

            Ans. $x_0 = 0.360$ in.

88 The collar A is given a harmonic oscillation along the fixed horizontal shaft with a frequency of 3 Hz. The ball is attached to the collar by the elastic rod of negligible mass and deflects 60 mm when a horizontal force of 40 N is applied statically to the ball with the collar clamped in place. If the double amplitude $2x_0$ of the ball is twice the double amplitude $2\delta_0$ of the collar with the two motions 180° out of phase, compute the mass m of the ball.

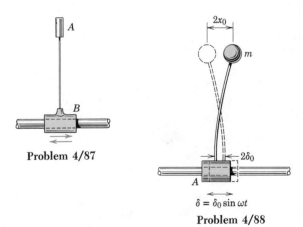

Problem 4/87

Problem 4/88

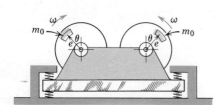

Problem 4/90

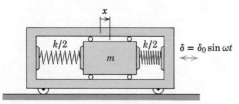

Problem 4/91

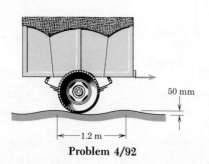

Problem 4/92

4/89 A spring-mounted machine with a mass of 24 kg is observed to vibrate harmonically in the vertical direction with an amplitude of 0.30 mm under the action of a vertical force which varies harmonically between F_0 and $-F_0$ with a frequency of 4 Hz. Damping is negligible. If a static force of magnitude F_0 causes a deflection of 0.60 mm, calculate the equivalent spring constant k for the springs which support the machine.   *Ans.* $k = 5050$ N/m

4/90 A device to produce vibrations consists of the two counter-rotating wheels, each carrying an eccentric mass $m_0 = 1$ kg with a center of mass at a distance $e = 12$ mm from its axis of rotation. The wheels are synchronized so that the vertical positions of the unbalanced masses are always identical. The total mass of the device is 10 kg. Determine the two possible values of the equivalent spring constant k for the mounting which will permit the amplitude of the periodic force transmitted to the fixed mounting to be 1500 N due to the unbalance of the rotors at a speed of 1800 rev/min. Neglect damping.
     Ans. $k = 823$ kN/m   or   227 kN/m

4/91 The seismic mass m is mounted in the frame between two compression springs each of stiffness $k/2$ (which is equivalent to the action of a single spring of stiffness k). The frame is given a horizontal vibration $\delta = \delta_0 \sin \omega t$. If x is the displacement of the mass relative to the frame, derive the expression for the amplification factor defined as the ratio of the amplitude of the relative displacement x_0 to δ_0. (Compare results with Eq. 4/28.)

$$Ans. \quad \frac{x_0}{\delta_0} = \frac{(\omega/p)^2}{1 - (\omega/p)^2}$$

▶**4/92** Determine the amplitude of vertical vibration of the spring-mounted trailer as it travels at a velocity of 25 km/h over the corduroy road whose contour may be expressed by a sine or cosine term. The mass of the trailer is 500 kg and that of the wheels alone may be neglected. During the loading each 75 kg added to the load caused the trailer to sag 3 mm on its springs. Assume that the wheels are in contact with the road at all times and neglect damping. At what critical speed v_c is the vibration of the trailer greatest?   *Ans.* $x_0 = 14.75$ mm, $v_c = 15.23$ km/h

KINETICS OF SYSTEMS OF PARTICLES 5

/1 **INTRODUCTION.** In the previous three chapters we have dealt with the principles of dynamics for the motion of a particle. Although attention was focused primarily on the kinetics of a single particle in Chapters 3 and 4, reference to the motion of two particles, considered together as a system, was made in the discussions of work-energy and impulse-momentum. Our next major step in the development of the subject of dynamics is to extend these principles which we applied to a single particle to describe the motion of a general system of particles. This extension provides a unity to the remaining sections of dynamics and enables us to treat the motion of rigid bodies and the motion of nonrigid systems as well. We recall that a rigid body is a solid system of particles wherein the distances between particles remain essentially unchanged. The overall motions found with machines, land and air vehicles, rockets and spacecraft, and many moving structures provide examples of rigid-body problems. A nonrigid body, on the other hand, may be a solid body where the object of investigation is the time dependence of the changes in shape due to elastic or nonelastic deformations. Or a nonrigid body may be a defined mass of liquid or gaseous particles which are flowing at a specified rate. Examples would be the air and fuel flowing through the turbine of an aircraft engine, the burned gases issuing from the nozzle of a rocket motor, or the water passing through a rotary pump.

Although the extension of the equations for single-particle motion to a general system of particles is accomplished without undue difficulty, it cannot be expected that the full generality and significance of these extended principles can be fully understood without considerable problem experience. For this reason we strongly recommend that the general results obtained in the following article be reviewed frequently during the remainder of the study of dynamics. In this way the unity contained in these broader principles of dynamics will be brought into sharper focus and a more fundamental view of the subject will be acquired.

/2 **DEFINING EQUATIONS.** Newton's second law of motion and the equations of work-energy and impulse-momentum will now be ex-

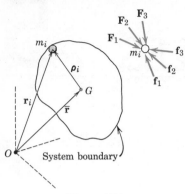

Figure 5/1

tended to cover a general mass system which we model by considering n mass particles bounded by a closed surface in space, Fig. 5/1. This bounding envelope, for example, might be the exterior surface of a given rigid body, the bounding surface of an arbitrary portion of the body, the exterior surface of a rocket containing both rigid and flowing particles, or a particular volume of fluid particles. In each case the system considered is the mass within the envelope, and the mass must be clearly defined and isolated.

(*a*) *Generalized Newton's Second Law.* Figure 5/1 shows a representative particle of mass m_i of the system isolated with forces $\mathbf{F}_1, \mathbf{F}_2, \mathbf{F}_3, \ldots$ acting on m_i from sources *external* to the envelope, and forces $\mathbf{f}_1, \mathbf{f}_2, \mathbf{f}_3, \ldots$ acting on m_i from sources *internal* to the envelope. The external forces are due to contact with external bodies or to external gravitational, electric, or magnetic forces. The internal forces are forces of reaction with other mass particles within the envelope. The particle m_i is located by its position vector $\mathbf{r}_i$ measured from a Newtonian set of reference axes.° The center of mass G of the system of particles isolated is located by the position vector $\bar{\mathbf{r}}$ which, from Varignon's theorem of moments in statics, is given by

$$m\bar{\mathbf{r}} = \Sigma m_i \mathbf{r}_i$$

where the total system mass is $m = \Sigma m_i$. The summation sign Σ represents the summation over all n particles. Newton's second law, Eq. 3/3, when applied to m_i gives

$$\mathbf{F}_1 + \mathbf{F}_2 + \mathbf{F}_3 + \cdots + \mathbf{f}_1 + \mathbf{f}_2 + \mathbf{f}_3 + \cdots = m_i \ddot{\mathbf{r}}_i$$

where $\ddot{\mathbf{r}}_i$ is the acceleration of m_i. A similar equation may be written for each of the particles of the system. If these equations written for *all* particles of the system are added together, there results

$$\Sigma \mathbf{F} + \Sigma \mathbf{f} = \Sigma m_i \ddot{\mathbf{r}}_i$$

The term $\Sigma \mathbf{F}$ then becomes the vector sum of *all* forces acting on all particles of the isolated system from sources external to the system, and $\Sigma \mathbf{f}$ becomes the vector sum of all forces on all particles produced by the internal actions and reactions between particles. This last sum is identically zero since all internal forces occur in pairs of equal and opposite actions and reactions. By differentiating the equation defining $\bar{\mathbf{r}}$ twice with time, we have $m\ddot{\bar{\mathbf{r}}} = \Sigma m_i \ddot{\mathbf{r}}_i$ where m has no time derivative as long as mass is not entering or leaving the system.°° Substitu-

---

° It was shown in Art. 4/4 that any nonrotating and nonaccelerating set of axes constitutes a Newtonian reference system in which the principles of Newtonian mechanics are valid.

°° If m is a function of time, a more complex situation develops; this situation is discussed in Art. 5/4 on variable mass.

tion into the summation of the equations of motion gives

$$\Sigma \mathbf{F} = m\ddot{\overline{\mathbf{r}}} \quad \text{or} \quad \Sigma \mathbf{F} = m\overline{\mathbf{a}} \tag{5/1}$$

where $\overline{\mathbf{a}}$ is the acceleration $\ddot{\overline{\mathbf{r}}}$ of the center of mass of the system.

Equation 5/1 is the generalized Newton's second law of motion for a mass system and is referred to as the *equation of motion* of m. The equation states that the resultant of the external forces on any system of mass equals the total mass of the system times the acceleration of the center of mass. This law expresses the so-called *principle of motion of the mass center*. It should be observed that $\overline{\mathbf{a}}$ is the acceleration of the mathematical point which represents instantaneously the position of the mass center for the given n particles. For a nonrigid body this acceleration need not represent the acceleration of any particular particle. It should be noted further that Eq. 5/1 holds for each instant of time and is therefore an instantaneous relationship. Equation 5/1 for the mass system had to be proved as it may not be inferred directly from Eq. 3/3 for the single particle. Equation 5/1 may be expressed in component form using x-y-z coordinates or whatever coordinate system is most convenient for the problem at hand. Thus

$$\Sigma F_x = m\overline{a}_x \qquad \Sigma F_y = m\overline{a}_y \qquad \Sigma F_z = m\overline{a}_z \tag{5/1a}$$

Although Eq. 5/1, as a vector equation, requires that the acceleration vector $\overline{\mathbf{a}}$ must have the same direction as the resultant external force $\Sigma \mathbf{F}$, it does not follow that $\Sigma \mathbf{F}$ necessarily passes through G. In general, in fact, $\Sigma \mathbf{F}$ does not pass through G, as will be shown later.

(b) *Work-Energy.* In Art. 3/6 the work-energy relation for a single particle was developed. Applicability to a system of two joined particles was also noted. We now turn our attention to the general system of Fig. 5/1, where the work-energy relation for the representative particle of mass m_i is $U_i = \Delta T_i$. Here U_i is the work done on m_i by all forces $\mathbf{F}_i = \mathbf{F}_1 + \mathbf{F}_2 + \mathbf{F}_3 + \cdots$ applied from sources external to the system and by all forces $\mathbf{f}_i = \mathbf{f}_1 + \mathbf{f}_2 + \mathbf{f}_3 + \cdots$ applied from sources internal to the sytem. The change in kinetic energy of m_i is $\Delta T_i = \Delta(\frac{1}{2}m_i v_i^2)$ where v_i is the magnitude of the particle velocity $\mathbf{v}_i = \dot{\mathbf{r}}_i$.

For the entire system the sum of the work-energy equations written for all particles is $\Sigma U_i = \Sigma \Delta T_i$ which may be represented by the same expression as Eq. 3/11 of Art. 3/6, namely,

$$U = \Delta T \tag{5/2}$$

where $U = \Sigma U_i$, the work done by all forces on all particles, and ΔT is the change in the total kinetic energy $T = \Sigma T_i$ of the system.

For a rigid body or a system of rigid bodies joined by ideal frictionless connections no net work is done by the internal interacting forces or moments in the connections. We see that the work done by

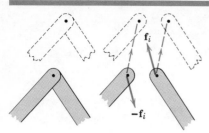

Figure 5/2

all pairs of internal forces $\mathbf{f}_i$ and $-\mathbf{f}_i$ at a typical connection, Fig. 5/2, in the system is zero since their points of application have identical displacement components in the direction of the forces. For this situation U becomes the work done on the system by the external forces only.

For a nonrigid mechanical system which has elastic members capable of storing energy, a part of the work done by the external forces goes into changing the internal elastic potential energy V_e. Also, if the works done by the gravity forces are *excluded* from the U-term and are accounted for by the changes in gravitational potential energy V_g, then we may write the work-energy equation for the nonrigid mechanical system

$$U = \Delta T + \Delta V_g + \Delta V_e \tag{5/3}$$

which is the same as Eq. 3/17 in Art. 3/7.

We now examine the expression $T = \Sigma \frac{1}{2} m_i v_i^2$ for the kinetic energy of the mass system in more detail. By our principle of relative motion discussed in Art. 2/8, we may write the velocity of the representative particle as

$$\mathbf{v}_i = \bar{\mathbf{v}} + \dot{\boldsymbol{\rho}}_i$$

where $\bar{\mathbf{v}}$ is the velocity of the mass center G and $\dot{\boldsymbol{\rho}}_i$ is the velocity of m_i with respect to a translating reference frame moving with the mass center G. We recall the identity $v_i^2 = \mathbf{v}_i \cdot \mathbf{v}_i$ and write the kinetic energy of the system as

$$T = \Sigma \tfrac{1}{2} m_i \mathbf{v}_i \cdot \mathbf{v}_i = \Sigma \tfrac{1}{2} m_i (\bar{\mathbf{v}} + \dot{\boldsymbol{\rho}}_i) \cdot (\bar{\mathbf{v}} + \dot{\boldsymbol{\rho}}_i)$$

$$= \Sigma \tfrac{1}{2} m_i \bar{v}^2 + \Sigma \tfrac{1}{2} m_i |\dot{\boldsymbol{\rho}}_i|^2 + \Sigma m_i \bar{\mathbf{v}} \cdot \dot{\boldsymbol{\rho}}_i$$

Since $\boldsymbol{\rho}_i$ is measured from the mass center, $\Sigma m_i \boldsymbol{\rho}_i = \mathbf{0}$ and the third term is $\bar{\mathbf{v}} \cdot \Sigma m_i \dot{\boldsymbol{\rho}}_i = \bar{\mathbf{v}} \cdot \dfrac{d}{dt} \Sigma (m_i \boldsymbol{\rho}_i) = \mathbf{0}$. Also $\Sigma \tfrac{1}{2} m_i \bar{v}^2 = \tfrac{1}{2} \bar{v}^2 \Sigma m_i$ $= \tfrac{1}{2} m \bar{v}^2$. Therefore the total kinetic energy becomes

$$T = \tfrac{1}{2} m \bar{v}^2 + \Sigma \tfrac{1}{2} m_i |\dot{\boldsymbol{\rho}}_i|^2 \tag{5/4}$$

This equation expresses the fact that the total kinetic energy of a mass system equals the energy of mass-center translation of the system as a whole plus the energy due to motion of all particles relative to the mass center.

(*c*) *Impulse-Momentum.* From our definition in Art. 3/8 the linear momentum of the representative particle of the system depicted in Fig. 5/1 is $\mathbf{G}_i = m_i \mathbf{v}_i$ where the velocity of m_i is $\mathbf{v}_i = \dot{\mathbf{r}}_i$. The linear momentum of the system is defined as the vector sum of the linear momenta of all of its particles or $\mathbf{G} = \Sigma m_i \mathbf{v}_i$. By substituting the relative velocity relation $\mathbf{v}_i = \bar{\mathbf{v}} + \dot{\boldsymbol{\rho}}_i$ and noting again that $\Sigma m_i \boldsymbol{\rho}_i = \mathbf{0}$ we get

$$\mathbf{G} = \Sigma m_i(\bar{\mathbf{v}} + \dot{\boldsymbol{\rho}}_i) = \Sigma m_i \bar{\mathbf{v}} + \frac{d}{dt}\Sigma m_i \boldsymbol{\rho}_i$$

$$= \bar{\mathbf{v}}\Sigma m_i + \frac{d}{dt}(\mathbf{0})$$

or
$$\boxed{\mathbf{G} = m\bar{\mathbf{v}}} \tag{5/5}$$

Thus the linear momentum of any mass system of constant mass is the product of the mass and the velocity of its center of mass.

The time derivative of $\mathbf{G}$ is $m\dot{\bar{\mathbf{v}}} = m\bar{\mathbf{a}}$, which by Eq. 5/1 is the resultant external force acting on the system. Thus we have

$$\boxed{\Sigma\mathbf{F} = \dot{\mathbf{G}}} \tag{5/6}$$

which has the same form as Eq. 3/21 for a single particle. Equation 5/6 states that the resultant of the external forces on any mass system equals the time rate of change of the linear momentum of the system and is an alternative form of the generalized second law of motion, Eq. 5/1. In Eq. 5/6 the total mass was assumed constant during differentiation with time, so that the equation does not apply to systems whose mass changes with time.

The angular momentum of the mass system of Fig. 5/1 about a fixed point O in the Newtonian reference system is defined as the vector sum of the moments of the linear momenta about O of all particles of the system and is

$$\mathbf{H}_O = \Sigma(\mathbf{r}_i \times m_i\mathbf{v}_i)$$

The time derivative of the vector product is $\dot{\mathbf{H}}_O = \Sigma(\dot{\mathbf{r}}_i \times m_i\mathbf{v}_i) + \Sigma(\mathbf{r}_i \times m_i\dot{\mathbf{v}}_i)$. The first summation vanishes since the cross product of two equal vectors $\dot{\mathbf{r}}_i$ and $\mathbf{v}_i$ is zero. The second summation is $\Sigma(\mathbf{r}_i \times m_i\mathbf{a}_i) = \Sigma(\mathbf{r}_i \times \mathbf{F}_i)$ which is the vector sum of the moments about O of all forces acting on all particles of the system. This moment sum $\Sigma\mathbf{M}_O$ represents only the moments of forces external to the system, since the internal forces cancel one another. Thus the moment sum becomes

$$\boxed{\Sigma\mathbf{M}_O = \dot{\mathbf{H}}_O} \tag{5/7}$$

which has the same form as Eq. 3/27 for a single particle. Equation 5/7 states that the resultant vector moment about any fixed point of all external forces on any system of mass equals the time rate of change of angular momentum of the system about the fixed point. As in the linear-momentum case, Eq. 5/7 does not apply if the total mass of the system is changing with time.

The principle expressed by Eq. 5/7 also holds when the mass center G is chosen as the reference point. The angular momentum about G is the sum of the moments of the linear momentum about G and becomes

$$\bar{\mathbf{H}} = \Sigma(\boldsymbol{\rho}_i \times m_i \mathbf{v}_i)$$

When we substitute the relative-velocity relation $\mathbf{v}_i = \bar{\mathbf{v}} + \dot{\boldsymbol{\rho}}_i$, the summation becomes

$$\Sigma(\boldsymbol{\rho}_i \times m_i[\bar{\mathbf{v}} + \dot{\boldsymbol{\rho}}_i]) = -\bar{\mathbf{v}} \times \Sigma m_i \boldsymbol{\rho}_i + \Sigma(\boldsymbol{\rho}_i \times m_i \dot{\boldsymbol{\rho}}_i)$$

Here we have inverted the first cross product with a change in sign so that $\bar{\mathbf{v}}$ can be factored out and reveal the sum $\Sigma m_i \boldsymbol{\rho}_i$, which is zero as seen previously. Thus

$$\bar{\mathbf{H}} = \Sigma(\boldsymbol{\rho}_i \times m_i \dot{\boldsymbol{\rho}}_i) \tag{5/8}$$

which shows that the angular momentum about the mass center may be written alternatively as the sum of the moments about G of the linear momenta relative to G where the relative velocity $\dot{\boldsymbol{\rho}}_i$ is used.

To find the relation between the moments of the external forces and the angular momentum, we differentiate Eq. 5/8 with time and get

$$\dot{\bar{\mathbf{H}}} = \Sigma(\dot{\boldsymbol{\rho}}_i \times m_i \dot{\boldsymbol{\rho}}_i) + \Sigma(\boldsymbol{\rho}_i \times m_i \ddot{\boldsymbol{\rho}}_i)$$

The first summation is zero since $\dot{\boldsymbol{\rho}}_i \times \dot{\boldsymbol{\rho}}_i = \mathbf{0}$. With the aid of the relative motion equation $\mathbf{a}_i = \bar{\mathbf{a}} + \ddot{\boldsymbol{\rho}}_i$ the second term becomes $\Sigma(\boldsymbol{\rho}_i \times m_i[\mathbf{a}_i - \bar{\mathbf{a}}]) = \Sigma(\boldsymbol{\rho}_i \times \mathbf{F}_i) + \bar{\mathbf{a}} \times \Sigma m_i \boldsymbol{\rho}_i = \Sigma\bar{\mathbf{M}} + \mathbf{0}$. Thus we have

$$\boxed{\Sigma\bar{\mathbf{M}} = \dot{\bar{\mathbf{H}}}} \tag{5/9}$$

The moment sum $\Sigma\bar{\mathbf{M}}$ about the mass center is due to external forces only. Equations 5/7 and 5/9 are among the most powerful of the governing relations in dynamics and form the basis for much of the analysis of three-dimensional problems.

(*d*) *Conservation of Energy and Momentum.* A mass system is said to be *conservative* if it does not lose energy by virtue of internal friction forces which do negative work or by virtue of nonelastic members which dissipate energy upon cycling. If no work is done on a conservative system during an interval of motion by external forces (other than gravity or other potential forces), then no part of the energy of the system is lost. In such a case Eq. 5/3 gives

$$\boxed{\Delta T + \Delta V_e + \Delta V_g = 0} \tag{5/10}$$

which expresses the *law of conservation of dynamical energy*. This law holds only in the ideal case where internal kinetic friction is sufficiently small to be neglected.

If, for a certain interval of time, the resultant external force $\Sigma\mathbf{F}$ acting on a conservative or nonconservative mass system is zero, Eq. 5/6 requires that $\dot{\mathbf{G}} = \mathbf{0}$, so that during this interval

$$\boxed{\Delta\mathbf{G} = \mathbf{0}} \tag{5/11}$$

which expresses the *principle of conservation of linear momentum.* Thus in the absence of an external impulse the linear momentum of a system remains unchanged.

Similarly, if the resultant moment about a fixed point O or about the mass center G of all external forces on any mass system is zero, Eq. 5/7 or 5/9 requires, respectively, that

$$\boxed{\Delta\mathbf{H}_O = \mathbf{0} \qquad \text{or} \qquad \Delta\bar{\mathbf{H}} = \mathbf{0}} \tag{5/12}$$

These relations express the *principle of conservation of angular momentum* for a general mass system in the absence of an angular impulse. Thus, if there is no angular impulse about a fixed point (or about the mass center), the angular momentum of the system about the fixed point (or about the mass center) remains unchanged. Either equation may hold without the other.

It was proved in Art. 4/4 that the basic laws of Newtonian mechanics hold for measurements made relative to a set of axes which translate with a constant velocity. Thus Eqs. 5/1 through 5/12 are valid provided all quantities are expressed relative to the translating axes.

Equations 5/1 through 5/12 are among the most important of the basic derived laws of mechanics. In this chapter these laws have been derived for the most general system of constant mass in order that the generality of the laws will be established. These laws find repeated use when applied to specific mass systems such as rigid and nonrigid solids and certain fluid systems which are discussed in the articles that follow. The reader is urged to study these laws carefully and to compare them with their more restricted forms encountered earlier in Chapter 3 on the kinetics of particles.

Sample Problem 5/1

Each of the three balls has a mass m and is welded to the rigid equiangular frame of negligible mass. The assembly rests on a smooth horizontal surface. If a force $\mathbf{F}$ is suddenly applied to the one bar as shown, determine (a) the acceleration of point O and (b) the angular acceleration $\ddot{\theta}$ of the frame.

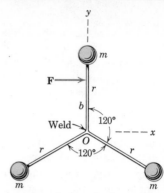

Solution. (a) Point O is the mass center of the system of the three balls, so that its acceleration is given by Eq. 5/1.

① $[\Sigma\mathbf{F} = m\bar{\mathbf{a}}]$ $F\mathbf{i} = 3m\bar{\mathbf{a}}$ $\bar{\mathbf{a}} = \mathbf{a}_O = \dfrac{F}{3m}\mathbf{i}$ *Ans.*

(b) We determine $\ddot{\theta}$ from the moment principle, Eq. 5/9. To find $\bar{H}$ we note that the velocity of each ball relative to the mass center O as measured in the nonrotating axes x-y is $r\dot{\theta}$ where $\dot{\theta}$ is the common angular velocity of the spokes. The angular momentum of the system about O is the sum of the moments of the relative linear momenta as shown by Eq. 5/8 so is expressed by

$$H_O = \bar{H} = 3(mr\dot{\theta})r = 3mr^2\dot{\theta}$$

② Equation 5/9 now gives

$[\Sigma\bar{M} = \dot{\bar{H}}]$ $Fb = \dfrac{d}{dt}(3mr^2\dot{\theta}) = 3mr^2\ddot{\theta}$ so $\ddot{\theta} = \dfrac{Fb}{3mr^2}$ *Ans.*

① We note that the result depends only on the magnitude and direction of $\mathbf{F}$ and not on b, which locates the line of action of $\mathbf{F}$.

② Although $\dot{\theta}$ is initially zero we need the expression for H_O in order to get $\dot{H}_O$. We observe also that $\ddot{\theta}$ is independent of the motion of O.

Sample Problem 5/2

Consider the same conditions as for Sample Problem 5/1 except that the spokes are freely hinged at O and so do not constitute a rigid system. Explain the difference between the two problems.

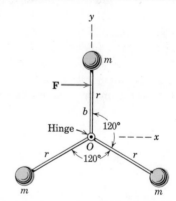

Solution. The generalized Newton's second law holds for any mass system, so that the acceleration $\bar{\mathbf{a}}$ of the mass center G is the same as with Sample Problem 5/1, namely,

$$\bar{\mathbf{a}} = \dfrac{F}{3m}\mathbf{i}$$ *Ans.*

Although G coincides with O at the instant represented, the motion of the hinge O is not the same as the motion of G since O will not remain the center of mass as the angles between the spokes change.

① Both $\Sigma\bar{M}$ and $\dot{\bar{H}}$ have the same values for the two problems at the instant represented. However the angular motions of the spokes in this problem are all different and are not easily determined.

① This present system could be dismembered and the motion equations written for each of the parts with the unknowns eliminated one by one. Or a more sophisticated method using the equations of Lagrange could be employed. (See the author's *Dynamics*, 2nd Edition 1971 for a discussion of this approach.)

Sample Problem 5/3

A shell with a mass of 20 kg is fired from point O with a velocity $u = 300$ m/s in the vertical x-z plane at the inclination shown. When it reaches the top of its trajectory at P it explodes into three fragments A, B, and C. Immediately after the explosion fragment A is observed to rise vertically a distance of 500 m above P, and fragment B is seen to have a horizontal velocity $\mathbf{v}_B$ and eventually land at point Q. When recovered the masses of the fragments A, B, and C are found to be 5, 9, and 6 kg, respectively. Calculate the velocity which fragment C has immediately after the explosion. Neglect atmospheric resistance.

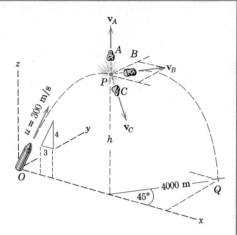

Solution. From our knowledge of projectile motion (Sample Problem 2/6) the time required for the shell to reach P and its vertical rise are

$$t = u_z/g = 300(4/5)/9.81 = 24.5 \text{ s}$$

$$h = \frac{u_z{}^2}{2g} = \frac{[(300)(4/5)]^2}{2(9.81)} = 2936 \text{ m}$$

The velocity of A has the magnitude

$$v_A = \sqrt{2gh_A} = \sqrt{2(9.81)(500)} = 99.0 \text{ m/s}$$

With no z-component of velocity initially fragment B requires 24.5 s to return to the ground. Thus its horizontal velocity, which remains constant, is

$$v_B = s/t = 4000/24.5 = 163.5 \text{ m/s}$$

Since the force of the explosion is internal to the system of the shell and its three fragments, the linear momentum of the system remains unchanged during the explosion. Thus

$$[\Delta \mathbf{G} = 0] \qquad m\mathbf{v} = m_A\mathbf{v}_A + m_B\mathbf{v}_B + m_C\mathbf{v}_C$$

$$20(300)(\tfrac{3}{5})\mathbf{i} = 5(99.0\mathbf{k}) + 9(163.5)(\mathbf{i} \cos 45° + \mathbf{j} \sin 45°) + 6\mathbf{v}_C$$

$$6\mathbf{v}_C = 2560\mathbf{i} - 1040\mathbf{j} - 495\mathbf{k}$$

$$\mathbf{v}_C = 427\mathbf{i} - 173\mathbf{j} - 82.5\mathbf{k} \text{ m/s}$$

$$v_C = \sqrt{(427)^2 + (173)^2 + (82.5)^2} = 468 \text{ m/s} \qquad Ans.$$

① The velocity $\mathbf{v}$ of the shell at the top of its trajectory is, of course, the constant horizontal component of its initial velocity $\mathbf{u}$, which becomes $u(3/5)$.

② We note that the mass center of the three fragments while still in flight continues to follow the same trajectory that the shell would have followed if it had not exploded.

Sample Problem 5/4

The 32.2-lb carriage A moves horizontally in its guide with a speed of 4 ft/sec and carries two assemblies of balls and light rods which rotate about a shaft at O in the carriage. Each of the four balls weighs 3.22 lb. The assembly on the front face rotates counterclockwise at a speed of 80 rev/min, and the assembly on the back side rotates clockwise at a speed of 100 rev/min. For the entire system calculate (a) the kinetic energy T, (b) the linear momentum G, and (c) the angular momentum H_O about point O.

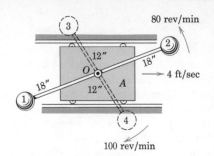

Solution. (a) *Kinetic energy.* The velocity of the balls with respect to O is

$$[|\dot{\boldsymbol{\rho}}_i| = v_{\text{rel}} = r\dot{\theta}] \qquad (v_{\text{rel}})_{1,2} = \frac{18}{12}\frac{80(2\pi)}{60} = 12.57 \text{ ft/sec}$$

$$(v_{\text{rel}})_{3,4} = \frac{12}{12}\frac{100(2\pi)}{60} = 10.47 \text{ ft/sec}$$

The kinetic energy of the system is given by Eq. 5/4. The translational part is

$$\tfrac{1}{2}m\bar{v}^2 = \frac{1}{2}\left(\frac{32.2}{32.2} + 4\frac{3.22}{32.2}\right)(4^2) = 11.20 \text{ ft-lb}$$

The rotational part of the kinetic energy depends on the squares of the relative velocities and is

$$\Sigma\tfrac{1}{2}m_i|\dot{\boldsymbol{\rho}}_i|^2 = 2\left[\frac{1}{2}\frac{3.22}{32.2}(12.57)^2\right]_{(1,2)} + 2\left[\frac{1}{2}\frac{3.22}{32.2}(10.47)^2\right]_{(3,4)}$$

$$= 15.80 + 10.96 = 26.76 \text{ ft-lb}$$

The total kinetic energy is

$$T = \tfrac{1}{2}m\bar{v}^2 + \Sigma\tfrac{1}{2}m_i|\dot{\boldsymbol{\rho}}_i|^2 = 11.20 + 26.76 = 37.96 \text{ ft-lb} \qquad Ans.$$

(b) *Linear momentum.* The linear momentum of the system by Eq. 5/5 is the total mass times v_O, the velocity of the center of mass. Thus

$$[\mathbf{G} = m\bar{\mathbf{v}}] \qquad G = \left(\frac{32.2}{32.2} + 4\frac{3.22}{32.2}\right)(4) = 5.6 \text{ lb-sec}$$

(c) *Angular momentum about O.* The angular momentum about O is due to the moments of the linear momenta of the balls. Taking counterclockwise as positive, we have

$$H_O = \Sigma|\mathbf{r}_i \times m_i\mathbf{v}_i|$$

$$H_O = \left[2\left(\frac{3.22}{32.2}\right)\left(\frac{18}{12}\right)(12.57)\right]_{(1,2)} - \left[2\left(\frac{3.22}{32.2}\right)\left(\frac{12}{12}\right)(10.47)\right]_{(3,4)}$$

$$= 3.77 - 2.09 = 1.68 \text{ ft-lb-sec} \qquad Ans.$$

① Note that the mass m is the total mass, carriage plus the four balls, and that $\bar{v}$ is the velocity of the mass center O which is the carriage velocity.

② Note that the direction of rotation, clockwise or counterclockwise, makes no difference in the calculation of kinetic energy, which depends on the square of the velocity.

③ There is a temptation to overlook the contribution of the balls since their linear momenta relative to O in each pair are in opposite directions and cancel. However, each ball also has a velocity component $\bar{v}$ and hence a momentum component $m_i\bar{v}$.

④ Contrary to the case of kinetic energy where the direction of rotation was immaterial, angular momentum is a vector quantity and the direction of rotation must be accounted for.

PROBLEMS

5/1 Three monkeys *A*, *B*, and *C* weighing 20, 30, and 16 lb, respectively, are climbing up and down the rope suspended from *D*. At the instant represented, *A* is descending the rope with an acceleration of 5 ft/sec², and *C* is pulling himself up with an acceleration of 3 ft/sec². Monkey *B* is climbing up with a constant speed of 1.8 ft/sec. Treat the rope and monkeys as a complete system and calculate the tension *T* in the rope at *D*.

Ans. T = 64.4 lb

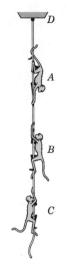

Problem 5/1

5/2 Each of the four systems slides on a smooth horizontal surface and has the same total mass *m*. The configuration of mass in each of the two pairs is identical. What can be said about the acceleration of the mass center for each system? Explain any difference in the accelerations of the members.

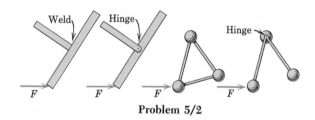

Problem 5/2

5/3 The two 3-kg spheres slide on a smooth horizontal surface in the *x-y* plane and are connected by the two identical springs of negligible mass. Calculate the *x*-component of the acceleration of each sphere due to the action of the 60-N force applied to the junction of the springs.

Ans. $a_x = 10$ m/s²

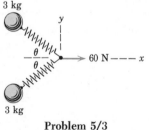

Problem 5/3

5/4 The system consists of two smooth spheres each of mass *m* connected by a light spring and by the two bars of negligible mass hinged freely at their ends. The spheres are confined to slide in the smooth horizontal guide. If a horizontal force *F* is applied to the one bar as shown, what is the acceleration of the center *C* of the spring? Why does the result not depend on *b*?

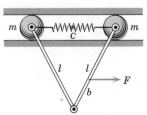

Problem 5/4

5/5 Calculate the acceleration of the center of mass of the system of the four 5-kg cylinders. Neglect friction and the mass of the pulleys and springs.

Ans. $\bar{a} = 2.69$ m/s²

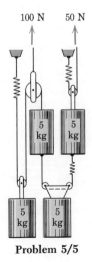

Problem 5/5

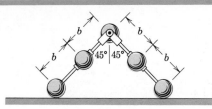

Problem 5/6

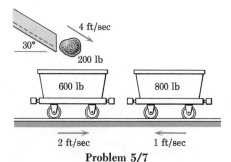

Problem 5/7

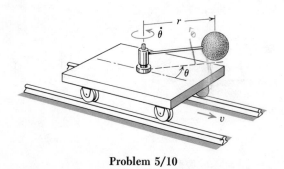

Problem 5/8

Problem 5/10

5/6 The configuration of five identical spheres is released from rest in the position shown and moves in the vertical plane under the influence of gravity. The lower two spheres on each side are securely fastened to the light rigid rods which are freely hinged to the upper sphere. Neglect friction and determine the velocity v with which the upper sphere strikes the horizontal surface. (Note that the velocity of the middle spheres is half that of the upper sphere upon impact.)

5/7 The 600-lb and 800-lb mine cars are rolling in opposite directions along the horizontal track with the respective speeds of 2 ft/sec and 1 ft/sec. Upon impact the cars become coupled together. Just prior to impact a 200-lb boulder leaves the delivery chute with a speed of 4 ft/sec in the direction shown and lands in the 600-lb car. Calculate the velocity of the system after the boulder has come to rest relative to the car. Would the final velocity be the same if the cars were coupled before the boulder dropped?
Ans. $v = 0.68$ ft/sec, yes

5/8 The man of mass m_1 and the woman of mass m_2 are standing on opposite ends of the platform of mass m_0 which moves with negligible friction and is initially at rest with $s = 0$. The man and woman begin to approach each other. Derive an expression for the displacement s of the platform when the two meet in terms of the displacement x_1 of the man relative to the platform.

5/9 Determine the expression for the velocity $\dot{s}$ of the platform of Prob. 5/8 if the man runs at a speed $\dot{x}_1$ and the woman walks at the speed $\dot{x}_2$ both relative to the platform.
$$Ans. \; \dot{s} = \frac{m_1\dot{x}_1 - m_2\dot{x}_2}{m_0 + m_1 + m_2}$$

5/10 The small car which has a mass of 20 kg rolls freely on the horizontal track and carries the 5-kg sphere mounted on the light rotating rod with $r = 0.4$ m. A geared motor drive maintains a constant angular speed $\dot{\theta} = 4$ rad/s of the rod. If the car has a velocity $v = 0.6$ m/s when $\theta = 0$, calculate v when $\theta = 60°$. Neglect the mass of the wheels and any friction.

5/11 Determine the maximum value of the velocity v of the car in Prob. 5/10 and the corresponding angle θ.

> *Ans.* $v = 0.920$ m/s, $\theta = 90°$

5/12 The girl A, the captain B, and the sailor C weigh 120, 180, and 160 lb, respectively, and are sitting in the 300-lb skiff which is gliding through the water with a speed of 1 knot. If the three people change their positions as shown in the second figure, find the distance x from the skiff to the position where it would have been if the people had not moved. Neglect any resistance to motion afforded by the water. Does the sequence or timing of the change in positions affect the final result?

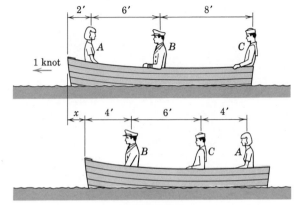

Problem 5/12

5/13 The three small spheres are welded to the light rigid frame which is rotating in a horizontal plane about a vertical axis through O with an angular velocity $\dot{\theta} = 20$ rad/s. If a couple $M_O = 30$ N·m is applied to the frame for 5 s, compute the new angular velocity $\dot{\theta}'$.

> *Ans.* $\dot{\theta}' = 80.7$ rad/s

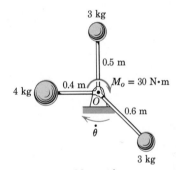

Problem 5/13

5/14 The angular momentum of a system of 6 particles about a fixed point O at time $t = 4$ s is $\mathbf{H}_4 = 3.65\mathbf{i} + 4.27\mathbf{j} - 5.36\mathbf{k}$ kg·m²/s. At time $t = 4.1$ s the angular momentum is $\mathbf{H}_{4.1} = 3.67\mathbf{i} + 4.30\mathbf{j} - 5.20\mathbf{k}$ kg·m²/s. Determine the average value of the resultant moment about point O of all forces acting on all particles during the 0.1-s interval.

5/15 At a certain instant particles A, B, and C with masses of 2, 3, and 4 kg, respectively, have position vectors $\mathbf{r}_A = 2\mathbf{i} + \mathbf{j} - 2\mathbf{k}$ m, $\mathbf{r}_B = 2\mathbf{i} + 2\mathbf{j} - \mathbf{k}$ m, and $\mathbf{r}_C = \mathbf{i} - 2\mathbf{j} - 2\mathbf{k}$ m. The velocities of the particles at this same instant are $\mathbf{v}_A = \dot{\mathbf{r}}_A = 3\mathbf{i} + 2\mathbf{j} - 2\mathbf{k}$ m/s, $\mathbf{v}_B = \dot{\mathbf{r}}_B = 2\mathbf{i} + 3\mathbf{j} + \mathbf{k}$ m/s, and $\mathbf{v}_C = \dot{\mathbf{r}}_C = -2\mathbf{i} - 2\mathbf{j} + \mathbf{k}$ m/s. Calculate the angular momentum of the system of three particles about the origin O.

> *Ans.* $\mathbf{H}_O = -5\mathbf{i} - 4\mathbf{j} - 16\mathbf{k}$ kg·m²/s

5/16 Prove that the linear momentum **G** of a system of mass m with respect to fixed axes is given by

$$\mathbf{G} = \mathbf{G}_r + m\mathbf{v}_0$$

where $\mathbf{G}_r$ is the linear momentum of the system with respect to a set of translating axes moving with a velocity $\mathbf{v}_0$.

5/17 For any system of mass m show that the angular momenta about a fixed point O and about the mass center are related by

$$\mathbf{H}_O = \bar{\mathbf{H}} + \bar{\mathbf{r}} \times \mathbf{G}$$

where $\bar{\mathbf{r}}$ is the position vector of the mass center with respect to O and **G** is the linear momentum of the system.

5/18 Show that the sum of the moments about a fixed point O of the external forces acting on a system of mass m is given by

$$\Sigma \mathbf{M}_O = \dot{\bar{\mathbf{H}}} + \bar{\mathbf{r}} \times \dot{\mathbf{G}}$$

where $\bar{\mathbf{H}}$ is the angular momentum about the mass center, $\bar{\mathbf{r}}$ is the position vector of the mass center with respect to O, and **G** is the linear momentum of the system.

5/19 Each of the bars A and B has a mass of 10 kg and slides in the horizontal guideways with negligible friction. Motion is controlled by the lever of negligible mass connected to the bars as shown. Calculate the acceleration of point C on the lever when the 200-N force is applied as indicated. To verify your result analyze the kinetics of each member separately and determine a_C by kinematic considerations from the calculated accelerations of the two bars.   *Ans.* $a_C = 10$ m/s^2

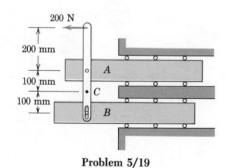

Problem 5/19

5/20 A test firing of two projectiles each weighing 20 lb takes place from the vehicle which weighs 2000 lb and is moving with an initial velocity $v_0 = 4$ ft/sec in the direction opposite to the firing. The muzzle velocity of each projectile (relative to the barrel) is $v_r = 800$ ft/sec. Calculate the velocity v' of the vehicle after the projectiles have been fired (*a*) simultaneously or (*b*) in sequence. Neglect the friction and mass of the wheels.

Ans. (*a*) $v' = 19.69$ ft/sec, (*b*) $v' = 19.76$ ft/sec

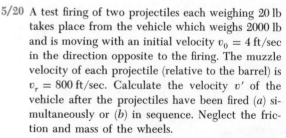

Problem 5/20

5/3 STEADY MASS FLOW.   The momentum relation developed in Art. 5/2 for a general system of mass provides us with a direct means of analyzing the action of mass flow where a change of momentum occurs. The dynamics of mass flow is of great importance in the description of fluid machinery of all types including turbines, pumps, nozzles, air-breathing jet engines, and rockets. The treatment of mass flow in this article is not intended to take the place of a study of fluid mechanics but merely to present the basic principles and equations of momentum which find important usage in fluid mechanics and in the general flow of mass whether the form be liquid, gaseous, or granular.

One of the most important cases of mass flow occurs during steady flow conditions where the rate at which mass enters a given volume equals the rate at which mass leaves the same volume. The volume in question may be enclosed by a rigid container, fixed or moving, such as the nozzle of a jet aircraft or rocket, the space between blades in a gas turbine, the volume within the casing of a centrifugal pump, or the volume within the bend of a pipe through which a fluid is flowing at a steady rate. The design of such fluid machines is dependent on the analysis of the forces and moments which are developed through the corresponding momentum changes of the flowing mass.

Consider a rigid container, shown in section in Fig. 5/3a, into which mass flows in a steady stream at the rate m' through the entrance section of area A_1. Mass leaves the container through the exit section of area A_2 at the same rate, so that there is no accumulation or depletion of the total mass within the container during the period of observation. The velocity of the entering stream is $\mathbf{v}_1$ normal to A_1 and that of the leaving stream is $\mathbf{v}_2$ normal to A_2. If ρ_1 and ρ_2 are the respective densities of the two streams, the continuity of flow requires that

$$\rho_1 A_1 v_1 = \rho_2 A_2 v_2 = m' \qquad (5/13)$$

The forces which act are described by isolating either the mass of fluid within the container or the entire container and the fluid within it. We would use the first approach if the forces between the container and the fluid were to be described, and we would adopt the second approach when the forces external to the container are desired. The latter situation is our primary interest in which case the *system isolated* consists of the fixed structure of the container and the fluid within it at a particular instant of time. This isolation is described by a free-body diagram of the mass within a closed space volume defined by the exterior surface of the container and the entrance and exit surfaces. We must account for all forces applied

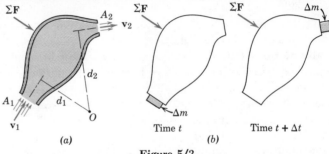

Figure 5/3

externally to this system, and in Fig. 5/3*a* the vector sum of this external force system is denoted by $\Sigma\mathbf{F}$. Included in $\Sigma\mathbf{F}$ are the forces exerted on the container at points of its attachment to other structures including attachments at A_1 and A_2, if present; the forces acting on the fluid within the container at A_1 and A_2 due to any static pressure which may exist in the fluid at these positions; and the weight of the fluid and structure if appreciable. The resultant $\Sigma\mathbf{F}$ of all of these external forces must equal $\dot{\mathbf{G}}$, the time rate of change of the linear momentum of the isolated system, according to Eq. 5/6 which was developed in Art. 5/2 for any system of constant mass, rigid or nonrigid.

The expression for $\dot{\mathbf{G}}$ may be obtained by an incremental analysis. Figure 5/3*b* illustrates the system at time t when the system mass is that of the container, the mass within it, and an increment Δm about to enter during time Δt. At time $t + \Delta t$ the same total mass is that of the container, the mass within it, and an equal increment Δm which leaves the container in time Δt. The linear momentum of the container and mass within it between the two sections A_1 and A_2 remains unchanged during Δt so that the change in momentum of the system in time Δt is

$$\Delta\mathbf{G} = (\Delta m)\mathbf{v}_2 - (\Delta m)\mathbf{v}_1 = \Delta m(\mathbf{v}_2 - \mathbf{v}_1)$$

Division by Δt and passage to the limit yield $\dot{\mathbf{G}} = m'\,\Delta\mathbf{v}$ where

$$m' = \lim_{\Delta t \to 0}\left(\frac{\Delta m}{\Delta t}\right) = \frac{dm}{dt}$$

Thus by Eq. 5/6

$$\boxed{\Sigma\mathbf{F} = m'\,\Delta\mathbf{v}} \tag{5/14}$$

Equation 5/14 establishes the relation between the resultant force on a steady-flow system and the corresponding mass flow rate and vector velocity increment.

Alternatively we may note that the time rate of change of linear momentum is the vector difference between the rate at which linear momentum leaves the system and the rate at which linear momentum enters the system. Thus we may write $\dot{\mathbf{G}} = m'\mathbf{v}_2 - m'\mathbf{v}_1 = m'\,\Delta\mathbf{v}$, which agrees with the foregoing result.

We can now see one of the powerful applications of our general force-momentum equation which we derived for any mass system. Our system here includes particles which are rigid (the structural container for the mass stream) and particles which are in motion (the flow of mass). By defining the boundary of the system, the mass within which is constant for steady flow conditions, we are able to utilize the generality of Eq. 5/6. However, we must be very careful to account for *all* external forces acting on the system, and they become clear if our free-body diagram is correct.

A similar formulation is obtained for the case of angular momentum in steady-flow systems. The resultant moment of all external forces about some fixed point O on or off the system, Fig. 5/3a, equals the time rate of change of angular momentum of the system about O. This fact was established in Eq. 5/7 which, for the case of steady flow in a single plane, becomes

$$\Sigma M_O = m'(v_2 d_2 - v_1 d_1) \qquad (5/15)$$

When the velocities of the incoming and outgoing flows are not in the same plane, the equation may be written in vector form as

$$\boxed{\Sigma \mathbf{M}_O = m'(\mathbf{d}_2 \times \mathbf{v}_2 - \mathbf{d}_1 \times \mathbf{v}_1)} \qquad (5/15a)$$

where $\mathbf{d}_1$ and $\mathbf{d}_2$ are the position vectors to the centers of A_1 and A_2 from the fixed reference O. In both relations the mass center G may be used alternatively as a moment center by virtue of Eq. 5/9.

Equations 5/14 and 5/15 are very simple relations which find important use in describing relatively complex fluid actions. It should be noted that these equations relate *external* forces to the resultant changes in momentum and are independent of the flow path and momentum changes *internal* to the system.

The foregoing analysis may be applied to systems moving with constant speed by noting that the basic relations $\Sigma\mathbf{F} = \dot{\mathbf{G}}$ and $\Sigma\mathbf{M}_O = \dot{\mathbf{H}}_O$ or $\Sigma\bar{\mathbf{M}} = \dot{\bar{\mathbf{H}}}$ apply to systems moving with constant speed as discussed in Arts. 4/4 and 5/2. The only restriction is that the mass within the system remain constant with respect to time.

Three examples of the analysis of steady mass flow are given in the following sample problems which illustrate the application of the principles embodied in Eqs. 5/14 and 5/15.

Sample Problem 5/5

The smooth vane shown diverts the open stream of fluid of cross-sectional area A, mass density ρ, and velocity v. (a) Determine the force components R and F required to hold the vane in a fixed position. (b) Find the forces when the vane is given a constant velocity u less than v and in the direction of v.

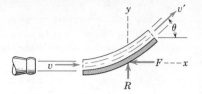

Fixed vane

Moving vane

Solution. *Part (a).* The free-body diagram of the vane together with the fluid portion undergoing the momentum change is shown. The momentum equation may be applied to the isolated system for the change in motion in both the x- and y-directions. With the vane stationary the magnitude of the exit velocity v' equals that of the entering velocity v with fluid friction neglected. The changes in the velocity components are, then,

$$\Delta v_x = v' \cos \theta - v = -v(1 - \cos \theta)$$

and

$$\Delta v_y = v' \sin \theta - 0 = v \sin \theta$$

The mass rate of flow is $m' = \rho A v$, and substitution into Eq. 5/14 gives

$$[\Sigma F_x = m' \, \Delta v_x] \qquad -F = \rho A v[-v(1 - \cos \theta)]$$
$$F = \rho A v^2 (1 - \cos \theta) \qquad \qquad Ans.$$

$$[\Sigma F_y = m' \, \Delta v_y] \qquad R = \rho A v[v \sin \theta]$$
$$R = \rho A v^2 \sin \theta \qquad \qquad Ans.$$

① Be careful with algebraic signs when using Eq. 5/14. The change in v_x is the final value minus the initial value measured in the positive x-direction. Also we must be careful to write $-F$ for ΣF_x.

Part (b). In the case of the moving vane the final velocity v' of the fluid upon exit is the vector sum of the velocity u of the vane plus the velocity of the fluid relative to the vane $v - u$. This combination is shown in the velocity diagram to the right of the figure for the exit conditions. The x-component of v' is the sum of the components of its two parts, so $v_x' = (v - u) \cos \theta + u$. The change in x-velocity of the stream is

$$\Delta v_x = (v - u) \cos \theta + (u - v) = -(v - u)(1 - \cos \theta)$$

The y-component of v' is $(v - u) \sin \theta$, so that the change in the y-velocity of the stream is $\Delta v_y = (v - u) \sin \theta$.

The mass rate of flow m' is the mass undergoing momentum change per unit of time. This rate is the mass flowing over the vane per unit time and *not* the rate of issuance from the nozzle. Thus

$$m' = \rho A(v - u)$$

The impulse-momentum principle of Eq. 5/14 applied in the positive coordinate direction gives

$$[\Sigma F_x = m' \, \Delta v_x] \qquad -F = \rho A(v - u)[-(v - u)(1 - \cos \theta)]$$
$$F = \rho A(v - u)^2 (1 - \cos \theta) \qquad Ans.$$

$$[\Sigma F_y = m' \, \Delta v_y] \qquad R = \rho A(v - u)^2 \sin \theta \qquad Ans.$$

② Observe that for given values of u and v the angle for maximum force F is $\theta = 180°$.

Sample Problem 5/6

For the moving vane of Sample Problem 5/5 determine the optimum speed u of the vane for the generation of maximum power by the action of the fluid on the vane.

Solution. The force R shown with the figure for Sample Problem 5/5 is normal to the velocity of the vane so does no work. The work done by the force F shown is negative, but the power developed by the force (equal and opposite to F) exerted by the fluid on the moving vane is

$$[P = Fu] \qquad P = \rho A(v - u)^2 u(1 - \cos \theta)$$

The velocity of the vane for maximum power for the one blade in the stream is specified by

$$\left[\frac{dP}{du} = 0\right] \qquad \rho A(1 - \cos\theta)(v^2 - 4uv + 3u^2) = 0$$

$$(v - 3u)(v - u) = 0 \qquad u = \frac{v}{3} \qquad \textit{Ans.}$$

The second solution $u = v$ gives a minimum condition of zero power. An angle $\theta = 180°$ completely reverses the direction of the fluid and clearly produces both maximum force and maximum power for any value of u.

① The result here applies to a single vane only. In the case of multiple vanes, such as the blades on a turbine disk, the rate at which fluid issues from the nozzles is the same rate at which fluid is undergoing momentum change. Thus $m' = \rho A v$ rather than $\rho A(v - u)$. With this change the optimum value of u turns out to be $u = v/2$.

Sample Problem 5/7

The offset nozzle has a discharge area A at B and an inlet area A_0 at C. A liquid enters the nozzle at a static gage pressure p through the fixed pipe and issues from the nozzle with a velocity v in the direction shown. If the constant density of the liquid is ρ, write expressions for the tension T, shear Q, and bending moment M in the pipe at C.

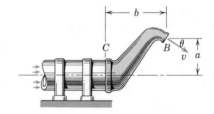

Solution. The free-body diagram of the nozzle and the fluid within it shows the tension T, shear Q, and bending moment M acting on the flange of the nozzle where it attaches to the fixed pipe. The force pA_0 on the fluid within the nozzle due to the static pressure is an additional external force.

Continuity of flow with constant density requires that

$$Av = A_0 v_0$$

where v_0 is the velocity of the fluid at entrance to the nozzle. The momentum principle of Eq. 5/14 applied to the system in the two coordinate directions gives

$$[\Sigma F_x = m' \Delta v_x] \qquad pA_0 - T = \rho Av(v \cos\theta - v_0)$$

$$T = pA_0 + \rho Av^2\left(\frac{A}{A_0} - \cos\theta\right) \qquad \textit{Ans.}$$

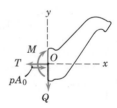

$$[\Sigma F_y = m' \Delta v_y] \qquad -Q = \rho Av(-v \sin\theta - 0)$$

$$Q = \rho Av^2 \sin\theta \qquad \textit{Ans.}$$

The moment principle of Eq. 5/15 applied in the clockwise sense gives

$$[\Sigma M_O = m'(v_2 d_2 - v_1 d_1)] \qquad M = \rho Av(va \cos\theta + vb \sin\theta - 0)$$

$$M = \rho Av^2(a \cos\theta + b \sin\theta) \qquad \textit{Ans.}$$

① Again, be careful to observe the correct algebraic signs of the terms on both sides of Eqs. 5/14 and 5/15.

② The forces and moment acting on the pipe are equal and opposite to those shown acting on the nozzle.

Sample Problem 5/8

An air-breathing jet aircraft of total mass m flying with a constant speed v consumes air at the mass rate m_a' and exhausts burned gas at the mass rate m_g' with a velocity u relative to the aircraft. Fuel is consumed at the constant rate m_f'. The total aerodynamic forces acting on the aircraft are the lift L, normal to the direction of flight, and the drag D opposite to the direction of flight. Any force due to the static pressure across the inlet and exhaust surfaces is assumed to be included in D. Write the equation for the motion of the aircraft and identify the thrust T.

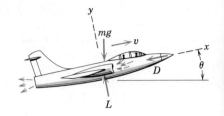

Solution. The free-body diagram of the aircraft together with the air, fuel, and exhaust gas within it is given and shows only the weight, lift, and drag forces as defined. We attach axes x-y to the aircraft and apply our momentum equation relative to the moving system.

The fuel will be treated as a steady stream entering the aircraft with no velocity relative to the system and leaving with a relative velocity u in the exhaust stream. We now apply Eq. 5/14 relative to the reference axes and treat the air and fuel flows separately. For the air flow the change in velocity in the x-direction relative to the moving system is

$$\Delta v_a = -u - (-v) = -(u - v)$$

and for the fuel flow the x-change in velocity relative to x-y is

$$\Delta v_f = -u - (0) = -u$$

Thus we have

$$[\Sigma F_x = m' \Delta v_x] \qquad -mg \sin \theta - D = -m_a'(u - v) - m_f'u$$
$$= -m_g'u + m_a'v$$

where the substitution $m_g' = m_a' + m_f'$ has been made. Changing signs gives

$$m_g'u - m_a'v = mg \sin \theta + D$$

which is simply an equation of equilibrium of the aircraft.

If we modify the boundaries of our system to expose the interior surfaces upon which the air and gas act, we will have the simulated model shown where the air exerts a force $m_a'v$ on the interior of the turbine and the exhaust gas reacts against the interior surfaces with the force $m_g'u$.

The commonly used model is shown in the final diagram where the net effect of air and exhaust momentum changes is replaced by a simulated thrust

$$T = m_g'u - m_a'v \qquad Ans.$$

applied to the aircraft from a presumed external source.

Inasmuch as m_f' is generally only 2 percent or less of m_a', we can use the approximation $m_g' \approx m_a'$ and express the thrust as

$$T \approx m_g'(u - v) \qquad Ans.$$

We have analyzed the case of constant velocity. Although our Newtonian principles do not generally hold relative to accelerating axes, it can be shown that we may use the $F = ma$ equation for the simulated model and write $T - mg \sin \theta - D = m\dot{v}$ with virtually no error.

① Note that the boundary of the system cuts across the air stream at the entrance to the air scoop and across the exhaust stream at the nozzle.

② We are permitted to use moving axes which translate with constant velocity. See Arts. 4/4 and 5/2.

③ Riding with the aircraft we observe the air entering our system with a velocity $-v$ measured in the plus x-direction and leaving the system with an x-velocity of $-u$. The final minus the initial values give the expression cited, namely, $-u - (-v) = -(u - v)$.

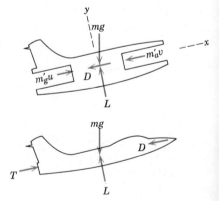

④ We now see that the "thrust" is, in reality, not a force external to the entire airplane shown in the first figure but can be modeled as an external force.

PROBLEMS

5/21 Fresh water issues from the nozzle *A* with a velocity of 100 ft/sec at the rate of 2 ft^3/sec and is split into the two equal streams by the fixed vane shown. Calculate the force *F* required to hold the vane in place. The specific weight of fresh water is 62.4 lb/ft^3. *Ans. F* = 388 lb

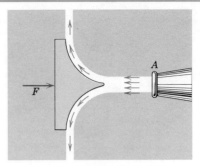

Problem 5/21

5/22 An experimental aircraft designed for vertical take-off and landing (VTOL) has a mass of 7 Mg. Movable vanes deflect the exhaust stream downward at takeoff. Each of the craft's two engines sucks in air at the rate of 60 kg/s at a density of 1.217 kg/m^3 and discharges the exhaust at a velocity of 600 m/s. Each engine uses fuel at the rate of 1 kg/s. Determine the initial vertical acceleration *a*.

Problem 5/22

5/23 A jet-engine noise suppressor consists of a movable duct which is secured directly behind the jet exhaust by cable *A* and deflects the blast directly upward. During ground test the engine sucks in air at the rate of 43 kg/s and burns fuel at the rate of 0.8 kg/s. The exhaust velocity is 720 m/s. Determine the tension *T* in the cable. *Ans. T* = 32.6 kN

Problem 5/23

5/24 The fire tug discharges a stream of salt water (specific weight 64.0 lb/ft^3) with a nozzle velocity of 120 ft/sec at the rate of 1200 gal/min. Calculate the propeller thrust *T* which must be developed by the tug to maintain a fixed position while pumping. (231 in.^{3} = 1 gal)

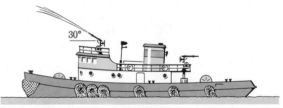

Problem 5/24

5/25 The pump shown draws air with a density ρ through the fixed duct *A* of diameter *d* with a velocity *u* and discharges it at high velocity *v* through the two outlets *B*. The pressure in the air streams at *A* and *B* is atmospheric. Determine the expression for the tension *T* exerted on the pump unit through the flange at *C*.

5/26 If the vane of Problem 5/21 has a velocity of 20 ft/sec to the left, compute the force *F* and the power *P* developed by the vane. *Ans. F* = 248 lb, *P* = 9.02 hp

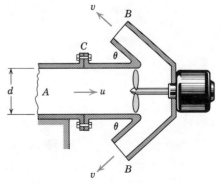

Problem 5/25

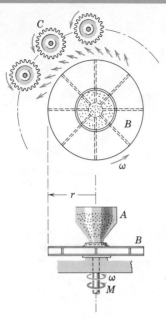

Problem 5/27

Problem 5/29

Problem 5/30

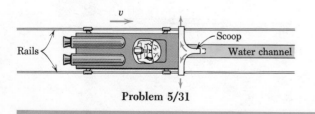

Problem 5/31

5/27 Shot peening is used to increase the fatigue strength of gear teeth and other machine parts. The device shown consists of a reservoir A which feeds the shot pellets into the center of a rotating impeller B that flings them at high speed and produces a spray of shot. The spray of high-velocity pellets impinges against the rotating gears C that are being treated. If m kg of shot are consumed per second and if the angular velocity is ω, determine the torque M required to drive the impeller.

5/28 An air-breathing jet aircraft has a constant speed of 1500 km/h in horizontal flight at an altitude of 12 km. The turbojet engine consumes air at the rate of 110 kg/s at this speed and uses fuel at the rate of 0.97 kg/s. The gases are exhausted at a relative nozzle velocity of 780 m/s at the atmospheric pressure corresponding to the flight altitude. Determine the total drag D (air resistance) on the exterior surface of the aircraft and the useful power P (thrust horsepower) of the engine at this speed.
Ans. $D = 40.7$ kN, $P = 16.97$ MW

5/29 The jet aircraft has a mass of 4 Mg and has a drag (air resistance) of 24 kN at a speed of 800 km/h at a particular altitude. The aircraft consumes air at the rate of 90 kg/s through its intake scoop and uses fuel at the rate of 0.90 kg/s. If the exhaust has a rearward velocity of 600 m/s relative to the exhaust nozzle, determine the maximum angle of elevation α at which the jet can fly with a constant speed of 800 km/h at the particular altitude in question.

5/30 The snow plow eats its way through a 1.5-m drift and throws the snow off to the side at the rate of 30 Mg per minute. The snow has a velocity of 12 m/s relative to the plow in the 45° direction shown. Determine the lateral force R between the tires and the road and the tractive force F required to drive the plow at 15 km/h.
Ans. $R = 4.24$ kN, $F = 2.08$ kN

5/31 The figure shows the top view of an experimental rocket sled which is traveling at a speed of 1000 ft/sec when its forward scoop enters a water channel to act as a brake. The water is diverted at right angles relative to the sled. If the frontal flow area of the scoop is 15 in.², calculate the initial braking force. The specific weight of water is 62.4 lb/ft³.

32 A 25-mm steel slab 1.2 m wide enters the rolls at the speed of 0.4 m/s and is reduced in thickness to 19 mm. Calculate the small horizontal thrust T on the bearings of each of the two rolls.

Ans. $T = 5.93$ N

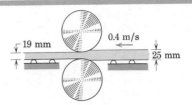

Problem 5/32

33 In a test of the operation of a "cherry picker" fire truck the equipment is free to roll with its brakes released. For the position shown the truck is observed to deflect the spring of stiffness $k = 15$ kN/m a distance of 150 mm because of the action of the horizontal stream of water issuing from the nozzle when the pump is activated. If the exit diameter of the nozzle is 30 mm, calculate the velocity v of the stream as it leaves the nozzle. Also determine the added moment M which the joint at A must resist when the pump is in operation with the nozzle in the position shown.

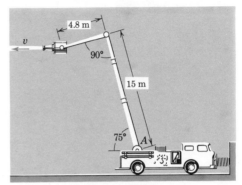

Problem 5/33

34 The experimental ground-effect machine has a total weight of 4200 lb. It hovers 1 or 2 ft off the ground by pumping air at atmospheric pressure through the circular intake duct at B and discharging it horizontally under the periphery of the skirt C. For an intake velocity v of 150 ft/sec calculate the average air pressure p under the 18-ft-diameter machine. The specific weight of the air is 0.076 lb/ft³.

Ans. $p = 0.156$ lb/in.²

Problem 5/34

35 The pipe bend shown has a cross-sectional area A and is supported in its plane by the tension T applied to its flanges by the adjacent connecting pipes (not shown). If the velocity of the liquid is v, its density ρ, and its static pressure p, determine T and show that it is independent of the bend angle θ.

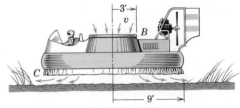

Problem 5/35

36 The 90° vane moves to the left with a constant velocity of 10 m/s against a stream of fresh water issuing from the 25-mm-diameter nozzle with a velocity of 20 m/s. Calculate the forces F_x and F_y on the vane required to support the motion.

Ans. $F_x = 442$ N, $F_y = 442$ N

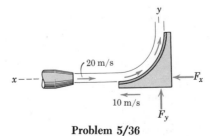

Problem 5/36

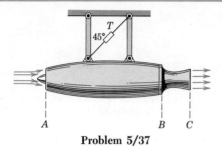

Problem 5/37

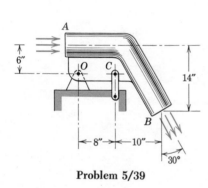

Problem 5/39

Problem 5/40

5/37 In the static test of a jet engine and exhaust nozzle assembly, air is sucked into the engine at the rate of 30 kg/s and fuel is burned at the rate of 0.6 kg/s. The flow area, static pressure, and axial-flow velocity for the three sections shown are as follows:

	Sec. A	Sec. B	Sec. C
Flow area, m^2	0.15	0.16	0.06
Static pressure, kPa	-14	140	14
Axial-flow velocity, m/s	120	315	600

Determine the tension T in the diagonal member of the supporting test stand and calculate the force F exerted on the nozzle flange at B by the bolts and gasket to hold the nozzle to the engine housing.

5/38 In the case of multiple vanes where each vane which enters the jet is followed immediately by another, as in a turbine or water wheel (see figures for Probs. 5/51 and 5/52) determine the maximum power P which can be developed for a given blade angle and the corresponding optimum peripheral speed u of the vanes in terms of the jet velocity v for maximum power. Modify Sample Problem 5/6 by assuming an infinite number of vanes so that the rate at which fluid leaves the nozzle equals the rate at which fluid passes over the vanes.

$$Ans. \ P = \tfrac{1}{4}\rho Av^3(1 - \cos\theta), \ u = v/2$$

5/39 The open pipe bend AB is welded to the flange which is supported by the pin at O and the link at C. A stream of fresh water (specific weight 62.4 lb/ft^3) is directed into the pipe at A with a velocity of 80 ft/sec and emerges at B with essentially the same speed. If the flow rate is 8 ft^3/sec, calculate the force C in the link.

5/40 A ball of mass m is supported in a vertical jet of water with density ρ. If the stream of water issuing from the nozzle has a diameter d and velocity u, determine the height h above the nozzle at which the ball is supported. Assume that the jet remains intact and that there is no energy loss in the jet.

$$Ans. \ h = \frac{u^2}{2g} - 8g\left(\frac{m}{\pi\rho ud^2}\right)^2$$

√41 Air is pumped through the stationary duct A with a velocity of 15 m/s and exhausted through an experimental nozzle section BC. The average static pressure across section B is 1050 kPa gage, and air density at this pressure and at the temperature prevailing is 13.5 kg/m^3. The average static pressure across the exit section C is measured to be 14 kPa gage, and the corresponding density of air is 1.217 kg/m^3. Calculate the force T exerted on the nozzle flange at B by the bolts and the gasket to hold the nozzle in place.　　　　*Ans. T* = 28.7 kN

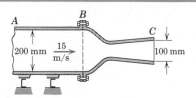

Problem 5/41

√42 Air enters the pipe at A at the rate of 12 lb/sec under a pressure of 200 lb/in.^{2} gage and leaves the whistle at atmospheric pressure through the opening at B. The entering velocity of the air at A is 150 ft/sec, and the exhaust velocity at B is 1200 ft/sec. Calculate the tension T, shear Q, and bending moment M in the pipe at A. The net flow area at A is 12 in.^{2}

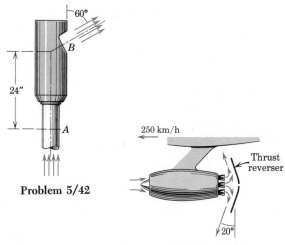

Problem 5/42

√43 Reverse thrust to decelerate a jet aircraft during landing is obtained by swinging the exhaust deflectors into place as shown, thus causing a partial reversal of the exhaust stream. For a jet engine which consumes 40 kg of air per second at a ground speed of 250 km/h and which uses 0.6 kg of fuel per second, determine the reverse thrust as a fraction n of the forward thrust with deflectors removed. The exhaust velocity relative to the nozzle is 600 m/s. Assume air enters the engine with a relative velocity equal to the ground speed of the aircraft.　　　　*Ans. n* = 0.515

Problem 5/43

√44 The sump pump has a net mass of 310 kg and pumps fresh water against a 6-m head at the rate of 0.125 m^3/s. Determine the vertical force R between the supporting base and the pump flange at A during operation. The mass of water in the pump may be taken to be the equivalent of a 200-mm-diameter column 6 m in height.

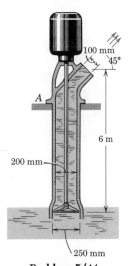

Problem 5/44

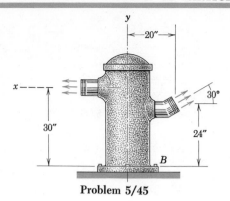

Problem 5/45

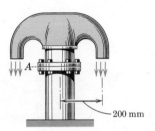

Problem 5/46

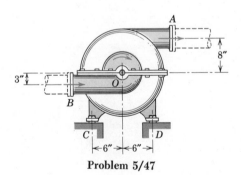

Problem 5/47

5/45 The fire hydrant is tested under a high standpipe pressure of 120 lb/in.² The total flow of 10 ft³/sec is divided equally between the two outlets each of which has a cross-sectional area of 0.040 ft². The inlet cross-sectional area at the base is 0.75 ft². Neglect the weight of the hydrant and water within it and compute the tension T, the shear Q, and the bending moment M in the base of the standpipe at B. The specific weight of water is 62.4 lb/ft³. The static pressure of the water as it enters the base at B is 120 lb/in.²

   Ans. $T = 12{,}610$ lb, $Q = 162$ lb, $M = 1939$ lb-ft

5/46 The 180° pipe return discharges salt water (density 1030 kg/m³) into the atmosphere at a constant rate of 0.05 m³/s. The static pressure in the water at section A is 70 kPa above atmospheric pressure. The flow area of the pipe at A is 12 500 mm² and that at each of the two outlets is 2000 mm². If each of the six flange bolts is tightened with a torque wrench so that it is under a tension of 750 N, determine the average pressure p on the gasket between the two flanges. The flange area in contact with the gasket is 10 000 mm². Also determine the bending moment M in the pipe at section A if the left-hand side of the tee is blocked off and the flow rate is cut in half.

5/47 The centrifugal pump handles 5000 gal of fresh water per minute with inlet and outlet velocities of 60 ft/sec. The impeller is turned clockwise through the shaft at O by a motor which delivers 50 hp at a pump speed of 900 rev/min. With the pump filled but not turning, the vertical reactions at C and D are each 50 lb. Calculate the forces exerted by the foundation on the pump at C and D while the pump is running. The tensions in the connecting pipes at A and B are exactly balanced by the respective forces due to the static pressure in the water. (*Suggestion:* Isolate the entire pump and water within it between sections A and B and apply the momentum principle to the entire system.)

   Ans. $C = 946$ lb up, $D = 846$ lb down

48 The helicopter shown has a mass m and hovers in position by imparting downward momentum to a column of air defined by the slip-stream boundary shown. Find the downward velocity v given to the air by the rotor at a section in the stream below the rotor where the pressure is atmospheric and the stream radius is r. Also find the power P required of the engine. Neglect the rotational energy of the air, any temperature rise due to air friction, and any change in air density ρ.

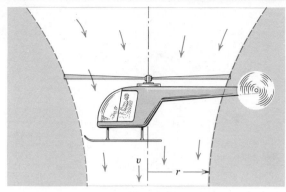

Problem 5/48

49 The sprinkler is made to rotate at the constant angular velocity ω and distributes water at the volume rate Q. Each of the four nozzles has an exit area A. Write an expression for the torque M on the shaft of the sprinkler necessary to maintain the given motion. For a given pressure and, hence, flow rate Q, at what speed ω_0 will the sprinkler operate with no applied torque? Let ρ be the density of the water.

$$Ans.\ \omega_0 = \frac{Qr}{4A(r^2 + b^2)}$$

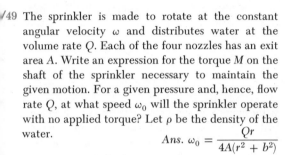

Problem 5/49

50 A test vehicle for impact studies has a mass of 1.6 Mg and is accelerated from rest by the action of a water jet which impinges against the deflector vane attached to the vehicle as shown. The freshwater jet is generated by the action of a piston which is activated by air released from a high-pressure chamber. The jet is 150 mm in diameter and has a velocity which is essentially constant at 180 m/s for the short duration of the test. If frictional resistance is 10 percent of the vehicle weight, determine the velocity u of the vehicle after the jet has acted for 4 s. Assume steady-flow conditions for the vane. $\qquad$ *Ans.* $u = 167.3$ m/s

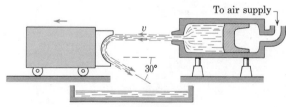

Problem 5/50

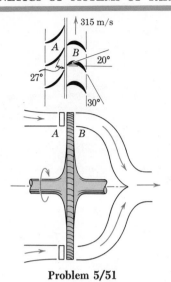

Problem 5/51

▶5/51 In the figure is shown a detail of the stationary nozzle diaphragm *A* and the rotating blades *B* of a gas turbine. The products of combustion pass through the fixed diaphragm blades at the 27° angle and impinge on the moving rotor blades. The angles shown are selected so that the velocity of the gas relative to the moving blade at entrance is at the 20° angle for minimum turbulence, corresponding to a mean blade velocity of 315 m/s at a radius of 375 mm. If gas flows past the blades at the rate of 15 kg/s, determine the theoretical power output *P* of the turbine. Neglect fluid and mechanical friction with the resulting heat-energy loss, and assume that all the gases are deflected along the surfaces of the blades with a velocity of constant magnitude relative to the blade.   *Ans. P* = 1.197 MW

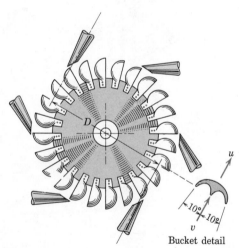

Bucket detail

Problem 5/52

▶5/52 In the figure is shown an impulse-turbine wheel for a hydroelectric power plant which is to operate with a static head of water of 300 m at each of its 6 nozzles and is to rotate at the speed of 270 rev/min. Each wheel and generator unit is to develop an output power of 22 000 kW. The efficiency of the generator may be taken to be 0.90, and an efficiency of 0.85 for the conversion of the kinetic energy of the water jets to energy delivered by the turbine may be expected. The mean peripheral speed of such a wheel for greatest efficiency will be about 0.47 times the jet velocity. If each of the buckets is to have the shape shown, determine the necessary jet diameter *d* and wheel diameter *D*. Assume that the water acts on the bucket which is at the tangent point of each jet stream.
        Ans. d = 165.3 mm, *D* = 2.55 m

5/4 VARIABLE MASS. In Art. 5/2 we extended the equations for the motion of a particle to include a system of particles. In the *c*-part of the article this extension led to the very general expressions $\Sigma \mathbf{F} = \dot{\mathbf{G}}$, $\Sigma \mathbf{M}_O = \dot{\mathbf{H}}_O$, and $\Sigma \bar{\mathbf{M}} = \dot{\bar{\mathbf{H}}}$, which are Eqs. 5/6, 5/7, and 5/9, respectively. In the derivation of these equations the summations were taken over a fixed number of particles, so that the mass of the system to be analyzed was constant. In Art. 5/3 these momentum principles were extended in Eqs. 5/14 and 5/15 to describe the action of forces on a system defined by a geometric volume through which passes a steady flow of mass. The amount of mass within the control volume was, therefore, constant with respect to time and we were, thus, permitted to use Eqs. 5/6, 5/7, and 5/9. When the mass within the boundary of a system under consideration is not constant, the foregoing relationships are no longer valid.[*]

We will now develop the equation of motion for a system whose mass varies with time. For this simplified case we consider, first, a body which gains mass by virtue of its overtaking and swallowing a stream of matter, Fig. 5/4*a*. The mass of the body and its velocity at any instant are m and v, respectively. The stream of matter is assumed to be moving in the same direction as m with a constant velocity v_0 less than v. By virtue of Eq. 5/14 the force exerted by m on the particles of the stream to accelerate them from a velocity v_0 to a greater velocity v is $R = m'(v - v_0) = \dot{m}u$ where the time rate of increase of m is $m' = \dot{m}$ and where u is the relative velocity with which the particles approach m. In addition to R, all other forces acting on m in the direction of its motion are designated by ΣF. The equation of motion

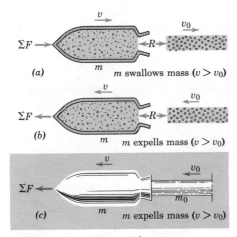

(a) m swallows mass ($v > v_0$)

(b) m expells mass ($v > v_0$)

(c) m expells mass ($v > v_0$)

Figure 5/4

of m from Newton's second law is, therefore, $\Sigma F - R = m\dot{v}$ or

$$\boxed{\Sigma F = m\dot{v} + \dot{m}u} \qquad (5/16)$$

Similarly, if the body loses mass by expelling it at a velocity v_0 less than v, Fig. 5/4b, the force R required to decelerate the particles from a velocity v to a lesser velocity v_0 is $R = m'(-v_0 - [-v]) = m'(v - v_0)$. But $m' = -\dot{m}$ since m is decreasing. Also the relative velocity with which the particles leave m is $u = v - v_0$. Thus the force R becomes $R = -\dot{m}u$. If ΣF denotes the resultant of all other forces acting on m in the direction of its motion, Newton's second law requires $\Sigma F + R = m\dot{v}$ or

$$\Sigma F = m\dot{v} + \dot{m}u$$

which is the same relationship as in the case where m is gaining mass. We may use Eq. 5/16, therefore, as the equation of motion of m whether it is gaining or losing mass.

A frequent error in the use of the force-momentum equation is to express the partial force sum ΣF as

$$\Sigma F = \frac{d}{dt}(mv) = m\dot{v} + \dot{m}v$$

From this expansion we see that the direct differentiation of the linear momentum gives the correct force ΣF *only* when the body picks up mass initially at rest or when it expels mass at a zero absolute velocity. In both instances $v_0 = 0$ and $u = v$.

We may also obtain Eq. 5/16 by a direct differentiation of the momentum from the basic relation $\Sigma F = \dot{G}$ provided a proper system of constant total mass is chosen. To illustrate this approach, we take the case where m is losing mass and use Fig. 5/4c which shows the system of m and an arbitrary portion m_0 of the stream of ejected mass. The mass of this system is $m + m_0$ and is constant. The ejected stream of mass is assumed to move undisturbed once separated from m, and the only force external to the entire system is ΣF which is applied directly to m as before. The reaction $R = -\dot{m}u$ is internal to the system and is not disclosed. With constant total mass the momentum principle $\Sigma F = \dot{G}$ is applicable and we have

$$\Sigma F = \frac{d}{dt}(mv + m_0 v_0) = m\dot{v} + \dot{m}v + \dot{m}_0 v_0 + m_0 \dot{v}_0$$

Clearly $\dot{m}_0 = -\dot{m}$, and the velocity of the ejected mass with respect to m is $u = v - v_0$. Also $\dot{v}_0 = 0$ since m_0 moves undisturbed with no acceleration once free of m. Thus the relation becomes

$$\Sigma F = m\dot{v} + \dot{m}u$$

which is identical with the result of the previous formulation, Eq. 5/16.

The case of m losing mass is clearly descriptive of rocket propulsion. Figure 5/5a shows a vertically ascending rocket, the system for which is the mass within the volume defined by the exterior surface of the rocket and the exit plane across the nozzle. External to this system the free-body diagram discloses the instantaneous values of gravitational attraction mg, aerodynamic resistance R, and the force pA due to the average static pressure p across the nozzle exit plane of area A. Also the rate of mass flow is $m' = -\dot{m}$. Thus we may write the equation of motion of the rocket $\Sigma F = m\dot{v} + \dot{m}u$ as

$$m'u + pA - mg - R = m\dot{v} \qquad (5/17)$$

Equation 5/17 is of the form "$\Sigma F = ma$" where the first term in "ΣF" is the thrust $T = m'u$. Thus the rocket may be simulated as a body to which an external thrust T is applied, Fig. 5/5b, and the problem may then be analyzed like any other "$F = ma$" problem except that m is a function of time.

It may be observed that, during the initial stages of motion when the magnitude of the velocity v of the rocket is less than the relative exhaust velocity u, the absolute velocity v_0 of the exhaust gases will be directed rearward. On the other hand, when the rocket reaches a velocity v whose magnitude is greater than u, the absolute velocity v_0 of the exhaust gases will be directed forward. For a given mass rate of flow the rocket thrust T depends only on the relative exhaust velocity u and not on the magnitude or direction of the absolute velocity v_0 of the exhaust gases.

In the foregoing treatment of bodies whose mass changes with time we have assumed that all elements of the mass m of the body were moving with the same velocity v at any instant of time and that the particles of mass which were added to or expelled from the body underwent an abrupt transition of velocity upon entering or leaving the body. Thus this velocity change has been modeled as a mathematical discontinuity. In reality this change in velocity can not be discontinuous even though the transition may be rapid. In the case of a rocket, for example, the velocity change occurs in the space between the combustion zone and the exit plane of the exhaust nozzle. A more general analysis[*] of variable-mass dynamics removes this restriction of discontinuous velocity change and introduces a slight correction to Eq. 5/16.

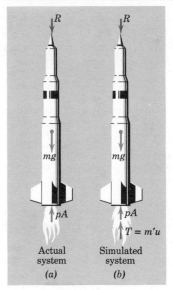

Figure 5/5

(a) Actual system (b) Simulated system

[*] For a development of the equations which describe the general motion of a time-dependent system of mass see Art. 53 of the author's *Dynamics, 2nd Edition* 1971 and *SI Version* 1975, John Wiley & Sons.

Sample Problem 5/9

The end of a chain of length L and mass ρ per unit length which is piled on a platform is lifted vertically with a constant velocity v by a variable force P. Find P as a function of the height x of the end above the platform. Also find the energy lost during the lifting of the chain.

Solution I (variable-mass approach). Equation 5/16 will be used and applied to the moving part of the chain of length x which is gaining mass. The force summation ΣF includes all forces acting on the moving part except the force exerted by the particles which are being attached. From the diagram we have

$$\Sigma F_x = P - \rho gx$$

The velocity is constant so that $\dot{v} = 0$. The rate of increase of mass is $\dot{m} = \rho v$, and the relative velocity with which the attaching particles approach the moving part is $u = v - 0 = v$. Thus Eq. 5/16 becomes

① $[\Sigma F = m\dot{v} + \dot{m}u]$ $\quad P - \rho gx = 0 + \rho v(v),$ $\quad P = \rho(gx + v^2)$ $\quad$ *Ans.*

① The model of Fig. 5/4a shows the mass being added to the leading end of the moving part. With the chain the mass is added to the trailing end, but the effect is the same.

Thus we see that the force P consists of the two parts, ρgx, which is the weight of the moving part of the chain, and ρv^2, which is the added force required to change the momentum of the links on the platform from a condition at rest to a velocity v.

Solution II (constant-mass approach). The principle of impulse and momentum for a system of particles expressed by Eq. 5/6 will be applied to the entire chain considered as the system of constant mass. The free-body diagram of the system shows the unknown force P, the total weight of all links ρgL, and the force $\rho g(L - x)$ exerted by the platform on those links which are at rest upon it. The momentum of the system at any position is $G_x = \rho xv$ and the momentum equation gives

② $\left[\Sigma F_x = \dfrac{dG_x}{dt}\right]$ $\quad P + \rho g(L - x) - \rho gL = \dfrac{d}{dt}(\rho xv),$ $\quad P = \rho(gx + v^2)$ $\quad$ *Ans.*

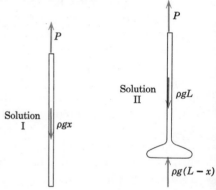

② We must be very careful not to use $\Sigma F = \dot{G}$ for a system whose mass is changing. Thus we have taken the total chain as the system since its mass is constant.

Again the force P is seen to be equal to the weight of the portion of the chain which is off the platform plus the added term which accounts for the time rate of increase of momentum of the chain.

Energy loss. Each link on the platform acquires its velocity abruptly through an impact with the link above it which lifts it off the platform. The succession of impacts gives rise to an energy loss ΔE. The work-energy equation may be written $U = \Delta T + \Delta V_g + \Delta E$.

$$U = \int P\, dx = \int_0^L (\rho gx + \rho v^2)dx = \tfrac{1}{2}\rho gL^2 + \rho v^2 L$$

$$\Delta T = \tfrac{1}{2}\rho Lv^2 \qquad \Delta V_g = \rho gL\frac{L}{2} = \tfrac{1}{2}\rho gL^2$$

Substituting into the work-energy equation gives

$$\tfrac{1}{2}\rho gL^2 + \rho v^2 L = \tfrac{1}{2}\rho Lv^2 + \tfrac{1}{2}\rho gL^2 + \Delta E, \quad \Delta E = \tfrac{1}{2}\rho Lv^2 \qquad \textit{Ans.}$$

Sample Problem 5/10

Replace the open-link chain of Sample Problem 5/9 by a flexible but inextensible rope or bicycle-type chain of length L and mass ρ per unit length. Determine the force P required to elevate the end of the rope with a constant velocity v and determine the corresponding reaction R between the coil and the platform.

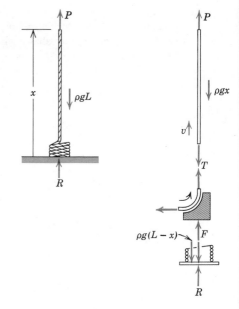

Solution. We shall neglect any horizontal motion of the rope as it uncoils. Since elements of the rope acquire their upward velocity in a smooth and continuous manner, there is no energy loss as was the case with the open-link chain. Therefore we apply the work-energy equation to the entire rope for an infinitesimal movement and get

$$[dU = dT + dV_g] \qquad P\,dx = d(\tfrac{1}{2}\rho x v^2) + d\left(\rho g x\,\frac{x}{2}\right)$$

$$= \tfrac{1}{2}\rho v^2\,dx + \rho g x\,dx$$

$$P = \tfrac{1}{2}\rho v^2 + \rho g x \qquad\qquad Ans.$$

Next we apply the force-momentum equation to the entire system and get

$$\left[\Sigma F = \frac{dG}{dt}\right] \qquad P + R - \rho g L = \frac{d}{dt}(\rho x v)$$

$$P + R = \rho g L + \rho v^2$$

Eliminating P between the two equations gives

$$R = \rho g L + \rho v^2 - (\tfrac{1}{2}\rho v^2 + \rho g x)$$

$$R = \rho g(L - x) + \tfrac{1}{2}\rho v^2 \qquad\qquad Ans.$$

To understand more fully the mechanism of the ideal rope and the difference between it and the open-link chain we have drawn free-body diagrams of the vertical portion, the transition section with a vane for supporting its change of momentum, and the coil and supporting plate. Equilibrium of the vertical section ($\dot{v} = 0$) requires

$$[\Sigma F = 0] \qquad T + \rho g x - P = 0, \quad T = P - \rho g x = \tfrac{1}{2}\rho v^2$$

The force-momentum equation for the transition section gives

$$[\Sigma F = m'\Delta v] \qquad F + T = m'(v - 0), \qquad F = -\tfrac{1}{2}\rho v^2 + \rho v(v - 0)$$

$$F = \tfrac{1}{2}\rho v^2$$

Equilibrium of the base plate gives

$$[\Sigma F = 0] \qquad R - F - \rho g(L - x) = 0, \quad R = \rho g(L - x) + \tfrac{1}{2}\rho v^2$$

which checks the previous calculation. Lacking a central diverting vane, the force F would come from the coil of rope with R remaining the same.

① Since P is a variable we must evaluate the work during an infinitesimal displacement rather than a finite displacement.

② The center of mass of the length x of rope is a distance $x/2$ above the base, so the potential energy relative to the base is the weight $\rho g x$ times $x/2$. Note that v is constant so is not differentiated.

③ Note that $\dot{x} = v$.

④ We see that the force required to hoist the rope is less than that required to hoist the open-link chain of Sample Problem 5/9 by the amount $\tfrac{1}{2}\rho v^2$. Also, the reaction under the rope exceeds that under the chain by the same amount $\tfrac{1}{2}\rho v^2$. A part of the force required to change the momentum of the rope comes from the tension T in the rope and a part from the platform. The total upward force $P + R$ in the two cases is the same.

Sample Problem   5/11

A rocket of initial total mass m_0 is fired vertically up from the north pole and accelerates until the fuel, which burns at a constant rate, is exhausted. The relative nozzle velocity of the exhaust gas has a constant value u, and the nozzle exhausts at atmospheric pressure throughout the flight. If the residual mass of the rocket structure and machinery is m_b when burnout occurs, determine the expression for the maximum velocity reached by the rocket. Neglect atmospheric resistance and the variation of gravity with altitude.

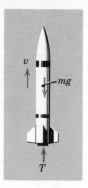

Solution I ($F = ma$ solution).   We adopt the approach illustrated with Fig. 5/5b and treat the thrust as an external force on the rocket. With the neglect of the back pressure p across the nozzle and the atmospheric resistance R, Eq. 5/17 or Newton's second law gives

$$T - mg = m\dot{v}$$

But the thrust is $T = m'u = -\dot{m}u$ so that the equation of motion becomes

$$-\dot{m}u - mg = m\dot{v}$$

Multiplication by dt, division by m, and rearrangement give

$$dv = -u\frac{dm}{m} - g\,dt$$

which is now in a form that can be integrated. The velocity v corresponding to the time t is given by the integration

$$\int_0^v dv = -u\int_{m_0}^m \frac{dm}{m} - g\int_0^t dt$$

or

$$v = u\ln\frac{m_0}{m} - gt$$

Since the fuel is burned at the constant rate $m' = -\dot{m}$, the mass at any time t is $m = m_0 + \dot{m}t$. If we let m_b stand for the mass of the rocket when burnout occurs, then the time at burnout becomes $t_b = (m_b - m_0)/\dot{m} = (m_0 - m_b)/(-\dot{m})$. This time gives the condition for maximum velocity which now becomes

$$v_{\max} = u\ln\frac{m_0}{m_b} + \frac{g}{\dot{m}}(m_0 - m_b) \qquad \textit{Ans.}$$

The quantity $\dot{m}$ is a negative number since the mass decreases with time.

Solution II (variable-mass solution).   If we use Eq. 5/16 then $\Sigma F = -mg$ and the equation becomes

$$[\Sigma F = m\dot{v} + \dot{m}u] \qquad -mg = m\dot{v} + \dot{m}u$$

But $\dot{m}u = -m'u = -T$ so that the equation of motion becomes

$$T - mg = m\dot{v}$$

which is the same as formulated with *Solution I.*

① The neglect of atmospheric resistance is not a bad assumption for a first approximation inasmuch as the velocity of the ascending rocket is smallest in the dense part of the atmosphere and greatest in the rarefied region. Also for an altitude of 320 km the acceleration due to gravity is 91 percent of the value at the surface of the earth.

② Vertical launch from the north pole is taken only to eliminate any complication due to the earth's rotation in figuring the absolute trajectory of the rocket.

PROBLEMS

53 The Saturn V rocket has a total launch weight of $6.32(10^6)$ lb, and each of its five F-1 engines develops a thrust of $1.5(10^6)$ lb. A total of $4.56(10^6)$ lb of fuel is burned by the first-stage F-1 engines in 160 sec at a constant rate. Calculate the initial vertical liftoff acceleration a of the rocket and the exhaust velocity u of the burned gases relative to the nozzle. Assume atmospheric pressure across the nozzle.

> Ans. $a = 6.01$ ft/sec^2, $u = 8470$ ft/sec

54 The tank of water is at rest on a horizontal surface when a 250-N force is applied to it as shown. If water issues from the rear discharge pipe at the rate of 20 kg/s with a velocity of 2.4 m/s relative to the opening in the direction shown, calculate the initial acceleration a of the tank if its total mass at the start is 300 kg. Neglect the rotational inertia of the wheels.

Problem 5/54

55 The solid-propellant rocket has a mass of 500 kg at launch. The rocket motor burns fuel at the rate of 4 kg/s with an exhaust velocity of 2500 m/s relative to the nozzle. Determine the acceleration a of the mass center of the rocket and the angle θ made by the acceleration with the horizontal as the rocket clears the launcher.

> Ans. $a = 10.83$ m/s^2, $\theta = 61.4°$

Problem 5/55

56 The mass m of a raindrop increases as it picks up moisture during its vertical descent through still air. If the air resistance to motion of the drop is R and its downward velocity is v, write the equation of motion for the drop and show that the relation $\Sigma F = d(mv)/dt$ is obeyed as a special case of the variable-mass equation.

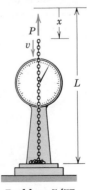

57 The upper end of the open-link chain of length L and mass ρ per unit length is lowered at a constant speed v by the force P. Determine the reading R of the platform scale in terms of x.

> Ans. $R = \rho g x + \rho v^2$

Problem 5/57

5/58 A small rocket of initial mass m_0 is fired vertically upward near the surface of the earth (g constant). If air resistance is neglected, determine the manner in which the mass m of the rocket must vary as a function of the time t after launching in order that the rocket may have a constant vertical acceleration a with a constant relative velocity u of the escaping gases with respect to the nozzle.

5/59 If Eq. 5/16 is applied to the moving part of the ideal rope of Sample Problem 5/10 using $\Sigma F = P - \rho g x$, $m = \rho x$, $\dot{m} = \rho v$, $\dot{v} = 0$, and $u = v$, explain why the result for P is invalid.

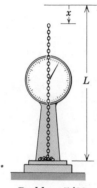

Problem 5/60

5/60 The upper end of the open-link chain of length L and mass ρ per unit length is released from rest with the lower end just touching the platform of the scale. Determine the expression for the force F read on the scale as a function of the distance x through which the upper end has fallen. (*Comment:* The chain acquires a free-fall velocity of $\sqrt{2gx}$ since the links on the scale exert no force on those above, which are still falling freely. Work the problem in two ways, first, by evaluating the time rate of change of momentum for the entire chain and, second, by considering the force F to be composed of the weight of the links at rest on the scale plus the force necessary to divert an equivalent stream of fluid.)   *Ans.* $F = 3\rho g x$

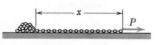

Problem 5/61

5/61 The end of a pile of loose-link chain of mass ρ per unit length is being pulled horizontally along the surface by a constant force P. If the coefficient of friction between the chain and the surface is μ, determine the acceleration a of the chain in terms of x and $\dot{x}$.

$$Ans.\ a = \frac{P}{\rho x} - \mu g - \frac{\dot{x}^2}{x}$$

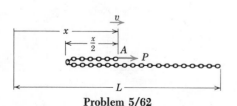

Problem 5/62

5/62 An open-link chain of length $L = 20$ ft weighing 100 lb is resting on a smooth horizontal surface when end A is doubled back on itself by a force P applied to end A. (*a*) Calculate the required value of P to give A a constant velocity of 4 ft/sec. (*b*) Calculate the acceleration of A if $P = 4$ lb and if $v = 4$ ft/sec when $x = 10$ ft.
                     Ans. (*a*) $P = 1.24$ lb, (*b*) $a = 3.55$ ft/sec^2

/63 An ideal rope or bicycle-type chain of length L and mass ρ per unit length is resting on a smooth horizontal surface when end A is doubled back on itself by a force P applied to end A. End B of the rope is secured to a fixed support. Determine the force P required to give A a constant velocity v. (*Hint:* The action of the loop can be modeled by inserting a circular disk of negligible mass as shown in the separate sketch and then taking the disk radius as zero. It is easily shown that the tensions in the rope at C, D, and B are all equal to P under the ideal conditions imposed and with constant velocity.)

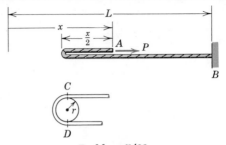

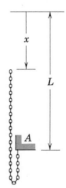

Problem 5/63

/64 The left end of the open-link chain of length L and mass ρ per unit length is released from rest at $x = 0$. Determine the expression for the tension T in the chain at its support at A in terms of x. Also determine the energy loss ΔE during the entire motion from $x = 0$ to $x = 2L$

$$\text{Ans. } T = \frac{3\rho g x}{2}, \ \Delta E = \rho g L^2$$

/65 The open-link chain of total length L and of mass ρ per unit length is released from rest at $x = 0$ at the same instant that the platform starts from rest at $y = 0$ and moves vertically up with a constant acceleration a. Determine the expression for the total force R exerted on the platform by the chain t seconds after the motion starts.

$$\text{Ans. } R = \frac{3}{2}\rho(a + g)^2 t^2$$

Problem 5/64

/66 In the figure is shown a system used to arrest the motion of an airplane landing on a field of restricted length. The plane of mass m rolling freely with a velocity v_0 engages a hook which pulls the ends of two heavy chains each of length L and mass ρ per unit length in the manner shown. A conservative calculation of the effectiveness of the device neglects the retardation of chain friction on the ground and any other resistance to the motion of the airplane. With these assumptions compute the velocity v of the airplane at the instant that the last link of each chain is put in motion. Also determine the relation between displacement x and the time t after contact with the chain. Assume each link of the chain acquires its velocity v suddenly upon contact with the moving links.

$$\text{Ans. } v = \frac{v_0}{1 + 2\rho L/m}, \quad x = \frac{m}{\rho}\left[\sqrt{1 + \frac{2v_0 t \rho}{m}} - 1\right]$$

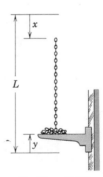

Problem 5/65

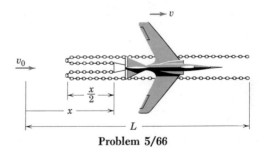

Problem 5/66

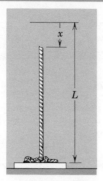

Problem 5/67

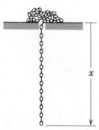

Problem 5/68

Problem 5/69

▶5/67 A rope or hinged-link bicycle-type chain of length L and mass ρ per unit length is released from rest with $x = 0$. Determine the expression for the total force R exerted on the fixed platform by the chain as a function of x. Note that the hinged-link chain is a conservative system during all but the last increment of motion. Compare the result with that of Prob. 5/65 if the upward motion of the platform in that problem is taken to be zero.

$$\text{Ans. } R = \rho g x \frac{4L - 3x}{2(L - x)}$$

▶5/68 One end of the pile of chain falls through a hole in its support and pulls the remaining links after it in a steady flow. If the links which are initially at rest acquire the velocity of the chain suddenly and without frictional resistance or interference from the support or from adjacent links, find the velocity v of the chain as a function of x if $v = 0$ when $x = 0$. Also find the acceleration a of the falling chain and the energy ΔE lost from the system as the last link leaves the platform. (*Hint.* Apply Eq. 5/16 and treat the product xv as the variable when solving the differential equation. Also note at the appropriate step that $dx = v\,dt$.) The total length of the chain is L, and its mass per unit length is ρ.

$$\text{Ans. } v = \sqrt{\frac{2gx}{3}}, \ a = \frac{g}{3}, \ \Delta E = \frac{\rho g L^2}{6}$$

▶5/69 Replace the pile of chain in Prob. 5/68 by a coil of rope of mass ρ per unit length and total length L as shown, and determine the velocity of the falling section in terms of x if it starts from rest at $x = 0$. Show that the acceleration is constant at $g/2$. The rope is considered to be perfectly flexible in bending but inextensible and constitutes a conservative system (no energy loss). Rope elements acquire their velocity in a continuous manner from zero to v in a small transition section of the rope at the top of the coil. For comparison with the chain of Prob. 5/68 this transition section may be considered to have negligible length without violating the requirement that there be no energy loss in the present problem. Also determine the force R exerted by the platform on the coil in terms of x and explain why R becomes zero when $x = 2L/3$. Neglect the dimensions of the coil compared with x.

$$\text{Ans. } v = \sqrt{gx}, \ R = \rho g\left(L - \frac{3x}{2}\right)$$

II
DYNAMICS OF
RIGID BODIES

PLANE KINEMATICS OF RIGID BODIES

6

6/1 INTRODUCTION. The description of the motion of rigid bodies is useful in two important ways. First, it is frequently necessary to generate, transmit, or control certain desired motions by the use of cams, gears, and linkages of various types. Here an analysis of the displacement, velocity, and acceleration of the motion is necessary to determine the design geometry of the mechanical parts. Furthermore, as a result of the motion generated, forces are frequently developed which must be accounted for in the design of the parts. Second, it is often necessary to determine the motion of a rigid body resulting from the forces applied to it. Calculation of the motion of a rocket under the influence of its jet thrust and gravitational attraction is an example of such a problem. In both types of problems we need to have command of the principles of rigid-body kinematics. In this chapter we will cover the kinematics of motion which may be analyzed as taking place in a single plane. In Chapter 8 we will present a brief introduction to the kinematics of motion which requires three spatial coordinates for its description.

A rigid body was defined in the previous chapter as a system of particles for which the distances between the particles remain unchanged. Thus if each particle of such a body is located by a position vector from reference axes attached to and rotating with the body, there would be no change in any position vector as measured from these axes. This formulation is, of course, an ideal one since all solid materials change shape to some exent when forces are applied to them. Nevertheless, if the movements associated with the changes in shape are very small compared with the overall movements of the body as a whole, then the ideal concept of rigidity is quite acceptable. As mentioned in Chapter 1 the displacements due to the flutter of an aircraft wing, for instance, are of no consequence in the description of the flight path of the aircraft as a whole, for which the rigid-body assumption is clearly in order. On the other hand, if the problem is one of describing, as a function of time, the internal wing stress due to wing flutter, then the relative motions of portions of the wing become of prime importance and cannot be neglected. In this case the wing may not be considered a rigid body. In the present chapter and in the two that follow essentially all of the material is based upon the assumption of rigidity.

The plane motion of a rigid body may be divided into several categories as represented in Fig. 6/1.

Translation is defined as any motion in which every line in the body remains parallel to its original position at all times. In translation there is *no rotation of any line in the body*. In *rectilinear translation*, part (*a*) of Fig. 6/1, all points in the body move in parallel straight lines. In *curvilinear translation*, part (*b*), all points move on congruent curves. We note that in each of the two cases of translation the motion of the body is completely specified by the motion of any point in the body, since all points have the same motion. Thus our earlier study of the motion of a point (particle) in Chapter 2 enables us to describe completely the translation of a rigid body.

Rotation about a fixed axis, part (*c*) of Fig. 6/1, is the angular motion about the axis. It follows that all particles move in circular paths about the axis of rotation, and all lines in the body (including those that do not pass through the axis) rotate through the same angle in the same time. Again, our discussion in Chapter 2 on the circular motion of a point enables us to describe the motion of a rotating rigid body, which is treated in the next article.

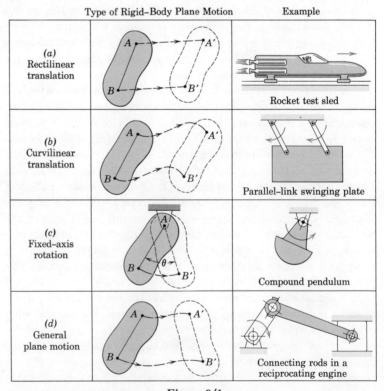

Type of Rigid–Body Plane Motion	Example
(*a*) Rectilinear translation	Rocket test sled
(*b*) Curvilinear translation	Parallel–link swinging plate
(*c*) Fixed–axis rotation	Compound pendulum
(*d*) General plane motion	Connecting rods in a reciprocating engine

Figure 6/1

General plane motion of a rigid body, part (*d*) of Fig. 6/1, is a combination of translation and rotation. We will utilize the principles of relative motion covered in Art. 2/8 in the description of general plane motion.

In each of the examples cited all particles in the body move in parallel planes. The motion, however, is represented by its projection onto a single plane parallel to the motion called the *plane of motion*. This plane is usually chosen as passing through the center of mass of the body.

Determination of the plane motion of rigid bodies is accomplished by either a direct calculation of the absolute displacements and their time derivatives from the absolute geometry involved or else by utilizing the principles of relative motion. Each method is important and useful and will be taken up in turn in the articles which follow.

6/2 ROTATION. The rotation of a rigid body is described by its angular motion. Figure 6/2 shows a rigid body which is rotating as it undergoes plane motion in the plane of the figure. The angular positions of any two lines 1 and 2 attached to the body are specified by θ_1 and θ_2 measured from any fixed reference direction which is convenient. Since the angle β is invariant, the relation $\theta_2 = \theta_1 + \beta$ upon differentiation with respect to time gives $\dot{\theta}_2 = \dot{\theta}_1$ and $\ddot{\theta}_2 = \ddot{\theta}_1$ or, during a finite interval, $\Delta\theta_2 = \Delta\theta_1$. Thus *all lines in a rigid body in its plane of motion have the same angular displacement, the same angular velocity, and the same angular acceleration.* It should be noted that the angular motion of a line depends only on its angular displacement with respect to any arbitrary fixed reference and on the time derivatives of the displacement. Angular motion does not require the presence of a fixed axis, normal to the plane of motion, about which the line and the body rotate.

(*a*) *Angular Motion Relations.* The angular velocity ω and angular acceleration α of a rigid body in plane rotation are, respectively, the first and second time derivatives of the angular position coordi-

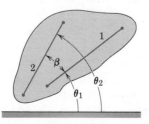

Figure 6/2

nate θ of any line in the body in the plane of motion. These definitions give

$$\omega = \frac{d\theta}{dt} = \dot{\theta}$$

$$\alpha = \frac{d\omega}{dt} = \dot{\omega} \qquad \text{or} \qquad \alpha = \frac{d^2\theta}{dt^2} = \ddot{\theta} \qquad (6/1)$$

$$\omega \, d\omega = \alpha \, d\theta \qquad \text{or} \qquad \dot{\theta} \, d\dot{\theta} = \ddot{\theta} \, d\theta$$

The third relation is obtained by eliminating dt from the first two. In each of these relations the positive direction for ω and α, clockwise or counterclockwise, is the same as that chosen for θ. Equations 6/1 should be recognized as analogous to the defining equations for the rectilinear motion of a particle expressed by Eqs. 2/1, 2/2, and 2/3. In fact all relations which were described for rectilinear motion in Art. 2/2 apply to the case of rotation in a plane if the linear quantities s, v, and a are replaced by their respective equivalent angular quantities θ, ω, and α.

For rotation with constant angular acceleration the integrals of Eqs. 6/1 become

$$\omega = \omega_0 + \alpha t$$

$$\omega^2 = \omega_0{}^2 + 2\alpha(\theta - \theta_0)$$

$$\theta = \theta_0 + \omega_0 t + \tfrac{1}{2}\alpha t^2$$

Here θ_0 and ω_0 are the values of the angular position coordinate and angular velocity, respectively, at $t = 0$ and t is the time of the interval of motion considered. The student should be able to carry out these integrations easily, as they are completely analogous to the corresponding equations for rectilinear motion with constant acceleration covered in Art. 2/2.

The graphical relationships described for s, v, a, and t in Figs. 2/3 and 2/4 may be used for θ, ω, and α merely by the substitution of corresponding symbols. The student should sketch these graphical relations for plane rotation. The mathematical procedures for obtaining velocity and displacement from acceleration described for rectilinear motion may be applied to rotation by merely replacing the linear quantities by their corresponding angular quantities.

(b) Rotation about a Fixed Axis. When a body rotates about a fixed axis, all points other than those on the axis move in concentric circles about the fixed axis. Thus for the rigid body in Fig. 6/3 rotating about a fixed axis normal to the plane of the figure through O, any point such as A moves in a circle of radius r. From our previous discussion in Art. 2/5 we are already familiar with the relationships between the linear motion of A and the angular motion of the line normal to its path, which is also the angular motion of the rigid body.

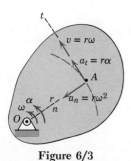

Figure 6/3

Using the notation $\omega = \dot{\theta}$ and $\alpha = \dot{\omega} = \ddot{\theta}$ for the angular velocity and angular acceleration, respectively, of the body we have Eqs. 2/11 rewritten as

$$v = r\omega$$
$$a_n = r\omega^2 = v^2/r = v\omega \qquad (6/2)$$
$$a_t = r\alpha$$

These quantities may be expressed alternatively using the cross-product relationship of vector notation. The vector formulation is especially important in three-dimensional analysis of motion. The angular velocity of the rotating body may be expressed by the vector $\boldsymbol{\omega}$ normal to the plane of rotation and having a sense governed by the right-hand rule, as shown in Fig. 6/4a. From the definition of the vector cross product we see that the vector $\mathbf{v}$ is obtained by crossing $\boldsymbol{\omega}$ into $\mathbf{r}$. This cross product gives the correct magnitude and direction for $\mathbf{v}$ and we write

$$\dot{\mathbf{r}} = \mathbf{v} = \boldsymbol{\omega} \times \mathbf{r}$$

The order of the vectors to be crossed must be retained. The reverse order gives $\mathbf{r} \times \boldsymbol{\omega} = -\mathbf{v}$.

The acceleration of A is obtained by differentiating the cross-product[*] expression for $\mathbf{v}$ which gives

$$\mathbf{a} = \dot{\mathbf{v}} = \boldsymbol{\omega} \times \dot{\mathbf{r}} + \dot{\boldsymbol{\omega}} \times \mathbf{r}$$
$$= \boldsymbol{\omega} \times (\boldsymbol{\omega} \times \mathbf{r}) + \dot{\boldsymbol{\omega}} \times \mathbf{r}$$
$$= \boldsymbol{\omega} \times \mathbf{v} + \boldsymbol{\alpha} \times \mathbf{r}$$

Here $\boldsymbol{\alpha} = \dot{\boldsymbol{\omega}}$ stands for the angular acceleration of the body. Thus the vector equivalents to Eqs. 6/2 are

$$\mathbf{v} = \boldsymbol{\omega} \times \mathbf{r}$$
$$\mathbf{a}_n = \boldsymbol{\omega} \times (\boldsymbol{\omega} \times \mathbf{r}) \qquad (6/3)$$
$$\mathbf{a}_t = \boldsymbol{\alpha} \times \mathbf{r}$$

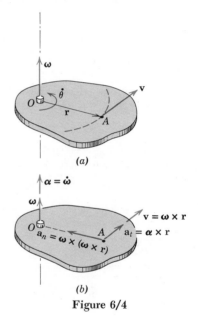

(a)

(b)

Figure 6/4

and are shown in Fig. 6/4b.

For three-dimensional motion of a rigid body the angular velocity vector $\boldsymbol{\omega}$ may change direction as well as magnitude, and in this case the angular acceleration, which is the time derivative of angular velocity $\boldsymbol{\alpha} = \dot{\boldsymbol{\omega}}$, will no longer be in the same direction as $\boldsymbol{\omega}$. An introduction to three-dimensional motion of rigid bodies is presented in Chapter 8.

[*] The rule for differentiating the cross product is covered in item 9 of Art. B7, Appendix B.

Sample Problem 6/1

A flywheel rotating freely at 1800 rev/min clockwise is subjected to a variable counterclockwise torque which is first applied at time $t = 0$. The torque produces a counterclockwise angular acceleration $\alpha = 4t$ rad/s² where t is the time in seconds during which the torque is applied. Determine (a) the time required for the flywheel to reduce its clockwise angular speed to 900 rev/min, (b) the time required for the flywheel to reverse its direction of rotation, and (c) the total number of revolutions, clockwise plus counterclockwise, turned by the flywheel during the first 14 seconds of torque application.

Solution. The counterclockwise direction will be taken arbitrarily as positive.

(a) Since α is a known function of the time, we may integrate it to obtain angular velocity. With the initial angular velocity of $-1800(2\pi)/60 = -60\pi$ rad/s we have

$$[d\omega = \alpha \, dt] \qquad \int_{-60\pi}^{\omega} d\omega = \int_{0}^{t} 4t \, dt, \qquad \omega = -60\pi + 2t^2$$

Substituting the clockwise angular speed of 900 rev/min or $\omega = -900(2\pi)/60 = -30\pi$ rad/s gives

$$-30\pi = -60\pi + 2t^2, \qquad t^2 = 15\pi, \qquad t = 6.86 \text{ s} \qquad Ans.$$

(b) The flywheel changes direction when its angular velocity is momentarily zero. Thus

$$0 = -60\pi + 2t^2, \qquad t^2 = 30\pi, \qquad t = 9.71 \text{ s} \qquad Ans.$$

(c) The total number of revolutions through which the flywheel turns during 14 seconds is the number of clockwise turns N_1 during the first 9.71 seconds plus the number of counterclockwise turns N_2 during the remainder of the interval. Integrating the expression for ω in terms of t gives us the angular displacement in radians. Thus for the first interval

$$[d\theta = \omega \, dt] \qquad \int_{0}^{\theta_1} d\theta = \int_{0}^{9.71} (-60\pi + 2t^2) \, dt$$

$$\theta_1 = [-60\pi t + \tfrac{2}{3}t^3]_{0}^{9.71} = -1220 \text{ rad}$$

or $N_1 = 1220/2\pi = 194.2$ revolutions clockwise.
For the second interval

$$\int_{0}^{\theta_2} d\theta = \int_{9.71}^{14} (-60\pi + 2t^2) \, dt$$

$$\theta_2 = [-60\pi t + \tfrac{2}{3}t^3]_{9.71}^{14} = 410 \text{ rad}$$

or $N_2 = 410/2\pi = 65.3$ revolutions counterclockwise. Thus the total number of revolutions turned during the 14 seconds is

$$N = N_1 + N_2 = 194.2 + 65.3 = 259.5 \text{ rev} \qquad Ans.$$

We have plotted ω versus t and we see that θ_1 is represented by the negative area and θ_2 by the positive area. If we had integrated over the entire interval in one step, we would have obtained $|\theta_2| - |\theta_1|$.

① We must be very careful to be consistent with our algebraic signs. The lower limit is the negative (clockwise) value of the initial angular velocity. Also we must convert revolutions to radians since α is in radian units.

② Again note that the minus sign signifies clockwise in this problem.

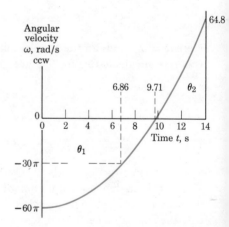

③ We could have converted the original expression for α into the units of rev/s², in which case our integrals would have come out directly in revolutions.

Sample Problem 6/2

The pinion A of the hoist motor drives gear B which is attached to the hoisting drum. The load L is lifted from its rest position and acquires an upward velocity of 3 ft/sec in a vertical rise of 4 ft with constant acceleration. As the load passes this position compute (a) the acceleration of point C on the cable in contact with the drum and (b) the angular velocity and angular acceleration of the pinion A.

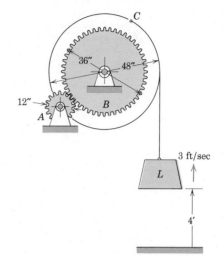

Solution. (a) If the cable does not slip on the drum, the vertical velocity and acceleration of the load L are, of necessity, the same as the tangential velocity v and tangential acceleration a_t of point C. For the rectilinear motion of L with constant acceleration we have

$$[v^2 = 2as] \qquad a = a_t = \frac{v^2}{2s} = \frac{3^2}{2(4)} = 1.125 \text{ ft/sec}^2$$

The normal component of acceleration of C is

$$[a_n = v^2/r] \qquad a_n = \frac{3^2}{24/12} = 4.5 \text{ ft/sec}^2$$

and the total acceleration of C has the magnitude

$$[a = \sqrt{a_n{}^2 + a_t{}^2}] \qquad a_C = \sqrt{(4.5)^2 + (1.125)^2} = 4.64 \text{ ft/sec}^2 \qquad \textit{Ans.}$$

with the direction indicated on the sketch.

    (b) The angular motion of gear A is determined from the angular motion of gear B by the velocity v_1 and tangential acceleration a_1 of their common point of contact. First the angular motion of gear B is determined from the motion of point C on the attached drum. Thus

$$[v = r\omega] \qquad \omega_B = v/r = \frac{3}{24/12} = 1.5 \text{ rad/sec}$$

$$[a_t = r\alpha] \qquad \alpha_B = a_t/r = \frac{1.125}{24/12} = 0.563 \text{ rad/sec}^2$$

Then from $v_1 = r_A\omega_A = r_B\omega_B$ and $a_1 = r_A\alpha_A = r_B\alpha_B$ we have

$$\omega_A = \frac{r_B}{r_A}\omega_B = \frac{18/12}{6/12}1.5 = 4.5 \text{ rad/sec CW} \qquad \textit{Ans.}$$

$$\alpha_A = \frac{r_B}{r_A}\alpha_B = \frac{18/12}{6/12}0.563 = 1.69 \text{ rad/sec}^2 \text{ CW} \qquad \textit{Ans.}$$

① Be sure to recognize that a point on the cable changes the direction of its velocity after it contacts the drum and therefore acquires a normal component of acceleration.

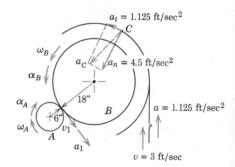

PROBLEMS

6/1 A braking torque causes a flywheel to slow down at the constant rate of 20 rev/min during each second. If the flywheel comes to rest from a speed of 1800 rev/min, find the number of revolutions N turned during this interval.   *Ans.* $N = 1350$ rev

6/2 Magnetic tape is being fed over and around the light pulleys mounted in a computer. If the speed v of the tape is constant and if the magnitude of the acceleration of point A on the tape is $\frac{4}{3}$ times that of point B, calculate the radius r of the smaller pulley.

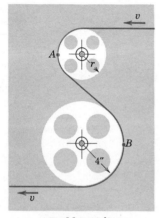

Problem 6/2

6/3 The clockwise angular coordinate of a radial line on a wheel is given by $\theta = 2t^3 + 3t^2 + 2$ where θ is in radians and t is in seconds. Find the angular displacement $\Delta\theta$ during the time $(t > 0)$ in which the angular velocity changes from 12 to 72 rad/s, both clockwise.   *Ans.* $\Delta\theta = 76$ rad

6/4 Experimental data for a rotating control element reveal the plotted relation between angular velocity and the angular coordinate θ as shown. Approximate the angular acceleration α of the element when $\theta = 4$ rad.

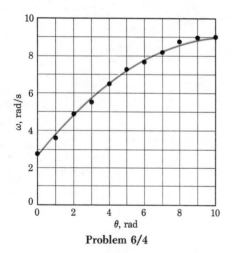

Problem 6/4

6/5 A wheel is rotating at a uniform clockwise speed of 2 rad/s when a variable torque is applied to its shaft at time $t = 0$. As a result there is a clockwise angular acceleration α which increases in direct proportion to the angle θ through which the wheel turns from the instant $t = 0$. After 16 complete turns from this time the angular acceleration is 4 rad/s². Determine the angular velocity ω of the wheel when this condition is reached.

Ans. $\omega = 20.2$ rad/s

6/6 The angular velocity of a gear is controlled according to $\omega = 12 - 3t^2$ where ω, in radians per second, is positive in the clockwise sense and where t is the time in seconds. Find the net angular displacement $\Delta\theta$ from the time $t = 0$ to $t = 3$ s. Also find the total number of revolutions N through which the gear turns during the 3 seconds.

7 The circular disk rotates about its center O. At a certain instant point A has a velocity $v_A = 0.8$ m/s in the direction shown, and at the same instant the tangent of the angle θ made by the total acceleration vector of any point B with its radial line to O is 0.6. For this instant compute the angular acceleration α of the disk.   *Ans.* $\alpha = 38.4$ rad/s²

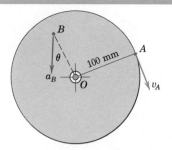

Problem 6/7

8 The circular disk rotates about its vertical z-axis. At a certain instant the velocity of point B on the disk is $v_B = 8i$ in./sec, and the tangential acceleration of A is $\mathbf{a}_{A_t} = 18\mathbf{j}$ in./sec². Determine the vector expression for the angular velocity $\boldsymbol{\omega}$ of the disk and the total acceleration $\mathbf{a}$ of B for this instant.

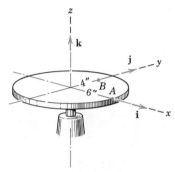

Problem 6/8

9 Refer to the pulley and tape unit shown in Prob. 6/2. At a certain instant the speed of the tape is 120 in./sec, and the magnitude of the total acceleration of point B is 4500 in./sec². Determine the magnitude of the total acceleration of point A for this same instant if $r = 3$ in.
         Ans. $a_A = 5510$ in./sec²

10 A radial line on a gear, which rotates with constant angular acceleration, has an angular coordinate $\theta = 4$ rad when $t = 0$. After 10 seconds its angular coordinate is zero. Also, when $t = 4$ s, the gear reverses the direction of its motion. Calculate the angular velocity of the gear when $t = 10$ s.

11 The frictional resistance to the rotation of a flywheel consists of a retardation due to air friction which varies as the square of the angular velocity and a constant frictional retardation in the bearing. As a result the angular acceleration of the flywheel while it is allowed to coast is given by $\alpha = -K - k\omega^2$ where K and k are constants. Determine an expression for the time required for the flywheel to come to rest from an initial angular velocity ω_0.

$$\text{Ans. } t = \frac{1}{\sqrt{Kk}} \tan^{-1}\left(\omega_0 \sqrt{\frac{k}{K}}\right)$$

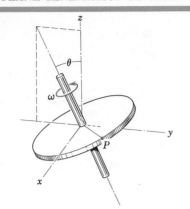

Problem 6/13

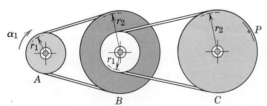

Problem 6/14

6/12 The motion of a rotating body is controlled so that the rate of change of its angular velocity ω with respect to its angular coordinate θ is a constant k. If the angular velocity of the body is ω_0 at the start of the interval where $\theta = 0$ and $t = 0$, derive the expressions for θ, ω, and the angular acceleration α as functions of the time t.

$$\text{Ans. } \theta = \frac{\omega_0}{k}(e^{kt} - 1), \ \omega = \omega_0 e^{kt}, \ \alpha = \omega_0 k e^{kt}$$

6/13 The circular disk rotates with a constant angular velocity $\omega = 50$ rad/s about its axis, which is inclined in the y-z plane at the angle $\theta = \tan^{-1} \frac{3}{4}$. Determine the vector expressions for the velocity and acceleration of point P, whose position vector at the instant shown is $\mathbf{r} = 60\mathbf{i} + 64\mathbf{j} + 48\mathbf{k}$ mm. (Check the magnitudes of your results from the scalar values $v = r\omega$ and $a_n = r\omega^2$.)

6/14 A V-belt speed-reduction drive is shown where pulley A drives the two integral pulleys B which in turn drive pulley C. If A starts from rest at time $t = 0$ and is given a constant angular acceleration α_1, derive expressions for the angular velocity of C and the magnitude of the acceleration of a point P on the belt both at time t.

$$\text{Ans. } \omega_C = \left(\frac{r_1}{r_2}\right)^2 \alpha_1 t, \ a_P = \frac{r_1^2}{r_2} \alpha_1 \sqrt{1 + \left(\frac{r_1}{r_2}\right)^4 \alpha_1^2 t^4}$$

6/3 ABSOLUTE MOTION. The determination of velocities and accelerations, both linear and angular, in rigid-body plane motion by direct differentiation of the equations for displacements is a straightforward and direct approach. The selection of this method in preference to solution by relative-motion methods will depend mainly on the relative simplicity of the geometry involved. Choice is best indicated after experience has been acquired with both approaches.

In the absolute-motion analysis of plane motion use is made of the differential relations, Eqs. 2/1, 2/2, and 2/3, developed in Art. 2/2 for the rectilinear motion of particles, and Eqs. 6/1 and 6/2 or 6/3, developed in the previous article for rotation. The procedure for carrying out an absolute-motion analysis consists, first, of writing an equation which satisfies the geometry of the problem and relates the appropriate linear and the angular coordinates. This geometric equation is then differentiated once with respect to time to obtain velocities and a second time to obtain accelerations. The details of this procedure are best illustrated through actual example. Sample Problem 6/3 contains an analysis of the motion of a rolling wheel, and the results of this basic problem should be mastered thoroughly as they find repeated use in much of the work which follows.

Sample Problem   6/3

A wheel of radius r rolls on a flat surface without slipping. Determine the angular motion of the wheel in terms of the linear motion of its center O. Also determine the acceleration of a point on the rim of the wheel as the point comes into contact with the surface upon which the wheel rolls.

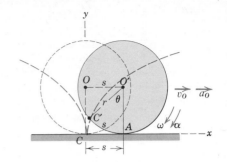

Solution.   The figure shows the wheel rolling to the right from the dotted to the full position without slipping. The linear displacement of the center O is s, which is also the arc length $C'A$ along the rim upon which the wheel rolls. The radial line CO rotates to the new position $C'O'$ through the angle θ, where θ is measured from the vertical direction. If the wheel does not slip, the arc $C'A$ must equal the distance s. Thus the displacement relationship and its two time derivatives give

$$s = r\theta$$

$$v_O = r\omega \qquad \qquad \textit{Ans.}$$

$$a_O = r\alpha$$

where $v_O = \dot{s}$, $a_O = \dot{v}_O = \ddot{s}$, $\omega = \dot{\theta}$, and $\alpha = \dot{\omega} = \ddot{\theta}$. The angle θ, of course, must be in radians. The acceleration a_O will be directed in the sense opposite to that of v_O if the wheel is slowing down. In this event the angular acceleration α will have the opposite sense to ω.

The origin of fixed coordinates is taken arbitrarily but conveniently at the point of contact between C on the rim of the wheel and the ground. When point C has moved along its cycloidal path to C', its new coordinates and their time derivatives become

$$x = s - r\sin\theta = r(\theta - \sin\theta) \qquad\qquad y = r - r\cos\theta = r(1 - \cos\theta)$$

$$\dot{x} = r\dot{\theta}(1 - \cos\theta) = v_O(1 - \cos\theta) \qquad\qquad \dot{y} = r\dot{\theta}\sin\theta = v_O\sin\theta$$

$$\ddot{x} = \dot{v}_O(1 - \cos\theta) + v_O\dot{\theta}\sin\theta \qquad\qquad \ddot{y} = \dot{v}_O\sin\theta + v_O\dot{\theta}\cos\theta$$

$$\quad = a_O(1 - \cos\theta) + r\omega^2\sin\theta \qquad\qquad\quad = a_O\sin\theta + r\omega^2\cos\theta$$

For the desired instant of contact, $\theta = 0$ and

$$\ddot{x} = 0 \quad \text{and} \quad \ddot{y} = r\omega^2 \qquad \textit{Ans.}$$

Thus the acceleration of the point C on the rim at the instant of contact with the ground depends only on r and ω and is directed toward the center of the wheel. If desired, the velocity and acceleration of C at any position θ may be obtained by writing the expressions $\mathbf{v} = \mathbf{i}\dot{x} + \mathbf{j}\dot{y}$ and $\mathbf{a} = \mathbf{i}\ddot{x} + \mathbf{j}\ddot{y}$.

① These three relations are not entirely unfamiliar at this point, and their application to the rolling wheel should be mastered thoroughly.

② Clearly when $\theta = 0$, the point of contact has zero velocity so that $\dot{x} = \dot{y} = 0$. The acceleration of the contact point on the wheel will also be obtained by the principles of relative motion in Art. 6/6.

Sample Problem 6/4

The load L is being hoisted by the pulley and cable arrangement shown. Each cable is wrapped securely around its respective pulley so it does not slip. The two pulleys to which L is attached are fastened together to form a single rigid body. Calculate the velocity and acceleration of the load L and the corresponding angular velocity ω and angular acceleration α of the double pulley under the following conditions:

Case (a) Pulley 1: $\omega_1 = \dot{\omega}_1 = 0$ (pulley at rest)

Pulley 2: $\omega_2 = 2$ rad/sec, $\alpha_2 = \dot{\omega}_2 = -3$ rad/sec^2

Case (b) Pulley 1: $\omega_1 = 1$ rad/sec, $\alpha_1 = \dot{\omega}_1 = 4$ rad/sec^2

$\omega_2 = 2$ rad/sec, $\alpha_2 = \dot{\omega}_2 = -2$ rad/sec^2

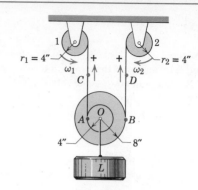

Solution. The tangential displacement, velocity, and acceleration of a point on the rim of pulley 1 or 2 equal the corresponding vertical motions of point A or B since the cables are assumed to be inextensible.

Case (a). With A momentarily at rest, line AB rotates to AB' through the angle $d\theta$ during time dt. From the diagram we see that the displacements and their time derivatives give

$$ds_B = \overline{AB}\,d\theta \qquad v_B = \overline{AB}\,\omega \qquad (a_B)_t = \overline{AB}\,\alpha$$
$$ds_O = \overline{AO}\,d\theta \qquad v_O = \overline{AO}\,\omega \qquad a_O = \overline{AO}\,\alpha$$

With $v_D = r_2\omega_2 = 4(2) = 8$ in./sec, $a_D = r_2\alpha_2 = 4(-3) = -12$ in./sec^2 we have for the angular motion of the double pulley

$$\omega = v_B/\overline{AB} = v_D/\overline{AB} = 8/12 = 2/3 \text{ rad/sec (CCW)} \qquad Ans.$$
$$\alpha = (a_B)_t/\overline{AB} = a_D/\overline{AB} = -12/12 = -1 \text{ rad/sec}^2 \text{ (CW)} \qquad Ans.$$

The corresponding motion of O and the load L is

$$v_O = \overline{AO}\,\omega = 4(2/3) = 8/3 \text{ in./sec} \qquad Ans.$$
$$a_O = \overline{AO}\,\alpha = 4(-1) = -4 \text{ in./sec}^2 \qquad Ans.$$

Case (b). With point C, and hence point A, in motion line AB moves to $A'B'$ during time dt. From the diagram for this case we see that the displacements and their time derivatives give

$$ds_B - ds_A = \overline{AB}\,d\theta \qquad v_B - v_A = \overline{AB}\,\omega \qquad (a_B)_t - (a_A)_t = \overline{AB}\,\alpha$$
$$ds_O - ds_A = \overline{AO}\,d\theta \qquad v_O - v_A = \overline{AO}\,\omega \qquad a_O - (a_A)_t = \overline{AO}\,\alpha$$

With $v_C = r_1\omega_1 = 4(1) = 4$ in./sec, $v_D = r_2\omega_2 = 4(2) = 8$ in./sec

$$a_C = r_1\alpha_1 = 4(4) = 16 \text{ in./sec}^2, \; a_D = r_2\alpha_2 = 4(-2) = -8 \text{ in./sec}^2$$

we have for the angular motion of the double pulley

$$\omega = \frac{v_B - v_A}{\overline{AB}} = \frac{v_D - v_C}{\overline{AB}} = \frac{8 - 4}{12} = 1/3 \text{ rad/sec (CCW)} \qquad Ans.$$

$$\alpha = \frac{(a_B)_t - (a_A)_t}{\overline{AB}} = \frac{a_D - a_C}{\overline{AB}} = \frac{-8 - 16}{12} = -2 \text{ rad/sec}^2 \text{ (CW)} \quad Ans.$$

The corresponding motion of O and the load L is

$$v_O = v_A + \overline{AO}\,\omega = v_C + \overline{AO}\,\omega = 4 + 4(1/3) = 16/3 \text{ in./sec} \qquad Ans.$$
$$a_O = (a_A)_t + \overline{AO}\,\alpha = a_C + \overline{AO}\,\alpha = 16 + 4(-2) = 8 \text{ in./sec}^2 \qquad Ans.$$

① Recognize that the inner pulley is a wheel rolling along the fixed line of the left-hand cable. Thus the expressions of Sample Problem 6/3 hold.

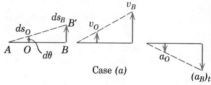

Case (a)

② Since B moves along a curved path, in addition to its tangential component of acceleration $(a_B)_t$, it will also have a normal component of acceleration toward O which does not affect the angular acceleration of the pulley.

③ The diagrams show these quantities and the simplicity of their linear relationships. The visual picture of the motion of O and B as AB rotates through the angle $d\theta$ should clarify the analysis.

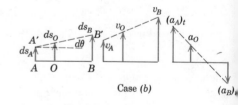

Case (b)

④ Again, as in case (a), the differential rotation of line AB as seen from the figure establishes the relation between the angular velocity of the pulley and the linear velocities of points A, O, and B. The negative sign for $(a_B)_t = a_D$ produces the acceleration diagram shown but does not destroy the linearity of the relationships.

Sample Problem 6/5

Motion of the equilateral triangular plate ABC in its plane is controlled by the hydraulic cylinder D. If the piston rod in the cylinder is moving upward at the constant rate of 0.3 m/s during an interval of its motion, calculate for the instant when $\theta = 30°$ the velocity and acceleration of the center of the roller B in the horizontal guide and the angular velocity and angular acceleration of edge CB.

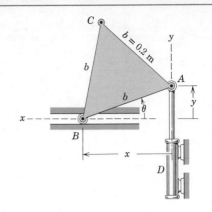

Solution. With the x-y coordinates chosen as shown the given motion of A is $v_A = \dot{y} = 0.3$ m/s and $a_A = \ddot{y} = 0$. The accompanying motion of B is given by x and its time derivatives which may be obtained from $x^2 + y^2 = b^2$. Differentiating gives

$$x\dot{x} + y\dot{y} = 0 \qquad \dot{x} = -\frac{y}{x}\dot{y}$$

$$x\ddot{x} + \dot{x}^2 + y\ddot{y} + \dot{y}^2 = 0 \qquad \ddot{x} = -\frac{\dot{x}^2 + \dot{y}^2}{x} - \frac{y}{x}\ddot{y}$$

① Observe that it is simpler to differentiate a product than a quotient. Thus differentiate $x\dot{x} + y\dot{y} = 0$ rather than $\dot{x} = -y\dot{y}/x$.

With $y = b \sin\theta$, $x = b \cos\theta$, and $\ddot{y} = 0$ the expressions become

$$v_B = \dot{x} = -v_A \tan\theta$$

$$a_B = \ddot{x} = -\frac{v_A^2}{b}\sec^3\theta$$

Substituting the numerical values $v_A = 0.3$ m/s and $\theta = 30°$ gives

$$v_B = -0.3\left(\frac{1}{\sqrt{3}}\right) = -0.1732 \text{ m/s} \qquad \text{Ans.}$$

$$a_B = -\frac{(0.3)^2(2/\sqrt{3})^3}{0.2} = -0.693 \text{ m/s}^2 \qquad \text{Ans.}$$

The negative signs indicate that the velocity and the acceleration of B are both to the right since x and its derivatives are positive to the left.

The angular motion of CB is the same as that of every line on the plate including AB. Differentiating $y = b \sin\theta$ gives

$$\dot{y} = b\dot{\theta}\cos\theta, \qquad \omega = \dot{\theta} = \frac{v_A}{b}\sec\theta$$

The angular acceleration is

$$\alpha = \dot{\omega} = \frac{v_A}{b}\dot{\theta}\sec\theta\tan\theta = \frac{v_A^2}{b^2}\sec^2\theta\tan\theta$$

Substitution of the numerical values gives

$$\omega = \frac{0.3}{0.2}\frac{2}{\sqrt{3}} = 1.732 \text{ rad/s} \qquad \text{Ans.}$$

$$\alpha = \frac{(0.3)^2}{(0.2)^2}\left(\frac{2}{\sqrt{3}}\right)^2\frac{1}{\sqrt{3}} = 1.732 \text{ rad/s}^2 \qquad \text{Ans.}$$

Both ω and α are counterclockwise since their signs are positive in the sense of positive measurement of θ.

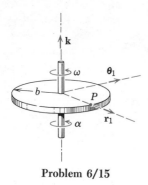

Problem 6/15

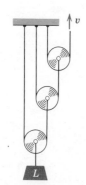

Problem 6/18

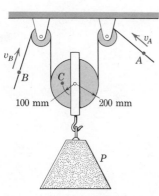

Problem 6/19

PROBLEMS

6/15 The circular disk of radius b revolves about its fixed axis with an angular velocity ω and an angular acceleration α as shown. Use cylindrical coordinates with unit vectors $\mathbf{r}_1$, $\boldsymbol{\theta}_1$, and $\mathbf{k}$ to express the velocity and acceleration of point P on the rim.

6/16 A point P located on a rigid body rotating about the fixed z-axis has a velocity $\mathbf{v} = 3\mathbf{i} - 2\mathbf{j}$ m/s at a certain instant. If the angular velocity of the body is $\boldsymbol{\omega} = 4\mathbf{k}$ rad/s at this same instant, calculate the coordinates of P.   *Ans.* $x = -\frac{1}{2}$ m, $y = -\frac{3}{4}$ m

6/17 A point P on a rigid body which rotates in the x-y plane about an axis through the origin O has coordinates $x = -3$ in. and $y = 4$ in. at a certain instant. At this same instant the angular velocity and angular acceleration of the body around the z-axis are $\boldsymbol{\omega} = 2\mathbf{k}$ rad/sec and $\boldsymbol{\alpha} = -3\mathbf{k}$ rad/sec^2, respectively. Determine the vector expression for the velocity $\mathbf{v}$ and acceleration $\mathbf{a}$ of P for this instant. Identify the normal and tangential components of the acceleration.
$$Ans.\ \mathbf{v} = -8\mathbf{i} - 6\mathbf{j} \text{ in./sec}$$
$$\mathbf{a}_n = 4(3\mathbf{i} - 4\mathbf{j}) \text{ in./sec}^2$$
$$\mathbf{a}_t = 3(4\mathbf{i} + 3\mathbf{j}) \text{ in./sec}^2$$

6/18 The load L is hoisted by the pulley-and-cable combination shown. If the system starts from rest and the upper cable acquires a velocity $v = 4$ m/s with constant acceleration when the load is 6 m above its starting position, calculate the acceleration of the load and find its velocity at this instant.

6/19 The concrete pier P is being lowered by the pulley and cable arrangement shown. If points A and B have velocities of 0.4 m/s and 0.2 m/s, respectively, compute the velocity of P, the velocity of point C for the instant represented, and the angular velocity of the pulley.
$$Ans.\ \omega = 0.5 \text{ rad/s CW}, \ v_P = 0.3 \text{ m/s},$$
$$v_C = 0.25 \text{ m/s}$$

6/20 The telephone-cable reel is rolled down the incline by the cable leading from the upper drum and wrapped around the inner hub of the reel. If the drum turns at the constant rate $\omega_1 = 2$ rad/sec, calculate the time required for the center of the reel to move 100 ft along the incline.

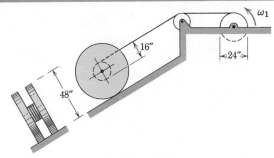

Problem 6/20

6/21 The cables at A and B are wrapped securely around the rims and the hub of the integral pulley as shown. If the cables at A and B are given upward velocities of 4 m/s and 3 m/s, respectively, calculate the velocity of the center O and the angular velocity of the pulley.

Ans. $\omega = 3.70$ rad/s CW, $v_O = 3.33$ m/s

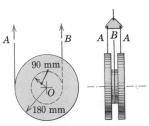

Problem 6/21

6/22 For the rolling telephone-cable reel of Prob. 6/20 if the upper drum starts from rest with an angular acceleration of 0.06 rad/sec², compute the time t required for the center of the reel to move 100 ft down the incline and specify its angular acceleration α.

6/23 The spool rolls on its hub up the inner cable A as the equalizer plate B pulls the outer cables down. The three cables are wrapped securely around their respective peripheries and do not slip. If, at the instant represented, B has moved down a distance of 800 mm from rest with a constant acceleration of 100 mm/s², determine the velocity of point C and the acceleration of the center O for this particular instant.

Ans. $a_O = 66.7$ mm/s²
$v_C = 933$ mm/s

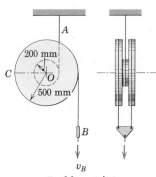

Problem 6/23

6/24 The cable from drum A turns the double wheel B which rolls on its hubs without slipping. Determine the angular velocity ω and the angular acceleration α of the drum C at the instant when the angular velocity and angular acceleration of A are 4 rad/s and 3 rad/s², respectively, both in the counterclockwise direction.

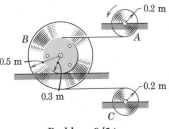

Problem 6/24

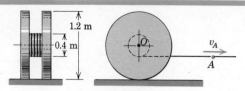

Problem 6/25

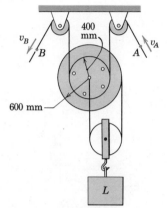

Problem 6/26

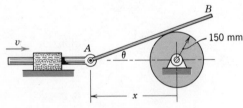

Problem 6/28

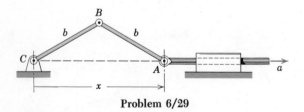

Problem 6/29

6/25 The telephone-cable reel rolls without slipping on the horizontal surface. If point A on the cable has a velocity $v_A = 0.8$ m/s to the right, compute the velocity of the center O and the angular velocity ω of the reel. (Be careful not to make the mistake of assuming that the reel rolls to the left.)

$\qquad$ *Ans.* $v_O = 1.2$ m/s, $\omega = 2$ rad/s CW

6/26 The double pulleys are fastened together to form a rigid unit. If $v_B = 0$ and $v_A = 4$ m/s, compute the velocity of the load L.

6/27 If $v_A = 9$ m/s and $v_B = 3$ m/s in Prob. 6/26, compute the velocity of the load L and the corresponding angular velocity of the double pulley which is a rigid unit. $\quad$ *Ans.* $\omega = 15$ rad/s CW, $v_L = 7.5$ m/s

6/28 The horizontal control rod has a constant velocity $v = 4$ in./sec. Calculate the angular velocity ω of rod AB when $x = 12$ in.

6/29 Point A is given a constant acceleration a to the right starting from rest with x essentially zero. Determine the angular velocity ω of link AB in terms of x and a.

$$\text{Ans. } \omega = \frac{\sqrt{2ax}}{\sqrt{4b^2 - x^2}} \text{ CCW}$$

6/30 Calculate the angular velocity ω of the slender bar AB as a function of the distance x and the constant angular velocity ω_0 of the drum.

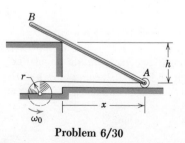

Problem 6/30

6/31 Determine an expression for the angular velocity ω of link AB in terms of the angle θ and the angular velocity $\omega_0 = \dot\theta$ of OA.

$$Ans. \ \omega = \frac{\cos\theta}{\sqrt{4 - \sin^2\theta}}\omega_0$$

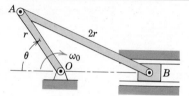

Problem 6/31

6/32 The crank OA starts from rest at $\theta = 0$ and turns clockwise with a constant angular acceleration α. Determine the acceleration a_x of the shaft B for the position $\theta = 90°$.

6/33 The circular cam is mounted eccentrically about its fixed bearing at O and turns counterclockwise at the constant angular velocity ω. The cam causes the fork A and attached control rod to oscillate in the horizontal x-direction. Write the expressions for the velocity v_x and acceleration a_x of the control rod in terms of the angle θ measured from the vertical. The contact surfaces of the fork are vertical.

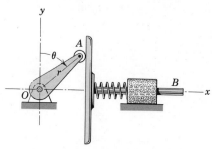

Problem 6/32

6/34 The slotted arm OA rotates with a constant angular velocity $\omega = \dot\theta$ during a limited interval of its motion and moves the pivoted slider block along the horizontal slot. Write the expressions for the velocity v_B and acceleration a_B of the pin B in the slider block in terms of θ.

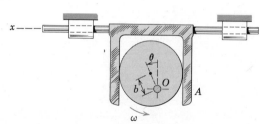

Problem 6/33

6/35 The elements of a wheel-and-disk mechanical integrator are shown in the figure. The integrator wheel A turns about its fixed shaft and is driven by friction from disk B with no slipping occurring tangent to its rim. The distance y is a variable and can be controlled at will. Show that the angular displacement of the integrator wheel is given by $z = (1/b)\int y \, dx$, where x is the angular displacement of the disk B.

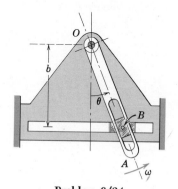

Problem 6/34

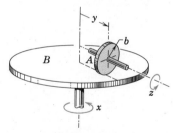

Problem 6/35

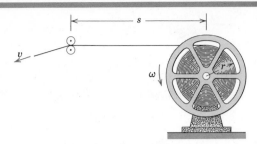

Problem 6/36

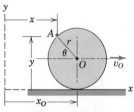

Problem 6/37

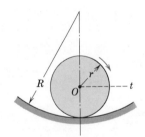

Problem 6/38

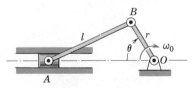

Problem 6/39

6/36 A reel of film is being unwound by pulling the film through the guide rollers with a constant velocity v. The reel speeds up as its radius diminishes. Determine the angular acceleration α of the reel if the thickness of the film is t and the diminishing radius of the roll of film is r. The distance s is very large compared with r.

$$\text{Ans. } \alpha = \frac{tv^2}{2\pi r^3}$$

6/37 The wheel rolls to the right without slipping, and its center O has a constant velocity v_0. Determine the velocity v and acceleration a of a point A on the rim of the wheel in terms of the angle θ measured clockwise from the horizontal. Make use of the coordinates shown.

$$\text{Ans. } v = v_0\sqrt{2(1 + \sin\theta)}$$
$$a = v_0{}^2/r \text{ toward } O$$

6/38 Show that the expressions $v = r\omega$ and $a_t = r\alpha$ hold for the motion of the center O of the wheel which rolls on the circular arc, where ω and α are the absolute angular velocity and acceleration, respectively, of the wheel. (*Hint:* Follow the example of Sample Problem 6/3 and allow the wheel to roll a small distance. Be very careful to identify the correct *absolute* angle through which the wheel turns in determining its angular velocity and angular acceleration.)

▶**6/39** One of the most common mechanisms is the slider-crank. Express the angular velocity ω_{AB} and angular acceleration α_{AB} of the connecting rod AB in terms of the crank angle θ for a given constant crank speed ω_0. Take ω_{AB} and α_{AB} to be positive counterclockwise.

$$\text{Ans. } \omega_{AB} = \frac{r\omega_0}{l}\frac{\cos\theta}{\sqrt{1 - \frac{r^2}{l^2}\sin^2\theta}}$$

$$\alpha_{AB} = \frac{r\omega_0{}^2}{l}\sin\theta\frac{\frac{r^2}{l^2} - 1}{\left(1 - \frac{r^2}{l^2}\sin^2\theta\right)^{3/2}}$$

6/4 RELATIVE VELOCITY. The second approach to rigid-body kinematics is to use the principles of relative motion. In Art. 2/8 we
developed these principles for measurements relative to translating
axes and applied the relative velocity equation

$$\mathbf{v}_A = \mathbf{v}_B + \mathbf{v}_{A/B} \qquad\qquad [2/20]$$

to the motions of two particles A and B.

   We now choose two points on the *same* rigid body for our two
particles. The consequence of this choice is that the motion of one
point as seen by an observer translating with the other point must be
circular since the radial distance to the observed point from the
reference point does not change. This observation is the *key* to the
successful understanding of a large majority of problems in the plane
motion of rigid bodies. This concept is illustrated in Fig. 6/5a, which
shows a rigid body moving in the plane of the figure from position AB
to $A'B'$ during time Δt. This movement may be visualized as occurring in two parts. First, the body translates to the parallel position
$A''B'$ with the displacement $\Delta\mathbf{r}_B$. Second, the body rotates about B'
through the angle $\Delta\theta$. From the nonrotating reference axes x'-y'
attached to the reference point B', it is seen that this remaining
motion of the body is one of simple rotation about B', giving rise to
the displacement $\Delta\mathbf{r}_{A/B}$ of A with respect to B. To the nonrotating
observer attached to B the body appears to undergo fixed-axis rotation about B with A executing circular motion as emphasized in Fig.
6/5b. Therefore the relationships developed for circular motion in
Arts. 2/5 and 6/2 and cited as Eqs. 2/11 and 6/2 (or 6/3) describe the
relative portion of A's motion.

   Point B was arbitrarily chosen as the reference point for attachment of our nonrotating reference axes x-y. Point A could have been
used just as well, in which case we observe B to have circular motion
about A considered fixed as shown in Fig. 6/5c. We see that the sense

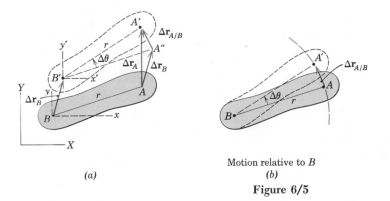

(a)

Motion relative to B
(b)

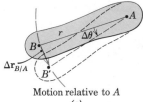

Motion relative to A
(c)

Figure 6/5

of the rotation, counterclockwise in this example, is the same whether we choose A or B as the reference, and we see that $\Delta\mathbf{r}_{B/A} = -\Delta\mathbf{r}_{A/B}$.

With B as the reference point the total displacement of A is seen from Fig. 6/5a to be

$$\Delta\mathbf{r}_A = \Delta\mathbf{r}_B + \Delta\mathbf{r}_{A/B}$$

where $\Delta\mathbf{r}_{A/B}$ has the magnitude $r\Delta\theta$ as $\Delta\theta$ approaches zero. We note that the *relative linear motion* $\Delta\mathbf{r}_{A/B}$ is accompanied by the *absolute angular motion* $\Delta\theta$, as seen from the translating axes x'-y'. Dividing the expression for $\Delta\mathbf{r}_A$ by the corresponding time interval Δt and passing to the limit give the relative-velocity equation

$$\boxed{\mathbf{v}_A = \mathbf{v}_B + \mathbf{v}_{A/B}} \tag{6/4}$$

This expression is the same as Eq. 2/20 with the one restriction that the distance r between A and B remains constant. The magnitude of the relative velocity is thus seen to be $v_{A/B} = \lim_{\Delta t \to 0} (|\Delta\mathbf{r}_{A/B}|/\Delta t) = \lim_{\Delta t \to 0} (r\Delta\theta/\Delta t)$ which, with $\omega = \dot{\theta}$, becomes

$$\boxed{v_{A/B} = r\omega} \tag{6/5}$$

With $\mathbf{r}$ representing the vector $\mathbf{r}_{A/B}$, from the first of Eqs. 6/3 we may write the relative velocity as the vector

$$\boxed{\mathbf{v}_{A/B} = \boldsymbol{\omega} \times \mathbf{r}} \tag{6/6}$$

where $\boldsymbol{\omega}$ is the angular-velocity vector normal to the plane of the motion in the sense determined by the right-hand rule. A critical observation seen from Figs. 6/5b and 6/5c is that the relative linear velocity is always perpendicular to the line joining the two points in question. This conclusion should be perfectly clear from Figs. 6/5b and 6/5c.

The application of Eq. 6/4 is clarified by visualizing the separate translation and rotation components of the equation. These components are emphasized in Fig. 6/6 which shows a rigid body in plane

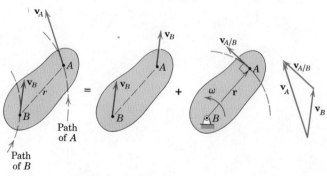

Figure 6/6

motion. With B chosen as the reference point, the velocity of A is the vector sum of the translational portion $\mathbf{v}_B$ plus the rotational portion $\mathbf{v}_{A/B} = \boldsymbol{\omega} \times \mathbf{r}$ which has the magnitude $v_{A/B} = r\omega$, where $|\boldsymbol{\omega}| = \dot{\theta}$, the *absolute* angular velocity of AB. The fact that the *relative linear velocity* is *always perpendicular* to the line joining the two points in question is an important key to the solution of many problems. The student should draw the equivalent diagram where point A is used as the reference point rather than B.

A second use of Eq. 6/4 for relative-velocity problems in plane motion may be made for constrained sliding contact between two links in a mechanism. In this case we choose points A and B as coincident points, one on each link, for the instant under consideration. This second use of the relative-velocity equation is illustrated in Sample Problem 6/9.

Solution of the relative-velocity equation may be carried out by scalar or vector algebra, or a graphical analysis may be employed. In any event a sketch of the vector polygon which represents the vector equation should be made to reveal the physical relationships involved. From this sketch, scalar component equations may be written by projecting the vectors along convenient directions. Usually a simultaneous solution may be avoided by a careful choice of the projections. Alternatively, each term in the relative-motion equation may be written in terms of its **i**- and **j**-components from which two scalar equations result when the equality is applied, separately, to the coefficients of the **i**- and **j**-terms. Many problems lend themselves to a graphical solution, particularly when the given geometry does not readily lend itself to mathematical expression. In this case the known vectors are first constructed in their correct positions using a convenient scale. Next, the unknown vectors, which complete the polygon and satisfy the vector equation, are measured directly from the drawing. The choice of method to be used depends upon the particular problem at hand, the accuracy required, and upon individual preference and experience. All three of the approaches are illustrated in the sample problems which follow.

Regardless of which method of solution we employ, we note that the single vector equation in two dimensions is equivalent to two scalar equations, so that two scalar unknowns can be solved. The unknowns, for instance, might be the magnitude of one vector and the direction of another.

Sample Problem 6/6

The wheel of radius $r = 300$ mm rolls to the right without slipping and has a velocity $v_O = 3$ m/s of its center O. Calculate the velocity of point A on the wheel for the instant represented.

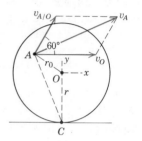

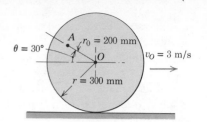

   Solution I (scalar—geometric). The center O is chosen as the reference point for the relative-velocity equation since its motion is given. We therefore write

$$\mathbf{v}_A = \mathbf{v}_O + \mathbf{v}_{A/O}$$

where the relative-velocity term is observed from the translating axes x-y attached to O. The angular velocity of AO is the same as that of the wheel which, from Sample Problem 6/3, is $\omega = v_O/r = 3/0.3 = 10$ rad/s. Thus from Eq. 6/5 we have

$$[v_{A/O} = r_0\dot{\theta}] \qquad v_{A/O} = 0.2(10) = 2 \text{ m/s}$$

which is normal to AO as shown. The vector sum $\mathbf{v}_A$ is shown on the diagram and may be calculated from the law of cosines. Thus

$$v_A^2 = 3^2 + 2^2 + 2(3)(2)\cos 60° = 19 \ (\text{m/s})^2, \ v_A = 4.36 \text{ m/s} \qquad Ans.$$

   The contact point C momentarily has zero velocity and can be used alternatively as the reference point, in which case the relative-velocity equation becomes $\mathbf{v}_A = \mathbf{v}_C + \mathbf{v}_{A/C} = \mathbf{v}_{A/C}$ where

$$v_{A/C} = \overline{AC}\omega = \frac{\overline{AC}}{\overline{OC}}v_O = \frac{0.436}{0.300}(3) = 4.36 \text{ m/s}, \qquad v_A = v_{A/C} = 4.36 \text{ m/s}$$

The distance $\overline{AC} = 436$ mm is calculated separately. We see that $\mathbf{v}_A$ is normal to AC since A is momentarily rotating about point C.

   Solution II (vector). We will now use Eq. 6/6 and write

$$\mathbf{v}_A = \mathbf{v}_O + \mathbf{v}_{A/O} = \mathbf{v}_O + \boldsymbol{\omega} \times \mathbf{r}_0$$

where

$$\boldsymbol{\omega} = -10\mathbf{k} \text{ rad/s}$$
$$\mathbf{r}_0 = 0.2(-\mathbf{i}\cos 30° + \mathbf{j}\sin 30°) = -0.173\mathbf{i} + 0.1\mathbf{j} \text{ m}$$
$$\mathbf{v}_O = 3\mathbf{i} \text{ m/s}$$

We now solve the vector equation

$$\mathbf{v}_A = 3\mathbf{i} + \begin{vmatrix} \mathbf{i} & \mathbf{j} & \mathbf{k} \\ 0 & 0 & -10 \\ -0.173 & 0.1 & 0 \end{vmatrix} = 3\mathbf{i} + 1.73\mathbf{j} + 1.0\mathbf{i}$$

$$= 4\mathbf{i} + 1.73\mathbf{j} \text{ m/s} \qquad\qquad Ans.$$

The magnitude $v_A = \sqrt{4^2 + (1.73)^2} = \sqrt{19} = 4.36$ m/s and direction agree with the previous solution.

① Be sure to visualize $\mathbf{v}_{A/O}$ as the velocity which A appears to have in its circular motion relative to O.

② The vectors may also be laid off to scale graphically and the magnitude and direction of v_A measured directly from the diagram.

③ The velocity of any point on the wheel is easily determined by using the contact point C as the reference point. The student should construct the velocity vectors for a number of points on the wheel for practice.

④ The vector $\boldsymbol{\omega}$ is directed into the paper by the right-hand rule whereas the positive z-direction is out from the paper, hence the minus sign.

Sample Problem 6/7

Crank CB oscillates about C through a limited arc causing crank OA to oscillate about O. When the linkage passes the position shown with CB horizontal and OA vertical, the angular velocity of CB is 2 rad/s counterclockwise. For this instant determine the angular velocities of OA and AB.

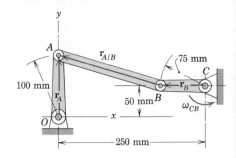

Solution I (vector). The relative-velocity equation $\mathbf{v}_A = \mathbf{v}_B + \mathbf{v}_{A/B}$ is rewritten as

① $$\boldsymbol{\omega}_{OA} \times \mathbf{r}_A = \boldsymbol{\omega}_{CB} \times \mathbf{r}_B + \boldsymbol{\omega}_{AB} \times \mathbf{r}_{A/B}$$

where $\boldsymbol{\omega}_{OA} = \omega_{OA}\mathbf{k}$, $\boldsymbol{\omega}_{CB} = 2\mathbf{k}$ rad/s, $\boldsymbol{\omega}_{AB} = \omega_{AB}\mathbf{k}$

$\mathbf{r}_A = 100\mathbf{j}$ mm, $\mathbf{r}_B = -75\mathbf{i}$ mm, $\mathbf{r}_{A/B} = -175\mathbf{i} + 50\mathbf{j}$ mm

① We are using here the first of Eqs. 6/3 and Eq. 6/6.

Substitution gives

$$\omega_{OA}\mathbf{k} \times 100\mathbf{j} = 2\mathbf{k} \times (-75\mathbf{i}) + \omega_{AB}\mathbf{k} \times (-175\mathbf{i} + 50\mathbf{j})$$

$$-100\,\omega_{OA}\mathbf{i} = -150\mathbf{j} - 175\,\omega_{AB}\mathbf{j} - 50\,\omega_{AB}\mathbf{i}$$

Matching coefficients of the respective $\mathbf{i}$- and $\mathbf{j}$-terms gives

$$-100\,\omega_{OA} + 50\,\omega_{AB} = 0 \qquad 25(6 + 7\,\omega_{AB}) = 0$$

the solutions of which are

② $$\omega_{AB} = -6/7 \text{ rad/s} \quad \text{and} \quad \omega_{OA} = -3/7 \text{ rad/s} \qquad Ans.$$

② The minus signs in the answers indicate that the vectors ω_{AB} and ω_{OA} are in the negative $\mathbf{k}$-direction. Hence, the angular velocities are clockwise.

Solution II (scalar-geometric). Solution by the scalar geometry of the vector triangle is particularly simple here since $\mathbf{v}_A$ and $\mathbf{v}_B$ are at right angles for this special position of the linkages. First we compute v_B, which is

$[v = r\omega]$ $$v_B = 0.075(2) = 0.150 \text{ m/s}$$

and represent it in its correct direction as shown. The vector $\mathbf{v}_{A/B}$ must be perpendicular to AB, and the angle θ between $\mathbf{v}_{A/B}$ and $\mathbf{v}_B$ is also the angle made by AB with the horizontal direction. This angle is given by

$$\tan\theta = \frac{100 - 50}{250 - 75} = \frac{2}{7}$$

③ The horizontal vector $\mathbf{v}_A$ completes the triangle for which we have

$$v_{A/B} = v_B/\cos\theta = 0.150/\cos\theta$$

$$v_A = v_B \tan\theta = 0.150(2/7) = 0.30/7 \text{ m/s}$$

③ Always make certain that the sequence of vectors in the vector polygon agrees with the equality of vectors specified by the vector equation.

The angular velocities become

$[\omega = v/r]$ $$\omega_{AB} = v_{A/B}/\overline{AB} = \frac{0.150}{\cos\theta}\frac{\cos\theta}{0.250 - 0.075}$$

$$= 6/7 \text{ rad/s CW} \qquad Ans.$$

$[\omega = v/r]$ $$\omega_{OA} = v_A/\overline{OA} = \frac{0.30}{7}\frac{1}{0.100} = 3/7 \text{ rad/s CW} \qquad Ans.$$

Sample Problem 6/8

The common configuration of a reciprocating engine is that of the slider-crank mechanism shown. If the crank OB has a clockwise rotational speed of 1500 rev/min, determine for the position where $\theta = 60°$ the velocity of the piston A, the velocity of point G on the connecting rod, and the angular velocity of the connecting rod.

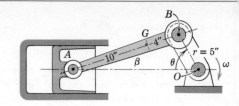

Solution. The velocity of the crank pin B as a point on AB is easily found, so that B will be used as the reference point for determining the velocity of A. The relative-velocity equation may now be written

$$\mathbf{v}_A = \mathbf{v}_B + \mathbf{v}_{A/B}$$

The crank-pin velocity is

① $[v = r\omega]$ $\qquad v_B = \dfrac{5}{12}\dfrac{1500(2\pi)}{60} = 65.4 \text{ ft/sec}$

and is normal to OB. The direction of $\mathbf{v}_A$ is, of course, along the horizontal cylinder axis. The direction of $\mathbf{v}_{A/B}$ must be perpendicular to the line AB as explained in the present article and as indicated on the diagram where the reference point B is shown as fixed. We obtain this direction by computing angle β from the law of sines which gives

$$\frac{5}{\sin \beta} = \frac{14}{\sin 60°}, \qquad \beta = \sin^{-1} 0.3093 = 18.0°$$

We now complete the sketch of the velocity triangle where the angle between $\mathbf{v}_{A/B}$ and $\mathbf{v}_A$ is $90° - 18.0° = 72.0°$ and the third angle is $180° - 30° - 72.0° = 78.0°$. Vectors $\mathbf{v}_A$ and $\mathbf{v}_{A/B}$ are shown with their proper sense such that the head-to-tail sum of $\mathbf{v}_B$ and $\mathbf{v}_{A/B}$ equals $\mathbf{v}_A$. The magnitudes of the unknowns are now calculated from the trigonometry of the vector triangle or are scaled from the diagram if a graphical solution is used. Solving for $\mathbf{v}_A$ and $\mathbf{v}_{A/B}$ by the law of sines gives

② $\qquad \dfrac{v_A}{\sin 78.0°} = \dfrac{65.4}{\sin 72.0°} \qquad v_A = 67.3 \text{ ft/sec} \qquad$ *Ans.*

$$\frac{v_{A/B}}{\sin 30°} = \frac{65.4}{\sin 72.0°} \qquad v_{A/B} = 34.4 \text{ ft/sec}$$

The angular velocity of AB is counterclockwise as revealed by the sense of $\mathbf{v}_{A/B}$ and is

$[\omega = v/r]$ $\qquad \omega_{AB} = \dfrac{v_{A/B}}{AB} = \dfrac{34.4}{14/12} = 29.5 \text{ rad/sec} \qquad$ *Ans.*

We now determine the velocity of G by writing

$$\mathbf{v}_G = \mathbf{v}_B + \mathbf{v}_{G/B}$$

where $\quad v_{G/B} = \overline{GB}\,\omega_{AB} = \dfrac{\overline{GB}}{\overline{AB}} v_{A/B} = \dfrac{4}{14}(34.4) = 9.83 \text{ ft/sec.}$ As seen from the diagram $\mathbf{v}_{G/B}$ has the same direction as $\mathbf{v}_{A/B}$. The vector sum is shown on the last diagram. We can calculate v_G with some geometric labor or simply measure its magnitude and direction from the velocity diagram drawn to scale. For simplicity we adopt the latter procedure here and obtain

$$v_G = 64.1 \text{ ft/sec} \qquad \textit{Ans.}$$

As seen, the diagram may be superposed directly on the first velocity diagram.

① Remember always to convert ω to radians per unit time when using $v = r\omega$.

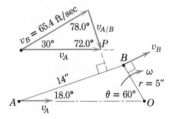

② A graphical solution to this problem is the quickest to achieve although its accuracy is limited. Solution by vector algebra can, of course, be used but would involve somewhat more labor in this problem.

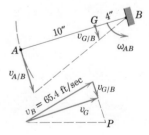

Sample Problem   6/9

The power screw turns at a speed which gives the threaded collar C a velocity of 0.8 ft/sec vertically down. Determine the angular velocity of the slotted arm when $\theta = 30°$.

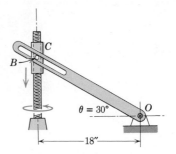

Solution.   The angular velocity of the arm can be found if the velocity of a point on the arm is known. We choose a point A on the arm coincident with the pin B of the collar for this purpose. If we use B as our reference point and write $\mathbf{v}_A = \mathbf{v}_B + \mathbf{v}_{A/B}$, we see from the diagram which shows the arm and points A and B an instant before and an instant after coincidence that $\mathbf{v}_{A/B}$ has a direction along the slot away from O.

The magnitudes of $\mathbf{v}_A$ and $\mathbf{v}_{A/B}$ are the only unknowns in the vector equation, so that it may now be solved. We draw the known vector $\mathbf{v}_B$ and then obtain the intersection P of the known directions of $\mathbf{v}_{A/B}$ and $\mathbf{v}_A$. The solution gives

$$v_A = v_B \cos \theta = 0.8 \cos 30° = 0.693 \text{ ft/sec}$$

$$[\omega = v/r] \qquad \omega = v_A / \overline{OA} = \frac{0.693}{(\frac{18}{12})/\cos 30°}$$

$$= 0.400 \text{ rad/sec CCW} \qquad\qquad Ans.$$

We note the difference between this problem of constrained sliding contact between two links and the three preceding sample problems of relative velocity where no sliding contact occurred and where the points A and B were located on the same rigid body in each case.

① Physically, of course, this point does not exist, but we can imagine such a point in the middle of the slot and attached to the arm.

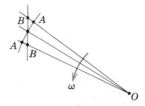

② Always identify the knowns and unknowns before attempting the solution of a vector equation.

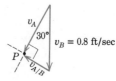

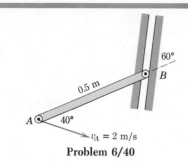

0.5 m

60°

40°

$v_A = 2$ m/s

Problem 6/40

A

θ

L

B

β

v_B

v_A

Problem 6/41

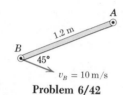

A

1.2 m

B

45°

$v_B = 10$ m/s

Problem 6/42

650 mm

A

O

v_0

Problem 6/43

660 mm

70 mm

160 mm

170 mm

Problem 6/44

PROBLEMS

6/40 End A of the link has the velocity shown at the instant depicted. End B is confined to move in the slot. For this instant calculate the velocity of B and the angular velocity of AB.

Ans. $v_B = 3.06$ m/s, $\omega_{AB} = 7.88$ rad/s CCW

6/41 If v_A and v_B are the instantaneous velocities of the respective ends of the rigid link, show that the angular velocity of the link is given by $(v_A \sin \theta + v_B \sin \beta)/L$ and that $v_A \cos \theta = v_B \cos \beta$.

6/42 At a particular instant end B of the bar AB has a velocity of 10 m/s in the direction shown. What is the minimum possible velocity which end A can have? What would be the corresponding angular velocity of the bar?

Ans. $(v_A)_{\min} = 7.07$ m/s, $\omega_{AB} = 5.89$ rad/s CCW

6/43 The magnitude of the absolute velocity of point A on the automobile tire is 10 m/s when A is in the position shown. What is the corresponding velocity v_O of the car and the angular velocity ω of the wheel? (The wheel rolls without slipping.)

6/44 The rider of the bicycle shown pumps steadily to maintain a constant speed of 16 km/h against a slight head wind. Calculate the maximum and minimum magnitudes of the absolute velocity of the pedal A.

Ans. $(v_A)_{\max} = 5.33$ m/s, $(v_A)_{\min} = 3.56$ m/s

45 End A of the link has a downward velocity v_A of 2 m/s during an interval of its motion. For the position where $\theta = 30°$ determine the angular velocity ω of AB and the velocity v_G of the midpoint G of the link.

46 Solve Prob. 6/45 by using vector algebra.
   Ans. $\omega = 11.55\mathbf{k}$ rad/s, $\mathbf{v}_G = 0.578\mathbf{i} - 1.0\mathbf{j}$ m/s

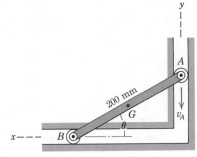

Problem 6/45

47 The wheel rolls to the left without slipping. If the velocity of B relative to A is 4 ft/sec, determine the velocity of point P for this position.

48 The two pulleys are fastened together to form a single rigid unit, and each of the two cables is wrapped securely around its respective pulley. If point A on the hoisting cable has a velocity $v_A = 3$ ft/sec, determine the velocity of point O and the velocity of point B on the large pulley for the position shown.
   Ans. $v_O = 2$ ft/sec, $v_B = 2.83$ ft/sec

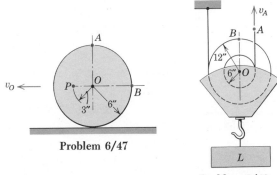

Problem 6/47

Problem 6/48

49 For the instant shown vertex A of the equi-angular plate has a velocity of 3 ft/sec in the positive x-direction. Also, the velocity of B with respect to C is 2 ft/sec in the negative y-direction. Determine the magnitude of the velocity of C for this position.

50 Determine the velocity of point D which will produce a counterclockwise angular velocity of 40 rad/s for link AB in the position shown for the four-bar linkage.

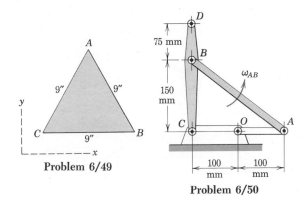

Problem 6/49

Problem 6/50

51 Determine the velocity of the piston and the angular velocity of the connecting rod AB in Sample Problem 6/8 for the position $\theta = 240°$.
   Ans. $v_A = 46.0$ ft/sec, $\omega_{AB} = 29.5$ rad/sec CW

52 For the linkage shown determine the velocity of C if OB has a counterclockwise angular velocity of 4 rad/sec at the position for which $x = 24$ in.

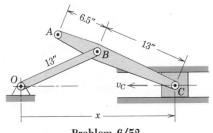

Problem 6/52

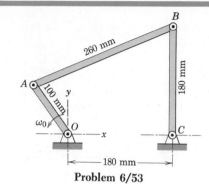

Problem 6/53

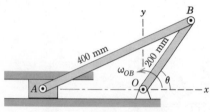

Problem 6/54

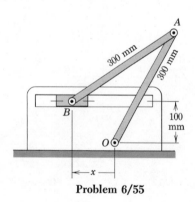

Problem 6/55

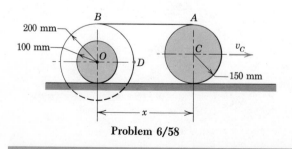

Problem 6/58

6/53 In the four-bar linkage shown control link OA has a counterclockwise angular velocity $\omega_0 = 10$ rad/s during a short interval of motion. When link CB passes the vertical position shown, point A has coordinates $x = -60$ mm and $y = 80$ mm. By means of vector algebra determine the angular velocity ω_{AB} of AB.   *Ans.* $\omega_{AB} = 2.5\mathbf{k}$ rad/s

6/54 The crank OB has a counterclockwise angular velocity $\omega_{OB} = 4$ rad/s. For the position in which $\theta = \tan^{-1}\frac{4}{3}$ determine the angular velocity of link AB. Solve by using vector algebra.

6/55 If the slider at B has a velocity $\dot{x} = 0.4$ m/s for the position where $x = 75$ mm, compute the corresponding angular velocity of OA.

Ans. $\omega_{OA} = 2.97$ rad/s CCW

6/56 If link OA in Prob. 6/55 has a clockwise angular velocity of 2 rad/s in the position for which $x = 75$ mm, determine the velocity of the slider at B.

6/57 If the slider C of the linkage in Prob. 6/52 has a velocity $v_C = 20$ in./sec at the position for which $x = 24$ in., for this instant determine the angular velocity of AC and the horizontal component of the velocity of A.

Ans. $\omega_{AC} = 2$ rad/sec CW
$(v_A)_x = 5$ in./sec to the left

6/58 The center C of the smaller wheel has a velocity $v_C = 0.4$ m/s in the direction shown. The cord which connects the two wheels is securely wrapped around the respective peripheries and does not slip. Calculate the velocity of point D when in the position shown. Also compute the change Δx which occurs per second if v_C is constant.

Ans. $v_D = 0.596$ m/s, $\Delta x = 0.133$ m

√59 The crank *OA* oscillates about the $\theta = 0$ position causing *CB*, in turn, to oscillate. If *OA* has a counterclockwise angular velocity of 6 rad/sec when $\theta = 30°$, determine the corresponding angular velocity of *CB* for this instant.

$$\text{Ans. } \omega_{CB} = 2.00 \text{ rad/sec CW}$$

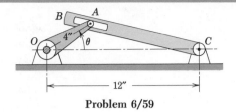

Problem 6/59

√60 The rotation of the gear is controlled by the horizontal motion of end *A* of the rack *AB*. If the piston rod has a constant velocity $\dot{x} = 300$ mm/s during a short interval of motion, determine the angular velocity ω_0 of the gear and the angular velocity ω_{AB} of *AB* at the instant when $x = 800$ mm.

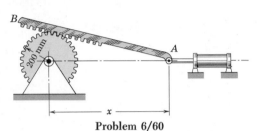

Problem 6/60

√61 At the instant represented the slotted arm is revolving counterclockwise about *O* with an angular velocity of 2 rad/s. Also for this instant, $\theta = 30°$ and $\beta = 45°$. The pin *A* is attached to arm *BC*. Use a graphical solution to determine the velocity of point *B* for this position. $\text{Ans. } v_B = 2.58 \text{ m/s}$

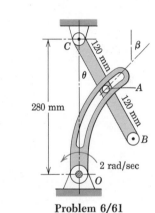

Problem 6/61

√62 For a short interval of motion the collars have velocities $v_A = 2$ m/s to the right and $v_B = 3$ m/s to the left. For the position $\theta = 60°$ determine the velocity of point *C*. $\text{Ans. } v_C = 1.528 \text{ m/s}$

√63 At the instant represented $x = 50$ mm and $\dot{s} = 1.6$ m/s. Determine the corresponding velocity of point *B*. $\text{Ans. } v_B = 1.029 \text{ m/s}$

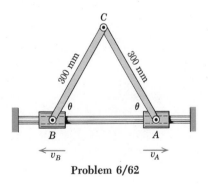

Problem 6/62

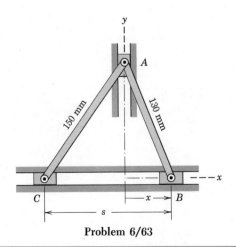

Problem 6/63

6/5 INSTANTANEOUS CENTER OF ZERO VELOCITY.

In the previous article we determined the velocity of a point on a rigid body in plane motion by adding the relative velocity due to rotation about a convenient reference point to the velocity of the reference point. In the present article we will solve the problem by choosing a unique reference point which momentarily has zero velocity. Insofar as velocities are concerned the body may be considered to be in pure rotation about an axis, normal to the plane of motion, passing through this point. This axis is called the *instantaneous axis* of zero velocity, and the intersection of this axis with the plane of motion is known as the *instantaneous center* of zero velocity. This approach provides us with a valuable means for visualizing and analyzing velocities in plane motion.

The existence of the instantaneous center is easily shown. For the body in Fig. 6/7a let us assume that the directions of the absolute velocities of any two points A and B on the body are known and are not parallel. If there is a point about which A has absolute circular motion at the instant considered, this point must lie on the normal to $\mathbf{v}_A$ through A. Similar reasoning applies to B, and the intersection C of these two perpendiculars fulfills the requirement for an absolute center of rotation *at the instant considered.* Point C is the instantaneous center of zero velocity and may lie on or off the body. If it lies off the body, it may be visualized as lying on the body extended. The instantaneous center is not a fixed point in the body nor a fixed point in the plane.

If the magnitude of the velocity of one of the points, say v_A, is also known, the angular velocity ω of the body and the linear velocity of every point in the body are easily obtained. Thus the angular velocity of the body, Fig. 6/7a, is

$$\omega = \frac{v_A}{r_A}$$

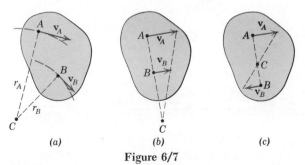

(a)   (b)   (c)

Figure 6/7

which is, of course, also the angular velocity of *every* line in the body. Therefore the velocity of B is $v_B = r_B\omega = (r_B/r_A)v_A$. Once the instantaneous center is located, the direction of the instantaneous velocity of every point in the body is readily found since it must be perpendicular to the radial line joining the point in question with C.

If the velocities of two points in a body having plane motion are parallel, Fig. 6/7*b* or 6/7*c*, the line joining the points is perpendicular to the direction of the velocities, and the instantaneous center C is located by direct proportion as shown. We can readily see from Fig. 6/7*b* that as the parallel velocities approach equality in magnitude, the instantaneous center C moves farther away from the body and approaches infinity in the limit as the body gives up its angular velocity and translates only.

As the body changes its position, the instantaneous center C also changes its position both in space and on the body. The locus of the instantaneous centers in space is known as the *space centrode,* and the locus of the positions of the instantaneous centers on the body is known as the *body centrode.* At the instant considered the two curves are tangent at the position of point C. It may be shown that the body centrode curve rolls on the space centrode curve during the motion of the body, as indicated schematically in Fig. 6/8.

Whereas the instantaneous center of zero velocity as a point on the body momentarily is at rest, its acceleration generally is *not* zero. Thus this point may *not* be used as an instantaneous center of zero acceleration in a manner analogous to its use for finding velocity. An instantaneous center of zero acceleration does exist for bodies in general plane motion, but its location and use represents a specialized topic in mechanism kinematics and will not be treated here.

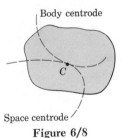

Body centrode

C

Space centrode

Figure 6/8

Sample Problem 6/10

The wheel of Sample Problem 6/6, shown again here, rolls to the right without slipping with its center O having a velocity $v_O = 3$ m/s. Locate the instantaneous center of zero velocity and use it to find the velocity of point A for the position indicated.

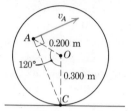

Solution. The point on the rim of the wheel in contact with the ground has no velocity if the wheel is not slipping; it is, therefore, the instantaneous center C of zero velocity. The angular velocity of the wheel becomes

$[\omega = v/r]$ $\qquad \omega = v_O/\overline{OC} = 3/0.300 = 10$ rad/s

The distance from A to C is

① $\overline{AC} = \sqrt{(0.300)^2 + (0.200)^2 - 2(0.300)(0.200)\cos 120°} = 0.436$ m

The velocity of A becomes

② $[v = r\omega]$ $\qquad v_A = \overline{AC}\omega = 0.436(10) = 4.36$ m/s $\qquad$ *Ans.*

The direction of $\mathbf{v}_A$ is perpendicular to AC as shown.

① Be sure to recognize that the cosine of $120°$ is itself negative.

② From the results of this problem you should be able to visualize and sketch the velocities of all points on the wheel.

Sample Problem 6/11

Arm OB of the linkage has a clockwise angular velocity of 10 rad/sec in the position shown where $\theta = 45°$. Determine the velocity of A, the velocity of D, and the angular velocity of link AB for the instant shown.

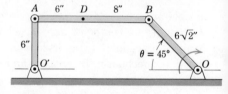

Solution. The directions of the velocities of A and B are tangent to their circular paths about the fixed centers O' and O as shown. The intersection of the two perpendiculars to the velocities from A and B ① locates the instantaneous center C for the link AB. The distances $\overline{AC}$, $\overline{BC}$, and $\overline{DC}$ shown on the diagram are computed or scaled from the drawing. The angular velocity of BC, considered as a line on the body extended, is equal to the angular velocity of AC, DC, and AB and is

$[\omega = v/r]$ $\qquad \omega_{BC} = \dfrac{v_B}{\overline{BC}} = \dfrac{\overline{OB}\,\omega_{OB}}{\overline{BC}} = \dfrac{6\sqrt{2}(10)}{14\sqrt{2}}$

$\qquad\qquad\qquad = 4.29$ rad/sec CCW $\quad$ *Ans.*

Thus the velocities of A and D are

$[v = r\omega]$ $\qquad v_A = \dfrac{14}{12}(4.29) = 5.00$ ft/sec $\qquad$ *Ans.*

$\qquad\qquad v_D = \dfrac{15.23}{12}(4.29) = 5.44$ ft/sec $\qquad$ *Ans.*

in the directions shown.

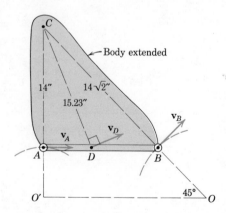

① For the instant depicted we should visualize link AB and its body extended to be rotating as a single unit about point C.

PROBLEMS

6/64 Solve Prob. 6/45 by the method of Art. 6/5.
Ans. $\omega = 11.55$ rad/s CW, $v_G = 1.155$ m/s

6/65 For the instant represented point A of the right-triangular plate has a velocity $v_A = 0.6$ m/s in the direction shown, and the plate has a clockwise angular velocity of 2 rad/s. For this condition find the distance b from A to the instantaneous center of zero velocity and find the velocity of point D.

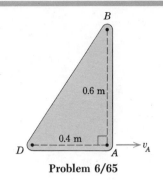

Problem 6/65

6/66 Solve Prob. 6/48 by the method of Art. 6/5.

6/67 End A of the slender pole is given a velocity v_A to the right along the horizontal surface. Show that the magnitude of the velocity of end B equals v_A when the midpoint M of the pole comes in contact with the semicircular obstruction.

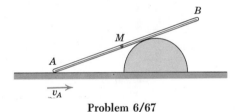

Problem 6/67

6/68 Solve Prob. 6/40 by the method of Art. 6/5.

6/69 The circular disk rolls without slipping on the two plates A and B, which move parallel to each other but in opposite directions. If $v_A = 2$ m/s and $v_B = 4$ m/s, locate the instantaneous center of zero velocity for the disk and determine the velocity of point D at the instant represented.

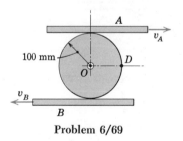

Problem 6/69

6/70 The two gears of 6-in. and 10-in. radius are fastened together and mesh with the two horizontal racks (teeth are not shown). If A has a velocity of 8 ft/sec to the right and B has a velocity of 4 ft/sec to the left, determine the velocity of the center O and of point P in the position shown.
Ans. $v_O = 3.50$ ft/sec, $v_P = 8.28$ ft/sec

6/71 Solve for the velocity of point D in Prob. 6/58 by the method of Art. 6/5.

6/72 The blade of a rotary power mower turns counterclockwise at the angular speed of 1800 rev/min. If the body centrode is a circle of radius 0.75 mm, compute the velocity v_O of the mower.
Ans. $v_O = 0.141$ m/s

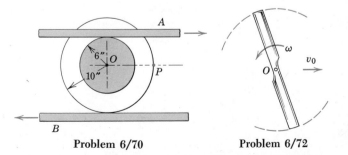

Problem 6/70   **Problem 6/72**

6/73 The rear wheels of a $\frac{1}{2}$-ton pickup truck have a diameter of 28 in. and are slipping on an icy road as torque is applied to them through the differential. If the body centrode for each wheel is a circle of diameter 2 in. and if each wheel is turning at the rate of 300 rev/min, determine the velocity of the truck.

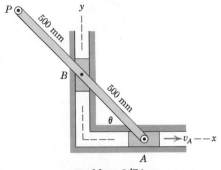

Problem 6/74

6/74 Motion of the bar is controlled by the constrained paths of points A and B. If the angular velocity of the bar is 2 rad/s counterclockwise as the position $\theta = 45°$ is passed, determine the corresponding velocities of points P and A.

6/75 Link OA has a counterclockwise angular velocity $\dot{\theta} = 4$ rad/sec during an interval of its motion. Determine the angular velocity of link AB and of sector BD for $\theta = 45°$ at which instant AB is horizontal and BD is vertical.

$$\textit{Ans. } \omega_{AB} = 1.414 \text{ rad/sec CCW}$$
$$\omega_{BD} = 3.77 \text{ rad/sec CW}$$

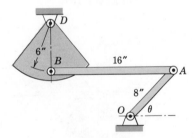

Problem 6/75

6/76 Solve Prob. 6/54 by the method of Art. 6/5.

6/77 Solve Sample Problem 6/8 by the method of Art. 6/5.

6/78 Solve for the velocity of the piston in Sample Problem 6/8 for $\theta = 210°$.   *Ans.* $v_A = 22.4$ ft/sec

6/79 Solve Prob. 6/55 by the method of Art. 6/5.

6/80 Solve Prob. 6/52 by the method of Art. 6/5.
$$\textit{Ans. } v_C = 3.33 \text{ ft/sec, } v_A = 6.05 \text{ ft/sec}$$

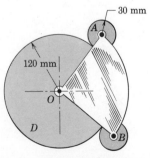

Problem 6/81

6/81 The gear D (teeth not shown) rotates counterclockwise about O with a constant angular velocity of 3 rad/s. The 90° sector AOB is mounted on an independent shaft at O, and each of the small gears at A and B meshes with gear D. If the sector has a clockwise angular velocity of 4 rad/s at the instant represented, determine the corresponding angular velocity ω of each of the small gears.

82 Solve Prob. 6/53 by the method of Art. 6/5. Also find the velocity of point B.

        Ans. $\omega_{AB} = 2.5$ rad/s CCW, $v_B = 1.05$ m/s

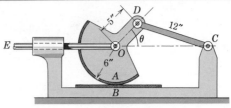

Problem 6/83

83 A device which tests the resistance to wear of two materials A and B is shown. If the link EO has a velocity of 4 ft/sec to the right when $\theta = 45°$, determine the rubbing velocity v_A.

        Ans. $v_A = 9.19$ ft/sec

84 The rectangular body B is pivoted to the crank OA at A and is supported by the wheel at D. If OA has a counterclockwise angular velocity of 2 rad/s, determine the velocity of point E and the angular velocity of body B when the crank OA passes the vertical position shown.

    Ans. $v_E = 0.139$ m/s, $\omega_B = 0.289$ rad/s CW

Problem 6/84

85 Vertical oscillation of the spring-loaded plunger F is controlled by a periodic change in pressure in the vertical hydraulic cylinder E. For the position $\theta = 60°$ determine the angular velocity of AD and the linear velocity of the roller A in its horizontal guide for a downward velocity of 2 m/s of the plunger F.

        Ans. $\omega_{AD} = 13.33$ rad/s
                $v_A = 2.31$ m/s

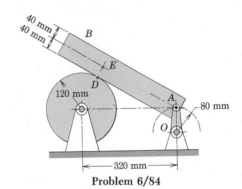

86 The shaft at O drives the arm OA at a clockwise speed of 90 rev/min about the fixed bearing at O. Use the method of the instantaneous center of zero velocity to determine the rotational speed of gear B (gear teeth not shown) if (*a*) ring gear D is fixed and (*b*) ring gear D rotates counterclockwise about O with a speed of 80 rev/min.

        Ans. (*a*) $\omega_B = 360$ rev/min
             (*b*) $\omega_B = 600$ rev/min

Problem 6/86

Problem 6/85

87 The large roller bearing rolls to the left on its outer race with a velocity of its center O of 0.9 m/s. At the same time the central shaft and inner race rotate counterclockwise with an angular speed of 240 rev/min. Determine the angular velocity ω of each of the rollers.   *Ans.* $\omega = 10.73$ rad/s

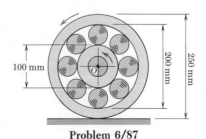

Problem 6/87

6/6 RELATIVE ACCELERATION.   To obtain the relative-acceleration relationship the equation $\mathbf{v}_A = \mathbf{v}_B + \mathbf{v}_{A/B}$ for relative velocities using nonrotating reference axes for two points A and B in plane motion may be differentiated with respect to time to get $\dot{\mathbf{v}}_A = \dot{\mathbf{v}}_B + \dot{\mathbf{v}}_{A/B}$ or

$$\mathbf{a}_A = \mathbf{a}_B + \mathbf{a}_{A/B} \tag{6/7}$$

In words Eq. 6/7 states that the acceleration of point A equals the acceleration of point B plus (vectorially) the acceleration which A appears to have to a nonrotating observer moving with B. If points A and B are located on the same rigid body in the plane of motion, the distance r between them remains constant so that the observer moving with B perceives A to have circular motion about B, as we saw in Art. 6/4 with the relative-velocity relationship. With the relative motion being circular it follows that the relative-acceleration term will have both a normal component directed from A toward B due to the change of direction of $\mathbf{v}_{A/B}$ and a tangential component perpendicular to AB due to the change in magnitude of $\mathbf{v}_{A/B}$. These acceleration components for circular motion, cited in Eqs. 6/2, were covered earlier in Art. 2/5 and should be thoroughly familiar by now. Thus we may write

$$\mathbf{a}_A = \mathbf{a}_B + (\mathbf{a}_{A/B})_n + (\mathbf{a}_{A/B})_t \tag{6/8}$$

where the magnitudes of the relative-acceleration components are

$$
\begin{aligned}
(a_{A/B})_n &= v_{A/B}{}^2/r = r\omega^2 \\
(a_{A/B})_t &= \dot{v}_{A/B} = r\alpha
\end{aligned}
\tag{6/9}
$$

or in vector notation the acceleration components become

$$
\begin{aligned}
(\mathbf{a}_{A/B})_n &= \boldsymbol{\omega} \times (\boldsymbol{\omega} \times \mathbf{r}) \\
(\mathbf{a}_{A/B})_t &= \boldsymbol{\alpha} \times \mathbf{r}
\end{aligned}
\tag{6/9a}
$$

In these relationships $\boldsymbol{\omega}$ is the angular velocity and $\boldsymbol{\alpha}$ is the angular acceleration of the body. The vector locating A from B is $\mathbf{r}$. It is important to observe that the *relative* acceleration terms depends on the respective *absolute* angular velocity and *absolute* angular acceleration.

The meaning of Eqs. 6/8 and 6/9 is illustrated in Fig. 6/9, which shows a rigid body in plane motion with points A and B moving along separate curved paths with absolute accelerations $\mathbf{a}_A$ and $\mathbf{a}_B$. Contrary to the case with velocities the accelerations $\mathbf{a}_A$ and $\mathbf{a}_B$ are, in general, not tangent to the paths described by A and B when these paths are curvilinear. The figure shows the acceleration of A to be composed of two parts, the acceleration of B and the acceleration of A with respect to B. A sketch showing the reference point as fixed is useful in disclosing the correct sense of each of the two components of the relative-acceleration term.

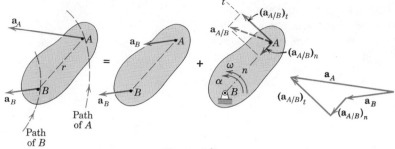

Figure 6/9

Alternatively we may write the acceleration equation in the reverse order, which puts the nonrotating reference axes on *A* rather than *B*. This order gives

$$\mathbf{a}_B = \mathbf{a}_A + \mathbf{a}_{B/A}$$

Here $\mathbf{a}_{B/A}$ and its *n*- and *t*-components are the negatives of $\mathbf{a}_{A/B}$ and its *n*- and *t*-components. The student should make a sketch corresponding to Fig. 6/9 for this inverse sequence of terms.

As in the case of the relative-velocity equation we can handle the solution to Eq. 6/8 in three different ways, namely, by scalar algebra and geometry, by vector algebra, or by graphical construction. The student will be well advised to become familiar with all three techniques. In any event a sketch of the vector polygon representing the vector equation should be made with close attention paid to the head-to-tail combination of vectors which agrees with the equation. Known vectors should be added first, and the unknown vectors will become the closing legs of the vector polygon. It is vital that we visualize the vectors in their geometrical sense, as only then can the full significance of the acceleration equation be perceived. Before attempting a solution it is well to identify the knowns and unknowns and to keep in mind that a solution to a vector equation in two dimensions can be carried out when the unknowns have been reduced to two scalar quantities. These quantities may be the magnitude or direction of any of the terms of the equation. We observe that when both points move on curved paths, there will, in general, be six scalar quantities to account for in Eq. 6/8.

Inasmuch as the normal acceleration components depend on velocities, it is generally necessary to solve for the velocities before the acceleration calculations can be made. The reference point in the relative-acceleration equation is chosen as some point on the body in question whose acceleration is either known or can be easily found. Care must be exercised *not* to use the instantaneous center of zero velocity as the reference point unless its acceleration is known and accounted for. An instantaneous center of zero acceleration exists for a rigid body in general plane motion but will not be discussed here since its use is somewhat specialized.

Sample Problem 6/12

The wheel of radius r rolls to the left without slipping and, at the instant considered, the center O has a velocity $\mathbf{v}_O$ and an acceleration $\mathbf{a}_O$ to the left. Determine the acceleration of points A and C on the wheel for the instant considered.

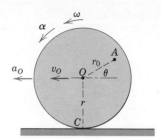

Solution. From our previous analysis of Sample Problem 6/3 we know that the angular velocity and angular acceleration of the wheel are

$$\omega = v_O/r \quad \text{and} \quad \alpha = a_O/r$$

The acceleration of A is written in terms of the given acceleration of O. Thus

$$\mathbf{a}_A = \mathbf{a}_O + \mathbf{a}_{A/O} = \mathbf{a}_O + (\mathbf{a}_{A/O})_n + (\mathbf{a}_{A/O})_t$$

The relative-acceleration terms are viewed as though O were fixed, and for this relative circular motion they have the magnitudes

$$(a_{A/O})_n = r_0 \omega^2 = r_0 \left(\frac{v_O}{r}\right)^2$$

$$(a_{A/O})_t = r_0 \alpha = r_0 \left(\frac{a_O}{r}\right)$$

① and the directions shown.

Adding the vectors head-to-tail gives $\mathbf{a}_A$ as shown. In a numerical problem we may obtain the combination algebraically or graphically. The algebraic expression for the magnitude of $\mathbf{a}_A$ is found from the square root of the sum of the squares of its components. If we use n- and t-directions we have

② $$a_A = \sqrt{(a_A)_n{}^2 + (a_A)_t{}^2} = \sqrt{[a_O \cos \theta + (a_{A/O})_n]^2 + [a_O \sin \theta + (a_{A/O})_t]^2}$$
$$= \sqrt{(r\alpha \cos \theta + r_0 \omega^2)^2 + (r\alpha \sin \theta + r_0 \alpha)^2} \qquad \textit{Ans.}$$

The direction of $\mathbf{a}_A$ can be computed if desired.

The acceleration of the instantaneous center C of zero velocity considered as a point on the wheel is obtained from the expression

$$\mathbf{a}_C = \mathbf{a}_O + \mathbf{a}_{C/O}$$

where the components of the relative-acceleration term are $(a_{C/O})_n = r\omega^2$ directed from C to O and $(a_{C/O})_t = r\alpha$ directed to the right on account of the counterclockwise angular acceleration of line CO about O. The terms are added together in the lower diagram, and it is seen that

③ $$a_C = r\omega^2 \qquad \textit{Ans.}$$

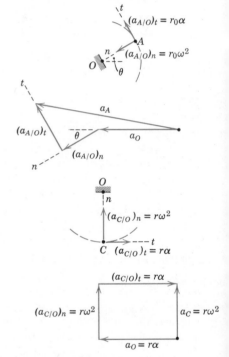

① The counterclockwise angular acceleration α of OA determines the positive direction of $(a_{A/O})_t$. The normal component $(a_{A/O})_n$ is, of course, directed toward the reference center O.

② If the wheel were rolling to the right with the same velocity v_O but still had an acceleration a_O to the left, note that the solution for a_A would be unchanged.

③ We note that the acceleration of the instantaneous center of zero velocity is independent of α and is directed toward the center of the wheel. This conclusion is a useful result to remember.

Sample Problem 6/13

The linkage of Sample Problem 6/7 is repeated here. Crank CB has a constant counterclockwise angular velocity of 2 rad/s in the position shown during a short interval of its motion. Determine the angular acceleration of links AB and OA for this position. Solve by using vector algebra.

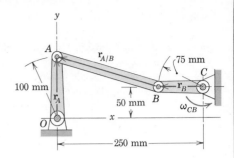

Solution. We first solve for the velocities which were obtained in Sample Problem 6/7. They are

$$\omega_{AB} = -6/7 \text{ rad/s} \quad \text{and} \quad \omega_{OA} = -3/7 \text{ rad/s}$$

where the counterclockwise direction ($+\mathbf{k}$-direction) is taken as positive. The acceleration equation is

$$\mathbf{a}_A = \mathbf{a}_B + (\mathbf{a}_{A/B})_n + (\mathbf{a}_{A/B})_t$$

where, from Eqs. 6/3 and 6/9a, we may write

$$\mathbf{a}_A = \boldsymbol{\alpha}_{OA} \times \mathbf{r}_A + \boldsymbol{\omega}_{OA} \times (\boldsymbol{\omega}_{OA} \times \mathbf{r}_A)$$
$$= \alpha_{OA}\mathbf{k} \times 100\mathbf{j} + (-\tfrac{3}{7}\mathbf{k}) \times (-\tfrac{3}{7}\mathbf{k} \times 100\mathbf{j})$$
$$= -100\,\alpha_{OA}\mathbf{i} - 100(\tfrac{3}{7})^2\mathbf{j} \text{ mm/s}^2$$

$$\mathbf{a}_B = \boldsymbol{\alpha}_{CB} \times \mathbf{r}_B + \boldsymbol{\omega}_{CB} \times (\boldsymbol{\omega}_{CB} \times \mathbf{r}_B)$$
$$= \mathbf{0} + 2\mathbf{k} \times (2\mathbf{k} \times [-75\mathbf{i}])$$
$$= 300\mathbf{i} \text{ mm/s}^2$$

$$(\mathbf{a}_{A/B})_n = \boldsymbol{\omega}_{AB} \times (\boldsymbol{\omega}_{AB} \times \mathbf{r}_{A/B})$$
$$= -\tfrac{6}{7}\mathbf{k} \times [(-\tfrac{6}{7}\mathbf{k}) \times (-175\mathbf{i} + 50\mathbf{j})]$$
$$= (\tfrac{6}{7})^2(175\mathbf{i} - 50\mathbf{j}) \text{ mm/s}^2$$

$$(\mathbf{a}_{A/B})_t = \boldsymbol{\alpha}_{AB} \times \mathbf{r}_{A/B}$$
$$= \alpha_{AB}\mathbf{k} \times (-175\mathbf{i} + 50\mathbf{j})$$
$$= -50\,\alpha_{AB}\mathbf{i} - 175\,\alpha_{AB}\mathbf{j} \text{ mm/s}^2$$

① Remember to preserve the order of the factors in the cross products.

We now substitute these results into the relative-acceleration equation and equate separately the coefficients of the $\mathbf{i}$-terms and the coefficients of the $\mathbf{j}$-terms to give

$$-100\,\alpha_{OA} = 428.6 - 50\,\alpha_{AB}$$
$$-18.37 = -36.73 - 175\,\alpha_{AB}$$

The solutions are

$$\alpha_{AB} = -0.105 \text{ rad/s}^2 \quad \text{and} \quad \alpha_{OA} = -4.34 \text{ rad/s}^2 \quad \textit{Ans.}$$

Since the unit vector $\mathbf{k}$ points out from the paper in the positive z-direction, we see that the angular accelerations of AB and OA are both clockwise (negative).

It is recommended that the student sketch each of the acceleration vectors in its proper geometric relationship according to the relative-acceleration equation to help clarify the meaning of the solution.

Sample Problem 6/14

The slider-crank mechanism of Sample Problem 6/8 is repeated here. The crank OB has a constant clockwise angular speed of 1500 rev/min. For the instant when the crank angle θ is $60°$ determine the acceleration of the piston A and the angular acceleration of the connecting rod AB.

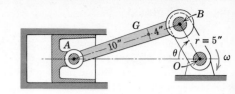

Solution. The acceleration of A may be expressed in terms of the acceleration of the crank pin B. Thus

$$\mathbf{a}_A = \mathbf{a}_B + (\mathbf{a}_{A/B})_n + (\mathbf{a}_{A/B})_t$$

Point B moves in a circle of 5-in. radius with a constant speed so that it has only a normal component of acceleration directed from B to O.

$$[a_n = r\omega^2] \qquad a_B = \frac{5}{12}\left(\frac{1500[2\pi]}{60}\right)^2 = 10,280 \text{ ft/sec}^2$$

The relative-acceleration terms are visualized with A rotating in a circle relative to B, which is considered fixed, as shown. From Sample Problem 6/8 the angular velocity of AB for these same conditions is $\omega_{AB} = 29.5$ rad/sec so that

$$[a_n = r\omega^2] \qquad (a_{A/B})_n = \frac{14}{12}(29.5)^2 = 1015 \text{ ft/sec}^2$$

directed from A to B. The tangential component $(\mathbf{a}_{A/B})_t$ is known in direction only since its magnitude depends on the unknown angular acceleration of AB. We also know the direction of $\mathbf{a}_A$ since the piston is confined to move along the horizontal axis of the cylinder. There are now only two remaining unknowns left in the equation, namely, the magnitudes of $\mathbf{a}_A$ and $(\mathbf{a}_{A/B})_t$, so that the solution may be carried out.

If we adopt an algebraic solution using the geometry of the acceleration polygon, we first compute the angle between AB and the horizontal. With the law of sines this angle becomes $18.0°$. Equating separately the horizontal components and the vertical components of the terms in the acceleration equation, as seen from the acceleration polygon, gives

$$a_A = 10{,}280 \cos 60° + 1015 \cos 18.0° - (a_{A/B})_t \sin 18.0°$$

$$0 = 10{,}280 \sin 60° - 1015 \sin 18.0° - (a_{A/B})_t \cos 18.0°$$

The solution to these equations gives the magnitudes

$$(a_{A/B})_t = 9030 \text{ ft/sec}^2 \qquad \text{and} \qquad a_A = 3310 \text{ ft/sec}^2 \qquad \textit{Ans.}$$

With the sense of $(\mathbf{a}_{A/B})_t$ also determined from the diagram, the angular acceleration of AB is seen from the figure representing rotation relative to B to be

$$[\alpha = a_t/r] \qquad \alpha_{AB} = 9030/(14/12) = 7740 \text{ rad/sec}^2 \text{ clockwise} \qquad \textit{Ans.}$$

If we adopt a graphical solution, we begin with the known vectors $\mathbf{a}_B$ and $(\mathbf{a}_{A/B})_n$ and add them head-to-tail using a convenient scale. Next we construct the direction of $(\mathbf{a}_{A/B})_t$ through the head of the last vector. The solution of the equation is obtained by the intersection P of this last line with a horizontal line through the starting point representing the known direction of the vector sum $\mathbf{a}_A$. Scaling the magnitudes from the diagram gives values which agree with the calculated results of

$$a_A = 3310 \text{ ft/sec}^2 \qquad \text{and} \qquad (a_{A/B})_t = 9030 \text{ ft/sec}^2 \qquad \textit{Ans.}$$

① If the crank OB had an angular acceleration, $\mathbf{a}_B$ would also have a tangential component of acceleration.

② Alternatively the relation $a_n = v^2/r$ may be used for calculating $(a_{A/B})_n$ provided the relative velocity $v_{A/B}$ is used for v. The equivalence is easily seen when it is recalled that $v_{A/B} = r\omega$.

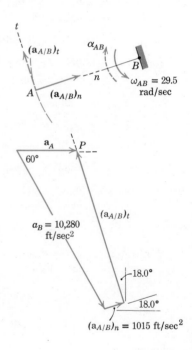

③ Except where extreme accuracy is required do not hesitate to use a graphical solution, as it is quick and reveals the physical relationships among the vectors. The known vectors, of course, may be added in any order as long as the governing equation is satisfied.

PROBLEMS

88 The blade of 2-ft radius rotates with a constant angular velocity ω about the shaft at O which is mounted in the sliding block. If the block has an acceleration $a_O = 6$ ft/sec^2 and the magnitude of the acceleration of the blade tip P is 10 ft/sec^2 in the vertical position shown, determine ω. (Can you determine the sense of rotation from the given information?)   *Ans.* $\omega = 2$ rad/sec, No

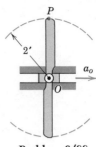

Problem 6/88

89 Assume that the rotating blade of Prob. 6/88 no longer has a constant angular velocity. If the bearing block O still has an acceleration $a_O = 6$ ft/sec^2 to the right, what must be the angular acceleration of the blade if the total acceleration of the tip P has no horizontal component for the position shown?

90 A car with tires of 700-mm diameter accelerates at a constant rate from rest to a velocity of 50 km/h in a distance of 25 m. Determine the magnitude of the acceleration of a point A on the top of the wheel as the car reaches a speed of 10 km/h.
              Ans. $a_A = 23.4$ m/s^2

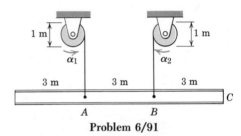

Problem 6/91

91 The 9-m steel beam is being hoisted from its horizontal position by the two cables attached at A and B. If the initial angular accelerations of the hoisting drums are $\alpha_1 = 0.2$ rad/s^2 and $\alpha_2 = 0.5$ rad/s^2 in the directions shown, determine the corresponding angular acceleration α of the beam, the acceleration of C, and the distance b from A to a point P on the beam center line that has no acceleration.
              Ans. $\alpha = 0.05$ rad/s^2 CCW, $a_C = 0.4$ m/s^2
                  $b = 2$ m left of A

92 The sheave with a diameter of 30 in. is elevated by the cable as shown. If the block B has a constant upward acceleration of 6 ft/sec^2, determine the acceleration of a point A which is on the bottom of the sheave at the time when B reaches an upward velocity of 4 ft/sec.

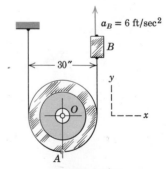

Problem 6/92

93 The circular disk rolls to the left without slipping. If $\mathbf{a}_{A/B} = -2.7\mathbf{j}$ m/s^2, determine the velocity and acceleration of the center O of the disk.
              Ans. $\mathbf{v}_O = -0.6\mathbf{i}$ m/s, $\mathbf{a}_O = -1.8\mathbf{i}$ m/s^2

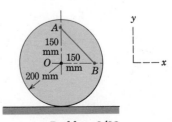

Problem 6/93

Problem 6/94

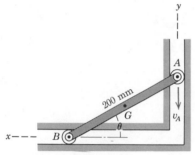

Problem 6/95

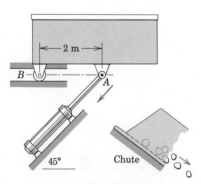

Problem 6/98

6/94 The bicycle of Prob. 6/44 is repeated here. If the cyclist starts from rest and acquires a velocity of 2 m/s in a distance of 3 m with constant acceleration, compute the magnitude of the acceleration of pedal A for this condition if it is in the top position.

6/95 For the link of Prob. 6/45 shown again here determine the acceleration of end B and the angular acceleration of AB when $\theta = 30°$ if end A has a constant downward velocity of 2 m/s.
Ans. $\mathbf{a}_B = -30.8\mathbf{i}$ m/s^2, $\boldsymbol{\alpha}_{AB} = -77.0\mathbf{k}$ rad/s^2

6/96 Use the information and answers cited for Prob. 6/95 and determine the acceleration of the mass center G of link AB.   *Ans.* $\mathbf{a}_G = -15.40\mathbf{i}$ m/s^2

6/97 Determine the acceleration of the piston of Sample Problem 6/14 for (*a*) $\theta = 0°$, (*b*) $\theta = 90°$, and (*c*) $\theta = 180°$. Take the positive x-direction to the right.
Ans. (*a*) $\mathbf{a}_A = 13{,}950\mathbf{i}$ ft/sec^2
(*b*) $\mathbf{a}_A = -3930\mathbf{i}$ ft/sec^2
(*c*) $\mathbf{a}_A = -6610\mathbf{i}$ ft/sec^2

6/98 A container for waste materials is dumped by the hydraulic-actuated linkage shown. If the piston rod starts from rest in the position indicated and has an acceleration of 0.5 m/s^2 in the direction shown, compute the initial angular acceleration of the container.

6/99 Assume that the piston rod of the dumping device of Prob. 6/98 has a velocity of 300 mm/s in the direction of the arrow as point A passes the horizontal through B. If A has an acceleration of 40 mm/s^2 in the same direction, determine the corresponding angular acceleration of the container.
Ans. $\alpha = 14.1(10^{-3})$ rad/s^2 CW

100 The slider at *B* has a constant horizontal velocity v_B to the left as link *AB* passes the vertical position and link *AO* becomes horizontal. For this instant determine the angular acceleration of *AO*.

101 Calculate the angular acceleration of the plate in the position shown where control link *AO* has a constant angular velocity $\omega_{OA} = 4$ rad/sec and $\theta = 60°$ for both links.

Ans. $\alpha_{AB} = 15.40$ rad/sec² CW

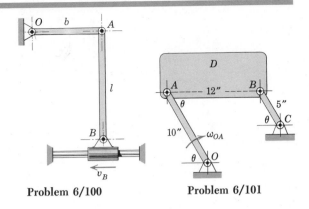

Problem 6/100   **Problem 6/101**

102 The center *O* of the spool starts from rest and acquires a velocity of 1.2 m/s up the incline with constant acceleration in a distance *s* of 2.4 m under the action of a steady pull *P* applied to point *A* of the cable. The cable is wrapped securely around the hub, and the wheel rolls without slipping. For the 2.4-m position shown, calculate the acceleration of point *A* on the cable and point *B* on the spool.

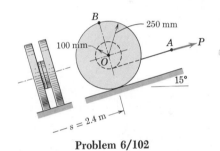

Problem 6/102

103 The two connected wheels of Prob. 6/58 are shown again here. Determine the magnitude of the acceleration of point *D* in the position shown if the center *C* of the smaller wheel has an acceleration to the right of 0.8 m/s² and has reached a velocity of 0.4 m/s at this instant.   *Ans.* $a_D = 1.388$ m/s²

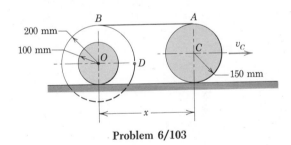

Problem 6/103

104 The two rigidly connected gears and two racks (teeth not shown) of Prob. 6/70 are repeated here. If rack *A* has an acceleration of 5.6 ft/sec² to the right and rack *B* has an acceleration of 4 ft/sec² to the left as they both start from rest, compute the magnitude of the acceleration of point *P* in the position shown.

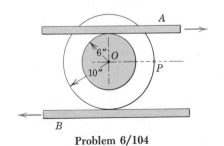

Problem 6/104

6/105 For Prob. 6/104 if, in addition to the specified accelerations, rack A has a velocity of 1.5 ft/sec to the right and rack B has a velocity of 2.5 ft/sec to the left, determine the magnitude of the acceleration of P.   *Ans.* $a_P = 8.14$ ft/sec^2

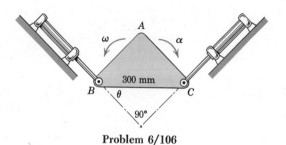

Problem 6/106

6/106 At the instant represented $\theta = 45°$, and the triangular plate ABC has a counterclockwise angular velocity of 20 rad/s and a clockwise angular acceleration of 100 rad/s^2. Determine the magnitudes of the corresponding velocity v and acceleration a of the piston rod of the hydraulic cylinder attached to C.

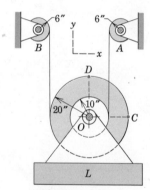

Problem 6/107

6/107 The load L is lowered by the two sheaves which are fastened together. For the instant shown drum A has a counterclockwise angular velocity of 4 rad/sec which is decreasing at the rate of 4 rad/sec each second. Simultaneously drum B has a clockwise angular velocity of 6 rad/sec which is increasing at the rate of 2 rad/sec each second. Calculate the acceleration of points C and D and the load L.
    Ans. $\mathbf{a}_C = 3\mathbf{j} - 0.267\mathbf{i}$ ft/sec^2
    $\mathbf{a}_D = 0.733\mathbf{j} - 2\mathbf{i}$ ft/sec^2
    $\mathbf{a}_L = 1.0\mathbf{j}$ ft/sec^2

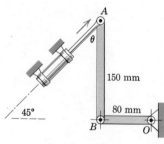

Problem 6/108

6/108 The piston rod has a constant velocity of 80 mm/s in the direction indicated, and at the instant represented $\theta = 45°$ and BO is horizontal. Compute the magnitude of the acceleration of B for this position.
    Ans. $a_B = 45.3$ mm/s^2

109 Use the information given in the solution to Sample Problem 6/14 and determine the acceleration of the mass center G of the connecting rod for the position described. Solve graphically.

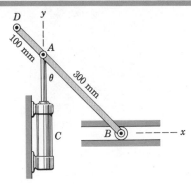

Problem 6/110

110 If the piston rod of the hydraulic cylinder C has a constant upward velocity of 0.5 m/s, calculate the acceleration of point D for the position where θ is 45°.

111 The slider-crank linkage of Prob. 6/54 is repeated here. If OB has a constant counterclockwise angular velocity of 4 rad/s, determine the acceleration of A when OB reaches the position $\theta = \tan^{-1}\frac{4}{3}$. Solve by vector algebra.

$$\text{Ans. } \mathbf{a}_A = -2.29 \text{ m/s}^2$$

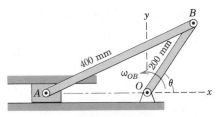

Problem 6/111

112 The linkage of Prob. 6/53 is shown again here. If OA has a constant counterclockwise angular velocity $\omega_0 = 10$ rad/s, calculate the angular acceleration of link AB for the position where the coordinates of A are $x = -60$ mm and $y = 80$ mm. Link BC is vertical for this position. Solve by vector algebra. (Use the results of Prob. 6/53 for the angular velocity of AB.)

$$\text{Ans. } \boldsymbol{\alpha}_{AB} = 10.42\mathbf{k} \text{ rad/s}^2, \ \boldsymbol{\alpha}_{BC} = -19.21\mathbf{k} \text{ rad/s}^2$$

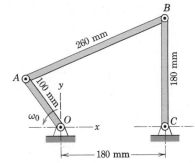

Problem 6/112

113 The elements of a power hacksaw are shown in the figure. The saw blade is mounted in a frame which slides along the horizontal guide. If the motor turns the flywheel at a constant counterclockwise speed of 60 rev/min, determine the acceleration of the blade for the position where $\theta = 90°$, and find the corresponding angular acceleration of the link AB.

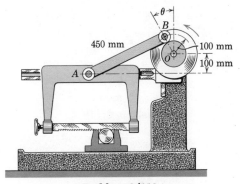

Problem 6/113

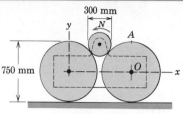

Problem 6/114

6/114 The small vehicle is driven by the 300-mm friction wheel, which acquires a rotational speed $N = 150$ rev/min while the vehicle moves through a distance of 6 m from rest. Calculate the constant acceleration of the vehicle during the motion and find the acceleration, expressed in vector form, of point A on the top of the front wheel as the vehicle reaches the 6-m position. No slipping occurs at any of the rolling surfaces.

$$Ans. \ a_0 = 0.463 \ \text{m/s}^2$$
$$\mathbf{a}_A = 0.925\mathbf{i} - 14.80\mathbf{j} \ \text{m/s}^2$$

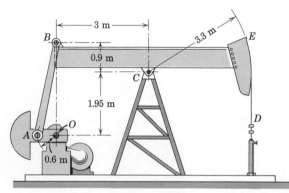

Problem 6/115

▶6/115 An oil pumping rig is shown in the figure. The flexible pump rod D is fastened to the sector at E and is always vertical as it enters the fitting below D. The link AB causes the beam BCE to oscillate as the weighted crank OA revolves. If OA has a constant clockwise speed of 1 rev every 3 s, determine the acceleration of the pump rod D when the beam and the crank OA are both in the horizontal position shown. $Ans. \ a_D = 0.568 \ \text{m/s}^2$ down

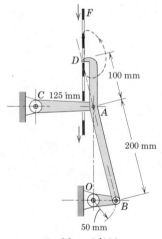

Problem 6/116

▶6/116 An intermittent-drive mechanism for perforated tape F consists of the link DAB driven by the crank OB. The trace of the motion of the finger at D is shown by the dotted line. Determine the acceleration of D at the instant shown when both OB and CA are horizontal if OB has a constant clockwise rotational velocity of 120 rev/min.

$$Ans. \ a_D = 1997 \ \text{mm/s}^2$$

/7 MOTION RELATIVE TO ROTATING AXES.   In our discussion of
the relative motion of particles in Art. 2/8 and in our use of the
relative-motion equations for the plane motion of rigid bodies in this
present chapter we have made all of our relative-velocity and relative-acceleration measurements from *nonrotating* reference axes. The
solution of many problems in kinematics where motion is generated
within or observed from a system which itself is rotating is greatly
facilitated by the use of rotating reference axes. An example of such a
motion would be the movement of a fluid particle along the curved
vane of a centrifugal pump where the path relative to the vanes of
the impeller becomes an important design consideration. A second
example is the measured motion of a rocket or an artificial satellite
which differs from the absolute motion on account of the rotating
frame of reference of the earth.

We begin the description of motion using rotating axes by considering the plane motion of two particles A and B in the fixed X-Y
plane, Fig. 6/10a. For the time being we will consider A and B to be
moving independently of one another for the sake of generality. The
motion of A will now be observed from a moving reference frame x-y
which has its origin attached to B and which rotates with an angular
velocity $\omega = \dot{\theta}$. We may write this angular velocity as the vector
$\boldsymbol{\omega} = \mathbf{k}\omega = \mathbf{k}\dot{\theta}$ where the vector is normal to the plane of motion and
where its positive sense is in the positive z-direction (out from the
paper) as established by the right-hand rule. The absolute position
vector of A is given by

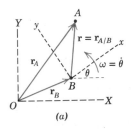

(a)

$$\mathbf{r}_A = \mathbf{r}_B + \mathbf{r} = \mathbf{r}_B + (\mathbf{i}x + \mathbf{j}y)$$

where $\mathbf{i}$ and $\mathbf{j}$ are unit vectors in the x-y frame and $\mathbf{r} = \mathbf{i}x + \mathbf{j}y$ stands
for $\mathbf{r}_{A/B}$, the position vector of A with respect to B. The velocity and
acceleration equations require successive differentiation of the position-vector equation with respect to time. In contrast to the case of
translating axes treated in Art 2/8, the unit vectors $\mathbf{i}$ and $\mathbf{j}$ are now
rotating with the x-y axes and, hence, have time derivatives which
must be evaluated. These derivatives may be seen from Fig. 6/10b
which shows the infinitesimal change in each unit vector during time
dt when the reference axes rotate through an angle $d\theta = \omega\,dt$. The
differential change in $\mathbf{i}$ is $d\mathbf{i}$, and it has the direction of $\mathbf{j}$ and a
magnitude equal to the angle $d\theta$ times the length of the vector $\mathbf{i}$,
which is unity. Thus $d\mathbf{i} = \mathbf{j}\,d\theta$. Similarly the unit vector $\mathbf{j}$ has an
infinitesimal change $d\mathbf{j}$ which points in the negative x-direction, so
that $d\mathbf{j} = -\mathbf{i}\,d\theta$. Dividing by dt and replacing $d\mathbf{i}/dt$ by $\dot{\mathbf{i}}$, $d\mathbf{j}/dt$ by $\dot{\mathbf{j}}$,
and $d\theta/dt$ by $\dot{\theta} = \omega$ gives us

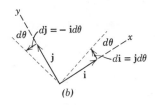

(b)

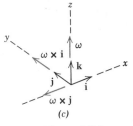

(c)

Figure 6/10

$$\dot{\mathbf{i}} = \mathbf{j}\omega \qquad \text{and} \qquad \dot{\mathbf{j}} = -\mathbf{i}\omega$$

When the cross product is introduced, we see from Fig. 6/10c that
$\boldsymbol{\omega} \times \mathbf{i} = \mathbf{j}\omega$ and $\boldsymbol{\omega} \times \mathbf{j} = -\mathbf{i}\omega$. Therefore the time derivatives of the

unit vectors may be written as

$$\dot{\mathbf{i}} = \boldsymbol{\omega} \times \mathbf{i} \qquad \text{and} \qquad \dot{\mathbf{j}} = \boldsymbol{\omega} \times \mathbf{j} \tag{6/10}$$

(a) Relative Velocity. We will now use the expressions of Eqs. 6/10 when we take the time derivative of the position-vector equation for A and B to obtain the relative-velocity relation. Thus

$$\dot{\mathbf{r}}_A = \dot{\mathbf{r}}_B + \frac{d}{dt}(\mathbf{i}x + \mathbf{j}y)$$

$$= \dot{\mathbf{r}}_B + (\dot{\mathbf{i}}x + \dot{\mathbf{j}}y) + (\mathbf{i}\dot{x} + \mathbf{j}\dot{y})$$

But $\dot{\mathbf{i}}x + \dot{\mathbf{j}}y = \boldsymbol{\omega} \times \mathbf{i}x + \boldsymbol{\omega} \times \mathbf{j}y = \boldsymbol{\omega} \times \mathbf{r}$. Also, since the observer in x-y measures velocity components $\dot{x}$ and $\dot{y}$, we see that $\mathbf{i}\dot{x} + \mathbf{j}\dot{y} = \mathbf{v}_{\text{rel}}$ which is the velocity relative to the x-y frame of reference. Thus the relative-velocity equation becomes

$$\mathbf{v}_A = \mathbf{v}_B + \boldsymbol{\omega} \times \mathbf{r} + \mathbf{v}_{\text{rel}} \tag{6/11}$$

Comparison of Eq. 6/11 with Eq. 2/20 for nonrotating reference axes shows that $\mathbf{v}_{A/B} = \boldsymbol{\omega} \times \mathbf{r} + \mathbf{v}_{\text{rel}}$ from which we conclude that the term $\boldsymbol{\omega} \times \mathbf{r}$ is the difference between the relative velocity as measured from nonrotating and rotating axes.

To illustrate further the meaning of the last two terms in Eq. 6/11, the motion of particle A relative to the x-y plane is shown in Fig. 6/11 as taking place in a curved slot in a plate which represents the rotating x-y reference system. The velocity of A as measured relative to the plate, $\mathbf{v}_{\text{rel}}$, would be tangent to the path fixed in the x-y plate and would have a magnitude $\dot{s}$ where s is measured along the path. This relative velocity may also be viewed as the velocity $\mathbf{v}_{A/P}$ relative to a point P attached to the plate and coincident with A at the instant under consideration. The term $\boldsymbol{\omega} \times \mathbf{r}$ has a magnitude $r\dot{\theta}$ and a direction normal to $\mathbf{r}$ and is the velocity relative to B of point P as seen from nonrotating axes attached to B.

The following comparison will help to establish the equivalence and clarify the difference between the relative-velocity equations written for rotating and nonrotating reference axes.

$$\mathbf{v}_A = \mathbf{v}_B + \boldsymbol{\omega} \times \mathbf{r} + \mathbf{v}_{\text{rel}}$$

$$\mathbf{v}_A = \underbrace{\mathbf{v}_B + \mathbf{v}_{P/B}} + \mathbf{v}_{A/P}$$

$$\mathbf{v}_A = \mathbf{v}_P \underbrace{\qquad\quad} + \mathbf{v}_{A/P} \tag{6/11a}$$

$$\mathbf{v}_A = \mathbf{v}_B + \mathbf{v}_{A/B}$$

In the second equation the term $\mathbf{v}_{P/B}$ is measured from a nonrotating position, otherwise it would be zero. The term $\mathbf{v}_{A/P}$ is the same as $\mathbf{v}_{\text{rel}}$ and is the velocity of A as measured in the x-y frame. In the third equation $\mathbf{v}_P$ is the absolute velocity of P and represents the effect of

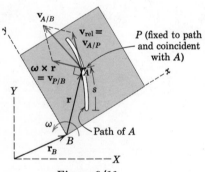

Figure 6/11

the moving coordinate system, both translational and rotational. The fourth equation is the same as that developed for nonrotating axes, Eq. 2/20, and it is seen that $v_{A/B} = v_{P/B} + v_{A/P} = \omega \times r + v_{rel}$.

(b) *Transformation of a Time Derivative.* Equation 6/11 represents a transformation of the time derivative of the position vector between rotating and nonrotating axes. We may easily generalize this result to apply to the time derivative of any vector quantity $\mathbf{V} = \mathbf{i}V_x + \mathbf{j}V_y$. Accordingly, the total time derivative as taken in X-Y is

$$\left(\frac{d\mathbf{V}}{dt}\right)_{XY} = (\mathbf{i}\dot{V}_x + \mathbf{j}\dot{V}_y) + (\dot{\mathbf{i}}V_x + \dot{\mathbf{j}}V_y)$$

Replacing $\mathbf{r}$ in the previous analysis by $\mathbf{V}$ gives us

$$\left(\frac{d\mathbf{V}}{dt}\right)_{XY} = \left(\frac{d\mathbf{V}}{dt}\right)_{xy} + \boldsymbol{\omega} \times \mathbf{V} \qquad (6/12)$$

Here $\boldsymbol{\omega} \times \mathbf{V}$ represents the difference between the time derivative of the vector in a fixed reference and its time derivative in the rotating reference. As we shall see in Art. 8/2 where three-dimensional motion is introduced, Eq. 6/12 is valid in three dimensions as well as in two dimensions.

The physical significance of Eq. 6/12 is illustrated in Fig. 6/12, which shows the vector $\mathbf{V}$ at time t as observed both in the fixed axes X-Y and in the rotating axes x-y. Since we are dealing with the effects of rotation only, the vector may be shown through the origin of coordinates without loss of generality. During time dt the vector swings to position V', and the observer in x-y measures the components dV due to its change in magnitude and $V\,d\beta$ due to its rotation $d\beta$ relative to x-y. To the rotating observer, then, the derivative $(d\mathbf{V}/dt)_{xy}$ which he measures has the components dV/dt and $V\,d\beta/dt = V\dot{\beta}$. The remaining part of the total time derivative not measured by the rotating observer has the magnitude $V\,d\theta/dt$ and, expressed as a vector, is $\boldsymbol{\omega} \times \mathbf{V}$. Thus we see physically from the diagram that

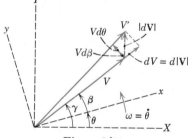

Figure 6/12

$$(\dot{\mathbf{V}})_{XY} = (\dot{\mathbf{V}})_{xy} + \boldsymbol{\omega} \times \mathbf{V}$$

which is Eq. 6/12.

(c) *Relative Acceleration.* The relative-acceleration equation may be obtained by differentiating the relative-velocity relation, Eq. 6/11. Thus

$$\mathbf{a}_A = \mathbf{a}_B + \dot{\boldsymbol{\omega}} \times \mathbf{r} + \boldsymbol{\omega} \times \dot{\mathbf{r}} + \dot{\mathbf{v}}_{rel}$$

From the derivation of Eq. 6/11 we see that the third term on the right of the acceleration equation becomes

$$\boldsymbol{\omega} \times \dot{\mathbf{r}} = \boldsymbol{\omega} \times \frac{d}{dt}(\mathbf{i}x + \mathbf{j}y) = \boldsymbol{\omega} \times (\boldsymbol{\omega} \times \mathbf{r}) + \boldsymbol{\omega} \times \mathbf{v}_{rel}$$

The last term on the right of the equation for $\mathbf{a}_A$ is

$$\dot{\mathbf{v}}_{\text{rel}} = \frac{d}{dt}(\mathbf{i}\dot{x} + \mathbf{j}\dot{y}) = (\dot{\mathbf{i}}\dot{x} + \dot{\mathbf{j}}\dot{y}) + (\mathbf{i}\ddot{x} + \mathbf{j}\ddot{y})$$

$$= \boldsymbol{\omega} \times (\mathbf{i}\dot{x} + \mathbf{j}\dot{y}) + (\mathbf{i}\ddot{x} + \mathbf{j}\ddot{y})$$

$$= \boldsymbol{\omega} \times \mathbf{v}_{\text{rel}} + \mathbf{a}_{\text{rel}}$$

Substitution into the expression for $\mathbf{a}_A$ and collection of terms give

$$\mathbf{a}_A = \mathbf{a}_B + \dot{\boldsymbol{\omega}} \times \mathbf{r} + \boldsymbol{\omega} \times (\boldsymbol{\omega} \times \mathbf{r}) + 2\boldsymbol{\omega} \times \mathbf{v}_{\text{rel}} + \mathbf{a}_{\text{rel}} \qquad (6/13)$$

Equation 6/13 is the general vector expression for the absolute acceleration of a particle A in terms of its acceleration $\mathbf{a}_{\text{rel}}$ measured relative to a moving coordinate system which rotates with an angular velocity $\boldsymbol{\omega}$. The terms $\dot{\boldsymbol{\omega}} \times \mathbf{r}$ and $\boldsymbol{\omega} \times (\boldsymbol{\omega} \times \mathbf{r})$ are shown in Fig. 6/13. They represent, respectively, the t- and n-components of the acceleration $\mathbf{a}_{P/B}$ of the coincident point P in its circular motion with respect to B. This motion would be observed from a set of nonrotating axes moving with B. The magnitude of $\dot{\boldsymbol{\omega}} \times \mathbf{r}$ is $r\ddot{\theta}$ and its direction is tangent to the circle. The magnitude of $\boldsymbol{\omega} \times (\boldsymbol{\omega} \times \mathbf{r})$ is $r\omega^2$ and its direction is from P to B along the normal to the circle. The acceleration of A relative to the path, $\mathbf{a}_{\text{rel}}$, may be expressed in rectangular, normal and tangential, or polar coordinates in the rotating system. Generally n- and t-components are most frequently used, and these components are depicted in Fig. 6/13. The tangential component would have the magnitude $(a_{\text{rel}})_t = \ddot{s}$ where s is the distance measured along the path to A. The normal component would have the magnitude v_{rel}^2/ρ where ρ is the radius of curvature of the path as measured in x-y. The sense of this vector would always be toward the center of curvature.

The term $2\boldsymbol{\omega} \times \mathbf{v}_{\text{rel}}$, shown in Fig. 6/13, is known as the *Coriolis* acceleration° and represents the difference between the acceleration of A relative to P as measured from nonrotating axes and from rotating axes. The direction is always normal to the term $\mathbf{v}_{\text{rel}}$, and the sense is established by the right-hand rule for the cross product.

The Coriolis acceleration is difficult to visualize because it is composed of two separate physical effects. To help with this visualization, we will consider the simplest possible motion in which this term appears. In Fig. 6/14a we have a rotating disk with a radial slot in which a small particle A is confined to slide. Let the disk turn with a constant angular velocity $\omega = \dot{\theta}$ and let the particle move along the slot with a constant speed $v_{\text{rel}} = \dot{x}$ relative to the slot. The velocity of A has the components $\dot{x}$ due to motion along the slot and $x\omega$ due to the rotation of the slot. The changes in these two velocity components

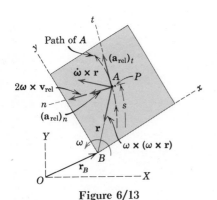

Figure 6/13

° Named after a French military engineer G. Coriolis (1792–1843) who was the first to call attention to this term.

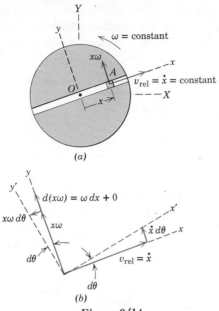

Figure 6/14

due to the rotation of the disk are shown in the *b*-part of the figure for the interval dt during which the x-y axes rotate with the disk through the angle $d\theta$ to x'-y'. The velocity increment due to the change in direction of $\mathbf{v}_{\text{rel}}$ is $\dot{x}\,d\theta$ and that due to the change in magnitude of $x\omega$ is $\omega\,dx$, both being in the y-direction normal to the slot. Dividing each increment by dt and adding gives the sum $\omega\dot{x} + \dot{x}\omega = 2\dot{x}\omega$, which is the magnitude of the Coriolis acceleration $2\boldsymbol{\omega} \times \mathbf{v}_{\text{rel}}$. Dividing the remaining velocity increment due to the change in direction of $x\omega$ by dt gives $x\omega\dot{\theta}$ or $x\omega^2$, which is the acceleration of a point P fixed to the slot and momentarily coincident with the particle A. We now see how Eq. 6/13 fits these results. With B taken as our origin, $\mathbf{a}_B = 0$. With constant angular velocity, $\dot{\boldsymbol{\omega}} \times \mathbf{r} = 0$. With $\mathbf{v}_{\text{rel}}$ constant in magnitude and no curvature to the slot, $\mathbf{a}_{\text{rel}} = 0$. We are left with

$$\mathbf{a}_A = \boldsymbol{\omega} \times (\boldsymbol{\omega} \times \mathbf{r}) + 2\boldsymbol{\omega} \times \mathbf{v}_{\text{rel}}$$

Replacing $\mathbf{r}$ by $\mathbf{i}x$, $\boldsymbol{\omega}$ by $\omega\mathbf{k}$, and $\mathbf{v}_{\text{rel}}$ by $\mathbf{i}\dot{x}$ gives

$$\mathbf{a}_A = -x\omega^2\mathbf{i} + 2\dot{x}\omega\mathbf{j}$$

which checks our analysis from Fig. 6/14. We also note that this same result is contained in our polar-coordinate analysis of plane curvilinear motion in Eq. 2/14 when we let $\ddot{r} = 0$ and $\ddot{\theta} = 0$ and replace r by x and $\dot{\theta}$ by ω. If the slot in the disk of Fig. 6/14 had been curved, we would have had a normal component of acceleration relative to the slot so that $\mathbf{a}_{\text{rel}}$ would not be zero.

The following comparison will help to establish the equivalence and clarify the difference between the relative-acceleration equations written for rotating and nonrotating reference axes.

$$
\begin{aligned}
\mathbf{a}_A &= \mathbf{a}_B + \underbrace{\dot{\boldsymbol{\omega}} \times \mathbf{r} + \boldsymbol{\omega} \times (\boldsymbol{\omega} \times \mathbf{r})}_{} + \underbrace{2\boldsymbol{\omega} \times \mathbf{v}_{\mathrm{rel}} + \mathbf{a}_{\mathrm{rel}}}_{} \\
\mathbf{a}_A &= \underbrace{\mathbf{a}_B + \underbrace{\mathbf{a}_{P/B}}_{}}_{} + \mathbf{a}_{A/P} \\
\mathbf{a}_A &= \underbrace{\mathbf{a}_P}_{} + \underbrace{\mathbf{a}_{A/P}}_{} \\
\mathbf{a}_A &= \mathbf{a}_B + \underbrace{\mathbf{a}_{A/B}}_{}
\end{aligned}
\tag{6/13a}
$$

The equivalence of $\mathbf{a}_{P/B}$ and $\dot{\boldsymbol{\omega}} \times \mathbf{r} + \boldsymbol{\omega} \times (\boldsymbol{\omega} \times \mathbf{r})$, as shown in the second equation, has already been described. From the third equation where $\mathbf{a}_B + \mathbf{a}_{P/B}$ has been combined to give $\mathbf{a}_P$, it is seen that the relative-acceleration term $\mathbf{a}_{A/P}$, unlike the corresponding relative-velocity term, is not equal to the relative acceleration $\mathbf{a}_{\mathrm{rel}}$ measured from the rotating x-y frame of reference. The Coriolis term is, therefore, the difference between the acceleration $\mathbf{a}_{A/P}$ of A relative to P as measured in a nonrotating system and the acceleration $\mathbf{a}_{\mathrm{rel}}$ of A relative to P as measured in a rotating system. From the fourth equation it is seen that the acceleration $\mathbf{a}_{A/B}$ of A with respect to B as measured in a nonrotating system, Eq. 2/21, is a combination of the last four terms in the first equation for the rotating system.

In the analysis of acceleration using a rotating frame of reference we frequently find it convenient to take the origin of the reference coordinates at the point P, coincident with the position of the particle A at the instant under observation. This choice eliminates the terms $\dot{\boldsymbol{\omega}} \times \mathbf{r}$ and $\boldsymbol{\omega} \times (\boldsymbol{\omega} \times \mathbf{r})$ since the vector $\mathbf{r}$ vanishes. Hence, from the foregoing discussion and comparisons, we see that the relative-acceleration equation may be written simply as

$$
\boxed{\mathbf{a}_A = \mathbf{a}_P + 2\boldsymbol{\omega} \times \mathbf{v}_{\mathrm{rel}} + \mathbf{a}_{\mathrm{rel}}}
\tag{6/13b}
$$

When this form is used, we must note that the point P may not be picked at random since it is the one point in the rotating reference system coincident with the particle A at the instant of analysis.

It is important to observe that the vector notation employed depends on the consistent use of a *right-handed* set of coordinate axes. Before the student endeavors to use Eqs. 6/11 or 6/13 it is important that he study the derivations carefully and understand the physical interpretation of each of the terms prior to attempting the problem work. We note that Eqs. 6/11 and 6/13, developed here for plane motion, also hold for space motion, and this extension of generality will be covered in Art. 8/6 on relative motion in space.

Sample Problem 6/15

The pin A of the hinged link AC is confined to move in the rotating slot of link BO. The angular velocity of BO is $\omega = 2$ rad/s clockwise and is constant for the interval of motion concerned. For the position where $\theta = 45°$ with AC horizontal, determine the angular velocity of AC and the velocity of A relative to the rotating slot in BO.

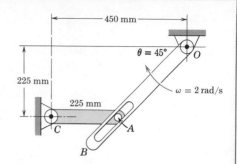

Solution. Motion of a point (pin A) along a rotating path (the slot) suggests rotating coordinates attached to the arm OB. With the origin at O the term v_B of Eq. 6/11 vanishes, and we have

$$\mathbf{v}_A = \boldsymbol{\omega} \times \mathbf{r} + \mathbf{v}_{\text{rel}}$$

where $\boldsymbol{\omega}$ is the angular-velocity vector of the slotted arm (into the paper) and $\mathbf{r}$ is the vector from O to a point P on the arm coincident with A. The velocity of P is $\mathbf{v}_P = \boldsymbol{\omega} \times \mathbf{r}$, and the relative velocity is $\mathbf{v}_{\text{rel}} = \mathbf{v}_{A/P}$. Alternatively, then, we may write

$$\mathbf{v}_A = \mathbf{v}_P + \mathbf{v}_{A/P}$$

The velocity of the point P on member BO is

$$[v = r\omega] \qquad v_P = \overline{OP}\omega = 225\sqrt{2}(2) = 450\sqrt{2} \text{ mm/s}$$

① This equivalence was noted in the third of Eqs. 6/11a.

The relative velocity $\mathbf{v}_{A/P}$, which is the same as $\mathbf{v}_{\text{rel}}$, is seen from the figure to be along the slot toward O. This conclusion becomes clear when it is observed that A is approaching P along the slot from below before coincidence and is receding from P upward along the slot following coincidence. The velocity of A is tangent to its circular arc about C. The vector equation may now be solved since there are only two remaining scalar unknowns, namely, the magnitude of $\mathbf{v}_{A/P}$ and the magnitude of $\mathbf{v}_A$. For the 45° position the figure requires

$$v_{A/P} = v_{\text{rel}} = 450\sqrt{2} \tan 45° = 450\sqrt{2} \text{ mm/s} \qquad \textit{Ans.}$$

$$v_A = 450\sqrt{2}(\sqrt{2}) = 900 \text{ mm/s}$$

each in its direction shown. The angular velocity of AC is now determined as

$$\left[\omega = \frac{v}{r}\right] \qquad \omega_{AC} = \frac{v_A}{AC} = \frac{900}{225} = 4 \text{ rad/s counterclockwise} \qquad \textit{Ans.}$$

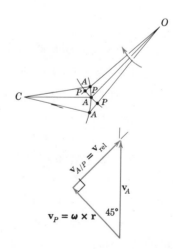

Sample Problem 6/16

At the instant represented the disk with the radial slot is rotating about O with a counterclockwise angular velocity of 4 rad/sec which is decreasing at the rate of 10 rad/sec². The motion of slider A is separately controlled, and at this instant $r = 6$ in., $\dot{r} = 5$ in./sec, and $\ddot{r} = 81$ in./sec². Determine the absolute velocity and acceleration of A for this position.

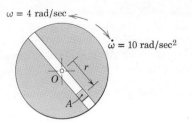

$\omega = 4$ rad/sec

$\dot{\omega} = 10$ rad/sec²

r

O

A

Solution. We have motion relative to a rotating path, so that a rotating coordinate system with origin at O is indicated. We attach x-y axes to the disk and use the unit vectors $\mathbf{i}$ and $\mathbf{j}$.

Velocity. With the origin at O the term $\mathbf{v}_B$ of Eq. 6/11 disappears and we have

①
②
$$\mathbf{v}_A = \boldsymbol{\omega} \times \mathbf{r} + \mathbf{v}_{rel}$$

The angular velocity as a vector is $\boldsymbol{\omega} = 4\mathbf{k}$ rad/sec where $\mathbf{k}$ is the unit vector normal to the x-y plane. Our relative-velocity equation becomes

$$\mathbf{v}_A = 4\mathbf{k} \times 6\mathbf{i} + 5\mathbf{i} = 24\mathbf{j} + 5\mathbf{i} \text{ in./sec} \qquad Ans.$$

in the direction indicated and has the magnitude

$$v_A = \sqrt{(24)^2 + 5^2} = 24.5 \text{ in./sec} \qquad Ans.$$

Acceleration. Equation 6/13 written for zero acceleration of the origin of the rotating coordinate system is

$$\mathbf{a}_A = \boldsymbol{\omega} \times (\boldsymbol{\omega} \times \mathbf{r}) + \dot{\boldsymbol{\omega}} \times \mathbf{r} + 2\boldsymbol{\omega} \times \mathbf{v}_{rel} + \mathbf{a}_{rel}$$

The terms become

③
$$\boldsymbol{\omega} \times (\boldsymbol{\omega} \times \mathbf{r}) = 4\mathbf{k} \times (4\mathbf{k} \times 6\mathbf{i}) = 4\mathbf{k} \times 24\mathbf{j} = -96\mathbf{i} \text{ in./sec}^2$$

$$\dot{\boldsymbol{\omega}} \times \mathbf{r} = -10\mathbf{k} \times 6\mathbf{i} = -60\mathbf{j} \text{ in./sec}^2$$

$$2\boldsymbol{\omega} \times \mathbf{v}_{rel} = 2(4\mathbf{k}) \times 5\mathbf{i} = 40\mathbf{j} \text{ in./sec}^2$$

$$\mathbf{a}_{rel} = 81\mathbf{i} \text{ in./sec}^2$$

The total acceleration is, therefore,

$$\mathbf{a}_A = (81 - 96)\mathbf{i} + (40 - 60)\mathbf{j} = -15\mathbf{i} - 20\mathbf{j} \text{ in./sec}^2 \qquad Ans.$$

in the direction indicated and has the magnitude

$$a_A = \sqrt{(15)^2 + (20)^2} = 25 \text{ in./sec}^2 \qquad Ans.$$

Vector notation is certainly not essential to the solution of this problem. The student should be able to work out the steps with scalar notation just as easily. The correct direction of the Coriolis-acceleration term may always be found by the direction in which the head of the $\mathbf{v}_{rel}$ vector would move if rotated about its tail in the sense of $\boldsymbol{\omega}$ as shown.

① This equation is the same as $\mathbf{v}_A = \mathbf{v}_P + \mathbf{v}_{A/P}$ where P is a point attached to the disk coincident with A at this instant.

② Note that the x-y-z axes chosen constitute a right-handed system.

③ Be sure to recognize that $\boldsymbol{\omega} \times (\boldsymbol{\omega} \times \mathbf{r})$ and $\dot{\boldsymbol{\omega}} \times \mathbf{r}$ represent the normal and tangential components of acceleration of a point P on the disk coincident with A. This description becomes that of Eq. 6/13b.

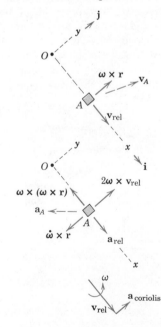

Sample Problem 6/17

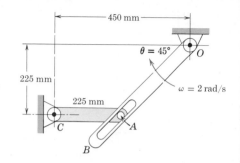

For the conditions of Sample Problem 6/15 determine the angular acceleration of AC and the acceleration of A relative to the rotating slot in arm OB.

Solution. We attach a rotating coordinate system to arm OB with origin at O and use Eq. 6/13b, which is

$$\mathbf{a}_A = \mathbf{a}_P + 2\boldsymbol{\omega} \times \mathbf{v}_{rel} + \mathbf{a}_{rel}$$

where P is the point on OB coincident with A. From the solution to Sample Problem 6/15 we use the values $\omega_{AC} = 4$ rad/s and $v_{rel} = 450\sqrt{2}$ mm/s and compute the following terms:

$(a_A)_n = \overline{AC}\omega_{AC}^2 = 225(4)^2 = 3600$ mm/s² directed toward C

$(a_A)_t = \overline{AC}\alpha_{AC} = 225\alpha_{AC}$ normal to AC, sense unknown

$(a_P)_n = \overline{OP}\omega^2 = 225\sqrt{2}(2)^2 = 900\sqrt{2}$ mm/s² directed toward O

$(a_P)_t = \overline{OP}\alpha = 0$ since $\alpha = \dot{\omega} = 0$

$|2\boldsymbol{\omega} \times \mathbf{v}_{rel}| = 2\omega v_{rel} = 2(2)(450\sqrt{2}) = 1800\sqrt{2}$ mm/s² directed as shown

$\mathbf{a}_{rel} = $ vector measured along slot

The vector equation may now be solved since there are only two remaining scalar unknowns, namely, the magnitudes of $(\mathbf{a}_A)_t$ and $\mathbf{a}_{rel}$. The vector polygon representing Eq. 6/13b and shown in the figure is begun at point R and ends at point S where the lines with the known directions of $(\mathbf{a}_A)_t$ and $\mathbf{a}_{rel}$ intersect.

By equating components normal to $\mathbf{a}_{rel}$, for example, the term a_{rel} is eliminated automatically from the equation, and $(a_A)_t$ is obtained immediately without resorting to the labor of a simultaneous solution. Thus

$$\frac{(a_A)_t}{\sqrt{2}} = \frac{(a_A)_n}{\sqrt{2}} + |2\boldsymbol{\omega} \times \mathbf{v}_{rel}|$$

$$(a_A)_t = 3600 + 1800\sqrt{2}(\sqrt{2}) = 7200 \text{ mm/s}^2$$

Similarly, equating components normal to $(\mathbf{a}_A)_t$ gives

$$\frac{a_{rel}}{\sqrt{2}} = (a_A)_n + \frac{(a_P)_n}{\sqrt{2}} + \frac{|2\boldsymbol{\omega} \times \mathbf{v}_{rel}|}{\sqrt{2}}$$

$$a_{rel} = (3600 + 900 + 1800)\sqrt{2} = 8910 \text{ mm/s}^2 \qquad Ans.$$

The angular acceleration of AC may be found since we now know $(a_A)_t$. Thus

$$\alpha_{AC} = \frac{(a_A)_t}{\overline{AC}} = \frac{7200}{225} = 32 \text{ rad/s}^2 \qquad Ans.$$

With $(\mathbf{a}_A)_t$ pointing down in the figure, α_{AC} is seen to be clockwise.

The foregoing results may also be obtained by expressing each of the terms in vector notation using **i**- and **j**-components corresponding to any convenient choice of the rotating reference axes x-y. If the x-axis is attached to OB with the y-axis perpendicular to OB, then equating the coefficients of the **i**-terms and then of the **j**-terms in the vector equation would produce two algebraic equations which could be solved simultaneously for the magnitudes of $(\mathbf{a}_A)_t$ and $\mathbf{a}_{rel}$.

① If OB had had an angular acceleration in addition to an angular velocity, the acceleration of the point P on OB would have had a tangential as well as a normal component.

② If the slot had been curved with a radius of curvature ρ, the term $\mathbf{a}_{rel}$ would have had a component v_{rel}^2/ρ normal to the slot and directed toward the center of curvature in addition to its component along the slot.

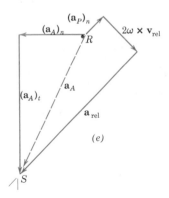

(e)

③ Alternatively, a graphical solution may also be employed where the acceleration polygon is constructed to a convenient scale and the magnitudes of $(\mathbf{a}_A)_t$ and $\mathbf{a}_{rel}$ are measured directly from the diagram.

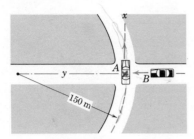

Problem 6/117

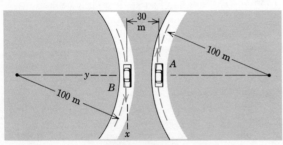

Problem 6/118

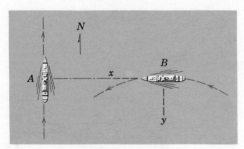

Problem 6/119

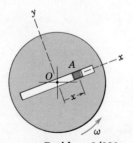

Problem 6/120

PROBLEMS

6/117 The cars of Prob. 2/172 are shown again here. Car A rounds the curve at a constant speed of 54 km/h, and car B is moving at a speed of 81 km/h. Determine the velocity which car B appears to have to an observer riding in and turning with car A. Is this apparent velocity the negative of the velocity which A appears to have to a nonrotating observer in car B? The distance separating the cars at the instant considered is 30 m.

Ans. $\mathbf{v}_{\text{rel}} = -18\mathbf{i} + 22.5\mathbf{j}$ m/s, No

6/118 Cars A and B are rounding the curves with equal speeds of 72 km/h. Determine the velocity which A appears to have to an observer riding in and turning with car B for the instant represented. Does the curvature of the road for car A affect the result? Axes x-y are attached to car B.

6/119 Ship A is proceeding north at a constant speed of 12 knots, while ship B is making a constant speed of 10 knots while turning to port (left) at the constant rate of 10 deg/min. When the ships are separated by a distance of 2 nautical miles in the relative positions shown, with B momentarily heading west and A crossing its bow, the navigator of B measures the apparent velocity of A. Find this velocity and specify the clockwise angle β which it makes with the north direction.

Ans. $|\mathbf{v}_{\text{rel}}| = 34.4$ knots, $\beta = 16.89°$

6/120 The disk with the radial slot rotates about O with an angular acceleration $\dot{\omega} = 15$ rad/s² while the slider A moves with a constant speed $\dot{x} = 100$ mm/s relative to the slot during a certain interval of its motion. If the angular velocity of the disk is $\omega = 12$ rad/s at the instant when the slider crosses the center O of rotation of the disk, determine the acceleration of the slider at this instant.

121 A particle is dropped down a vertical 30-m pipe situated on the equator. What are the magnitude and direction of the particle's absolute acceleration component normal to the pipe axis after it has fallen 25 m from rest? Assume frictionless vertical fall. The angular velocity of the earth is $0.729(10^{-4})$ rad/s and the gravitational acceleration relative to the earth at the equator is 9.781 m/s². (See Fig. 1/2.)

Ans. $a_n = 3.22$ mm/s² west

122 A vehicle *A* travels west at high speed on a perfectly straight road *B* which is tangent to the earth's surface at the equator. The road has no curvature whatsoever in the vertical plane. Determine the necessary speed v_{rel} of the vehicle relative to the road which will give rise to zero acceleration of the vehicle in the vertical direction. Assume that the center of the earth has no acceleration.

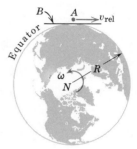

Problem 6/122

123 If the road of Prob. 6/122, instead of being straight with no curvature, followed the curvature of the earth's surface, determine the necessary speed v_{rel} of the vehicle to the west relative to the road which will give rise to zero vertical acceleration of the vehicle. Assume that the center of the earth has no acceleration.

Ans. $v_{\text{rel}} = R\omega = 1672$ km/h or 1038 mi/hr

124 Cars *A* and *B* are both moving with a constant speed of 50 km/h in the directions shown. Calculate the acceleration which *A* appears to have to an observer riding in *B* and turning with it as it rounds the curve of 150-m radius.

125 If the cars of Prob. 6/118 both have a constant speed of 72 km/h as they round the curves, determine the acceleration which *A* appears to have to an observer riding in and turning with car *B* for the instant represented. Axes *x-y* are attached to car *B*.

Ans. $\mathbf{a}_{\text{rel}} = 9.2\mathbf{j}$ m/s²

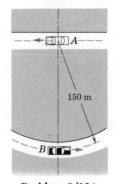

Problem 6/124

126 For the two ships described in Prob. 6/119 and for their relative positions shown where the distance between them is 2 nautical miles, determine the acceleration which *A* appears to have to an observer moving with ship *B*. Express the result in **i**- and **j**-components where the *x-y* axes are attached to *B* as indicated.   *Ans.* $\mathbf{a}_{\text{rel}} = -471\mathbf{i} + 105\mathbf{j}$ n.mi/h²

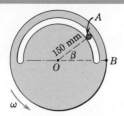

Problem 6/127

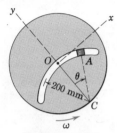

Problem 6/128

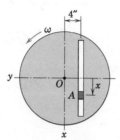

Problem 6/129

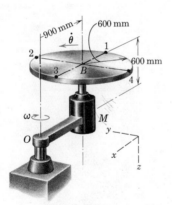

Problem 6/130

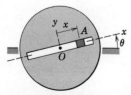

Problem 6/131

6/127 The disk with the circular slot rotates about O with a constant counterclockwise angular velocity $\omega = 10$ rad/s. The pin A moves in the slot so that its radial line AO rotates relative to the line OB, fixed to the disk, with a constant rate $\dot{\beta} = 5$ rad/s for a certain interval of its motion. Solve for the acceleration of A by both Eqs. 2/14 and 6/13.

6/128 The disk with the circular slot of 200-mm radius rotates about O with a constant angular velocity $\omega = 15$ rad/s. Determine the acceleration of the slider A at the instant when it passes the center of the disk if, at that moment, $\dot{\theta} = 12$ rad/s and $\ddot{\theta} = 0$.

6/129 The slider A moves in the slot at the same time that the disk rotates about its center O with an angular speed ω positive in the counterclockwise sense. Determine the x- and y-components of the absolute acceleration of A if, at the instant represented, $\omega = 5$ rad/sec, $\dot{\omega} = -10$ rad/sec^2, $x = 4$ in., $\dot{x} = 6$ in./sec, and $\ddot{x} = 20$ in./sec^2.

   Ans. $a_x = -40$ in./sec^2, $a_y = 80$ in./sec^2

6/130 The shaft and attached disk of the motor M turn counterclockwise when viewed from above at a constant rate $\dot{\theta} = 3$ rad/s *relative* to the motor housing and attached arm OM. Simultaneously the arm is set into clockwise rotation with a constant angular speed $\omega = 2$ rad/s. Determine the absolute acceleration of each of the four points on the disk when in the position shown.

   Ans. $\mathbf{a}_1 = 4.2\mathbf{i}$ m/s^2, $\mathbf{a}_2 = 3.6\mathbf{i} - 0.6\mathbf{j}$ m/s^2
   $\mathbf{a}_3 = 3.0\mathbf{i}$ m/s^2, $\mathbf{a}_4 = 3.6\mathbf{i} + 0.6\mathbf{j}$ m/s^2

6/131 The slider A oscillates in the slot about the neutral position O with a frequency of 2 cycles per second and an amplitude x_{max} of 2 in. so that its displacement in inches may be written $x = 2 \sin 4\pi t$ where t is the time in seconds. The disk, in turn, is set into angular oscillation about O with a frequency of 4 cycles per second and an amplitude $\theta_{max} = 0.20$ rad. The angular displacement is thus given by $\theta = 0.20 \sin 8\pi t$. Calculate the acceleration of A for the positions (*a*) $x = 0$ with $\dot{x}$ positive and (*b*) $x = 2$ in.   *Ans.* (*a*) $\mathbf{a}_A = 253\mathbf{j}$ in./sec^2
   (*b*) $\mathbf{a}_A = -366\mathbf{i}$ in./sec^2

132 Two boys A and B are sitting on opposite sides of a horizontal turntable which rotates at a constant counterclockwise angular velocity ω as seen from above. Boy A throws a ball toward B by giving it a horizontal velocity $\mathbf{u}$ relative to the turntable toward B. Assume that the ball has no horizontal acceleration once released and write an expression for the acceleration $\mathbf{a}_{\text{rel}}$ which B would observe the ball to have in the plane of the turntable just after it is thrown. Sketch the path of the ball on the turntable as observed by B.

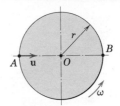

Problem 6/132

133 The figure shows the vanes of a centrifugal-pump impeller which turns with a constant clockwise speed of 200 rev/min. The fluid particles are observed to have an absolute velocity whose component in the r-direction is 10 ft/sec at discharge from the vane. Furthermore, the magnitude of the velocity of the particles measured relative to the vane is increasing at the rate of 80 ft/sec² just before they leave the vane. Determine the magnitude of the total acceleration of a fluid particle an instant before it leaves the impeller. The radius of curvature ρ of the vane at its end is 8 in.

         Ans. $a = 156$ ft/sec²

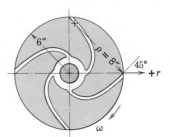

Problem 6/133

134 The slotted disk sector rotates with a constant counterclockwise angular velocity $\omega = 3$ rad/s. Simultaneously the slotted arm OC oscillates about the line OB (fixed to the disk) so that θ changes at the constant rate of 2 rad/s except at the extremities of the oscillation during reversal of direction. Determine the total acceleration of the pin A when $\theta = 30°$ and $\dot{\theta}$ is positive (clockwise).

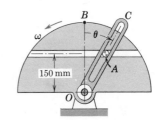

Problem 6/134

135 In the position shown the bar DC is rotating counterclockwise at the constant rate $N = 2$ rad/sec. Determine the angular velocity ω and the angular acceleration α of EBO at this instant.

         Ans. $\omega = 2$ rad/sec CCW
         $\alpha = 8$ rad/sec² CW

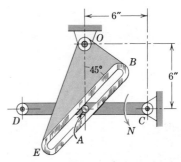

Problem 6/135

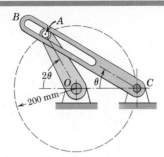

Problem 6/136

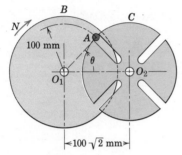

Problem 6/137

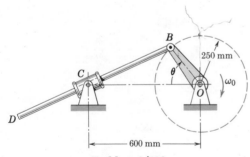

Problem 6/138

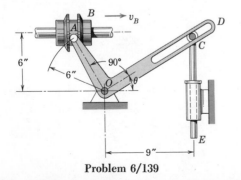

Problem 6/139

6/136 The crank OA revolves clockwise with a constant angular velocity of 10 rad/s within a limited arc of its motion. For the position $\theta = 30°$ determine the angular velocity of the slotted link CB and the acceleration of A as measured relative to the slot in CB. *Ans.* $\omega_{BC} = 5$ rad/s CW
$a_{\text{rel}} = 8.66$ m/s² toward C

6/137 Intermittent rotary motion can be obtained with the Geneva mechanism shown, where the driving pin of disk B engages the radial slots of disk C in the 45° position and turns the disk through 90° before disengaging itself from the slot. If driver B has a constant clockwise angular velocity $N = 60$ rev/min, determine the angular velocity ω and the angular acceleration α of C for the instant when $\theta = 30°$. *Ans.* $\omega = 2.57$ rad/s CCW
$\alpha = 92.1$ rad/s² CCW

6/138 The crank OB revolves clockwise at the constant rate ω_0 of 5 rad/s. For the instant when $\theta = 90°$ determine the angular acceleration α of the rod BD, which slides through the pivoted collar at C. *Ans.* $\alpha = 6.25$ rad/s² CW

▶**6/139** The pin A in the bell crank AOD is guided by the flanges of the collar B, which slides with a constant velocity v_B of 3 ft/sec along the fixed shaft. For the position $\theta = 30°$ determine the acceleration of the plunger CE, whose upper end is positioned by the radial slot in the bell crank. *Ans.* $a_C = 83.1$ ft/sec² up

▶**6/140** Determine the angular acceleration of link OB in the position shown if $\omega = 2$ rad/s, $\dot{\omega} = 6$ rad/s², and $\theta = \beta = 60°$. Pin A is a part of link OB, and the curved slot has a radius of curvature of 125 mm. *Ans.* $\alpha_{OB} = 9.96$ rad/s² CCW

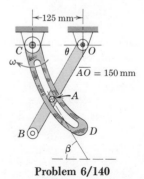

Problem 6/140

6/8 PROBLEM FORMULATION AND REVIEW. In Chapter 6 we have applied our knowledge of basic kinematics which we developed in Chapter 2 to the plane motion of rigid bodies. We approached the problem essentially in two ways. First we wrote an equation which described the general geometric configuration of a given problem in terms of knowns and variables. Then we differentiated the equation with respect to time to obtain velocities and accelerations both linear and angular. We called this approach an absolute-motion analysis, and it is satisfactory when the geometric expression and its time derivatives can be easily written.

In our second approach we applied the principles of relative motion to rigid bodies and found that this approach enabled us to solve many problems which would be too awkward to handle by mathematical differentiation. The relative-velocity equation, the instantaneous center of zero velocity, and the relative-acceleration equation all require that we visualize clearly and analyze correctly the case of circular motion of one point around another point as viewed from nonrotating axes. Recognition of the relationships for circular motion is certainly an essential key without which little progress can be made.

The relative-velocity and relative-acceleration relationships are vector equations which we may solve in any one of three ways: by a scalar-geometric analysis of the vector polygon, by vector algebra, or by a graphical construction of the vector polygon. Each method has its advantages, and the student is urged to gain experience with all three. If one uses a vector-algebra approach, he should sketch the geometry of the vectors in any event, because the ability to visualize physical-geometrical relationships of force and motion is probably the most important experience of all in the study of mechanics.

In Chapter 7 we shall study the kinetics of rigid bodies in plane motion. Here we find that the ability to solve for the linear and angular accelerations of rigid bodies is absolutely necessary in order to apply the force and moment equations which relate the applied forces to the resulting motions. Thus the material of Chapter 6 is essential to that in Chapter 7.

Finally, in Chapter 6 we introduced rotating coordinate systems which enable us to solve problems where the motion is observed relative to a rotating frame of reference. Whenever a point moves along a path which itself is turning, analysis by rotating axes is indicated if a relative-motion approach is used. In deriving Eq. 6/11 for relative velocity and Eq. 6/13 for relative acceleration, where the relative terms are measured from a rotating reference system, it was necessary for us to account for the time derivatives of the unit vectors **i** and **j**. Equations 6/11 and 6/13 also apply to spatial motion, as will be shown in Chapter 8.

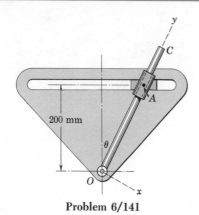

Problem 6/141

REVIEW PROBLEMS

6/141 The slider blocks are pinned together at point A with one block confined to move in the horizontal slot of the fixed plate while the other block slides along the rotating rod OC. If the rod has an angular velocity $\dot{\theta} = 3$ rad/s when the position $\theta = 30°$ is passed, determine the corresponding velocity of point A as measured by an observer attached to the rotating axes x-y. Let P be a point attached to OC and coincident with point A at the instant considered. Visualize the proper direction and sense of $v_{A/P}$ by constructing the positions of P and A just prior to, at, and just after the instant considered.

Ans. $v_{A/P} = 400$ mm/s

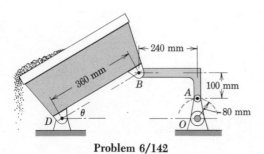

Problem 6/142

6/142 The tilting device maintains a sloshing water bath for washing vegetable produce. If the crank OA oscillates about the vertical and has a clockwise angular velocity of 4π rad/s when OA is vertical, determine the angular velocity of the basket in the position shown where $\theta = 30°$.

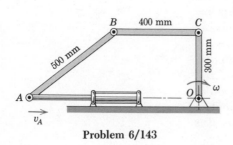

Problem 6/143

6/143 In the linkage shown OC has a constant clockwise angular velocity $\omega = 2$ rad/s during an interval of motion while the hydraulic cylinder gives pin A a constant velocity of 1.2 m/s to the right. For the position shown where OC is vertical and BC is horizontal calculate the angular velocity of BC. Solve by drawing the necessary velocity polygon.

Ans. $\omega_{BC} = 2$ rad/s CW

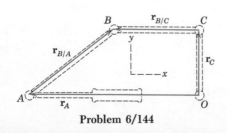

Problem 6/144

6/144 Solve Prob. 6/143 by vector algebra starting with the position-vector equation $r_C + r_{B/C} = r_A + r_{B/A}$ as shown in the separate diagram here.

145 Determine the angular velocity ω of the ram head
    AE of the rock crusher in the position for which
    $\theta = 60°$. The crank OB has an angular speed of
    60 rev/min. When B is at the bottom of its circle, D
    and E are on a horizontal line through F and lines
    BD and AE are vertical. The dimensions are
    $\overline{OB} = 4$ in., $\overline{BD} = 30$ in., and $\overline{AE} = \overline{ED} = \overline{DF} =$
    15 in. Construct the given configuration graphi-
    cally and use the method of instantaneous center of
    zero velocity. *Ans.* $\omega = 1.10$ rad/sec CW

146 Show that the expressions $v_O = r\omega$ and $a_t = r\alpha$ hold
    for the motion of the center O of the wheel which
    rolls on the fixed circular surface, where ω and α are
    the absolute angular velocity and angular accelera-
    tion, respectively, of the wheel.

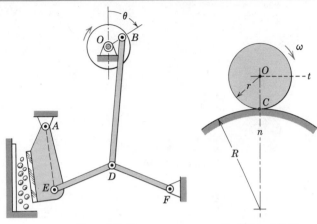

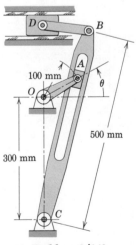

Problem 6/145   **Problem 6/146**

147 If the center O of the wheel shown in Prob. 6/146
    has a velocity v_O, write an expression for the accel-
    eration of point C, the instantaneous center of zero
    velocity on the wheel.

$$\text{Ans. } a_C = \frac{v_O{}^2}{r(1 + r/R)}, \ -n\text{-direction}$$

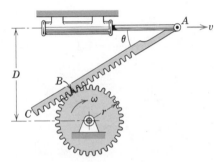

Problem 6/148

148 The hydraulic cylinder moves pin A to the right
    with a constant velocity v. Use the fact that the
    distance from A to B is invariant, where B is the
    point on AC momentarily in contact with the gear,
    and write an expression for the angular velocity ω of
    the gear and the angular velocity of the rack AC.

$$\text{Ans. } \omega = \frac{v \cos \theta}{r}, \ \omega_{AC} = \frac{v \sin^2 \theta}{D - r \cos \theta}$$

149 The figure illustrates a commonly used quick-return
    mechanism which produces a slow cutting stroke of
    the tool (attached to D) and a rapid return stroke. If
    the driving crank OA is turning at the constant rate
    $\dot{\theta} = 3$ rad/s, determine the velocity of point B for
    the instant when $\theta = 30°$. *Ans.* $v_B = 288$ mm/s

Problem 6/149

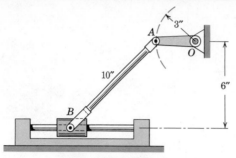

Problem 6/150

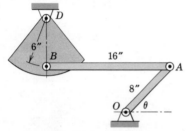

Problem 6/151

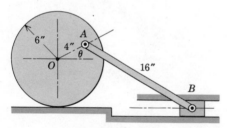

Problem 6/152

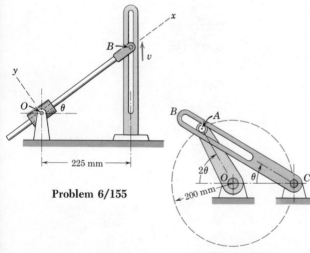

Problem 6/155

Problem 6/156

6/150 If the collar at B is given a constant velocity of 12 in./sec to the right, determine the angular acceleration of crank AO when it reaches the horizontal position shown.

6/151 The mechanism of Prob. 6/75 is repeated here. If link OA has a constant counterclockwise angular velocity $\dot{\theta} = 4$ rad/sec in the position where $\theta = 45°$, and if AB is horizontal and BD is vertical, determine the angular acceleration of the sector BD.   *Ans.* $\alpha_{BD} = 9.75$ rad/sec² CW

6/152 The wheel rolls without slipping, and its position is controlled by the motion of the slider B. If B has a constant velocity of 10 in./sec to the left, determine the angular velocity of AB and the velocity of the center O of the wheel when $\theta = 0$.

6/153 If the center O of the wheel of Prob. 6/152 has a constant velocity of 6 in./sec to the left, calculate the acceleration of the slider B for the position $\theta = 0$.   *Ans.* $a_B = 5.25$ in./sec²

6/154 Solve for the acceleration of pin A in Prob. 6/141 by using Eq. 6/13b and the results of Prob. 6/141. Assume that $\dot{\theta}$ is constant.   *Ans.* $a_A = 2.77$ m/s²

6/155 The shaft slides through the guide bearing which is pivoted at O, and pin B at the end of the shaft has a constant upward velocity $v = 450$ mm/s in the fixed slot for an interval of motion. For the instant when $\theta = 30°$, determine the acceleration a_A of a point A which is on the center line of the shaft and is coincident with the point O at this instant. (*Hint:* Express the acceleration of A in terms of the known motion of B and also in terms of the motion of A relative to O.)   *Ans.* $a_A = 893$ mm/s²

6/156 The figure for Prob. 6/136 is repeated here. If the slotted link BC has a constant clockwise angular velocity of 10 rad/s in the position for which $\theta = 30°$, determine the angular velocity of AO and the acceleration of A as measured relative to the slot in CB.   *Ans.* $\omega_{AO} = 20$ rad/s clockwise
$a_{rel} = 34.6$ m/s² toward C

PLANE KINETICS OF RIGID BODIES

7

7/1 INTRODUCTION. The kinetics of rigid bodies treats the relationships between the forces which act upon the bodies from sources external to their boundaries and the corresponding translational and rotational motions of the bodies. In Chapter 6 we developed the kinematical relationships for the plane motion of rigid bodies, and we will use these relationships extensively in this present chapter where the effects of forces on the two-dimensional motion of rigid bodies are examined.

In Chapter 3 we found that two force equations of motion were required to define the plane motion of a particle whose motion has two linear components. For the plane motion of a rigid body an additional equation is needed to specify the state of rotation of the body. Thus two force equations and one moment equation or their equivalent are required to determine the state of rigid-body plane motion.

The relationships which form the basis for most of the analysis of rigid-body motion were developed in Art. 5/2 of Chapter 5 for a general system of mass particles. Frequent reference will be made to these equations as they are further developed in Chapter 7 and applied specifically to the plane motion of rigid bodies. The reader is advised to refer to the developments of Chapter 5 frequently as he studies Chapter 7. Readers will also be well advised not to proceed further until they have a firm grasp of the calculation of velocities and accelerations as developed in Chapter 6 for rigid-body plane motion. Without the ability to determine accelerations correctly from the principles of kinematics, it is frequently useless to attempt application of the force and moment principles of motion. Consequently, readers are urged to master the necessary kinematics, including the calculation of relative accelerations, before proceeding.

Basic to the approach to kinetics is the isolation of the body or system to be analyzed. This isolation was illustrated and used in Chapter 3 for particle kinetics and will be employed consistently in the present chapter. For problems involving the instantaneous relationships among force, mass, and acceleration or momentum, the body or system should be explicitly defined by isolating it with its *free-body diagram*. When the principles of work and energy are employed, an *active-force diagram* which shows only those external forces that do work on the system may be used in lieu of the free-body diagram. *No solution of a problem should be attempted without first defining the complete external boundary of the body or system and identifying all external forces that act on it.*

Chapter 7 is organized into the same three sections in which we treated the kinetics of particles in Chapter 3. Section A relates the forces and moments to the instantaneous linear and angular accelerations. Section B treats the solution of problems by the method of work and energy. Section C covers the methods of impulse and momentum. Virtually all of the basic concepts and approaches covered in these three sections were treated in Chapter 3 on particle kinetics. This repetition will result in accelerated progress in Chapter 7 provided the kinematics of rigid-body plane motion is well in hand. In each of the three sections the three types of motion, namely, translation, fixed-axis rotation, and general plane motion, will be separately identified.

In the kinetics of rigid bodies which have angular motion it is necessary to introduce a property of the body which accounts for the radial distribution of its mass with respect to a particular axis of rotation normal to the plane of motion. This property is known as the *mass moment of inertia* of the body, and it is essential that we be able to calculate this property in order to solve rotational problems. Familiarity with the calculation of mass moments of inertia is assumed of the reader at this point. Appendix A treats this topic and is included for those who need instruction or review.

SECTION A. FORCE, MASS, AND ACCELERATION

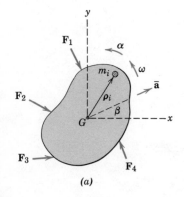

(a)

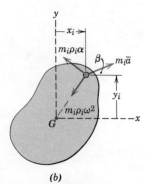

(b)

Figure 7/1

7/2 GENERAL EQUATIONS OF MOTION. Specifying the position of a rigid body in its plane of motion requires the values of three scalar coordinates. The two coordinates of the mass center, or some other convenient reference point, and the angular position of the body about the reference point selected will determine uniquely the position of each point in the body. Therefore we see that plane motion requires three independent scalar equations for its description. Two of these equations, the x- and y-components of Eq. 5/1a for motion in the x-y plane, have already been described for a general mass system, which includes a rigid body.

We may obtain the third equation by simplifying Eq. 5/9, the moment equation about the mass center $\Sigma \overline{\mathbf{M}} = \dot{\overline{\mathbf{H}}}$, which we derived in Art. 5/2 for a general system of mass. For this purpose consider the rigid body shown in Fig. 7/1a, which illustrates plane motion in the x-y plane under the action of forces $\mathbf{F}_1$, $\mathbf{F}_2$, $\mathbf{F}_3$, . . . applied to the body from external sources. The acceleration of the mass center G is represented by the vector $\overline{\mathbf{a}}$, and the angular velocity and angular acceleration, shown arbitrarily in the counterclockwise sense, have the magnitudes ω and α, respectively. For our rigid body in plane motion $\Sigma \bar{\mathbf{M}}$ represents the sum of the moments of all external forces about the z-axis through the mass center G. The angular momentum about the mass center for the general system was expressed in Eq. 5/8 as $\bar{\mathbf{H}} = \Sigma \boldsymbol{\rho}_i \times m_i \dot{\boldsymbol{\rho}}_i$ where $\boldsymbol{\rho}_i$ is the position vector relative to G of the representative particle of mass m_i. For our rigid body the velocity of m_i with respect to G is $\dot{\boldsymbol{\rho}}_i = \boldsymbol{\omega} \times \boldsymbol{\rho}_i$ and lies in the plane of motion normal to $\boldsymbol{\rho}_i$. The product $\boldsymbol{\rho}_i \times \dot{\boldsymbol{\rho}}_i$ is, then, a vector normal to the x-y plane and in the sense of $\boldsymbol{\omega}$, which, as a vector, is taken here to be in the positive z-direction by the right-hand rule. Thus the magnitude of $\bar{\mathbf{H}}$ becomes $\bar{H} = \Sigma \rho_i^2 m_i \omega = \omega \Sigma \rho_i^2 m_i$. The summation, which may also be written as $\int \rho^2 \, dm$, is defined as the *mass moment of inertia* $\bar{I}$ of the body about the z-axis through G. Hence we write

$$\bar{H} = \bar{I}\omega$$

where $\bar{I}$ is a constant property of the body and is a measure of the radial distribution of mass around the z-axis through G. With this substitution our moment equation becomes

$$\Sigma \bar{M} = \dot{\bar{H}} = \bar{I}\dot{\omega} = \bar{I}\alpha$$

where $\alpha = \dot{\omega}$ is the angular acceleration of the body.

The moment equation and the two scalar components of the generalized Newton's second law of motion, Eq. 5/1a, are

$$\boxed{\begin{aligned} \Sigma F_x &= m\bar{a}_x \\ \Sigma F_y &= m\bar{a}_y \\ \Sigma \bar{M} &= \bar{I}\alpha \end{aligned}} \qquad (7/1)$$

Equations 7/1 are the general equations of motion for a rigid body in plane motion.

As an alternative approach it is instructive to derive the moment equation by referring directly to the forces which act on the representative particle of mass m_i, as shown in Fig. 7/1b. The acceleration of m_i equals $\bar{a}$ plus the relative terms $\rho_i \omega^2$ and $\rho_i \alpha$ where the mass center G is used as the reference point. It follows that the resultant of all forces on m_i has the components $m_i \bar{a}$, $m_i \rho_i \omega^2$, and $m_i \rho_i \alpha$ in the directions shown. The sum of the moments of these force components about G in the sense of α becomes

$$\bar{M}_i = m_i \rho_i^2 \alpha + (m_i \bar{a} \sin \beta) x_i - (m_i \bar{a} \cos \beta) y_i$$

Similar moment expressions exist for all particles in the body, and the sum of these moments about G for the resultant forces acting on all particles may be written as

$$\Sigma \bar{M} = \Sigma m_i \rho_i^2 \alpha + \bar{a} \sin \beta \, \Sigma m_i x_i - \bar{a} \cos \beta \, \Sigma m_i y_i$$

But the origin of coordinates is taken at the mass center, so that $\Sigma m_i x_i = m\bar{x} = 0$ and $\Sigma m_i y_i = m\bar{y} = 0$. Thus the moment sum becomes

$$\Sigma \bar{M} = \Sigma m_i \rho_i^2 \alpha = \bar{I}\alpha$$

as before. The contribution to $\Sigma \bar{M}$ of the forces internal to the body is, of course, zero since they occur in pairs of equal and opposite forces of action and reaction between neighboring particles. Thus $\Sigma \bar{M}$, as before, represents the sum of moments about G of only the external forces acting on the body as disclosed by the free-body diagram. We note that the force component $m_i \rho_i \omega^2$ has no moment about G and conclude, therefore, that the angular velocity ω has no influence on the moment equation about the mass center G.

Figure 7/2a illustrates schematically the free-body diagram of a rigid body in plane motion with angular acceleration α and mass-center acceleration $\bar{a}$ at the instant considered. Figure 7/2b illustrates the equivalent resultants of the applied force system which are a resultant force $m\bar{a}$ through G in the direction of $\bar{a}$ and a resultant couple $\bar{I}\alpha$ in the sense of α. This *resultant-force diagram* illustrates one of the most important and useful of the conclusions in dynamics, namely, the equivalence between the external force system, Fig. 7/2a, which produces plane motion of a rigid body, and the resultants of this system, Fig. 7/2b. This equivalence makes it possible for us to write the necessary equations that establish the instantaneous relations between the forces and the accelerations for any given problem. Representation of the resultants $m\bar{a}$ and $\bar{I}\alpha$ for every problem in plane motion will ensure that the force and moment sums as disclosed from the free-body diagram are equated to their proper resultants. This representation permits complete freedom of choice of a convenient moment center. If, for example, we choose point O in

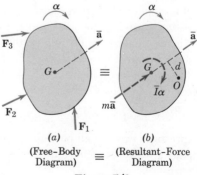

<table>
<tr><td>(a)</td><td></td><td>(b)</td></tr>
<tr><td>(Free–Body
Diagram)</td><td>≡</td><td>(Resultant–Force
Diagram)</td></tr>
</table>

Figure 7/2

Fig. 7/2*b* as such a convenient point, the summation of moments of the external forces about O would give

$$\boxed{\Sigma M_O = \bar{I}\alpha + m\bar{a}d} \qquad\qquad (7/2)$$

If the moment center is chosen on the opposite side of $m\bar{a}$, we see that the sign of the $m\bar{a}d$ term would be negative for a clockwise summation in the sense of α. Equation 7/2 is merely an expression of the principle of moments. It states that the sum of the moments about some point O of all external forces acting on the body in the plane of motion equals the moment of their resultant. Their resultant is expressed as the couple $\bar{I}\alpha$ and the force $m\bar{\mathbf{a}}$ through G.

　　The motion of a rigid body may be unconstrained or constrained. The rocket moving in a vertical plane, Fig. 7/3*a*, is an example of unconstrained motion as there are no physical confinements to its motion. The two components $\bar{a}_x$ and $\bar{a}_y$ of the mass-center acceleration and the angular acceleration α may be determined independently of one another by direct application of Eqs. 7/1. The bar in Fig. 7/3*b*, on the other hand, represents a constrained motion where the vertical and horizontal guides for the ends of the bar impose a kinematical relationship between the acceleration components of the mass center and the angular acceleration of the bar. Thus it is necessary to determine this kinematic relationship from the principles established in Chapter 6 and to combine it with the force and moment equations of motion before a solution can be carried out. In general, dynamics problems which involve physical constraints to motion require a kinematical analysis relating linear to angular acceleration before the force and moment equations of motion can be solved. It is for this reason that an understanding of the principles and methods of Chapter 6 is so vital to the work of Chapter 7.

　　Upon occasion when dealing with two or more connected rigid bodies whose motions are related kinematically, it is convenient to

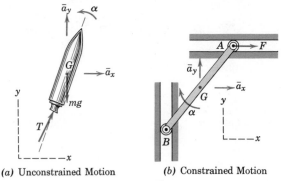

(a) Unconstrained Motion　　　　(b) Constrained Motion

Figure 7/3

analyze the bodies as an entire system. Figure 7/4 illustrates two
rigid bodies hinged at A and subjected to the external forces shown.
The forces in the connection at A are internal to the system and are
not disclosed. The resultant of all external forces must equal the
vector sum of the two resultants $m_1\bar{\mathbf{a}}_1$ and $m_2\bar{\mathbf{a}}_2$, and the sum of the
moments about some arbitrary point such as O of all external forces
must equal the moment of the resultants, $\bar{I}_1\alpha_1 + \bar{I}_2\alpha_2 + m_1\bar{a}_1d_1 +$
$m_2\bar{a}_2d_2$. Thus we may state

$$
\begin{aligned}
\Sigma F &= \Sigma m\bar{\mathbf{a}} \\
\Sigma M_O &= \Sigma \bar{I}\alpha + \Sigma m\bar{a}d
\end{aligned}
\tag{7/3}
$$

where the summations on the right-hand side of the equations represent as many terms as there are separate bodies. If there are more
than three remaining unknowns in a system, however, the three
independent scalar equations of motion, when applied to the system,
are not sufficient to solve the problem. In this case more advanced
methods such as virtual work (Art. 7/7) or Lagrange's equations (not
discussed in this book*) could be employed or else the system dismembered and each part analyzed separately with the resulting
equations solved simultaneously.

   Dynamic equilibrium. An alternative viewpoint for the solution
of problems involving the relationship between forces and accelerations is contained in an approach generally referred to as
D'Alembert's principle, which was described briefly in Art. 4/4 on
the kinetics of particles as viewed from reference axes moving with
the particle. This principle may easily be extended to describe the
plane motion of rigid bodies. By attaching reference axes to the rigid
body of Fig. 7/2 which executes general plane motion, an observer
who is fixed to these axes can measure no motion of the body with

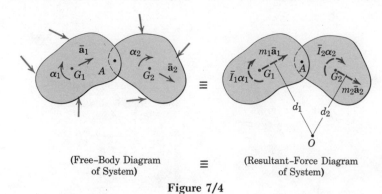

(Free–Body Diagram $\equiv$ (Resultant-Force Diagram
   of System)   of System)

Figure 7/4

---

*See the author's *Dynamics, 2nd Edition* and *SI Version*, for a treatment of
Lagrange's equations.

respect to his moving reference frame. Relative to these moving axes, then, the observer can conclude that the body is in "equilibrium." Clearly, the only way in which the applied forces can be made to have no unbalanced resultant and be consistent with this "equilibrium" is to impose a fictitious force $-m\overline{a}$ on the body to counterbalance the resultant or *effective force* which equals $m\overline{a}$. This fictitious force is referred to as the *inertia force* or *reversed effective force,* and it must be applied through the mass center G in the direction *opposite* to the acceleration $\overline{a}$ of G. Similarly, the only way in which the applied forces can be made to have no unbalanced moment is to impose a fictitious *inertia couple* $-\overline{I}\alpha$ on the body in the rotational sense *opposite* to the angular acceleration α.

Thus the applied system of actual forces $\mathbf{F}_1$, $\mathbf{F}_2$, $\mathbf{F}_3$, . . . along with the fictitious inertia force and inertia couple would constitute an equilibrium system and would create an artificial state known as *dynamic equilibrium*. The familiar equilibrium equations $\Sigma\mathbf{F} = \mathbf{0}$ and $\Sigma M = 0$ may then be applied using any convenient reference directions for the zero force summations and any convenient moment center for the zero moment summation.

If the method known as D'Alembert's principle is used to create dynamic equilibrium, the inertia force and the inertia couple should be added to the free-body diagram and the principles of equilibrium should be stated and applied. If the dynamic-equilibrium method is not used, the free-body diagram should disclose only the actual applied forces which should then be equated to their resultant force and their resultant couple with the aid of a resultant-force diagram and by direct application of the equations of motion, Eqs. 7/1. This latter and more direct description is employed in this book in preference to the method of dynamic equilibrium.

The foregoing developments will now be applied to the three cases of motion in a plane, namely, translation, fixed-axis rotation, and general plane motion in the three articles which follow.

7/3 TRANSLATION. Rigid-body translation in plane motion was described in Art. 6/1 and illustrated in Figs. 6/1*a* and 6/1*b* where we saw that every line in a translating body remains parallel to its original position at all times. In rectilinear translation all points move in straight lines, whereas in curvilinear translation all points move on congruent curved paths. In either case there can be no angular motion of the translating body. With zero angular acceleration Eqs. 7/1 become

$$\left.\begin{array}{l} \Sigma F_x = m\overline{a}_x \\[4pt] \Sigma F_y = m\overline{a}_y \\[4pt] \Sigma \overline{M} = 0 \end{array}\right\} \tag{7/4}$$

which eliminates all reference to the moment of inertia. Figure 7/5*a* represents the free-body diagram and its corresponding resultant-force diagram for a body with rectilinear translation where the mass center G and every point such as C have the same acceleration $\bar{\mathbf{a}}$. With the equivalence established between the two diagrams, we readily see that the moment equation may be written in any of the following ways:

$$\Sigma \bar{M} = 0 \qquad \Sigma M_A = 0 \qquad \Sigma M_O = m\bar{a}d \qquad (7/5)$$

where O is any point not on the line through G in the direction of $\bar{\mathbf{a}}$ and where A is any point on this line.

Figure 7/5*b* represents the free-body diagram and its equivalent resultant-force diagram for a body with curvilinear translation. Here the mass center G moves along a curved path, and the resultant $m\bar{\mathbf{a}}$ is represented by its normal and tangential components $m\bar{a}_n$ and $m\bar{a}_t$ for convenience. Thus we may write the three motion equations as

$$\boxed{\begin{aligned} \Sigma F_n &= m\bar{a}_n \\ \Sigma F_t &= m\bar{a}_t \\ \Sigma \bar{M} &= 0 \end{aligned}} \qquad (7/6)$$

Again, we see from the equivalence between the free-body diagram and the resultant-force diagram that three alternative forms of the moment equation may be written as

$$\Sigma \bar{M} = 0 \qquad \Sigma M_A = m\bar{a}_n d_A \qquad \Sigma M_B = m\bar{a}_t d_B \qquad (7/7)$$

For our particular diagram the second equation would be positive in a clockwise sense and the third equation would be positive in a counterclockwise sense.

The most expedient choice of a moment center is one which eliminates as many unwanted unknowns as possible from the moment equation. This choice is revealed to us by inspection of the resultant-force diagram and the free-body diagram.

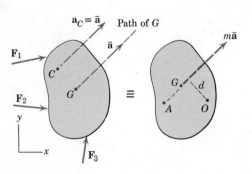

(Free–Body Diagram) ≡ (Resultant–Force Diagram)

(*a*) Rectilinear Translation

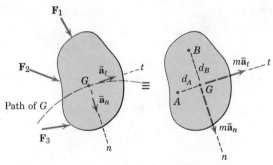

(Free–Body Diagram) ≡ (Resultant–Force Diagram)

(*b*) Curvilinear Translation

Figure 7/5

Sample Problem 7/1

The car weighs 3220 lb and reaches a speed of 30 mi/hr from rest in a distance of 120 ft up the 10-percent incline with constant acceleration. Calculate the normal force under each pair of wheels and the friction force under the rear driving wheels. The coefficient of friction between the tires and the road is known to be at least 0.8.

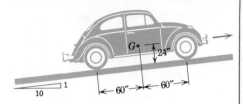

① Without this assumption we would be obliged to account for the relatively small additional forces which produce moments that give the wheels their angular acceleration.

Solution. We will assume that the mass of the wheels is negligible compared with the total mass of the car. The car may now be simulated by a single rigid body in rectilinear translation with an acceleration of

$$[v^2 = 2as] \qquad \bar{a} = \frac{(44)^2}{2(120)} = 8.07 \text{ ft/sec}^2$$

The free-body diagram of the complete car shows the normal forces N_1 and N_2, the friction force F which is in the direction to oppose the slipping of the driving wheels, and the weight W represented by its two components. With $\theta = \tan^{-1} 1/10 = 5.71°$ these components are $W \cos \theta = 3220 \cos 5.71° = 3204$ lb and $W \sin \theta = 3220 \sin 5.71° = 320$ lb. The resultant-force diagram shows the resultant which passes through the mass center and is in the direction of its acceleration. Its magnitude is

$$m\bar{a} = \frac{3220}{32.2}(8.07) = 807 \text{ lb}$$

A moment equation about B eliminates two of the unknowns and gives

$$[\Sigma M_B = m\bar{a}d_B] \qquad 3204(5) - 320(2) - 10N_1 = 807(2)$$

$$N_1 = 1377 \text{ lb} \qquad \textit{Ans.}$$

The force equations now give

$$[\Sigma F_y = 0] \qquad N_2 + 1377 - 3204 = 0 \qquad N_2 = 1827 \text{ lb} \qquad \textit{Ans.}$$

$$[\Sigma F_x = m\bar{a}_x] \qquad F - 320 = \frac{3220}{32.2}(8.07) \qquad F = 1127 \text{ lb} \qquad \textit{Ans.}$$

In order to support this much friction force a coefficient of friction of at least $F/N_2 = 1127/1827 = 0.62$ is required. Since our coefficient of friction is at least 0.8, the surfaces are rough enough to support the calculated value of F so that our result is correct.

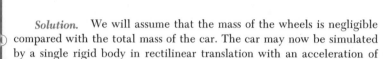

Free–Body Diagram

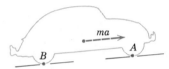

Resultant–Force Diagram

② Recall that 30 mi/hr is 44 ft/sec.

③ What other sequence of equations could have been used just as effectively? What would be the consequence of starting the solution with $\Sigma \bar{M} = 0$?

④ We must be careful not to use the friction equation $F = \mu N$ here since we do not have a case of slipping or impending slipping. If the given coefficient of friction were less than 0.62, the friction force would be μN_2, and the car would be unable to attain the acceleration of 8.07 ft/sec². In this case the unknowns would be N_1, N_2, and a.

⑤ The left-hand side of the equation is evaluated from the free-body diagram, and the right-hand side from the resultant-force diagram. The positive sense for the moment sum is arbitrary but must be the same for both sides of the equation. In this problem we have taken the clockwise sense as positive for the moment of the resultant force about B.

Sample Problem 7/2

The vertical bar AB has a mass of 150 kg with center of mass G midway between the ends. The bar is elevated from rest at $\theta = 0$ by means of the parallel links of negligible mass with a constant couple $M = 5$ kN·m applied to the lower link at C. Determine the angular acceleration α of the links as a function of θ, and find the force B in the link DB at the instant when $\theta = 30°$.

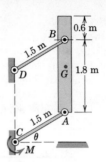

Solution. The motion of the bar is seen to be curvilinear translation since the bar itself does not rotate during the motion. With negligible mass of the links the tangential component A_t of the force at A is $A_t = M/\overline{AC} = 5/1.5 = 3.33$ kN, and the force at B is along the link. All applied forces are shown on the free-body diagram, and the resultant-force diagram is also indicated where the resultant force is shown in terms of its two components.

A summation of forces in the t-direction will eliminate all unknowns other than α, so that for the general position shown

$$[\Sigma F_t = m\bar{a}_t] \quad 3.33 - 0.15(9.81)\cos\theta = 0.15(1.5\alpha)$$

$$\alpha = 14.81 - 6.54\cos\theta \text{ rad/s}^2 \quad Ans.$$

With α a known function of θ, the angular velocity ω of the links is obtained from

$$[\omega\, d\omega = \alpha\, d\theta] \quad \int_0^\omega \omega\, d\omega = \int_0^\theta (14.81 - 6.54\cos\theta)d\theta$$

$$\omega^2 = 29.6\,\theta - 13.08\sin\theta$$

Substitution for $\theta = 30°$ gives

$$(\omega^2)_{30°} = 8.97 \text{ (rad/s)}^2 \quad \alpha_{30°} = 9.15 \text{ rad/s}^2$$

and

$$m\bar{r}\omega^2 = 0.15(1.5)(8.97) = 2.02 \text{ kN}$$

$$m\bar{r}\alpha = 0.15(1.5)(9.15) = 2.06 \text{ kN}$$

The force B may be obtained by a moment summation about A, which eliminates A_n and A_t and the weight. Or a moment summation may be taken about the intersection of A_n and the line of action of $m\bar{r}\alpha$, which eliminates A_n and $m\bar{r}\alpha$. Using A as a moment center gives

$$[\Sigma M_A = m\bar{a}d] \quad 1.8\frac{\sqrt{3}}{2}B = 2.02(1.2)\frac{\sqrt{3}}{2} + 2.06(0.6)$$

$$B = 2.14 \text{ kN} \quad Ans.$$

② The component A_n could be obtained from a force summation in the n-direction or from a moment summation about G or about the intersection of B and the line of action of $m\bar{r}\alpha$.

① The force and moment equations for a body of negligible mass become the same as the equations of equilibrium. Link BD, therefore, acts as a two-force member in equilibrium.

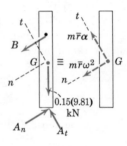

② Generally speaking the best choice of reference axes is to make them coincide with the directions in which the components of the mass-center acceleration are expressed. Examine the consequences of choosing horizontal and vertical axes in this problem.

PROBLEMS

7/1 The right-angled bar with uniform legs of equal length b is pinned to a support at O and hangs in the dotted position while the support is at rest. Calculate the steady horizontal acceleration a of the support which will cause the angle bar to assume the position shown. *Ans. $a = 3g$*

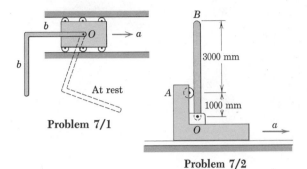

Problem 7/1

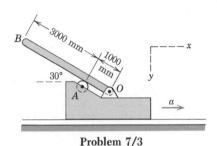

Problem 7/2

7/2 The uniform 30-kg bar OB is secured in the vertical position to the accelerating frame by the hinge at O and the roller at A. If the horizontal acceleration of the frame is $a = 20$ m/s^2, compute the force F_A on the roller and the horizontal component of the force supported by the pin at O.

7/3 The uniform 30-kg bar OB is secured to the accelerating frame in the 30° position from the horizontal by the hinge at O and roller at A. If the horizontal acceleration of the frame is $a = 20$ m/s^2, compute the force F_A on the roller and the x- and y-components of the force supported by the pin at O.
 Ans. $F_A = 1.11$ kN, $O_x = 45$ N, $O_y = 667$ N

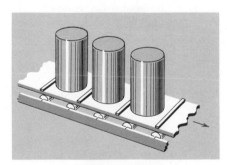

Problem 7/3

7/4 Solid homogeneous cylinders 400 mm high and 250 mm in diameter are supported by a flat conveyor belt which moves horizontally. If the speed of the belt increases according to $v = 1.2 + 0.9t^2$ m/s, where t is the time in seconds measured from the instant the increase begins, calculate the value of t for which the cylinders begin to tip over. Cleats on the belt prevent the cylinders from slipping.

Problem 7/4

7/5 The homogeneous crate of mass m is mounted on small wheels at its front and back edges. Determine the maximum force P which can be applied without overturning the crate about (a) its front edge with $h = b$ and (b) its back edge with $h = 0$.
 Ans. (a) $P = mgc/b$, (b) $P = mgc/b$

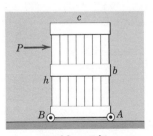

Problem 7/5

7/6 Determine the value of h in Prob. 7/5 for which the reaction under the wheel at A is twice that under the wheel at B if $P = mg/2$.

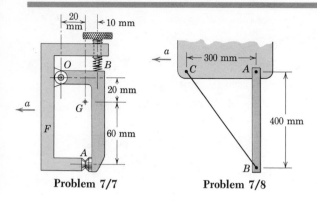

Problem 7/7   **Problem 7/8**

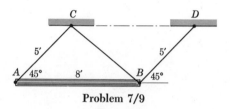

Problem 7/9

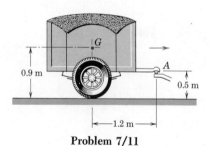

Problem 7/11

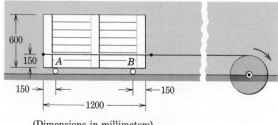

(Dimensions in millimeters)

Problem 7/12

7/7 The arm *OA* of the classifying accelerometer has a mass of 0.2 kg with center of mass at *G*. The adjusting screw and spring are preset to a force of 8 N at *B*. At what acceleration *a* would the electrical contacts at *A* be on the verge of opening? Motion is in the vertical plane of the figure.   *Ans. a = 7.12g*

7/8 The uniform 5-kg bar *AB* is suspended in a vertical position from an accelerating vehicle and restrained by the wire *BC*. If the acceleration is *a* = 0.6*g*, determine the tension *T* in the wire and the magnitude of the total force supported by the pin at *A*.

7/9 The uniform pole *AB* weighs 100 lb and is suspended in the horizontal position by the three wires shown. If wire *CB* breaks, calculate the tension in wire *BD* immediately after the break. (*Suggestion:* By a thoughtful choice of moment center solve by using only one equation of motion.)

    Ans. T_B = 35.4 lb

7/10 Calculate the tension in wire *BD* in Prob. 7/9 when the 45° angle has become 60° with the 100-lb pole moving under the action of its weight after *CB* breaks. At the 60° position wires *BD* and AC have an angular velocity of 1.43 rad/sec.

7/11 The loaded trailer has a mass of 900 kg with center of mass at *G* and is attached at *A* to a rear-bumper hitch. If the car and trailer reach a velocity of 60 km/h on a level road in a distance of 30 m from rest with constant acceleration, compute the vertical component of the force supported by the hitch at *A*. Neglect the small friction force exerted on the relatively light wheels.   *Ans. A_y = 1389 N*

7/12 The 600-kg crate is supported by rollers at *A* and *B* and is being moved along the floor by the horizontal cable. If the initial cable tension is 3000 N as the winch takes hold, determine the corresponding forces under the rollers. The center of mass of the crate is located at its geometric center.

    Ans. F_A = 3440 N, F_B = 2440 N

13 The force P is applied to the homogeneous crate of mass m. If the coefficient of friction between the crate and the horizontal platform is μ, determine the limiting values of h so that the crate will slide without tipping about either the front edge or the rear edge.

$$Ans.\ h_{\substack{(\max \\ \min)}} = \tfrac{1}{2}\left[b - \frac{mg}{P}(\mu b \mp c)\right]$$

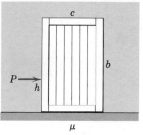

Problem 7/13

14 The 50-kg plate with center of mass at G is suspended by the light parallel links AD and BC and is free to swing in the vertical plane. At what angle θ of the links can the plate be released from rest and have zero force in link BC immediately after release? At this instant what is the force in link AD?

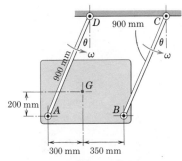

Problem 7/14

15 The vertical 1610-lb plate with its mass center at G is held in the position shown by the parallel cables A and B and the horizontal cable C. If the cable C is suddenly released, calculate the tension in cable B immediately after release.

$$Ans.\ T_B = 368\ \text{lb}$$

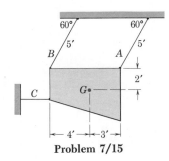

Problem 7/15

16 The block A and attached rod have a combined mass of 50 kg and are confined to move along the 60° guide under the action of the 600-N applied force. The uniform horizontal rod has a mass of 15 kg and is welded to the block at B. Friction in the guide is negligible. Compute the bending moment M exerted by the weld on the rod at B.

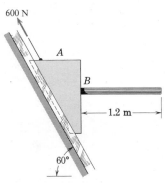

Problem 7/16

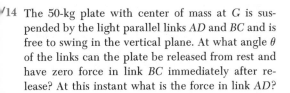

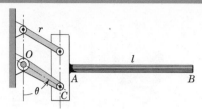

Problem 7/17

7/17 The uniform rod AB of mass m and length l is welded at A to the vertical member of the parallelogram linkage and always remains horizontal. If link OC is given a constant counterclockwise angular velocity ω in the vertical plane, determine expressions for the bending moment M, the tension T, and the shear V applied to the beam at A by the weld in terms of θ.

$$\text{Ans. } M = m\frac{l}{2}(g + r\omega^2 \cos \theta)$$
$$T = mr\omega^2 \sin \theta$$
$$V = m(g + r\omega^2 \cos \theta)$$

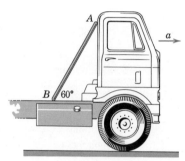

Problem 7/19

7/18 If the driving link OC in Prob. 7/17 starts from rest in the position shown and is given a counterclockwise angular acceleration α, calculate the corresponding bending moment M, shear force V, and compression P applied to the beam at A by the weld.

7/19 The coefficient of friction at both ends of the uniform bar is 0.40. Determine the maximum horizontal acceleration a which the truck may have without causing the bar to slip. (*Suggestion:* The problem can be solved by using only one equation, a moment equation. The location of the moment center may be determined graphically.)

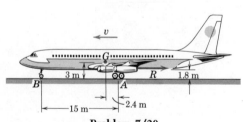

Problem 7/20

7/20 A jet transport with a landing speed of 200 km/h reduces its speed to 50 km/h with a negative thrust R from its jet thrust reversers in a distance of 450 m along the runway with constant deceleration. The total mass of the aircraft is 125 Mg with mass center at G. Compute the reaction N under the nose wheel B toward the end of the braking interval and prior to the application of mechanical braking. At the lower speed aerodynamic forces on the aircraft are small and may be neglected.   *Ans.* $N = 228$ kN

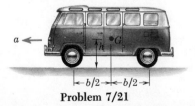

Problem 7/21

7/21 Determine the expression for the maximum velocity v which the microbus can reach in a distance s from rest without slipping its rear driving wheels if the coefficient of friction between the tires and the pavement is μ. Neglect the mass of the wheels.

√22 The car seen from the rear is traveling at a speed v around a turn of mean radius r banked inward at an angle θ. The coefficient of friction between the tires and the road is μ. Determine (a) the proper bank angle for a given v to eliminate any tendency to slip or tip, and (b) the maximum speed v before the car tips or slips for a given θ. Note that the forces and the acceleration lie in the plane of the figure so that the problem may be treated as one of plane motion even though the velocity is normal to this plane.

Problem 7/22

√23 The rocket sled shown is used to test human tolerance to high accelerations. The sled and rocket motors have an initial weight of 2700 lb with mass center at G_1. The man and instrumentation together weigh 600 lb with mass center at G_2. If the rocket motors produce a total thrust T of 24,000 lb and the initial frictional resistance at each runner is 30 percent of the vertical load supported by the runner, calculate the vertical reaction N_A under the front runner.   *Ans.* $N_A = 3250$ lb

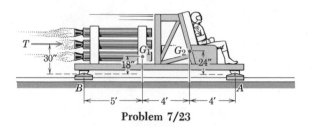

Problem 7/23

/24 The mine skip has a loaded mass of 2000 kg and is attached to the towing vehicle by the light hinged link CD. If the towing vehicle has an acceleration of 3 m/s², calculate the corresponding reactions under the small wheels at A and B.

Ans. $F_A = 11.37$ kN, $F_B = 7.05$ kN

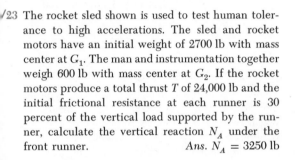

(Dimensions in millimeters)

Problem 7/24

√25 Determine the maximum counterweight W for which the loaded 4000-lb coal car will not overturn about the rear wheels B. Neglect the mass of all pulleys and wheels. (Note that the tension in the cable at C is not $2W$.)

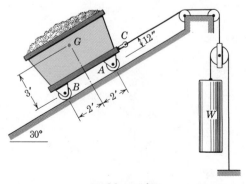

Problem 7/25

7/26 If the counterweight W in Prob. 7/25 is 1500 lb, compute the force under the rear wheels B.

Ans. $F_B = 2932$ lb

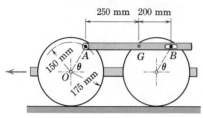

Problem 7/27

7/27 The two identical wheels are mounted on the central frame and roll to the left without slipping with a constant velocity $v = 2.4$ m/s. The 7.5-kg connecting rod AB is pinned to the forward wheel at A, and the pin B of the second wheel fits into a smooth horizontal slot in AB. Determine the total forces exerted on the rod by pins A and B for the position $\theta = 60°$.   Ans. $A = 116.5$ N, $B = 60.9$ N

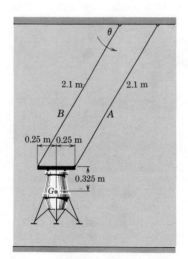

Problem 7/28

7/28 Model tests of the landing of the lunar excursion module (LEM) are conducted using the pendulum which consists of the model suspended by the two parallel wires A and B. If the model has a mass of 10 kg with center of mass at G, and if $\dot{\theta} = 2$ rad/s when $\theta = 60°$, calculate the tension in each of the wires at this instant.

Ans. $T_A = 147.9$ N, $T_B = 21.1$ N

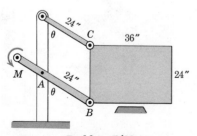

Problem 7/29

▶**7/29** The homogeneous rectangular plate weighs 50 lb and is supported in the vertical plane by the light parallel links shown. If a couple $M = 120$ lb-ft is applied to the end of link AB with the system initially at rest, calculate the force supported by the pin at C as the plate lifts off its support with $\theta = 60°$.   Ans. $C = 51.0$ lb

/4 FIXED-AXIS ROTATION.   Equations 7/1, which we developed in Art. 7/2 for the general case of plane motion, may now be applied to the important special case of fixed-axis rotation. Figure 7/6*a* represents a rigid body rotating about a fixed point O (i.e., about a fixed axis through O normal to the plane of rotation) with an angular velocity ω and an angular acceleration α at the instant considered. The mass center G has the familiar acceleration components for circular motion, namely, $\bar{a}_n = \bar{r}\omega^2$ and $\bar{a}_t = \bar{r}\alpha$. Figure 7/6*b* represents the free-body diagram which discloses all forces acting on the body including the force exerted by the bearing at O on the body. If the motion is not in the horizontal plane, the weight $W = mg$ will also be shown. We learned from Art. 7/2 that this external force system is equivalent to the resultants $m\bar{a}$ and $\bar{I}\alpha$ which are represented in the *c*-part of the figure. Here the resultant force is replaced by its components $m\bar{r}\omega^2$ and $m\bar{r}\alpha$ which act through G in the direction of its normal and tangential accelerations. The equivalence between the free-body diagram and the resultant-force diagram of Fig. 7/6*c* is sufficient to solve any problem of fixed-axis rotation.

Alternatively we may combine the couple $\bar{I}\alpha$ and the force $m\bar{r}\alpha$ by moving the force to a parallel position through a point Q a distance q from the point of rotation O and in line with G and O, Fig. 7/6*d*. The distance q is calculated to preserve the same resultant moment about O. Thus $m\bar{r}^2\alpha + \bar{I}\alpha = m\bar{r}\alpha q$. Substituting $\bar{I} = m\bar{k}^2$ and cancelling $m\alpha$ give $\bar{r}^2 + \bar{k}^2 = \bar{r}q$ so that $q = (\bar{r}^2 + \bar{k}^2)/\bar{r} = k_0^2/\bar{r}$ where k_0 is the radius of gyration of the body about O. Point Q is a unique point for a given body with a given fixed point O and is called the *center of percussion.* If the resultant is represented by a single resultant force $m\bar{a}$, then this resultant force must pass through the center of percussion.

The sum of moments about point O may now be written $\Sigma M_O = m\bar{r}\alpha(k_0^2/\bar{r}) = I_0\alpha$ where $I_0 = mk_0^2$. The three equations of

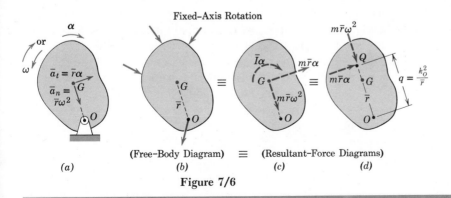

Fixed–Axis Rotation

(Free–Body Diagram) $\equiv$ (Resultant–Force Diagrams)

(a)   (b)   (c)   (d)

Figure 7/6

motion for fixed-axis rotation, written in their most useful form, are, therefore

$$\Sigma F_n = m\bar{r}\omega^2$$
$$\Sigma F_t = m\bar{r}\alpha \qquad (7/8)$$
$$\Sigma M_O = I_O\alpha$$

It is important to note that, whereas the moment equation about O does not involve the bearing force exerted on the body at O, each of the force equations must include any component of this bearing force in its respective n- or t-direction. Thus the bearing force must not be omitted from the free-body diagram.

With the equivalence established between the free-body diagram and either of the two resultant-force diagrams we have complete freedom in choosing a moment center which is most convenient in any particular problem. It is readily apparent from Fig. 7/6d that the center of percussion Q provides a unique moment center since a moment sum about Q must be zero. We now observe the alternative moment equations about the three special points O, G, and Q. They are

$$\Sigma M_O = I_O\alpha \qquad \Sigma\bar{M} = \bar{I}\alpha \qquad \Sigma M_Q = 0$$

These equations are not independent, but any two of them may be used to replace the second and third of Eqs. 7/8.

For the special but common case of rotation of a rigid body about a fixed axis through the mass center G, Fig. 7/7a, the resultant force $m\bar{a}$ is zero since the mass center has no acceleration. The resultant of the force system, then, is the couple $\bar{I}\alpha$, and we may write

$$\Sigma F_x = 0$$
$$\Sigma F_y = 0 \qquad (7/9)$$
$$\Sigma\bar{M} = \bar{I}\alpha$$

Here x and y are any mutually perpendicular directions which are convenient. The free-body diagram is represented in Fig. 7/7b and shows the force components F_x and F_y of the bearing reactions at the mass center G and all other external forces on the body. The resultant-force diagram in the c-part of the figure consists of the resultant couple $\bar{I}\alpha$ only. Since the resultant of the applied forces is a couple, we can take moments about *any* point O and have the same result. Therefore, we may write $\Sigma M_O = \bar{I}\alpha$ for the case of rotation about a fixed mass center.

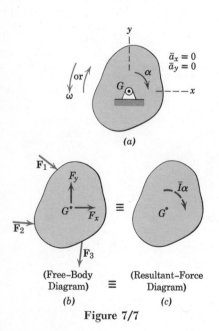

$\bar{a}_x = 0$
$\bar{a}_y = 0$

(a)

(Free-Body Diagram) ≡ (Resultant-Force Diagram)

(b) ≡ *(c)*

Figure 7/7

Sample Problem 7/3

The concrete block weighing 644 lb is elevated by the hoisting mechanism shown where the cables are securely wrapped around the respective drums. The drums, which are fastened together and turn as a single unit about their mass center at O, have a combined weight of 322 lb and a radius of gyration about O of 18 in. If a constant tension P of 400 lb is maintained by the power unit at A, determine the vertical acceleration of the block and the resultant force supported by the bearing at O.

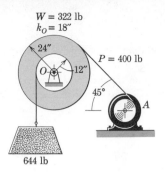

Solution I. The free-body and resultant-force diagrams of the drums and concrete block are drawn showing all forces which act including the components O_x and O_y of the bearing reaction. The resultant of the force system on the drums for centroidal rotation is the couple $\bar{I}\alpha = I_O\alpha$ where

$$[I = k^2 m] \qquad \bar{I} = I_0 = \left(\frac{18}{12}\right)^2 \frac{322}{32.2} = 22.5 \text{ lb-ft-sec}^2$$

Taking moments about the mass center O for the pulley in the sense of the angular acceleration α gives

$$[\Sigma\bar{M} = \bar{I}\alpha] \qquad 400\left(\frac{24}{12}\right) - T\left(\frac{12}{12}\right) = 22.5\alpha \qquad (a)$$

The acceleration of the block is described by

$$[\Sigma F_y = ma_y] \qquad T - 644 = \frac{644}{32.2}a \qquad (b)$$

The angular acceleration α and the acceleration a are related by

$$[a_t = r\alpha] \qquad a = \frac{12}{12}\alpha$$

With this substitution Eqs. (a) and (b) are combined to give

$$T = 717 \text{ lb} \qquad \alpha = 3.67 \text{ rad/sec}^2 \qquad a = 3.67 \text{ ft/sec}^2 \qquad Ans.$$

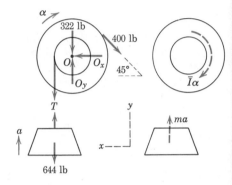

The bearing reaction is computed from its components. Since $\bar{a} = 0$ we use the equilibrium equations

$$[\Sigma F_x = 0] \qquad O_x - 400 \cos 45° = 0 \qquad O_x = 283 \text{ lb}$$

$$[\Sigma F_y = 0] \qquad O_y - 322 - 717 - 400 \sin 45° = 0 \qquad O_y = 1322 \text{ lb}$$

$$O = \sqrt{(283)^2 + (1322)^2} = 1352 \text{ lb} \qquad Ans.$$

Solution II. If we are not interested in calculating the tension T, we may use a more condensed approach by drawing the free-body diagram of the entire system, thus eliminating reference to T, which becomes internal to the new system. From the resultant-force diagram for the system we see that the moment sum about O must equal the resultant moment $\bar{I}\alpha$ for the drums and the moment of the resultant ma for the block. Thus from the principle of Eq. 7/3 we have

$$[\Sigma M_O = \bar{I}\alpha + mad] \qquad 400\left(\frac{24}{12}\right) - 644\left(\frac{12}{12}\right) = 22.5\alpha + \frac{644}{32.2}a\left(\frac{12}{12}\right)$$

With $a = (12/12)\alpha$ the solution gives, as before, $a = 3.67 \text{ ft/sec}^2$.

We may equate the force sums on the entire system to the sums of the resultants. Thus

$$[\Sigma F_y = \Sigma m\bar{a}_y] \qquad O_y - 322 - 644 - 400 \sin 45° = \frac{322}{32.2}(0) + \frac{644}{32.2}(3.67)$$

$$O_y = 1322 \text{ lb}$$

$$[\Sigma F_x = \Sigma m\bar{a}_x] \qquad O_x - 400 \cos 45° = 0$$

$$O_x = 283 \text{ lb}$$

① Be alert to the fact that the tension T is not $300 (9.81)$ N. If it were, the block would not accelerate.

② Don't overlook the need to express k_o in meters when using g in m/s².

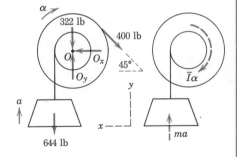

Sample Problem 7/4

The pendulum has a mass of 7.5 kg with center of mass at G and has a radius of gyration about the pivot O of 295 mm. If the pendulum is released from rest at $\theta = 0$, determine the total force supported by the bearing at the instant when $\theta = 60°$. Friction in the bearing is negligible.

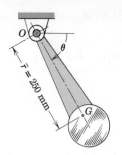

Solution. The free-body diagram of the pendulum in a general position is shown along with the corresponding resultant-force diagram where the components of the resultant force have been drawn through G. Alternatively, as shown in the lower figure, the resultant forces can be drawn through the center of percussion Q without the couple.

The normal component O_n is found from a force equation in the n-direction which involves the normal acceleration $\bar{r}\omega^2$. Since the angular velocity ω of the pendulum is found from the integral of the angular acceleration and since O_t depends on the tangential acceleration $\bar{r}\alpha$, it follows that α should be obtained first. To this end with $I_O = k_O^2 m$, the moment equation about O gives

$[\Sigma M_O = I_O \alpha]$ $7.5(9.81)(0.25) \cos \theta = (0.295)^2(7.5)\alpha$

$$\alpha = 28.2 \cos \theta \text{ rad/s}^2$$

and for $\theta = 60°$

$[\omega \, d\omega = \alpha \, d\theta]$ $\displaystyle\int_0^\omega \omega \, d\omega = \int_0^{\pi/3} 28.2 \cos \theta \, d\theta$ $\omega^2 = 48.8 \text{ (rad/s)}^2$

The remaining two equations of motion applied to the 60° position yield

$[\Sigma F_n = m\bar{r}\omega^2]$ $O_n - 7.5(9.81) \sin 60° = 7.5(0.25)(48.8)$

$$O_n = 155.2 \text{ N}.$$

$[\Sigma F_t = m\bar{r}\alpha]$ $-O_t + 7.5(9.81) \cos 60° = 7.5(0.25)(28.2) \cos 60°$

$$O_t = 10.37 \text{ N}$$

$$O = \sqrt{(155.2)^2 + (10.37)^2} = 155.6 \text{ N} \qquad Ans.$$

The proper sense for O_t may be observed at the outset by applying the alternative moment equation $\Sigma \bar{M} = \bar{I}\alpha$ where the moment about G due to O_t must be clockwise to agree with α. The force O_t may also be obtained initially by a moment equation about Q, which avoids the necessity of computing α. First, we must obtain the distance q, which is

$[q = k_O^2/\bar{r}]$ $\displaystyle q = \frac{(0.295)^2}{0.250} = 0.348 \text{ m}$

$[\Sigma M_Q = 0]$ $O_t(0.348) - 7.5(9.81)(\cos 60°)(0.348 - 0.250) = 0$

$$O_t = 10.37 \text{ N} \qquad Ans.$$

① The acceleration components of G are, of course, $\bar{a}_n = \bar{r}\omega^2$ and $\bar{a}_t = \bar{r}\alpha$.

② Review the theory again and satisfy yourself that $\Sigma M_O = I_O\alpha = \bar{I}\alpha + m\bar{r}^2\alpha = m\bar{r}\alpha q$.

③ Note especially here that the force summations are taken in the positive directions of the acceleration components of the mass center G.

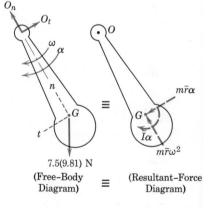

(Free–Body Diagram) $\equiv$ (Resultant–Force Diagram)

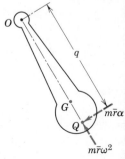

PROBLEMS

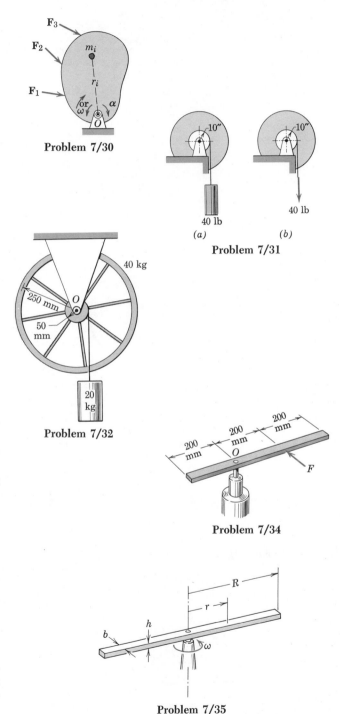

/30 Derive the moment equation of motion $\Sigma M_O = I_O \alpha$ for the plane rotation of a general rigid body about the fixed axis O. Start with the equation of motion for a representative particle of mass m_i of the body.

Problem 7/30

/31 Each of the two drums and connected hubs of 10-in. radius weighs 200 lb and has a radius of gyration about its center of 15 in. Calculate the angular acceleration of each drum. Friction in each bearing is negligible.

        Ans. $\alpha_a = 3.15 \text{ rad/sec}^2$, $\alpha_b = 3.43 \text{ rad/sec}^2$

Problem 7/31

/32 The 40-kg flywheel is turned by the action of the 20-kg cylinder supported by a cord wrapped around the 50-mm-radius hub. In the absence of friction determine the downward velocity of the cylinder after it has fallen 2 m from rest and find the force R on the bearing O during the motion. Neglect the mass of the hub, shaft, and spokes compared with the total mass of the flywheel. Also neglect the radial thickness of the flywheel rim compared with its radius.

Problem 7/32

/33 If the 20-kg cylinder in Prob. 7/32 acquires a downward velocity of 0.5 m/s after falling 2 m from rest, compute the constant resisting moment M_f due to friction in the bearing O of the flywheel.

        Ans. $M_f = 6.62 \text{ N·m}$

Problem 7/34

/34 The 600-mm slender bar has a mass of 10 kg and is pivoted freely about a vertical axis at O. A horizontal force F, which has an initial value of 150 N, is applied to the bar when it is at rest. Calculate the initial value of the horizontal component of the reaction at O. (*Suggestion:* The problem can be solved by using only one moment equation.)

/35 The uniform slender bar of density ρ rotates about a vertical axis through its mass center with a constant angular velocity ω. Determine an expression for the tensile stress σ in the bar as a function of r.

Problem 7/35

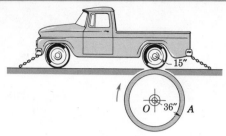

Problem 7/36

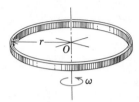

Problem 7/37

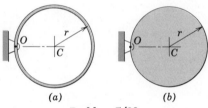

(a)   *(b)*

Problem 7/38

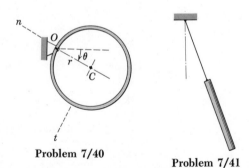

Problem 7/40

Problem 7/41

Problem 7/42

7/36 The automotive dynamometer is able to simulate road conditions for an acceleration of 0.6g for the loaded pickup truck with a gross weight of 5600 lb. Calculate the required moment of inertia of the dynamometer drum about its center O assuming that the drum turns freely during the acceleration phase of the test.   *Ans.* $I_O = 1565$ lb-ft-sec^2

7/37 The ring of mean radius r and small cross-sectional dimensions is rotating in a horizontal plane with a constant angular velocity ω about its center O. Determine the tension T in the ring. Solve, first, by analyzing one half of the ring as a free body and, second, by analyzing a differential element of arc as a free body.

7/38 Determine the angular acceleration and the force on the bearing at O for (a) the narrow ring of mass m and (b) the flat circular disk of mass m immediately after each is released from rest in the vertical plane with OC horizontal.

Ans. (a) $\alpha = g/(2r)$, $O = mg/2$
(b) $\alpha = 2g/(3r)$, $O = mg/3$

7/39 Solve for the bearing reactions in Prob. 7/38 using only one motion equation without finding the angular acceleration.

7/40 The narrow ring of mass m is free to rotate in the vertical plane about O. If the ring is released from rest at $\theta = 0$, determine expressions for the n- and t-components of the force at O in terms of θ.

Ans. $O_t = -\dfrac{mg}{2} \cos \theta$, $O_n = 2\,mg \sin \theta$

7/41 The uniform slender bar is suspended by a light cord from its end and swings as a pendulum. With the aid of a free-body diagram show that the bar cannot remain in line with the cord during the motion.

7/42 Gear A weighs 48 lb and has a centroidal radius of gyration of 6 in. Gear B weighs 16 lb and has a centroidal radius of gyration of 3 in. Calculate the angular acceleration of gear B when a torque M of 60 lb-in. is applied to the shaft of gear A. (Gear teeth are not shown.)   *Ans.* $\alpha_B = 20.1$ rad/sec^2

/43 The uniform slender bar AB has a mass of 8 kg and swings in a vertical plane about the pivot at A. If $\dot{\theta} = 2$ rad/s when $\theta = 30°$, compute the force supported by the pin at A at that instant.

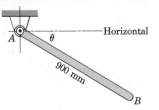

Horizontal

Problem 7/43

/44 For the conditions of Prob. 7/43 calculate the component of the pin force at A normal to the bar using only one equation of motion.   *Ans.* $A_t = 16.99$ N

/45 The uniform 72-ft mast weighs 600 lb and is hinged at its lower end to a fixed support at O. If the winch C develops a starting torque of 900 lb-ft, calculate the total force supported by the pin at O as the mast begins to lift off its support at B. Also find the corresponding angular acceleration α of the mast. The cable at A is horizontal, and the mass of the pulleys and winch is negligible.

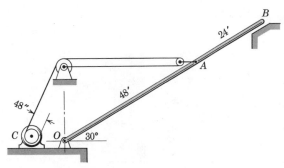

Problem 7/45

/46 For the conditions of Prob. 7/45 calculate the component of the pin reaction at O normal to the mast using only one motion equation not involving the angular acceleration of the mast.
Ans. $O_t = 129.9$ lb

/47 The solid circular disk of mass m with center at G is attached to the vertical shaft at O and rotates in the horizontal plane under the action of an applied couple M as shown. Determine the required length of time t, starting from rest, for the n- and t-components of the force exerted by the disk on the shaft to become momentarily equal.

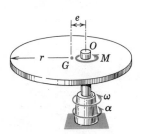

Problem 7/47

/48 The solid cylindrical rotor B has a mass of 43 kg and is mounted on its central axis C–C. The frame A rotates about the fixed vertical axis O–O under the applied torque $M = 30$ N·m. The rotor may be unlocked from the frame by withdrawing the locking pin P. Calculate the angular acceleration α of the frame A if the locking pin is (a) in place and (b) withdrawn. Neglect all friction and the mass of the frame.   *Ans.* (a) $\alpha = 8.46$ rad/s^2
(b) $\alpha = 11.16$ rad/s^2

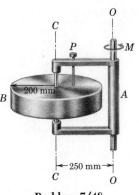

Problem 7/48

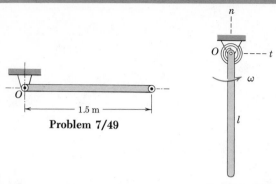

Problem 7/49

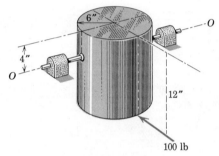

Problem 7/50

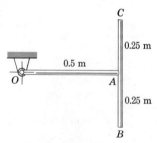

Problem 7/51

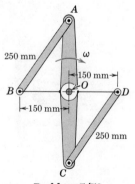

Problem 7/52

Problem 7/53

7/49 The uniform slender bar has a mass of 20 kg and is pivoted about a horizontal axis through O. The bearing at O fits closely on its shaft and exerts a constant frictional moment of 6 N·m on the bar opposing rotation. If the bar is released from rest in the horizontal position, compute the initial reaction on the bar at O.

7/50 The slender bar of mass m and length l is acted upon by a torsion spring at O which exerts a counterclockwise couple M on the bar as it swings past the vertical position. If the angular velocity of the bar is ω in the vertical position, write the expressions for the n- and t-components of the bearing reaction on the bar at this instant.

$$Ans.\ O_n = mg\left(1 + \frac{l\omega^2}{2g}\right),\ O_t = 3M/(2l)$$

7/51 The solid homogeneous cylinder weighs 300 lb and is free to rotate about the horizontal axis O–O. If the cylinder, initially at rest, is acted upon by the 100-lb force shown, calculate the horizontal component R of the force supported by each of the two symmetrically placed bearings when the 100-lb force is first applied.

7/52 Each of the two uniform slender bars OA and BC has a mass of 8 kg. The bars are welded at A to form a T-shaped member and are rotating freely about a horizontal axis through O. If the bars have an angular velocity ω of 4 rad/s as OA passes the horizontal position shown, calculate the total force R supported by the bearing at O.

$$Ans.\ R = 101.3\ N$$

7/53 Link AC is made to revolve with a constant angular velocity $\omega = 10$ rad/s about a fixed vertical axis through its center O. Each of the uniform links AB and CD has a mass of 4 kg and is held in the configuration shown by a cord leading perpendicularly to the rotating link AC. Calculate the tension T in BO and DO.

$$Ans.\ T = 30\ N$$

7/54 The bar A of mass m is formed into a 90° circular arc of radius r and attached to the hub by the light rods. The curved bar oscillates about the vertical axis under the action of a torsion spring B. At the instant under consideration the angular velocity is ω and the angular acceleration is α. Write expressions for the moment M exerted by the spring on the hub and the horizontal force R exerted by the shaft on the hub.

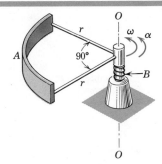

Problem 7/54

7/55 The 16-ft I-beam weighs 2000 lb and is held in the horizontal position by the pin at O and by the vertical cable which passes around the pulley at A and around the 500-lb motorized winch at B. If the winch motor has an output starting torque of 600 lb-ft, calculate the initial vertical force supported by the pin at O. Treat the beam as a slender bar and the winch unit as a mass concentrated at the center of the pulley. (Is the horizontal component of the force on the pin zero?)

Ans. $O_y = 1469$ lb, No

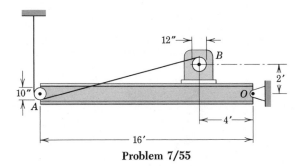

Problem 7/55

7/56 The moment supported by the torsion spring B of Prob. 7/54 is proportional to the angle through which the sector turns from its equilibrium position. If the sector is released from rest in a position where the moment is M_0 and the angle is θ_0, determine the horizontal force R on the bearing (a) as the sector is released and (b) when the sector has a maximum angular velocity.

7/57 The rim of the wheel weighs 100 lb and has a mean radius r of 18 in. The three spokes are spaced 90° apart, and each spoke is a uniform 15-lb rod whose length may be taken to be 18 in. If a torque M of 400 lb-in. is applied to the wheel through its vertical shaft at O, calculate the horizontal component of the bearing reaction at O as the wheel starts from rest. Neglect the mass of the hub.

Ans. $O_t = 1.45$ lb

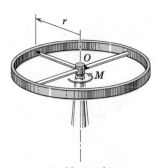

Problem 7/57

7/58 A uniform metal rod with a mass of 0.60 kg per meter of length is bent into the shape shown in the vertical plane and pivoted freely about a horizontal axis at O normal to the plane. If the rod is released from rest in the position shown with the straight section horizontal, compute the initial x- and y-components of the reaction **R** supported by the bearing at O. *Ans.* $R_x = 0.339$ N, $R_y = 4.79$ N

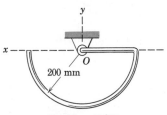

Problem 7/58

Problem 7/59

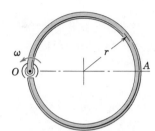

Problem 7/60

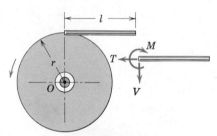

Problem 7/61

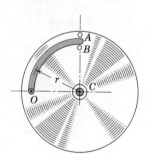

Problem 7/62

7/59 The uniform semicircular bar of mass m and radius r is hinged freely about a horizontal axis through A. If the bar is released from rest in the position shown where AB is horizontal, determine the initial angular acceleration α of the bar and the expression for the force exerted on the bar by the pin at A. (Note carefully that the initial tangential acceleration of the mass center is not vertical.)

▶**7/60** The split ring of mass m and radius r is rotating in its plane with an angular velocity ω about a fixed vertical axis through one end at O. Determine the bending moment M and shear force V in the ring at A if ω is constant.

$$Ans.\ M = \frac{2}{\pi}\,mr^2\omega^2,\ V = mr\omega^2/2$$

▶**7/61** The uniform slender rod of length l and mass m is welded at its end tangent to the rim of the circular disk of radius r which rotates about a vertical axis through O. Determine the bending moment M, the shear force V, and the axial force T which the weld exerts on the rod (*a*) for a constant angular velocity ω of the disk and (*b*) as the disk starts from rest with a counterclockwise angular acceleration α.

$$Ans.\ (a)\ M = \frac{mrl\omega^2}{2},\ V = mr\omega^2,\ T = \frac{ml\omega^2}{2}$$

$$(b)\ M = -\frac{ml^2\alpha}{3},\ V = -\frac{ml\alpha}{2},\ T = mr\alpha$$

▶**7/62** The curved bar of mass m is hinged to the rotating disk at O and bears against one of the smooth pins A and B which are fastened to the disk. If the disk rotates about its vertical axis C, determine the force exerted on the bar by the hinge at O and the reaction A or B on the bar (*a*) if the disk has a constant angular velocity ω and (*b*) as the disk starts from rest with a counterclockwise angular acceleration α.

$$Ans.\ (a)\ O = \frac{2mr\omega^2}{\pi},\ A = \frac{2mr\omega^2}{\pi}$$

$$(b)\ O = mr\alpha\sqrt{1 + \frac{4}{\pi^2}},\ B = mr\alpha\left(1 - \frac{2}{\pi}\right)$$

7/5 GENERAL PLANE MOTION. The dynamics of general plane motion of a rigid body combines translation and rotation. In Art. 7/2 we represented such a body with its free-body diagram and its force and moment resultants in Fig. 7/2, which is repeated here for convenient reference. The equations of motion which apply to this general case, Eqs. 7/1, are also repeated here.

$$\boxed{\begin{aligned} \Sigma F_y &= m\bar{a}_y \\ \Sigma F_x &= m\bar{a}_x \\ \Sigma \bar{M} &= \bar{I}\alpha \end{aligned}} \qquad [7/1]$$

The x- and y-directions are any two mutually perpendicular directions which are convenient. As in the previous two articles, our procedure is again to establish the equivalence between the externally applied forces, as disclosed by the free-body diagram, and their force and moment resultants, as represented by the resultant-force diagram.

    In working a problem in general plane motion we first observe whether the motion is unconstrained or constrained, as illustrated in the examples of Fig. 7/3. If the motion is constrained, we must account for the kinematic relationship between the linear and the angular accelerations and incorporate it into our force and moment equations of motion. If the motion is unconstrained, the accelerations can be determined independently of one another by direct application of the three motion equations, Eqs. 7/1.

    In Art. 7/2 we noted from Fig. 7/2 that our resultant-force diagram enabled us to choose any convenient moment center, such as O, and write

$$\boxed{\Sigma M_O = \bar{I}\alpha + m\bar{a}d} \qquad [7/2]$$

Our experience with mechanics has already shown us the advantages of choosing convenient moment centers, both in statics and in dy-

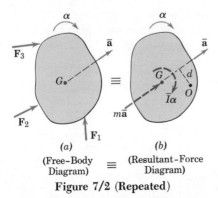

(*a*)		(*b*)
(Free–Body	$\equiv$	(Resultant–Force
Diagram)		Diagram)

Figure 7/2 (Repeated)

namics, and the kinetics of general plane motion provides us with further example. As before the equivalence between the free-body diagram and the resultant-force diagram enables us to select the most convenient moment center for the moment equation. The advantage of such a selection is generally a reduction in the number of separate motion equations which must be solved.

Interconnected bodies in plane motion may often be analyzed by considering them together as a single system, thus eliminating reference to the forces in their connections. The forces external to the system as disclosed by the free-body diagram of the system must be equivalent to the resultant forces and the resultant couples as illustrated in Fig. 7/4. Treating interconnected bodies as a system, however, does not relieve us of the necessity to establish the kinematic relationships between the linear and the angular accelerations of the members.[°]

Mention should be made of two special cases in plane motion which occur with sufficient frequency to warrant special attention. The first case occurs when the moment center O, as a point on the body or body extended, has no acceleration. The moment equation about O then becomes

$$\Sigma M_O = I_O \alpha \qquad (7/10)$$

which meets the same conditions as for a body rotating about a fixed axis through O. Point O need not be fixed but could have a constant velocity.

The second case of frequent occurrence exists when a moment center O is chosen which has an acceleration directed toward or away from G, Fig. 7/8a. The acceleration of G, written in terms of the acceleration of O, has the components a_O, $\bar{r}\omega^2$, and $\bar{r}\alpha$ so that the resultant force $m\bar{a}$ has the components ma_O, $m\bar{r}\omega^2$, and $m\bar{r}\alpha$ as shown in the b-part of the figure. The moment sum about O becomes $\Sigma M_O = \bar{I}\alpha + m\bar{r}^2\alpha$. Substitution of $I_O = \bar{I} + m\bar{r}^2$ gives

$$\Sigma M_O = I_O \alpha \qquad (7/10a)$$

Figure 7/8c shows the frequently-encountered example of the foregoing situation which occurs for a rolling wheel with mass center G at the geometric center. Here the instantaneous center C of zero velocity has an acceleration directed toward the mass center, and therefore the equation

$$\Sigma M_C = I_C \alpha \qquad (7/11)$$

may be used for the moment equation as long as the wheel is not

---

[°] When an interconnected system has more than one degree of freedom, that is, requires more than one coordinate to specify completely the configuration of the system, the more advanced equations of Lagrange are generally used.

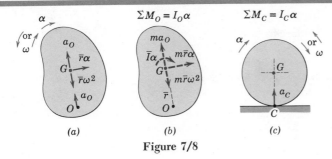

$$\Sigma M_O = I_O \alpha \qquad \Sigma M_C = I_C \alpha$$

Figure 7/8

slipping. If the wheel slips or if the mass center is not the geometric center, then the acceleration of the contact point C would not in general pass through G and Eq. 7/11 would not hold.

Strong emphasis should be placed on the importance of a clear choice of the body to be isolated and the representation of this isolation by a correct free-body diagram. Only after this vital step has been completed can we properly evaluate the equivalence between the external forces and their resultants. Of equal importance in the analysis of plane motion is a clear understanding of the kinematics involved. Very often the difficulties experienced at this point have to do with kinematics, and a thorough review of the relative acceleration relations for plane motion will be most helpful. In formulating the solution to a problem we recognize that the directions of certain forces or accelerations may not be known at the outset, so that it may be necessary to make initial assumptions whose validity will be proved or disproved when the solution is carried out. It is essential, however, that all assumptions made be consistent with the principle of action and reaction and with any kinematical requirements, which are also called conditions of constraint. Thus if a wheel is rolling on a horizontal surface, its center is constrained to move on a horizontal line. Furthermore, if the unknown linear acceleration a of the center of the wheel is assumed positive to the right, the unknown angular acceleration α must be positive in a clockwise sense in order that $a = +r\alpha$, assuming the wheel does not slip. Also, we note that for a wheel which rolls without slipping $a = r\alpha$, but the friction force F between the wheel and its supporting surface is generally less than its maximum value, so that $F \neq \mu N$. But if the wheel slips as it rolls, $a \neq r\alpha$ although the friction force has reached its limiting value so that $F = \mu N$. It may be necessary to test the validity of either assumption in a given problem.

Sample Problem 7/5

The drum A is given a constant angular acceleration α_0 of 3 rad/s² and causes the 70-kg spool B to roll on the horizontal surface by means of the connecting cable which wraps around the inner hub of the spool. The radius of gyration $\bar{k}$ of the spool about its mass center G is 250 mm, and the coefficient of friction between the spool and the horizontal surface is 0.25. Determine the tension T in the cable and the friction force F exerted by the horizontal surface on the spool.

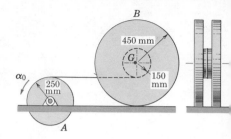

Solution. The free-body diagram and the resultant-force diagram of the spool are drawn as shown. The correct direction of the friction force may be assigned in this problem by observing from both diagrams that with counterclockwise angular acceleration a moment sum about point G (and also about point D) must be counterclockwise. A point on the connecting cable has an acceleration $a_t = r\alpha = 0.25(3) = 0.75$ m/s², which is also the horizontal component of the acceleration of point D on the spool. It will be assumed initially that the spool rolls without slipping, in which case it would have a counterclockwise angular acceleration
① $\alpha = (a_D)_x/\overline{DC} = 0.75/0.30 = 2.5$ rad/s². The acceleration of the mass center G is, therefore, $\bar{a} = r\alpha = 0.45(2.5) = 1.125$ m/s².

With the kinematics determined, the forces may be found from the equivalence between the free-body diagram and the resultant-force diagram. Thus

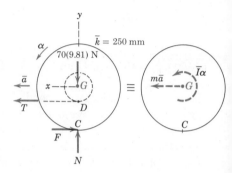

② $[\Sigma M_C = \bar{I}\alpha + m\bar{a}r]$ $0.3T = 70(0.25)^2(2.5) + 70(1.125)(0.45)$

$$T = 154.6 \text{ N} \qquad \textit{Ans.}$$

The friction force may be obtained by the force summation

$[\Sigma F_x = m\bar{a}_x]$ $154.6 - F = 70(1.125)$ $F = 75.8$ N   *Ans.*

Establishing the validity of the answer requires a check of the assumption that no slipping occurs. With

$[\Sigma F_y = 0]$ $N = 687$ N

the surfaces are capable of supporting a maximum friction force equal to $F = \mu N = 0.25(687) = 171.7$ N. Thus the assumption that the spool rolls without slipping is valid since a friction force of only 75.8 N is required to maintain pure rolling.

If the coefficient of friction had been 0.1, for example, then the friction force would have been limited to $0.1(687) = 68.7$ N which is less than 75.8 N, and the spool would slip. In this event the kinematical relation $\bar{a} = r\alpha$ would no longer hold. With $(a_D)_x$ known, the angular
③ acceleration would be $\alpha = [\bar{a} - (a_D)_x]/\overline{GD}$. The correct acceleration $\bar{a}$ of G would then be obtained from a moment equation about point D, and a force equation in the x-direction would determine the new value of T.

Alternatively, the given problem with the assumption of no slipping could be solved by using Eq. 7/11 directly. This equation is identical with the moment equation about C which was obtained from the resultant-force diagram. Also, an equation for moments about point D could be written initially to give F after which T could be obtained from a force equation.

① The relation between $\bar{a}$ and α is the kinematical requirement which accompanies the constraint to the motion imposed by the assumption that the spool rolls without slipping.

② The choice of C for a moment center eliminates the unknowns F and N. The moment arm of T is $0.450 - 0.150 = 0.3$ m. Be careful not to make the mistake of using $\frac{1}{2}mr^2$ for $\bar{I}$ of the spool, which is not a uniform circular disk.

③ Our principles of relative acceleration are a necessity here. Hence the relation $(a_{G/D})_t = \overline{GD}\alpha$ should be recognized.

Sample Problem 7/6

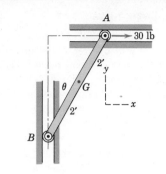

The slender bar AB weighs 60 lb and moves in the vertical plane with its ends constrained to follow the smooth horizontal and vertical guides. If the 30-lb force is applied at A with the bar initially at rest in the position for which $\theta = 30°$, calculate the resulting angular acceleration of the bar and the forces on the small end rollers at A and B.

Solution. The bar undergoes constrained motion, so that we must establish the relationship between the mass-center acceleration and the angular acceleration. The relative-acceleration equation $\mathbf{a}_A = \mathbf{a}_B + \mathbf{a}_{A/B}$ must be solved first, and then the equation $\bar{\mathbf{a}} = \mathbf{a}_G = \mathbf{a}_B + \mathbf{a}_{G/B}$ is next solved to obtain expressions relating $\bar{a}$ and α. The corresponding acceleration polygons are shown, and their solution gives

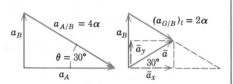

$$\bar{a}_x = \bar{a} \cos 30° = 2\alpha \cos 30° = 1.732\alpha \text{ ft/sec}^2$$

$$\bar{a}_y = \bar{a} \sin 30° = 2\alpha \sin 30° = 1.0\alpha \text{ ft/sec}^2$$

Next we construct the free-body diagram and the resultant-force diagram as shown. With $\bar{a}_x$ and $\bar{a}_y$ now known in terms of α, the remaining unknowns are α and the forces A and B. We could apply Eqs. 7/1 as they stand and solve the three equations simultaneously for the three unknowns. However, we may avoid this labor by a careful choice of moment center. If we choose point C, we automatically eliminate two of the unknowns, A and B, which allows us to solve for α directly. Our equation for moments about C is

$$[\Sigma M_C = \bar{I}\alpha + \Sigma m\bar{a}d] \quad 30(4 \cos 30°) - 60(2 \sin 30°) = \frac{1}{12}\frac{60}{32.2}(4^2)\alpha$$

$$+ \frac{60}{32.2}(1.732\alpha)(2 \cos 30°) + \frac{60}{32.2}(1.0\alpha)(2 \sin 30°)$$

$$43.9 = 9.94\alpha, \qquad \alpha = 4.42 \text{ rad/sec}^2 \qquad Ans.$$

With α determined we can now apply the force equations independently and get

$$[\Sigma F_y = m\bar{a}_y] \qquad A - 60 = \frac{60}{32.2}(1.0)(4.42) \qquad A = 68.2 \text{ lb} \qquad Ans.$$

$$[\Sigma F_x = m\bar{a}_x] \qquad 30 - B = \frac{60}{32.2}(1.732)(4.42) \qquad B = 15.74 \text{ lb} \qquad Ans.$$

If we had applied Eqs. 7/1 directly following acquisition of the kinematic equations of constraint, we would have

$$[\Sigma\bar{M} = \bar{I}\alpha] \quad 30(2 \cos 30°) - A(2 \sin 30°) + B(2 \cos 30°) = \frac{1}{12}\frac{60}{32.2}(4^2)\alpha$$

$$[\Sigma F_x = m\bar{a}_x] \qquad 30 - B = \frac{60}{32.2}(1.732\alpha)$$

$$[\Sigma F_y = ma_y] \qquad A - 60 = \frac{60}{32.2}(1.0\alpha)$$

Solving the three equations simultaneously gives us the results obtained previously.

① If the application of the relative-acceleration equations is not perfectly clear at this point, then review Art. 6/6. Note that the relative normal acceleration term is absent since there is no angular velocity of the bar.

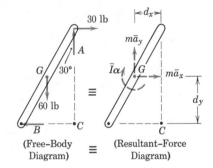

(Free–Body Diagram) $\equiv$ (Resultant–Force Diagram)

② Recall that the moment of inertia of a slender rod about its center is $\frac{1}{12}ml^2$. From the resultant-force diagram the terms $\Sigma m\bar{a}d$ are $m\bar{a}_x d_y + m\bar{a}_y d_x$. Since both are clockwise in the sense of $\bar{I}\alpha$, they are positive.

Sample Problem 7/7

A car door is inadvertently left slightly open when the brakes are applied to give the car a constant rearward acceleration a. Derive expressions for the angular velocity of the door as it swings past the 90° position and the components of the hinge reactions for any value of θ. The mass of the door is m, its mass center is a distance $\bar{r}$ from the hinge axis O, and the radius of gyration about O is k_O.

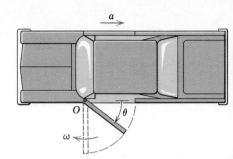

Solution. Because the angular velocity ω increases with θ, we need to find how the angular acceleration α varies with θ so that we may integrate it over the interval to obtain ω. We obtain α from a moment equation about O. First we draw the free-body diagram of the door in the horizontal plane for a general position θ. The only forces in this plane are the components of the hinge reaction shown here in the x- and y-directions. In addition to the resultant couple $\bar{I}\alpha$ shown in the sense of α, we represent the resultant force $m\bar{a}$ in terms of its components by using an equation of relative acceleration with respect to O. This equation becomes the kinematical equation of constraint and is

$$\bar{a} = a_G = a_O + (a_{G/O})_n + (a_{G/O})_t$$

The magnitudes of the $m\bar{a}$ components are, then,

$$ma_O = ma, \qquad m(a_{G/O})_n = m\bar{r}\omega^2, \qquad m(a_{G/O})_t = m\bar{r}\alpha$$

where $\omega = \dot{\theta}$ and $\alpha = \ddot{\theta}$.

For a given angle θ the three unknowns are α, O_x, and O_y. We can eliminate O_x and O_y by a moment equation about O, which gives

$$[\Sigma M_O = \bar{I}\alpha + \Sigma m\bar{a}d] \qquad 0 = m(k_O{}^2 - \bar{r}^2)\alpha + m\bar{r}\alpha(\bar{r}) - ma(\bar{r}\sin\theta)$$

Solving for α gives

$$\alpha = \frac{a\bar{r}}{k_O{}^2}\sin\theta$$

Now we integrate α first to a general position and get

$$[\omega\, d\omega = \alpha\, d\theta] \qquad \int_0^\omega \omega\, d\omega = \int_0^\theta \frac{a\bar{r}}{k_O{}^2}\sin\theta\, d\theta$$

$$\omega^2 = \frac{2a\bar{r}}{k_O{}^2}(1 - \cos\theta)$$

For $\theta = \pi/2$

$$\omega = \frac{1}{k_O}\sqrt{2a\bar{r}} \qquad\qquad Ans.$$

To find O_x and O_y for any given value of θ the force equations give

$$[\Sigma F_x = m\bar{a}_x] \qquad O_x = ma - m\bar{r}\omega^2\cos\theta - m\bar{r}\alpha\sin\theta$$

$$= m\left(a - \frac{2a\bar{r}^2}{k_O{}^2}[1 - \cos\theta]\cos\theta - \frac{a\bar{r}^2}{k_O{}^2}\sin^2\theta\right)$$

$$= ma\left(1 - \frac{\bar{r}^2}{k_O{}^2}[1 + 2\cos\theta - 3\cos^2\theta]\right) \qquad Ans.$$

$$[\Sigma F_y = m\bar{a}_y] \qquad O_y = m\bar{r}\alpha\cos\theta - m\bar{r}\omega^2\sin\theta$$

$$= m\bar{r}\frac{a\bar{r}}{k_O{}^2}\sin\theta\cos\theta - m\bar{r}\frac{2a\bar{r}}{k_O{}^2}(1 - \cos\theta)\sin\theta$$

$$= \frac{ma\bar{r}^2}{k_O{}^2}(3\cos\theta - 2)\sin\theta \qquad Ans.$$

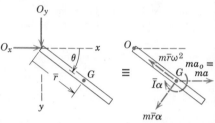

(Free-Body Diagram) $\equiv$ (Resultant-Force Diagram)

① Point O is chosen because it is the only point on the door whose acceleration is known.

② Be careful to place $m\bar{r}\alpha$ in the sense of positive α with respect to rotation about O.

③ The free-body diagram shows that there is zero moment about O. We use the transfer-of-axis theorem here and substitute $k_O{}^2 = \bar{k}^2 + \bar{r}^2$. If this relation is not totally familiar, review Art. A1 in Appendix A.

④ The resultant-force diagram shows clearly the terms which make up $m\bar{a}_x$ and $m\bar{a}_y$.

PROBLEMS

63 The 60-lb wheel has a centroidal radius of gyration of 6 in. If friction is sufficient to prevent slipping, compute the friction force F acting on the wheel during its downhill roll. What is the minimum coefficient of friction to prevent slipping?

Ans. $F = 7.10$ lb, $\mu_{\text{min}} = 0.13$

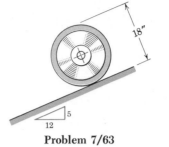

Problem 7/63

64 The solid circular cylinder with a cord wrapped around its periphery and fastened to the fixed support is released from rest on the incline. The minimum coefficient of friction μ to maintain static equilibrium is $\frac{1}{2} \tan \theta$. If μ is less than this value, find the acceleration a of the center of the cylinder.

Ans. $a = \frac{2}{3}(\sin \theta - 2\mu \cos \theta)g$

Problem 7/64

65 A long cable of length L and mass ρ per unit length is wrapped around the periphery of a spool of negligible mass. One end of the cable is fixed, and the spool is released from rest in the position shown. Show that the center of the spool has a constant acceleration during the motion.

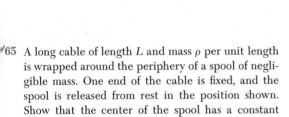

Problem 7/65

66 Calculate the acceleration of the center G of the attached wheels for the system shown. Each cable is wrapped securely around its respective periphery.

Ans. $a = 0.665$ m/s² up

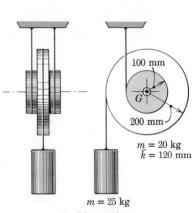

$m = 20$ kg
$\bar{k} = 120$ mm

$m = 25$ kg

Problem 7/66

67 For a given coefficient of friction μ determine the maximum angle θ which will permit the solid circular disk to roll down the incline without slipping.

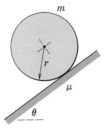

Problem 7/67

68 If the coefficient of friction between the wheel of Prob. 7/63 and the incline is 0.10, calculate the acceleration of the center of the wheel and its angular acceleration. Do the results depend on the weight of the wheel?

Ans. $a = 9.41$ ft/sec², $\alpha = 8.92$ rad/sec², No

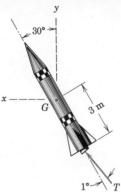

Problem 7/69

7/69 Above the earth's atmosphere at an altitude of 400 km where the acceleration due to gravity is 8.69 m/s² a certain rocket has a total remaining mass of 300 kg and is directed 30° from the vertical. If the thrust T from the rocket motor is 4 kN and if the rocket nozzle is tilted through an angle of 1° as shown, calculate the angular acceleration α and the x- and y-components of the acceleration of the mass center G. The rocket has a centroidal radius of gyration of 1.5 m.

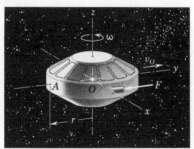

Problem 7/70

7/70 The spacecraft is spinning with a constant angular velocity ω about the z-axis at the same time that its mass center O is traveling with a velocity v_O in the y-direction. If a tangential hydrogen-peroxide jet is fired when the craft is in the position shown, determine the expression for the absolute acceleration of point A on the spacecraft rim at the instant the jet force is F. The radius of gyration of the craft about the z-axis is k, and its mass is m.

$$\text{Ans. } \mathbf{a}_A = -\frac{Fr^2}{mk^2}\mathbf{i} - \left(\frac{F}{m} - r\omega^2\right)\mathbf{j}$$

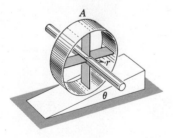

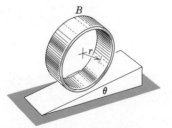

Problem 7/71

7/71 Determine the angular acceleration of each of the two wheels as they roll without slipping down the inclines. For wheel A investigate the case where the mass of the rim and spokes is negligible and the mass of the bar is concentrated along its center line. For wheel B assume that the thickness of the rim is negligible compared with its radius so that all of the mass is concentrated in the rim. Also specify the minimum coefficient of friction μ required to prevent each wheel from slipping.

$$\text{Ans. } \alpha_A = \frac{g \sin \theta}{r}, \ \mu_A = 0$$

$$\alpha_B = \frac{g \sin \theta}{2r}, \ \mu_B = \tfrac{1}{2} \tan \theta$$

72 The mass center G of the 20-lb wheel is off center by 0.50 in. If G is in the position shown as the wheel rolls without slipping through the bottom of the circular path of 6-ft radius with an angular velocity ω of 10 rad/sec, compute the force P exerted by the path on the wheel. (Be careful to use the correct mass-center acceleration.)   *Ans.* $P = 20.2$ lb

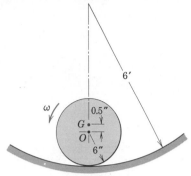

Problem 7/72

73 What should be the radius r_0 of the circular groove in order that there be no friction force acting between the wheel and the horizontal surface regardless of the magnitude of the force P applied to the cord? The centroidal radius of gyration of the wheel is $\overline{k}$.

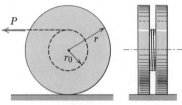

Problem 7/73

74 The circular disk of 200-mm radius has a mass of 25 kg with centroidal radius of gyration of 175 mm and has a concentric circular groove of 75 mm radius cut into it. If a steady horizontal force of 20 N is applied to a cord wrapped around the groove as shown, calculate the angular acceleration α of the disk as it starts from rest. The coefficient of friction between the disk and the horizontal surface is 0.10. (Be sure to recognize that the wheel rolls clockwise and *not* counterclockwise. Assume first that the wheel does not slip and then verify your assumption with the results.)   *Ans.* $\alpha = 1.416$ rad/s^2

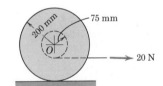

Problem 7/74

75 In the design of a space station, remote from gravitational attraction, a long tubular antenna is to be deployed in the following way. Thin spring-steel tape of length L and mass ρ per unit of length is preformed into a single split tube, after which it is flattened and wound around a spool of radius r. The bending moment in the tape as it leaves the spool tangentially is M. This moment acts to unroll the tape, and the spool is pushed away from the clamped end A of the tube. The guiding bracket contains the tape and keeps it tangent to the spool as it unwinds. Neglect the mass of the spool and bracket compared with the mass of the tape and determine the acceleration a of the center of the spool and the compressive force P in the tube as the roll accelerates away from A. Neglect any variation in r and assume that the clamp A is fixed in space. The roll may be analyzed with the free-body diagram shown.

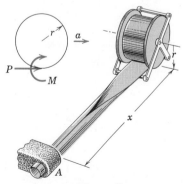

Problem 7/75

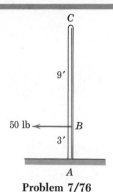

Problem 7/76

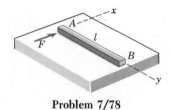

Problem 7/78

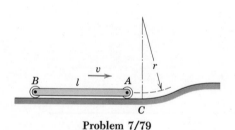

Problem 7/79

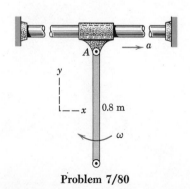

Problem 7/80

7/76 The uniform pole AC weighs 64.4 lb and is balanced in the vertical position when a horizontal force of 50 lb is suddenly applied to the rope at B. In the absence of friction between the pole and the horizontal surface determine the initial acceleration of end C.

7/77 In Prob. 7/76 if the coefficient of friction between the vertical pole and the horizontal surface is 0.30, calculate the initial acceleration of end C.

Ans. $a_C = 6.82 \text{ ft/sec}^2$ to the left

7/78 The uniform slender bar of length l and mass m is resting on a smooth horizontal surface. If a horizontal force F is applied normal to the bar at end A, determine the expression for the acceleration a_B of end B during application of F.

7/79 The uniform heavy bar AB of mass m is moving on its light end rollers along the horizontal with a velocity v when end A passes point C and begins to move on the curved portion of the path with radius r. Determine the force exerted by the path on the roller A immediately after it passes C.

Ans. $A = mg\left(\dfrac{1}{2} + \dfrac{v^2}{3gr}\right)$

7/80 End A of the uniform 5-kg bar is pinned freely to the collar which has an acceleration $a = 4 \text{ m/s}^2$ along the fixed horizontal shaft. If the bar has a clockwise angular velocity $\omega = 2 \text{ rad/s}$ as it swings past the vertical, determine the components of the force on the bar at A for this instant.

/81 A 15-kg roll of wrapping paper in the form of a solid circular cylinder with a diameter of 300 mm is resting on a horizontal surface. If a force of 40 N is applied to the paper at the angle $\theta = 30°$, determine the initial acceleration a of the center of the roll and the corresponding angular acceleration α. The coefficient of friction between the paper and the horizontal surface is 0.2.

Ans. $a = 0.614$ m/s², $\alpha = 12.95$ rad/s²

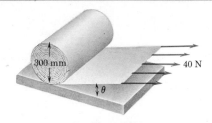

Problem 7/81

/82 Each of the solid circular disk wheels has a mass of 2 kg, and the inner solid cylinder has a mass of 3 kg. The disks and cylinder are mounted on the small central shaft so that each can rotate independently of the other with negligible friction in the bearings. Calculate the acceleration of the center of the wheels when the 20-N force is applied as shown. The coefficient of friction between the wheels and the horizontal surface is 0.30.

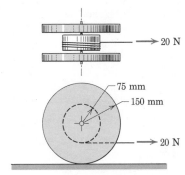

Problem 7/82

/83 The truck, initially at rest with a solid cylindrical roll of paper in the position shown, moves forward with a constant acceleration a. Find the distance s which the truck goes before the paper rolls off the edge of its horizontal bed. Friction is sufficient to prevent slipping. *Ans.* $s = 3d/2$

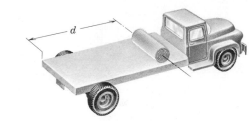

Problem 7/83

/84 The uniform steel beam is 5 m long and has a mass of 500 kg. If the supporting cable CB breaks, determine the tension T in the remaining cable AC an instant after the break occurs. The beam may be treated as a slender bar.

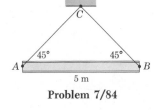

Problem 7/84

/85 A uniform rod of length l and mass m is secured to a circular hoop of radius l as shown. The mass of the hoop is negligible. If the rod and hoop are released from rest on a horizontal surface in the position illustrated, determine the initial values of the friction force F and normal force N under the hoop if friction is sufficient to prevent slipping.

Ans. $F = \frac{3}{8}mg$, $N = \frac{13}{16}mg$

Problem 7/85

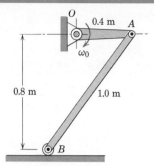

Problem 7/86

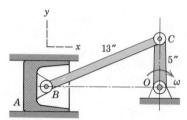

Problem 7/87

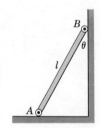

Problem 7/88

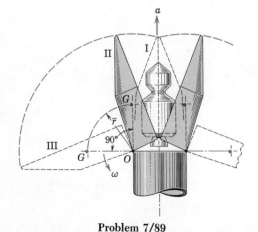

Problem 7/89

7/86 The crank OA rotates in the vertical plane with a constant clockwise angular velocity ω_0 of 4.5 rad/s. For the position where OA is horizontal calculate the force under the light roller B of the 10-kg slender bar AB.   *Ans.* $B = 36.4$ N

7/87 Calculate the x- and y-components of the force on the piston pin B of the slider-crank mechanism for the position shown where the crank OC passes the perpendicular to OB. The crank turns with a constant angular speed of 1200 rev/min. The connecting rod weighs 2 lb and may be treated as a uniform slender bar. The piston weighs 1.8 lb, and any gas pressure and friction on it are negligible. Also neglect the weight of BC compared with the other forces acting.   *Ans.* $B_x = 153.3$ lb, $B_y = 13.2$ lb

7/88 The uniform slender bar of mass m and length l is released from rest in the position shown. If friction against the vertical and horizontal surfaces is negligible, determine the expression for the initial angular acceleration α of the bar.

7/89 The fairing which covers the spacecraft package in the nose of the booster rocket is jettisoned when the rocket is in space where gravitational attraction is negligible. A mechanical actuator moves the two halves slowly from the closed position I to position II at which point the fairings are released to rotate freely about their hinges at O under the influence of a constant acceleration a of the rocket. When position III is reached, the hinge at O is released and the fairings drift away from the rocket. Determine the angular velocity ω of the fairings at the 90° position. The mass of each fairing is m with center of mass at G and radius of gyration k_O about O.

90 The unbalanced wheel weighs 64.4 lb with center of mass located 3 in. from the center O. The radius of gyration about G is 8 in. Compute the normal component of the force of contact at C for the position shown as the wheel rolls without slipping down the 15° incline. The wheel has an angular velocity $\omega = 2$ rad/sec at this particular instant.

Ans. $N = 55.3$ lb

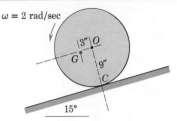

Problem 7/90

91 A uniform square plate with a mass of 40 kg is suspended by the two vertical wires attached to corners A and B. If the wire at B suddenly breaks, compute the tension induced in the wire at A immediately after the break occurs.

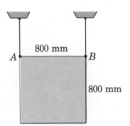

Problem 7/91

92 The circular disk of mass m and radius r is released from rest with θ essentially zero and rolls without slipping on the circular guide of radius R. Determine the expression for the normal force N between the disk and the guide in terms of θ. (Be careful to use the correct acceleration of the mass center.)

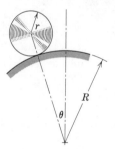

Problem 7/92

93 A bowling ball with a circumference of 27 in. weighs 14 lb and has a radius of gyration of 3.28 in. If the ball is released with a velocity of 20 ft/sec but with no angular velocity as it touches the alley floor, compute the distance traveled by the ball before it begins to roll without slipping. The coefficient of friction between the ball and the floor is 0.20

Ans. $s = 18.7$ ft

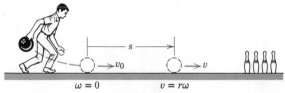

Problem 7/93

94 The slender bar of mass m is released from rest in the position shown. Determine the normal force N exerted on the roller by the supporting horizontal surface an instant after release. Neglect the mass of the small roller.

Problem 7/94

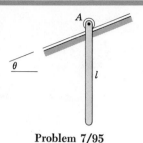

Problem 7/95

7/95 The slender rod of mass m and length l is released from rest in the vertical position with the small roller at end A resting on the incline. Determine the initial acceleration of A.

$$Ans. \; a_A = \frac{g \sin \theta}{1 - \frac{3}{4} \cos^2 \theta}$$

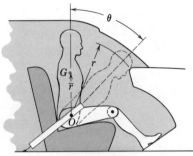

Problem 7/96

▶7/96 In a study of head injury against the instrument panel of a car during sudden or crash stops where lap belts without shoulder straps are used, the segmented human model shown in the figure is analyzed. The hip joint O is assumed to remain fixed relative to the car, and the torso above the hip is treated as a rigid body of mass m freely pivoted at O. The center of mass of the torso is at G with the initial position of OG taken as vertical. The radius of gyration of the torso about O is k_O. If the car is brought to a sudden stop with a constant deceleration a, determine the velocity v relative to the car with which the model's head strikes the instrument panel. Substitute the values $m = 50$ kg, $\bar{r} = 450$ mm, $r = 800$ mm, $k_O = 550$ mm, $\theta = 45°$, and $a = 10g$ and compute v.

Ans. $v = 11.73$ m/s

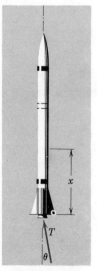

Problem 7/97

▶7/97 A rocket is given a thrust T at an angle θ with its axis in order to change the direction of its motion. Treat the rocket as a uniform slender bar of length l and mass m, and write the expression for the bending moment M in the rocket as a function of x when the rocket is in the vertical position shown.

$$Ans. \; M = \left(\frac{l - x}{l}\right)^2 xT \sin \theta$$

SECTION B. WORK AND ENERGY

/6 WORK-ENERGY RELATIONS.   In our study of the kinetics of
particles in Arts. 3/6 and 3/7 of Chapter 3 we developed and
applied the principles of work and energy to the motion of a particle
and to selected cases of connected particles. We found that these
principles were especially useful in describing motion resulting from
the cumulative effect of forces acting through distances. Furthermore, when the forces were conservative, we were able to determine
velocity changes by analyzing the energy conditions at the beginning
and end of the motion interval. For finite displacements the work-
energy method eliminates the necessity for determining the acceleration and integrating it over the interval to obtain the velocity change.
These same advantages are realized when we extend the work-energy
principles to describe rigid-body motion. Before carrying out this
extension it is strongly recommended that the definitions and concepts of work, kinetic energy, gravitational and elastic potential
energy, conservative forces, and power treated in Arts. 3/6 and 3/7
be carefully reviewed. We shall assume familiarity with these quantities as we apply them to rigid-body problems. In part (b) of Art. 5/2
in Chapter 5 on the kinetics of systems of particles we extended the
principles of Arts. 3/6 and 3/7 to encompass any general system of
mass particles, which includes rigid bodies, and this discussion should
also be reviewed.

   (a) *Work of Forces and Couples.*   The work done by a force **F**
has been treated in detail in Art. 3/6 and is given by

$$U = \int \mathbf{F} \cdot d\boldsymbol{r} \qquad \text{or} \qquad U = \int (F \cos \alpha) \, ds$$

where $d\mathbf{r}$ is the infinitesimal vector displacement of the point of
application of **F** during time dt. In the equivalent scalar form of the
integral α is the angle between **F** and the direction of the displacement and ds is the magnitude of the vector displacement $d\boldsymbol{r}$. We
frequently have occasion to evaluate the work done by a couple M
which acts on a rigid body during its motion. Figure 7/9 shows a
couple $M = Fb$ acting on a rigid body which moves in the plane of
the couple. During time dt the body rotates through an angle $d\theta$, and
line AB moves to $A'B'$. We may consider this motion in two parts,
first a translation to $A'B''$ and then a rotation $d\theta$ about A'. We see
immediately that during the translation the work done by one of the
forces cancels that due to the other force, so that the net work done is
$dU = F(b \, d\theta) = M \, d\theta$ due to the rotational part of the motion. If the
couple acts in the opposite sense to the rotation, the work done is
negative. During a finite rotation the work done by a couple M whose
plane is parallel to the plane of motion is, therefore,

$$U = \int M \, d\theta$$

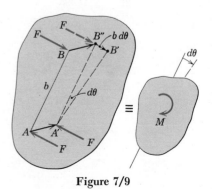

Figure 7/9

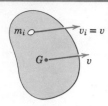

(a) Translation

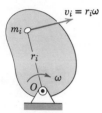

(b) Fixed–Axis
Rotation

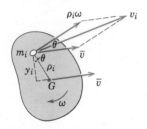

(c) General Plane
Motion

Figure 7/10

(b) **Kinetic Energy.** We now use the familiar expression for the kinetic energy of a particle to develop expressions for the kinetic energy of a rigid body for each of the three classes of rigid-body plane motion illustrated in Fig. 7/10.

(Translation) The translating rigid body of Fig. 7/10*a* has a mass m and all of its particles have a common velocity v. The kinetic energy of any particle of mass m_i of the body is $T_i = \frac{1}{2}m_i v^2$, so for the entire body $T = \Sigma \frac{1}{2}m_i v^2 = \frac{1}{2}v^2 \Sigma m_i$ or

$$\boxed{T = \tfrac{1}{2}mv^2} \tag{7/12}$$

This expression holds for both rectilinear and curvilinear translation.

(Fixed-axis rotation) The rigid body in Fig. 7/10*b* rotates with an angular velocity ω about the fixed axis through O. The kinetic energy of a representative particle of mass m_i is $T_i = \frac{1}{2}m_i(r_i\omega)^2$. Thus for the entire body $T = \frac{1}{2}\omega^2 \Sigma m_i r_i{}^2$. But the moment of inertia of the body about O is $I_O = \Sigma m_i r_i{}^2$, so

$$\boxed{T = \tfrac{1}{2}I_O\omega^2} \tag{7/13}$$

The similarity in form of the kinetic energy expressions for translation and rotation should be noted. The reader should verify that the dimensions of the two expressions are identical.

(General plane motion) The rigid body in Fig. 7/10*c* executes plane motion where, at the instant considered, the velocity of its mass center G is $\bar{v}$ and the angular velocity is ω. The velocity v_i of a representative particle of mass m_i may be expressed in terms of the mass-center velocity $\bar{v}$ and the velocity $\rho_i\omega$ relative to the mass center as shown. With the aid of the law of cosines we write the kinetic energy of the body as the sum ΣT_i of the kinetic energies of all its particles. Thus

$$T = \Sigma \tfrac{1}{2}m_i v_i{}^2 = \Sigma \tfrac{1}{2}m_i(\bar{v}^2 + \rho_i{}^2\omega^2 + 2\bar{v}\rho_i\omega\cos\theta)$$

Since ω and $\bar{v}$ are common to all terms in the third summation, we may factor them out. Thus the third term in the expression for T becomes

$$\omega\bar{v}\Sigma m_i\rho_i\cos\theta = \omega\bar{v}\Sigma m_i y_i = 0$$

since $\Sigma m_i y_i = m\bar{y} = 0$. The kinetic energy of the body is, then, $T = \frac{1}{2}\bar{v}^2\Sigma m_i + \frac{1}{2}\omega^2\Sigma m_i\rho_i{}^2$ or

$$\boxed{T = \tfrac{1}{2}m\bar{v}^2 + \tfrac{1}{2}\bar{I}\omega^2} \tag{7/14}$$

where $\bar{I}$ is the moment of inertia of the body about its mass center. This expression for kinetic energy clearly shows the separate contributions to the total kinetic energy resulting from the translational velocity $\bar{v}$ of the mass center and the rotational velocity ω about the mass center.

The kinetic energy of plane motion may also be expressed in terms of the rotational velocity about the instantaneous center C of zero velocity. Since C momentarily has zero velocity, the proof leading to Eq. 7/13 for the fixed point O holds equally well for point C, so that, alternatively, we may write the kinetic energy of a rigid body in plane motion as

$$T = \tfrac{1}{2}I_C\omega^2 \qquad (7/15)$$

In Art. 5/2 we derived Eq. 5/4 for the kinetic energy of any system of mass. We now see that this expression becomes equivalent to Eq. 7/14 when the mass system is rigid. For a rigid body the quantity $\dot{\rho}_i$ in Eq. 5/4 is the velocity of the representative particle relative to the mass center and is the vector $\boldsymbol{\omega} \times \boldsymbol{\rho}_i$ which has the magnitude $\rho_i\omega$. The summation term in Eq. 5/4 becomes $\Sigma\tfrac{1}{2}m_i(\rho_i\omega)^2 = \tfrac{1}{2}\omega^2\Sigma m_i\rho_i{}^2 = \tfrac{1}{2}\bar{I}\omega^2$, which brings Eq. 5/4 into agreement with Eq. 7/14.

(c) *Potential Energy.* Gravitational potential energy V_g and elastic potential energy V_e were covered in detail in Art. 3/7. The only additional comment is a reminder that, if the potential energy term ΔV_g is used, then the work of the weight $W = mg$ must *not* be included also in the expression for U.

Work-energy equation. The work-energy relation, Eq. 3/17, was introduced in Art. 3/7 of Chapter 3 for particle motion and was generalized in Art. 5/2 of Chapter 5 to include the motion of a general system of particles. This equation,

$$U = \Delta T + \Delta V_g + \Delta V_e \qquad [5/3]$$

applies to any conservative mechanical system. For application to the motion of a single rigid body the term ΔV_e disappears, and the total work U done on the body by the external forces (other than gravity forces) equals the corresponding change ΔT in the kinetic energy of the body plus the change ΔV_g in its potential energy of position in the gravitational field. Alternatively the equation may be written $U = \Delta T$ provided that the work of gravitational forces is included in the expression for U.

When applied to an interconnected conservative system of rigid bodies, Eq. 5/3 will include the change ΔV_e in the stored elastic energy in the connections. The term U will include the work of all forces external to the system (other than gravitational forces) including the negative work of internal friction forces if any. The term ΔT is the sum of the changes in kinetic energy of all moving parts during the interval of motion in question, and ΔV_g is the sum of the changes in gravitational potential energy for the various members. Alternatively, if the work of gravitational forces is included in the U-term, then the ΔV_g-term must be omitted.

When the work-energy principle is applied to a single rigid body, either a *free-body diagram* or an *active-force diagram* should be used. In the case of an interconnected system of rigid bodies, an active-force diagram of the entire system should be drawn in order to isolate the system and disclose all forces which do work on the system. Diagrams should also be drawn which disclose the initial and final positions of the system for the given interval of motion.

The work-energy equation provides a direct relationship between the forces which do work and the corresponding changes in the motion of a mechanical system. However, if there is appreciable internal mechanical friction, then the system must be dismembered in order to disclose the kinetic-friction forces and account for the negative work that they do. When the system is dismembered, one of the primary advantages of the work-energy approach is automatically lost. The work-energy method is most useful for analyzing conservative systems of interconnected bodies where energy loss due to the negative work of friction forces is negligible.

(*d*) *Power.* The concept of power was discussed in Art. 3/6 on work-energy for particle motion. It is recalled that power is the time rate at which work is performed. For a rigid body of mass m having a motion of translation with a velocity $\mathbf{v}$ and an acceleration $\mathbf{a}$, the power P developed by the resultant force $\mathbf{R}$ which acts on the body equals the time rate of increase of the translational kinetic energy, which becomes

$$P = \dot{T} = \frac{d}{dt}(\tfrac{1}{2}m\mathbf{v}\cdot\mathbf{v}) = \tfrac{1}{2}m(\mathbf{a}\cdot\mathbf{v} + \mathbf{v}\cdot\mathbf{a})$$

$$= m\mathbf{a}\cdot\mathbf{v} = \mathbf{R}\cdot\mathbf{v}$$

The dot product accounts for the case of curvilinear translation where the velocity and the acceleration are not in the same direction.

For the plane rotation of a rigid body with mass moment of inertia $\bar{I}$ about an axis normal to the plane and passing through the mass center, the power P developed by the resultant couple $\bar{M}$ which acts on the body equals the time rate of increase of the rotational kinetic energy, which becomes

$$P = \dot{T} = \frac{d}{dt}(\tfrac{1}{2}\bar{I}\omega^2) = \bar{I}\dot{\omega}\omega = \bar{M}\omega$$

If the body is slowing down in either translational or rotational motion, the respective power become negative with the body giving up kinetic energy and doing work on another body to which it is connected. For a rigid body which changes both its translational and rotational velocities, the net power supplied to it or taken from it is the sum of the powers for translation and rotation.

Power must also be supplied to a body to increase its gravitational potential energy at a certain rate, in which case $P = \dot{V}_g$. Power is also required to increase the elastic potential energy of a spring at a certain rate, so that $P = \dot{V}_e$. In general, if kinetic and both types of potential energy are changing, then

$$P = \dot{T} + \dot{V}_g + \dot{V}_e$$

Sample Problem 7/8

The wheel rolls up the incline on its hubs without slipping and is pulled by the 100-N force applied to the cord wrapped around its outer rim. If the wheel starts from rest, compute its angular velocity ω after its center has moved a distance of 3 m up the incline. The wheel has a mass of 40 kg with center of mass at O and has a centroidal radius of gyration of 150 mm.

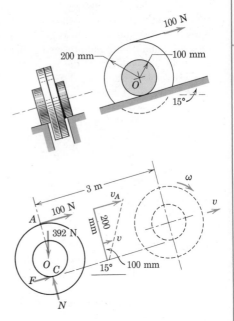

Solution. Of the four forces shown on the free-body diagram of the wheel only the 100-N pull and the weight of $40(9.81) = 392$ N do work. The friction force does no work as long as the wheel does not slip. By use of the concept of the instantaneous center C of zero velocity we see that a point A on the cord to which the 100-N force is applied has a velocity $v_A = [(200 + 100)/100]v$. Hence point A on the cord moves a distance of $(200 + 100)/100$ times as far as the center O. Thus, with the effect of the weight included in the U-term, the work done on the wheel becomes

$$U = 100 \frac{200 + 100}{100}(3) - (392 \sin 15°)(3) = 595 \text{ J}$$

The wheel has general plane motion, so that the change in its kinetic energy is

$$[T = \tfrac{1}{2}m\bar{v}^2 + \tfrac{1}{2}\bar{I}\omega^2] \qquad \Delta T = \left[\frac{1}{2}40(0.10\omega)^2 + \frac{1}{2}40(0.15)^2\omega^2\right] - 0$$
$$= 0.650\omega^2$$

The work-energy equation gives

$$[U = \Delta T] \qquad 595 = 0.650\omega^2 \qquad \omega = 30.3 \text{ rad/s}$$

Alternatively the kinetic energy of the wheel may be written

$$[T = \tfrac{1}{2}I_C\omega^2] \qquad T = \frac{1}{2}40[(0.15)^2 + (0.10)^2]\omega^2 = 0.650\omega^2$$

① Since the velocity of the instantaneous center C on the wheel is zero, it follows that the rate at which the friction force does work is continuously zero. Hence F does no work as long as the wheel does not slip. If the wheel were rolling on a moving platform, however, the friction force would do work even though the wheel were not slipping.

② Note that the component of the weight down the plane does negative work.

③ Be careful to use the correct radius in the expression $v = r\omega$ for the velocity of the center of the wheel.

④ Recall that $I_C = \bar{I} + m\overline{OC}^2$ where $\bar{I} = I_O = mk_o^2$.

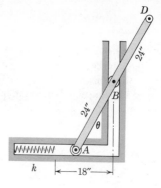

Sample Problem 7/9

The 4-ft slender bar weighs 40 lb with mass center at B and is released from rest in the position for which θ is essentially zero. Point B is confined to move in the smooth vertical guide, while end A moves in the smooth horizontal guide and compresses the spring as the bar falls. Determine (a) the angular velocity of the bar as the position $\theta = 30°$ is passed and (b) the velocity with which B strikes the horizontal surface if the stiffness of the spring is 30 lb/in.

Solution.   With the friction and mass of the small rollers at A and B neglected, the system may be treated as being conservative.

Part (a).   For the first interval of motion from $\theta = 0$ to $\theta = 30°$ the spring is not engaged, so that there is no V_e term in the energy equation. If we adopt the alternative of treating the work of the weight in the V_g term, then there are no other forces which do work and $U = 0$.

Since we have a constrained plane motion, there is a kinematic relation between the velocity v_B of the center of mass and the angular velocity ω of the bar. This relation is easily obtained by using the instantaneous center C of zero velocity and noting that $v_B = \overline{CB}\omega$. Thus the kinetic energy of the bar in the 30° position becomes

$$[T = \tfrac{1}{2}m\overline{v}^2 + \tfrac{1}{2}\overline{I}\omega^2] \quad T = \frac{1}{2}\frac{40}{32.2}\left(\frac{12}{12}\omega\right)^2 + \frac{1}{2}\left(\frac{1}{12}\frac{40}{32.2}4^2\right)\omega^2 = 1.449\,\omega^2$$

The change in gravitational potential energy is the weight times the change in height of the mass center and equals

$$[\Delta V_g = W\Delta h] \qquad \Delta V_g = 40(2\cos 30° - 2) = -10.72 \text{ ft-lb}$$

We now substitute into the energy equation and get

$$[U = \Delta T + \Delta V_g] \qquad 0 = 1.449\omega^2 - 10.72, \quad \omega = 2.72 \text{ rad/sec} \qquad Ans.$$

Part (b).   For the entire interval of motion we include the spring as a part of the system where

$$[V_e = \tfrac{1}{2}kx^2] \qquad \Delta V_e = \frac{1}{2}(30)(24-18)^2\frac{1}{12} - 0 = 45 \text{ ft-lb}$$

In the final horizontal position point A has no velocity, so that the bar is, in effect, rotating about A. Hence its kinetic energy is

$$[T = \tfrac{1}{2}I_A\omega^2] \qquad T = \frac{1}{2}\left(\frac{1}{3}\frac{40}{32.2}4^2\right)\left(\frac{v_B}{24/12}\right)^2 = 0.828v_B^2$$

The change in gravitational potential energy is

$$[\Delta V_g = W\Delta h] \qquad \Delta V_g = 40(-2) = -80 \text{ ft-lb}$$

Substituting into the energy equation gives

$$[U = \Delta T + \Delta V_g + \Delta V_e] \qquad 0 = (0.828v_B^2 - 0) - 80 + 45$$

$$v_B = 6.50 \text{ ft/sec} \qquad Ans.$$

Alternatively, if the bar alone constitutes the system, the active-force diagram shows the weight which does positive work and the spring force kx which does negative work. We would then write

$$[U = \Delta T] \qquad 80 - 45 = 0.828v_B^2$$

which is identical with the previous result.

① We recognize that the forces acting on the bar at A and B are normal to the respective directions of motion and, hence, do no work.

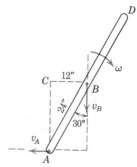

② If we convert k to lb/ft we have $\Delta V_e = \frac{1}{2}\left(30\,\frac{\text{lb}}{\text{in.}}\right)\left(12\,\frac{\text{in.}}{\text{ft}}\right)\left(\frac{24-18}{12}\text{ft}\right)^2 = 45$ ft-lb. Always check the consistency of your units.

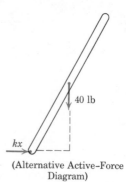

(Alternative Active–Force Diagram)

Sample Problem 7/10

In the mechanism shown each of the two wheels has a mass of 30 kg and a centroidal radius of gyration of 100 mm. Each link OB has a mass of 10 kg and may be treated as a slender bar. The 7-kg collar at B slides on the fixed vertical shaft with negligible friction. The spring has a stiffness $k = 30$ kN/m and is contacted by the bottom of the collar when the links reach the horizontal position. If the collar is released from rest at the position $\theta = 45°$ and if friction is sufficient to prevent the wheels from slipping, determine (a) the velocity v_B of the collar as it first strikes the spring, and (b) the maximum deformation x of the spring.

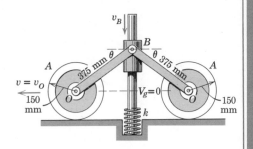

Solution. The mechanism executes plane motion and is conservative with the neglect of kinetic friction losses. The datum for zero gravitational potential energy V_g is conveniently taken through O as shown.

(a) For the interval from $\theta = 45°$ to $\theta = 0$, it is noted that ΔT_{wheels} is zero since each wheel starts from rest and momentarily comes to rest at $\theta = 0$. Also, at the lower position each link is merely rotating about its point O so that

$$\Delta T = [2(\tfrac{1}{2}I_O\omega^2) - 0]_{\text{links}} + [\tfrac{1}{2}mv^2 - 0]_{\text{collar}}$$

$$= \frac{1}{3}10(0.375)^2 \left(\frac{v_B}{0.375}\right)^2 + \frac{1}{2}7v_B{}^2 = 6.83v_B{}^2$$

The collar at B drops a distance $0.375/\sqrt{2} = 0.265$ m so that

$$\Delta V = \Delta V_g = 0 - 2(10)(9.81)\frac{0.265}{2} - 7(9.81)(0.265) = -44.2 \text{ J}$$

Also, $U = 0$. Hence,

$$[U = \Delta T + \Delta V] \qquad 0 = 6.83v_B{}^2 - 44.2 \qquad v_B = 2.54 \text{ m/s} \qquad Ans.$$

(b) At the condition of maximum deformation x of the spring, all parts are momentarily at rest which makes $\Delta T = 0$. Thus

$$[U = \Delta T + \Delta V_g + \Delta V_e]$$

$$0 = 0 - 2(10)(9.81)\left(\frac{0.265}{2} + \frac{x}{2}\right) - 7(9.81)(0.265 + x) + \tfrac{1}{2}(30)(10^3)x^2$$

Solution for the positive value of x gives

$$x = 60.1 \text{ mm} \qquad Ans.$$

It should be noted that the results of parts (a) and (b) involve a very simple net energy change despite the fact that the mechanism has undergone a fairly complex sequence of motions. Solution of this and similar problems by other than a work-energy approach is not an inviting prospect.

① With the work of the weight of the collar B included in the ΔV_g term, there are no other forces external to the system which do work. The friction force acting under each wheel does no work since the wheel does not slip, and, of course, the normal force does no work here. Hence $U = 0$.

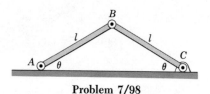

Problem 7/98

PROBLEMS

7/98 The uniform slender bars are hinged at B and are confined to move in the vertical plane. If they are released from rest in the positions shown, determine the expression for the velocity v with which B strikes the horizontal plane. Neglect all friction.

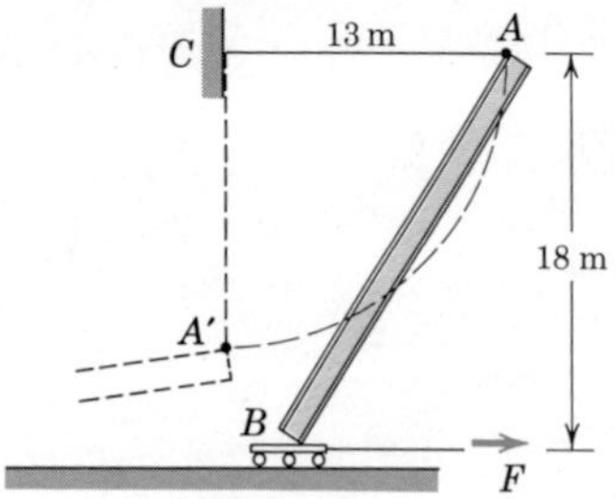

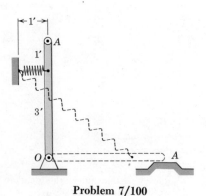

Problem 7/99

7/99 The steel I-beam AB is being positioned by the horizontal force F in the cable at B. If this cable breaks in the position shown where the cable AC is horizontal, determine the velocity of end A as it reaches A'. Neglect friction.   *Ans.* $v = 11.29$ m/s

Problem 7/100

7/100 The uniform slender bar weighs 60 lb and is released from rest in the near-vertical position shown where the spring of stiffness 10 lb/ft is unstretched. Calculate the velocity with which end A strikes the horizontal surface.

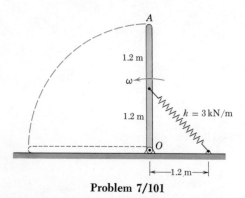

Problem 7/101

7/101 Calculate the initial angular velocity of the 30-kg slender bar OA in the vertical position so that the bar will reach the dotted horizontal position with zero velocity under the action of the spring. In the initial position the spring is unstretched.

Ans. $\omega = 3.67$ rad/s

102 A 1500-kg flywheel with a radius of gyration of 500 mm has its speed reduced from 4000 to 3600 rev/min during a 3-min interval. Calculate the average power supplied by the flywheel.

103 The dump truck carries 6 yd³ of dirt which weighs 110 lb/ft³, and the elevating mechanism rotates the dump about the pivot *A* at a constant angular rate of 4 deg/sec. The mass center of the dump and load is at *G*. Determine the maximum power *P* required during the tilting of the load.   *Ans. P = 11.31 hp*

Problem 7/103

104 The drum of 375-mm radius and its shaft have a mass of 41 kg and a radius of gyration of 300 mm about the axis of rotation. A total of 18 m of flexible steel cable with a mass of 3.08 kg per meter of length is wrapped around the drum with one end secured to the surface of the drum. The free end of the cable has an initial overhang $x = 0.6$ m as the drum is released from rest. Determine the angular velocity ω of the drum for the instant when $x = 6$ m. Assume that the center of mass of the portion of cable remaining on the drum at any time lies on the shaft axis. Neglect friction.

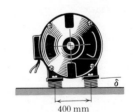

Problem 7/104

105 The electric motor shown is delivering 4.5 kW at 1725 rev/min to a pump which it drives. Calculate the angle δ through which the motor deflects under load if the stiffness of each of its four spring mounts is 10 kN/m. In what direction does the motor turn?

   Ans. δ = 0.892°, motor turns clockwise

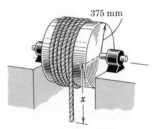

Problem 7/105

106 The uniform bar *ABC* weighs 6 lb and is initially at rest with end *A* bearing against the stop in the horizontal guide. When a constant couple $M = 72$ lb-in. is applied to end *C*, the bar rotates causing end *A* to strike the side of the vertical guide with a velocity of 10 ft/sec. Calculate the loss of energy ΔQ due to friction in the guides and rollers. The mass of the rollers may be neglected.

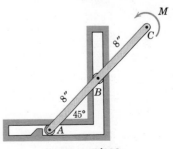

Problem 7/106

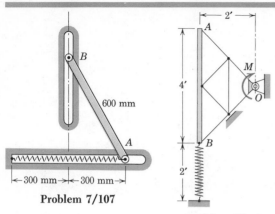

Problem 7/107

Problem 7/108

7/107 The uniform slender bar AB has a mass of 10 kg and is released from rest in the position shown where the spring with a stiffness of 360 N/m has been stretched 200 mm. Determine the velocity of end A as the bar passes the vertical position. Since the spring is secured at both ends, it is capable of supporting both tension and compression. Friction is assumed negligible. *Ans.* $v = 0.935$ m/s

7/108 The figure shows the cross section AB of a 200-lb door which is a 4 ft by 6 ft panel of uniform thickness. The door is supported by a framework of negligible weight hinged about a horizontal shaft at O. In the position shown, the spring, which has a stiffness $k = 30$ lb/ft, is unstretched. If a constant torque $M = 650$ lb-ft is applied to the frame through its shaft at O starting from the rest position shown, determine the angular velocity of the door when it reaches the horizontal position.

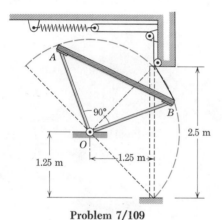

Problem 7/109

7/109 The figure shows the cross section AB of a garage door which is a rectangular 2.5 by 5 m panel of uniform thickness with a mass of 200 kg. The door is supported by the struts having negligible mass and hinged at O. Two spring-and-cable assemblies, one on each side of the door, control the movement. When the door is in the horizontal open position, each spring is unextended. If the door is given a slight unbalance from the open position and allowed to fall, determine the value of the spring constant k for each spring which will limit the angular velocity of the door to 1.5 rad/s when edge B strikes the floor. *Ans.* $k = 1.270$ kN/m

Problem 7/110

7/110 The figure shows the cross section of a garage door which is a uniform rectangular panel 8 by 8 ft and weighing 200 lb. The door carries two spring assemblies, one on each side of the door, like the one shown. Each spring has a stiffness of 50 lb/ft and is unstretched when the door is in the open position shown. If the door is released from rest in this position, calculate the velocity of the edge at A as it strikes the garage floor.

111 The sheave of 400-mm radius has a mass of 50 kg and a radius of gyration of 300 mm. The sheave and its 100-kg load are suspended by the cable and the spring, which has a stiffness of 1.5 kN/m. If the system is released from rest with the spring initially stretched 100 mm, determine the velocity of O after it has dropped 50 mm.   *Ans. v = 0.757 m/s*

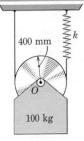

Problem 7/111

Problem 7/112

112 A slender metal rod of length l and mass m is welded to the rim of a hoop of radius l. If the hoop is released from rest in the position shown where the rod is normal to the supporting surface, determine the speed v of the center of the hoop after it has made one complete revolution. Assume no slipping and continuous contact between the hoop and its supporting surface. Also, neglect the mass of the hoop.

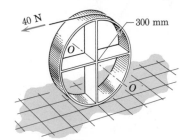

Problem 7/113

113 The wheel is composed of a 10-kg hoop stiffened by four thin spokes each with a mass of 2 kg. A horizontal force of 40 N is applied to the wheel initially at rest. Calculate the angular velocity of the wheel after its center has moved 3 m. Friction is sufficient to prevent slipping.   *Ans. $\omega = 13.19$ rad/s*

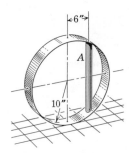

Problem 7/114

114 The slender rod A weighs 8 lb and is welded to the inside of the 6-lb thin metal hoop which rolls on the horizontal surface without slipping. If the hoop is released from rest with the rod in the vertical position, calculate the maximum angular velocity reached by the hoop.

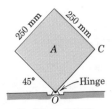

115 The solid square block is hinged about a horizontal axis at O and is released from rest in the position shown. Calculate the velocity v_C with which corner C hits the horizontal surface.   *Ans. $v_C = 1.234$ m/s*

Problem 7/115

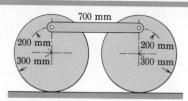

Problem 7/116

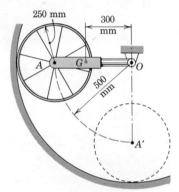

Problem 7/117

Problem 7/118

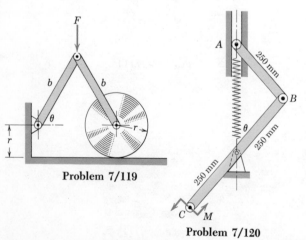

Problem 7/119

Problem 7/120

7/116 The 700-mm horizontal bar has a mass of 24 kg and is freely pinned to the two identical wheels. Each wheel is a solid circular disk with a mass of 16 kg. If the system is released from rest with the bar in essentially the top position, compute the maximum angular velocity ω of the wheels during the subsequent motion. The wheels roll without slipping.

7/117 The wheel consists of a 4-kg rim of 250-mm radius with hub and spokes of negligible mass. The wheel is mounted on the 3-kg yoke OA with mass center at G and with a radius of gyration about O of 350 mm. If the assembly is released from rest in the horizontal position shown and if the wheel rolls on the circular surface without slipping, compute the velocity of point A when it reaches A'.

Ans. $v_A = 2.45$ m/s

7/118 Motive power for the experimental 10-Mg bus comes from the energy stored in a rotating flywheel which it carries. The flywheel has a mass of 1500 kg and a radius of gyration of 500 mm and is brought up to a maximum speed of 4000 rev/min. If the bus starts from rest and acquires a speed of 72 km/h at the top of a hill 20 m above the starting position, compute the reduced speed N of the flywheel. Assume that 10 percent of the energy taken from the flywheel is lost. Neglect the rotational energy of the wheels of the bus.

7/119 A constant force F is applied in the vertical direction to the symmetrical linkage starting from the rest position shown. Determine the angular velocity ω which the links acquire as they reach the position $\theta = 0$. Each link has a mass m_0. The wheel is a solid circular disk of mass m and rolls on the horizontal surface without slipping.

Ans. $\omega = \sqrt{\dfrac{3(F + m_0 g)\sin\theta}{m_0 b}}$

7/120 A couple $M = 12$ N·m is applied at C to the spring toggle mechanism which is released from rest in the position $\theta = 45°$. In this position the spring, which has a stiffness of 140 N/m, is stretched 150 mm. Bar AB has a mass of 3 kg and BC a mass of 6 kg. Calculate the angular velocity ω of BC as it crosses the position $\theta = 0$. Motion is in the vertical plane, and friction is negligible.

121 The solid semicircular disk of radius r is released from rest in the position shown. If no slipping occurs between the disk and the horizontal surface, determine the expression for the angular velocity ω reached by the disk when its kinetic energy is a maximum.

$$Ans. \ \omega = 4\sqrt{\frac{g/r}{9\pi - 16}}$$

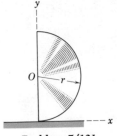

Problem 7/121

122 A hemispherical shell of radius r is released from rest with its axis of symmetry inclined an angle θ from the vertical. Determine the angular velocity ω of the shell as it rocks past the equilibrium position $\theta = 0$. Assume no slipping.

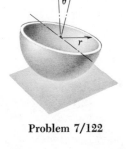

Problem 7/122

123 The center of the 200-lb wheel with a centroidal radius of gyration of 4 in. has a velocity of 2 ft/sec down the incline in the position shown. Calculate the normal reaction N under the wheel as it rolls past position A. Assume no slipping occurs.

$$Ans. \ N = 346 \ lb$$

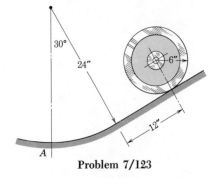

Problem 7/123

124 The horizontal platform A has a mass of 15 kg, and each of its uniform slender legs has a mass of 3 kg. When the legs are vertical at $\theta = 0$, each of the two springs of stiffness $k = 700$ N/m is unstretched. If a constant torque $M = 18$ N·m is applied to the one leg as shown, starting from rest in the position $\theta = 0$, determine the angular velocity ω of the legs when the position $\theta = 45°$ is passed.

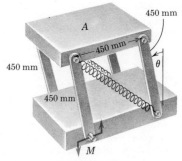

Problem 7/124

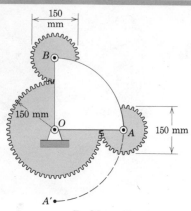

Problem 7/125

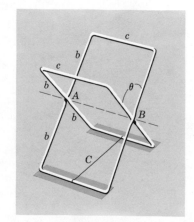

Problem 7/126

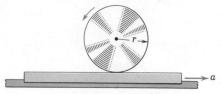

Problem 7/127

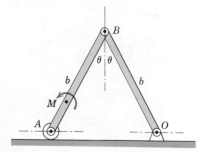

Problem 7/128

7/125 The large gear is fixed and does not rotate. Each of the small gears has a mass of 2 kg and may be treated as a uniform circular disk. The sector is a uniform quarter-circular plate with a mass of 6 kg and is pivoted freely about O. If the assembly is released from rest in the position shown with OB vertical, calculate the angular velocity ω of the small gears when A reaches the bottom position A'.

Ans. $\omega = 28.2$ rad/s

7/126 The two identical steel frames with the dimensions shown are fabricated from the same bar stock and are hinged at the midpoints A and B of their sides. If the frame is resting in the position shown on a horizontal surface with negligible friction, determine the velocity v with which each of the upper ends of the frame hits the horizontal surface if the cord at C is cut.

Ans. $v = \sqrt{12\, gb\, \dfrac{c + 2b}{3c + 4b} \cos \dfrac{\theta}{2}}$

7/127 The solid circular cylinder of mass m and radius r is resting on a horizontal surface which is given a constant acceleration a to the right from rest. Determine the work done on the cylinder during the interval in which it has rotated through 360°. The cylinder rolls without slipping. (*Hint:* The cylinder rolls in the direction shown relative to the moving surface.)

7/128 The two slender bars each of mass m and length b are pinned together and move in the vertical plane. If the bars are released from rest in the position shown and move together under the action of a couple M of constant magnitude applied to AB, determine the velocity of A as it strikes O.

Ans. $v_A = \sqrt{3\left[\dfrac{M\theta}{m} - gb(1 - \cos \theta)\right]}$

/129 The small vehicle is designed for high-speed travel over the snow. The endless tread for each side of the vehicle has a mass ρ per unit length and is driven by the front wheels. Determine that portion M of the constant front-axle torque required to give both vehicle treads their motion corresponding to a vehicle velocity v achieved with constant acceleration in a distance s from rest on level terrain.

$$\text{Ans. } M = \frac{4\rho r v^2}{s}(\pi r + b)$$

Problem 7/129

/130 The figure shows the cross section of a uniform 200-lb ventilator door hinged about its upper horizontal edge at O. The door is controlled by the spring-loaded cable which passes over the small pulley at A. The spring has a stiffness of 15 lb per foot of stretch and is undeformed when $\theta = 0$. If the door is released from rest in the horizontal position, determine the maximum angular velocity ω reached by the door and the corresponding angle θ.

/131 The solid square block is supported on the horizontal plane by a small roller with negligible friction. The block is released from rest in the near-vertical position shown. Calculate the angular velocity ω of the block and the linear velocity of corner O as corner C hits the horizontal surface. (Compare with Prob. 7/115.) *Ans.* $\omega = 6.25$ rad/s, $v_O = 0.781$ m/s

/132 For the conditions of Sample Problem 7/10 calculate the velocity of the center O of either wheel for the instant when the position $\theta = 30°$ is passed.
Ans. $v = 0.439$ m/s

/133 A small experimental vehicle has a total mass m of 500 kg including wheels and driver. Each of the four wheels has a mass of 40 kg and a centroidal radius of gyration of 400 mm. Total frictional resistance R to motion is 400 N and is measured by towing the vehicle at a constant speed on a level road with engine disengaged. Determine the power output of the engine for a speed of 72 km/h when going up the 10 percent grade (a) with zero acceleration and (b) with an acceleration of 3 m/s². (*Hint:* Power equals the time rate of increase of the total energy of the vehicle plus the rate at which frictional work is overcome.)
Ans. (a) $P = 17.76$ kW, (b) $P = 52.0$ kW

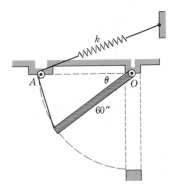

Problem 7/130

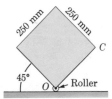

Problem 7/131

Problem 7/133

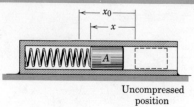

Uncompressed
position

Problem 7/134

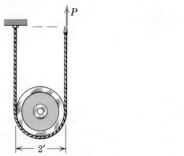

Problem 7/135

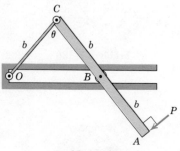

Problem 7/136

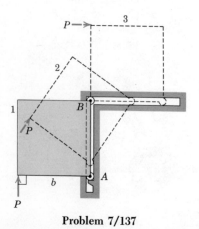

Problem 7/137

7/134 Slider A of mass m is placed in the smooth horizontal tube and released from rest in the position for which the spring of stiffness k has been compressed a distance x_0. Determine the maximum power P developed by the spring in moving the slider and the displacement x at which this condition occurs.

▶7/135 Calculate the constant force P required to give the center of the pulley a velocity of 4 ft/sec in an upward movement of the center of 3 ft from the rest position shown. The pulley weighs 30 lb with a radius of gyration of 10 in., and the cable has a total length of 15 ft with a weight of 2 lb/ft.

Ans. $P = 38.6$ lb

▶7/136 The uniform bar ABC has a mass m and starts from rest with $\theta = 180°$ where A, B, C, and O are collinear. If the applied force P is constant in magnitude, determine the angular velocity ω of the bar as B reaches O with $\theta = 0$. The mass of the roller at B and the mass of the strut OC are negligible.

Ans. $\omega = \sqrt{\dfrac{6P\pi}{13mb}}$

▶7/137 The two corners A and B of the uniform square plate of mass m are confined to move in the vertical and horizontal slots, respectively. If the plate is subjected to a force P greater than the weight, constant in magnitude, and always perpendicular to the edge of the plate, determine the velocity v_A with which corner A strikes the top of the vertical slot. Friction in the small guide rollers is negligible. (*Hint:* Replace P by an equivalent force at A and a couple.)

Ans. $v_A = \sqrt{3b}\sqrt{\dfrac{P}{2m}(1 + \pi) - g}$

7/7 ACCELERATION FROM WORK-ENERGY; VIRTUAL WORK.

In addition to the determination of velocities resulting from the action of forces acting over finite intervals of motion, we may use the work-energy equation to establish the instantaneous accelerations of the members of a system of interconnected bodies as a result of the active forces applied. Or we may modify the equation to establish the configuration of such a system when it undergoes a constant acceleration.

Equation 5/3 when written for an infinitesimal interval of motion becomes

$$dU = dT + dV$$

The term dU represents the total work done by all active nonpotential forces acting on the system under consideration during the infinitesimal displacement of the system. The work of potential forces is included in the dV-term. If we use the subscript i to denote a representative body of the interconnected system, the differential change in kinetic energy T for the entire system becomes

$$dT = d(\Sigma \tfrac{1}{2} m_i \bar{v}_i{}^2 + \Sigma \tfrac{1}{2} \bar{I}_i \omega_i{}^2) = \Sigma m_i \bar{v}_i \, d\bar{v}_i + \Sigma \bar{I}_i \omega_i \, d\omega_i$$

where $d\bar{v}_i$ and $d\omega_i$ are the respective changes in the magnitudes of the velocities and where the summation is taken over all bodies of the system. But for each body $m_i \bar{v}_i \, d\bar{v}_i = m_i \bar{\mathbf{a}}_i \cdot d\bar{\mathbf{s}}_i$ and $\bar{I}_i \omega_i \, d\omega_i = \bar{I}_i \alpha_i \, d\theta_i$, where $d\bar{\mathbf{s}}_i$ represents the infinitesimal linear displacement of the center of mass and where $d\theta_i$ represents the infinitesimal angular displacement of the body in the plane of motion. We note that $\bar{\mathbf{a}}_i \cdot d\bar{\mathbf{s}}_i$ is identical to $(\bar{a}_i)_t \, d\bar{s}_i$ where $(\bar{a}_i)_t$ is the component of $\bar{\mathbf{a}}_i$ along the tangent to the curve described by the mass center of the body in question. Also α_i represents $\ddot{\theta}_i$, the angular acceleration of the representative body. Consequently for the entire system

$$dT = \Sigma m_i \bar{\mathbf{a}}_i \cdot d\bar{\mathbf{s}}_i + \Sigma \bar{I}_i \alpha_i \, d\theta_i$$

This change may also be written as

$$dT = \Sigma \mathbf{R}_i \cdot d\bar{\mathbf{s}}_i + \Sigma \bar{\mathbf{M}}_i \cdot d\boldsymbol{\theta}_i$$

where $\mathbf{R}_i$ and $\bar{\mathbf{M}}_i$ are the resultant force and resultant couple acting on body i and where $d\boldsymbol{\theta}_i = d\theta_i \mathbf{k}$. These last two equations merely show us that the differential change in kinetic energy equals the differential work done on the system by the resultant forces and resultant couples acting on all the bodies of the system.

The term dV represents the differential change in the total gravitational potential energy V_g and the total elastic potential energy V_e and has the form

$$dV = d(\Sigma m_i g h_i + \Sigma \tfrac{1}{2} k_j x_j{}^2) = \Sigma m_i g \, dh_i + \Sigma k_j x_j \, dx_j$$

where h_i represents the vertical distance of the center of mass of the representative body of mass m_i above any convenient datum plane and where x_j stands for the deformation, tensile or compressive, of a representative elastic member of the system (spring) whose stiffness is k_j.

The complete expression for dU may now be written as

$$dU = \Sigma m_i \bar{\mathbf{a}}_i \cdot d\bar{s}_i + \Sigma \bar{I}_i \alpha_i \, d\theta_i + \Sigma m_i g \, dh_i + \Sigma k_j x_j \, dx_j \qquad (7/16)$$

In applying Eq. 7/16 to a system of one degree of freedom, which is any system whose configuration or position is uniquely determined by the value of a single coordinate, the terms $m_i \bar{\mathbf{a}}_i \cdot d\bar{s}_i$ and $\bar{I}_i \alpha_i \, d\theta_i$ will be positive if the accelerations are in the same direction as the respective displacements and negative if in the opposite direction. Equation 7/16 has the advantage of relating the accelerations to the active forces directly, which eliminates the need for dismembering the system and then eliminating the internal forces and reactive forces by simultaneous solution of the force-mass-acceleration equations written for each separate member of the system.

Virtual work. In Eq. 7/16 the differential motions are differential changes in the real or actual displacements which occur. For a mechanical system which assumes a steady-state configuration during constant acceleration we often find it convenient to introduce the concept of *virtual work*. The concepts of virtual work and virtual displacement have been introduced and used to establish equilibrium configurations for static systems of interconnected bodies (see Chapter 7 of Vol. 1, *Statics*). A virtual displacement is any assumed and arbitrary displacement, linear or angular, away from the natural or actual position. For a system of connected bodies the virtual displacements must be consistent with the constraints of the system. For example, when one end of a link is hinged about a fixed pivot, the virtual displacement of the other end must be normal to the line joining the two ends. Or, if two links are freely pinned together, any virtual displacement of the joint considered as a point on one link must be identical with the virtual displacement of the joint considered as a point on the other link. Such requirements for displacements consistent with the constraints are purely kinematical, that is, depend solely on the geometry of possible motions, and provide what are known as the *equations of constraint*. If a set of virtual displacements satisfying the equations of constraint and therefore consistent with the constraints is given to a mechanical system, the proper relationship between the coordinates which specify the configuration of the system will be established by applying the work-energy relationship of Eq. 7/16 expressed in terms of virtual changes. Thus

$$\delta U = \Sigma m_i \bar{\mathbf{a}}_i \cdot \delta \bar{s}_i + \Sigma \bar{I}_i \alpha_i \, \delta \theta_i + \Sigma m_i g \, \delta h_i + \Sigma k_j x_j \, \delta x_j \qquad (7/16a)$$

It is customary to use the differential symbol d to refer to differential changes in the *real* displacements, whereas the symbol δ is used to signify differential changes which are assumed or *virtual* changes.

Sample Problem 7/11

The movable rack A has a mass of 3 kg, and rack B is fixed. The gear has a mass of 2 kg and a radius of gyration of 60 mm. In the position shown the spring, which has a stiffness of 1.2 kN/m, is stretched a distance of 40 mm. For the instant represented determine the acceleration a of rack A under the action of the 80-N force. The plane of the figure is vertical.

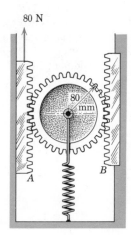

Solution. The given figure represents the active-force diagram for the entire system, which is conservative.

During an infinitesimal upward displacement dx of rack A the work done on the system is $80\ dx$, where x is in meters, and this work equals the sum of the corresponding changes in the total energy of the system. These changes which appear in Eq. 7/16 are as follows:

$$[dT = \Sigma m_i \bar{a}_i \cdot d\bar{s}_i + \Sigma \bar{I}_i \alpha_i d\theta_i]$$

$$dT_{\text{rack}} = 3a\ dx$$

$$dT_{\text{gear}} = 2\frac{a}{2}\frac{dx}{2} + 2(0.06)^2 \frac{a/2}{0.08}\frac{dx/2}{0.08} = 0.781a\ dx$$

The change in potential energies of the system from Eq. 7/16 becomes

$$[dV = \Sigma m_i g\ dh_i + \Sigma k_j x_j dx_j]$$

$$dV_{\text{rack}} = 3g\ dx = 3(9.81)dx = 29.4\ dx$$

$$dV_{\text{gear}} = 2g(dx/2) = g\ dx = 9.81\ dx$$

$$dV_{\text{spring}} = kx_j\ dx_j = 1200(0.04)dx/2 = 24\ dx$$

Substitution into Eq. 7/16 gives us

$$80\ dx = 3a\ dx + 0.781a\ dx + 29.4\ dx + 9.81\ dx + 24\ dx$$

Cancelling dx and solving for a give

$$a = 16.76/3.781 = 4.43\ \text{m/s}^2 \qquad\qquad \textit{Ans.}$$

We see that using the work-energy method for an infinitesimal displacement has given us the direct relation between the applied force and the resulting acceleration. It was unnecessary to dismember the system, draw two free-body diagrams, apply $\Sigma F = m\bar{a}$ twice, apply $\Sigma M = \bar{I}\alpha$ and $F = kx$, eliminate unwanted terms, and finally solve for a.

① Note that none of the remaining forces external to the system do any work. The work done by the weight and by the spring is accounted for in the potential energy terms.

② Note that $\bar{a}_i$ for the gear is its mass-center acceleration, which is half that for the rack A. Also its displacement is $dx/2$. For the rolling gear the angular acceleration from $a = r\alpha$ is $\alpha_i = (a/2)/0.08$, and the angular displacement from $ds = r\ d\theta$ is $d\theta_i = (dx/2)/0.08$.

③ Note here that the displacement of the spring is one half that of the rack. Hence $x_i = x/2$.

Sample Problem 7/12

A constant force P is applied to end A of the two identical and uniform links and causes them to move to the right in their vertical plane with a horizontal acceleration a. Determine the steady-state angle θ made by the bars with one another.

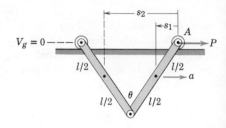

Solution. The figure constitutes the active-force diagram for the system. To find the steady-state configuration, consider a virtual displacement of each bar from the natural position assumed during the acceleration. Measurement of the displacement with respect to end A eliminates any work done by force P during the virtual displacement. Thus

①
$$\delta U = 0$$

The virtual change in kinetic energy from Eq. 7/16a is

$$\delta T = \Sigma m\bar{a} \cdot \delta\bar{s} = ma(-\delta s_1) + ma(-\delta s_2)$$

$$= -ma\left[\delta\left(\frac{l}{2}\sin\frac{\theta}{2}\right) + \delta\left(\frac{3l}{2}\sin\frac{\theta}{2}\right)\right]$$

②
$$= -ma\left(l\cos\frac{\theta}{2}\,\delta\theta\right)$$

We choose the horizontal line through A as the datum for zero potential energy. Thus the potential energy of the links is

$$V_g = 2mg\left(-\frac{l}{2}\cos\frac{\theta}{2}\right)$$

and the virtual change in potential energy becomes

$$\delta V_g = \delta\left(-2mg\frac{l}{2}\cos\frac{\theta}{2}\right) = \frac{mgl}{2}\sin\frac{\theta}{2}\,\delta\theta$$

Substitution into the work-energy equation for virtual changes, Eq. 7/16a, gives

$$[\delta U = \delta T + \delta V_g] \qquad 0 = -mal\cos\frac{\theta}{2}\,\delta\theta + \frac{mgl}{2}\sin\frac{\theta}{2}\,\delta\theta$$

from which
$$\theta = 2\tan^{-1}\frac{2a}{g} \qquad\qquad Ans.$$

Again, in this problem we see that the work-energy approach obviated the necessity for dismembering the system, drawing separate free-body diagrams, applying motion equations, eliminating unwanted terms, and solving for θ.

① Note that we use the symbol δ to refer to an assumed or virtual differential change rather than the symbol d, which refers to an infinitesimal change in the real displacement.

② We have chosen to use the angle θ to describe the configuration of the links, although we could have used the distance between the two ends of the links just as well.

PROBLEMS

138 Determine the initial angular acceleration of the links in Prob. 7/124, shown again here, as the mechanism starts from rest at $\theta = 0$ where the springs are unstretched. *Ans.* $\alpha = 4.68 \text{ rad/s}^2$

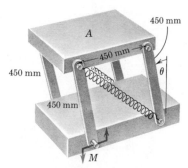

Problem 7/138

139 Links A and B each weigh 8 lb, and bar C weighs 12 lb. Calculate the angle θ assumed by the links if the body to which they are pinned is given a steady horizontal acceleration a of 4 ft/sec².

Ans. $\theta = 7.1°$

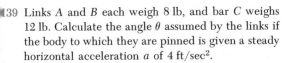

Problem 7/139

140 The cargo box of the food-delivery truck for aircraft servicing has a loaded mass m and is elevated by the application of a couple M on the lower end of the link which is hinged to the truck frame. The horizontal slots allow the linkage to unfold as the cargo box is elevated. Determine the upward acceleration of the box in terms of h for a given value of M. Neglect the mass of the links.

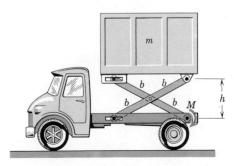

Problem 7/140

141 The hydraulic cylinder A exerts an upward force P on pin B of the uniform bar BC of mass m. Neglect the mass of link OG and determine the angular acceleration α of BC as the bar starts from rest in the position shown. Motion occurs in the vertical plane. *Ans.* $\alpha = \dfrac{3}{4b}\left(\dfrac{2P}{m} - g\right)\cos\theta$

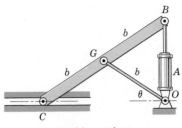

Problem 7/141

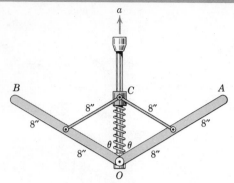

Problem 7/142

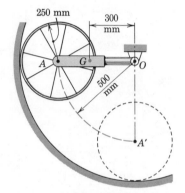

Problem 7/143

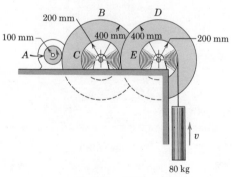

Problem 7/144

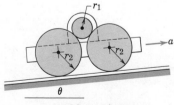

Problem 7/145

7/142 Each of the uniform bars OA and OB weighs 4 lb and is freely hinged at O to the vertical shaft, which is given an upward acceleration $a = g/2$. The links which connect the light collar C to the bars have negligible weight, and the collar slides freely on the shaft. The spring has a stiffness $k = 0.75$ lb/in. and is uncompressed for the position equivalent to $\theta = 0$. Calculate the angle θ assumed by the bars under steady accelerating conditions.

7/143 The wheel and yoke of Prob. 7/117 are repeated here. Determine the initial angular acceleration of OA when the assembly is released from rest with OA horizontal. The wheel consists of a 4-kg rim with spokes and hub of negligible mass. Friction is sufficient to prevent slipping. The mass center of the 3-kg yoke OA is at G, and its radius of gyration about O is 350 mm.   *Ans.* $\alpha = 12.02$ rad/s^2

7/144 Pinion A of the electric motor turns gear B and its attached pinion C. This unit meshes with gear D and its attached hoisting drum E. The motor rotor and pinion A have a combined mass of 30 kg and a radius of gyration of 150 mm. The unit B and C, as well as the unit D and E, has a mass of 95 kg with a radius of gyration of 300 mm. The motor receives 2 kW of electrical power, 94 percent of which is converted into mechanical power. For an instant when the upward velocity of the 80-kg mass is 0.9 m/s, calculate the upward acceleration a of the load.   *Ans.* $a = 0.585$ m/s^2

7/145 The small vehicle has a total mass m, and each of its four wheels has a moment of inertia I about its center. The vehicle is propelled by an electric motor which supplies a torque M to the frictional-drive pinion of radius r_1. The rotor of the electric motor and the attached pinion have a combined moment of inertia I_O about the shaft axis. Determine the torque M required to give the vehicle an acceleration a up the incline. No slipping occurs between any of the frictional surfaces at the wheels or pinion.

146 The uniform arm *OA* weighs 8 lb, and gear *D* weighs 10 lb with a radius of gyration about its center of 2.4 in. The large gear *B* is fixed and cannot rotate, but the arm and small gear are free to rotate in the vertical plane about their respective bearings. Find the angular acceleration α of *OA* just after it is released from rest in the horizontal position.

Ans. $\alpha = 27.7$ rad/sec²

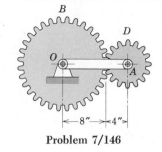

Problem 7/146

147 The two uniform slender bars, each of mass *m*, are suspended in the vertical plane by the small rollers which bear on the horizontal surface. The bars are prevented from collapse by the cord *A*. If the cord is cut, determine the initial downward acceleration of the pin at *O* which joins the bars.

$$\text{Ans. } a = \frac{3(1 + \cos\theta)}{5 + 3\cos\theta} g$$

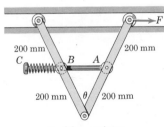

Problem 7/147

148 Each of the two uniform links has a mass of 3 kg. The rod *AC* has a mass of 2.5 kg and is free to slide through the pivoted collar at *B*. The spring has negligible mass, so that the total mass of the assembly is 8.5 kg. Determine the steady-state angle θ assumed by the links under a constant accelerating force $F = 50$ N. The links are suspended in the vertical plane by the rollers, and the spring has a stiffness $k = 350$ N/m. Also, the spring is uncompressed when θ is essentially zero for $F = 0$.

Ans. $\theta = 27.0°$

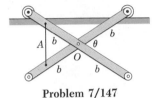

Problem 7/148

149 The sliding bar of mass *m* in Prob. 7/136 is repeated here. Determine the angular acceleration of *AC* due to the action of the force *P* as the bar starts from rest with any value θ. Link *OC* has negligible mass, and the smooth guide is horizontal.

$$\text{Ans. } \alpha = \left(\frac{3P}{mb}\right)\frac{2 + \cos\theta}{7 + 6\cos\theta}$$

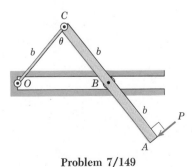

Problem 7/149

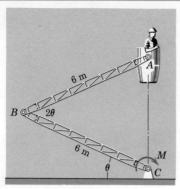

Problem 7/150

▶7/150 The aerial tower shown is designed to elevate a workman in a vertical direction. An internal mechanism at B maintains the angle between AB and BC at twice the angle θ between BC and the ground. If the combined mass of the man and the cab is 200 kg and if all other masses are neglected, determine the torque M applied to BC at C and the torque M_B in the joint at B required to give the cab an initial vertical acceleration of 1.2 m/s² when started from rest in the position $\theta = 30°$.

Ans. $M_B = 11.44$ kN · m
$M = 0$

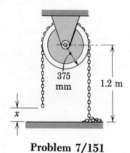

Problem 7/151

▶7/151 The 36-kg pulley has a radius of gyration about its center of 250 mm. The chain has a mass of 7.5 kg/m and has a length of 2.4 m plus the half-circumferential portion over the pulley. For a very slight unbalance starting from rest, the pulley turns clockwise and the chain piles up on the platform. As each link strikes the platform and comes to rest, it is unable to transmit any force to the links above it. Compute the angular velocity ω of the pulley at the instant that $x = 1.2$ m. (*Hint:* Analyze the energy change during a differential interval of motion noting the energy loss of each increment of chain which comes to rest on the platform.)   *Ans.* $\omega = 4.53$ rad/s

SECTION C. IMPULSE AND MOMENTUM

7/8 IMPULSE-MOMENTUM EQUATIONS. The principles of impulse and momentum were developed and used in Art. 3/8 of Chapter 3 in the description of particle motion. In this treatment we observed that these principles were of particular importance when the applied forces were expressible as functions of the time and when interactions between particles occurred during short periods of time, such as with impact. Similar advantages result when the impulse-momentum principles are applied to the motion of rigid bodies.

In Art. 5/2 of Chapter 5 the impulse-momentum principles were extended to cover any defined system of mass particles without restriction as to the connections between the particles of the system. These extended relations all apply to the motion of a rigid body, which is merely a special case of a general system of mass. We will now apply these equations for rigid-body motion in two dimensions.

(a) Linear Momentum. In Art. 5/2 we defined the linear momentum of a mass system as the vector sum of the linear momenta of all of its particles and wrote $\mathbf{G} = \Sigma m_i \mathbf{v}_i$. With $\mathbf{r}_i$ representing the position vector to m_i we have $\mathbf{v}_i = \dot{\mathbf{r}}_i$ and $\mathbf{G} = \Sigma m_i \dot{\mathbf{r}}_i$ which, for a system whose total mass is constant, may be written as $\mathbf{G} = d(\Sigma m_i \mathbf{r}_i)/dt$. When we substitute the principle of moments $m\bar{\mathbf{r}} = \Sigma m_i \mathbf{r}_i$ to locate the mass center, the momentum becomes $\mathbf{G} = d(m\bar{\mathbf{r}})/dt = m\dot{\bar{\mathbf{r}}}$ where $\dot{\bar{\mathbf{r}}}$ is the velocity $\bar{\mathbf{v}}$ of the mass center. Therefore, as before, we find that the linear momentum of any mass system, rigid or nonrigid, is

$$\boxed{\mathbf{G} = m\bar{\mathbf{v}}} \qquad [5/5]$$

In the derivation of Eq. 5/5 we note that it was unnecessary to employ the kinematical condition for a rigid body, Fig. 7/11, which is $\mathbf{v}_i = \bar{\mathbf{v}} + \boldsymbol{\omega} \times \boldsymbol{\rho}_i$. In that case we obtain the same result by writing $\mathbf{G} = \Sigma m_i(\bar{\mathbf{v}} + \boldsymbol{\omega} \times \boldsymbol{\rho}_i)$. The first sum is $\bar{\mathbf{v}}\Sigma m_i = m\bar{\mathbf{v}}$, and the second sum becomes $\boldsymbol{\omega} \times \Sigma m_i \boldsymbol{\rho}_i = \boldsymbol{\omega} \times m\bar{\boldsymbol{\rho}} = \mathbf{0}$ since $\boldsymbol{\rho}_i$ is measured from the mass center making $\bar{\boldsymbol{\rho}}$ zero.

Next in Art. 5/2 we rewrote Newton's generalized second law as Eq. 5/6. This equation and its integrated form are

$$\boxed{\Sigma \mathbf{F} = \dot{\mathbf{G}}} \quad \text{and} \quad \boxed{\int_{t_1}^{t_2} \Sigma \mathbf{F}\, dt = \mathbf{G}_2 - \mathbf{G}_1} \qquad (7/17)$$

Equation 7/17 may be written in its scalar component form which, for plane motion in the *x-y* plane, gives

$$\boxed{\begin{aligned} \Sigma F_x &= \dot{G}_x \\[6pt] \Sigma F_y &= \dot{G}_y \end{aligned}} \quad \text{and} \quad \boxed{\begin{aligned} \int_{t_1}^{t_2} \Sigma F_x\, dt &= G_{x_2} - G_{x_1} \\[6pt] \int_{t_1}^{t_2} \Sigma F_y\, dt &= G_{y_2} - G_{y_1} \end{aligned}} \qquad (7/17a)$$

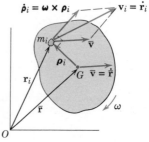

Figure 7/11

In words the first of Eqs. 7/17 and 7/17a states that the resultant force equals the time rate of change of momentum. The integrated form of Eqs. 7/17 and 7/17a states that the linear impulse on the body during the interval $t_2 - t_1$ equals the corresponding change in linear momentum. As in the force-mass-acceleration formulation, the force summations in Eqs. 7/17 and 7/17a must include *all* forces acting externally on the body considered. We emphasize, therefore, that when using the impulse-momentum equations, it is essential to construct the complete free-body diagram so as to disclose all forces which appear in the force summation. In contrast to the method of work and energy, all forces exert impulses whether they do work or not.

(b) *Angular Momentum.* Angular momentum is defined as the moment of linear momentum. In Art. 5/2 we expressed the angular momentum about the mass center of any prescribed system of mass as $\bar{\mathbf{H}} = \Sigma \boldsymbol{\rho}_i \times m_i \mathbf{v}_i$, which is merely the vector sum of the moments about G of the linear momenta of all particles. We showed in Eq. 5/8 that this sum could also be written as $\bar{\mathbf{H}} = \Sigma \boldsymbol{\rho}_i \times m_i \dot{\boldsymbol{\rho}}_i$ where $\dot{\boldsymbol{\rho}}_i$ is the velocity of m_i with respect to G. Although we have simplified this expression in Art. 7/2 in the course of deriving the moment equation of motion, we will pursue this same expression again for sake of emphasis by using the rigid body in plane motion represented in Fig. 7/11. The relative velocity becomes $\dot{\boldsymbol{\rho}}_i = \boldsymbol{\omega} \times \boldsymbol{\rho}_i$ where the angular velocity of the body is $\boldsymbol{\omega} = \omega \mathbf{k}$. The unit vector $\mathbf{k}$ is directed into the paper for the sense of $\boldsymbol{\omega}$ shown. Since $\boldsymbol{\rho}_i$, $\dot{\boldsymbol{\rho}}_i$, and $\boldsymbol{\omega}$ are at right angles to one another, the magnitude of $\dot{\boldsymbol{\rho}}_i$ is $\rho_i \omega$, and the magnitude of $\boldsymbol{\rho}_i \times m_i \dot{\boldsymbol{\rho}}_i$ is $\rho_i^2 \omega m_i$. Thus we may write $\bar{\mathbf{H}} = \Sigma \rho_i^2 m_i \omega \mathbf{k} = \bar{I} \omega \mathbf{k}$ where $\bar{I} = \Sigma m_i \rho_i^2$ is the mass moment of inertia of the body about its mass center. Because the angular-momentum vector is always normal to the plane of motion, vector notation is generally unnecessary, and we may write the angular momentum about the mass center as the scalar

$$\boxed{\bar{H} = \bar{I} \omega} \qquad (7/18)$$

This angular momentum appears in the moment-angular-momentum relation, Eq. 5/9, which in scalar notation for plane motion, along with its integrated form, is

$$\boxed{\Sigma \bar{M} = \dot{\bar{H}}} \quad \text{and} \quad \boxed{\int_{t_1}^{t_2} \Sigma \bar{M} \, dt = \bar{H}_2 - \bar{H}_1} \qquad (7/19)$$

In words the first of Eqs. 7/19 states that the sum of the moments about the mass center of *all* forces acting on the body equals the time rate of change of angular momentum about the mass center. The integrated form of Eq. 7/19 states that the angular impulse about the mass center of all forces acting on the body during the interval $t_2 - t_1$ equals the corresponding change in the angular momentum about G. The sense for positive rotation must be clearly established, and the algebraic signs of $\Sigma \bar{M}$, $\bar{H}_2$, and $\bar{H}_1$ must be consistent with this choice. Again, a free-body diagram is essential.

With the moments about G of the linear momenta of all particles accounted for by $\bar{H} = \bar{I}\omega$, it follows that we may represent the linear momentum $\mathbf{G} = m\bar{\mathbf{v}}$ as a vector through the mass center G, as shown in Fig. 7/12a. Thus $\mathbf{G}$ and $\bar{\mathbf{H}}$ have vector properties analogous to those of the resultant force and couple.

With the establishment of the linear and angular momentum resultants in Fig. 7/12a, which represents the momentum diagram, the angular momentum H_O about any point O is easily written as

$$H_O = \bar{I}\omega + m\bar{v}d \qquad (7/20)$$

This expression holds at any particular instant of time about O, which may be a fixed or moving point on or off the body.

When a body rotates about a fixed point O on the body or body extended, as shown in Fig. 7/12b, the relations $\bar{v} = \bar{r}\omega$ and $d = \bar{r}$ may be substituted into the expression for H_O, giving $H_O = (\bar{I}\omega + m\bar{r}^2\omega)$. But $\bar{I} + m\bar{r}^2 = I_O$ so that

$$H_O = I_O\omega \qquad (7/21)$$

In Art. 5/2 we derived Eq. 5/7, which is the moment-angular-momentum equation about a fixed point O. This equation, written in scalar notation for plane motion, along with its integrated form is

$$\Sigma M_O = I_O\dot{\omega} \quad \text{and} \quad \int_{t_1}^{t_2} \Sigma M_O \, dt = I_O(\omega_2 - \omega_1) \qquad (7/22)$$

In addition to drawing a complete free-body diagram in order that the force and moment summations may be correctly evaluated when applying the linear and angular impulse-momentum equations for rigid-body motion, we find it useful to draw the momentum diagram which indicates the resultant linear-momentum vector and the angular-momentum couple. We caution the reader here not to add linear momentum and angular momentum for the same reason that force and moment cannot be added directly.

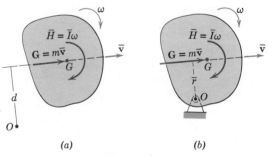

(a)   (b)

Figure 7/12

(*c*) *Interconnected Rigid Bodies.* The equations of impulse and momentum may also be used for a system of interconnected rigid bodies since the momentum principles are applicable to any general system of constant mass. In Fig. 7/13 are shown the combined free-body diagram and momentum diagram for two interconnected bodies. Equations 5/6 and 5/7, which are $\Sigma \mathbf{F} = \dot{\mathbf{G}}$ and $\Sigma \mathbf{M}_O = \dot{\mathbf{H}}_O$ where O is a fixed reference point, may be written for each member of the system and added. The sums would be

$$\Sigma \mathbf{F} = \dot{\mathbf{G}}_1 + \dot{\mathbf{G}}_2 + \cdots$$
$$\Sigma \mathbf{M}_O = \dot{\mathbf{H}}_{O_1} + \dot{\mathbf{H}}_{O_2} + \cdots \tag{7/23}$$

In integrated form for a finite time interval, these expressions are

$$\int_{t_1}^{t_2} \Sigma \mathbf{F}\, dt = (\Delta \mathbf{G})_{\text{system}} \qquad \int_{t_1}^{t_2} \Sigma \mathbf{M}_O\, dt = (\Delta \mathbf{H}_O)_{\text{system}} \tag{7/24}$$

We note that the equal and opposite actions and reactions in the connections are internal to the system and cancel one another so will not be involved in the force and moment summations. Also point O is one fixed reference point for the entire system.

(*d*) *Conservation of Momentum.* In Part (*d*) of Art. 5/2 the principles of conservation of momentum for a general mass system were expressed by Eqs. 5/11 and 5/12. These principles are applicable to either a single rigid body or to a system of interconnected rigid bodies. Thus, if $\Sigma \mathbf{F} = \mathbf{0}$ for a given interval of time, then

$$\boxed{\Delta \mathbf{G} = \mathbf{0}} \tag{5/11}$$

which says that the linear momentum vector undergoes no change in the absence of a resultant linear impulse. For the system of interconnected rigid bodies there may be linear momentum changes of individual parts of the system during the interval, but there will be no resultant momentum change for the system if there is no resultant linear impulse.

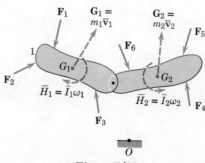

Figure 7/13

Similarly, if the resultant moment about a given fixed point O or about the mass center is zero during a particular interval of time for a single rigid body or for a system of interconnected rigid bodies, then

$$\boxed{\Delta \mathbf{H}_O = \mathbf{0}} \quad \text{or} \quad \boxed{\Delta \bar{\mathbf{H}} = \mathbf{0}} \qquad [5/12]$$

which says that the angular momentum either about the fixed point or about the mass center undergoes no change in the absence of a corresponding resultant angular impulse. Again, in the case of the interconnected system, there may be angular-momentum changes of individual components during the interval, but there will be no resultant angular-momentum change for the system if there is no resultant angular impulse about the fixed point or the mass center. Either of Eqs. 5/12 may hold without the other. In the case of an interconnected system the use of the center of mass for the system is generally inconvenient. As was illustrated previously in Art. 3/9 in the chapter on particle motion, the use of momentum principles greatly facilitates the analysis of situations where forces and couples act for very short periods of time.

(*e*) *Impact of Rigid Bodies.* Impact phenomena involve a fairly complex interrelationship of energy and momentum transfer, energy dissipation, elastic and plastic deformation, relative impact velocity, and body geometry. In Art. 4/3 we treated the impact of bodies modeled as particles and considered only the case of central impact where the contact forces of impact passed through the mass centers of the bodies, as would always happen with colliding smooth spheres, for example. To relate the conditions after impact to those before impact required the introduction of the so-called coefficient of restitution e or impact coefficient, which compares the relative separation velocity to the relative approach velocity measured along the direction of the contact forces. Although in the classical theory of impact e was considered a constant for given materials, more modern investigations show that e is highly dependent on geometry and impact velocity as well as on materials. At best, even for spheres and rods under direct central and longitudinal impact, the coefficient of restitution is a complex and variable factor of limited use.

Any attempt to extend this simplified theory of impact utilizing a coefficient of restitution for the noncentral impact of rigid bodies of varying shape is a gross oversimplification which has little practical value. For this reason we shall not include such an exercise in this book, even though such a theory is easily developed and appears in certain references. We can and do, however, make full use of the principles of conservation of linear and angular momentum where applicable in discussing the impact and other interactions of rigid bodies.

Sample Problem 7/13

The force P, which is applied to the cable wrapped around the central hub of the symmetrical wheel, is increased slowly according to $P = 1.50t$, where P is in pounds and t is the time in seconds after P is first applied. Determine the angular velocity ω of the wheel 10 sec after P is applied if the wheel is rolling to the left with a velocity of its center of 3 ft/sec at time $t = 0$. The wheel weighs 120 lb with a radius of gyration about its center of 10 in. and rolls without slipping.

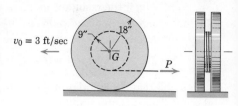

Solution. The free-body diagram of the wheel for any position within the interval is shown. Also indicated are the initial linear and angular momenta at time $t = 0$ and the final linear and angular momenta at time $t = 10$ sec. The correct direction of the friction force F is that to oppose the slipping which would occur without friction. ①

Application of the linear impulse-momentum equation and the angular impulse-momentum equation over the *entire* interval gives

② $\left[\int_{t_1}^{t_2} \Sigma F_x \, dt = G_{x_2} - G_{x_1} \right]$ $\quad \int_0^{10} (1.5t - F) \, dt = \dfrac{120}{32.2} \left[\dfrac{18}{12} \omega - (-3) \right]$

③ $\left[\int_{t_1}^{t_2} \Sigma \bar{M} \, dt = \bar{H}_2 - \bar{H}_1 \right]$

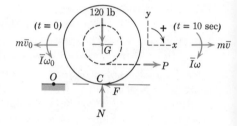

$$\int_0^{10} \left(\frac{18}{12} F - \frac{9}{12}[1.5t] \right) dt = \frac{120}{32.2} \left(\frac{10}{12} \right)^2 \left[\omega - \left(-\frac{3}{18/12} \right) \right]$$

Since the force F is variable, it must remain under the integral sign. We eliminate F between the two equations by multiplying the second one by $\frac{12}{18}$ and adding to the first one. Integrating and solving for ω give

$$\omega = 3.13 \text{ rad/sec clockwise} \qquad \qquad \textit{Ans.}$$

Alternative solution. We could avoid the necessity of a simultaneous solution by applying the second of Eqs. 7/22 about a fixed point O on the horizontal surface. The moments of the 120-lb weight and the equal and opposite force N cancel one another, and F is eliminated since its moment about O is zero. Thus the angular momentum about O becomes $H_O = \bar{I}\omega + m\bar{v}r = m\bar{k}^2\omega + mr^2\omega = m(\bar{k}^2 + r^2)\omega$ where $\bar{k}$ is the centroidal radius of gyration and r is the 18-in. rolling radius. Thus we see that $H_O = H_C$ since $\bar{k}^2 + r^2 = k_C^2$ and $H_C = I_C\omega = mk_C^2\omega$. Equation 7/22 now gives

$$\left[\int_{t_1}^{t_2} \Sigma M_O \, dt = H_{O_2} - H_{O_1} \right]$$

$$\int_0^{10} 1.5t \left(\frac{18 - 9}{12} \right) dt = \frac{120}{32.2} \left[\left(\frac{10}{12} \right)^2 + \left(\frac{18}{12} \right)^2 \right] \left(\omega - \left[-\frac{3}{18/12} \right] \right)$$

Solution of this one equation is equivalent to the simultaneous solution of the two previous equations.

① Also, we note the clockwise unbalance of moments about C which causes a clockwise angular acceleration as the wheel rolls without slipping. Since the moment sum about G must be in the clockwise sense of α, the friction force must act to the left to provide it.

② Note carefully the signs of the momentum terms. The final linear velocity is assumed in the positive x-direction, so G_{x_2} is positive. The initial linear velocity is negative, so G_{x_1} is negative. When we subtract the negative term, we get the double minus sign.

③ Since the wheel rolls without slipping, a positive x-velocity requires a clockwise angular velocity, and vice versa. Again we subtract a minus quantity.

Sample Problem 7/14

The sheave E of the hoisting rig shown has a mass of 30 kg and a centroidal radius of gyration of 250 mm. The 40-kg load D which is carried by the sheave has an initial downward velocity $v_0 = 1.2$ m/s at the instant when a clockwise torque is applied to the hoisting drum A to maintain essentially a constant force $F = 380$ N in the cable at B. Compute the angular velocity ω of the sheave 5 s after the torque is applied to the drum and find the tension T in the cable at O during the interval. Neglect all friction.

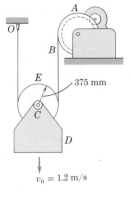

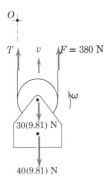

Solution. The load and the sheave taken together constitute the system, and its free-body diagram is shown. The tension T in the cable at O and the final angular velocity ω of the sheave are the two unknowns. We eliminate T initially by applying the moment-angular-momentum equation about the fixed point O, taking counterclockwise as positive.

$$\left[\int_{t_1}^{t_2} \Sigma M_O \, dt = H_{O_2} - H_{O_1}\right]$$

$$\int_{t_1}^{t_2} \Sigma M_O \, dt = \int_0^5 [380(0.750) - (30 + 40)(9.81)(0.375)]\, dt$$

$$= 137.4 \text{ N} \cdot \text{m} \cdot \text{s}$$

$$(H_{O_2} - H_{O_1})_D = mv_2 d - mv_1 d = md(v_2 - v_1)$$

$$= 40(0.375)(v - [-1.2]) = 15(0.375\omega + 1.2)$$

$$= 5.63\omega + 18 \text{ N} \cdot \text{m} \cdot \text{s}$$

$$(H_{O_2} - H_{O_1})_E = \bar{I}(\omega_2 - \omega_1) + md(\bar{v}_2 - \bar{v}_1)$$

$$= 30(0.250)^2(\omega - [-1.2/0.375]) + 30(0.375)(0.375\omega - [-1.2])$$

$$= 6.09\omega + 19.50 \text{ N} \cdot \text{m} \cdot \text{s}$$

Substituting into the momentum equation gives

$$137.4 = 5.63\omega + 18 + 6.09\omega + 19.50$$

$$\omega = 8.53 \text{ rad/s counterclockwise} \qquad \textit{Ans.}$$

① Watch for the double minus sign in subtracting a negative quantity. Also, the units of angular momentum, which are those of angular impulse, may be written as kg·m²·s⁻¹.

The force-linear-momentum equation is now applied to the system to determine T. With the positive direction up we have

$$\left[\int_{t_1}^{t_2} \Sigma F \, dt = G_2 - G_1\right]$$

$$\int_0^5 [T + 380 - 70(9.81)]\, dt = 70[0.375(8.53) - (-1.2)]$$

$$5T = 1841 \qquad T = 368 \text{ N} \qquad \textit{Ans.}$$

If we had taken our moment equation around the center C of the sheave instead of point O, it would contain both unknowns T and ω, and we would be obliged to solve it simultaneously with the foregoing force equation which would also contain the same two unknowns.

Sample Problem 7/15

The uniform rectangular block of dimensions shown is sliding to the left on the horizontal surface with a velocity v when it strikes the small step in the surface. Assume negligible rebound at the step, and compute the minimum value of v which will permit the block to pivot about the edge of the step and just reach the standing position A with no velocity. Compute the percentage energy loss $\Delta E/E$ for $b = c$.

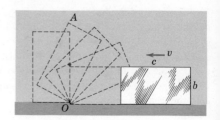

Solution. It will be assumed that the edge of the step O acts as a latch on the corner of the block, so that the block pivots about O. Furthermore, the height of the step is assumed negligible compared with the dimensions of the block. During impact the only force which exerts a moment about O is the weight mg, but the angular impulse due to the weight is extremely small since the time of impact is negligible. Thus we ① may assume that the angular momentum about O is conserved.

The initial angular momentum of the block about O just before impact is the moment of its linear momentum and is $H_O = mv(b/2)$. The velocity ② of the center of mass G immediately after impact is $\bar{v}$, and the angular velocity is $\omega = \bar{v}/\bar{r}$. The angular momentum about O just after impact when the block is starting its rotation about O is

③ $$[H_O = I_O\omega] \qquad H_O = \left[\frac{1}{12}m(b^2 + c^2) + m\left(\left[\frac{c}{2}\right]^2 + \left[\frac{b}{2}\right]^2\right)\right]\omega$$

$$= \frac{m}{3}(b^2 + c^2)\omega$$

Conservation of angular momentum gives

$$[\Delta H_O = 0] \qquad \frac{m}{3}(b^2 + c^2)\omega = mv\frac{b}{2} \qquad \omega = \frac{3vb}{2(b^2 + c^2)}$$

This angular velocity will be sufficient to raise the block just past position A if the kinetic energy of rotation equals the increase in potential energy. Thus

$$[\Delta T + \Delta V_g = 0] \qquad \frac{1}{2}I_O\omega^2 - mg\left(\sqrt{\left(\frac{b}{2}\right)^2 + \left(\frac{c}{2}\right)^2} - \frac{b}{2}\right) = 0$$

$$\frac{1}{2}\frac{m}{3}(b^2 + c^2)\left[\frac{3vb}{2(b^2 + c^2)}\right]^2 - \frac{mg}{2}(\sqrt{b^2 + c^2} - b) = 0$$

$$v = 2\sqrt{\frac{g}{3}\left(1 + \frac{c^2}{b^2}\right)}(\sqrt{b^2 + c^2} - b) \qquad \text{Ans.}$$

The percentage loss of energy is

$$\frac{\Delta E}{E} = \frac{\frac{1}{2}mv^2 - \frac{1}{2}I_O\omega^2}{\frac{1}{2}mv^2} = 1 - \frac{k_O^2\omega^2}{v^2} = 1 - \left(\frac{b^2 + c^2}{3}\right)\left[\frac{3b}{2(b^2 + c^2)}\right]^2$$

$$= 1 - \frac{3}{4\left(1 + \frac{c^2}{b^2}\right)}, \quad \Delta E/E = 62.5 \text{ percent for } b = c \qquad \text{Ans.}$$

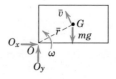

① If the corner of the block struck a spring instead of the rigid step, then the time of the interaction during compression of the spring could become appreciable, and the angular impulse about the fixed point at the end of the spring due to the moment of the weight would have to be accounted for.

② Be sure to use the transfer theorem $I_O = \bar{I} + m\bar{r}^2$ correctly here.

③ Notice the abrupt change in direction and magnitude of the velocity of G during the impact.

PROBLEMS

152 The center of the homogeneous solid cylinder is given an initial velocity of 2 ft/sec up the incline. Determine the time t required for it to reach a velocity of 4 ft/sec down the incline if it rolls without slipping.   *Ans.* $t = 2.81$ sec

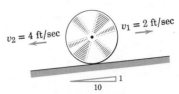

Problem 7/152

153 The constant tensions of 200 N and 160 N are applied to the hoisting cable as shown. If the velocity v of the load is 2 m/s down and the angular velocity ω of the pulley is 8 rad/s counterclockwise at time $t = 0$, determine v and ω after the cable tensions have been applied for 5 s. Note the independence of the results.

$$\text{Ans. } v = 0.379 \text{ m/s up}$$
$$\omega = 56.0 \text{ rad/s CW}$$

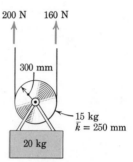

Problem 7/153

154 The frictional moment M_f acting on a rotating turbine disk and its shaft is given by $M_f = k\omega^2$ where ω is the angular velocity of the turbine. If the source of power is cut off while the turbine is running with an angular velocity ω_0, determine the time t for the speed of the turbine to drop to half of its initial value. The moment of inertia of the turbine disk and shaft is I.

155 The 30-g bullet has a horizontal velocity of 500 m/s as it strikes the 10-kg slender bar OA, which is suspended from point O and is initially at rest. Calculate the angular velocity ω which the bar with its embedded bullet has acquired immediately after impact. (Analyze the bar and bullet together as the system.)   *Ans.* $\omega = 2.81$ rad/s

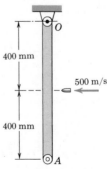

Problem 7/155

156 If the bullet of Prob. 7/155 takes 10^{-4} s to embed itself in the bar, calculate the time average of the horizontal force O_x exerted by the pin on the bar at O during the interaction between the bullet and the bar. Use the results cited for Prob. 7/155.

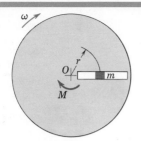

Problem 7/157

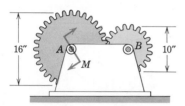

Problem 7/158

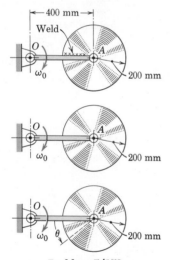

Problem 7/159

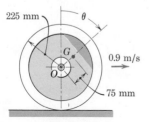

Problem 7/160

7/157 The small block of mass m slides along the radial slot of the disk while the disk rotates in the horizontal plane about its center O. The block is released from rest relative to the disk and moves outward with an increasing velocity $\dot{r}$ along the slot as the disk turns. Determine the expression in terms of r and $\dot{r}$ for the torque M which must be applied to the disk to maintain a constant angular velocity ω of the disk.

7/158 With the gears initially at rest and the couple M equal to zero, the forces exerted by the frame on the shafts of the gears at A and B are 30 and 16 lb, respectively, both upward to support the weights of the two gears. A couple $M = 60$ lb-in. is now applied to the larger gear through its shaft at A. After 4 sec the larger gear has a clockwise angular momentum of 12 ft-lb-sec, and the smaller gear has a counterclockwise angular momentum of 4 ft-lb-sec. Calculate the new values of the forces R_A and R_B exerted by the frame on the shafts during the 4-sec interval. Isolate the two gears together as the system. *Ans.* $R_A = 27.23$ lb, $R_B = 18.77$ lb

7/159 The uniform circular disk of 200-mm radius has a mass of 25 kg and is mounted on the rotating bar OA in three different ways. In each case the bar rotates about its vertical shaft at O with a clockwise angular velocity $\omega = 4$ rad/s. In case (a) the disk is welded to the bar. In case (b) the disk, which is pinned freely at A, moves with curvilinear translation and therefore has no rigid-body rotation. In case (c) the relative angle between the disk and the bar is increasing at the rate $\dot{\theta} = 8$ rad/s. Calculate the angular momentum of the disk about point O for each case.

Ans. (a) $H_O = 18$ kg·m²/s
(b) $H_O = 16$ kg·m²/s
(c) $H_O = 14$ kg·m²/s

7/160 The unbalanced wheel is made to roll to the right without slipping with a constant velocity of 0.9 m/s of its center O. The wheel has a mass of 8 kg with center of mass at G and has a radius of gyration about O of 150 mm. Determine the angular momentum H_O of the wheel about O at the instant (a) when G passes directly over O with $\theta = 0$ and (b) when G passes the horizontal line through O where $\theta = 90°$.

161 The large rotor has a mass of 60 kg and a radius of gyration about its vertical axis of 200 mm. The small rotor is a solid circular disk with a mass of 8 kg and is initially rotating with an angular velocity $\omega_1 = 80$ rad/s with the large rotor at rest. A spring-loaded pin P which rotates with the large rotor is released and bears against the periphery of the small disk bringing it to a stop relative to the large rotor. Neglect any bearing friction and calculate the final angular velocity of the assembly.

Ans. $\omega = 1.22$ rad/s

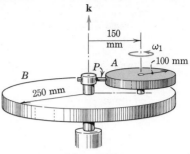

Problem 7/161

162 The uniform circular disk D weighs 8 lb and is free to turn about the bearing axis C–C. The arm B weighs 5 lb and is fastened to the vertical shaft O–O. The arm may be approximated as a slender bar 12 in. long, and the moment of inertia of the vertical shaft O–O may be neglected. The rod A weighs 6 lb and is fastened securely to the arm. If the initial angular velocity of D is $\omega_0 = 7$ rad/sec in the direction shown and the arm B is at rest, determine the angular velocity ω_B of the arm after a torque $M = 15$ lb-in. has been applied to the shaft for 4 sec.

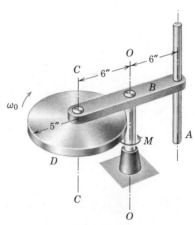

Problem 7/162

163 Each of the two 300-mm uniform rods A has a mass of 1.5 kg and is hinged at its end to the rotating base B. The 4-kg base has a radius of gyration of 40 mm and is initially rotating freely about its vertical axis with a speed of 300 rev/min and with the rods latched in the vertical positions. If the latches are released and the rods assume the horizontal dotted positions, calculate the new rotational speed N of the assembly.

Ans. $N = 32.0$ rev/min

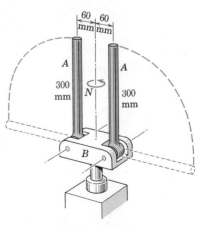

Problem 7/163

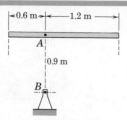

Problem 7/164

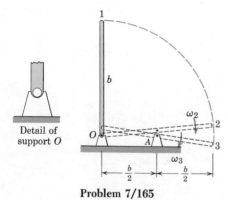

Detail of support O

Problem 7/165

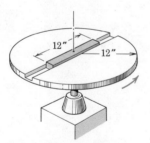

Problem 7/166

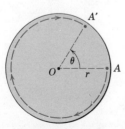

Problem 7/168

7/164 The 24-kg slender bar is released from rest in the horizontal position shown. If point A of the bar becomes attached to the pivot at B upon impact after dropping through the 0.9-m distance, calculate the angular velocity ω of the bar immediately after impact.

7/165 The slender bar of mass m and length b is pivoted at its lower end at O in the manner shown in the separate detail of the support O. The bar is released from rest in the vertical position 1. When the middle of the bar strikes the pivot at A in horizontal position 2, it becomes latched to the pivot, and simultaneously the connection at O becomes disengaged. Determine the angular velocity ω_3 of the bar just after it engages the pivot at A in position 3.

Ans. $\omega_3 = \sqrt{3g/b}$

7/166 The 12-in. slender bar weighs 3 lb and is initially centered in the radial slot of the circular disk which is rotating freely about its vertical axis with a speed of 400 rev/min. The disk weighs 30 lb and has a radius of gyration of 9 in. If the bar is given a slight nudge, it will slide radially from its central position. Calculate the rotational velocity of the disk when the end of the bar reaches the periphery of the disk.

Ans. $N = 383$ rev/min

7/167 The small block of mass m in Prob. 7/157 slides outward along the radial slot as the disk is rotating freely in the horizontal plane about its bearing at O with zero moment M. If the angular velocity of the disk was ω_1 when the radial distance to m was r_1, determine ω as a function of r. The moment of inertia of the disk about its axis of rotation is I_O.

7/168 A man of mass m stands at point A on the horizontal turntable initially at rest. The turntable is free to rotate about O with negligible friction and has a moment of inertia I about O. If the man starts to walk clockwise in a circle of radius r, determine the angle θ through which the turntable rotates until the man reaches his starting point in the new position A'. If the man walks at the speed v_r relative to the surface of the turntable, what is the corresponding angular velocity ω of the turntable?

Ans. $\theta = \dfrac{2\pi}{1 + (I/mr^2)}$, $\omega = \dfrac{v_r/r}{1 + (I/mr^2)}$

169 The slotted circular disk whose mass is 6 kg has a radius of gyration about O of 175 mm. The disk carries the four steel balls, each of mass 0.15 kg and located as shown, and rotates freely about a vertical axis through O with an angular speed of 120 rev/min. Each of the small balls is held in place by a latching device not shown. If the balls are released while the disk is rotating and come to rest in the dotted positions relative to the slots, compute the new angular velocity ω of the disk. Also find the magnitude $|\Delta E|$ of the energy loss due to the impact of the balls with the ends of the slots. Neglect the diameter of the balls and discuss this approximation.

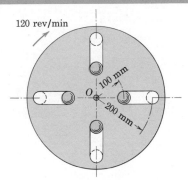

120 rev/min

100 mm

200 mm

Problem 7/169

170 Refer to the rotors of Prob. 7/161 and introduce the unit vector **k** vertically up. Assume initially that disk A has an angular velocity $\boldsymbol{\omega}_1 = 80\mathbf{k}$ rad/s *relative* to rotor B while B has an angular velocity $-10\mathbf{k}$ rad/s. Calculate the common angular velocity $\boldsymbol{\omega}$ of the disks after the braking pin P has been released and all relative rotation between the disks has ceased.   *Ans.* $\boldsymbol{\omega} = -8.78\mathbf{k}$ rad/s

171 The 9-in.-radius wheel with rigidly attached 6-in.-radius hub weighs 128.8 lb and is released from rest on the 60° incline. The cord is securely wrapped around the hub and fastened to the fixed point A. Calculate the velocity of the center O for the position reached 3 seconds after release.

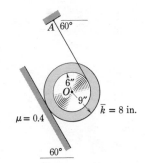

A 60°

6″

O

9″

$\bar{k} = 8$ in.

$\mu = 0.4$

60°

Problem 7/171

172 The solid circular cylinder of radius r is at rest on the flat belt when a force P is applied to the belt. If P is sufficient to cause slipping between the belt and the cylinder at all times, determine the time t required for the cylinder to reach the dotted position. Also determine the angular velocity ω of the cylinder in this same position. The coefficient of friction between the cylinder and the belt is μ.

$$\textit{Ans. } t = \sqrt{\frac{2s}{\mu g}}, \ \omega = \frac{2\sqrt{2\mu g s}}{r}$$

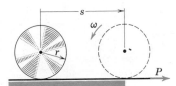

s

ω

r

P

Problem 7/172

173 A cylindrical shell of 400-mm diameter and mass m is rotating about its central horizontal axis with an angular velocity of 30 rad/s when it is released onto a horizontal surface with no velocity of its center. If slipping between the shell and the surface occurs for 1.5 s, calculate the coefficient of friction μ and the maximum velocity v reached by the center of the shell.

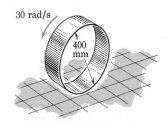

30 rad/s

400 mm

Problem 7/173

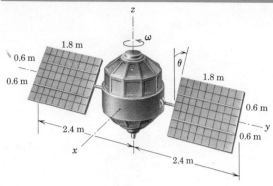

Problem 7/174

7/174 The body of the spacecraft has a mass of 160 kg and a radius of gyration about its z-axis of 0.45 m. Each of the two solar panels may be treated as a uniform flat plate of 8-kg mass. If the spacecraft is rotating about its z-axis at the angular rate of 1.0 rad/s with $\theta = 0$, determine the angular rate ω after the panels are rotated to the position $\theta = \pi/2$ by an internal mechanism. Neglect the small momentum change of the body about the y-axis.

Ans. $\omega = 0.974$ rad/s

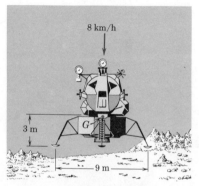

Problem 7/175

7/175 The 17.5-Mg lunar landing module with center of mass at G has a radius of gyration of 1.8 m about G. The module is designed to contact the lunar surface with a vertical free-fall velocity of 8 km/h. If one of the four legs hits the lunar surface on a small incline and suffers no rebound, compute the angular velocity ω of the module immediately after impact as it pivots about the contact point. The 9-m dimension is the distance across the diagonal of the square formed by the four feet as corners.

7/176 A uniform pole of length L is dropped at an angle θ with the vertical, and both ends have a velocity v as end A hits the ground. If end A pivots about its contact point during the remainder of the motion, determine the velocity v' with which end B hits the ground.

Ans. $v' = \sqrt{\dfrac{9v^2}{4}\sin^2\theta + 3gL\cos\theta}$

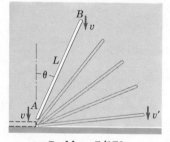

Problem 7/176

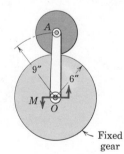

Problem 7/177

7/177 The small gear is made to rotate in a horizontal plane about the large stationary gear by means of the torque M applied to the arm OA. The small gear weighs 6 lb and may be treated as a circular disk. The arm OA weighs 4 lb and has a radius of gyration about the fixed bearing at O of 6 in. Determine the constant torque M required to give the arm OA an absolute angular velocity of 20 rad/sec in 3 sec, starting from rest. Neglect friction.

Ans. $M = 1.255$ lb-ft

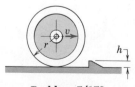

Problem 7/178

7/178 Determine the minimum velocity v which the wheel may have and just roll over the obstruction. The centroidal radius of gyration of the wheel is k, and it is assumed that the wheel does not slip.

179 The uniform stone block with $b = 1.2$ m and $h = 0.9$ m is released from rest with its center of mass G almost directly above the supporting corner. Determine the angular velocity ω' of the block about corner A immediately after impact assuming that A remains in contact with the ground. Also assume that contact occurs at the corners only and that no slipping takes place. What fraction $\Delta E/E$ of the energy is lost due to impact?

Ans. $\omega' = 0.112$ rad/s, $\Delta E/E = 0.998$

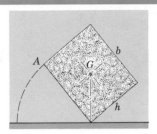

Problem 7/179

180 Determine the maximum value of b/h for the stone block of Prob. 7/179 for which any rotation about corner A is possible following impact. Assume corner contact and no slipping.

181 The motor M drives disk A which turns disk B with no slipping. Disk A and its attached shaft and motor armature weigh 36 lb and have a combined radius of gyration of 3.4 in. Disk B weighs 10 lb and has a radius of gyration of 5.6 in. The motor housing and attached arm C together weigh 48 lb and have a radius of gyration about the axis O–O of 18 in. Before the motor is turned on, the entire assembly is rotating as a unit about O–O with a rotational speed $\omega_0 = 30$ rev/min in the direction shown. The motor M has an operating speed of 1720 rev/min in the direction shown as measured with C fixed. Determine the new rotational speed of arm C if the motor is turned on.   *Ans.* $N = 26.2$ rev/min clockwise

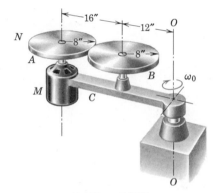

Problem 7/181

182 A mechanical device for reducing the spin of a satellite consists of two masses m attached to wires or thin metal tapes which unwind when released and acquire additional angular momentum at the expense of the angular momentum of the satellite. The masses are then released when unwound. Each tape has a length $R\phi$ and is fastened to its point A. If the satellite is spinning at an initial speed ω_0 when the masses are released with $\beta = 0$, determine the rate $\dot{\beta}$ of unwinding and find the new angular velocity ω_ϕ of the satellite when $\beta = \phi$. The moment of inertia of the satellite exclusive of the m's is I. (*Hint.* Equate the angular momentum and the kinetic energy of the system to their respective initial values.)   *Ans.* $\dot{\beta} = \omega_0$ constant

$$\omega_\phi = \frac{I + 2mR^2(1 - \phi^2)}{I + 2mR^2(1 + \phi^2)}\omega_0$$

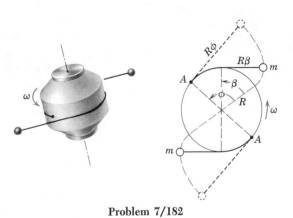

Problem 7/182

7/9 PROBLEM FORMULATION AND REVIEW. In Chapter 7 we have made use of essentially all of the elements of dynamics studied so far. We have found that a knowledge of kinematics, using both absolute- and relative-motion analysis, is an essential part of the solution to problems in rigid-body kinetics. Our approach in Chapter 7 paralleled that used in Chapter 3 where we developed the kinetics of particles using force-mass-acceleration, work-energy, and impulse-momentum methods.

With this previous background Chapter 7 should bring a good many of the principles and methods of dynamics into focus. The following outline will help to summarize the most important considerations in solving rigid-body problems in plane motion.

(*a*) *Identification of Body or System.* As we have seen throughout mechanics, both statics and dynamics, it is essential that we make an unambiguous decision as to which body or system of bodies we are analyzing and then isolate the body or system selected by drawing its free-body diagram, resultant-force diagram, or active-force diagram, whichever is appropriate.

(*b*) *Type of Motion.* Next identify the category of motion as rectilinear translation, curvilinear translation, fixed-axis rotation, or general plane motion. Always see that the kinematics of the problem is properly described before attempting to solve the kinetic equations. Remember that the principles of relative motion find application whenever a rigid body has any rotation.

(*c*) *Coordinate System.* Choose an appropriate coordinate system. The geometry of the particular kinematics involved is usually the deciding factor. Designate the positive sense for moment and force summations and be consistent with the choice.

(*d*) *Principle and Method.* If the instantaneous relationship between the applied forces and the acceleration is desired, then the equivalence between the forces and their $m\bar{a}$ and $\bar{I}\alpha$ resultants, as disclosed by the free-body and resultant-force diagrams, will yield the solution and indicate the most direct approach to be followed.

When motion occurs over an interval of displacement, the work-energy approach is indicated, and we are able to relate initial to final velocities without calculating the acceleration. We have seen the advantage of this approach for interconnected mechanical systems with negligible internal friction (conservative systems).

If the interval of motion is specified in terms of time rather than displacement, the impulse-momentum approach is indicated. When the angular motion of a rigid body is suddenly generated or changed, we use the principle of conservation of angular momentum.

(*e*) *Assumptions and Approximations.* By now we should have acquired a feel for the practical significance of certain assumptions and approximations such as treating a rod as an ideal slender bar and neglecting friction where minimal. These and other idealizations are important to the process of obtaining solutions to real problems.

REVIEW PROBLEMS

183 A total length of 120 m of cable having a mass of 0.863 kg/m is wrapped around the drum A, which turns with negligible friction about its horizontal axis O–O. If the drum is released from rest with $x = 15$ m of cable unwound from the drum, calculate the initial angular acceleration α of the drum. The drum alone has a mass of 110 kg with radius of gyration about O–O of 546 mm.

Ans. $\alpha = 1.087$ rad/s^2

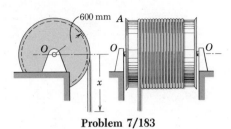

Problem 7/183

184 The two hinged links are released from rest with OA in the horizontal position shown. Calculate the velocity of end B along the horizontal surface for the instant when OA reaches the vertical position. Link OA has twice the mass of link AB, and both may be treated as slender bars. Neglect all friction and the mass of the small roller at B.

Ans. $v = 3.99$ m/s

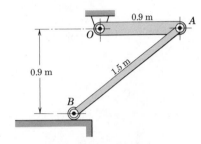

Problem 7/184

185 Calculate the upward acceleration a of the center O of the 16-kg solid circular disk of 200-mm radius when the 10-kg counterweight is allowed to fall.

186 What initial clockwise angular velocity ω must the uniform and slender 15-lb bar have as it crosses the vertical position $(\theta = 0)$ in order that it just reach the horizontal position $(\theta = 90°)$? The spring has a stiffness of 3 lb/ft and is unstretched when $\theta = 0$.

Ans. $\omega = 2.94$ rad/sec

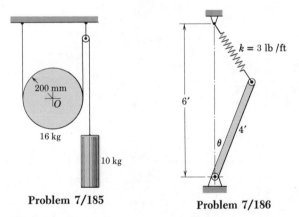

Problem 7/185

Problem 7/186

187 The wheel shown rolls on its 150-mm-diameter hubs on inclined rails without slipping. If the wheel has a clockwise angular velocity of 4 rad/s at time $t = 0$, calculate its angular velocity ω at $t = 16$ s. The radius of gyration of the wheel about its center is 200 mm.

Ans. $\omega = 62.8$ rad/s

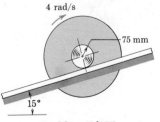

Problem 7/187

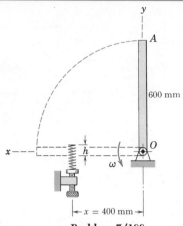

Problem 7/188

7/188 The 15-kg slender bar *OA* is released from rest in the vertical position and compresses the spring of stiffness $k = 20$ kN/m as the horizontal position is passed. Determine the proper setting of the spring by specifying the distance h which will result in the bar having an angular velocity $\omega = 4$ rad/s as it crosses the horizontal position. What is the effect of x on the dynamics of the problem?

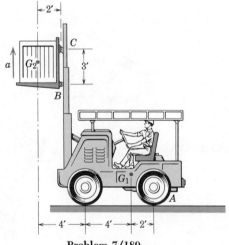

Problem 7/189

7/189 The forklift truck with center of mass at G_1 has a weight of 3200 lb including the vertical mast. The fork and load have a combined weight of 1800 lb with center of mass at G_2. The roller guide at *B* is capable of supporting horizontal force only, whereas the connection at *C*, in addition to supporting horizontal force, also transmits the vertical elevating force. If the fork is given an upward acceleration which is sufficient to reduce the force under the rear wheels at *A* to zero, calculate the corresponding reaction at *B*. *Ans. B* = 2133 lb

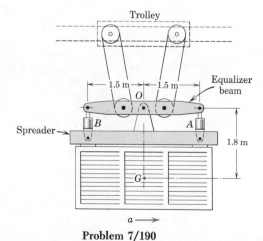

Problem 7/190

7/190 An antiswing system for unloading cargo containers from ships consists of the moving trolley and its suspended equalizer beam and spreader, to which the loaded container is secured. Rotation of the spreader relative to the equalizer beam is controlled by the hydraulic cylinders *A* and *B* which have a piston diameter of 100 mm. During the horizontal acceleration of the system with a load of 4 Mg the oil pressure in cylinder *A* is 400 kPa and that in cylinder *B* is 500 kPa. Compute the acceleration a of the system. The center of mass of the load is at *G*.

191 The figure shows the Saturn V mobile launch platform *A* together with the umbilical tower *B*, unfueled rocket *C*, and crawler-transporter *D* which carries the system to the launch site. The approximate dimensions of the structure and locations of the mass centers *G* are given. The approximate masses are $m_A = 3$ Gg, $m_B = 3.3$ Gg, $m_C = 0.23$ Gg, $m_D = 3$ Gg. The minimum stopping distance from the top speed of 1.5 km/h is 0.1 m. Compute the vertical component of the reaction under the front crawler unit *F* during the period of maximum deceleration.   *Ans. F* = 59.5 MN

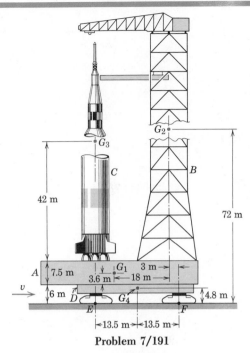

Problem 7/191

192 The parallel links *OE* and *CD* are given a constant counterclockwise angular acceleration of 10 rad/s² by the torque *M*. As the angle θ reaches 60°, the angular velocity of the links is $\dot{\theta} = 4$ rad/s. For this instant calculate the bending moment supported by the weld at *A*. The vertical bar has a mass of 2.4 kg.   *Ans. M* = 4.00 N · m

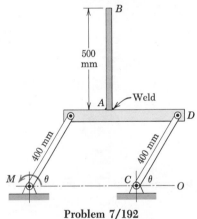

Problem 7/192

193 The riding power mower has a weight of 280 lb with center of mass at G_1. The operator weighs 180 lb with center of mass at G_2. Calculate the minimum coefficient of friction μ which will permit the front wheels of the mower to lift off the ground as the mower starts to move forward.

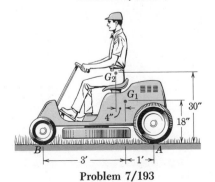

Problem 7/193

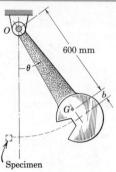

600 mm

Specimen

Problem 7/194

7/194   The pendulum for the impact-testing machine has a mass of 35 kg with center of mass at G and has a radius of gyration about O of 650 mm. The pendulum is designed so that the force on the bearing at O has no horizontal component during impact with the specimen at the bottom of the swing. Determine the distance b. Also calculate the total force on the bearing at O for the instant after the pendulum is released from rest at $\theta = 60°$.

Ans. $b = 104.2$ mm, $O = 177.2$ N

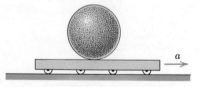

Problem 7/195

7/195   A uniform circular disk which rolls without slipping with a velocity v encounters an abrupt change in the direction of its motion as it rolls onto the incline θ. Determine the new velocity v' of the center of the disk as it starts up the incline, and find the fraction n of the initial energy which is lost due to contact with the incline if $\theta = 10°$.

Ans. $v' = \dfrac{v}{3}(1 + 2\cos\theta)$, $n = 0.020$

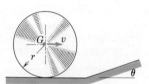

Problem 7/196

7/196   The solid sphere is resting on a platform which is given a horizontal acceleration $a = 2g$. Determine the acceleration $\bar{a}$ of the center of the sphere if the coefficient of friction between the sphere and the platform is (*a*) 0.80 and (*b*) 0.40.

Ans. (*a*) $\bar{a} = 4g/7$, (*b*) $\bar{a} = 2g/5$

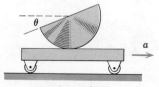

Problem 7/197

7/197   The horizontal acceleration of the cart is gradually increased until the solid half-cylinder tilts through a steady angle θ. Determine the corresponding acceleration a. Also specify the minimum value of the coefficient of friction to prevent slipping.

198 The two uniform slender bars, each weighing 10 lb, are released from rest in the vertical plane from the positions shown where OA is horizontal. Friction in the hinges at O and A is negligible. Calculate the force exerted on the bar by the bearing O (a) just after release and (b) as OA swings through the vertical position OA'.

$\qquad$ *Ans.* (a) $F_O = 3.12$ lb, (b) $F_O = 53.8$ lb

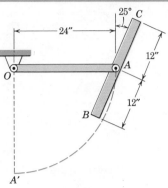

Problem 7/198

199 The toggle mechanism is used to move the mass m in the vertical guide. Determine the upward acceleration of m due to the action of P for a given angle θ by analyzing the system as a whole. Neglect the mass of the links.

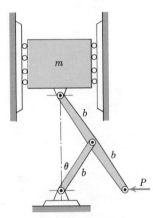

Problem 7/199

200 The rectangular block, which is solid and homogeneous, is supported at its corners by small rollers resting on horizontal surfaces. If the supporting surface at B is suddenly removed, determine the expression for the initial acceleration of corner A.

$$\textit{Ans. } a_A = \frac{3g}{4\dfrac{b}{h} + \dfrac{h}{b}}$$

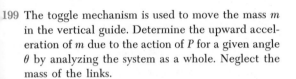

Problem 7/200

201 The light circular hoop carries a uniform heavy band of mass m around half of its circumference. If the hoop is released from rest in the position shown, write an expression for the normal force N under the hoop for the position where the kinetic energy of the hoop is greatest. Friction is sufficient to prevent slipping.

$$\textit{Ans. } N = mg\left[1 + \frac{8}{\pi(\pi - 2)}\right]$$

Problem 7/201

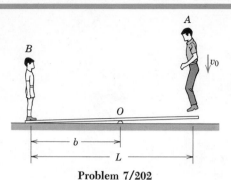

Problem 7/202

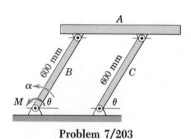

Problem 7/203

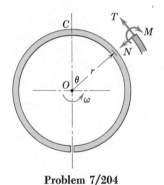

Problem 7/204

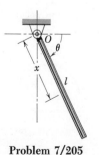

Problem 7/205

7/202 In an acrobatic stunt man A of mass m_A drops from a raised platform onto the end of the light but strong beam with a velocity v_0. The boy of mass m_B is propelled upward with a velocity v_B. For a given ratio $n = m_B/m_A$ determine b in terms of L which will maximize the upward velocity of the boy. Assume that both man and boy act as rigid bodies.

$$\text{Ans. } b = \frac{L}{1 + \sqrt{n}}$$

7/203 The bar A of mass $m = 30$ kg is elevated by the action of the couple $M = 270$ N·m applied to one end of the two parallel links B and C as shown. Each link is a uniform slender bar and has a mass of 10 kg and a length of 600 mm. Determine the angular acceleration α of the links as a function of the angle θ.   Ans. $\alpha = 20.5 - 17.8 \cos \theta$ rad/s²

▶**7/204** The split ring of radius r is rotating about a vertical axis through its center O with a constant angular velocity ω. Use a differential element of the ring and derive expressions for the shear force N and rim tension T in the ring in terms of the angle θ. Determine the bending moment M_C at point C by using one half of the ring as a free body. The mass of the ring per unit length of rim is ρ.

$$\text{Ans. } N = \rho r^2 \omega^2 \sin \theta$$
$$T = \rho r^2 \omega^2 (1 + \cos \theta)$$
$$M_C = 2\rho r^3 \omega^2$$

▶**7/205** The uniform slender bar of mass m and length l is pivoted freely at its end about a horizontal axis through O and released from rest at $\theta = 0$. Write the expression for the bending moment M in the bar in terms of x and θ. For a given θ, find the maximum value of M and the value of x at which it occurs.

$$\text{Ans. } M = \frac{mgx}{4}\left(1 - \frac{x}{l}\right)^2 \cos \theta$$

$$M_{\max} = \frac{mgl}{27} \cos \theta \text{ at } x = \frac{l}{3}$$

INTRODUCTION TO THREE-DIMENSIONAL DYNAMICS OF RIGID BODIES

8

8/1 INTRODUCTION. Although a large percentage of dynamics problems in engineering lend themselves to solution by means of the principles of plane motion, modern developments have focused increasing attention upon problems which call for the analysis of motion in three dimensions. Inclusion of the third dimension adds considerable complexity to the kinematic and kinetic relationships. Not only does the added dimension introduce a third component to vectors which represent force, linear velocity, linear acceleration, and linear momentum, but the introduction of the third dimension adds the possibility of two additional components for vectors representing angular quantities including moments of forces, angular velocity, angular acceleration, and angular momentum. It is in three-dimensional motion that the full power of vector analysis is utilized.

A good background in the dynamics of plane motion is extremely useful in the study of three-dimensional dynamics, as the approach to problems and many of the terms are the same as or analogous to those in two dimensions. If the study of three-dimensional dynamics is undertaken without the benefit of prior study of plane-motion dynamics, more time will be required to master the principles and to become familiar with the approach to problems.

The treatment presented in Chapter 8 is not intended as a complete development of the three-dimensional motion of rigid bodies but merely as a basic introduction to the subject. This introduction should, however, be sufficient to solve many of the more common problems in three-dimensional motion and also to lay the foundation for more advanced study. We shall proceed as we did for particle motion and for rigid-body plane motion by first examining the necessary kinematics and then proceeding to the kinetics.

SECTION A. KINEMATICS

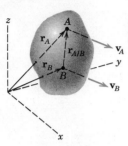

Figure 8/1

8/2 TRANSLATION. Figure 8/1 shows a rigid body translating in three-dimensional space. Any two points in the body, such as A and B, will move along parallel straight lines if the motion is one of *rectilinear translation* or will move along congruent curves if the motion is one of *curvilinear translation*. In either case every line in the body, such as AB, remains parallel to its original position. The position vectors and their first and second time derivatives are

$$\mathbf{r}_A = \mathbf{r}_B + \mathbf{r}_{A/B} \qquad \mathbf{v}_A = \mathbf{v}_B \qquad \mathbf{a}_A = \mathbf{a}_B$$

where $\mathbf{r}_{A/B}$ remains constant and therefore has no time derivative. Thus all points in the body have the same velocity and the same acceleration. The kinematics of translation presents no special difficulty, and further elaboration is unnecessary.

8/3 FIXED-AXIS ROTATION. Consider now the *rotation* of a rigid body about a fixed axis n-n in space with an angular velocity $\boldsymbol{\omega}$, as shown in Fig. 8/2. The angular velocity is a vector in the direction of the rotation axis with a sense established by the familiar right-hand rule. For fixed-axis rotation $\boldsymbol{\omega}$ does not change its direction since it lies along the axis. We choose the origin O of the fixed coordinate system on the rotation axis for convenience. Any point such as A that is not on the axis moves in a circular arc in a plane normal to the axis and has a velocity

$$\boxed{\mathbf{v} = \boldsymbol{\omega} \times \mathbf{r}} \tag{8/1}$$

which may be seen by replacing $\mathbf{r}$ by $\mathbf{h} + \mathbf{b}$ and noting that $\boldsymbol{\omega} \times \mathbf{h} = \mathbf{0}$. The acceleration of A is given by the time derivative of Eq. 8/1. Thus,

$$\boxed{\mathbf{a} = \dot{\boldsymbol{\omega}} \times \mathbf{r} + \boldsymbol{\omega} \times (\boldsymbol{\omega} \times \mathbf{r})} \tag{8/2}$$

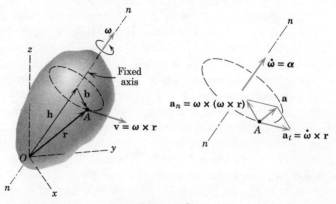

Figure 8/2

where $\dot{\mathbf{r}}$ has been replaced by its equal, $\mathbf{v} = \boldsymbol{\omega} \times \mathbf{r}$. The normal and tangential components of $\mathbf{a}$ for the circular motion have the familiar magnitudes $a_n = |\boldsymbol{\omega} \times (\boldsymbol{\omega} \times \mathbf{r})| = b\omega^2$ and $a_t = |\dot{\boldsymbol{\omega}} \times \mathbf{r}| = b\alpha$, where $\alpha = \dot{\omega}$. Inasmuch as both $\mathbf{v}$ and $\mathbf{a}$ are perpendicular to $\boldsymbol{\omega}$ and $\dot{\boldsymbol{\omega}}$, it follows that $\mathbf{v} \cdot \boldsymbol{\omega} = 0$, $\mathbf{v} \cdot \dot{\boldsymbol{\omega}} = 0$, $\mathbf{a} \cdot \boldsymbol{\omega} = 0$, and $\mathbf{a} \cdot \dot{\boldsymbol{\omega}} = 0$ for fixed-axis rotation.

8/4 PARALLEL-PLANE MOTION. When all points in a rigid body move in planes which are parallel to a fixed plane P, Fig. 8/3, we have a general form of plane motion. The reference plane is customarily taken through the mass center G and is referred to as the plane of motion. Since each point in the body, such as A', has a motion identical with the motion of point A in plane P, it follows that the kinematics of plane motion covered in Chapter 6 provides a complete description of the motion when applied to the reference plane.

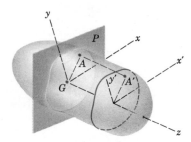

Figure 8/3

8/5 ROTATION ABOUT A FIXED POINT. When a body rotates about a fixed point, the angular velocity vector no longer remains fixed in direction, and this change calls for a more general concept of rotation.

We must first examine the conditions under which rotation vectors obey the parallelogram law of addition and may, therefore, be treated as proper vectors. Consider a solid sphere, Fig. 8/4, cut from a rigid body confined to rotate about the fixed point O.

The x-y-z axes here are taken fixed in space and do not rotate with the body. In the a-part of the figure two successive 90° rotations of the sphere about, first, the x-axis and, second, the y-axis result in the motion of a point which is initially on the y-axis in position 1, to positions 2 and 3, successively. On the other hand, if the order of the rotations is reversed, the point suffers no motion during the y-rotation but moves to point 3 during the 90° rotation about the x-axis. Thus the two cases do not yield the same final position, and it is evident from

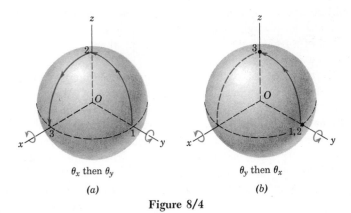

θ_x then θ_y θ_y then θ_x

(a)   (b)

Figure 8/4

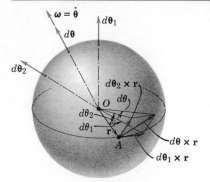

Figure 8/5

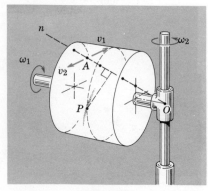

Figure 8/6

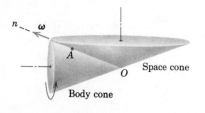

Figure 8/7

this one special example that finite rotations do not generally obey the parallelogram law of vector addition and are not commutative. Thus finite rotations may *not* be treated as proper vectors.

Infinitesimal rotations, however, do obey the parallelogram law of vector addition. This fact is shown in Fig. 8/5 which represents the combined effect of two infinitesimal rotations $d\boldsymbol{\theta}_1$ and $d\boldsymbol{\theta}_2$ of a rigid body about the respective axes through the fixed point O. As a result of $d\boldsymbol{\theta}_1$ point A has a displacement $d\boldsymbol{\theta}_1 \times \mathbf{r}$, and likewise $d\boldsymbol{\theta}_2$ causes a displacement $d\boldsymbol{\theta}_2 \times \mathbf{r}$ of point A. Either order of addition of these infinitesimal displacements clearly produces the same resultant displacement, which is $d\boldsymbol{\theta}_1 \times \mathbf{r} + d\boldsymbol{\theta}_2 \times \mathbf{r} = (d\boldsymbol{\theta}_1 + d\boldsymbol{\theta}_2) \times \mathbf{r}$. Hence the two rotations are equivalent to the single rotation $d\boldsymbol{\theta} = d\boldsymbol{\theta}_1 + d\boldsymbol{\theta}_2$. It follows that the angular velocities $\boldsymbol{\omega}_1 = \dot{\boldsymbol{\theta}}_1$ and $\boldsymbol{\omega}_2 = \dot{\boldsymbol{\theta}}_2$ may be added vectorially to give $\boldsymbol{\omega} = \dot{\boldsymbol{\theta}} = \boldsymbol{\omega}_1 + \boldsymbol{\omega}_2$. We conclude, therefore, that at any instant of time a body with one fixed point is rotating instantaneously about a particular axis passing through the fixed point.

To aid in visualizing the concept of the instantaneous axis of rotation we will cite a specific example. Figure 8/6 represents a solid cylindrical rotor made of clear plastic containing many black particles embedded in the plastic. The rotor is spinning about its shaft axis at the steady rate ω_1, and its shaft, in turn, is rotating about the fixed vertical axis at the steady rate ω_2, with rotations in the directions indicated. If the rotor is photographed at a certain instant during its motion, the resulting picture would show one line of black dots in sharp focus indicating that, momentarily, their velocity was zero. This line of points with no velocity establishes the instantaneous position of the axis of rotation O-n. Any dot on this line, such as A, would have equal and opposite velocity components, v_1 due to ω_1 and v_2 due to ω_2. All other dots, such as the one at P, would appear out of focus, and their movements would show as short blurred streaks in the form of small circular arcs in planes normal to the axis O-n. Thus all particles of the body, except those on line O-n, are momentarily rotating in circular arcs about the instantaneous axis of rotation. If a succession of photographs were taken, we would observe that the rotation axis would be defined by a new series of dots in focus and that the axis would change position both in space and relative to the body. For rotation of a rigid body about a fixed point, then, it is seen that the rotation axis is in general not a line fixed in the body.

(a) Body and Space Cones. Relative to the plastic cylinder of Fig. 8/6 the instantaneous axis of rotation O-A-n generates a right-circular cone about the cylinder axis called the *body cone*. As the two rotations continue and the cylinder swings around the vertical axis, the instantaneous axis of rotation also generates a right-circular cone about the vertical axis called the *space cone*. These cones are shown in Fig. 8/7 for this particular example, and we see that the body cone rolls on the space cone. For a more general case where the rotations

are not steady, the space and body cones will not be right-circular cones, Fig. 8/8, but the body cone still rolls on the space cone.

(*b*) *Angular Acceleration.* The angular acceleration $\boldsymbol{\alpha}$ of a rigid body in three-dimensional motion is the time derivative of its angular velocity, $\boldsymbol{\alpha} = \dot{\boldsymbol{\omega}}$. In contrast to the case of rotation in a single plane where the scalar α measures only the change in magnitude of the angular velocity, in three-dimensional motion the vector $\boldsymbol{\alpha}$ reflects the change in direction of $\boldsymbol{\omega}$ as well as its change in magnitude. Thus in Fig. 8/8 where the tip of the angular velocity vector $\boldsymbol{\omega}$ follows the space curve p and changes both in magnitude and direction, the angular acceleration $\boldsymbol{\alpha}$ becomes a vector tangent to this curve in the direction of the change in $\boldsymbol{\omega}$.

When the magnitude of $\boldsymbol{\omega}$ remains constant, the angular acceleration $\boldsymbol{\alpha}$ becomes normal to $\boldsymbol{\omega}$. For this case if we let $\boldsymbol{\Omega}$ stand for the angular velocity with which the vector $\boldsymbol{\omega}$ itself rotates (*precesses*) as it forms the space cone, the angular acceleration may be written

$$\boldsymbol{\alpha} = \boldsymbol{\Omega} \times \boldsymbol{\omega} \qquad (8/3)$$

This relation is easily seen from Fig. 8/9 where the vectors $\boldsymbol{\alpha}$, $\boldsymbol{\omega}$, and $\boldsymbol{\Omega}$ in the lower figure bear exactly the same relationship to each other as do the vectors $\mathbf{v}$, $\mathbf{r}$, and $\boldsymbol{\omega}$ in the upper figure for relating the velocity of a point A on a rigid body to its position vector from O and the angular velocity of the body.

If we use Fig. 8/2 to represent a rigid body rotating about a fixed point O with the instantaneous axis of rotation n-n, we see that the velocity $\mathbf{v}$ and acceleration $\mathbf{a} = \dot{\mathbf{v}}$ of any point A in the body are given by the same expressions as apply to the case where the axis is fixed, namely,

$$\mathbf{v} = \boldsymbol{\omega} \times \mathbf{r} \qquad [8/1]$$

$$\mathbf{a} = \dot{\boldsymbol{\omega}} \times \mathbf{r} + \boldsymbol{\omega} \times (\boldsymbol{\omega} \times \mathbf{r}) \qquad [8/2]$$

The one difference between the case of rotation about a fixed axis and rotation about a fixed point lies in the fact that for rotation about a fixed point the angular acceleration $\boldsymbol{\alpha} = \dot{\boldsymbol{\omega}}$ will have a component normal to $\boldsymbol{\omega}$ due to the change in direction of $\boldsymbol{\omega}$ as well as a component in the direction of $\boldsymbol{\omega}$ to reflect any change in the magnitude of $\boldsymbol{\omega}$. Although any point on the rotation axis n-n momentarily will have zero velocity, it will *not* have zero acceleration as long as $\boldsymbol{\omega}$ is changing its direction. On the other hand, for rotation about a fixed axis $\boldsymbol{\alpha} = \dot{\boldsymbol{\omega}}$ has only the one component along the fixed axis to reflect the change in the magnitude of $\boldsymbol{\omega}$. Furthermore, points which lie on the fixed rotation axis clearly have no velocity or acceleration.

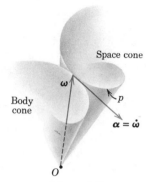

Figure 8/8

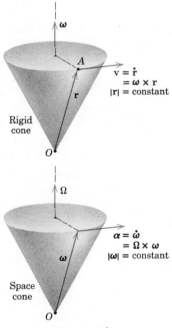

Figure 8/9

Sample Problem 8/1

The 0.8-m arm OA for a remote-control mechanism is pivoted about the horizontal x-axis of the clevis, and the entire assembly rotates about the z-axis with a constant speed $N = 60$ rev/min. Simultaneously the arm is being raised at the constant rate $\dot{\beta} = 4$ rad/s. For the position where $\beta = 30°$ determine (a) the angular velocity of OA, (b) the angular acceleration of OA, (c) the velocity of point A, and (d) the acceleration of point A.

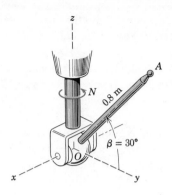

Solution. (a) Since the arm OA is rotating about both the x- and the z-axes, it has the components $\omega_x = \dot{\beta} = 4$ rad/s and $\omega_z = 2\pi N/60 = 2\pi(60)/60 = 6.283$ rad/s. The angular velocity is

$$\omega = \omega_x + \omega_z = 4\mathbf{i} + 6.283\mathbf{k} \text{ rad/s} \qquad Ans.$$

(b) The angular acceleration of OA is

$$\alpha = \dot{\omega} = \dot{\omega}_x + \dot{\omega}_z$$

Since ω_z is not changing in magnitude or direction, $\dot{\omega}_z = 0$. But ω_x is changing direction and thus has a derivative which, from Eq. 8/3, is

$$\dot{\omega}_x = \omega_z \times \omega_x = 6.283\mathbf{k} \times 4\mathbf{i} = 25.13\mathbf{j} \text{ rad/s}^2$$

① Therefore $\qquad \alpha = 25.13\mathbf{j} + 0 = 25.13\mathbf{j} \text{ rad/s}^2 \qquad Ans.$

(c) With the position vector of A given by $\mathbf{r} = 0.693\mathbf{j} + 0.4\mathbf{k}$ m, the velocity of A from Eq. 8/1 becomes

$$\mathbf{v} = \omega \times \mathbf{r} = \begin{vmatrix} \mathbf{i} & \mathbf{j} & \mathbf{k} \\ 4 & 0 & 6.283 \\ 0 & 0.693 & 0.4 \end{vmatrix} = -4.35\mathbf{i} - 1.60\mathbf{j} + 2.77\mathbf{k} \text{ m/s} \qquad Ans.$$

(d) The acceleration of A from Eq. 8/2 is

$$\mathbf{a} = \dot{\omega} \times \mathbf{r} + \omega \times (\omega \times \mathbf{r})$$

$$= \alpha \times \mathbf{r} + \omega \times \mathbf{v}$$

$$= \begin{vmatrix} \mathbf{i} & \mathbf{j} & \mathbf{k} \\ 0 & 25.13 & 0 \\ 0 & 0.693 & 0.4 \end{vmatrix} + \begin{vmatrix} \mathbf{i} & \mathbf{j} & \mathbf{k} \\ 4 & 0 & 6.283 \\ -4.35 & -1.60 & 2.77 \end{vmatrix}$$

$$= (10.05\mathbf{i}) + (10.05\mathbf{i} - 38.44\mathbf{j} - 6.40\mathbf{k})$$

② $\qquad = 20.11\mathbf{i} - 38.44\mathbf{j} - 6.40\mathbf{k} \text{ m/s}^2 \qquad Ans.$

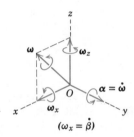

① Alternatively consider axes x-y-z to be attached to the vertical shaft and clevis so that they rotate. The derivative of ω_x becomes $\dot{\omega}_x = 4\mathbf{i}$. But from Eq. 6/10 we have $\dot{\mathbf{i}} = \omega_z \times \mathbf{i} = 6.283\mathbf{k} \times \mathbf{i} = 6.283\mathbf{j}$. Thus $\alpha = \dot{\omega}_x = 4(6.283)\mathbf{j} = 25.13\mathbf{j} \text{ rad/s}^2$ as before.

② To compare methods it is suggested that these results for $\mathbf{v}$ and $\mathbf{a}$ be obtained by applying Eqs. 2/18 and 2/19 for particle motion in spherical coordinates, changing symbols as necessary.

Sample Problem 8/2

The electric motor with attached disk is running at a constant low speed of 120 rev/min in the direction shown. Its housing and mounting base are initially at rest. The entire assembly is next set in rotation about the vertical Z-axis at the constant rate $N = 60$ rev/min with a fixed angle γ of 30°. Determine (a) the angular velocity and angular acceleration of the disk, (b) the space and body cones, and (c) the velocity and acceleration of point A at the top of the disk for the instant shown.

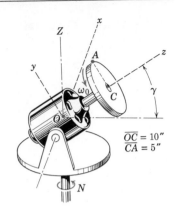

$\overline{OC} = 10''$
$\overline{CA} = 5''$

Solution. The axes x-y-z with unit vectors $\mathbf{i}, \mathbf{j}, \mathbf{k}$ are attached to the motor frame with the z-axis coinciding with the rotor axis and the x-axis coinciding with the horizontal axis through O about which the motor tilts. The Z-axis is vertical and carries the unit vector $\mathbf{K} = \mathbf{j} \cos \gamma + \mathbf{k} \sin \gamma$.

(a) The rotor and disk have two components of angular velocity, $\omega_0 = 120(2\pi)/60 = 4\pi$ rad/sec about the z-axis and $\Omega = 60(2\pi)/60 = 2\pi$ rad/sec about the Z-axis. Thus the angular velocity becomes

$$\omega = \omega_0 + \Omega = \omega_0 \mathbf{k} + \Omega \mathbf{K}$$

$$= \omega_0 \mathbf{k} + \Omega(\mathbf{j} \cos \gamma + \mathbf{k} \sin \gamma) = (\Omega \cos \gamma)\mathbf{j} + (\omega_0 + \Omega \sin \gamma)\mathbf{k}$$

$$= (2\pi \cos 30°)\mathbf{j} + (4\pi + 2\pi \sin 30°)\mathbf{k} = \pi(\sqrt{3}\mathbf{j} + 5.0\mathbf{k}) \text{ rad/sec} \qquad Ans.$$

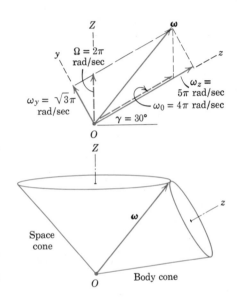

The angular acceleration of the motor and disk from Eq. 8/3 is

$$\alpha = \dot{\omega} = \Omega \times \omega$$

$$= \Omega(\mathbf{j} \cos \gamma + \mathbf{k} \sin \gamma) \times [(\Omega \cos \gamma)\mathbf{j} + (\omega_0 + \Omega \sin \gamma)\mathbf{k}]$$

$$= \Omega(\omega_0 \cos \gamma + \Omega \sin \gamma \cos \gamma)\mathbf{i} - (\Omega^2 \sin \gamma \cos \gamma)\mathbf{i}$$

$$= (\Omega \omega_0 \cos \gamma)\mathbf{i} = \mathbf{i}(2\pi)(4\pi) \cos 30° = 68.4\mathbf{i} \text{ rad/sec}^2 \qquad Ans.$$

(b) The angular velocity vector ω is the common element of the space and body cones which may now be constructed as shown.

Space cone

Body cone

(c) The position vector of point A for the instant considered is

$$\mathbf{r} = 5\mathbf{j} + 10\mathbf{k} \text{ in.}$$

From Eq. 8/1 the velocity of A is

$$\mathbf{v} = \omega \times \mathbf{r} = \begin{vmatrix} \mathbf{i} & \mathbf{j} & \mathbf{k} \\ 0 & \sqrt{3}\pi & 5\pi \\ 0 & 5 & 10 \end{vmatrix} = -7.68\pi\mathbf{i} \text{ in./sec} \qquad Ans.$$

From Eq. 8/2 the acceleration of point A is

$$\mathbf{a} = \dot{\omega} \times \mathbf{r} + \omega \times (\omega \times \mathbf{r}) = \alpha \times \mathbf{r} + \omega \times \mathbf{v}$$

$$= 68.4\mathbf{i} \times (5\mathbf{j} + 10\mathbf{k}) + \pi(\sqrt{3}\mathbf{j} + 5\mathbf{k}) \times (-7.68\pi\mathbf{i})$$

$$= -1063\mathbf{j} + 473\mathbf{k} \text{ in./sec}^2 \qquad Ans.$$

① Note that $\omega_0 + \Omega = \omega = \omega_y + \omega_z$ as shown on the vector diagram.

② Remember that Eq. 8/3 gives the complete expression for α only for steady precession where $|\omega|$ is constant, which applies to this problem.

③ Since the magnitude of ω is constant, α must be tangent to the base circle of the space cone which puts it in the plus x-direction in agreement with our calculated conclusion.

PROBLEMS

8/1 A rigid body rotates about a fixed axis with an angular velocity $\boldsymbol{\omega} = \pi(2\mathbf{i} + 3\mathbf{j} + 3\mathbf{k})$ rad/s, and at a certain instant a point A on the body has a velocity whose x- and y-components are 6 and 4 m/s, respectively. Find the magnitude v of the velocity of point A.   *Ans. v = 10.77 m/s*

8/2 A rigid body rotates about a fixed axis and has an angular acceleration $\boldsymbol{\alpha} = 3\pi^2(\mathbf{i} + 1.5\mathbf{j} + 1.5\mathbf{k})$ rad/s². At a certain instant the velocity of a point A on the body has x- and z-components equal to 3 and -4 m/s, respectively. Find the magnitude v of the velocity of point A.

8/3 A timing mechanism consists of the rotating distributor arm AB and the fixed contact C. If the arm rotates about the fixed axis OA with a constant angular velocity $\boldsymbol{\omega} = 3(3\mathbf{i} + 2\mathbf{j} + 6\mathbf{k})$ rad/sec, and if the coordinates of the contact C expressed in inches are $(3, 3, 6)$, determine the magnitude of the acceleration of the tip B of the distributor arm as it passes point C.   *Ans. |$\mathbf{a}_B$| = 423 in./sec²*

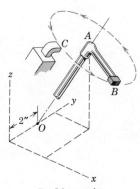

Problem 8/3

8/4 A rigid body rotates about a fixed axis with a constant angular velocity of 14 rad/s. A point O on the rotation axis is selected as the origin of coordinates, and a point B on the rotation axis has x-y-z coordinates of 0.2, 0.3, 0.6 meters. Determine the magnitudes of the velocity and acceleration of a point A on the body whose x-y-z coordinates are 200, 200, 100 millimeters.

              Ans. v = 2.72 m/s, a = 38.1 m/s²

8/5 The disk rotates with a spin velocity of 15 rad/s about its horizontal z-axis first in the direction (a) and second in the direction (b). The assembly rotates with the velocity $N = 10$ rad/s about the vertical axis. Construct the space and body cones for each case.

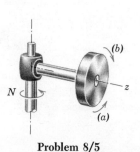

Problem 8/5

8/6 The rod A is hinged about the axis O-O of the clevis, which is attached to the end of the rotating vertical shaft. The shaft rotates with an angular velocity Ω as shown, and the x-y axes are attached to the vertical shaft. If θ is changing at the constant rate $\dot{\theta} = -p$, write the expressions for the angular velocity and angular acceleration of the rod.

Ans. $\omega = p\mathbf{j} + \Omega\mathbf{k}, \ \boldsymbol{\alpha} = -\Omega p\mathbf{i}$

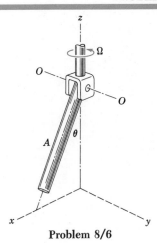

Problem 8/6

8/7 The vertical shaft and clevis of Prob. 8/6 have a constant angular velocity about the z-axis. If the angle θ varies with the time t according to $\theta = bt^3$ where b is a constant, determine expressions for the angular velocity ω and angular acceleration $\boldsymbol{\alpha}$ of the rod A as functions of the time t.

8/8 If the motor of Sample Problem 8/2 pivots about the x-axis at the constant rate $\dot{\gamma} = 3\pi$ rad/sec with no rotation about the Z-axis $(N = 0)$, determine the angular acceleration $\boldsymbol{\alpha}$ of the rotor and disk as the position $\gamma = 30°$ is passed. The constant speed of the motor is 120 rev/min. Also find the velocity and acceleration of a point A, which is on the top of the disk for this position.

Ans. $\mathbf{v} = 5\pi(-4\mathbf{i} + 6\mathbf{j} - 3\mathbf{k})$ in./sec
$\mathbf{a} = -5\pi^2(25\mathbf{j} + 18\mathbf{k})$ in./sec^2

8/9 For the figure for Sample Problem 8/2 assume that the motor shaft and attached disk are rotating about the z-axis with a spin ω_0 of 120 rev/min which is increasing at the rate of 10 rad/sec^2 at the position $\gamma = 30°$. At the same time $\dot{\gamma}$ is 12 rad/sec and is increasing at the rate of 15 rad/sec^2. Determine the vector expression for the angular acceleration $\boldsymbol{\alpha}$ of the rotor at this instant. There is no rotation about the Z-axis.

8/10 The revolving crane of Prob. 2/157 is shown again here. The boom OP has a length of 24 m, and the crane is revolving about the vertical axis at the constant rate of 2 rev/min in the direction shown. Simultaneously the boom is being lowered at the constant rate $\dot{\beta} = 0.10$ rad/s. Calculate the magnitudes of the velocity and acceleration of the end P of the boom for the instant when it passes the position $\beta = 30°$.

Ans. $v = 3.48$ m/s, $a = 1.104$ m/s^2

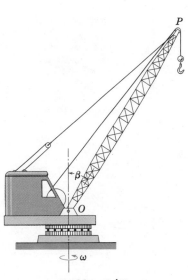

Problem 8/10

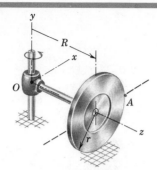

Problem 8/11

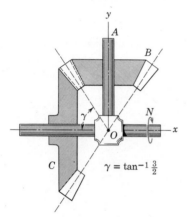

Problem 8/13

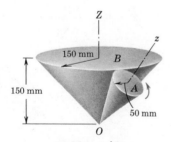

Problem 8/15

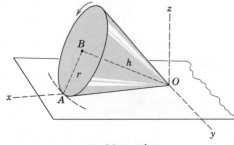

Problem 8/16

8/11 The wheel rolls without slipping in a circular arc of radius R and makes one complete turn about the vertical y-axis with constant speed in time τ. Determine the vector expression for the angular acceleration $\boldsymbol{\alpha}$ of the wheel and construct the space and body cones.

8/12 Determine the velocity $\mathbf{v}$ and acceleration $\mathbf{a}$ of point A on the wheel of Prob. 8/11 for the position shown where A crosses the horizontal line through the center of the wheel.

$$Ans.\ \mathbf{v} = \frac{2\pi R}{\tau}\left(\mathbf{i} - \mathbf{j} - \frac{r}{R}\mathbf{k}\right),$$

$$\mathbf{a} = -\left(\frac{2\pi}{\tau}\right)^2 R\left[\left(\frac{R}{r} + \frac{r}{R}\right)\mathbf{i} + \mathbf{k}\right]$$

8/13 The shaft OA of the bevel gear B rotates about the fixed axis O-x with a constant speed $N = 60$ rev/min in the direction shown. Gear B meshes with the bevel gear C along its pitch cone of semivertex angle $\gamma = \tan^{-1}\frac{3}{2}$ as shown. Determine the angular velocity $\boldsymbol{\omega}$ and angular acceleration $\boldsymbol{\alpha}$ of gear B if gear C is fixed and does not rotate. The y-axis revolves with the shaft OA.

8/14 If gear C of Prob. 8/13 has a constant rotational velocity of 20 rev/min about the axis O-x in the same sense as N while OA maintains its constant rotational speed $N = 60$ rev/min, calculate the angular velocity $\boldsymbol{\omega}$ and angular acceleration $\boldsymbol{\alpha}$ of gear B. $\quad Ans.\ \boldsymbol{\omega} = 2\pi(-\mathbf{i} + \mathbf{j})$ rad/s
$$\boldsymbol{\alpha} = -4\pi^2\mathbf{k}\ \text{rad/s}^2$$

▶**8/15** The right circular cone A rolls on the fixed right circular cone B at a constant rate and makes one complete trip around B every 4 s. Compute the magnitude of the angular acceleration $\boldsymbol{\alpha}$ of cone A.
$$Ans.\ \alpha = 6.32\ \text{rad/s}^2$$

▶**8/16** The solid right circular cone of base radius r and height h rolls on a flat surface without slipping. The center B of the circular base moves in a circular path around the z-axis with a constant speed v. Determine the angular velocity $\boldsymbol{\omega}$ and the angular acceleration $\boldsymbol{\alpha}$ of the solid cone.

$$Ans.\ \boldsymbol{\omega} = v\sqrt{\frac{1}{r^2} + \frac{1}{h^2}}\,\mathbf{i}$$

$$\boldsymbol{\alpha} = -\frac{v^2}{h^2}\left(\frac{r}{h} + \frac{h}{r}\right)\mathbf{j}$$

8/6 GENERAL MOTION.   The kinematic analysis of a rigid body which has general three-dimensional motion is best accomplished with the aid of our principles of relative motion. These principles have been applied to problems in plane motion and will now be extended to space motion. We will make use of both translating reference axes and rotating reference axes.

(a) *Translating Reference Axes.* In Fig. 8/10 is shown a rigid body which has an angular velocity ω. We choose any convenient point B as the origin of a translating reference system x-y-z. The velocity $\mathbf{v}$ and acceleration $\mathbf{a}$ of any other point A in the body is given by the relative-velocity and relative-acceleration expressions

$$\mathbf{v}_A = \mathbf{v}_B + \mathbf{v}_{A/B} \quad \text{and} \quad \mathbf{a}_A = \mathbf{a}_B + \mathbf{a}_{A/B} \quad [6/4, 6/7]$$

which were developed in Arts. 6/4 and 6/6 for the plane motion of rigid bodies and which hold equally well for any plane in space in which the three vectors for each of the equations must lie.

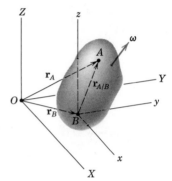

Figure 8/10

In applying these relations to rigid-body motion in space we note from Fig. 8/10 that the distance $\overline{AB}$ remains constant. Thus, from an observer's position on x-y-z, the body appears to rotate about the point B and point A appears to lie on a spherical surface with B as the center. Consequently, we may view the general motion as a translation of the body with the motion of B plus a rotation of the body about B.

The relative motion terms represent the effect of the rotation about B and are identical to the velocity and acceleration expressions discussed in the previous article for rotation of a rigid body about a fixed point. Therefore the relative-velocity and relative-acceleration equations may be written

$$\boxed{\begin{aligned} \mathbf{v}_A &= \mathbf{v}_B + \omega \times \mathbf{r}_{A/B} \\ \mathbf{a}_A &= \mathbf{a}_B + \dot{\omega} \times \mathbf{r}_{A/B} + \omega \times (\omega \times \mathbf{r}_{A/B}) \end{aligned}} \quad (8/4)$$

where ω is the instantaneous angular velocity of the body.

The selection of the reference point B is quite arbitrary in theory. In practice point B is chosen for convenience as some point in the body whose motion is known in whole or in part. If point A is chosen as the reference point, the relative motion equations become

$$\mathbf{v}_B = \mathbf{v}_A + \omega \times \mathbf{r}_{B/A}$$

$$\mathbf{a}_B = \mathbf{a}_A + \dot{\omega} \times \mathbf{r}_{B/A} + \omega \times (\omega \times \mathbf{r}_{B/A})$$

where $\mathbf{r}_{B/A} = -\mathbf{r}_{A/B}$. It should be clear that ω and, hence, $\dot{\omega}$ are the same vectors for either formulation since the absolute angular motion of the body is independent of the choice of reference point. When we come to the kinetic equations for general motion, we will see that the mass center of a body is frequently the most convenient reference point to choose.

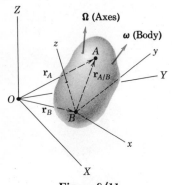

Figure 8/11

(*b*) *Rotating Reference Axes.* A more general formulation of the motion of a rigid body in space calls for the use of reference axes which rotate as well as translate. The description of Fig. 8/10 is modified in Fig. 8/11 to show reference axes with origin attached to the reference point *B* as before but which rotate with an absolute angular velocity $\boldsymbol{\Omega}$ which may be different from the absolute angular velocity $\boldsymbol{\omega}$ of the body.

We now make use of Eqs. 6/10, 6/11, 6/12, and 6/13 developed in Art. 6/7 for describing the plane motion of a rigid body using rotating axes. The extension of these relations from two to three dimensions is easily accomplished by merely including the *z*-component of the vectors, and this step is left to the student to carry out. Replacing $\boldsymbol{\omega}$ in these equations by the angular velocity $\boldsymbol{\Omega}$ of our rotating *x-y-z* axes gives us

$$\dot{\mathbf{i}} = \boldsymbol{\Omega} \times \mathbf{i} \qquad \dot{\mathbf{j}} = \boldsymbol{\Omega} \times \mathbf{j} \qquad \dot{\mathbf{k}} = \boldsymbol{\Omega} \times \mathbf{k} \tag{8/5}$$

for the time derivatives of the rotating unit vectors attached to *x-y-z*. The expressions for the velocity and acceleration of point *A* become

$$\mathbf{v}_A = \mathbf{v}_B + \boldsymbol{\Omega} \times \mathbf{r}_{A/B} + \mathbf{v}_{\text{rel}}$$
$$\mathbf{a}_A = \mathbf{a}_B + \dot{\boldsymbol{\Omega}} \times \mathbf{r}_{A/B} + \boldsymbol{\Omega} \times (\boldsymbol{\Omega} \times \mathbf{r}_{A/B}) + 2\boldsymbol{\Omega} \times \mathbf{v}_{\text{rel}} + \mathbf{a}_{\text{rel}} \tag{8/6}$$

where $\mathbf{v}_{\text{rel}} = \mathbf{i}\dot{x} + \mathbf{j}\dot{y} + \mathbf{k}\dot{z}$ and $\mathbf{a}_{\text{rel}} = \mathbf{i}\ddot{x} + \mathbf{j}\ddot{y} + \mathbf{k}\ddot{z}$ are, respectively, the velocity and acceleration of point *A* measured relative to *x-y-z* by an observer attached to *x-y-z*. Again we note that $\boldsymbol{\Omega}$ is the angular velocity of the axes and that $\mathbf{r}_{A/B}$ remains constant in magnitude for a rigid body. We observe further that, if *x-y-z* are rigidly attached to the body, $\boldsymbol{\Omega} = \boldsymbol{\omega}$ and $\mathbf{v}_{\text{rel}}$ and $\mathbf{a}_{\text{rel}}$ are both zero, which makes the equations identical with Eqs. 8/4.

In Art. 6/7 we also developed the relationship (Eq. 6/12) between the time derivative of a vector $\mathbf{V}$ as measured in the fixed *X-Y* system and the time derivative of $\mathbf{V}$ as measured relative to the rotating *x-y* system. For our three-dimensional case this relation becomes

$$\left(\frac{d\mathbf{V}}{dt}\right)_{XYZ} = \left(\frac{d\mathbf{V}}{dt}\right)_{xyz} + \boldsymbol{\Omega} \times \mathbf{V} \tag{8/7}$$

When we apply this transformation to the relative position vector $\mathbf{r}_{A/B} = \mathbf{r}_A - \mathbf{r}_B$ for our rigid body of Fig. 8/11, we get

$$\left(\frac{d\mathbf{r}_A}{dt}\right)_{XYZ} = \left(\frac{d\mathbf{r}_B}{dt}\right)_{XYZ} + \left(\frac{d\mathbf{r}_{A/B}}{dt}\right)_{xyz} + \boldsymbol{\Omega} \times \mathbf{r}_{A/B}$$

or
$$\mathbf{v}_A = \mathbf{v}_B + \mathbf{v}_{\text{rel}} + \boldsymbol{\Omega} \times \mathbf{r}_{A/B}$$

which gives us the first of Eqs. 8/6.

Equations 8/6 are particularly useful when the reference axes are attached to a moving body within which relative motion occurs.

Sample Problem 8/3

Crank CB rotates about the horizontal axis with an angular velocity $\omega_1 = 6$ rad/s which is constant for a short interval of motion which includes the position shown. The link AB has a ball-and-socket fitting on each end and connects crank DA with CB. For the instant shown, determine the angular velocity ω_2 of crank DA and the angular velocity $\boldsymbol{\omega}_n$ of link AB.

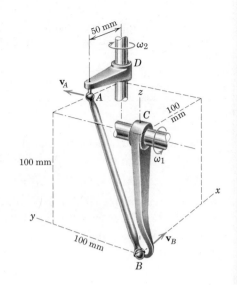

Solution. The relative-velocity relation, Eq. 8/4, will be solved first ① using translating reference axes attached to B. The equation is

$$\mathbf{v}_A = \mathbf{v}_B + \boldsymbol{\omega}_n \times \mathbf{r}_{A/B}$$

② where $\boldsymbol{\omega}_n$ is the angular velocity of link AB taken normal to AB. The velocities of A and B are

$$[v = r\omega] \qquad \mathbf{v}_A = 50\omega_2 \mathbf{j} \qquad \mathbf{v}_B = 100(6)\mathbf{i} = 600\mathbf{i} \text{ mm/s}$$

Also $\mathbf{r}_{A/B} = 50\mathbf{i} + 100\mathbf{j} + 100\mathbf{k}$ mm. Substitution into the velocity relation gives

$$50\omega_2 \mathbf{j} = 600\mathbf{i} + \begin{vmatrix} \mathbf{i} & \mathbf{j} & \mathbf{k} \\ \omega_{n_x} & \omega_{n_y} & \omega_{n_z} \\ 50 & 100 & 100 \end{vmatrix}$$

Expanding the determinant and equating the coefficients of the $\mathbf{i}$, $\mathbf{j}$, $\mathbf{k}$ terms give

$$-6 = \qquad + \omega_{n_y} - \omega_{n_z}$$
$$\omega_2 = -2\omega_{n_x} \qquad + \omega_{n_z}$$
$$0 = 2\omega_{n_x} - \omega_{n_y}$$

These equations may be solved for ω_2, which becomes

$$\omega_2 = 6 \text{ rad/s} \qquad\qquad Ans.$$

As they stand the three equations incorporate the fact that $\boldsymbol{\omega}_n$ is normal to $\mathbf{v}_{A/B}$, but they cannot be solved until the requirement that $\boldsymbol{\omega}_n$ be normal ③ to $\mathbf{r}_{A/B}$ is included. Thus,

$$[\boldsymbol{\omega}_n \cdot \mathbf{r}_{A/B} = 0] \qquad 50\omega_{n_x} + 100\omega_{n_y} + 100\omega_{n_z} = 0$$

Combination with two of the three previous equations yields the solutions

$$\omega_{n_x} = -\tfrac{4}{3} \text{ rad/s} \qquad \omega_{n_y} = -\tfrac{8}{3} \text{ rad/s} \qquad \omega_{n_z} = \tfrac{10}{3} \text{ rad/s}$$

Thus

$$\boldsymbol{\omega}_n = \tfrac{2}{3}(-2\mathbf{i} - 4\mathbf{j} + 5\mathbf{k}) \text{ rad/s}$$

with

$$\omega_n = \tfrac{2}{3}\sqrt{2^2 + 4^2 + 5^2} = 2\sqrt{5} \text{ rad/s} \qquad\qquad Ans.$$

① We select B as the reference point since its motion can easily be determined from the given angular velocity ω_1 of CB.

② The angular velocity $\boldsymbol{\omega}$ of AB is taken as a vector $\boldsymbol{\omega}_n$ normal to AB since any rotation of the link about its own axis AB has no influence on the behavior of the linkage.

③ The relative-velocity equation may be written as $\mathbf{v}_A - \mathbf{v}_B = \mathbf{v}_{A/B} = \boldsymbol{\omega}_n \times \mathbf{r}_{A/B}$ which requires that $\mathbf{v}_{A/B}$ be perpendicular to both $\boldsymbol{\omega}_n$ and $\mathbf{r}_{A/B}$. This equation alone does not incorporate the additional requirement that $\boldsymbol{\omega}_n$ be perpendicular to $\mathbf{r}_{A/B}$. Thus we must also satisfy $\boldsymbol{\omega}_n \cdot \mathbf{r}_{A/B} = 0$.

Sample Problem 8/4

Determine the angular acceleration $\dot{\omega}_2$ of crank AD in Sample Problem 8/3 for the conditions cited. Also find the angular acceleration of link AB.

Solution. The accelerations of the links may be found from the second of Eqs. 8/4, which may be written

$$\mathbf{a}_A = \mathbf{a}_B + \dot{\boldsymbol{\omega}}_n \times \mathbf{r}_{A/B} + \boldsymbol{\omega}_n \times (\boldsymbol{\omega}_n \times \mathbf{r}_{A/B})$$

where $\boldsymbol{\omega}_n$, as in Sample Problem 8/3, is the angular velocity of AB taken normal to AB. The angular acceleration of AB is written as $\dot{\boldsymbol{\omega}}_n$.

In terms of their normal and tangential components, the accelerations of A and B are

$$\mathbf{a}_A = 50\omega_2{}^2\mathbf{i} + 50\dot{\omega}_2\mathbf{j} = 1800\mathbf{i} + 50\dot{\omega}_2\mathbf{j} \text{ mm/s}^2$$

$$\mathbf{a}_B = 100\omega_1{}^2\mathbf{k} + (0)\mathbf{i} = 3600\mathbf{k} \text{ mm/s}^2$$

Also

$$\boldsymbol{\omega}_n \times (\boldsymbol{\omega}_n \times \mathbf{r}_{A/B}) = -\omega_n{}^2\mathbf{r}_{A/B} = -20(50\mathbf{i} + 100\mathbf{j} + 100\mathbf{k}) \text{ mm/s}^2$$

$$\dot{\boldsymbol{\omega}}_n \times \mathbf{r}_{A/B} = \mathbf{i}(100\dot{\omega}_{n_y} - 100\dot{\omega}_{n_z}) + \mathbf{j}(50\dot{\omega}_{n_z} - 100\dot{\omega}_{n_x}) + \mathbf{k}(100\dot{\omega}_{n_x} - 50\dot{\omega}_{n_y})$$

Substitution into the relative acceleration equation and equating respective coefficients of $\mathbf{i}$, $\mathbf{j}$, $\mathbf{k}$ give

$$28 = \dot{\omega}_{n_y} - \dot{\omega}_{n_z}$$

$$\dot{\omega}_2 + 40 = -2\dot{\omega}_{n_x} + \dot{\omega}_{n_z}$$

$$-32 = 2\dot{\omega}_{n_x} - \dot{\omega}_{n_y}$$

Solution of these equations for $\dot{\omega}_2$ gives

$$\dot{\omega}_2 = -36 \text{ rad/s}^2 \qquad\qquad Ans.$$

The vector $\dot{\boldsymbol{\omega}}_n$ is normal to $\mathbf{r}_{A/B}$ but is not normal to $\mathbf{v}_{A/B}$ as was the case with $\boldsymbol{\omega}_n$.

$$[\dot{\boldsymbol{\omega}}_n \cdot \mathbf{r}_{A/B} = 0] \qquad 2\dot{\omega}_{n_x} + 4\dot{\omega}_{n_y} + 4\dot{\omega}_{n_z} = 0$$

which, when combined with the preceding relations for these same quantities, gives

$$\dot{\omega}_{n_x} = -8 \text{ rad/s}^2 \qquad \dot{\omega}_{n_y} = 16 \text{ rad/s}^2 \qquad \dot{\omega}_{n_z} = -12 \text{ rad/s}^2$$

Thus

$$\dot{\boldsymbol{\omega}}_n = 4(-2\mathbf{i} + 4\mathbf{j} - 3\mathbf{k}) \text{ rad/s}^2 \qquad\qquad Ans.$$

and

$$|\dot{\boldsymbol{\omega}}_n| = 4\sqrt{2^2 + 4^2 + 3^2} = 4\sqrt{29} \text{ rad/s}^2 \qquad\qquad Ans.$$

① If the link AB had an angular velocity component along AB, then a change in both magnitude and direction of this component could occur which would contribute to the actual angular acceleration of the link as a rigid body. However, since any rotation about its own axis AB has no influence on the motion of the cranks at C and D, we shall concern ourselves only with $\dot{\boldsymbol{\omega}}_n$.

② The component of $\dot{\boldsymbol{\omega}}_n$ which is not normal to $\mathbf{v}_{A/B}$ gives rise to the change in direction of $\mathbf{v}_{A/B}$.

Sample Problem 8/5

The motor housing and its bracket rotate about the Z-axis at the constant rate $\Omega = 3$ rad/s. The motor shaft and disk have a constant angular velocity of spin $p = 8$ rad/s with respect to the motor housing in the direction shown. If γ is constant at 30°, determine the velocity and acceleration of point A at the top of the disk and the angular acceleration $\boldsymbol{\alpha}$ of the disk.

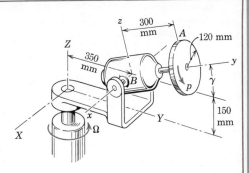

Solution. The rotating reference axes x-y-z are attached to the motor housing, and the rotating base for the motor has the momentary orientation shown with respect to the fixed axes X-Y-Z. We will use both X-Y-Z components with unit vectors $\mathbf{I}, \mathbf{J}, \mathbf{K}$ and x-y-z components with unit vectors $\mathbf{i}, \mathbf{j}, \mathbf{k}$. The angular velocity of the x-y-z axes becomes $\boldsymbol{\Omega} = \Omega\mathbf{K} = 3\mathbf{K}$ rad/s.

① This choice for the reference axes provides a simple description for the motion of the disk relative to these axes.

Velocity. The velocity of A is given by the first of Eqs. 8/6

$$\mathbf{v}_A = \mathbf{v}_B + \boldsymbol{\Omega} \times \mathbf{r}_{A/B} + \mathbf{v}_{rel}$$

where $\mathbf{v}_B = \boldsymbol{\Omega} \times \mathbf{r}_B = 3\mathbf{K} \times 0.350\mathbf{J} = -1.05\mathbf{I} = -1.05\mathbf{i}$ m/s

$$\boldsymbol{\Omega} \times \mathbf{r}_{A/B} = 3\mathbf{K} \times (0.300\mathbf{j} + 0.120\mathbf{k})$$
$$= (-0.9\cos 30°)\mathbf{i} + (0.36\sin 30°)\mathbf{i} = -0.599\mathbf{i}$ m/s$$

$$\mathbf{v}_{rel} = \mathbf{p} \times \mathbf{r}_{A/B} = 8\mathbf{j} \times (0.300\mathbf{j} + 0.120\mathbf{k}) = 0.960\mathbf{i}$ m/s$$

Thus $\mathbf{v}_A = -1.05\mathbf{i} - 0.599\mathbf{i} + 0.960\mathbf{i} = -0.689\mathbf{i}$ m/s   *Ans.*

② Note that $\mathbf{K} \times \mathbf{i} = \mathbf{J} = \mathbf{j}\cos\gamma - \mathbf{k}\sin\gamma$, $\mathbf{K} \times \mathbf{j} = -\mathbf{i}\cos\gamma$, and $\mathbf{K} \times \mathbf{k} = \mathbf{i}\sin\gamma$.

Acceleration. The acceleration of A is given by the second of Eqs. 8/6

$$\mathbf{a}_A = \mathbf{a}_B + \dot{\boldsymbol{\Omega}} \times \mathbf{r}_{A/B} + \boldsymbol{\Omega} \times (\boldsymbol{\Omega} \times \mathbf{r}_{A/B}) + 2\boldsymbol{\Omega} \times \mathbf{v}_{rel} + \mathbf{a}_{rel}$$

where $\mathbf{a}_B = \boldsymbol{\Omega} \times (\boldsymbol{\Omega} \times \mathbf{r}_B) = 3\mathbf{K} \times (3\mathbf{K} \times 0.350\mathbf{J}) = -3.15\mathbf{J}$

$$= 3.15(-\mathbf{j}\cos 30° + \mathbf{k}\sin 30°) = -2.73\mathbf{j} + 1.58\mathbf{k}$ m/s^2$

$$\dot{\boldsymbol{\Omega}} = \mathbf{0}$$

$$\boldsymbol{\Omega} \times (\boldsymbol{\Omega} \times \mathbf{r}_{A/B}) = 3\mathbf{K} \times (3\mathbf{K} \times [0.300\mathbf{j} + 0.120\mathbf{k}])$$
$$= 3\mathbf{K} \times (-0.599\mathbf{i}) = -1.557\mathbf{j} + 0.899\mathbf{k}$ m/s^2$

$$2\boldsymbol{\Omega} \times \mathbf{v}_{rel} = 2(3\mathbf{K}) \times 0.960\mathbf{i} = 5.76\mathbf{J}$$
$$= 5.76(\mathbf{j}\cos 30° - \mathbf{k}\sin 30°) = 4.99\mathbf{j} - 2.88\mathbf{k}$ m/s^2$

$$\mathbf{a}_{rel} = \mathbf{p} \times (\mathbf{p} \times \mathbf{r}_{A/B}) = 8\mathbf{j} \times (8\mathbf{j} \times [0.300\mathbf{j} + 0.120\mathbf{k}]) = -7.68\mathbf{k}$ m/s^2$

Substitution into the expression for $\mathbf{a}_A$ and collecting terms give us

$$\mathbf{a}_A = 0.703\mathbf{j} - 8.086\mathbf{k}$ m/s$^2 \quad \text{and}$$
$$a_A = \sqrt{(0.703)^2 + (8.086)^2} = 8.12$ m/s$^2 \quad \text{*Ans.*}$$

Angular acceleration. Since the precession is steady, we may use Eq. 8/3 to give us

$$\boldsymbol{\alpha} = \dot{\boldsymbol{\omega}} = \boldsymbol{\Omega} \times \boldsymbol{\omega} = 3\mathbf{K} \times (3\mathbf{K} + 8\mathbf{j})$$
$$= 0 + (-24\cos 30°)\mathbf{i} = -20.8\mathbf{i}$ rad/s$^2 \quad \text{*Ans.*}$$

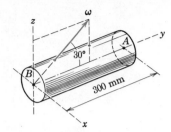

Problem 8/17

Problem 8/18

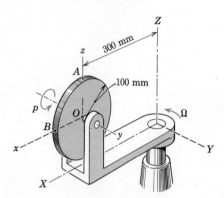

Problem 8/19

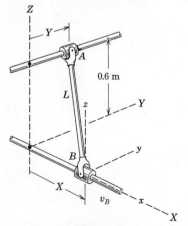

Problem 8/21

PROBLEMS

8/17 The solid cylinder shown has an angular velocity ω whose magnitude is 40 rad/s. Calculate the velocity of A with respect to B. Represent $\mathbf{v}_{A/B}$ on a sketch. Also determine the rate p at which the cylinder is spinning about its central axis.

Ans. $\mathbf{v}_{A/B} = -6\mathbf{i}$ m/s, $p = 34.6$ rad/s

8/18 The helicopter is nosing over at the constant rate q rad/s. If the rotor blades revolve at the constant speed p rad/s, write the expression for the angular acceleration $\boldsymbol{\alpha}$ of the rotor. Take the y-axis to be attached to the fuselage and pointing forward perpendicular to the rotor axis.   *Ans.* $\boldsymbol{\alpha} = pq\mathbf{j}$

8/19 The circular disk is spinning about its own axis (y-axis) at the constant rate $p = 10\pi$ rad/s. Simultaneously the frame is rotating about the Z-axis at the constant rate $\Omega = 4\pi$ rad/s. Calculate the angular acceleration α of the disk and the acceleration of point A at the top of the disk. Axes x-y-z are attached to the frame, which has the momentary orientation shown with respect to the fixed axes X-Y-Z.

Ans. $\boldsymbol{\alpha} = -40\pi^2\mathbf{i}$ rad/s^2
$\mathbf{a}_A = 2\pi^2(-2.4\mathbf{i} + 4\mathbf{j} - 5\mathbf{k})$ m/s^2

8/20 Determine the acceleration of point B on the disk of Prob. 8/19 for the conditions stated.

Ans. $\mathbf{a}_B = -16.4\pi^2\mathbf{i}$ m/s^2

8/21 For the instant represented collar B is moving along the fixed shaft in the X-direction with a constant velocity $v_B = 4$ m/s. Also at this instant $X = 0.3$ m and $Y = 0.2$ m. Calculate the velocity of collar A which moves along the fixed shaft parallel to the Y-axis. Solve, first, by differentiating the relation $X^2 + Y^2 + Z^2 = L^2$ with respect to time and, second, by using the first of Eqs. 8/4 with translating axes attached to B. Each clevis is free to rotate about the axis of the rod.   *Ans.* $\mathbf{v}_A = -6\mathbf{j}$ m/s

8/22 End A of the rigid link is confined to move in the $-x$-direction while end B is confined to move along the z-axis. Determine the component ω_n normal to AB of the angular velocity of the link as it passes the position shown with $v_A = 3$ ft/sec.

          Ans. $\omega_n = \frac{1}{49}(-3\mathbf{i} + 20\mathbf{j} + 9\mathbf{k})$ rad/sec

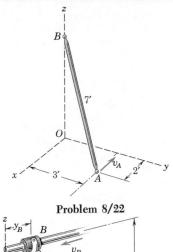

Problem 8/22

8/23 The collars at the ends of the telescoping link AB slide along the parallel fixed shafts shown. During an interval of motion $v_A = 125$ mm/s and $v_B = 50$ mm/s. Determine the vector expression for the component ω_n normal to AB of the angular velocity of the link for the position where $y_A = 100$ mm and $y_B = 50$ mm.

          Ans. $\omega_n = \frac{9}{49}(2\mathbf{i} + \mathbf{k})$ rad/s

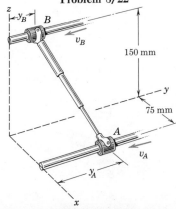

Problem 8/23

8/24 The spacecraft is revolving about its z-axis, which has a fixed space orientation, at the constant rate $p = \frac{1}{10}$ rad/s. Simultaneously its solar panels are unfolding at the rate $\dot{\beta}$ which is programmed to produce the variation with β shown in the graph. Determine the angular acceleration α of panel A an instant (a) before and an instant (b) after it reaches the position $\beta = 18°$.

      Ans. (a) $\alpha = -0.00388\mathbf{i} - 0.00349\mathbf{j}$ rad/s^2
           (b) $\alpha = -0.00349\mathbf{j}$ rad/s^2

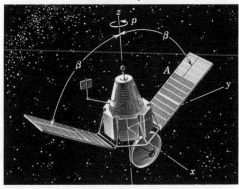

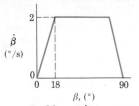

Problem 8/24

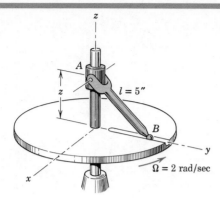

Problem 8/25

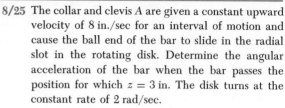

8/25 The collar and clevis A are given a constant upward velocity of 8 in./sec for an interval of motion and cause the ball end of the bar to slide in the radial slot in the rotating disk. Determine the angular acceleration of the bar when the bar passes the position for which $z = 3$ in. The disk turns at the constant rate of 2 rad/sec.

Ans. $\boldsymbol{\alpha} = -3\mathbf{i} - 4\mathbf{j}$ rad/sec^2

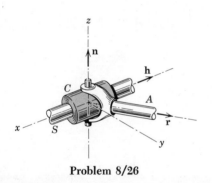

Problem 8/26

8/26 The end of the link A is attached to the collar C by the yoke which permits rotation of A about the z-axis. The collar C in turn may rotate about the fixed shaft S as well as slide along it. Irrespective of the motion of the other end of link A, show that the angular velocity $\boldsymbol{\omega}$ of the link must obey the relation $\boldsymbol{\omega} \cdot \mathbf{h} \times (\mathbf{r} \times \mathbf{h}) = 0$. Vectors $\mathbf{r}$ and $\mathbf{h}$ are any vectors, respectively, along the link and along the fixed shaft. The axis of the link A is normal to the yoke axis and, hence, lies in the x-y plane.

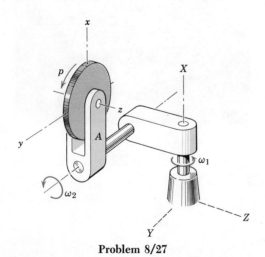

Problem 8/27

8/27 The disk has a constant angular velocity p about its z-axis, and the yoke A has a constant angular velocity ω_2 about its shaft as shown. Simultaneously the entire assembly revolves about the fixed X-axis with a constant angular velocity ω_1. Determine the expression for the angular acceleration of the disk as the yoke brings it into the vertical plane in the position shown. Solve by picturing the vector changes in the angular-velocity components.

Ans. $\boldsymbol{\alpha} = p\omega_2\mathbf{i} - p\omega_1\mathbf{j} + \omega_1\omega_2\mathbf{k}$

8/28 A simulator for perfecting the docking procedure for spacecraft consists of the frame A which is mounted on four air-bearing pads so that it can translate and rotate freely on the horizontal surface. Mounted in the frame is a drum B which can rotate about the horizontal axis of the frame A. The coordinate axes x-y-z are attached to the drum B, and the z-axis of the drum makes an angle β with the horizontal. Inside the drum is the simulated command module C which can rotate within the drum about its z-axis at a rate p. For a certain test run frame A is rotating on the horizontal surface in the direction shown with a constant angular velocity of 0.2 rad/s. Simultaneously the drum B is rotating about the x-axis at the constant rate $\dot\beta = 0.15$ rad/s, and the module C is turning inside the drum at the constant rate $p = 0.9$ rad/s in the direction indicated. For these conditions determine the angular velocity ω and the angular acceleration α of the simulator C as it passes the position $\beta = 0$.

Ans. $\omega = 0.15\mathbf{i} + 0.2\mathbf{j} + 0.9\mathbf{k}$ rad/s
$\alpha = 0.18\mathbf{i} - 0.135\mathbf{j} - 0.03\mathbf{k}$ rad/s^2

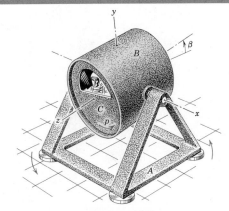

Problem 8/28

8/29 The test chamber of a flight simulator consists of a drum which spins about the a-a axis at a constant angular rate p relative to the cylindrical housing A. The housing in turn is mounted on transverse horizontal bearings and rotates about the axis b-b at the constant rate $\dot\theta$. The entire assembly is made to rotate about the fixed vertical z-axis at the constant rate Ω. Write the expression for the angular acceleration α of the spinning drum during the compounded motion for a given value of θ.

Ans. $\alpha = \dot\theta(p\sin\theta - \Omega)\mathbf{i} - (p\Omega\cos\theta)\mathbf{j}$
$+ (p\dot\theta\cos\theta)\mathbf{k}$

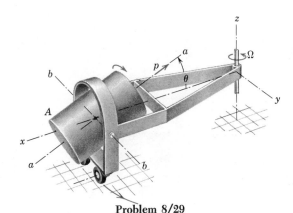

Problem 8/29

8/30 The center O of the spacecraft is moving through space with a constant velocity. During the period of motion prior to stabilization, the spacecraft has a constant rotational rate $\Omega = \frac12$ rad/s about its z-axis. The x-y-z axes are attached to the body of the craft, and the solar panels rotate about the y-axis at the constant rate $\dot\theta = \frac14$ rad/sec with respect to the spacecraft. If ω is the absolute angular velocity of the solar panels, determine $\dot\omega$. Also find the acceleration of point A when $\theta = 30°$.

Ans. $\dot\omega = \frac18\mathbf{i}$ rad/sec^2
$\mathbf{a}_A = 0.313\mathbf{i} - 2.43\mathbf{j} - 0.108\mathbf{k}$ ft/sec^2

Problem 8/30

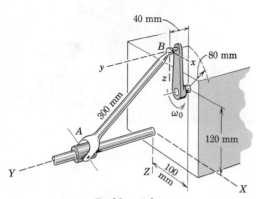

Problem 8/32

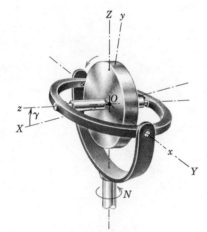

Problem 8/33

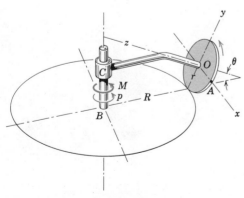

Problem 8/34

▶8/31 For the conditions specified with Sample Problem 8/2, except that γ is increasing at the steady rate of 3π rad/sec, determine the angular velocity $\boldsymbol{\omega}$ and the angular acceleration $\boldsymbol{\alpha}$ of the rotor when the position $\gamma = 30°$ is passed. (*Suggestion:* Apply Eq. 8/7 to the vector $\boldsymbol{\omega}$ to find $\boldsymbol{\alpha}$. Note that Ω in Sample Problem 8/2 is no longer the complete angular velocity of the axes.)

Ans. $\boldsymbol{\omega} = \pi(-3\mathbf{i} + \sqrt{3}\mathbf{j} + 5\mathbf{k})$ rad/sec
$\boldsymbol{\alpha} = \pi^2(4\sqrt{3}\mathbf{i} + 9\mathbf{j} + 3\sqrt{3}\mathbf{k})$ rad/sec^2

▶8/32 The crank with a radius of 80 mm turns with a constant angular velocity $\omega_0 = 4$ rad/s and causes the collar A to oscillate along the fixed shaft. Determine the velocity of the collar A and the angular velocity of the rigid-body link AB as the crank crosses the vertical position shown. (*Hint:* Use the relation cited in Prob. 8/26.)

Ans. $\mathbf{v}_A = 0.160\mathbf{j}$ m/s
$\boldsymbol{\omega} = 0.32(-2\mathbf{i} + 4\mathbf{j} - \mathbf{k})$ rad/s

▶8/33 The gyro rotor shown is spinning at the constant rate of 100 rev/min relative to the *x-y-z* axes in the direction indicated. If the angle γ between the gimbal ring and the horizontal *X-Y* plane is made to increase at the constant rate of 4 rad/sec and if the unit is forced to precess about the vertical at the constant rate $N = 20$ rev/min, calculate the magnitude of the angular acceleration $\boldsymbol{\alpha}$ of the rotor when $\gamma = 30°$. Solve by using Eq. 8/7 applied to the angular velocity of the rotor.

Ans. $\alpha = 42.8$ rad/sec^2

▶8/34 The wheel of radius r is free to rotate about the bent axle CO which turns about the vertical axis at the constant rate p rad/s. If the wheel rolls without slipping on the horizontal circle of radius R, determine the expressions for the angular velocity $\boldsymbol{\omega}$ and angular acceleration $\boldsymbol{\alpha}$ of the wheel. The *x*-axis is always horizontal.

Ans. $\boldsymbol{\omega} = p\left[\mathbf{j}\cos\theta + \mathbf{k}\left(\sin\theta + \dfrac{R}{r}\right)\right]$

$\boldsymbol{\alpha} = \mathbf{i}\dfrac{Rp^2}{r}\cos\theta$

SECTION B. KINETICS

8/7 ANGULAR MOMENTUM. The force equation for a mass system, rigid or nonrigid, Eq. 5/1 or 5/6, is the generalization of Newton's second law for the motion of a particle and should require no further explanation. The moment equation for three-dimensional motion, however, is not nearly as simple as the third of Eqs. 7/1 for plane motion since the change of angular momentum has a number of additional components which are absent in plane motion.

We consider now a rigid body moving with any general motion in space, Fig. 8/12a. Axes x-y-z are *attached* to the body with origin at the mass center G. Thus the angular velocity $\boldsymbol{\omega}$ of the body becomes the angular velocity of the x-y-z axes as observed from the fixed reference axes X-Y-Z. The absolute angular momentum $\bar{\mathbf{H}}$ of the body about its mass center G is the sum of the moments about G of the linear momenta of all elements of the body and was expressed in part (c) of Art. 5/2 as $\bar{\mathbf{H}} = \Sigma(\boldsymbol{\rho}_i \times m_i\mathbf{v}_i)$, where $\mathbf{v}_i$ is the absolute velocity of the mass element m_i. But for the rigid body $\mathbf{v}_i = \bar{\mathbf{v}} + \boldsymbol{\omega} \times \boldsymbol{\rho}_i$ where $\boldsymbol{\omega} \times \boldsymbol{\rho}_i$ is the relative velocity of m_i with respect to G as seen from nonrotating axes. Thus we may write

$$\bar{\mathbf{H}} = -\bar{\mathbf{v}} \times \Sigma m_i\boldsymbol{\rho}_i + \Sigma(\boldsymbol{\rho}_i \times m_i[\boldsymbol{\omega} \times \boldsymbol{\rho}_i])$$

where we have factored out $\bar{\mathbf{v}}$ from the first summation terms by reversing the order of the cross-product and changing the sign. With origin at the mass center G the first term in $\bar{\mathbf{H}}$ is zero since $\Sigma m_i\boldsymbol{\rho}_i = m\bar{\boldsymbol{\rho}} = 0$. The second term with the substitution of dm for m_i and $\boldsymbol{\rho}$ for $\boldsymbol{\rho}_i$ gives

$$\bar{\mathbf{H}} = \int (\boldsymbol{\rho} \times [\boldsymbol{\omega} \times \boldsymbol{\rho}]) \, dm \tag{8/8}$$

Before expanding the integrand of Eq. 8/8, we consider also the case of a rigid body rotating about a fixed point O, Fig. 8/12b. The x-y-z axes are attached to the body, and both body and axes have an angular velocity $\boldsymbol{\omega}$. The angular momentum about O was expressed in part (c) of Art. 5/2 and is $\mathbf{H}_O = \Sigma(\mathbf{r}_i \times m_i\mathbf{v}_i)$ where, for the rigid body,

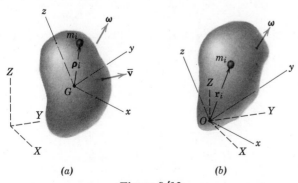

(a)   (b)

Figure 8/12

$\mathbf{v}_i = \boldsymbol{\omega} \times \mathbf{r}_i$. Thus, with the substitution of dm for m_i and $\mathbf{r}$ for $\mathbf{r}_i$, the angular momentum is

$$\mathbf{H}_O = \int (\mathbf{r} \times [\boldsymbol{\omega} \times \mathbf{r}]) \, dm \qquad (8/9)$$

We observe now that for the two cases of Figs. 8/12a and 8/12b, the position vector $\boldsymbol{\rho}_i$ or $\mathbf{r}_i$ is given by the same expression $\mathbf{i}x + \mathbf{j}y + \mathbf{k}z$. Thus Eqs. 8/8 and 8/9 are identical in form, and the symbol $\mathbf{H}$ will be used here for either case. We now carry out the expansion of the integrand in the two expressions for angular momentum with recognition of the fact that the components of $\boldsymbol{\omega}$ are invariant with respect to the integrals over the body and, hence, become constant multipliers of the integrals. The cross-product expansion applied to the triple vector product upon collection of terms gives

$$
\begin{aligned}
d\mathbf{H} = \ &\mathbf{i}[(y^2 + z^2)\omega_x && -xy\,\omega_y && -xz\omega_z]\,dm \\
&+\mathbf{j}[&& -yx\omega_x + (z^2 + x^2)\omega_y && -yz\omega_z]\,dm \\
&+\mathbf{k}[&& -zx\omega_x && -zy\omega_y + (x^2 + y^2)\omega_z]\,dm
\end{aligned}
$$

Now let

$$
\begin{aligned}
I_{xx} &= \int (y^2 + z^2)\,dm & I_{xy} &= \int xy\,dm \\
I_{yy} &= \int (z^2 + x^2)\,dm & I_{xz} &= \int xz\,dm \\
I_{zz} &= \int (x^2 + y^2)\,dm & I_{yz} &= \int yz\,dm
\end{aligned}
\qquad (8/10)
$$

The quantitites I_{xx}, I_{yy}, I_{zz} are known as the *moments of inertia* of the body about the respective axes, and I_{xy}, I_{xz}, I_{yz} are known as the *products of inertia* with respect to the coordinate axes. These quantities describe the manner in which the mass of a rigid body is distributed with respect to the chosen axes. The calculation of moments and products of inertia is explained fully in Appendix A. The double subscripts for the moments and products of inertia preserve a symmetry of notation which has special meaning in their description by tensor notation.[°] It is observed that $I_{xy} = I_{yx}$, $I_{xz} = I_{zx}$, $I_{yz} = I_{zy}$. With the substitutions of Eqs. 8/10, the expression for $\mathbf{H}$ becomes

$$
\begin{aligned}
\mathbf{H} = \ &\mathbf{i}(\ I_{xx}\omega_x - I_{xy}\omega_y - I_{xz}\omega_z) \\
&+\mathbf{j}(-I_{yx}\omega_x + I_{yy}\omega_y - I_{yz}\omega_z) \\
&+\mathbf{k}(-I_{zx}\omega_x - I_{zy}\omega_y + I_{zz}\omega_z)
\end{aligned}
\qquad (8/11)
$$

and the components of $\mathbf{H}$ are clearly

$$
\begin{aligned}
H_x &= \ \ I_{xx}\omega_x - I_{xy}\omega_y - I_{xz}\omega_z \\
H_y &= -I_{yx}\omega_x + I_{yy}\omega_y - I_{yz}\omega_z \\
H_z &= -I_{zx}\omega_x - I_{zy}\omega_y + I_{zz}\omega_z
\end{aligned}
\qquad (8/12)
$$

[°] See, for example, the author's *Dynamics, 2nd Edition* or *SI Version*, Art. 41.

Equation 8/11 is the general expression for the angular momentum about either the mass center G or about a fixed point O for a rigid body rotating with an instantaneous angular velocity $\boldsymbol{\omega}$.

Attention is called to the fact that in each of the two cases represented, the reference axes x-y-z are *attached* to the rigid body. This attachment makes the moment-of-inertia integrals and the product-of-inertia integrals of Eqs. 8/10 invariant with time. If the x-y-z axes were to rotate with respect to an irregular body, then these inertia integrals would be functions of the time which would introduce an undesirable complexity into the angular-momentum relations. An important exception occurs when a rigid body is spinning about an axis of symmetry, in which case the inertia integrals are not affected by the angular position of the body about its spin axis. Thus it is frequently convenient to permit a body with axial symmetry to rotate relative to the reference system about one of the coordinate axes. In addition to the momentum components due to the angular velocity $\boldsymbol{\Omega}$ of the reference axes, then, an added angular momentum component along the spin axis due to the relative spin about the axis would have to be accounted for.

In Eq. 8/12 the array of moments and products of inertia

$$\begin{bmatrix} I_{xx} & -I_{xy} & -I_{xz} \\ -I_{yx} & I_{yy} & -I_{yz} \\ -I_{zx} & -I_{zy} & I_{zz} \end{bmatrix}$$

is known as the *inertia matrix* or *inertia tensor*. As we change the orientation of the axes the moments and products of inertia will also change in value. It can be shown° that there is one unique orientation of axes x-y-z for a given origin for which the products of inertia vanish and the moments of inertia I_{xx}, I_{yy}, I_{zz} take on stationary values. For this orientation the inertia matrix takes the form

$$\begin{bmatrix} I_{xx} & 0 & 0 \\ 0 & I_{yy} & 0 \\ 0 & 0 & I_{zz} \end{bmatrix}$$

and is said to be diagonalized. The axes x-y-z for which the products of inertia vanish are called the *principal axes of inertia*, and I_{xx}, I_{yy}, and I_{zz} are called the *principal moments of inertia*. The principal moments of inertia for a given origin represent the maximum, minimum, and an intermediate value of the moments of inertia.

If the coordinate axes coincide with the principal axes of inertia, Eq. 8/11 for the angular momentum about the mass center or about a fixed point becomes

$$\boxed{\mathbf{H} = \mathbf{i}I_{xx}\omega_x + \mathbf{j}I_{yy}\omega_y + \mathbf{k}I_{zz}\omega_z} \tag{8/13}$$

° See, for example, the author's *Dynamics, 2nd Edition* or *SI Version*, Art. 41.

It is always possible to locate the principal axes of inertia for a general three-dimensional rigid body. Thus we can express its angular momentum by Eq. 8/13, although it may not always be convenient to do so for geometric reasons. Except when the body rotates about one of the principal axes of inertia or when I_{xx}, I_{yy}, and I_{zz} are in the same ratios as ω_x, ω_y, and ω_z, the vectors $\mathbf{H}$ and $\boldsymbol{\omega}$ have different directions.

The momentum properties of a rigid body may be represented by the resultant linear momentum vector $\mathbf{G} = m\bar{\mathbf{v}}$ through the mass center and the resultant angular momentum vector $\bar{\mathbf{H}}$ about the mass center, as shown in Fig. 8/13. Although $\bar{\mathbf{H}}$ has the properties of a free vector, we represent it through G for convenience. These vectors have properties analogous to those of a force and a couple. Thus the angular momentum about any point A which is *not fixed* to the body equals the free vector $\bar{\mathbf{H}}$ plus the moment of the linear momentum vector about A. Therefore we may write

$$\boxed{\mathbf{H}_A = \bar{\mathbf{H}} + \bar{\mathbf{r}} \times \mathbf{G} = \bar{\mathbf{H}} + \bar{\mathbf{r}} \times m\bar{\mathbf{v}}} \qquad (8/14)$$

Equation 8/14 constitutes a transfer theorem for angular momentum. Attention is called to the fact that we can use Eq. 8/11 to calculate $\mathbf{H}_A$ only when A is either the mass center or a fixed point in space.

8/8 KINETIC ENERGY. In part (*b*) of Art. 5/2 in the chapter on the dynamics of systems of particles we developed the expression for the kinetic energy T of any general system of mass, rigid or nonrigid, and obtained the result

$$T = \tfrac{1}{2}m\bar{v}^2 + \Sigma\tfrac{1}{2}m_i|\dot{\boldsymbol{\rho}}_i|^2 \qquad [5/4]$$

where $\bar{v}$ is the velocity of the mass center and $\boldsymbol{\rho}_i$ is the position vector of a representative element of mass m_i with respect to the mass center. We identified the first term as the kinetic energy due to the translation of the system and the second term as the kinetic energy associated with the motion relative to the mass center. The translational term may be written alternatively as

$$\tfrac{1}{2}m\bar{v}^2 = \tfrac{1}{2}m\dot{\bar{\mathbf{r}}}\cdot\dot{\bar{\mathbf{r}}} = \tfrac{1}{2}\bar{v}\cdot\mathbf{G}$$

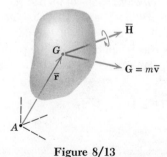

Figure 8/13

where $\dot{\bar{\mathbf{r}}}$ is the velocity $\bar{\mathbf{v}}$ of the mass center and $\mathbf{G}$ is the linear momentum of the body.

For a rigid body the relative term becomes the kinetic energy due to rotation about the mass center. Since $\dot{\boldsymbol{\rho}}_i$ is the velocity of the representative particle with respect to the mass center, then for the rigid body we may write it as $\dot{\boldsymbol{\rho}}_i = \boldsymbol{\omega} \times \boldsymbol{\rho}_i$ where $\boldsymbol{\omega}$ is the angular velocity of the body. With this substitution the relative term in the kinetic energy expression becomes

$$\Sigma \tfrac{1}{2} m_i |\dot{\boldsymbol{\rho}}_i|^2 = \Sigma \tfrac{1}{2} m_i (\boldsymbol{\omega} \times \boldsymbol{\rho}_i) \cdot (\boldsymbol{\omega} \times \boldsymbol{\rho}_i)$$

If we use the fact that the dot and the cross may be interchanged in the triple vector product, that is, $\mathbf{P} \times \mathbf{Q} \cdot \mathbf{R} = \mathbf{P} \cdot \mathbf{Q} \times \mathbf{R}$, we may write

$$(\boldsymbol{\omega} \times \boldsymbol{\rho}_i) \cdot (\boldsymbol{\omega} \times \boldsymbol{\rho}_i) = \boldsymbol{\omega} \cdot \boldsymbol{\rho}_i \times (\boldsymbol{\omega} \times \boldsymbol{\rho}_i)$$

Since $\boldsymbol{\omega}$ is the same factor in all terms of the summation, it may be factored out to give

$$\Sigma \tfrac{1}{2} m_i |\dot{\boldsymbol{\rho}}_i|^2 = \tfrac{1}{2} \boldsymbol{\omega} \cdot \Sigma \boldsymbol{\rho}_i \times m_i (\boldsymbol{\omega} \times \boldsymbol{\rho}_i) = \tfrac{1}{2} \boldsymbol{\omega} \cdot \bar{\mathbf{H}}$$

where $\bar{\mathbf{H}}$ is the same as the integral expressed by Eq. 8/8. Thus the general expression for the kinetic energy of a rigid body moving with mass-center velocity $\bar{\mathbf{v}}$ and angular velocity $\boldsymbol{\omega}$ is

$$\boxed{T = \tfrac{1}{2} \bar{\mathbf{v}} \cdot \mathbf{G} + \tfrac{1}{2} \boldsymbol{\omega} \cdot \bar{\mathbf{H}}} \qquad (8/15)$$

Expansion of this vector equation by substitution of the expression for $\bar{\mathbf{H}}$ written from Eq. 8/11 yields

$$T = \tfrac{1}{2} m \bar{v}^2 + \tfrac{1}{2}(\bar{I}_{xx}\omega_x{}^2 + \bar{I}_{yy}\omega_y{}^2 + \bar{I}_{zz}\omega_z{}^2)$$
$$- (\bar{I}_{xy}\omega_x\omega_y + \bar{I}_{xz}\omega_x\omega_z + \bar{I}_{yz}\omega_y\omega_z) \qquad (8/16)$$

If the axes coincide with the principal axes of inertia, the kinetic energy is merely

$$T = \tfrac{1}{2} m \bar{v}^2 + \tfrac{1}{2}(\bar{I}_{xx}\omega_x{}^2 + \bar{I}_{yy}\omega_y{}^2 + \bar{I}_{zz}\omega_z{}^2) \qquad (8/17)$$

When a rigid body is pivoted about a fixed point O or when there is a point O in the body which momentarily has zero velocity, the kinetic energy is $T = \Sigma \tfrac{1}{2} m_i \dot{\mathbf{r}}_i \cdot \dot{\mathbf{r}}_i$. This expression reduces to

$$\boxed{T = \tfrac{1}{2} \boldsymbol{\omega} \cdot \mathbf{H}_O} \qquad (8/18)$$

where $\mathbf{H}_O$ is the angular momentum about O, as may be seen by replacing $\boldsymbol{\rho}_i$ in the previous derivation by $\mathbf{r}_i$, the position vector from O. Equations 8/15 and 8/18 are the three-dimensional counterparts of Eqs. 7/13 and 7/14 for plane motion.

Sample Problem 8/6

The bent plate has a mass of 70 kg per square meter of surface area and revolves about the z-axis at the rate $\omega = 30$ rad/s. Determine (a) the angular momentum $\mathbf{H}$ of the plate about point O and (b) the kinetic energy T of the plate. Neglect the mass of the hub at the rotation axis and the thickness of the plate compared with its surface dimensions.

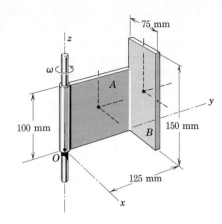

Solution. The moments and products of inertia are written with the aid of Eqs. A3 and A9 in Appendix A by transfer from the parallel centroidal axes for each part.

① First, the mass of each part is

$$m_A = (0.100)(0.125)(70) = 0.875 \text{ kg}, \quad m_B = (0.075)(0.150)(70) = 0.788 \text{ kg}$$

Part A

$$[I_{xx} = \bar{I}_{xx} + md^2] \quad I_{xx} = \frac{0.875}{12}[(0.100)^2 + (0.125)^2]$$
$$+ 0.875[(0.050)^2 + (0.0625)^2] = 0.007\ 47 \text{ kg} \cdot \text{m}^2$$

$$[I_{yy} = \tfrac{1}{3}ml^2] \quad I_{yy} = \frac{0.875}{3}(0.100)^2 = 0.002\ 92 \text{ kg} \cdot \text{m}^2$$

$$[I_{zz} = \tfrac{1}{3}ml^2] \quad I_{zz} = \frac{0.875}{3}(0.125)^2 = 0.004\ 56 \text{ kg} \cdot \text{m}^2$$

$$[I_{xy} = \int xy\ dm, \ I_{xz} = \int xz\ dm] \quad I_{xy} = 0, \quad I_{xz} = 0$$

$$[I_{yz} = \bar{I}_{yz} + md_y\, d_z] \quad I_{yz} = 0 + 0.875(0.0625)(0.050) = 0.002\ 73 \text{ kg} \cdot \text{m}^2$$

Part B

$$[I_{xx} = \bar{I}_{xx} + md^2] \quad I_{xx} = \frac{0.788}{12}(0.150)^2 + 0.788[(0.125)^2 + (0.075)^2]$$
$$= 0.018\ 21 \text{ kg} \cdot \text{m}^2$$

$$[I_{yy} = \bar{I}_{yy} + md^2] \quad I_{yy} = \frac{0.788}{12}[(0.075)^2 + (0.150)^2]$$
$$+ 0.788[(0.0375)^2 + (0.075)^2] = 0.007\ 38 \text{ kg} \cdot \text{m}^2$$

$$[I_{zz} = \bar{I}_{zz} + md^2] \quad I_{zz} = \frac{0.788}{12}(0.075)^2 + 0.788[(0.125)^2 + (0.0375)^2]$$
$$= 0.013\ 78 \text{ kg} \cdot \text{m}^2$$

$$[I_{xy} = \bar{I}_{xy} + md_x\, d_y] \quad I_{xy} = 0 + 0.788(0.0375)(0.125) = 0.003\ 69 \text{ kg} \cdot \text{m}^2$$

$$[I_{xz} = \bar{I}_{xz} + md_x\, d_z] \quad I_{xz} = 0 + 0.788(0.0375)(0.075) = 0.002\ 21 \text{ kg} \cdot \text{m}^2$$

$$[I_{yz} = \bar{I}_{yz} + md_y\, d_z] \quad I_{yz} = 0 + 0.788(0.125)(0.075) = 0.007\ 38 \text{ kg} \cdot \text{m}^2$$

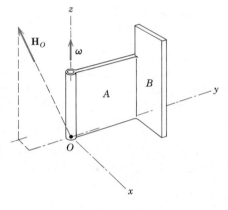

① The parallel-axis theorems for transferring moments and products of inertia from centroidal axes to parallel axes are explained in Appendix A and are most useful relations.

The sum of the respective inertia terms gives for the two plates together

$$I_{xx} = 0.0257 \text{ kg} \cdot \text{m}^2 \qquad I_{xy} = 0.003\ 69 \text{ kg} \cdot \text{m}^2$$
$$I_{yy} = 0.010\ 30 \text{ kg} \cdot \text{m}^2 \qquad I_{xz} = 0.002\ 21 \text{ kg} \cdot \text{m}^2$$
$$I_{zz} = 0.018\ 34 \text{ kg} \cdot \text{m}^2 \qquad I_{yz} = 0.010\ 12 \text{ kg} \cdot \text{m}^2$$

(a) The angular momentum of the body is given by Eq. 8/11 where $\omega_z = 30$ rad/s and ω_x and ω_y are zero. Thus

② $$\mathbf{H}_O = 30(-0.002\ 21\mathbf{i} - 0.010\ 12\mathbf{j} + 0.018\ 34\mathbf{k}) \text{ N} \cdot \text{m} \cdot \text{s} \qquad Ans.$$

(b) The kinetic energy from Eq. 8/18 becomes

$$T = \tfrac{1}{2}\boldsymbol{\omega} \cdot \mathbf{H}_O = \tfrac{1}{2}(30\mathbf{k}) \cdot 30(-0.002\ 21\mathbf{i} - 0.010\ 12\mathbf{j} + 0.018\ 34\ \mathbf{k}) = 8.25 \text{ J}$$

② Recall that the units of angular momentum may also be written in the base units as kg $\cdot$ m²/s.

PROBLEMS

8/35 The slender rod of mass m and length l is attached to the disk which rotates about the y-axis with an angular velocity ω. By inspection write the expression for the angular momentum **H** of the rod about the origin O of the x-y-z axes for the position shown. Verify the result by applying Eq. 8/11.

$$\text{Ans. } \mathbf{H} = mb^2\omega\left(\mathbf{j} - \frac{l}{2b}\mathbf{k}\right)$$

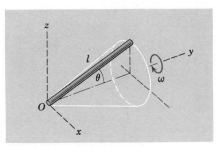

Problem 8/35

8/36 The slender rod of mass m and length l rotates about the y-axis as the element of a right circular cone. If the angular velocity about the y-axis is ω, determine the expression for the angular momentum of the rod with respect to the x-y-z axes for the particular position shown.

$$\text{Ans. } \mathbf{H} = \tfrac{1}{3}ml^2\omega\sin\theta\,(\mathbf{j}\sin\theta - \mathbf{k}\cos\theta)$$

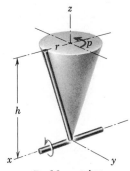

Problem 8/36

8/37 The slender bar of mass m and length $l = \sqrt{h^2 + r^2}$ is attached as an element of the cone which revolves about the z-axis with an angular rate p. Simultaneously the entire cone revolves about the x-axis with an angular rate ω. Determine the x-component of the angular momentum of the bar at the instant the bar crosses the x-z plane as shown.

$$\text{Ans. } H_x = \frac{mh}{3}(h\omega - rp)$$

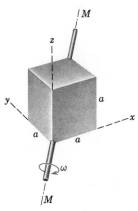

Problem 8/37

8/38 Determine the z-component of angular momentum for the bar of Prob. 8/37 for the position illustrated.

8/39 Determine the kinetic energy of the bar of Prob. 8/37 when it is in the position shown.

$$\text{Ans. } T = \frac{m}{6}(h\omega - rp)^2$$

8/40 The solid cube of mass m and side a revolves about an axis M-M through a diagonal with an angular velocity ω. Write the expression for the angular momentum **H** of the cube with respect to the axes indicated.

$$\text{Ans. } \mathbf{H} = \frac{ma^2\omega}{6\sqrt{3}}(\mathbf{i} + \mathbf{j} + \mathbf{k})$$

Problem 8/40

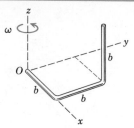

Problem 8/41

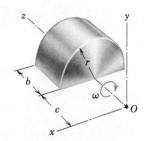

Problem 8/42

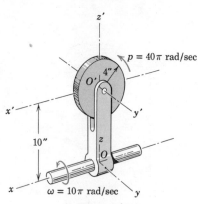

Problem 8/43

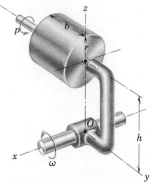

Problem 8/45

8/41 The bent rod has a mass m and revolves about the z-axis with an angular velocity ω. Determine the angular momentum of the rod about the origin O of the coordinates for the position shown. Also find the kinetic energy of the rod.

$$Ans. \ \mathbf{H}_O = \frac{m\omega b^2}{3}\left(-\tfrac{1}{2}\mathbf{i} - \tfrac{1}{2}\mathbf{j} + \tfrac{11}{3}\mathbf{k}\right)$$

$$T = \frac{11}{18}mb^2\omega^2$$

8/42 The half-circular cylinder revolves about the z-axis with an angular velocity ω as shown. Evaluate the angular momentum $\mathbf{H}$ with respect to the x-y-z axes. The mass of the cylinder is m.

$$Ans. \ \mathbf{H} = mr\omega\left[\frac{-2(2c + b)}{3\pi}\mathbf{j} + \frac{r}{2}\mathbf{k}\right]$$

8/43 The 4-in.-radius wheel weighs 6 lb and turns about its y'-axis with an angular velocity $p = 40\pi$ rad/sec in the direction shown. Simultaneously the fork rotates about its x-axis shaft with an angular velocity $\omega = 10\pi$ rad/sec as indicated. Calculate the angular momentum of the wheel about its center O'. Also compute the kinetic energy of the wheel.
$$Ans. \ \mathbf{H}_{O'} = 0.163(\mathbf{i} + 8\mathbf{j}) \ \text{lb-ft-sec}$$
$$T = 148.1 \ \text{ft-lb}$$

8/44 Calculate the angular momentum of the wheel of Prob. 8/43 about the origin O of the x-y-z axes.
$$Ans. \ \mathbf{H}_O = 4.23\mathbf{i} + 1.30\mathbf{j} \ \text{lb-ft-sec}$$

8/45 The solid circular cylinder of mass m, radius r, and length b revolves at an angular rate p about its geometric axis. Simultaneously the bracket and attached shaft axis revolve at the rate ω about the x-axis. Write the expressions for the x-, y-, z-components of angular momentum $\mathbf{H}$ of the cylinder about O.

$$Ans. \ H_x = \left(\frac{b^2}{3} + \frac{r^2}{4} + h^2\right)m\omega$$
$$H_y = \tfrac{1}{2}mr^2p$$
$$H_z = 0$$

8/46 The solid circular disk of mass $m = 2$ kg and radius $r = 100$ mm rolls in a circle of radius $b = 200$ mm on the horizontal plane without slipping. If the center line OC of the axle of the wheel rotates about the z-axis with an angular velocity $\omega = 4\pi$ rad/s, determine the expression for the angular momentum of the disk with respect to the fixed point O. Also compute the kinetic energy of the wheel.

$$Ans. \; \mathbf{H}_O = 0.251(-\mathbf{j} + 4.25\mathbf{k}) \, \text{N} \cdot \text{m} \cdot \text{s}$$
$$T = 9.87 \, \text{J}$$

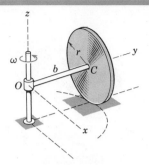

Problem 8/46

8/47 The rectangular plate, with a mass of 3 kg and a uniform small thickness, is welded at the 45° angle to the vertical shaft which rotates with the angular velocity of 20π rad/s. Determine the angular momentum $\mathbf{H}$ of the plate about O and find the kinetic energy of the plate.

$$Ans. \; \mathbf{H} = \pi(-0.4\mathbf{j} + 0.6\mathbf{k}) \, \text{N} \cdot \text{m} \cdot \text{s}, \; T = 59.2 \, \text{J}$$

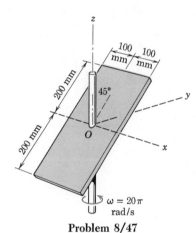

Problem 8/47

8/48 The uniform circular disk of mass m and radius r is mounted on its shaft which is pivoted at O about the vertical z_0-axis. If the disk rolls without slipping and makes one complete trip around the large fixed disk in time τ, write the expression for the angular momentum of the disk with respect to the x-y-z axes through O. (*Hint.* The spin relative to x-y-z is $R\omega/r$ where $\omega = 2\pi/\tau$.)

$$Ans. \; \mathbf{H}_O = \frac{2\pi mr^2}{\tau} \left[\frac{1}{2}\left(\frac{r}{R} - \frac{R}{r}\right)\mathbf{j} \right.$$
$$\left. + \left(\frac{R^2}{r^2} - \frac{3}{4}\right)\sqrt{1 - \frac{r^2}{R^2}}\mathbf{k} \right]$$

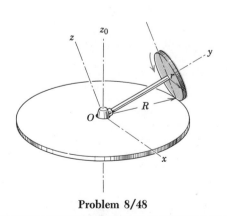

Problem 8/48

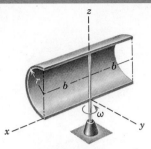

Problem 8/49

▶8/49   The half-cylindrical shell of radius r, length $2b$, and mass m revolves about the z-axis with an angular velocity ω as shown. Determine the expression for the angular momentum $\mathbf{H}$ of the shell with respect to the x-y-z axes.

$$Ans. \ \mathbf{H} = m\omega\left(\frac{2r^2}{\pi}\mathbf{j} + \left[\frac{r^2}{2} + \frac{b^2}{3}\right]\mathbf{k}\right)$$

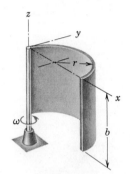

Problem 8/50

▶8/50   The half-cylindrical shell of mass m, radius r, and length b revolves about one edge along the z-axis with a constant rate ω as shown. Determine the angular momentum $\mathbf{H}$ of the shell with respect to the x-y-z axes.

$$Ans. \ \mathbf{H} = mr\omega\left(\frac{b}{2}\mathbf{i} + \frac{b}{\pi}\mathbf{j} + 2r\mathbf{k}\right)$$

8/9 MOMENTUM AND ENERGY EQUATIONS OF MOTION.

With the description of angular momentum, inertial properties, and kinetic energy of a rigid body established in the previous two articles we are ready to apply the general momentum and energy equations of motion.

(*a*) *Momentum Equations.* In Chapter 5 on the kinetics of systems of particles we established in part (*c*) of Art. 5/2 the general linear and angular momentum equations for a system of constant mass. These equations are

$$\Sigma \mathbf{F} = \dot{\mathbf{G}} \qquad\qquad [5/6]$$

$$\Sigma \mathbf{M} = \dot{\mathbf{H}} \qquad\qquad [5/7] \text{ or } [5/9]$$

The general moment relation, Eq. 5/7 or 5/9, is expressed here by the single equation $\Sigma \mathbf{M} = \dot{\mathbf{H}}$ where the terms are taken about either a fixed point O or about the mass center G. In the derivation of the moment principle the derivative of $\mathbf{H}$ was taken with respect to an absolute coordinate system. When $\mathbf{H}$ is expressed in terms of components measured relative to a moving coordinate system x-y-z which has an angular velocity $\boldsymbol{\Omega}$, then by Eq. 8/7 the moment relation becomes

$$\Sigma \mathbf{M} = \left(\frac{d\mathbf{H}}{dt}\right)_{xyz} + \boldsymbol{\Omega} \times \mathbf{H}$$

$$= (\mathbf{i}\dot{H}_x + \mathbf{j}\dot{H}_y + \mathbf{k}\dot{H}_z) + \boldsymbol{\Omega} \times \mathbf{H}$$

The terms in parentheses represent that part of $\dot{\mathbf{H}}$ due to the change in magnitude of the components of $\mathbf{H}$, and the cross-product term represents that part due to the changes in direction of the components of $\mathbf{H}$. Expansion of the cross product and rearrangement of terms give

$$\begin{aligned}\Sigma \mathbf{M} = {}& \mathbf{i}(\dot{H}_x - H_y\Omega_z + H_z\Omega_y) \\ {}& + \mathbf{j}(\dot{H}_y - H_z\Omega_x + H_x\Omega_z) \\ {}& + \mathbf{k}(\dot{H}_z - H_x\Omega_y + H_y\Omega_x)\end{aligned} \qquad (8/19)$$

Equation 8/19 is the most general form of the moment equation about a fixed point O or about the mass center G. The Ω's are the angular velocity components of rotation of the reference axes, and the H-components in the case of a rigid body are as defined in Eq. 8/12 where the ω's are the components of the angular velocity of the body. We now apply Eq. 8/19 to a rigid body where the coordinate axes are *attached* to the body. Under these conditions when expressed in the x-y-z coordinates, the *moments and products of inertia are invariant with time*, and $\boldsymbol{\Omega} = \boldsymbol{\omega}$. Thus for axes attached to the body, the three scalar components of Eq. 8/19 become

$$\Sigma M_x = \dot{H}_x - H_y\omega_z + H_z\omega_y$$
$$\Sigma M_y = \dot{H}_y - H_z\omega_x + H_x\omega_z \qquad (8/20)$$
$$\Sigma M_z = \dot{H}_z - H_x\omega_y + H_y\omega_x$$

Equations 8/20 are the general moment equations for rigid-body motion with reference axes *attached to the body,* and they hold with respect to axes through a fixed point O or through the mass center G.

In Art. 8/7 it was mentioned that, in general, for any origin fixed to a rigid body, there are three principal axes of inertia with respect to which the products of inertia vanish. If the reference axes coincide with the principal axes of inertia with origin at the mass center G or at a point O fixed to the body and fixed in space, the factors I_{xy}, I_{yz}, I_{xz} will be zero, and Eqs. 8/20 become

$$\Sigma M_x = I_{xx}\dot{\omega}_x - (I_{yy} - I_{zz})\omega_y\omega_z$$
$$\Sigma M_y = I_{yy}\dot{\omega}_y - (I_{zz} - I_{xx})\omega_z\omega_x \qquad (8/21)$$
$$\Sigma M_z = I_{zz}\dot{\omega}_z - (I_{xx} - I_{yy})\omega_x\omega_y$$

These relations are known as *Euler's equations*[°] and are among the most useful of the motion equations of three-dimensional dynamics.

(b) Energy Equations.   The resultant of all external forces acting on a rigid body may be replaced by the resultant force $\Sigma\mathbf{F}$ acting through the mass center and a resultant couple $\Sigma\overline{\mathbf{M}}$ acting about the mass center. Work is done by the resultant force and the resultant couple at the respective rates $\Sigma\mathbf{F}\cdot\overline{\mathbf{v}}$ and $\Sigma\overline{\mathbf{M}}\cdot\boldsymbol{\omega}$ where $\overline{\mathbf{v}}$ is the linear velocity of the mass center and $\boldsymbol{\omega}$ is the angular velocity of the body. Integration over the time from condition 1 to condition 2 gives the total work done during the time interval. Equating the works done to the respective changes in kinetic energy as expressed in Eq. 8/15 gives

$$\int_{t_1}^{t_2}\Sigma\mathbf{F}\cdot\overline{\mathbf{v}}\,dt = \tfrac{1}{2}\overline{\mathbf{v}}\cdot\mathbf{G}\,\bigg]_1^2$$
$$\int_{t_1}^{t_2}\Sigma\overline{\mathbf{M}}\cdot\boldsymbol{\omega}\,dt = \tfrac{1}{2}\boldsymbol{\omega}\cdot\overline{\mathbf{H}}\,\bigg]_1^2 \qquad (8/22)$$

These equations express the change in translational kinetic energy and the change in rotational kinetic energy, respectively, for the interval during which $\Sigma\mathbf{F}$ or $\Sigma\overline{\mathbf{M}}$ acts, and the sum of the two expressions equals ΔT.

The work-energy relationship, developed in Chapter 5 for a general system of particles and given by

$$U = \Delta T + \Delta V_e + \Delta V_g \qquad [5/3]$$

[°]Named after Leonhard Euler (1707–1783), a Swiss mathematician.

was used in Chapter 7 for rigid bodies in plane motion. The equation is equally applicable to rigid-body motion in three dimensions. As we have seen previously, the work-energy approach is of great advantage when analyzing end-point conditions of motion (initial and final). Here the work U done during the interval by all active forces external to the body or system is equated to the sum of the corresponding changes in kinetic energy ΔT, elastic potential energy ΔV_e, and gravitational potential energy ΔV_g. The potential-energy changes are determined in the usual way, as described previously in Art. 3/7.

We shall limit our application of the equations developed in this article to two problems of special interest, parallel-plane motion and gyroscopic motion, which are discussed in the two articles which follow.

8/10 PARALLEL-PLANE MOTION. When all particles of a rigid body move in planes which are parallel to a fixed plane, the body has a general form of plane motion, as described in Art. 8/4 and pictured in Fig. 8/3. Every line in such a body which is normal to the fixed plane remains parallel to itself at all times. We take the mass center G as the origin of coordinates x-y-z which are attached to the body with the x-y plane coinciding with the plane of motion P. The components of the angular velocity of both body and attached axes become $\omega_x = \omega_y = 0$, $\omega_z \neq 0$. For this case the angular momentum components from Eq. 8/12 become

$$H_x = -I_{xz}\omega_z \qquad H_y = -I_{yz}\omega_z \qquad H_z = I_{zz}\omega_z$$

and the moment relations of Eqs. 8/20 reduce to

$$\boxed{\begin{aligned} \Sigma M_x &= -I_{xz}\dot{\omega}_z + I_{yz}\omega_z{}^2 \\ \Sigma M_y &= -I_{yz}\dot{\omega}_z - I_{xz}\omega_z{}^2 \\ \Sigma M_z &= I_{zz}\dot{\omega}_z \end{aligned}} \qquad (8/23)$$

It is seen that the third moment equation is equivalent to the third of Eqs. 7/1 where the z-axis passes through the mass center or to the third of Eqs. 7/8 if the z-axis passes through a fixed point.

Equations 8/23 hold for an origin of coordinates at the mass center, as shown in Fig. 8/3, or for any origin on a fixed axis of rotation. The three independent force equations of motion which also apply to parallel-plane motion are clearly

$$\Sigma F_x = m\bar{a}_x \qquad \Sigma F_y = m\bar{a}_y \qquad \Sigma F_z = 0$$

Equations 8/23 find special use in describing the effect of dynamic unbalance in rotating shafts and in rolling bodies. As an alternative to the use of Eqs. 8/23 it is often just as easy to form the equivalence between the external forces and the $m\bar{a}$ and $\bar{I}\alpha$ resultants of the individual parts of the rigid body which has parallel-plane motion.

Sample Problem 8/7

The two circular disks each of mass m_1 are connected by the curved bar bent into quarter-circular arcs and welded to the disks. The bar has a mass m_2. The total mass of the assembly is $m = 2m_1 + m_2$. If the disks roll without slipping on a horizontal plane with a constant velocity v of the disk centers, determine the value of the friction force under each disk at the instant represented when the plane of the curved bar is horizontal.

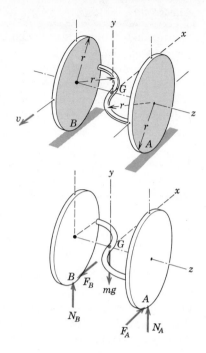

Solution. The motion is identified as parallel-plane motion since the planes of motion of all parts of the system are parallel. The free-body diagram shows the normal forces and friction forces at A and B and the total weight mg acting through the mass center G, which we take as the origin of coordinates which rotate with the body.

We now apply Eqs. 8/23 where $I_{yz} = 0$ and $\dot{\omega}_z = 0$. The moment equation about the y-axis requires determination of I_{xz}. From the diagram showing the geometry of the curved rod and with ρ standing for the mass of the rod per unit length, we have

① $$\left[I_{xz} = \int xz\, dm \right] \qquad I_{xz} = \int_0^{\pi/2} (r \sin \theta)(-r + r \cos \theta)\rho r\, d\theta$$

$$+ \int_0^{\pi/2} (-r \sin \theta)(r - r \cos \theta)\rho r\, d\theta$$

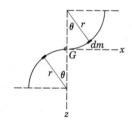

Evaluating the integrals gives

$$I_{xz} = -\rho r^3/2 - \rho r^3/2 = -\rho r^3 = -\frac{m_2 r^2}{\pi}$$

The second of Eqs. 8/23 with $\omega_z = v/r$ and $\dot{\omega}_z = 0$ gives

$$[\Sigma M_y = -I_{xz}\omega_z{}^2] \qquad F_A r + F_B r = -\left(-\frac{m_2 r^2}{\pi} \right)\frac{v^2}{r^2}$$

$$F_A + F_B = \frac{m_2 v^2}{\pi r}$$

But with $\bar{v} = v$ constant, $\bar{a}_x = 0$ so that

$$[\Sigma F_x = 0] \qquad F_A - F_B = 0, \qquad F_A = F_B$$

Thus

$$F_A = F_B = \frac{m_2 v^2}{2\pi r} \qquad \text{Ans.}$$

We also note for the given position that with $I_{yz} = 0$ and $\dot{\omega}_z = 0$ the moment equation about the x-axis gives

② $$[\Sigma M_x = 0] \qquad -N_A r + N_B r = 0, \qquad N_A = N_B = mg/2$$

① We must be very careful to observe the correct signs for each of the coordinates of the mass element dm which make up the product xz.

② When the plane of the curved bar is not horizontal, the normal forces under the disks are no longer equal.

PROBLEMS

8/51 Each of the two rods of mass m is welded to the face of the disk which rotates about the vertical axis with a constant angular velocity ω. Determine the bending moment M acting on each rod at its base.
Ans. $M = \frac{1}{2}mbl\omega^2$

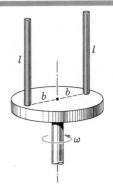

Problem 8/51

8/52 The 6-kg circular disk and attached shaft rotate at a constant speed $\omega = 10\,000$ rev/min. If the center of mass of the disk is 0.05 mm off center, determine the magnitudes of the horizontal forces A and B supported by the bearings because of the rotational unbalance.
Ans. $A = 576$ N, $B = 247$ N

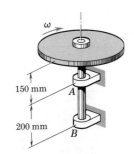

Problem 8/52

8/53 The irregular rod has a mass ρ per unit length and rotates about the shaft axis at the constant rate ω. Determine the expression for the bending moment M in the rod at A. Neglect the small moment caused by the weight of the rod.
Ans. $M = \sqrt{13}\rho b^3\omega^2$

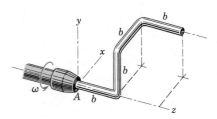

Problem 8/53

8/54 Each of the two right-angle rods has a mass of 120 g and is welded to the shaft which rotates at a steady speed $N = 3600$ rev/min. Compute the force supported by the bearing at A due to the dynamic unbalance of the shaft.
Ans. $A = 953$ N

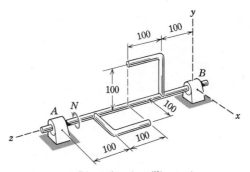

(Dimensions in millimeters)

Problem 8/54

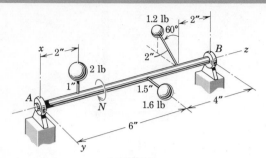

Problem 8/55

8/55 A shaft carries three unbalanced disk cams which are represented by the equivalent concentrated masses shown. Calculate the magnitudes of the reactions at the bearings A and B resulting from the dynamic unbalance at a shaft speed $N = 3600$ rev/min.   *Ans.* $A = 705$ lb, $B = 507$ lb

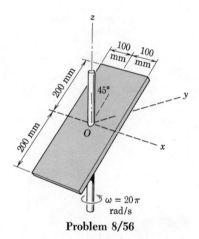

Problem 8/56

8/56 The rectangular plate of Prob. 8/47 is shown again here. The plate has a mass of 3 kg and is welded to the fixed vertical shaft which rotates at the constant speed of 20π rad/s. Compute the moment $\mathbf{M}$ applied *to* the shaft *by* the plate due to dynamic unbalance.   *Ans.* $\mathbf{M} = -79.0\mathbf{i}$ N·m

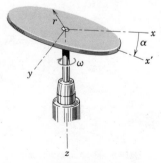

Problem 8/57

8/57 The circular disk of mass m and radius r is mounted on the vertical shaft with a small angle α between its plane and the plane of rotation of the shaft. Determine the expression for the bending moment $\mathbf{M}$ acting *on* the shaft due to the wobble of the disk at a shaft speed of ω rad/s.

Ans. $\mathbf{M} = (\tfrac{1}{8}mr^2\omega^2 \sin 2\alpha)\mathbf{j}$

8/58 The slender rod OA of mass m and length l is pivoted freely about a horizontal axis through O. The pivot at O and attached shaft rotate with a constant angular speed ω about the vertical z-axis. Write the expression for the angle θ assumed by the rod. What minimum value must ω attain before the rod will assume other than a vertical position?

$$\text{Ans. } \theta = \cos^{-1}\frac{3g}{2l\omega^2}, \ \omega_{\min} = \sqrt{\frac{3g}{2l}}$$

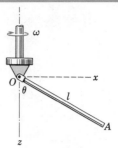

Problem 8/58

8/59 If the rod of Prob. 8/58 is welded to O at an angle θ with the vertical, determine the expression for the bending moment M in the rod at O due to the combined effect of the angular velocity ω of the shaft and the weight of the rod.

$$\text{Ans. } M_y = \frac{mgl}{2}\sin\theta\left(1 - \frac{2l}{3g}\omega^2\cos\theta\right)$$

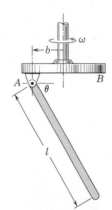

8/60 The uniform slender rod of length l is freely hinged to the bracket A on the under side of the disk B. The disk rotates about a vertical axis with a constant angular velocity ω. Determine the value of ω which will permit the rod to maintain the position $\theta = 60°$ with $b = l/4$.

$$\text{Ans. } \omega = 2\sqrt{\frac{\sqrt{3}\,g}{l}}$$

Problem 8/60

8/61 The partial ring has a mass ρ per unit length of rim and is welded to the shaft at A. Determine the bending moment M in the ring at A due to the effect of rotation of the ring for an angular velocity ω.

$$\text{Ans. } M = \tfrac{1}{2}\rho r^3 \omega^2$$

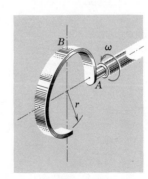

Problem 8/61

8/62 For the ring section of Prob. 8/61 determine the torsional moment T in the ring at B (moment about an axis tangent to the rim) as the ring starts from rest under the action of a torque M transmitted by the shaft to the ring at A. The mass per unit length of rim is ρ.

$$\text{Ans. } T = \tfrac{2}{3}M$$

Problem 8/63

8/63 Determine the bending moment M at the tangency point A in the semicircular rod of radius r and mass m as it rotates about the tangent axis with a constant angular velocity ω. Neglect the moment mgr produced by the weight compared with that caused by the rotation of the rod.

$$\text{Ans. } M = \frac{2mr^2\omega^2}{\pi}$$

8/64 Determine the normal forces under the two disks of Sample Problem 8/7 for the position where the plane of the curved bar is vertical. Take the curved bar to be at the top of disk A and at the bottom of disk B.

$$\text{Ans. } N_A = \frac{mg}{2} - \frac{m_2v^2}{2\pi r}$$

$$N_B = \frac{mg}{2} + \frac{m_2v^2}{2\pi r}$$

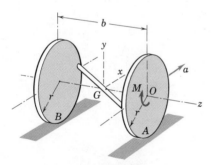

Problem 8/65

▶**8/65** Each of the two circular disks has a mass m and is welded to the end of the rigid rod of mass m_0 so that the disks have a common z-axis and are separated by a distance b. A couple M, applied to one of the disks with the assembly initially at rest, gives the centers of the disks an acceleration $\mathbf{a} = +a\mathbf{i}$. Friction is sufficient to prevent slipping. Derive expressions for the normal forces N_A and N_B exerted by the horizontal surface on the disks as they begin to roll. Express the results in terms of the acceleration a rather than the moment M.

$$\text{Ans. } N_A = mg + \frac{m_0g}{2}\left(1 + \frac{a}{3g}\right)$$

$$N_B = mg + \frac{m_0g}{2}\left(1 - \frac{a}{3g}\right)$$

▶**8/66** If a couple $\mathbf{M} = M\mathbf{k}$ is applied to one of the disks of Sample Problem 8/7 as they start from the rest position shown, determine the momentary values of the reactions N_A and N_B.

$$\text{Ans. } N_A = \frac{mg}{2} - \frac{Mm_2}{3\pi mr}$$

$$N_B = \frac{mg}{2} + \frac{Mm_2}{3\pi mr}$$

8/11 GYROSCOPIC MOTION: STEADY PRECESSION. One of the most interesting of all problems in dynamics is that of gyroscopic motion. This motion occurs whenever the axis about which a body is spinning is itself rotating about another axis. Whereas the complete description of this motion involves considerable complexity, the most common and useful examples of gyroscopic motion occur when the axis of a rotor spinning at constant speed turns (precesses) about another axis at a steady rate. Our discussion in this article will focus on this special case.

The gyroscope has important engineering applications. With a mounting in gimbal rings (see Fig. 8/19*b*) the gyro is free from external moments, and its axis will retain a fixed direction in space irrespective of the rotation of the structure to which it is attached. In this way the gyro is used for inertial guidance systems and other directional control devices. With addition of a pendulous mass to the inner gimbal ring the earth's rotation causes the gyro to precess so that the spin axis will always point north, and this action forms the basis of the gyro compass. The gyroscope has also found important use as a stabilizing device. The controlled precession of a large gyro mounted in a ship is used to produce a gyroscopic moment to counteract the rolling of a ship at sea. The gyroscopic effect is also an extremely important consideration in the design of bearings for the shafts of rotors which are subjected to forced precessions.

We shall describe gyroscopic action, first, with a simple physical approach which relies on our previous experience with the vector changes encountered in plane dynamics. This approach will help us gain a direct physical insight into gyroscopic action. Next we will make use of the general momentum relation, Eq. 8/19, for a more complete description.

(a) Simplified Approach. Figure 8/14 shows a symmetrical rotor spinning about the *z*-axis with a large angular velocity **p**, known as the *spin velocity*. If we should apply two forces *F* to the rotor axle to form a couple **M** whose vector is directed along the *x*-axis, we will find that the rotor shaft will rotate in the *x-z* plane about the *y*-axis in the sense indicated with a relatively slow angular velocity $\Omega = \dot{\psi}$ known as the *precession velocity*. Thus we identify the spin axis (**p**), the torque axis (**M**), and the precession axis (Ω) where the usual right-hand rule identifies the sense of the rotation vectors. The rotor shaft does *not* rotate about the *x*-axis in the sense of **M** as would be the case if the rotor were not spinning. To help understand this phenomenon a direct analogy may be made between the rotation vectors and the familiar vectors which describe the curvilinear motion of a particle.

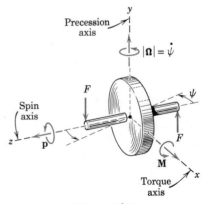

Figure 8/14

(a)

(b)

(c)

Figure 8/15

Figure 8/15a shows a particle of mass m moving in the x-z plane. The application of a force $\mathbf{F}$ normal to its linear momentum $\mathbf{G} = m\mathbf{v}$ causes a change $d\mathbf{G} = d(m\mathbf{v})$ in its momentum. We see that $d\mathbf{G}$, and hence $d\mathbf{v}$, is a vector in the direction of the force $\mathbf{F}$ according to Newton's second law $\mathbf{F} = \dot{\mathbf{G}}$ which may be written as $\mathbf{F}\,dt = d\mathbf{G}$. We have covered these relations extensively in Chapter 3. Recall now the analogous equation $\mathbf{M} = \dot{\mathbf{H}}$ which we developed for any prescribed mass system, rigid or nonrigid, about its mass center (Eq. 5/9) or about a fixed point O (Eq. 5/7). We now apply this relation to our symmetrical rotor as shown in Fig. 8/15b. For a high rate of spin $\mathbf{p}$ and a low precession rate $\mathbf{\Omega}$ about the y-axis the angular momentum is represented by the vector $\mathbf{H} = I\mathbf{p}$ where $I = I_{zz}$ is the moment of inertia of the rotor about the spin axis. Initially we are neglecting the small component of angular momentum about the y-axis which accompanies the slow precession. The application of the couple $\mathbf{M}$ normal to $\mathbf{H}$ causes a change $d\mathbf{H} = d(I\mathbf{p})$ in the angular momentum. We see that $d\mathbf{H}$, and hence $d\mathbf{p}$, is a vector in the direction of the couple $\mathbf{M}$ since $\mathbf{M} = \dot{\mathbf{H}}$ which may also be written $\mathbf{M}\,dt = d\mathbf{H}$. Just as the change in the linear momentum vector of the particle is in the direction of the applied force, so is the change in the angular momentum vector of the gyro in the direction of the couple. Thus we see that the vectors $\mathbf{M}$, $\mathbf{H}$, and $d\mathbf{H}$ are analogous to the vectors $\mathbf{F}$, $\mathbf{G}$, and $d\mathbf{G}$. With this insight it is no longer strange to see the rotation vector undergo a change in the direction of $\mathbf{M}$, thereby causing the axis of the rotor to precess about the y-axis.

In Fig. 8/15c we see that during time dt the angular momentum vector $I\mathbf{p}$ has swung through the angle $d\psi$, so that in the limit with $\tan d\psi = d\psi$ we have

$$d\psi = \frac{M\,dt}{Ip} \qquad \text{or} \qquad M = I\frac{d\psi}{dt}p$$

Substituting $\Omega = d\psi/dt$ for the magnitude of the precession velocity gives us

$$\boxed{M = I\Omega p} \tag{8/24}$$

We note that $\mathbf{M}$, $\mathbf{\Omega}$, and $\mathbf{p}$ as vectors are mutually perpendicular, and that their vector relationship may be represented by writing the equation in the cross-product form

$$\boxed{\mathbf{M} = I\mathbf{\Omega} \times \mathbf{p}} \tag{8/24a}$$

Equations 8/24 and 8/24a apply to moments taken about the mass center or about a fixed point on the axis of rotation.

The correct spatial relationship among the three vectors may be remembered from the fact that $d\mathbf{H}$, and hence $d\mathbf{p}$, is in the direction of $\mathbf{M}$, which establishes the correct sense for the precession $\mathbf{\Omega}$. Thus the spin vector **p** always tends to rotate toward the torque vector **M**. Figure 8/16 represents three orientations of the three vectors which are consistent with their correct order. Unless we establish this order correctly in a given problem, we are likely to arrive at a conclusion directly opposite to the correct one. It is emphasized that Eq. 8/24, like $\mathbf{F} = m\mathbf{a}$ and $M = I\alpha$, is an equation of motion, so that the couple **M** represents the couple due to *all* forces acting *on* the rotor as disclosed by a correct *free-body diagram of the rotor*. Also it is pointed out that, when a rotor is forced to precess, as occurs with the turbine in a ship which is executing a turn, the motion will generate a *gyroscopic couple* **M** which obeys Eq. 8/24a both in magnitude and sense.

In the foregoing discussion of gyroscopic motion it was assumed that the spin was large and the precession was small. Although we can see from Eq. 8/24 that for given values of I and M the precession Ω must be small if p is large, let us now examine the influence of Ω on the momentum relations. Again, we restrict our attention to steady precession where Ω has a constant magnitude. Figure 8/17 shows our same rotor again. Because it has a moment of inertia about the y-axis and an angular velocity of precession about this axis, there will be an additional component of angular momentum about the y-axis. Thus we have the two components $H_z = Ip$ and $H_y = I_0\Omega$ where I_0 stands for I_{yy} and, again, I stands for I_{zz}. The total angular momentum is **H** as shown. The change in **H** remains $d\mathbf{H} = \mathbf{M}\,dt$ as previously, and the precession during time dt is the angle $d\psi = M\,dt/H_z = M\,dt/(Ip)$ as before. Thus Eq. 8/24 is still valid and for steady precession is an exact description of the motion as long as the spin axis is perpendicular to the axis around which precession occurs.

Consider now the steady precession of a symmetrical top, Fig. 8/18, spinning about its axis with a high angular velocity **p** and supported at its point O. Here the spin axis makes an angle θ with the vertical Z-axis around which precession occurs. Again, we will neglect the small angular-momentum component due to the precession and consider **H** equal to Ip, the angular momentum about the axis of the top associated with the spin only. The moment about O is due to the weight and is $mg\bar{r}\sin\theta$ where $\bar{r}$ is the distance from O to the mass center G. From the diagram we see that the angular momentum vector $\mathbf{H}_O$ has a change $d\mathbf{H}_O = \mathbf{M}_O\,dt$ in the direction of **M** during time dt and that θ is unchanged. The increment in precessional angle around the Z-axis is

$$d\psi = \frac{M_O\,dt}{Ip\sin\theta}$$

Substituting the value of M_O and $\Omega = d\psi/dt$ gives

$$mg\bar{r}\sin\theta = I\Omega p\sin\theta \qquad \text{or} \qquad mg\bar{r} = I\Omega p$$

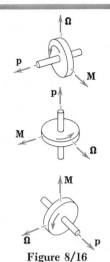

Figure 8/16

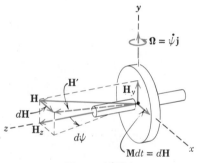

Figure 8/17

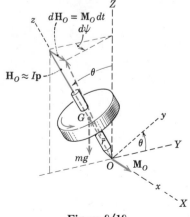

Figure 8/18

which is independent of θ. Introducing the radius of gyration so that $I = mk^2$ and solving for the precessional velocity give

$$\Omega = \frac{g\bar{r}}{k^2 p} \tag{8/25}$$

Unlike Eq. 8/24 which is an exact description for the rotor of Fig. 8/17 with precession confined to the x-z plane, Eq. 8/25 is an approximation based on the assumption that the angular momentum associated with Ω is negligible compared with that associated with p. We shall see the amount of the error in part (*b*) of this article. On the basis of our analysis the top will have a steady precession at the constant angle θ only if it is set in motion with a value of Ω which satisfies Eq. 8/25. When these conditions are not met, the precession becomes unsteady, and θ may oscillate with an amplitude which increases as the spin velocity decreases. The corresponding rise and fall of the rotation axis is known as *nutation*.

(*b*) *More Complete Analysis.* We will now make direct use of Eq. 8/19, which is the general angular-momentum equation for a rigid body, by applying it to a body spinning about its axis of rotational symmetry. This equation is valid for rotation about a fixed point or for rotation about the mass center. A spinning top, the rotor of a gyroscope, and a space capsule are examples of bodies whose motions can be described by the equations for rotation about a point. The general moment equations for this class of problems are fairly complex, and their complete solutions involve the use of elliptic integrals and somewhat lengthy computations. A large fraction of engineering problems where the motion is one of rotation about a point involve the steady precession of bodies of revolution which are spinning about their axis of symmetry. These conditions introduce simplifications which greatly facilitate solution of the equations.

Consider a body with axial symmetry, Fig. 8/19*a*, rotating about

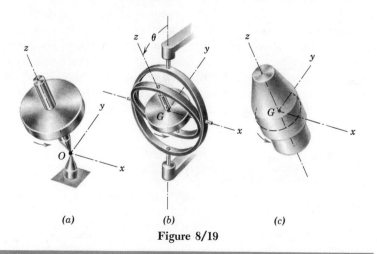

(*a*)   (*b*)   (*c*)

Figure 8/19

a fixed point O on its axis which is taken to be the z-direction. With O as origin, the x- and y-axes automatically become principal axes of inertia along with the z-axis. This same description may be used for the rotation of a similar symmetrical body about its center of mass G which is taken as the origin of coordinates, as shown with the gimbaled gyroscope rotor of Fig. 8/19b. Again the x- and y-axes are principal axes of inertia for point G. The same description may also be used to represent the rotation about its mass center of an axially symmetric body in space, such as the space capsule in Fig. 8/19c. In each case it is noted that, regardless of the rotation of the axes or of the rotation of the body relative to the axes (spin about the z-axis), the moments of inertia about the x- and y-axes remain constant with time. The principal moments of inertia are again designated $I_{zz} = I$ and $I_{xx} = I_{yy} = I_0$. The products of inertia are, of course, zero.

Before applying Eq. 8/19 we introduce a set of coordinates which provide a natural description for our problem. These coordinates are shown in Fig. 8/20 for the example of rotation about a fixed point O. The axes X-Y-Z are fixed in space, and plane A contains the X-Y axes and the fixed point O on the rotor axis. Plane B contains point O and is always normal to the rotor axis. Angle θ measures the inclination of the rotor axis from the vertical Z-axis and is also a measure of the angle between planes A and B. The intersection of the two planes is the x-axis which is located by the angle ψ from the X-axis. The y-axis lies in plane B, and the z-axis coincides with the rotor axis. The angles θ and ψ completely specify the position of the rotor axis. The angular displacement of the rotor with respect to axes x-y-z is specified by the angle ϕ measured from the x-axis to the x'-axis which is attached to the rotor. The spin velocity becomes $p = \dot{\phi}$.

The components of the angular velocity $\boldsymbol{\omega}$ of the rotor and the angular velocity $\boldsymbol{\Omega}$ of the axes x-y-z from Fig. 8/20 become

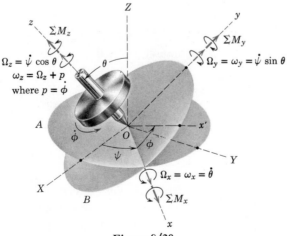

$$\Omega_z = \dot{\psi}\cos\theta$$
$$\omega_z = \Omega_z + p$$
where $p = \dot{\phi}$

$$\Omega_y = \omega_y = \dot{\psi}\sin\theta$$

$$\Omega_x = \omega_x = \dot{\theta}$$

Figure 8/20

$$\Omega_x = \dot{\theta} \qquad\qquad \omega_x = \dot{\theta}$$

$$\Omega_y = \dot{\psi}\sin\theta \qquad \omega_y = \dot{\psi}\sin\theta$$

$$\Omega_z = \dot{\psi}\cos\theta \qquad \omega_z = \dot{\psi}\cos\theta + p$$

It is important to note that the axes and the body have identical x- and y-components of angular velocity but that the z-components differ by the relative angular velocity p.

The angular-momentum components from Eq. 8/12 become

$$H_x = I_{xx}\omega_x = I_0\dot{\theta}$$

$$H_y = I_{yy}\omega_y = I_0\dot{\psi}\sin\theta$$

$$H_z = I_{zz}\omega_z = I(\dot{\psi}\cos\theta + p)$$

Substitution of the angular-velocity and angular-momentum components into Eq. 8/19 yields

$$\Sigma M_x = I_0(\ddot{\theta} - \dot{\psi}^2\sin\theta\cos\theta) + I\dot{\psi}(\dot{\psi}\cos\theta + p)\sin\theta$$

$$\Sigma M_y = I_0(\ddot{\psi}\sin\theta + 2\dot{\psi}\dot{\theta}\cos\theta) - I\dot{\theta}(\dot{\psi}\cos\theta + p) \qquad (8/26)$$

$$\Sigma M_z = I\frac{d}{dt}(\dot{\psi}\cos\theta + p)$$

Equations 8/26 are the general equations of rotation of a symmetrical body about either a fixed point O or the mass center G. In a given problem the solution to the equations will depend on the moment sums applied to the body about the three coordinate axes. We will confine our use of these equations to two particular cases of rotation about a point which are described in the following sections.

(c) *Steady-State Precession.* We now examine the conditions under which the rotor precesses at a steady rate $\dot{\psi}$ at a constant angle θ and with constant spin velocity p. Thus

$$\dot{\psi} = \text{constant}, \quad \ddot{\psi} = 0$$

$$\theta = \text{constant}, \quad \dot{\theta} = \ddot{\theta} = 0$$

$$p = \text{constant}, \quad \dot{p} = 0$$

and Eqs. 8/26 become

$$\Sigma M_x = \dot{\psi}\sin\theta\,[I(\dot{\psi}\cos\theta + p) - I_0\dot{\psi}\cos\theta]$$

$$\Sigma M_y = 0 \qquad (8/27)$$

$$\Sigma M_z = 0$$

From these results it is seen that the required moment acting on the rotor about O (or about G) must be in the x-direction since the y- and z-components are zero. Furthermore, with the constant values of θ, $\dot{\psi}$, and p the moment is seen to be of constant magnitude. It is also important to observe that the moment axis is perpendicular to the plane defined by the precession axis (Z-axis) and the spin axis (z-axis).

We may also obtain Eqs. 8/27 by observing that the components of **H** remain constant as observed in x-y-z so that $(\dot{\mathbf{H}})_{xyz} = 0$. Since in general $\Sigma\mathbf{M} = (\dot{\mathbf{H}})_{xyz} + \boldsymbol{\Omega} \times \mathbf{H}$, we have for the case of steady precession

$$\boxed{\Sigma\mathbf{M} = \boldsymbol{\Omega} \times \mathbf{H}} \qquad (8/28)$$

which reduces to Eqs. 8/27 upon substitution of the values of $\boldsymbol{\Omega}$ and **H**.

By far the most common engineering examples of gyroscopic motion occur when precession takes place about an axis which is normal to the rotor axis, as in Fig. 8/14. Thus with the substitution $\theta = \pi/2$, $\omega_z = p$, $\dot{\psi} = \Omega$, and $\Sigma M_x = M$, we have from Eqs. 8/27

$$M = I\Omega p \qquad [8/24]$$

which we derived initially in this article from a direct analysis of this special case.

Now let us examine the steady precession of the rotor (symmetrical top) of Fig. 8/20 for any constant value of θ other than $\pi/2$. The moment ΣM_x about the x-axis is due to the weight of the rotor and is $mg\bar{r}\sin\theta$. Substitution into Eqs. 8/27 and rearrangement of terms give us

$$mg\bar{r} = I\dot{\psi}p - (I_0 - I)\dot{\psi}^2\cos\theta$$

We see that $\dot{\psi}$ is small when p is large, so that the second term on the right-hand side of the equation becomes very small compared with $I\dot{\psi}p$. If we neglect this smaller term we have $\dot{\psi} = mg\bar{r}/(Ip)$ which, upon making the previous substitutions $\Omega = \dot{\psi}$ and $mk^2 = I$, becomes

$$\Omega = \frac{g\bar{r}}{k^2 p} \qquad [8/25]$$

We derived this same relation earlier by assuming that the angular momentum was entirely along the spin axis.

(d) *Steady Precession with Zero Moment.* Consider now the motion of a symmetrical rotor with no external moment about its mass center. Such motion is encountered with spacecraft and projectiles which both spin and precess during flight. Figure 8/21 represents such a body. Here the Z-axis, which has a fixed direction in space, is chosen to coincide with the direction of the angular momentum $\overline{\mathbf{H}}$ which is constant since $\Sigma\overline{\mathbf{M}} = \mathbf{0}$. The x-y-z axes are attached in the manner described in Fig. 8/20. From Fig. 8/21 the three components of momentum are $\overline{H}_x = 0$, $\overline{H}_y = \overline{H}\sin\theta$, $\overline{H}_z = \overline{H}\cos\theta$. From the defining relations, Eqs. 8/12, with the notation of this

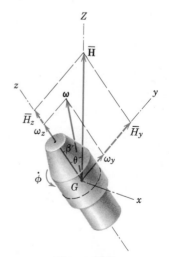

Figure 8/21

article these components are also given by $\bar{H}_x = I_0\omega_x$, $\bar{H}_y = I_0\omega_y$, $\bar{H}_z = I\omega_z$. Thus $\omega_x = \Omega_x = 0$ so that θ is constant. This result means that the motion is one of steady precession about the constant $\bar{\mathbf{H}}$ vector. With no x-component the angular velocity $\boldsymbol{\omega}$ of the rotor lies in the y-z plane along with the Z-axis and makes an angle β with the z-axis. The relationship between β and θ is obtained from $\tan\theta = \bar{H}_y/\bar{H}_z = I_0\omega_y/(I\omega_z)$, which is

$$\tan\theta = \frac{I_0}{I}\tan\beta \qquad (8/29)$$

Thus the angular velocity $\boldsymbol{\omega}$ makes a constant angle β with the spin axis.

The rate of precession is easily obtained from Eq. 8/27 with $M = 0$ which gives

$$\dot{\psi} = \frac{Ip}{(I_0 - I)\cos\theta} \qquad (8/30)$$

It is clear from this relation that the direction of the precession depends on the relative magnitudes of the two moments of inertia.

If $I_0 > I$, then $\beta < \theta$, as indicated in Fig. 8/22a, and the precession is said to be *direct*. Here the body cone rolls on the outside of the space cone.

If $I > I_0$, then $\theta < \beta$, as indicated in Fig. 8/22b, and the precession is said to be *retrograde*. In this instance the space cone is internal to the body cone, and $\dot{\psi}$ and p have opposite signs.

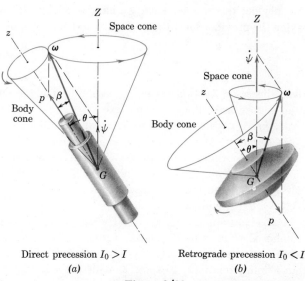

Direct precession $I_0 > I$ Retrograde precession $I_0 < I$

(a)   (b)

Figure 8/22

Sample Problem 8/8

The turbine rotor in a ship's power plant has a mass of 1000 kg with center of mass at G and a radius of gyration of 200 mm. The rotor shaft is mounted in bearings A and B with its axis in the horizontal fore-and-aft direction and turns counterclockwise at a speed of 5000 rev/min when viewed from the stern. Determine the vertical components of the bearing reactions at A and B if the ship is making a turn to port (left) of 400-m radius at a speed of 25 knots (1 knot = 0.514 m/s). Does the bow of the ship tend to rise or fall because of the gyroscopic action?

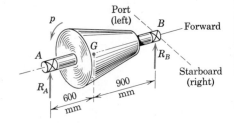

Solution. The vertical component of the bearing reactions will equal the static reactions R_1 and R_2 due to the weight of the rotor plus or minus the increment ΔR due to the gyroscopic effect. The moment principle from statics easily gives $R_1 = 5886$ N and $R_2 = 3924$ N. The given directions of the spin velocity $\mathbf{p}$ and the precession velocity $\mathbf{\Omega}$ are shown with the free-body diagram of the rotor. Since the spin axis always tends to rotate toward the torque axis, we see that the torque axis $\mathbf{M}$ points in the starboard direction as shown. The sense of the ΔR's is, therefore, up at B and down at A to produce the couple $\mathbf{M}$. Thus the bearing reactions at A and B are

$$R_A = R_1 - \Delta R \quad \text{and} \quad R_B = R_2 + \Delta R$$

The precession velocity Ω is the speed of the ship divided by the radius of its turn.

$$[v = \rho\Omega] \qquad \Omega = \frac{25(0.514)}{400} = 0.0321 \text{ rad/s}$$

Equation 8/24 is now applied around the mass center G of the rotor to give

$$[M = I\Omega p] \quad 1.500\,\Delta R = 1000(0.200)^2(0.0321)\left(\frac{5000[2\pi]}{60}\right)$$

$$\Delta R = 449 \text{ N}$$

The required bearing reactions become

$$R_A = 5886 - 449 = 5437 \text{ N} \quad \text{and} \quad R_B = 3924 + 449 = 4373 \text{ N} \qquad Ans.$$

We now observe from the principle of action and reaction that the forces just computed are those exerted *on* the rotor shaft *by* the structure of the ship. Consequently, the equal and opposite reactions are applied to the ship *by* the rotor shaft, as shown in the bottom sketch. Therefore the effect of the gyroscopic couple is to generate the increments ΔR shown, and the bow will tend to fall and the stern to rise (but only slightly).

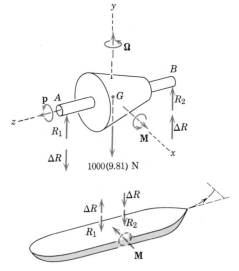

① If the ship is making a left turn, the rotation is counterclockwise when viewed from above, and the precession vector $\mathbf{\Omega}$ is up by the right-hand rule.

② After figuring the correct sense of $\mathbf{M}$ *on* the rotor, the common mistake is to apply it to the ship in the same sense, forgetting the action-and-reaction principle. Clearly the results are then reversed. (Be certain not to make this mistake when operating a vertical gyro stabilizer in your yacht to counteract its roll!)

Sample Problem   8/9

A proposed space station is closely approximated by four uniform spherical shells each of mass m and radius r. The mass of the connecting
structure and internal equipment may be neglected as a first approximation. If the station is designed to rotate about its z-axis at the rate of one
revolution every 4 seconds, determine (a) the number n of complete cycles
of precession for each revolution about the z-axis if the plane of rotation
deviates only slightly from a fixed orientation, and (b) find the period τ of
precession if the spin axis z makes an angle of $20°$ with respect to the axis
of fixed orientation about which precession occurs. Draw the space and
body cones for this latter condition.

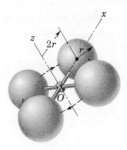

Solution. (a)   The number of precession cycles or wobbles for each
revolution of the station about the z-axis would be the ratio of the precessional velocity $\dot{\psi}$ to the spin velocity p which, from Eq. 8/30, is

$$\frac{\dot{\psi}}{p} = \frac{I}{(I_0 - I)\cos\theta}$$

The moments of inertia are

$$I_{zz} = I = 4[\tfrac{2}{3}mr^2 + m(2r)^2] = \tfrac{56}{3}\,mr^2$$

$$I_{xx} = I_0 = 2(\tfrac{2}{3})mr^2 + 2[\tfrac{2}{3}mr^2 + m(2r)^2] = \tfrac{32}{3}\,mr^2$$

With θ very small, $\cos\theta \approx 1$, and the ratio of angular rates becomes

$$n = \frac{\dot{\psi}}{p} = \frac{\tfrac{56}{3}}{\tfrac{32}{3} - \tfrac{56}{3}} = -\frac{7}{3} \qquad Ans.$$

The minus sign indicates retrograde precession where, in the present case,
$\dot{\psi}$ and p are essentially of opposite sense. Thus the station will make 7
wobbles for every 3 revolutions.

(b) For $\theta = 20°$ and $p = 2\pi/4$ rad/s, the period of precession or
wobble is $\tau = 2\pi/|\dot{\psi}|$, so that from Eq. 8/30

$$\tau = \frac{2\pi}{2\pi/4}\left|\frac{I_0 - I}{I}\cos\theta\right| = 4(\tfrac{3}{7})\cos 20° = 1.61 \text{ s} \qquad Ans.$$

The precession is retrograde, and the body cone is external to the
space cone as shown in the illustration where the body-cone angle, from
Eq. 8/29, is

$$\tan\beta = \frac{I}{I_0}\tan\theta = \frac{56/3}{32/3}(0.3640) = 0.6370 \qquad \beta = 32.50°$$

① Our theory is based on the assumption that
$I_{xx} = I_{yy}$ = the moment of inertia about
any axis through G perpendicular to the
z-axis. Such is the case here, and students
should prove it to their own satisfaction.

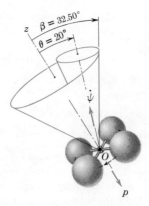

PROBLEMS

8/67 If the ship of Sample Problem 8/8 is on a straight course but its bow is rising as a wave passes under it, determine the direction of the gyroscopic moment exerted *by* the turbine rotor *on* the hull structure and its effect on the motion of the ship.

Ans. Bow tends to swing to port

8/68 A car makes a turn to the right on a level road. Determine whether the normal reaction under the right rear wheel is increased or decreased as a result of the gyroscopic effect of the precessing wheels.

8/69 The wheel spins at the rate p about its shaft OA, and the shaft in turn rotates about the vertical at the rate N. Determine whether the end A of the shaft deflects up or down as the shaft bends due to the gyroscopic effect during rotation.

Ans. End A deflects up

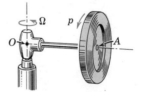

Problem 8/69

8/70 An airplane has just cleared the runway with a takeoff speed v. Each of its freely spinning wheels has a mass m with a radius of gyration k about its axle. As seen from the front of the airplane the wheel precesses at the angular rate Ω as the landing strut is folded into the wing about its pivot O. As a result of the gyroscopic action the supporting member A exerts a torsional moment M on B to prevent the tubular member from rotating in the sleeve at B. Determine M and identify whether it is in the sense of M_1 or M_2.

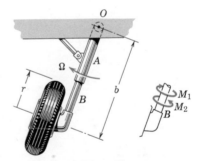

Problem 8/70

8/71 A small air compressor for an aircraft cabin consists of the 3.50-kg turbine A which drives the 2.40-kg blower B at a speed of 20 000 rev/min. The shaft of the assembly is mounted transversely to the direction of flight and is viewed from the rear of the aircraft in the figure. The radii of gyration of A and B are 79.0 and 71.0 mm, respectively. Calculate the radial forces exerted on the shaft by the bearings at C and D if the aircraft executes a clockwise roll (rotation about the longitudinal flight axis) of 2 rad/s viewed from the rear of the aircraft. Neglect the small moments caused by the weights of the rotors. *Ans.* $C = D = 948$ N

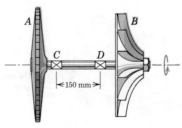

Problem 8/71

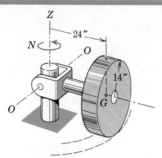

Problem 8/72

8/72 The wheel is a solid circular disk weighing 96.6 lb and rolls on the horizontal plane in a circle of 24-in. radius. The wheel shaft is pivoted about the axis *O-O* and is driven by the vertical shaft at the constant rate $N = 48$ rev/min about the Z-axis. Determine the normal force R between the wheel and the horizontal surface. Neglect the weight of the horizontal shaft.

8/73 The 225-kg rotor for a turbojet engine has a radius of gyration of 250 mm and rotates counterclockwise at 18 000 rev/min when viewed from the front of the airplane. If the airplane is traveling at 1000 km/h and making a turn to the left of 3-km radius, compute the gyroscopic moment M which the rotor bearings must support. Does the nose of the airplane tend to rise or fall as a result of the gyroscopic action?

Ans. $M = 2450$ N·m   Nose tends to rise

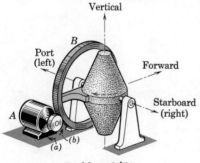

Problem 8/74

8/74 The figure shows a gyro mounted with vertical axis and used to stabilize a hospital ship against rolling. The motor A turns the pinion which precesses the gyro by rotating the large precession gear B and attached rotor assembly about a horizontal transverse axis in the ship. The rotor turns inside the housing at a clockwise speed of 960 rev/min when viewed from the top and has a mass of 80 Mg with radius of gyration of 1.45 m. Calculate the moment exerted on the hull structure by the gyro if the motor turns the precession gear B at the rate of 0.320 rad/s. In which of the two directions, (*a*) or (*b*), should the motor turn in order to counteract a roll of the ship to port?

8/75 An experimental car is equipped with a gyro stabilizer to counteract completely the tendency of the car to tip when rounding a curve (no change in normal force between tires and road). The rotor of the gyro has a mass m_0 and a radius of gyration k, and is mounted in fixed bearings on a shaft which is parallel to the rear axle of the car. The center of mass of the car is a distance h above the road, and the car is rounding an unbanked level turn at a speed v. At what speed p should the rotor turn and in what direction to counteract completely the tendency of the car to overturn for either a right or a left turn? The total mass of car and rotor is m.

Ans. $p = \dfrac{mvh}{m_0 k^2}$ direction opposite to car wheels

8/76 The electric motor has a total mass of 10 kg and is supported by the mounting brackets A and B attached to the rotating disk. The armature of the motor has a mass of 2.5 kg and a radius of gyration of 35 mm and turns counterclockwise at a speed of 1725 rev/min as viewed from A to B. The turntable revolves about its vertical axis at the constant rate of 48 rev/min in the direction shown. Determine the vertical components of the forces supported by the mounting brackets at A and B.

Ans. $R_A = 37.5$ N, $R_B = 60.6$ N

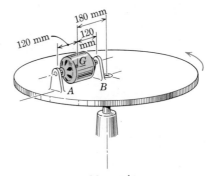

Problem 8/76

8/77 Show that the rate of steady precession for an axially symmetrical rotor under zero moment about its mass center G is given by $\dot{\psi} = I\omega_z/(I_0 \cos \theta)$ as well as by Eq. 8/30.

Problem 8/78

8/78 The cylindrical shell is rotating in space about its geometric axis. If the axis has a slight wobble, for what ratios of l/r will the motion be direct or retrograde precession?

Ans. Direct $l/r > \sqrt{6}$,   retrograde $l/r < \sqrt{6}$

8/79 The thin ring is projected into the air with a spin velocity of 300 rev/min. If its geometric axis is observed to have a very slight precessional wobble, determine the frequency f of the wobble.

Ans. $f = 10$ Hz (cycles/second), retrograde precession

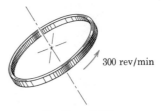

Problem 8/79

Problem 8/81

Problem 8/82

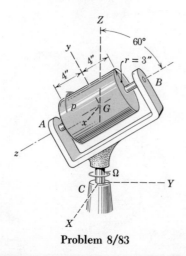

Problem 8/83

8/80 If the radius of gyration k of the rotor of Fig. 8/18 about its axis of symmetry is 150 mm, what must be the radius of gyration of the rotor about the y-axis so that Eq. 8/25 will become exact for steady precession about the Z-axis?

8/81 The spacecraft shown is symmetrical about its z-axis and has a radius of gyration of 720 mm about this axis. The radii of gyration about the x- and y-axes through the mass center are both equal to 540 mm. When moving in space the z-axis is observed to generate a cone with a total vertex angle of 4° as it precesses about the axis of total angular momentum. If the spacecraft has a spin velocity $\dot\phi$ about its z-axis of 1.5 rad/s, compute the period τ of each full precession. Is the spin vector in the positive or negative z-direction?

> *Ans.* $\tau = 1.83$ s, spin vector in negative z-direction

8/82 A solid right-circular cone of mass m, base radius $r = 100$ mm, and altitude $H = 3r = 300$ mm is suspended freely at its vertex O and is spinning about its geometric axis at the rate $p = 2700$ rev/min. Determine the period for steady slow precession of the cone about the vertical. Does the result depend upon the angle θ made by the cone axis with the vertical?

8/83 The solid cylindrical rotor weighs 64.4 lb and is mounted in bearings A and B of the frame which rotates about the vertical Z-axis. If the rotor spins at the constant rate $p = 50$ rad/sec relative to the frame and if the frame itself rotates at the constant rate $\Omega = 30$ rad/sec, compute the bending moment **M** in the shaft at C which the lower portion of the shaft exerts on the upper portion. Also compute the kinetic energy T of the rotor. Neglect the mass of the frame.

> *Ans.* **M** $= 97.9\mathbf{i}$ lb-ft, $T = 73.8$ ft-lb

/84 Let the rotor of Fig. 8/18 have a spin velocity of 3600 rev/min and execute steady precession about the vertical Z-axis at the constant angle $\theta = 30°$. Furthermore, $\bar{r} = 200$ mm, $k = 150$ mm, and the moment of inertia about the x-axis is five times that about the spin axis. Calculate the precession $\dot{\psi} = \Omega$ using Eq. 8/25 and obtain a close approximation to the percentage error e in this value.

Ans. $\dot{\psi} = 0.231$ rad/s, $e = 0.21$ percent too low

3/85 The 8-lb rotor with radius of gyration of 3 in. rotates on ball bearings at a speed of 3000 rev/min about its shaft OG. The shaft is free to pivot about the X-axis as well as to rotate about the Z-axis. Calculate the vector Ω for precession about the Z-axis. Neglect the mass of shaft OG and compute the gyroscopic couple M exerted by the shaft on the rotor at G.

Ans. $\Omega = -1.23K$ rad/s, $M = 67.7i$ lb-in.

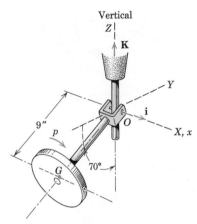

Problem 8/85

8/86 Derive Eq. 8/24 by relating the forces to the accelerations for a differential element of the thin ring of mass m. The ring has a constant angular velocity p about the z-axis and is given an additional constant angular velocity Ω about the y-axis by the application of an external moment M (not shown). (*Hint:* The acceleration of the element in the z-direction is due to (*a*) the change in magnitude of its velocity component in this direction resulting from Ω and (*b*) the change in direction of its x-component of velocity.)

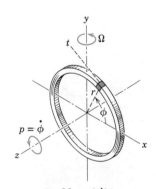

Problem 8/86

8/87 A projectile moving through the atmosphere with a velocity $\bar{v}$ which makes a small angle θ with its geometric axis is subjected to a resultant aerodynamic force R essentially opposite in direction to $\bar{v}$ as shown. If R passes through a point C slightly ahead of the mass center G, determine the expression for the minimum spin velocity p for which the projectile will be spin-stabilized with $\dot{\theta} = 0$. The moment of inertia about the spin axis is I and that about a transverse axis through G is I_0. (*Hint:* Determine M_x and substitute into Eq. 8/27. Express the result as a quadratic equation in $\dot{\psi}$ and determine the minimum value of p for which the expression under the radical is positive.)

Ans. $p > \dfrac{2}{I} \sqrt{R\bar{r}(I_0 - I)} \cos \theta$

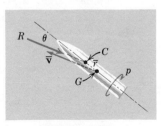

Problem 8/87

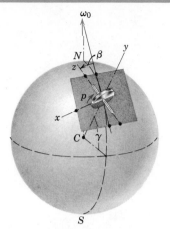

Problem 8/88

▶8/88 The elements of a gyrocompass are shown in the figure where the rotor, at north latitude γ on the surface of the earth, is mounted in a single gimbal ring which is free to rotate about the fixed vertical y-axis. The rotor axis is, then, able to rotate in the horizontal x-z plane as measured by the angle β from the north direction. Assume that the gyro spins at the rate p and that it has mass moments of inertia about the spin axis and transverse axis through G of I and I_0, respectively. Show that the gyro axis oscillates about the north direction according to the equation $\ddot{\beta} + K^2\beta = 0$ where $K^2 = I\omega_0 p \cos\gamma/I_0$, and that the period of oscillation about the north direction for small values of β is $\tau = 2\pi\sqrt{I_0/(I\omega_0 p \cos\gamma)}$. (*Hint:* The components of the angular velocity Ω of the axes in terms of the angular velocity ω_0 of the earth are

$$\Omega_x = -\omega_0 \cos\gamma \sin\beta$$

$$\Omega_y = \omega_0 \sin\gamma + \dot{\beta}$$

$$\Omega_z = \omega_0 \cos\gamma \cos\beta$$

which may be used in the y-component of the moment equations, Eqs. 8/19, to determine β as a function of time. Note that the square of the angular velocity ω_0 of the earth is small and may be neglected compared with the product $\omega_0 p$.)

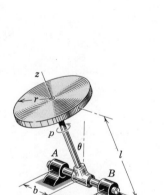

Problem 8/89

▶8/89 The solid circular disk of mass m and small thickness is spinning freely on its shaft at the rate p. If the assembly is released in the vertical position at $\theta = 0$ with $\dot{\theta} = 0$, determine the horizontal components of the forces A and B exerted by the respective bearings on the horizontal shaft as the position $\theta = \pi/2$ is passed. Neglect the mass of the two shafts compared with m, and neglect all friction. Solve by using the appropriate moment equations.

Ans. $A_z = -\dfrac{m\dot{\theta}}{2}\left(\dfrac{r^2}{2b}p + l\dot{\theta}\right)$

$B_z = \dfrac{m\dot{\theta}}{2}\left(\dfrac{r^2}{2b}p - l\dot{\theta}\right)$

where $\dot{\theta} = 2\sqrt{\dfrac{2gl}{r^2 + 4l^2}}$

AREA MOMENTS OF INERTIA

A/1 INTRODUCTION. When forces are distributed continuously over an area upon which they act, it is often necessary for us to calculate the moment of these forces about some axis either in or perpendicular to the plane of the area. Frequently the intensity of the force (pressure or stress) is proportional to the distance of the force from the moment axis. The elemental force acting on an element of area, then, is proportional to distance times differential area, and the elemental moment is proportional to distance squared times differential area. We see, therefore, that the total moment involves an integral of the form $\int (\text{distance})^2 \, d(\text{area})$. This integral is known as the *moment of inertia* of the area. The integral is a function of the geometry of the area and occurs so frequently in the applications of mechanics that we find it useful to develop its properties in some detail and to have these properties available for ready use when the integral arises.

Figure A/1 illustrates the physical origin of these integrals. In the *a*-part of the figure the surface area *ABCD* is subjected to a distributed pressure *p* whose intensity is proportional to the distance *y* from the axis *AB*. This situation was treated in Art. 5/8 of Chapter 5 where we described the action of liquid pressure on a plane surface. The moment about *AB* that is due to the pressure on the element of area dA is $py \, dA = ky^2 \, dA$. Thus the integral in question appears when the total moment $M = k \int y^2 \, dA$ is evaluated.

In Fig. A/1*b* we show the distribution of stress acting on a transverse section of a simple elastic beam bent by equal and opposite couples applied to its ends. At any section of the beam a linear distribution of force intensity or stress σ, given by $\sigma = ky$, is present, the stress being positive (tensile) below the axis *O–O* and negative (compressive) above the axis. We see that the elemental moment about the axis *O–O* is $dM = y(\sigma \, dA) = ky^2 \, dA$. Thus the same integral appears when the total moment $M = k \int y^2 \, dA$ is evaluated.

A third example is given in Fig. A/1*c*, which shows a circular shaft subjected to a twist or torsional moment. Within the elastic limit of the material this moment is resisted at each cross section of the shaft by a distribution of tangential or shear stress τ which is proportional to the radial distance r from the center. Thus $\tau = kr$,

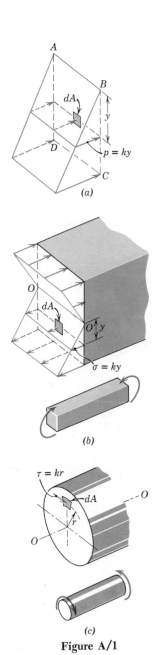

Figure A/1

and the total moment about the central axis is $M = \int r(\tau\, dA) = k \int r^2\, dA$. Here the integral differs from that in the preceding two examples in that the area is normal instead of parallel to the moment axis and in that r is a radial coordinate instead of a rectangular one.

Although the integral illustrated in the preceding examples is generally called the *moment of inertia* of the area about the axis in question, a more fitting term is the *second moment of area*, since the first moment $y\, dA$ is multiplied by the moment arm y to obtain the second moment for the element dA. The word *inertia* appears in the terminology by reason of the similarity between the mathematical form of the integrals for second moments of areas and those for the resultant moments of the so-called inertia forces in the case of rotating bodies. The moment of inertia of an area is a purely mathematical property of the area and in itself has no physical significance.

A/2   DEFINITIONS.

The following definitions of terms form the basis for the analysis of area moments of inertia.

(*a*) *Rectangular and Polar Moments of Inertia.*   Consider the area A in the x-y plane, Fig. A/2. The moments of inertia of the element dA about the x- and y-axes are, by definition, $dI_x = y^2\, dA$ and $dI_y = x^2\, dA$, respectively. Therefore the moments of inertia of A about the same axes are

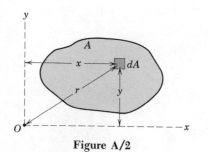

Figure A/2

$$\boxed{\begin{aligned} I_x &= \int y^2\, dA \\ I_y &= \int x^2\, dA \end{aligned}} \tag{A/1}$$

where we carry out the integration over the entire area.

The moment of inertia of dA about the pole O (z-axis) is, by similar definition, $dJ_z = r^2\, dA$, and the moment of inertia of the entire area about O is

$$\boxed{J_z = \int r^2\, dA} \tag{A/2}$$

The expressions defined by Eqs. A/1 are known as *rectangular* moments of inertia, whereas the expression of Eq. A/2 is known as the *polar* moment of inertia. Since $x^2 + y^2 = r^2$, it is clear that

$$\boxed{J_z = I_x + I_y} \tag{A/3}$$

A polar moment of inertia for an area whose boundaries are more simply described in rectangular coordinates than in polar coordinates is easily calculated with the aid of Eq. A/3.

We note that the moment of inertia of an element involves the square of the distance from the inertia axis to the element. An element whose coordinate is negative contributes as much to the moment of inertia as does an equal element with a positive coordinate of the same magnitude. Consequently we see that the area moment of inertia about any axis is always a positive quantity. In contrast, the first moment of the area, which was involved in the computations of centroids, could be either positive, negative, or zero.

The dimensions of moments of inertia of areas are clearly L^4, where L stands for the dimension of length. Thus the SI units for area moments of inertia are expressed as quartic meters (m^4) or quartic millimeters (mm^4). The U.S. customary units for area moments of inertia are quartic feet (ft^4) or quartic inches ($in.^4$).

The choice of the coordinates to use for the calculation of moments of inertia is important. Rectangular coordinates should be used for shapes whose boundaries are most easily expressed in these coordinates. Polar coordinates will usually simplify problems involving boundaries that are easily described in r and θ. The choice of an element of area which simplifies the integration as much as possible is also important. These considerations are quite analogous to those we discussed and illustrated in Chapter 5 for the calculation of centroids.

(b) Radius of Gyration. Consider an area A, Fig. A/3*a*, which has rectangular moments of inertia I_x and I_y and a polar moment of inertia J_z about O. We now visualize this area to be concentrated into a long narrow strip of area A a distance k_x from the x-axis, Fig. A/3*b*. By definition the moment of inertia of the strip about the x-axis will be the same as that of the original area if $k_x{}^2 A = I_x$. The distance k_x is known as the *radius of gyration* of the area about the x-axis. A similar relation for the y-axis is written by considering the area to be concentrated into a narrow strip parallel to the y-axis as shown in Fig. A/3*c*. Also, if we visualize the area to be concentrated into a narrow ring of radius k_z as shown in Fig. A/3*d*, we may express the polar moment of inertia as $k_z{}^2 A = J_z$. In summary we write

$$\boxed{\begin{array}{lll} I_x = k_x{}^2 A & & k_x = \sqrt{I_x/A} \\ I_y = k_y{}^2 A & \text{or} & k_y = \sqrt{I_y/A} \\ J_z = k_z{}^2 A & & k_z = \sqrt{J_z/A} \end{array}} \qquad \text{(A/4)}$$

The radius of gyration, then, is a measure of the distribution of the area from the axis in question. A rectangular or polar moment of inertia may be expressed by specifying the radius of gyration and the area.

When we substitute Eqs. A/4 into Eq. A/3 we have

$$\boxed{k_z{}^2 = k_x{}^2 + k_y{}^2} \qquad \text{(A/5)}$$

Thus the square of the radius of gyration about a polar axis equals the sum of the squares of the radii of gyration about the two corresponding rectangular axes.

It is imperative that there be no confusion between the coordinate to the centroid C of an area and the radius of gyration. In Fig. A/3*a* the square of the centroidal distance from the x-axis, for example, is $\bar{y}^2$, which is the square of the mean value of the distances from the

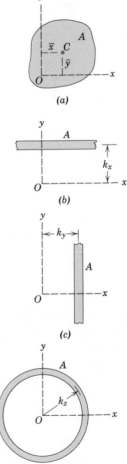
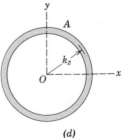

Figure A/3

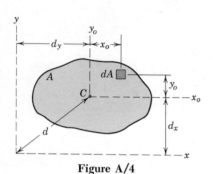

Figure A/4

elements of the area to the x-axis. The quantity $k_x{}^2$, on the other hand, is the mean of the squares of these distances. The moment of inertia is *not* equal to $A\bar{y}^2$, since the square of the mean is less than the mean of the squares.

(*c*) *Transfer of Axes.* The moment of inertia of an area about a noncentroidal axis may be easily expressed in terms of the moment of inertia about a parallel centroidal axis. In Fig. A/4 the x_0-y_0 axes pass through the centroid C of the area. Let us now determine the moments of inertia of the area about the parallel x-y axes. By definition the moment of inertia of the element dA about the x-axis is

$$dI_x = (y_o + d_x)^2 \, dA$$

Expanding and integrating give us

$$I_x = \int y_o{}^2 \, dA + 2d_x \int y_o \, dA + d_x{}^2 \int dA$$

We see that the first integral is by definition the moment of inertia $\bar{I}_x$ about the centroidal x_o-axis. The second integral is zero, since $\int y_o \, dA = A\bar{y}_o$ and $\bar{y}_o$ is automatically zero with the centroid on the x_o-axis. The third term is simply $Ad_x{}^2$. Thus the expression for I_x and the similar expression for I_y become

$$\boxed{\begin{aligned} I_x &= \bar{I}_x + Ad_x{}^2 \\ I_y &= \bar{I}_y + Ad_y{}^2 \end{aligned}} \tag{A/6}$$

By Eq. A/3 the sum of these two equations gives

$$\boxed{J_z = \bar{J}_z + Ad^2} \tag{A/6a}$$

Equations A/6 and A/6a are the so-called *parallel-axis theorems*. Two points in particular should be noted. First, the axes between which the transfer is made *must be parallel*, and, second, one of the axes *must pass through the centroid* of the area.

If a transfer is desired between two parallel axes neither one of which passes through the centroid, it is first necessary for us to transfer from one axis to the parallel centroidal axis and then to transfer from the centroidal axis to the second axis.

The parallel-axis theorems also hold for radii of gyration. With substitution of the definition of k into Eqs. A/6, the transfer relation becomes

$$\boxed{k^2 = \bar{k}^2 + d^2} \tag{A/6b}$$

where $\bar{k}$ is the radius of gyration about a centroidal axis parallel to the axis about which k applies and d is the distance between the two axes. The axes may be either in the plane or normal to the plane of the area.

A summary of the moment-of-inertia relations for some of the common plane figures is given in Table C/3, Appendix C.

Sample Problem A/1

Determine the moments of inertia of the rectangular area about the centroidal x_o-y_o axes, the centroidal polar axis z_o through C, the x-axis, and the polar axis z through O.

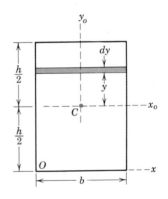

Solution. For the calculation of the moment of inertia $\bar{I}_x$ about the x_o-axis a horizontal strip of area $b\,dy$ is chosen so that all elements of the strip have the same y-coordinate. Thus

$$[I_x = \int y^2\, dA] \qquad \bar{I}_x = \int_{-h/2}^{h/2} y^2 b\, dy = \tfrac{1}{12}bh^3 \qquad \text{Ans.}$$

By interchanging symbols the moment of inertia about the centroidal y_o-axis is

$$\bar{I}_y = \tfrac{1}{12}hb^3 \qquad \text{Ans.}$$

The centroidal polar moment of inertia is

$$[J_z = I_x + I_y] \qquad \bar{J}_z = \tfrac{1}{12}(bh^3 + hb^3) = \tfrac{1}{12}A(b^2 + h^2) \qquad \text{Ans.}$$

By the parallel-axis theorem the moment of inertia about the x-axis is

$$[I_x = \bar{I}_x + Ad_x{}^2] \qquad I_x = \tfrac{1}{12}bh^3 + bh\left(\frac{h}{2}\right)^2 = \tfrac{1}{3}bh^3 = \tfrac{1}{3}Ah^2 \qquad \text{Ans.}$$

We also obtain the polar moment of inertia about O by the parallel-axis theorem which gives us

$$[J_z = \bar{J}_z + Ad^2] \qquad J_z = \tfrac{1}{12}A(b^2 + h^2) + A\left[\left(\frac{b}{2}\right)^2 + \left(\frac{h}{2}\right)^2\right]$$

$$J_z = \tfrac{1}{3}A(b^2 + h^2) \qquad \text{Ans.}$$

① If we had started with the second-order element $dA = dx\,dy$, integration with respect to x holding y constant amounts simply to multiplication by b and gives us the expression $y^2 b\,dy$ which we chose at the outset.

Sample Problem A/2

Determine the moments of inertia of the triangular area about its base and about parallel axes through its centroid and vertex.

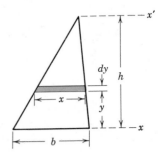

Solution. A strip of area parallel to the base is selected as shown in the figure, and it has the area $dA = x\,dy = [(h - y)b/h]\,dy$. By definition

$$[I_x = \int y^2\, dA] \quad I_x = \int_0^h y^2 \frac{h - y}{h} b\, dy = b\left[\frac{y^3}{3} - \frac{y^4}{4h}\right]_0^h = \frac{bh^3}{12} \qquad \text{Ans.}$$

By the parallel-axis theorem the moment of inertia $\bar{I}$ about an axis through the centroid, a distance $h/3$ above the x-axis, is

$$[\bar{I} = I - Ad^2] \qquad \bar{I} = \frac{bh^3}{12} - \left(\frac{bh}{2}\right)\left(\frac{h}{3}\right)^2 = \frac{bh^3}{36} \qquad \text{Ans.}$$

A transfer from the centroidal axis to the x'-axis through the vertex gives

$$[I = \bar{I} + Ad^2] \qquad I_{x'} = \frac{bh^3}{36} + \left(\frac{bh}{2}\right)\left(\frac{2h}{3}\right)^2 = \frac{bh^3}{4} \qquad \text{Ans.}$$

① Here again we choose the simplest possible element. If we had chosen $dA = dx\,dy$, we would have to integrate $y^2\,dx\,dy$ with respect to x first. This gives us $y^2x\,dy$, which is the expression we chose at the outset.

② Expressing x in terms of y should cause no difficulty if we observe the proportional relationship between the similar triangles.

Sample Problem A/3

Calculate the moments of inertia of the area of a circle about a diametral axis and about the polar axis through the center. Specify the radii of gyration.

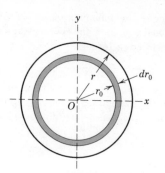

Solution. A differential element of area in the form of a circular ring
① may be used for the calculation of the moment of inertia about the polar z-axis through O since all elements of the ring are equidistant from O. The elemental area is $dA = 2\pi r_o\, dr_o$, and thus

$$[J_z = \int r^2\, dA] \qquad J_z = \int_0^r r_o^{\,2}(2\pi r_o\, dr_o) = \frac{\pi r^4}{2} = \tfrac{1}{2}Ar^2 \qquad \textit{Ans.}$$

The polar radius of gyration is

$$\left[k = \sqrt{\frac{J}{A}}\right] \qquad k_z = \frac{r}{\sqrt{2}} \qquad \textit{Ans.}$$

By symmetry $I_x = I_y$, so that from Eq. A3

$$[J_z = I_x + I_y] \qquad I_x = \tfrac{1}{2}J_z = \frac{\pi r^4}{4} = \tfrac{1}{4}Ar^2 \qquad \textit{Ans.}$$

The radius of gyration about the diametral axis is

$$\left[k = \sqrt{\frac{I}{A}}\right] \qquad k_x = \frac{r}{2} \qquad \textit{Ans.}$$

The foregoing determination of I_x is the simplest possible. The result may also be obtained by direct integration, using the element of area $dA = r_o\, dr_o\, d\theta$ shown in the lower figure. By definition

$$[I_x = \int y^2\, dA] \qquad I_x = \int_0^{2\pi}\int_0^r (r_o \sin\theta)^2 r_o\, dr_o\, d\theta$$

$$= \int_0^{2\pi} \frac{r^4 \sin^2\theta}{4}\, d\theta = \frac{r^4}{4}\frac{1}{2}\left[\theta - \frac{\sin 2\theta}{2}\right]_0^{2\pi} = \frac{\pi r^4}{4} \qquad \textit{Ans.}$$

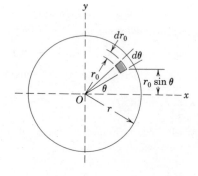

① Polar coordinates are certainly indicated here. Also, as before, we choose the simplest and lowest-order element possible, which is the differential ring. It should be evident immediately from the definition that the polar moment of inertia of the ring is its area $2\pi r_o dr_o$ times $r_o^{\,2}$.

② This integration is straightforward, but the use of Eq. A/3 along with the result for J_z is certainly simpler.

Sample Problem A/4

Determine the moment of inertia of the area under the parabola about the x-axis. Solve by using (a) a horizontal strip of area and (b) by using a vertical strip of area.

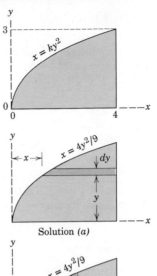

Solution. The constant $k = \frac{4}{9}$ is obtained first by substituting $x = 4$ and $y = 3$ into the equation for the parabola.

(a) *Horizontal strip.* Since all parts of the horizontal strip are the same distance from the x-axis, the moment of inertia of the strip about the x-axis is simply $y^2\, dA$ where $dA = (4 - x)\, dy = 4(1 - y^2/9)\, dy$. Integrating with respect to y gives us

$$[I_x = \int y^2\, dA] \quad I_x = \int_0^3 4y^2\left(1 - \frac{y^2}{9}\right) dy = \frac{72}{5} = 14.4 \text{ (units)}^4 \qquad Ans.$$

(b) *Vertical strip.* Here all parts of the element are at different distances from the x-axis, so we must use the correct expression for the moment of inertia of the elemental rectangle about its base which, from Sample Problem A/1, is $bh^3/3$. For the width dx and the height y the expression becomes

$$dI_x = \tfrac{1}{3}(dx)y^3$$

To integrate with respect to x we must express y in terms of x, which gives $y = 3\sqrt{x}/2$, and the integral becomes

$$I_x = \tfrac{1}{3}\int_0^4 \left(\frac{3\sqrt{x}}{2}\right)^3 dx = \frac{72}{5} = 14.4 \text{ (units)}^4 \qquad Ans.$$

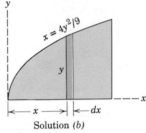

Solution (b)

① There is very little preference between Solutions (a) and (b). Solution (b), of course, requires that we know the moment of inertia for a rectangular area about its base, which we should remember because of its frequent use.

Sample Problem A/5

Determine the moment of inertia about the x-axis of the semicircular area.

Solution. The moment of inertia of the semicircular area about the x'-axis is one half of that for a complete circle about the same axis. Thus from the results of Sample Problem A/3

$$I_{x'} = \frac{1}{2}\frac{\pi r^4}{4} = \frac{20^4\pi}{8} = 2\pi(10^4) \text{ mm}^4$$

We obtain the moment of inertia $\bar{I}$ about the parallel centroidal axis x_o next. Transfer is made through the distance $\bar{r} = 4r/3\pi = (4)(20)/3\pi = 80/3\pi$ mm by the parallel-axis theorem. Hence

$$[\bar{I} = I - Ad^2] \quad \bar{I} = 2(10^4)\pi - \left(\frac{20^2\pi}{2}\right)\left(\frac{80}{3\pi}\right)^2 = 1.755(10^4) \text{ mm}^4$$

Finally, we transfer from the centroidal x_o-axis to the x-axis, which gives

$$[I = \bar{I} + Ad^2] \quad I_x = 1.755(10^4) + \left(\frac{20^2\pi}{2}\right)\left(30 + \frac{80}{3\pi}\right)^2$$

$$= 1.755(10^4) + 93.08(10^4) = 94.8(10^4) \text{ mm}^4 \qquad Ans.$$

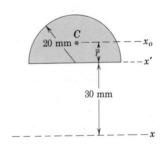

① This problem illustrates the caution we should observe in using a double transfer of axes since neither the x'- nor the x-axis passes through the centroid C of the area. If the circle were complete with the centroid on the x'-axis, only one transfer would be needed.

Sample Problem   A/6

Calculate the moment of inertia about the x-axis of the area enclosed between the y-axis and the circular arcs of radius a whose centers are at O and A.

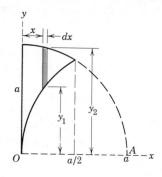

Solution. The choice of a vertical differential strip of area permits one integration to cover the entire area. A horizontal strip would require two integrations with respect to y by virtue of the discontinuity. The moment of inertia of the strip about the x-axis is that of a strip of height y_2 minus that of a strip of height y_1. Thus, from the results of Sample Problem A1 we write

$$dI_x = \tfrac{1}{3}(y_2\,dx)y_2{}^2 - \tfrac{1}{3}(y_1\,dx)y_1{}^2 = \tfrac{1}{3}(y_2{}^3 - y_1{}^3)\,dx$$

The values of y_2 and y_1 are obtained from the equations of the two curves, which are $x^2 + y_2{}^2 = a^2$ and $(x - a)^2 + y_1{}^2 = a^2$ and which give ① $y_2 = \sqrt{a^2 - x^2}$ and $y_1 = \sqrt{a^2 - (x-a)^2}$. Thus

① We choose the positive signs for the radicals here since both y_1 and y_2 lie above the x-axis.

$$I_x = \tfrac{1}{3}\int_0^{a/2} \{(a^2 - x^2)\sqrt{a^2 - x^2} - [a^2 - (x-a)^2]\sqrt{a^2 - (x-a)^2}\}\,dx$$

Simultaneous solution of the two equations which define the two circles gives the x-coordinate of the intersection of the two curves which, by inspection, is $a/2$. Evaluation of the integrals gives

$$\int_0^{a/2} a^2\sqrt{a^2 - x^2}\,dx = \frac{a^4}{4}\left(\frac{\sqrt{3}}{2} + \frac{\pi}{3}\right)$$

$$-\int_0^{a/2} x^2\sqrt{a^2 - x^2}\,dx = \frac{a^4}{16}\left(\frac{\sqrt{3}}{4} - \frac{\pi}{3}\right)$$

$$-\int_0^{a/2} a^2\sqrt{a^2 - (x-a)^2}\,dx = \frac{a^4}{4}\left(\frac{\sqrt{3}}{2} - \frac{2\pi}{3}\right)$$

$$\int_0^{a/2} (x-a)^2\sqrt{a^2 - (x-a)^2}\,dx = \frac{a^4}{8}\left(\frac{\sqrt{3}}{8} + \frac{\pi}{3}\right)$$

Collection of the integrals with the factor of $\tfrac{1}{3}$ gives

$$I_x = \frac{a^4}{96}(9\sqrt{3} - 2\pi) = 0.0969a^4 \qquad\qquad \textit{Ans.}$$

If we had started from a second-order element $dA = dx\,dy$, we would write $y^2\,dx\,dy$ for the moment of inertia of the element about the x-axis. Integrating from y_1 to y_2 holding x constant produces for the vertical strip

$$dI_x = \left[\int_{y_1}^{y_2} y^2\,dy\right]dx = \tfrac{1}{3}(y_2{}^3 - y_1{}^3)\,dx$$

which is the expression we started with by having the moment-of-inertia result for a rectangle in mind.

PROBLEMS

A/1 From the results of Sample Problem A/3 write the expression for the polar moment of inertia of the area of the circular sector about its center O.

Problem A/1

A/2 The narrow rectangular strip has an area of 6 mm², and its moment of inertia about the y-axis is 170 mm⁴. Obtain a close approximation to the radius of gyration about point O.

Ans. $k_O = 7.30$ mm

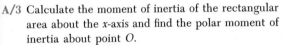

Problem A/2

A/3 Calculate the moment of inertia of the rectangular area about the x-axis and find the polar moment of inertia about point O.

Ans. $I_x = 21.06(10^6)$ mm⁴, $J_O = 35.64(10^6)$ mm⁴

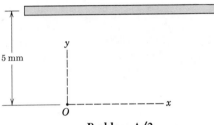

Problem A/3

A/4 From the results of Sample Problem A/1, state without calculation the moment of inertia of the area of the parallelogram about the x-axis through its base and about a parallel axis through its centroid.

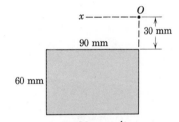

Problem A/4

A/5 The moments of inertia of the area A about the y- and y'-axes differ by 0.032 m⁴. Compute the area A, which has its centroid at C.

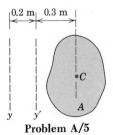

Problem A/5

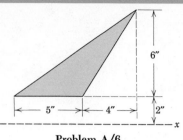

Problem A/6

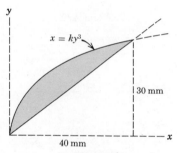

Problem A/7

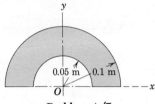

Problem A/8

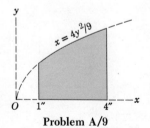

Problem A/9

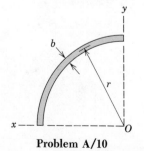

Problem A/10

A/6 Calculate the moment of inertia of the triangular area about the *x*-axis.   *Ans.* $I_x = 270$ in.^{4}

A/7 Obtain the polar moment of inertia of the area of the semicircular ring about point *O* by direct integration and use the result to find the moment of inertia about the *x*-axis.

$$Ans.\ I_x = 0.368(10^{-4})\ m^4$$

A/8 Calculate the moment of inertia of the shaded area about the *x*-axis.

A/9 Calculate the moment of inertia of the shaded area about the *x*-axis.   *Ans.* $I_x = 13.95$ in.^{4}

A/10 The quarter-circular strip of area has a width *b* which is small compared with its radius *r*. Determine the polar moment of inertia of the area about its centroid.

A/11 Determine the polar moments of inertia of the semicircular area about points A and B.

$$\text{Ans. } J_A = \tfrac{3}{4}\pi r^4, \; J_B = r^4\left(\frac{3\pi}{4} - \frac{4}{3}\right)$$

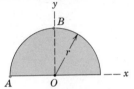

Problem A/11

A/12 Calculate by direct integration the moment of inertia of the shaded area about the x-axis. Solve, first, by using a horizontal strip having differential area and, second, by using a vertical strip of differential area.

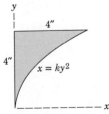

Problem A/12

A/13 Determine the polar radius of gyration of the area of the equilateral triangle of side b about its centroid C.

Problem A/13

A/14 From considerations of symmetry show that $I_{x'} = I_{y'} = I_x = I_y$ for the semicircular area regardless of the angle α.

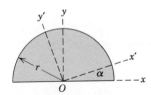

Problem A/14

A/15 Show that the moment of inertia of the area of the square about any axis x' through its center is the same as that about a central axis x parallel to a side.

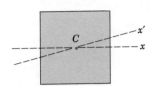

Problem A/15

A/16 Determine the moment of inertia about the x-axis of the area under the sine curve.

$$\text{Ans. } I_x = \frac{4ab^3}{9\pi}$$

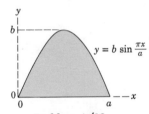

Problem A/16

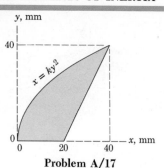

Problem A/17

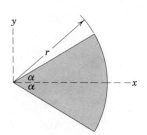

Problem A/19

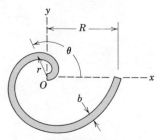

Problem A/20

A/17 Determine the moment of inertia of the shaded area about the *y*-axis.   *Ans.* $I_y = 27.8(10^4)$ mm^4

A/18 The area of a circular ring of inside radius r and outside radius $r + \Delta r$ is approximately equal to the circumference at the mean radius times the thickness Δr. The polar moment of inertia of the ring may be approximated by multiplying this area by the square of the mean radius. What percent error is involved if $\Delta r = r/10$?   *Ans.* Error = 0.226%

A/19 Determine the moments of inertia of the area of the circular sector about the *x*- and *y*-axes.

$$Ans. \ I_x = \frac{r^4}{4}\left(\alpha - \frac{\sin 2\alpha}{2}\right)$$

$$I_y = \frac{r^4}{4}\left(\alpha + \frac{\sin 2\alpha}{2}\right)$$

▶ A/20 A narrow strip of area of constant width b has the form of a spiral $r = k\theta$. After one complete turn from $\theta = 0$ to $\theta = 2\pi$, the end radius of the spiral is R. Determine the polar moment of inertia and the radius of gyration of the area about O.

Ans. $J_O = 1.609R^3b$, $k_O = 0.690R$

A/3 COMPOSITE AREAS. It is frequently necessary to calculate the moment of inertia of an area that is composed of a number of distinct parts of simple and calculable geometric shape. Since a moment of inertia is the integral or sum of the products of distance squared times element of area, it follows that the moment of inertia of a positive area is always a positive quantity. Therefore the moment of inertia of a composite area about a particular axis is simply the sum of the moments of inertia of its component parts about the same axis. It is often convenient to regard a composite area as being composed of positive and negative parts. We may then treat the moment of inertia of a negative area as a negative quantity.

When a composite area is composed of a large number of parts, it is convenient to tabulate the results for each of the parts in terms of its area A, its centroidal moment of inertia $\bar{I}$, the distance d from its centroidal axis to the axis about which the moment of inertia of the entire section is being computed, and the product Ad^2. For any one of the parts the moment of inertia about the desired axis by the transfer-of-axis theorem is $\bar{I} + Ad^2$. Thus for the entire section the desired moment of inertia becomes $I = \Sigma \bar{I} + \Sigma Ad^2$.

For such an area in the x-y plane, for example, and with the notation of Fig. A/4 where I_{x_0} is the same as $\bar{I}_x$ and I_{y_0} is the same as $\bar{I}_y$, the tabulation would include

Part	Area, A	d_x	d_y	$Ad_x^{\,2}$	$Ad_y^{\,2}$	I_{x_0}	I_{y_0}
			Sums	$\Sigma Ad_x^{\,2}$	$\Sigma Ad_y^{\,2}$	ΣI_{x_0}	ΣI_{y_0}

From the sums of the four columns, then, the moments of inertia for the composite area about the x- and y-axes become

$$I_x = \Sigma I_{x_0} + \Sigma Ad_x^{\,2}$$
$$I_y = \Sigma I_{y_0} + \Sigma Ad_y^{\,2}$$

Although we may add the moments of inertia of the individual parts of a composite area about a given axis, we may not add their radii of gyration. The radius of gyration for the composite area about the axis in question is given by $k = \sqrt{I/A}$ where I is the total moment of inertia and A is the total area of the composite figure.

Sample Problem   A/7

Calculate the moment of inertia and radius of gyration about the x-axis for the shaded area shown.

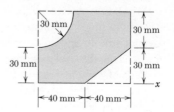

Solution. The composite area is composed of the positive area of the rectangle (1) and the negative areas of the quarter circle (2) and triangle (3). For the rectangle the moment of inertia about the x-axis, from Sample Problem A1 (or Table C3), is

$$I_x = \tfrac{1}{3}Ah^2 = \tfrac{1}{3}(80)(60)(60)^2 = 5.76(10^6) \text{ mm}^4$$

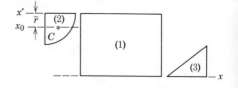

From Sample Problem A3 (or Table C3), the moment of inertia of the negative quarter-circular area about its base axis x' is

$$I_{x'} = -\frac{1}{4}\left(\frac{\pi r^4}{4}\right) = -\frac{\pi}{16}(30)^4 = -0.1590(10^6) \text{ mm}^4$$

We now transfer this result through the distance $\bar{r} = 4r/3\pi = 4(30)/3\pi = 12.73$ mm by the transfer-of-axis-theorem to get the centroidal moment of inertia of part (2) (or use Table C3 directly)

① $[\bar{I} = I - Ad^2]$ $\bar{I}_x = -0.1590(10^6) - \left[-\frac{\pi(30)^2}{4}(12.73)^2\right]$

$$= -0.0445(10^6) \text{ mm}^4$$

① Note that we must transfer the moment of inertia for the quarter-circular area to its centroidal axis x_0 before we can transfer it to the x-axis, as was done in Sample Problem A/5.

② The moment of inertia of the quarter-circular part about the x-axis is now

$[I = \bar{I} + Ad^2]$ $I_x = -0.0445(10^6) + \left[-\frac{\pi(30)^2}{4}\right](60 - 12.73)^2$

$$= -1.624(10^6) \text{ mm}^4$$

Finally, the moment of inertia of the negative triangular area (3) about its base, from Sample Problem A2 (or Table C3), is

② We watch our signs carefully here. Since the area is negative, both $\bar{I}$ and A carry negative signs.

$$I_x = -\tfrac{1}{12}bh^3 = -\tfrac{1}{12}(40)(30)^3 = -0.09(10^6) \text{ mm}^4$$

The total moment of inertia about the x-axis of the composite area is, consequently,

③ $I_x = 5.76(10^6) - 1.624(10^6) - 0.09(10^6) = 4.046(10^6) \text{ mm}^4$ *Ans.*

The net area of the figure is $A = 60(80) - \tfrac{1}{4}\pi(30)^2 - \tfrac{1}{2}(40)(30) = 3493$ mm^2 so that the radius of gyration about the x-axis is

$$k_x = \sqrt{I_x/A} = \sqrt{4.046(10^6)/3493} = 34.0 \text{ mm} \qquad Ans.$$

③ If there had been more than the three parts to the composite area, we would have arranged a tabulation of the $\bar{I}$ terms and the Ad^2 terms so as to keep systematic account of the terms and obtain $I = \Sigma\bar{I} + \Sigma Ad^2$.

PROBLEMS

A/21 Calculate the polar moment of inertia of the shaded area about point A.   *Ans.* $J_A = 138.9(10^4)$ mm^4

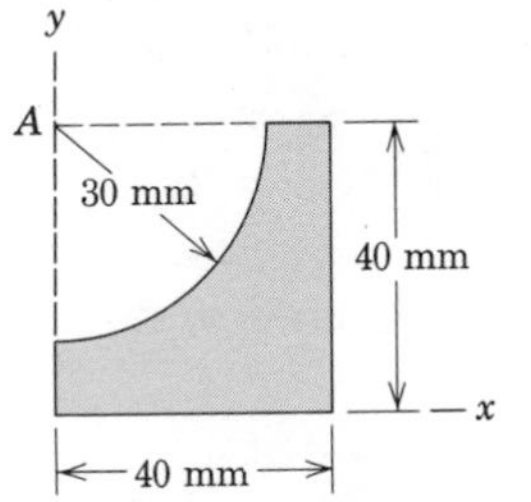

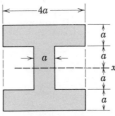

Problem A/21

A/22 Determine the moment of inertia of the shaded area about the x-axis in two different ways.

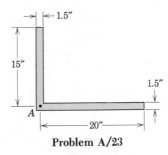

Problem A/22

A/23 Calculate the radius of gyration of the area of the angle section about point A. Note that the width of the legs is small compared to the length of each leg.   *Ans.* $k_A = 10.4$ in.

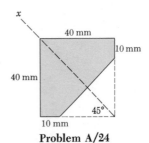

Problem A/23

A/24 Find the moment of inertia of the shaded area about the $45°$ x-axis of symmetry.

Problem A/24

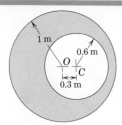

Problem A/25

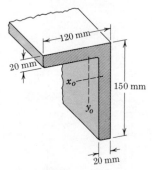

Problem A/26

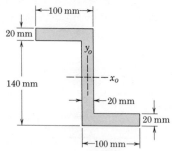

Problem A/27

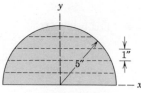

Problem A/28

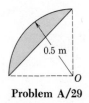

Problem A/29

A/25 Calculate the polar moment of inertia of the shaded area about point O.   *Ans.* $J_O = 1.265 \text{ m}^4$

A/26 Determine the moments of inertia of the Z-section about its centroidal x_o- and y_o-axes.

A/27 Calculate the moment of inertia of the cross section of the beam about its centroidal x_o-axis.
          Ans. $\bar{I}_x = 10.76(10^6) \text{ mm}^4$

A/28 Approximate the moment of inertia about the x-axis of the semicircular area by dividing it into five horizontal strips of equal width. Treat the moment of inertia of each strip as its area (width times length of its horizontal midline) times the square of the distance from its midline to the x-axis. Compare your result with the exact value.

A/29 Compute the polar radius of gyration of the shaded area about point O.   *Ans.* $k_O = 0.445 \text{ m}$

A/30 Derive the expression for the moment of inertia of the trapezoidal area about its base.

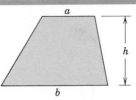

Problem A/30

A/31 Develop a formula for the moment of inertia of the regular hexagonal area of side b about its central x-axis.

$$Ans. \ I_x = \frac{5\sqrt{3}}{16}b^4$$

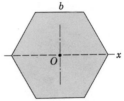

Problem A/31

A/32 Determine the expression for the radius of gyration of the hexagonal area of Prob. A/31 about a polar axis through its center O.

A/33 Calculate the moment of inertia of the shaded area about the x-axis. $Ans. \ I_x = 15.64 \ in.^4$

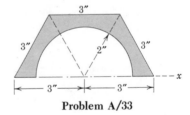

Problem A/33

A/34 Determine the moment of inertia of the rectangular area about its diagonal x-axis.

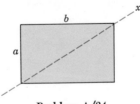

Problem A/34

A/35 Calculate the moment of inertia of the area of Prob. A/21 about the x-axis. $Ans. \ I_x = 28.3(10^4) \ mm^4$

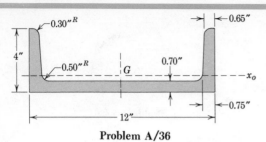

Problem A/36

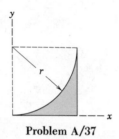

Problem A/37

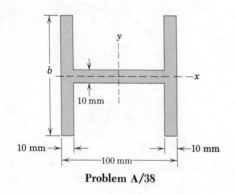

Problem A/38

A/36 Calculate the moment of inertia of the standard 12×4 in. channel section about the centroidal x_o-axis. Neglect the fillets and radii and compare with the handbook value of $I_x = 16.0$ in.[4]

A/37 Determine the moment of inertia of the shaded area about the x-axis.

$$Ans. \ I_x = \left(1 - \frac{5\pi}{16}\right)r^4$$

▶**A/38** For the H-beam section, determine the flange width b that will make the moments of inertia about the central x- and y-axes equal. (*Hint:* The solution of a cubic equation is required here. Refer to Appendix B, section B/4 for solving a cubic equation or approximate the answer with a numerical-graphical solution.)   *Ans.* $b = 161$ mm

A/4 PRODUCTS OF INERTIA AND ROTATION OF AXES.

(a) Definition. In certain problems involving unsymmetrical cross sections and in the calculation of moments of inertia about rotated axes, an expression

$$dI_{xy} = xy \, dA$$

occurs which has the integrated form

$$\boxed{I_{xy} = \int xy \, dA} \qquad (A/7)$$

where x and y are the coordinates of the element of area $dA = dx \, dy$. The quantity I_{xy} is called the *product of inertia* of the area A with respect to the x-y axes. Unlike moments of inertia, which are always positive for positive areas, the product of inertia may be positive, negative, or zero.

The product of inertia is zero whenever either one of the reference axes is an axis of symmetry, such as the x-axis for the area of Fig. A/5. Here we see that the sum of the terms $x(-y) \, dA$ and $x(+y) \, dA$ due to symmetrically placed elements vanishes. Since the entire area may be considered composed of pairs of such elements, it follows that the product of inertia I_{xy} for the entire area is zero.

(b) Transfer of Axes. A transfer-of-axis theorem similar to that for moments of inertia also exists for products of inertia. By definition the product of inertia of the area A in Fig. A/4 with respect to the x- and y-axes in terms of the coordinates x_o, y_o to the centroidal axes is

$$I_{xy} = \int (x_o + d_y)(y_o + d_x) \, dA$$

$$= \int x_o y_o \, dA + d_x \int x_o \, dA + d_y \int y_o \, dA + d_x d_y \int dA$$

The first integral is by definition the product of inertia about the centroidal axes, which we write $\bar{I}_{xy}$. The middle two integrals are both zero since the first moment of the area about its own centroid is necessarily zero. The third integral is merely $d_x d_y A$. Thus the transfer-of-axis theorem for products of inertia becomes

$$\boxed{I_{xy} = \bar{I}_{xy} + d_x d_y A} \qquad (A/8)$$

(c) Rotation of Axes. The product of inertia is useful when we need to calculate the moment of inertia of an area about inclined axes. This consideration leads directly to the important problem of determining the axes about which the moment of inertia is a maximum and a minimum.

In Fig. A/6 the moments of inertia of the area about the x'- and y'-axes are

$$I_{x'} = \int y'^2 \, dA = \int (y \cos \theta - x \sin \theta)^2 \, dA$$

$$I_{y'} = \int x'^2 \, dA = \int (y \sin \theta + x \cos \theta)^2 \, dA$$

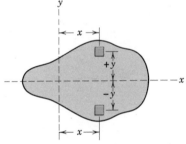

Figure A/5

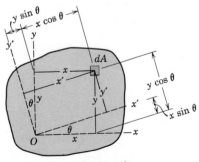

Figure A/6

where x' and y' have been replaced by their equivalent expressions as seen from the geometry of the figure.

Expanding and substituting the trigonometric identities,

$$\sin^2 \theta = \frac{1 - \cos 2\theta}{2} \qquad \cos^2 \theta = \frac{1 + \cos 2\theta}{2}$$

and the defining relations for I_x, I_y, I_{xy} give us

$$
\begin{aligned}
I_{x'} &= \frac{I_x + I_y}{2} + \frac{I_x - I_y}{2} \cos 2\theta - I_{xy} \sin 2\theta \\[2mm]
I_{y'} &= \frac{I_x + I_y}{2} - \frac{I_x - I_y}{2} \cos 2\theta + I_{xy} \sin 2\theta
\end{aligned}
\tag{A/9}
$$

In a similar manner we write the product of inertia about the inclined axes as

$$I_{x'y'} = \int x'y' \, dA = \int (y \sin \theta + x \cos \theta)(y \cos \theta - x \sin \theta) \, dA$$

Expanding and substituting the trigonometric identities

$$\sin \theta \cos \theta = \tfrac{1}{2} \sin 2\theta, \qquad \cos^2 \theta - \sin^2 \theta = \cos 2\theta$$

and the defining relations for I_x, I_y, I_{xy} give us

$$I_{x'y'} = \frac{I_x - I_y}{2} \sin 2\theta + I_{xy} \cos 2\theta \tag{A/9a}$$

Adding Eqs. A/9 gives $I_{x'} + I_{y'} = I_x + I_y = J_z$, the polar moment of inertia about O, which checks the results of Eq. A/3.

The angle which makes $I_{x'}$ and $I_{y'}$ a maximum or a minimum may be determined by setting the derivative of either $I_{x'}$ or $I_{y'}$ with respect to θ equal to zero. Thus

$$\frac{dI_{x'}}{d\theta} = (I_y - I_x) \sin 2\theta - 2I_{xy} \cos 2\theta = 0$$

Denoting this critical angle by α gives

$$\tan 2\alpha = \frac{2I_{xy}}{I_y - I_x} \tag{A/10}$$

Equation A/10 gives two values for 2α which differ by π since $\tan 2\alpha = \tan (2\alpha + \pi)$. Consequently the two solutions for α will differ by $\pi/2$. One value defines the axis of maximum moment of inertia, and the other value defines the axis of minimum moment of inertia. These two rectangular axes are known as the *principal axes of inertia*.

When we substitute Eq. A/10 for the critical value of 2θ in Eq. A/9a, we see that the product of inertia is zero for the principal axes of inertia. Substitution of $\sin 2\alpha$ and $\cos 2\alpha$, obtained from Eq. A/10, for sin 2θ and cos 2θ in Eqs. A/9 gives the magnitudes of the principal moments of inertia as

$$I_{max} = \frac{I_x + I_y}{2} + \frac{1}{2} \sqrt{(I_x - I_y)^2 + 4I_{xy}^2}$$

$$I_{min} = \frac{I_x + I_y}{2} - \frac{1}{2} \sqrt{(I_x - I_y)^2 + 4I_{xy}^2}$$

(A/11)

(d) **Mohr's Circle of Inertia.** We may represent the relations in Eqs. A/9, A/9a, A/10, and A/11 graphically by a diagram known as Mohr's circle. For given values of I_x, I_y, and I_{xy} the corresponding values of $I_{x'}$, $I_{y'}$, and $I_{x'y'}$ may be determined from the diagram for any desired angle θ. A horizontal axis for the measurement of moments of inertia and a vertical axis for the measurement of products of inertia are first selected, Fig. A/7. Next, point A, which has the

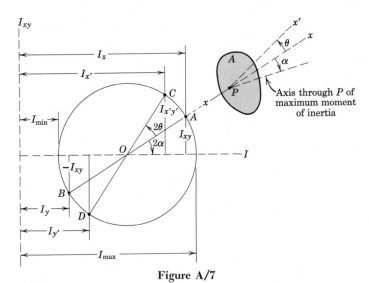

Figure A/7

coordinates (I_x, I_{xy}), and point B, which has the coordinates $(I_y, -I_{xy})$, are located. We now draw a circle with these two points as the extremities of a diameter. The angle from the radius OA to the horizontal axis is 2α or twice the angle from the x-axis of the area in question to the axis of maximum moment of inertia. The angle on the diagram and the angle on the area are both measured in the same sense as shown. The coordinates of any point C are $(I_{x'}, I_{x'y'})$, and those of the corresponding point D are $(I_{y'}, -I_{x'y'})$. Also the angle between OA and OC is 2θ or twice the angle from the x-axis to the x'-axis. Again we measure both angles in the same sense as shown. We may verify from the trigonometry of the circle that Eqs. A/9, A/9a, and A/10 agree with the statements made.

Sample Problem A/8

Write the expression for the product of inertia of the rectangular area with centroid at C with respect to the x-y axes parallel to its sides.

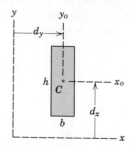

Solution. Since the product of inertia $\bar{I}_{xy}$ about the axes x_o-y_o is zero by symmetry, the transfer-of-axis theorem gives us

$$[I_{xy} = \bar{I}_{xy} + d_x d_y A] \qquad\qquad I_{xy} = d_x d_y bh \qquad\qquad Ans.$$

In this example both d_x and d_y are shown positive. We must be careful to be consistent with the positive directions of d_x and d_y as defined so that their proper signs are observed.

Sample Problem A/9

Determine the product of inertia about the x-y axes for the area under the parabola.

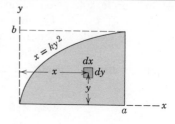

Solution. With the substitution of $x = a$ when $y = b$ the equation of the curve becomes $x = ay^2/b^2$.

Solution I. If we start with the second-order element $dA = dx\, dy$, we have $dI_{xy} = xy\, dx\, dy$. The integral over the entire area then gives us

$$I_{xy} = \int_0^b \int_{ay^2/b^2}^a xy\, dx\, dy = \int_0^b \frac{1}{2}\left(a^2 - \frac{a^2 y^4}{b^4}\right) y\, dy = \tfrac{1}{6}a^2 b^2 \quad Ans.$$

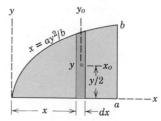

Solution II. Alternatively we can start with a first-order elemental strip and save one integration by using the results of Sample Problem A8. Taking a vertical strip $dA = y\, dx$ gives $dI_{xy} = 0 + (\tfrac{1}{2}y)(x)(y\, dx)$ where the distances to the centroidal axes of the elemental rectangle are $d_x = y/2$ and $d_y = x$. Now we have

$$I_{xy} = \int_0^a \frac{y^2}{2} x\, dx = \int_0^a \frac{xb^2}{2a} x\, dx = \frac{b^2}{6a} x^3 \Big]_0^a = \tfrac{1}{6}a^2 b^2 \qquad Ans.$$

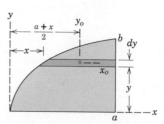

① If we had chosen a horizontal strip, our expression would have become $dI_{xy} = y\tfrac{1}{2}(a + x)[(a - x)\, dy]$ which when integrated, of course, gives us the same result as before.

Sample Problem A/10

Determine the product of inertia of the semicircular area with respect to the x-y axes.

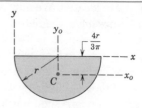

Solution. We may use the transfer-of-axis theorem, Eq. A8, to write

$$[I_{xy} = \bar{I}_{xy} + d_x d_y A] \qquad I_{xy} = 0 + \left(-\frac{4r}{3\pi}\right)(r)\left(\frac{\pi r^2}{2}\right) = -\frac{2r^4}{3} \qquad Ans.$$

where the x- and y-coordinates of the centroid C are $d_y = +r$ and $d_x = -4r/3\pi$. Since one of the centroidal axes is an axis of symmetry, $\bar{I}_{xy} = 0$.

① Proper use of the transfer-of-axis theorem saves a great deal of labor in computing products of inertia.

Sample Problem A/11

Determine the orientation of the principal axes of inertia through the centroid of the angle section and determine the corresponding maximum and minimum moments of inertia.

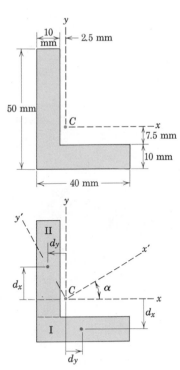

Solution. The location of the centroid C is easily calculated, and its position is shown on the diagram.

Products of inertia. The product of inertia for each rectangle about its own centroidal axes parallel to the x- and y-axes is zero by symmetry. Thus the product of inertia about the x-y axes for part I is

$$[I_{xy} = \bar{I}_{xy} + d_x d_y A] \quad I_{xy} = 0 + (-12.5)(+7.5)(400) = -3.75(10^4) \text{ mm}^4$$

where
$$d_x = -(7.5 + 5) = -12.5 \text{ mm}$$
and
$$d_y = +(20 - 10 - 2.5) = 7.5 \text{ mm}$$

Likewise for part II,

$$[I_{xy} = \bar{I}_{xy} + d_x d_y A] \quad I_{xy} = 0 + (12.5)(-7.5)(400) = -3.75(10^4) \text{ mm}^4$$

where
$$d_x = +(20 - 7.5) = 12.5 \text{ mm}$$
and
$$d_y = -(5 + 2.5) = -7.5 \text{ mm}$$

For the complete angle
$$I_{xy} = -3.75(10^4) - 3.75(10^4) = -7.50(10^4) \text{ mm}^4$$

Moments of inertia. The moments of inertia about the x- and y-axes for part I are

$$[I = \bar{I} + Ad^2] \quad I_x = \tfrac{1}{12}(40)(10)^3 + (400)(12.5)^2 = 6.583(10^4) \text{ mm}^4$$
$$I_y = \tfrac{1}{12}(10)(40)^3 + (400)(7.5)^2 = 7.583(10^4) \text{ mm}^4$$

and the moments of inertia for part II about these same axes are

$$[I = \bar{I} + Ad^2] \quad I_x = \tfrac{1}{12}(10)(40)^3 + (400)(12.5)^2 = 11.583(10^4) \text{ mm}^4$$
$$I_y = \tfrac{1}{12}(40)(10)^3 + (400)(7.5)^2 = 2.583(10^4) \text{ mm}^4$$

Thus for the entire section we have

$$I_x = 6.583(10^4) + 11.583(10^4) = 18.167(10^4) \text{ mm}^4$$
$$I_y = 7.583(10^4) + 2.583(10^4) = 10.167(10^4) \text{ mm}^4$$

Principal axes. The inclination of the principal axes of inertia is given by Eq. A10, so we have

$$\left[\tan 2\alpha = \frac{2I_{xy}}{I_y - I_x}\right] \quad \tan 2\alpha = \frac{-2(7.50)}{10.167 - 18.167} = 1.875$$

$$2\alpha = 61.9° \qquad \alpha = 31.0° \qquad\qquad \textit{Ans.}$$

We now compute the principal moments of inertia from Eqs. A9 using α for θ and get I_{max} from $I_{x'}$ and I_{min} from $I_{y'}$. Thus

$$I_{max} = \left(\frac{18.167 + 10.167}{2} + \frac{18.167 - 10.167}{2}(0.4705) + (7.50)(0.8824)\right)(10^4)$$
$$= 22.67(10^4) \text{ mm}^4 \qquad\qquad \textit{Ans.}$$
$$I_{min} = \left(\frac{18.167 + 10.167}{2} - \frac{18.167 - 10.167}{2}(0.4705) - (7.50)(0.8824)\right)(10^4)$$
$$= 5.67(10^4) \text{ mm}^4 \qquad\qquad \textit{Ans.}$$

Mohr's circle. Alternatively we could use Eqs. A11 to obtain the results for I_{max} and I_{min}, or we could construct the Mohr circle from the calculated values of I_x, I_y, and I_{xy}. These values are spotted on the diagram to locate points A and B, which are the extremities of the diameter of the circle. The angle 2α and I_{max} and I_{min} are obtained from the figure as shown.

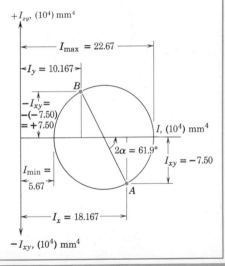

PROBLEMS

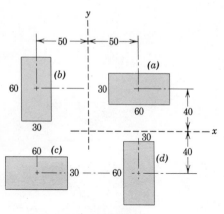

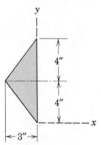

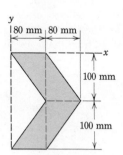

(Dimensions in Millimeters)

Problem A/39

A/39 Determine the product of inertia of each of the four areas about the x-y axes.

Ans. (*a*) and (*c*): $I_{xy} = 360(10^4)$ mm^4
(*b*) and (*d*): $I_{xy} = -360(10^4)$ mm^4

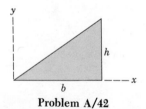

Problem A/40

A/40 Calculate the product of inertia of the triangular area about the x-y axes.

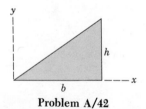

Problem A/41

A/41 Calculate the product of inertia of the shaded area about the x-y axes.   *Ans.* $I_{xy} = -128.0(10^6)$ mm^4

Problem A/42

A/42 Determine the product of inertia of the area of the right triangle about the x-y axes.

A/43 Obtain the product of inertia of the quarter-circular area with respect to the *x*-*y* axes and use this result to obtain the product of inertia with respect to the parallel centroidal axes.

$$\text{Ans. } I_{xy} = r^4/8, \; \bar{I}_{xy} = -0.01647r^4$$

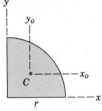

Problem A/43

A/44 Determine the moments and product of inertia of the area of the square with respect to the *x'*-*y'* axes.

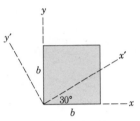

Problem A/44

A/45 An area has moments of inertia $I_x = 28$ in.⁴ and $I_y = 12$ in.⁴ about a set of *x*-*y* axes. The angle measured clockwise from the *x*-axis to the axis of maximum moment of inertia through the origin O is $20°$. Determine the minimum moment of inertia of the area about an axis through O.

$$\text{Ans. } I_{min} = 9.56 \text{ in.}^4$$

A/46 The products of inertia of the shaded area with respect to the *x*-*y* and *x'*-*y'* axes are $8(10^6)$ mm⁴ and $-42(10^6)$ mm⁴, respectively. Compute the area of the figure, whose centroid is C.

$$\text{Ans. } A = 1.316(10^4) \text{ mm}^2$$

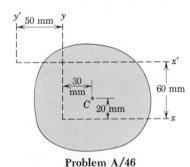

Problem A/46

A/47 Where $I_x = I_y$ for an area which is symmetrical about either the *x*- or the *y*-axis, prove that the moment of inertia is the same for all axes through the origin.

A/48 Determine the proportions of the rectangular area for which the moment of inertia about an *x'*-axis through the center point C of the base is a constant value regardless of $θ$. (See Prob. A/47.)

$$\text{Ans. } a = 2b$$

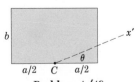

Problem A/48

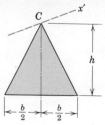

Problem A/49

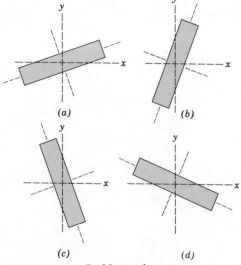

(a)   (b)

(c)   (d)

Problem A/51

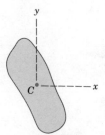

Problem A/52

A/49 Determine the relationship between the base b and altitude h of the isosceles triangle which will make the moments of inertia about all axes through the vertex C the same. (See Prob. A/47.)

A/50 Prove that the magnitude of the product of inertia can be computed from the relation
$$I_{xy} = \sqrt{I_x I_y - I_{max} I_{min}}.$$

A/51 Sketch the Mohr circle of inertia for each of the four rectangular areas with the proportions and positions shown. Indicate on each diagram point A which has coordinates I_x, I_{xy}, and the angle 2α where α is the angle from the x-axis to the axis of maximum moment of inertia.

A/52 The maximum and minimum moments of inertia of the shaded area are $25(10^6)$ mm^4 and $5(10^6)$ mm^4, respectively, about axes passing through the centroid C, and the product of inertia with respect to the x-y axes is $-8(10^6)$ mm^4. From the appropriate equations calculate I_x and the angle α measured counterclockwise from the x-axis to the axis of maximum moment of inertia.

$$\text{Ans. } I_x = 21(10^6) \text{ mm}^4, \ \alpha = 26.6°$$

A/53 Solve Prob. A/52 by constructing the Mohr circle of inertia.

A/54 The moments and product of inertia of an area with respect to x-y axes are $I_x = 14$ in.^{4}, $I_y = 24$ in.^{4}, and $I_{xy} = 12$ in.^{4} Construct the Mohr circle of inertia and use it to determine the principal moments of inertia and the angle α from the x-axis to the axis of maximum moment of inertia.

$$\text{Ans. } I_{max} = 32 \text{ in.}^4, \ I_{min} = 6 \text{ in.}^4,$$
$$\alpha = 56.3° \text{ clockwise}$$

A/55 Determine the maximum and minimum moments of inertia with respect to centroidal axes through C for the composite of the four square areas shown. Find the angle α measured from the x-axis to the axis of maximum moment of inertia.

Ans. $I_{\max} = a^4(\frac{10}{3} + \sqrt{5})$
$I_{\min} = a^4(\frac{10}{3} - \sqrt{5})$
$\alpha = 76.7°$

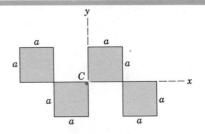

Problem A/55

A/56 Calculate the product of inertia of the rectangular area with respect to the x-y axes.

Ans. $I_{xy} = 1225$ mm^4

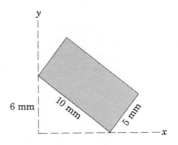

Problem A/56

A/57 Calculate the maximum and minimum moments of inertia of the structural angle about axes through its corner A and find the angle α measured counterclockwise from the x-axis to the axis of maximum inertia. Neglect the small radii and fillet.

Ans. $I_{\max} = 1.782(10^6)$ mm^4
$I_{\min} = 0.684(10^6)$ mm^4
$\alpha = -13.4°$

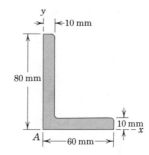

Problem A/57

A/58 Calculate the maximum and minimum moments of inertia about centroidal axes for the structural Z-section. Indicate the angle α measured counterclockwise from the x_o-axis to the axis of maximum moment of inertia.

Ans. $I_{\max} = 1.820(10^6)$ mm^4
$I_{\min} = 0.207(10^6)$ mm^4
$\alpha = 30.1°$

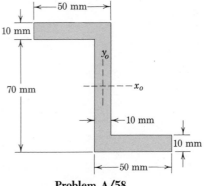

Problem A/58

MASS MOMENTS OF INERTIA

A/1 MASS MOMENTS OF INERTIA ABOUT AN AXIS. The equation of moments about an axis normal to the plane of motion for a rigid body in plane motion contains an integral which depends on the distribution of mass with respect to the moment axis. This integral occurs whenever a rigid body has an angular acceleration about its axis of rotation.

Consider a body of mass m, Fig. A/1, rotating about an axis O-O with an angular acceleration α. All particles of the body move in parallel planes which are normal to the rotation axis O-O. We may choose any one of the planes as the plane of motion, although the one containing the center of mass is usually the one so designated. An element of mass dm has a component of acceleration tangent to its circular path equal to $r\alpha$, and by Newton's second law of motion the resultant tangential force on this element equals $r\alpha\, dm$. The moment of this force about the axis O-O is $r^2\alpha\, dm$, and the sum of the moments of these forces for all elements is $\int r^2\alpha\, dm$. For a rigid body α is the same for all radial lines in the body and we may take it outside the integral sign. The remaining integral is known as the moment of inertia I of the mass m about the axis O-O and is

$$I = \int r^2 \, dm \qquad \text{(A/1)}$$

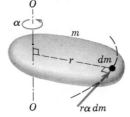

Figure A/1

This integral represents an important property of a body and is involved in the force analysis of any body that has rotational acceleration about a given axis. Just as the mass m of a body is a measure of the resistance to translational acceleration, the moment of inertia is a measure of resistance to rotational acceleration of the body.

The moment-of-inertia integral may be expressed alternatively as

$$I = \Sigma r_i^2 m_i \qquad \text{(A/1a)}$$

where r_i is the radial distance from the inertia axis to the representative particle of mass m_i and where the summation is taken over all particles of the body.

If the density ρ is constant throughout the body, the moment of inertia becomes

$$I = \rho \int r^2 \, dV$$

where dV is the element of volume. In this case the integral by itself defines a purely geometrical property of the body. When the density is not constant but is expressed as a function of the coordinates of the body, it must be left within the integral sign and its effect accounted for in the integration process.

In general the coordinates which best fit the boundaries of the body should be used in the integration. It is particularly important that we make a good choice of the element of volume dV. An element of lowest possible order should be chosen, and the correct expression for the moment of inertia of the element about the axis involved should be used. For example, in finding the moment of inertia of a solid right circular cone about its central axis, we may choose an element in the form of a circular slice of infinitesimal thickness, Fig. A/2a. The differential moment of inertia for this element is the expression for the moment of inertia of a circular cylinder of infinitesimal altitude about its central axis. (This expression will be obtained in Sample Problem A/1.) Alternatively we could choose an element in the form of a cylindrical shell of infinitesimal thickness as shown in Fig. A/2b. Since all of the mass of the element is at the same distance r from the inertia axis, the differential moment of inertia for this element is merely $r^2\,dm$ where dm is the differential mass of the elemental shell.

The dimensions of mass moments of inertia are (mass) (distance)^2 and are expressed in the units $kg \cdot m^2$ in SI units and in lb-ft-sec^2 in U.S. customary units.

(a) *Radius of Gyration.* The radius of gyration k of a mass m about an axis for which the moment of inertia is I is

$$k = \sqrt{\frac{I}{m}} \quad \text{or} \quad I = k^2 m \qquad (A/2)$$

Thus k is a measure of the distribution of mass of a given body about the axis in question, and its definition is analogous to the definition of the radius of gyration for second moments of area. If all the mass m could be concentrated at a distance k from the axis, the correct moment of inertia would be $k^2 m$. The moment of inertia of a body about a particular axis is frequently indicated by specifying the mass of the body and the radius of gyration of the body about the axis. The moment of inertia is then calculated from Eq. A/2.

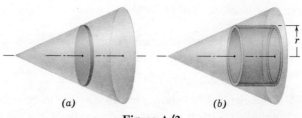

(a)   (b)

Figure A/2

(b) *Transfer of Axes.* If the moment of inertia of a body is known about a centroidal axis, it may be determined easily about any parallel axis. To prove this statement consider the two parallel axes in Fig. A/3, one being a centroidal axis through the mass center G and the other a parallel axis through some other point C. The radial distances from the two axes to any element of mass dm are r_o and r, and the separation of the axes is d. Substituting the law of cosines $r^2 = r_o^2 + d^2 + 2r_o d \cos\theta$ into the definition for the moment of inertia about the noncentroidal axis through C gives

$$I = \int r^2 \, dm = \int (r_o^2 + d^2 + 2r_o d \cos\theta)\, dm$$
$$= \int r_o^2 \, dm + d^2 \int dm + 2d \int u \, dm$$

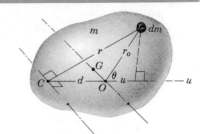

Figure A/3

The first integral is the moment of inertia $\bar{I}$ about the mass-center axis, the second term is md^2, and the third integral equals zero, since the u-coordinate of the mass center with respect to the axis through G is zero. Thus the parallel-axis theorem is

$$\boxed{I = \bar{I} + md^2} \qquad (A/3)$$

It must be remembered that the transfer cannot be made unless one axis passes through the center of mass and unless the axes are parallel. When the expressions for the radii of gyration are substituted in Eq. A/3, there results

$$\boxed{k^2 = \bar{k}^2 + d^2} \qquad (A/3a)$$

which is the parallel-axis theorem for obtaining the radius of gyration k about an axis a distance d from a parallel centroidal axis for which the radius of gyration is $\bar{k}$.

For plane-motion problems where rotation occurs about an axis normal to the plane of motion, a single subscript for I is sufficient to designate the inertia axis. Thus, if the plate of Fig. A/4 has plane motion in the x-y plane, the moment of inertia of the plate about the z-axis through O is designated I_O. For three-dimensional motion, however, where components of rotation may occur about more than one axis, we use a double subscript to preserve notational symmetry with product-of-inertia terms, which are described in Art. A/2. Thus the moments of inertia about the x-, y-, and z-axes are labeled I_{xx}, I_{yy}, and I_{zz}, respectively, and from Fig. A/5 we see that they become

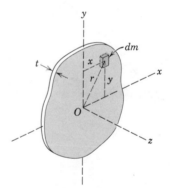

Figure A/4

$$\boxed{\begin{aligned} I_{xx} &= \int r_x^2 \, dm = \int (y^2 + z^2)\, dm \\ I_{yy} &= \int r_y^2 \, dm = \int (z^2 + x^2)\, dm \\ I_{zz} &= \int r_z^2 \, dm = \int (x^2 + y^2)\, dm \end{aligned}} \qquad (A/4)$$

These integrals are cited in Eqs. 8/10 of Art. 8/7 on angular momentum in three-dimensional rotation.

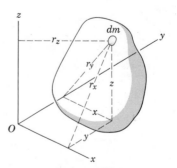

Figure A/5

The similarity between the defining expressions for mass moments of inertia and area moments of inertia is easily observed. An exact relationship between the two moment-of-inertia expressions exists in the case of flat plates. Consider the flat plate of uniform thickness in Fig. A/4. If the constant thickness is t and the density is ρ, the mass moment of inertia I_{zz} of the plate about the z-axis normal to the plate is

$$I_{zz} = \int r^2 \, dm = \rho t \int r^2 \, dA = \rho t J_z \qquad (A/5)$$

Thus the mass moment of inertia about the z-axis equals the mass per unit area, ρt, times the polar moment of inertia J_z of the plate area about the z-axis. If t is small compared with the dimensions of the plate in its plane, the mass moments of inertia I_{xx} and I_{yy} of the plate about the x- and y-axes are closely approximated by

$$\begin{aligned} I_{xx} &= \int y^2 \, dm = \rho t \int y^2 \, dA = \rho t I_x \\ I_{yy} &= \int x^2 \, dm = \rho t \int x^2 \, dA = \rho t I_y \end{aligned} \qquad (A/6)$$

Hence the mass moments of inertia equal the mass per unit area ρt times the corresponding area moments of inertia. The double subscripts for mass moments of inertia distinguish these quantities from area moments of inertia.

Inasmuch as $J_z = I_x + I_y$ for area moments of inertia, we have

$$I_{zz} = I_{xx} + I_{yy} \qquad (A/7)$$

which holds *only* for a thin flat plate. This restriction is observed from Eqs. A6, which do not hold true unless the thickness t or z-coordinate of the element is negligible compared with the distance of the element from the corresponding x- or y-axis. Equation A/7 is very useful when dealing with a differential mass element taken as a flat slice of differential thickness, say dz. In this case Eq. A/7 holds exactly and becomes

$$dI_{zz} = dI_{xx} + dI_{yy} \qquad (A/7a)$$

(c) *Composite Bodies.* The defining integral, Eq. A/1, involves the square of the distance from the axis to the element and so is always positive. Thus, as in the case of area moments of inertia, the mass moment of inertia of a composite body is the sum of the moments of inertia of the individual parts about the same axis. It is often convenient to consider a composite body as defined by positive volumes and negative volumes. The moment of inertia of a negative element, such as the material removed to form a hole, must be considered a negative quantity.

Thorough familiarity with the calculation of mass moments of inertia for rigid bodies is an absolute necessity for the study of the dynamics of rotation.

A summary of some of the more useful formulas for mass moments of inertia of various masses of common shape is given in Table C/4, Appendix C.

Sample Problem   A/1

Determine the moment of inertia and radius of gyration of a homogeneous
right circular cylinder of mass m and radius r about its central axis O–O.

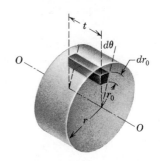

Solution.   An element of mass in cylindrical coordinates is
$dm = \rho \, dV = \rho t r_0 \, dr_0 \, d\theta$ where ρ is the density of the cylinder. The moment of inertia about the axis of the cylinder is

① If we had started with a cylindrical shell of radius r_0 and axial length t as our mass element dm, then $dI = r_0{}^2 \, dm$ directly. The student should evaluate the integral.

$$I = \int r_0{}^2 \, dm = \rho t \int_0^{2\pi} \int_0^r r_0{}^3 \, dr_0 \, d\theta = \rho t \frac{\pi r^4}{2} = \tfrac{1}{2}mr^2 \qquad Ans.$$

The radius of gyration is

$$k = \sqrt{\frac{I}{m}} = \frac{r}{\sqrt{2}} \qquad Ans.$$

② The result $I = \tfrac{1}{2}mr^2$ applies *only* to a solid homogeneous circular cylinder and cannot be used for any other wheel of circular periphery.

Sample Problem   A/2

Determine the moment of inertia and radius of gyration of a homogeneous
solid sphere of mass m and radius r about a diameter.

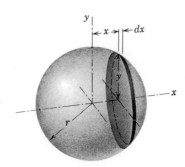

Solution.   A circular slice of radius y and thickness dx is chosen as
the volume element. From the results of Sample Prob. A1 the moment of
inertia about the x-axis of the elemental cylinder is

$$dI_{xx} = \tfrac{1}{2}(dm)y^2 = \tfrac{1}{2}(\pi \rho y^2 \, dx)y^2 = \frac{\pi \rho}{2}(r^2 - x^2)^2 \, dx$$

where ρ is the constant density of the sphere. The total moment of inertia
about the x-axis is

$$I_{xx} = \frac{\pi \rho}{2} \int_{-r}^{r} (r^2 - x^2)^2 \, dx = \tfrac{8}{15}\pi \rho r^5 = \tfrac{2}{5}mr^2 \qquad Ans.$$

The radius of gyration is

$$k = \sqrt{\frac{I}{m}} = \sqrt{\frac{2}{5}}\, r \qquad Ans.$$

① Here is an example where we utilize a previous result to express the moment of inertia of the chosen element, which in this case is a right-circular cylinder of differential axial length dx. It would be foolish to start with a third order element, such as $\rho \, dx \, dy \, dz$, when we can easily solve the problem with a first-order element.

Sample Problem A/3

Determine the moments of inertia of the homogeneous rectangular paral-lelepiped of mass m about the centroidal x_o- and z-axes and about the x-axis through one end.

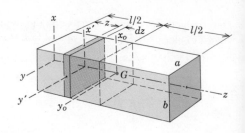

Solution. A transverse slice of thickness dz is selected as the element of volume. The moment of inertia of this slice of infinitesimal thickness equals the moment of inertia of the area of the section times the mass per unit area $\rho \, dz$. Thus the moment of inertia of the transverse slice about the y'-axis is

$$dI_{y'y'} = (\rho \, dz)(\tfrac{1}{12}ab^3)$$

and that about the x'-axis is

$$dI_{x'x'} = (\rho \, dz)(\tfrac{1}{12}a^3b)$$

As long as the element is a plate of differential thickness, the principle given by Eq. A/7a may be applied to give

① Refer to Eqs. A/6 and recall the expression for the area moment of inertia of a rectangle about an axis through its center parallel to its base.

$$dI_{zz} = dI_{x'x'} + dI_{y'y'} = (\rho \, dz)\frac{ab}{12}(a^2 + b^2)$$

These expressions may now be integrated to obtain the desired results.
The moment of inertia about the z-axis is

$$I_{zz} = \int dI_{zz} = \frac{\rho \, ab}{12}(a^2 + b^2)\int_0^l dz = \tfrac{1}{12}m(a^2 + b^2) \qquad Ans.$$

where m is the mass of the block. By interchanging symbols the moment of inertia about the x_o-axis is

$$I_{x_o x_o} = \tfrac{1}{12}m(a^2 + l^2) \qquad Ans.$$

The moment of inertia about the x-axis may be found by the parallel-axis theorem. Eq. A/3. Thus

$$I_{xx} = I_{x_o x_o} + m\left(\frac{l}{2}\right)^2 = \tfrac{1}{12}m(a^2 + 4l^2) \qquad Ans.$$

This last result may be obtained by expressing the moment of inertia of the elemental slice about the x-axis and integrating the expression over the length of the bar. Again by the parallel-axis theorem

$$dI_{xx} = dI_{x'x'} + z^2 \, dm = (\rho \, dz)(\tfrac{1}{12}a^3b) + z^2\rho ab \, dz = \rho ab\left(\frac{a^2}{12} + z^2\right)dz$$

Integrating gives the result obtained previously,

$$I_{xx} = \rho ab \int_0^l \left(\frac{a^2}{12} + z^2\right)dz = \frac{\rho abl}{3}\left(l^2 + \frac{a^2}{4}\right) = \tfrac{1}{12}m(a^2 + 4l^2)$$

The expression for I_{xx} may be simplified for a long prismatical bar or slender rod whose transverse dimensions are small compared with the length. In this case a^2 may be neglected compared with $4l^2$, and the moment of inertia of such a slender bar about an axis through one end normal to the bar becomes $I = \tfrac{1}{3}ml^2$. By the same approximation the moment of inertia about a centroidal axis normal to the bar is $I = \tfrac{1}{12}ml^2$.

PROBLEMS

A/1 What is the maximum diameter d of the circular cross section for the uniform rod of length l for which the approximate formula $\frac{1}{3}ml^2$ may be used for I_{xx} without exceeding a 1-percent error?

Ans. $d_{\max} = 0.232l$

Problem A/1

A/2 The moment of inertia of a solid homogeneous cylinder of radius r about an axis parallel to the central axis of the cylinder may be obtained approximately by multiplying the mass of the cylinder by the square of the distance d between the two axes. What percent error e results if (*a*) $d = 10r$, (*b*) $d = 2r$?

Ans. (a) $e = 0.498$ percent
(b) $e = 11.1$ percent

Problem A/2

A/3 From the results of Sample Prob. A/2 state without computation the moments of inertia of the solid homogeneous hemisphere of mass m about the x- and z-axes.

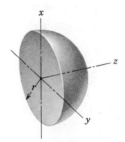

Problem A/3

A/4 State without calculation the moment of inertia about the z-axis of the thin conical shell of mass m and radius r from the results of Sample Problem A/1 applied to a circular disk. Observe the radial distribution of mass by viewing the cone along the z-axis.

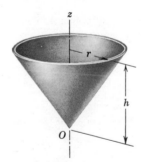

Problem A/4

A/5 The pattern is cut from sheet metal which has a mass of 20.5 kg/m². If the area moments of inertia of its face about the x- and y-axes are $1.40(10^8)$ mm⁴ and $3.05(10^8)$ mm⁴, respectively, determine the mass moment of inertia of the pattern about the z-axis normal to its surface.

Ans. $I_{zz} = 9.12(10^{-3})$ kg $\cdot$ m²

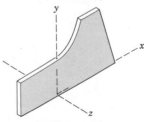

Problem A/5

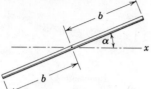

Problem A/6

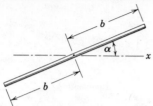

Problem A/7

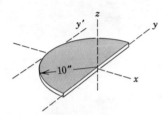

Problem A/8

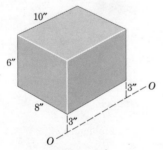

Problem A/9

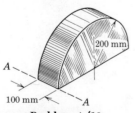

Problem A/10

A/6 Determine the moment of inertia of the inclined uniform slender rod of length $2b$ and mass m about the x-axis through its center.

A/7 The semicircular disk weighs 5 lb, and its small thickness may be neglected compared with its 10-in. radius. Compute the moment of inertia of the disk about the x-, y-, y'-, and z-axes.

$$Ans. \ I_{xx} = I_{yy} = 0.0270 \ \text{lb-ft-sec}^2$$
$$I_{zz} = 0.0539 \ \text{lb-ft-sec}^2$$
$$I_{y'y'} = 0.0433 \ \text{lb-ft-sec}^2$$

A/8 The rectangular metal plate has a mass of 15 kg. Compute its moment of inertia about the y-axis. What is the magnitude of the percentage error e introduced by using the approximate relation $\frac{1}{3}ml^2$ for I_{xx}? *Ans.* $I_{yy} = 0.5 \ \text{kg} \cdot \text{m}^2$, $e = 0.25$ percent

A/9 Calculate the mass moment of inertia about the axis O–O for the uniform 10-in. block of steel with cross-section dimensions of 6 and 8 in.

A/10 The semicircular disk is made of cast iron and has a mass of 45 kg. Calculate its moment of inertia about the axis A–A. *Ans.* $I_{AA} = 2.70 \ \text{kg} \cdot \text{m}^2$

A/11 Calculate the radius of gyration about axis O–O for the steel disk with the hole.

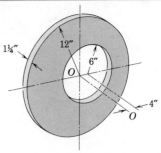

Problem A/11

A/12 A uniform brass rod having a mass of 0.6 kg is bent into the shape shown. Calculate the moment of inertia of the rod about an axis through O normal to the plane of the figure.

$$\text{Ans. } I_{OO} = 11.54(10^{-3}) \text{ kg} \cdot \text{m}^2$$

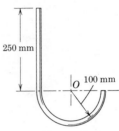

Problem A/12

A/13 The slender metal rods are welded together in the configuration shown. Each 6-in. segment weighs 0.30 lb. Compute the moment of inertia of the assembly about the y-axis.

$$\text{Ans. } I_{yy} = 0.01398 \text{ lb-ft-sec}^2$$

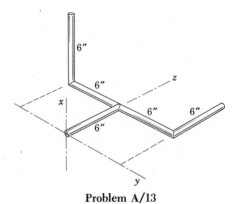

Problem A/13

A/14 Determine the moment of inertia of the half-ring of mass m about its diametral axis a-a and about axis b-b through the midpoint of the arc normal to the plane of the ring. The radius of the circular cross section is small compared with r.

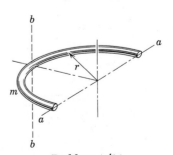

Problem A/14

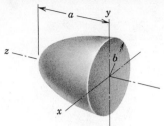

Problem A/15

A/15 Determine the moment of inertia about the z-axis of the homogeneous solid paraboloid of revolution of mass m.   Ans. $I_{zz} = \frac{1}{3}mb^2$

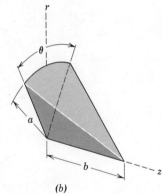

Problem A/16

A/16 Calculate the moment of inertia of the homogeneous right circular cone of mass m, base radius r, and altitude h about the cone axis x and about the y-axis through its vertex.

$$\text{Ans. } I_{xx} = \tfrac{3}{10}mr^2, \ I_{yy} = \tfrac{3}{5}m\left(\frac{r^2}{4} + h^2\right)$$

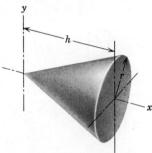

(a)   (b)

Problem A/17

A/17 Without integrating determine from the results of Sample Problem A/2 and Prob. A/16 the moments of inertia about the z-axis for (a) the spherical wedge and (b) the conical wedge. Each wedge has a mass m.   Ans. (a) $I_{zz} = \frac{2}{5}ma^2$, (b) $I_{zz} = \frac{3}{10}mr^2$

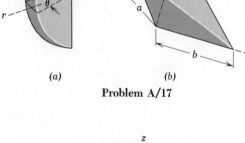

Problem A/18

A/18 A homogeneous solid of mass m is formed by revolving the $45°$ right triangle about the z-axis. Determine the moment of inertia of the solid about the z-axis.

A/19 Determine the moments of inertia of the half spherical shell with respect to the *x*- and *z*-axes. The mass of the shell is *m*, and its thickness is negligible compared with the radius *r*.

$$\text{Ans. } I_{xx} = I_{zz} = \tfrac{2}{3}mr^2$$

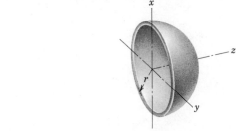

Problem A/19

A/20 In the study of high-speed reentry into the earth's atmosphere small solid cones are fired at high velocities into low-density gas. A condition of critical stability occurs when the moment of inertia of the cone about its axis of generation *a-a* equals that about a transverse axis *b-b* through the mass center. Determine the critical value of the cone angle *α* for this condition. Ans. *α* = 26.6°

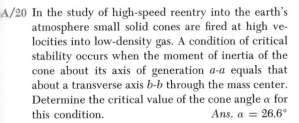

Problem A/20

A/21 Determine by integration the moment of inertia of the thin conical shell of Prob. A/4 about an axis through the vertex *O* normal to the *z*-axis. The mass of the shell is *m* and its altitude is *h*.

$$\text{Ans. } I_{OO} = \frac{m}{4}(r^2 + 2h^2)$$

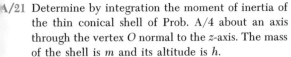

A/22 Determine the moment of inertia of the circular sector of mass *m* and radius *r* about the tangent line *O–O*. The thickness of the sector is small compared with *r*.

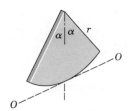

Problem A/22

A/23 Find the moment of inertia of the tetrahedron of mass *m* about the *z*-axis.

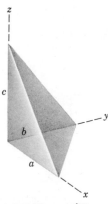

Problem A/23

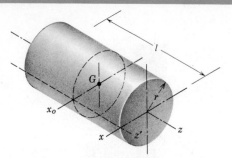

Problem A/24

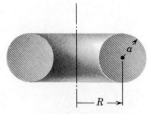

Problem A/25

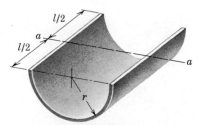

Problem A/26

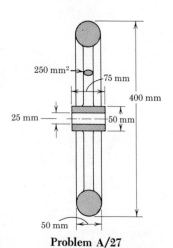

250 mm²

75 mm

400 mm

25 mm

50 mm

50 mm

Problem A/27

A/24 Determine the moments of inertia of the homoge-neous right circular cylinder of mass m about the x_o-, x-, and z'-axes shown.

A/25 Determine by integration the moment of inertia of the half-cylindrical shell of mass m about the axis a-a. The thickness of the shell is small compared with r.

$$Ans.\ I_{aa} = \frac{m}{2}\left(r^2 + \frac{l^2}{6}\right)$$

▶A/26 Determine the moment of inertia about the gener-ating axis of a complete ring of circular section (torus) with the dimensions shown in the sectional view. $Ans.\ I = m(R^2 + \frac{3}{4}a^2)$

▶A/27 Calculate the moment of inertia of the steel hand-wheel about its axis. There are six spokes, each of which has a uniform cross-sectional area of 250 mm². $Ans.\ I = 0.539\ kg \cdot m^2$

▶A/28 A shell of mass m is obtained by revolving the circular section about the z-axis. If the thickness of the shell is small compared with a and if $r = a/3$, determine the radius of gyration of the shell about the z-axis.   Ans. $k_z = 0.890a$

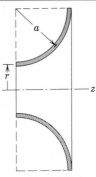

Problem A/28

▶A/29 Compute the moment of inertia of the mallet about the O–O axis. The mass of the head is 0.8 kg, and the mass of the handle is 0.5 kg.   Ans. $I_{OO} = 67.1$ g·m²

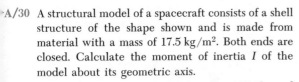

Problem A/29

▶A/30 A structural model of a spacecraft consists of a shell structure of the shape shown and is made from material with a mass of 17.5 kg/m². Both ends are closed. Calculate the moment of inertia I of the model about its geometric axis.
          Ans. $I = 1.594$ kg·m²

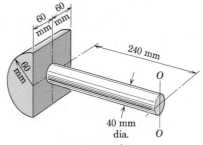

Problem A/30

▶A/31 The cube with semicircular grooves in two opposite faces is cast of lead. Calculate the moment of inertia of the solid about the a-a axis.
          Ans. $I_{aa} = 0.367$ kg·m²

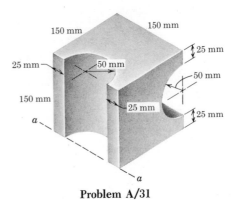

Problem A/31

A/2 PRODUCTS OF INERTIA.

For problems in the rotation of three-dimensional rigid bodies the expression for angular momentum contains, in addition to the moment-of-inertia terms, *product-of-inertia* terms defined as

$$I_{xy} = I_{yx} = \int xy \, dm$$
$$I_{xz} = I_{zx} = \int xz \, dm \qquad (A/8)$$
$$I_{yz} = I_{zy} = \int yz \, dm$$

These expressions were cited in Eqs. 8/10 in the expansion of the expression for angular momentum, Eq. 8/9.

Unlike moments of inertia which are always positive quantities, products of inertia may be positive or negative. The calculation of products of inertia involves the same basic procedure which we have followed in calculating moments of inertia and in evaluating other volume integrals insofar as the choice of element and the limits of integration are concerned. The only special precaution we need to observe is to be doubly watchful of the algebraic signs in the expressions. The units of products of inertia are the same as those of moments of inertia.

We have seen that the calculation of moments of inertia is often simplified by using the transfer-of-axis theorem. A similar theorem exists for transferring products of inertia, and we prove it easily as follows. In Fig. A/6 is shown the x-y view of a rigid body with parallel axes x_0-y_0 passing through the mass center G and located from the x-y axes by the distances d_x and d_y. The product of inertia about the x-y axes by definition is

$$I_{xy} = \int xy \, dm = \int (x_0 + d_x)(y_0 + d_y) \, dm$$
$$= \int x_0 y_0 dm + d_x d_y \int dm + d_x \int y_0 \, dm + d_y \int x_0 \, dm$$
$$= I_{x_0 y_0} + m d_x d_y$$

The last two integrals vanish since the first moments of mass about the mass center are necessarily zero. Similar relations exist for the remaining two product-of-inertia terms. If we drop the zero subscripts and use the bar to designate the mass-center quantity, we have

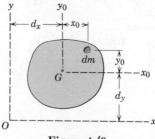

Figure A/6

$$\begin{aligned}
I_{xy} &= \bar{I}_{xy} + m d_x d_y \\
I_{xz} &= \bar{I}_{xz} + m d_x d_z \\
I_{yz} &= \bar{I}_{yz} + m d_y d_z
\end{aligned} \qquad \text{(A/9)}$$

These transfer-of-axis relations are valid *only* for transfer to or from *parallel axes* through the *mass center*.

With the aid of the product-of-inertia terms we can calculate the moment of inertia of a rigid body about any prescribed axis through the origin of coordinates. For the rigid body of Fig. A/7 let it be required to determine the moment of inertia about axis *OM*. The direction cosines of *OM* are *l*, *m*, *n*, and a unit vector $\boldsymbol{\lambda}$ along *OM* may be written $\boldsymbol{\lambda} = \mathbf{i}l + \mathbf{j}m + \mathbf{k}n$. The moment of inertia about *OM* is

$$I_M = \int h^2 \, dm = \int (\mathbf{r} \times \boldsymbol{\lambda}) \cdot (\mathbf{r} \times \boldsymbol{\lambda}) \, dm$$

where $|\mathbf{r} \times \boldsymbol{\lambda}| = r \sin \theta = h$. The cross product is

$$(\mathbf{r} \times \boldsymbol{\lambda}) = \mathbf{i}(yn - zm) + \mathbf{j}(zl - xn) + \mathbf{k}(xm - yl)$$

and, after collecting terms, the dot-product expansion gives

$$(\mathbf{r} \times \boldsymbol{\lambda}) \cdot (\mathbf{r} \times \boldsymbol{\lambda}) = h^2 = (y^2 + z^2)l^2 + (x^2 + z^2)m^2 + (x^2 + y^2)n^2$$
$$- 2xylm - 2xzln - 2yzmn$$

Thus, with the substitution of the expressions of Eqs. A/4 and A/8, we have

$$I_M = I_{xx}l^2 + I_{yy}m^2 + I_{zz}n^2 - 2I_{xy}lm - 2I_{xz}ln - 2I_{yz}mn \qquad \text{(A/10)}$$

This expression gives the moment of inertia about any axis *OM* in terms of the direction cosines of the axis and the moments and products of inertia about the coordinate directions.

(*a*) *Principal Axes of Inertia.* Expansion of the angular momentum expression, Eq. 8/11, for a rigid body with attached axes yields the array

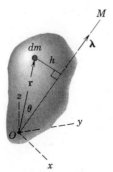

Figure A/7

$$\begin{bmatrix} I_{xx} & -I_{xy} & -I_{xz} \\ -I_{yx} & I_{yy} & -I_{yz} \\ -I_{zx} & -I_{zy} & I_{zz} \end{bmatrix}$$

which is known as the *inertia matrix* or *inertia tensor*. If we examine the moment- and product-of-inertia terms for all possible orientations of the axes with respect to the body for a given origin, we will find in the general case one unique orientation *x-y-z* for which the product-of-inertia terms vanish and the array takes the diagonalized form

$$\begin{bmatrix} I_{xx} & 0 & 0 \\ 0 & I_{yy} & 0 \\ 0 & 0 & I_{zz} \end{bmatrix}$$

Axes *x-y-z* are called the *principal axes of inertia*, and I_{xx}, I_{yy}, and I_{zz} are called the *principal moments of inertia* and represent the maximum, minimum, and an intermediate value for the moments of inertia for the particular origin chosen.

It may be shown° that for any given orientation of axes *x-y-z* the solution of the determinant equation

$$\begin{vmatrix} I_{xx} - I & -I_{xy} & -I_{xz} \\ -I_{yx} & I_{yy} - I & -I_{yz} \\ -I_{zx} & -I_{zy} & I_{zz} - I \end{vmatrix} = 0 \qquad (A/11)$$

for *I* yields three roots I_1, I_2, and I_3 of the resulting cubic equation which are the three principal moments of inertia. Also, the direction cosines *l*, *m*, and *n* of a principal inertia axis are given by

$$(I_{xx} - I)l - I_{xy}m - I_{xz}n = 0$$

$$-I_{yx}l + (I_{yy} - I)m - I_{yz}n = 0 \qquad (A/12)$$

$$-I_{zx}l - I_{zy}m + (I_{zz} - I)n = 0$$

These equations along with $l^2 + m^2 + n^2 = 1$ will enable a solution for the direction cosines to be made for each of the three *I*'s separately.

° See, for example, the author's *Dynamics, 2nd Edition,* or *SI Version*, Art. 41.

Sample Problem A/4

The bent plate has a uniform thickness t which is negligible compared with its other dimensions. The density of the plate material is ρ. Determine the products of inertia of the plate with respect to the axes as chosen.

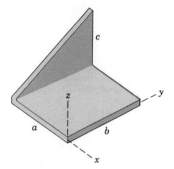

Solution. Each of the two parts is analyzed separately.

① **Rectangular part.** In the separate view of this part we introduce parallel axes x_0-y_0 through the mass center G and use the transfer-of-axis theorem. By symmetry we see that $\bar{I}_{xy} = I_{x_0 y_0} = 0$ so that

$$[I_{xy} = \bar{I}_{xy} + m d_x d_y] \qquad I_{xy} = 0 + \rho tab\left(-\frac{a}{2}\right)\left(\frac{b}{2}\right) = -\frac{1}{4}\rho t a^2 b^2$$

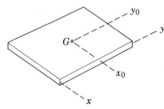

Since the z-coordinate of all elements of the plate is zero, it follows that $I_{xz} = I_{yz} = 0$.

Triangular part. In the separate view of this part we locate the mass center G and construct x_0-, y_0-, and z_0-axes through G. Since the x_0-coordinate of all elements is zero, it follows that $\bar{I}_{xy} = I_{x_0 y_0} = 0$ and $\bar{I}_{xz} = I_{x_0 z_0} = 0$. The transfer-of-axis theorems then give us

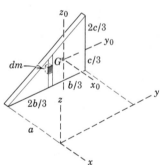

$$[I_{xy} = \bar{I}_{xy} + m d_x d_y] \qquad I_{xy} = 0 + \rho t \frac{b}{2} c(-a)\left(\frac{2b}{3}\right) = -\frac{1}{3}\rho t a b^2 c$$

$$[I_{xz} = \bar{I}_{xz} + m d_x d_z] \qquad I_{xz} = 0 + \rho t \frac{b}{2} c(-a)\left(\frac{c}{3}\right) = -\frac{1}{6}\rho t a b c^2$$

We obtain I_{yz} by direct integration, noting that the distance a of the plane of the triangle from the y-z plane in no way affects the y- and z-coordinates. With the mass element $dm = \rho t \, dy \, dz$ we have

② $$\left[I_{yz} = \int yz \, dm\right] \qquad I_{yz} = \rho t \int_0^b \int_0^{cy/b} yz \, dz \, dy = \rho t \int_0^b y\left[\frac{z^2}{2}\right]_0^{cy/b} dy$$

$$= \frac{\rho t c^2}{2b^2}\int_0^b y^3 \, dy = \frac{1}{8}\rho t b^2 c^2$$

Adding the expressions for the two parts gives

$$I_{xy} = -\frac{1}{4}\rho t a^2 b^2 - \frac{1}{3}\rho t a b^2 c = -\frac{1}{12}\rho t a b^2 (3a + 4c) \qquad \textit{Ans.}$$

$$I_{xz} = \quad -\frac{1}{6}\rho t a b c^2 = -\frac{1}{6}\rho t a b c^2 \qquad \textit{Ans.}$$

$$I_{yz} = \quad +\frac{1}{8}\rho t b^2 c^2 = +\frac{1}{8}\rho t b^2 c^2 \qquad \textit{Ans.}$$

① We must be careful to preserve the same sense of the coordinates. Thus plus x_0 and y_0 must agree with plus x and y.

② We choose to integrate with respect to z first, where the upper limit is the variable height $z = cy/b$. If we were to integrate first with respect to y, the limits of the first integral would be from the variable $y = bz/c$ to b.

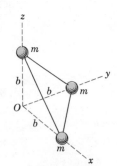

Problem A/32

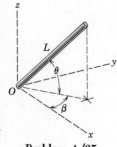

Problem A/35

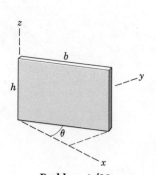

Problem A/33

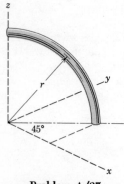

Problem A/37

PROBLEMS

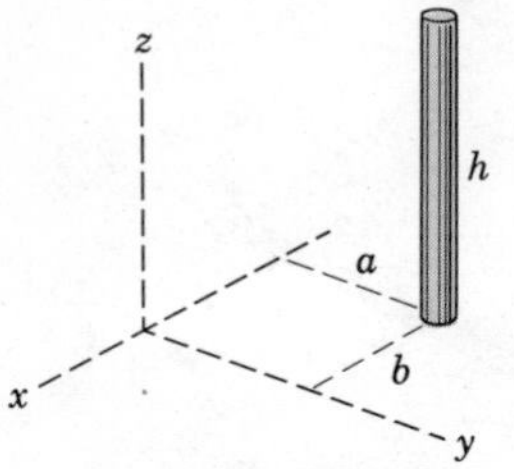

A/32 The slender rod of mass m is parallel to the z-axis. Determine the products of inertia of the rod with respect to the axes given.

Ans. $I_{xy} = -mab$, $I_{yz} = \frac{1}{2}mah$, $I_{xz} = -\frac{1}{2}mbh$

A/33 Prove that the moment of inertia of the rigid assembly of three identical balls each of mass m and radius r has the same value for all axes through O.

A/34 Prove that the moments of inertia about all axes passing through the center of a homogeneous cube are identical.

A/35 Determine the product of inertia I_{xz} of the uniform slender bar of mass m and length L for the orientation given.

Ans. $I_{xz} = \frac{1}{6}mL^2 \sin 2\theta \cos \beta$

A/36 Determine the three products of inertia with respect to the given axes for the uniform rectangular plate of mass m.

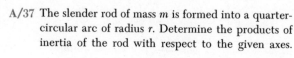

Problem A/36

A/37 The slender rod of mass m is formed into a quarter-circular arc of radius r. Determine the products of inertia of the rod with respect to the given axes.

Ans. $I_{xy} = \frac{1}{4}mr^2$, $I_{yz} = I_{xz} = \frac{1}{\pi\sqrt{2}}mr^2$

A/38 Determine the three products of inertia of the solid homogeneous half-cylinder of mass m for the axes shown.

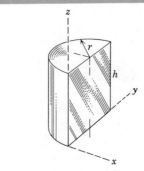

Problem A/38

A/39 Determine the moment of inertia I_M of the cube of mass m about its diagonal axis OM.

$$\text{Ans. } I_M = \tfrac{1}{6}ma^2$$

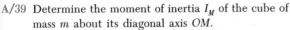

A/40 The bent rod has a mass m. Determine its moment of inertia about the diagonal axis OM.

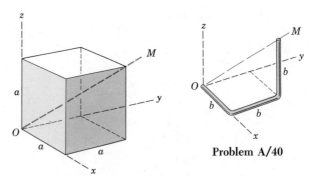

Problem A/40

Problem A/39

A/41 The rectangular block with the central hole is machined from steel. Compute the products of inertia of the block with respect to the axes shown.

$$\text{Ans. } I_{xy} = -0.0756 \text{ kg} \cdot \text{m}^2$$
$$I_{yz} = 0.0189 \text{ kg} \cdot \text{m}^2$$
$$I_{xz} = -0.0236 \text{ kg} \cdot \text{m}^2$$

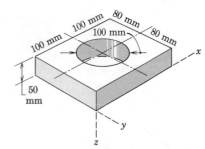

Problem A/41

A/42 The bracket is made from flat plate with a mass of 40 kg per square meter of plate area. Calculate the products of inertia of the bracket with respect to the given axes.

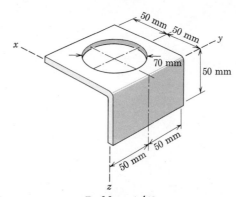

Problem A/42

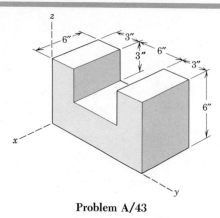

Problem A/43

A/43 The cast-iron block is machined to the dimensions shown. Compute the products of inertia of the block with respect to the coordinates shown.

$$\text{Ans. } I_{xy} = -0.328 \text{ lb-ft-sec}^2$$
$$I_{yz} = 0.273 \text{ lb-ft-sec}^2$$
$$I_{xz} = -0.136 \text{ lb-ft-sec}^2$$

Problem A/44

A/44 The steel plate with two right-angle bends has a thickness of approximately 15 mm and a total mass of exactly 33 kg. Calculate its moment of inertia about the diagonal axis through the corners A and B.

$$\text{Ans. } I = 0.670 \text{ kg} \cdot \text{m}^2$$

▶A/45 Each of the metal spheres has a mass m and a diameter which is small compared with the dimension b. Compute the values of the principal moments of inertia and determine the direction cosines for the axis of maximum moment of inertia.

$$\text{Ans. } I_1 = 5.532mb^2 \quad l_1 = -0.293$$
$$I_2 = 4.347mb^2 \quad m_1 = 0.844$$
$$I_3 = 2.121mb^2 \quad n_1 = -0.449$$

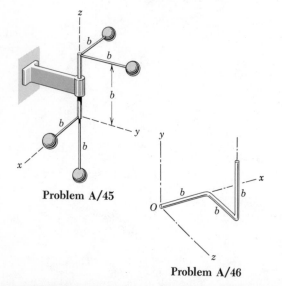

Problem A/45

Problem A/46

▶A/46 Determine the principal moments of inertia for the bent rod of total mass m.

$$\text{Ans. } I_1 = 0.145mb^2 \text{ minimum}$$
$$I_2 = 1.299mb^2 \text{ maximum}$$
$$I_3 = 1.223mb^2 \text{ intermediate}$$

▶A/47 Use the results cited for the principal moments of inertia of the bent rod in Prob. A/46 and specify the direction cosines of the axis of the intermediate value of principal moment of inertia.

SELECTED TOPICS OF MATHEMATICS

B

B/1 INTRODUCTION.

Appendix B contains an abbreviated summary and reminder of selected topics in basic mathematics which find frequent use in mechanics. The relationships are cited without proof. The student of mechanics will have frequent occasion to use many of these relations, and he will be handicapped if they are not well in hand. Other topics not listed will also be needed from time to time.

As the reader reviews and applies his mathematics, he should bear in mind that mechanics is an applied science descriptive of real bodies and actual motions. Therefore the geometric and physical interpretation of the applicable mathematics should be kept clearly in mind during the development of theory and the formulation and solution of problems.

B/2 PLANE GEOMETRY

1. When two intersecting lines are, respectively, perpendicular to two other lines, the angles formed by each pair are equal.

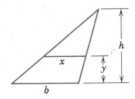

$\theta_1 = \theta_2$

2. Similar triangles

$$\frac{x}{b} = \frac{h - y}{h}$$

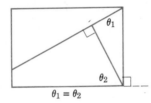

3. Any triangle

Area $= \frac{1}{2}bh$

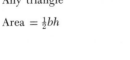

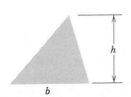

4. Circle

Circumference $= 2\pi r$
Area $= \pi r^2$
Arc length $s = r\theta$
Sector area $= \frac{1}{2}r^2\theta$

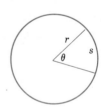

5. Every triangle inscribed within a semicircle is a right triangle.

$\theta_1 + \theta_2 = \pi/2$

6. Angles of a triangle

$\theta_1 + \theta_2 + \theta_3 = 180°$
$\theta_4 = \theta_1 + \theta_2$

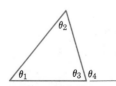

B/3 SOLID GEOMETRY

1. Sphere

   Volume $= \frac{4}{3}\pi r^3$

   Surface Area $= 4\pi r^2$

2. Spherical wedge

   Volume $= \frac{2}{3}r^3\theta$

3. Right-circular cone

   Volume $= \frac{1}{3}\pi r^2 h$

   Lateral area $= \pi r L$

   $L = \sqrt{r^2 + h^2}$

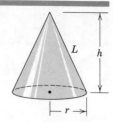

4. Any pyramid or cone

   Volume $= \frac{1}{3}Bh$

   where B = area of base

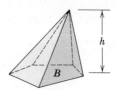

B/4 ALGEBRA

1. Quadratic equation

   $ax^2 + bx + c = 0$

   $x = \dfrac{-b \pm \sqrt{b^2 - 4ac}}{2a}$, $b^2 > 4ac$ for real roots

2. Logarithms

   $b^x = y$, $x = \log_b y$

   Natural logarithms

   $b = e = 2.718\ 282$

   $e^x = y$, $x = \log_e y = \ln y$

   $\log(ab) = \log a + \log b$

   $\log(a/b) = \log a - \log b$

   $\log(1/n) \doteq -\log n$

   $\log a^n = n \log a$

   $\log 1 = 0$

   $\log_{10} x = 0.4343 \ln x$

3. Determinants
   2nd order

   $\begin{vmatrix} a_1 & b_1 \\ a_2 & b_2 \end{vmatrix} = a_1 b_2 - a_2 b_1$

   3rd order

   $\begin{vmatrix} a_1 & b_1 & c_1 \\ a_2 & b_2 & c_2 \\ a_3 & b_3 & c_3 \end{vmatrix} = \begin{array}{l} +a_1 b_2 c_3 + a_2 b_3 c_1 + a_3 b_1 c_2 \\ -a_3 b_2 c_1 - a_2 b_1 c_3 - a_1 b_3 c_2 \end{array}$

4. Cubic equation

   $x^3 = Ax + B$

   Let $p = A/3$, $q = B/2$

   Case I: $q^2 - p^3$ negative (three roots real and distinct)

   $$\cos u = q/(p\sqrt{p}),\ 0 < u < 180°$$

   $$x_1 = 2\sqrt{p} \cos(u/3)$$

   $$x_2 = 2\sqrt{p} \cos(u/3 + 120°)$$

   $$x_3 = 2\sqrt{p} \cos(u/3 + 240°)$$

   Case II: $q^2 - p^3$ positive (one root real, two roots imaginary).

   $$x_1 = (q + \sqrt{q^2 - p^3})^{1/3} + (q - \sqrt{q^2 - p^3})^{1/3}$$

   Case III: $q^2 - p^3 = 0$ (three roots real, two roots equal)

   $$x_1 = 2q^{1/3},\ x_2 = x_3 = -q^{1/3}$$

   For general cubic equation

   $$x^3 + ax^2 + bx + c = 0$$

   Substitute $x = x_0 - a/3$ and get $x_0^3 = Ax_0 + B$. Then proceed as above to find values of x_0 from which $x = x_0 - a/3$.

B/5 ANALYTIC GEOMETRY

1. Straight line

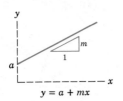

$$y = a + mx$$

$$\frac{x}{a} + \frac{y}{b} = 1$$

2. Circle

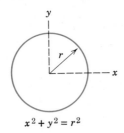

$$x^2 + y^2 = r^2$$

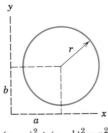

$$(x - a)^2 + (y - b)^2 = r^2$$

3. Parabola

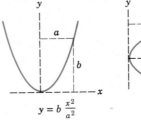

$$y = b\,\frac{x^2}{a^2}$$

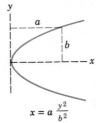

$$x = a\,\frac{y^2}{b^2}$$

4. Ellipse

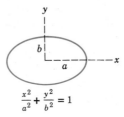

$$\frac{x^2}{a^2} + \frac{y^2}{b^2} = 1$$

5. Hyperbola

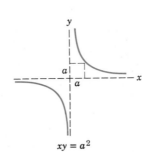

$$xy = a^2$$

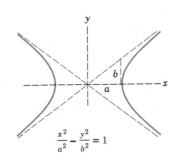

$$\frac{x^2}{a^2} - \frac{y^2}{b^2} = 1$$

B/6 TRIGONOMETRY

1. Definitions

$$\sin\theta = a/c \qquad \csc\theta = c/a$$
$$\cos\theta = b/c \qquad \sec\theta = c/b$$
$$\tan\theta = a/b \qquad \operatorname{ctn}\theta = b/a$$

2. Signs in the four quadrants

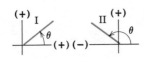
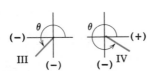

	I	II	III	IV
$\sin\theta$	$+$	$+$	$-$	$-$
$\cos\theta$	$+$	$-$	$-$	$+$
$\tan\theta$	$+$	$-$	$+$	$-$
$\csc\theta$	$+$	$+$	$-$	$-$
$\sec\theta$	$+$	$-$	$-$	$+$
$\operatorname{ctn}\theta$	$+$	$-$	$+$	$-$

3. Miscellaneous relations

$$\sin^2 \theta + \cos^2 \theta = 1$$
$$1 + \tan^2 \theta = \sec^2 \theta$$
$$1 + \mathrm{ctn}^2 \theta = \csc^2 \theta$$

$$\sin \frac{\theta}{2} = \sqrt{\tfrac{1}{2}(1 - \cos \theta)}$$

$$\cos \frac{\theta}{2} = \sqrt{\tfrac{1}{2}(1 + \cos \theta)}$$

$$\sin 2\theta = 2 \sin \theta \cos \theta$$
$$\cos 2\theta = \cos^2 \theta - \sin^2 \theta$$
$$\sin (a \pm b) = \sin a \cos b \pm \cos a \sin b$$
$$\cos (a \pm b) = \cos a \cos b \mp \sin a \sin b$$

4. Law of sines

$$\frac{a}{b} = \frac{\sin A}{\sin B}$$

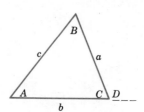

5. Law of cosines

$$c^2 = a^2 + b^2 - 2ab \cos C$$
$$c^2 = a^2 + b^2 + 2ab \cos D$$

B/7 VECTOR OPERATIONS

1. *Notation.* Vector quantities are printed in boldface type, and scalar quantities appear in lightface italic type. Thus, the vector quantity **V** has a scalar magnitude V. In longhand work vector quantities should always be consistently indicated by a symbol such as $\underline{V}$ or $\vec{V}$ to distinguish them from scalar quantities.

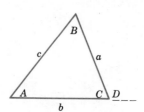

2. *Addition*

   Triangle addition $\mathbf{P} + \mathbf{Q} = \mathbf{R}$

   Parallelogram addition $\mathbf{P} + \mathbf{Q} = \mathbf{R}$

   Commutative law $\mathbf{P} + \mathbf{Q} = \mathbf{Q} + \mathbf{P}$

   Associative law $\mathbf{P} + (\mathbf{Q} + \mathbf{R}) = (\mathbf{P} + \mathbf{Q}) + \mathbf{R}$

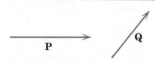

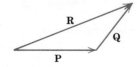

3. *Subtraction*

$$\mathbf{P} - \mathbf{Q} = \mathbf{P} + (-\mathbf{Q})$$

4. *Unit vectors* **i, j, k**

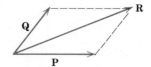

$$\mathbf{V} = \mathbf{i} V_x + \mathbf{j} V_y + \mathbf{k} V_z$$

where
$$|\mathbf{V}| = V = \sqrt{V_x{}^2 + V_y{}^2 + V_z{}^2}$$

5. *Direction cosines* l, m, n are the cosines of the angles between **V** and the x-, y-, z-axes. Thus

$$l = V_x/V \qquad m = V_y/V \qquad n = V_z/V$$

so that
$$\mathbf{V} = V(\mathbf{i}l + \mathbf{j}m + \mathbf{k}n)$$

and
$$l^2 + m^2 + n^2 = 1$$

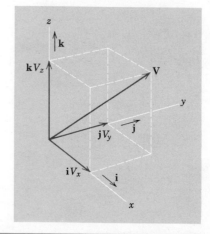

6. *Dot or scalar product*

$$\mathbf{P} \cdot \mathbf{Q} = PQ \cos \theta$$

This product may be viewed as the magnitude of **P** multiplied by the component $Q \cos \theta$ of **Q** in the direction of **P**, or as the magnitude of **Q** multiplied by the component $P \cos \theta$ of **P** in the direction of **Q**.

Commutative law $\mathbf{P} \cdot \mathbf{Q} = \mathbf{Q} \cdot \mathbf{P}$

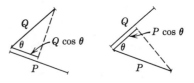

From the definition of the dot product

$$\mathbf{i} \cdot \mathbf{i} = \mathbf{j} \cdot \mathbf{j} = \mathbf{k} \cdot \mathbf{k} = 1$$

$$\mathbf{i} \cdot \mathbf{j} = \mathbf{j} \cdot \mathbf{i} = \mathbf{i} \cdot \mathbf{k} = \mathbf{k} \cdot \mathbf{i} = \mathbf{j} \cdot \mathbf{k} = \mathbf{k} \cdot \mathbf{j} = 0$$

$$\mathbf{P} \cdot \mathbf{Q} = (\mathbf{i}P_x + \mathbf{j}P_y + \mathbf{k}P_z) \cdot (\mathbf{i}Q_x + \mathbf{j}Q_y + \mathbf{k}Q_z)$$
$$= P_x Q_x + P_y Q_y + P_z Q_z$$

$$\mathbf{P} \cdot \mathbf{P} = P_x^{\,2} + P_y^{\,2} + P_z^{\,2}$$

It follows from the definition of the dot product that two vectors **P** and **Q** are perpendicular when their dot product vanishes, $\mathbf{P} \cdot \mathbf{Q} = 0$.

The angle θ between two vectors $\mathbf{P}_1$ and $\mathbf{P}_2$ may be found from their dot product expression $\mathbf{P}_1 \cdot \mathbf{P}_2 = P_1 P_2 \cos \theta$, which gives

$$\cos \theta = \frac{\mathbf{P}_1 \cdot \mathbf{P}_2}{P_1 P_2} = \frac{P_{1_x} P_{2_x} + P_{1_y} P_{2_y} + P_{1_z} P_{2_z}}{P_1 P_2} = l_1 l_2 + m_1 m_2 + n_1 n_2$$

where l, m, n stand for the respective direction cosines of the vectors. It is also observed that two vectors are perpendicular when their direction cosines obey the relation $l_1 l_2 + m_1 m_2 + n_1 n_2 = 0$.

Distributive law $\mathbf{P} \cdot (\mathbf{Q} + \mathbf{R}) = \mathbf{P} \cdot \mathbf{Q} + \mathbf{P} \cdot \mathbf{R}$

7. *Cross or vector product.* The cross product $\mathbf{P} \times \mathbf{Q}$ of the two vectors **P** and **Q** is defined as a vector with a magnitude

$$|\mathbf{P} \times \mathbf{Q}| = PQ \sin \theta$$

and a direction specified by the right-hand rule as shown. Reversing the order of the vectors and using the right-hand rule give $\mathbf{Q} \times \mathbf{P} = -\mathbf{P} \times \mathbf{Q}$.

Distributive law $\mathbf{P} \times (\mathbf{Q} + \mathbf{R}) = \mathbf{P} \times \mathbf{Q} + \mathbf{P} \times \mathbf{R}$

From the definition of the cross product, using a *right-handed coordinate system*

$$\mathbf{i} \times \mathbf{j} = \mathbf{k} \qquad \mathbf{j} \times \mathbf{k} = \mathbf{i} \qquad \mathbf{k} \times \mathbf{i} = \mathbf{j}$$
$$\mathbf{j} \times \mathbf{i} = -\mathbf{k} \qquad \mathbf{k} \times \mathbf{j} = -\mathbf{i} \qquad \mathbf{i} \times \mathbf{k} = -\mathbf{j}$$
$$\mathbf{i} \times \mathbf{i} = \mathbf{j} \times \mathbf{j} = \mathbf{k} \times \mathbf{k} = 0$$

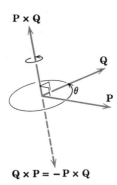

P × **Q**

$\mathbf{Q} \times \mathbf{P} = -\mathbf{P} \times \mathbf{Q}$

With the aid of these identities and the distributive law the vector product may be written

$$\mathbf{P} \times \mathbf{Q} = (\mathbf{i}P_x + \mathbf{j}P_y + \mathbf{k}P_z) \times (\mathbf{i}Q_x + \mathbf{j}Q_y + \mathbf{k}Q_z)$$
$$= \mathbf{i}(P_yQ_z - P_zQ_y) + \mathbf{j}(P_zQ_x - P_xQ_z) + \mathbf{k}(P_xQ_y - P_yQ_x)$$

The cross product may also be expressed by the determinant

$$\mathbf{P} \times \mathbf{Q} = \begin{vmatrix} \mathbf{i} & \mathbf{j} & \mathbf{k} \\ P_x & P_y & P_z \\ Q_x & Q_y & Q_z \end{vmatrix}$$

8. *Additional relations*

Triple scalar product $(\mathbf{P} \times \mathbf{Q}) \cdot \mathbf{R} = \mathbf{R} \cdot (\mathbf{P} \times \mathbf{Q})$. The dot and cross may be interchanged as long as the order of the vectors is maintained. Parentheses are unnecessary since $\mathbf{P} \times (\mathbf{Q} \cdot \mathbf{R})$ is meaningless since a vector $\mathbf{P}$ cannot be crossed into a scalar $\mathbf{Q} \cdot \mathbf{R}$. Thus the expression may be written

$$\mathbf{P} \times \mathbf{Q} \cdot \mathbf{R} = \mathbf{P} \cdot \mathbf{Q} \times \mathbf{R}$$

The triple scalar product has the determinant expansion

$$\mathbf{P} \times \mathbf{Q} \cdot \mathbf{R} = \begin{vmatrix} P_x & P_y & P_z \\ Q_x & Q_y & Q_z \\ R_x & R_y & R_z \end{vmatrix}$$

Triple vector product $(\mathbf{P} \times \mathbf{Q}) \times \mathbf{R} = -\mathbf{R} \times (\mathbf{P} \times \mathbf{Q})$
$$= \mathbf{R} \times (\mathbf{Q} \times \mathbf{P})$$

Here the parentheses must be used since an expression $\mathbf{P} \times \mathbf{Q} \times \mathbf{R}$ would be ambiguous because it would not identify the vector to be crossed. It may be shown that the triple vector product is equivalent to

$$(\mathbf{P} \times \mathbf{Q}) \times \mathbf{R} = \mathbf{R} \cdot \mathbf{PQ} - \mathbf{R} \cdot \mathbf{QP}$$

or

$$\mathbf{P} \times (\mathbf{Q} \times \mathbf{R}) = \mathbf{P} \cdot \mathbf{RQ} - \mathbf{P} \cdot \mathbf{QR}$$

The first term in the first expression, for example, is the dot product $\mathbf{R} \cdot \mathbf{P}$, a scalar, multiplied by the vector $\mathbf{Q}$.

9. *Derivatives of vectors* obey the same rules as for scalars.

$$\frac{d\mathbf{P}}{dt} = \dot{\mathbf{P}} = \mathbf{i}\dot{P}_x + \mathbf{j}\dot{P}_y + \mathbf{k}\dot{P}_z$$

$$\frac{d(\mathbf{P}u)}{dt} = \mathbf{P}\dot{u} + \dot{\mathbf{P}}u$$

$$\frac{d(\mathbf{P} \cdot \mathbf{Q})}{dt} = \mathbf{P} \cdot \dot{\mathbf{Q}} + \dot{\mathbf{P}} \cdot \mathbf{Q}$$

$$\frac{d(\mathbf{P} \times \mathbf{Q})}{dt} = \mathbf{P} \times \dot{\mathbf{Q}} + \dot{\mathbf{P}} \times \mathbf{Q}$$

10. *Integration of vectors.* If **V** is a function of x, y, and z and an element of volume is $d\tau = dx\,dy\,dz$, the integral of **V** over the volume may be written as the vector sum of the three integrals of its components. Thus

$$\int \mathbf{V}\,d\tau = \mathbf{i}\int V_x\,d\tau + \mathbf{j}\int V_y\,d\tau + \mathbf{k}\int V_z\,d\tau$$

B/8 SERIES

(Expression in brackets following series indicates range of convergence)

$$(1 \pm x)^n = 1 \pm nx + \frac{n(n-1)}{2!}x^2 \pm \frac{n(n-1)(n-2)}{3!}x^3 + \cdots \quad [x^2 < 1]$$

$$\sin x = x - \frac{x^3}{3!} + \frac{x^5}{5!} - \frac{x^7}{7!} + \cdots \qquad\qquad [x^2 < \infty]$$

$$\cos x = 1 - \frac{x^2}{2!} + \frac{x^4}{4!} - \frac{x^6}{6!} + \cdots \qquad\qquad [x^2 < \infty]$$

$$\sinh x = \frac{e^x - e^{-x}}{2} = x + \frac{x^3}{3!} + \frac{x^5}{5!} + \frac{x^7}{7!} + \cdots \qquad [x^2 < \infty]$$

$$\cosh x = \frac{e^x + e^{-x}}{2} = 1 + \frac{x^2}{2!} + \frac{x^4}{4!} + \frac{x^6}{6!} + \cdots \qquad [x^2 < \infty]$$

$$f(x) = \frac{a_0}{2} + \sum_{n=1}^{\infty} a_n \cos \frac{n\pi x}{l} + \sum_{n=1}^{\infty} b_n \sin \frac{n\pi x}{l}$$

where $a_n = \dfrac{1}{l}\displaystyle\int_{-l}^{l} f(x) \cos \frac{n\pi x}{l}\,dx,\ b_n = \dfrac{1}{l}\displaystyle\int_{-l}^{l} f(x) \sin \frac{n\pi x}{l}\,dx$

[Fourier expansion for $-l < x < l$]

B/9 DERIVATIVES

$$\frac{dx^n}{dx} = nx^{n-1}, \quad \frac{d(uv)}{dx} = u\frac{dv}{dx} + v\frac{du}{dx}, \quad \frac{d\left(\dfrac{u}{v}\right)}{dx} = \frac{v\dfrac{du}{dx} - u\dfrac{dv}{dx}}{v^2}$$

$$\lim_{\Delta x \to 0} \sin \Delta x = \sin dx = \tan dx = dx$$

$$\lim_{\Delta x \to 0} \cos \Delta x = \cos dx = 1$$

$$\frac{d \sin x}{dx} = \cos x, \quad \frac{d \cos x}{dx} = -\sin x, \quad \frac{d \tan x}{dx} = \sec^2 x$$

$$\frac{d \sinh x}{dx} = \cosh x, \quad \frac{d \cosh x}{dx} = \sinh x, \quad \frac{d \tanh x}{dx} = \text{sech}^2 x$$

B/10 INTEGRALS

$$\int x^n \, dx = \frac{x^{n+1}}{n+1}$$

$$\int \frac{dx}{x} = \ln x$$

$$\int \sqrt{a+bx} \, dx = \frac{2}{3b} \sqrt{(a+bx)^3}$$

$$\int x\sqrt{a+bx} \, dx = \frac{2}{15b^2}(3bx - 2a)\sqrt{(a+bx)^3}$$

$$\int \frac{dx}{\sqrt{a+bx}} = \frac{2\sqrt{a+bx}}{b}$$

$$\int \frac{x \, dx}{a+bx} = \frac{1}{b^2}[a + bx - a\ln(a+bx)]$$

$$\int \frac{x \, dx}{(a+bx)^n} = \frac{(a+bx)^{1-n}}{b^2}\left(\frac{a+bx}{2-n} - \frac{a}{1-n}\right)$$

$$\int \frac{dx}{a+bx^2} = \frac{1}{\sqrt{ab}} \tan^{-1} \frac{x\sqrt{ab}}{a} \quad \text{or} \quad \frac{1}{\sqrt{-ab}} \tanh^{-1} \frac{x\sqrt{-ab}}{a}$$

$$\int \frac{x \, dx}{a+bx^2} = \frac{1}{2b} \ln \frac{a+bx^2}{b}$$

$$\int \sqrt{x^2 \pm a^2} \, dx = \tfrac{1}{2}[x\sqrt{x^2 \pm a^2} \pm a^2 \ln(x + \sqrt{x^2 \pm a^2})]$$

$$\int \sqrt{a^2 - x^2} \, dx = \tfrac{1}{2}\left(x\sqrt{a^2 - x^2} + a^2 \sin^{-1}\frac{x}{a}\right)$$

$$\int x\sqrt{a^2 - x^2} \, dx = -\tfrac{1}{3}\sqrt{(a^2 - x^2)^3}$$

$$\int x^2 \sqrt{a^2 - x^2} \, dx = -\frac{x}{4}\sqrt{(a^2 - x^2)^3} + \frac{a^2}{8}\left(x\sqrt{a^2 - x^2} + a^2 \sin^{-1}\frac{x}{a}\right)$$

$$\int x^3 \sqrt{a^2 - x^2} \, dx = -\tfrac{1}{5}(x^2 + \tfrac{2}{3}a^2)\sqrt{(a^2 - x^2)^3}$$

$$\int \frac{dx}{\sqrt{a+bx+cx^2}} = \frac{1}{\sqrt{c}} \ln\left(\sqrt{a+bx+cx^2} + x\sqrt{c} + \frac{b}{2\sqrt{c}}\right) \quad \text{or} \quad \frac{-1}{\sqrt{-c}} \sin^{-1}\left(\frac{b+2cx}{\sqrt{b^2 - 4ac}}\right)$$

$$\int \frac{dx}{\sqrt{x^2 \pm a^2}} = \ln(x + \sqrt{x^2 \pm a^2}) \qquad\qquad \int \frac{x \, dx}{\sqrt{x^2 - a^2}} = \sqrt{x^2 - a^2}$$

$$\int \frac{dx}{\sqrt{a^2 - x^2}} = \sin^{-1}\frac{x}{a} \qquad\qquad\qquad \int \frac{x \, dx}{\sqrt{a^2 \pm x^2}} = \pm\sqrt{a^2 \pm x^2}$$

$$\int x\sqrt{x^2 \pm a^2} \, dx = \tfrac{1}{3}\sqrt{(x^2 \pm a^2)^3}$$

$$\int x^2\sqrt{x^2 \pm a^2} \, dx = \frac{x}{4}\sqrt{(x^2 \pm a^2)^3} \mp \frac{a^2}{8}x\sqrt{x^2 \pm a^2} - \frac{a^4}{8}\ln(x + \sqrt{x^2 \pm a^2})$$

$$\int \sin x \, dx = -\cos x$$

$$\int \cos x \, dx = \sin x$$

$$\int \sec x \, dx = \frac{1}{2} \ln \frac{1 + \sin x}{1 - \sin x}$$

$$\int \sin^2 x \, dx = \frac{x}{2} - \frac{\sin 2x}{4}$$

$$\int \cos^2 x \, dx = \frac{x}{2} + \frac{\sin 2x}{4}$$

$$\int \sin x \cos x \, dx = \frac{\sin^2 x}{2}$$

$$\int \sinh x \, dx = \cosh x$$

$$\int \cosh x \, dx = \sinh x$$

$$\int \tanh x \, dx = \ln \cosh x$$

$$\int \ln x \, dx = x \ln x - x$$

$$\int e^{ax} \, dx = \frac{e^{ax}}{a}$$

$$\int x e^{ax} \, dx = \frac{e^{ax}}{a^2} (ax - 1)$$

$$\int e^{ax} \sin px \, dx = \frac{e^{ax}(a \sin px - p \cos px)}{a^2 + p^2}$$

$$\int e^{ax} \cos px \, dx = \frac{e^{ax}(a \cos px + p \sin px)}{a^2 + p^2}$$

$$\int e^{ax} \sin^2 x \, dx = \frac{e^{ax}}{4 + a^2}\left(a \sin^2 x - \sin 2x + \frac{2}{a}\right)$$

$$\int e^{ax} \cos^2 x \, dx = \frac{e^{ax}}{4 + a^2}\left(a \cos^2 x + \sin 2x + \frac{2}{a}\right)$$

$$\int e^{ax} \sin x \cos x \, dx = \frac{e^{ax}}{4 + a^2}\left(\frac{a}{2} \sin 2x - \cos 2x\right)$$

$$\int \sin^3 x \, dx = -\frac{\cos x}{3}(2 + \sin^2 x)$$

$$\int \cos^3 x \, dx = \frac{\sin x}{3}(2 + \cos^2 x)$$

$$\int x \sin x \, dx = \sin x - x \cos x$$

$$\int x \cos x \, dx = \cos x + x \sin x$$

$$\int x^2 \sin x \, dx = 2x \sin x - (x^2 - 2) \cos x$$

$$\int x^2 \cos x \, dx = 2x \cos x + (x^2 - 2) \sin x$$

Radius of
curvature
$$\begin{cases} \rho_{xy} = \dfrac{\left[1 + \left(\dfrac{dy}{dx}\right)^2\right]^{3/2}}{\dfrac{d^2 y}{dx^2}} \\[4em] \rho_{r\theta} = \dfrac{\left[r^2 + \left(\dfrac{dr}{d\theta}\right)^2\right]^{3/2}}{r^2 + 2\left(\dfrac{dr}{d\theta}\right)^2 - r\dfrac{d^2 r}{d\theta^2}} \end{cases}$$

USEFUL TABLES

C

Table C/1   Properties

A. *Density,* ρ

	kg/m³	lbm/ft³		kg/m³	lbm/ft³
Aluminum	2 690	168	Iron (cast)	7 210	450
Concrete (av.)	2 400	150	Lead	11 370	710
Copper	8 910	556	Mercury	13 570	847
Earth (wet, av.)	1 760	110	Oil (av.)	900	56
(dry, av.)	1 280	80	Steel	7 830	489
Glass	2 590	162	Titanium	3 080	192
Gold	19 300	1205	Water (fresh)	1 000	62.4
Ice	900	56	(salt)	1 030	64
			Wood (soft pine)	480	30
			(hard oak)	800	50

B. *Coefficients of Friction,* μ

(The coefficients in the following table represent typical values under normal working conditions. Actual coefficients for a given situation will depend on the exact nature of the contacting surfaces. A variation of 25 to 100 per cent or more from these values could be expected in an actual application, depending on prevailing conditions of cleanliness, surface finish, pressure, lubrication, and velocity.)

CONTACTING SURFACE	TYPICAL VALUES OF COEFFICIENT OF FRICTION, μ	
	STATIC	KINETIC
Steel on steel (dry)	0.6	0.4
Steel on steel (greasy)	0.1	0.05
Teflon on steel	0.04	0.04
Steel on babbitt (dry)	0.4	0.3
Steel on babbitt (greasy)	0.1	0.07
Brass on steel (dry)	0.5	0.4
Brake lining on cast iron	0.4	0.3
Rubber tires on smooth pavement (dry)	0.9	0.8
Wire rope on iron pulley (dry)	0.2	0.15
Hemp rope on metal	0.3	0.2
Metal on ice	—	0.02

Table C/2   Solar System Constants

Universal gravitational constant $K = 6.673(10^{-11}) \text{ m}^3/(\text{kg} \cdot \text{s}^2)$
                                                  $= 3.439(10^{-8}) \text{ ft}^4/(\text{lbf-s}^4)$

Mass of Earth $m = 5.976(10^{24}) \text{ kg}$
                                                  $= 4.095(10^{23}) \text{ lbf-s}^2/\text{ft}$

Period of Earth's rotation (1 sidereal day) $= 23 \text{ h } 56 \text{ min } 4 \text{ s}$
                                                  $= 23.9344 \text{ h}$

Angular velocity of Earth $\omega = 0.7292(10^{-4}) \text{ rad/s}$
Angular velocity of Earth-Sun line $\omega' = 0.1991(10^{-6}) \text{ rad/s}$
Mean velocity of Earth's center about Sun $= 107\ 200 \text{ km/h}$
                                                  $= 66,610 \text{ mi/h}$

BODY	MEAN DISTANCE TO SUN km (mi)	ECCENTRICITY OF ORBIT e	PERIOD OF ORBIT SOLAR DAYS	MEAN DIAMETER km (mi)	MASS RELATIVE TO EARTH	SURFACE GRAVITATIONAL ACCELERATION m/s² (ft/s²)	ESCAPE VELOCITY km/s (mi/s)
Sun	—	—	—	1 392 000 (865 000)	333 000	274 (898)	616 (383)
Moon	384 398° (238 854)°	0.055	27.32	3 476 (2 160)	0.0123	1.62 (5.32)	2.37 (1.47)
Mercury	57.3 × 10⁶ (35.6 × 10⁶	0.206	87.97	5 000 (3 100)	0.054	3.47 (11.4)	4.17 (2.59)
Venus	108 × 10⁶ (67.2 × 10⁶)	0.0068	224.70	12 400 (7 700)	0.815	8.44 (27.7)	10.24 (6.36)
Earth	149.6 × 10⁶ (92.96 × 10⁶)	0.0167	365.26	12 742† (7 917)†	1.000	9.821‡ (32. 22)‡	11.18 (6.95)
Mars	227.9 × 10⁶ (141.6 × 10⁶)	0.093	686.98	6 788 (4 218)	0.107	3.73 (12.3)	5.03 (3.13)

° Mean distance to Earth (center-to-center)
† Diameter of sphere of equal volume
    polar diameter = 12 713 km
                   = 7 900 mi
    equatorial diameter = 12 755 km
                         = 7 926 mi
‡ For nonrotating spherical Earth, equivalent to absolute value at sea level and latitude 37.5°

Table C/3　Properties of Plane Figures

FIGURE	CENTROID	AREA MOMENTS OF INERTIA
Arc Segment 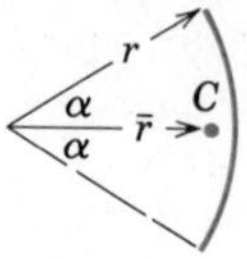	$\bar{r} = \dfrac{r \sin \alpha}{\alpha}$	—
Quarter and Semicircular Arcs 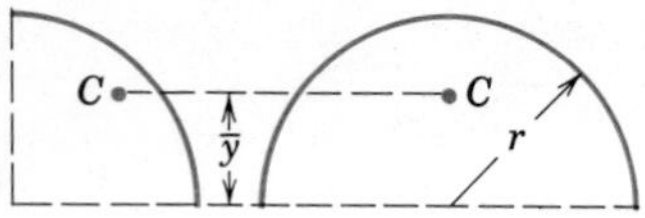	$\bar{y} = \dfrac{2r}{\pi}$	—
Triangular Area 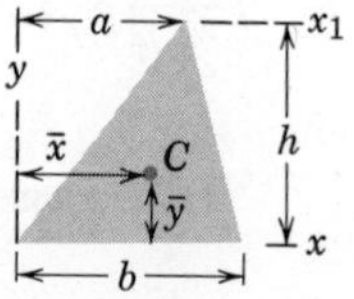	$\bar{x} = \dfrac{a + b}{3}$ $\bar{y} = \dfrac{h}{3}$	$I_x = \dfrac{bh^3}{12}$ $\bar{I}_x = \dfrac{bh^3}{36}$ $I_{x_1} = \dfrac{bh^3}{4}$
Rectangular Area 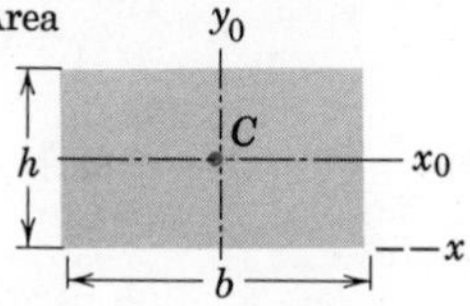	—	$I_x = \dfrac{bh^3}{3}$ $\bar{I}_x = \dfrac{bh^3}{12}$ $\bar{J} = \dfrac{bh}{12}(b^2 + h^2)$
Area of Circular Sector 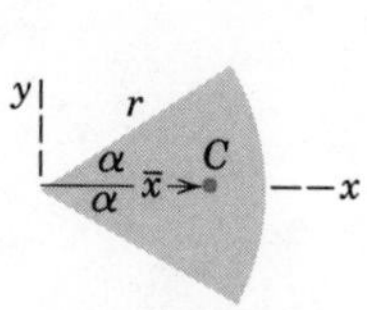	$\bar{x} = \dfrac{2}{3} \dfrac{r \sin \alpha}{\alpha}$	$I_x = \dfrac{r^4}{4}\left(\alpha - \tfrac{1}{2}\sin 2\alpha\right)$ $I_y = \dfrac{r^4}{4}\left(\alpha + \tfrac{1}{2}\sin 2\alpha\right)$ $J = \tfrac{1}{2}r^4\alpha$
Quarter Circular Area 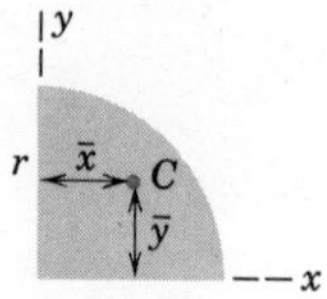	$\bar{x} = \bar{y} = \dfrac{4r}{3\pi}$	$I_x = I_y = \dfrac{\pi r^4}{16}$ $\bar{I}_x = \bar{I}_y = \left(\dfrac{\pi}{16} - \dfrac{4}{9\pi}\right)r^4$ $J = \dfrac{\pi r^4}{8}$
Area of Elliptical Quadrant Area $A = \dfrac{\pi ab}{4}$ 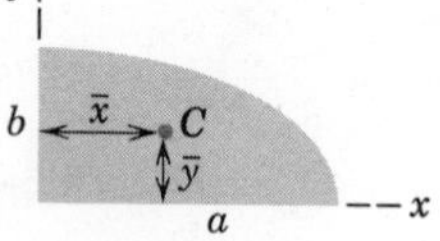	$\bar{x} = \dfrac{4a}{3\pi}$ $\bar{y} = \dfrac{4b}{3\pi}$	$I_x = \dfrac{\pi ab^3}{16},\ \bar{I}_x = \left(\dfrac{\pi}{16} - \dfrac{4}{9\pi}\right)ab^3$ $I_y = \dfrac{\pi a^3 b}{16},\ \bar{I}_y = \left(\dfrac{\pi}{16} - \dfrac{4}{9\pi}\right)a^3 b$ $J = \dfrac{\pi ab}{16}(a^2 + b^2)$

Table C/4   Properties of Homogeneous Solids

(m = mass of body shown)

BODY	MASS CENTER	MOMENTS OF INERTIA
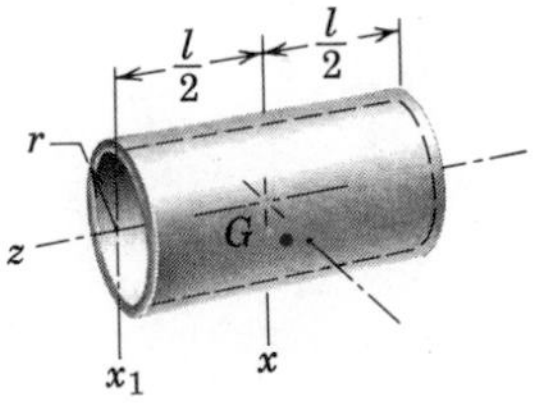 Circular Cylindrical Shell	—	$I_{xx} = \frac{1}{2}mr^2 + \frac{1}{12}ml^2$ $I_{x_1x_1} = \frac{1}{2}mr^2 + \frac{1}{3}ml^2$ $I_{zz} = mr^2$
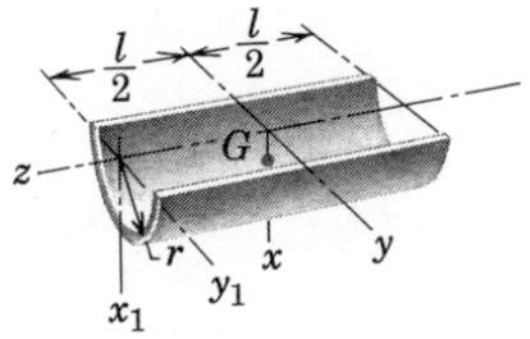 Half Cylindrical Shell	$\bar{x} = \dfrac{2r}{\pi}$	$I_{xx} = I_{yy}$ $\quad = \frac{1}{2}mr^2 + \frac{1}{12}ml^2$ $I_{x_1x_1} = I_{y_1y_1}$ $\quad = \frac{1}{2}mr^2 + \frac{1}{3}ml^2$ $I_{zz} = mr^2$ $\bar{I}_{zz} = \left(1 - \dfrac{4}{\pi^2}\right)mr^2$
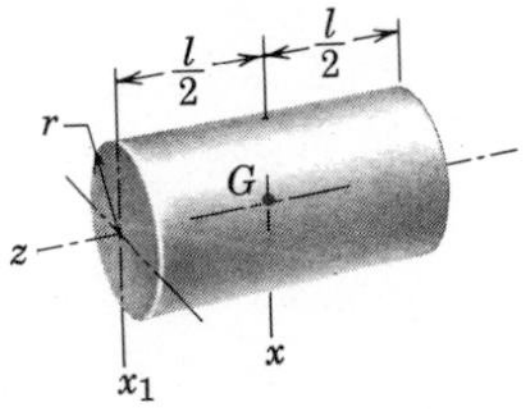 Circular Cylinder	—	$I_{xx} = \frac{1}{4}mr^2 + \frac{1}{12}ml^2$ $I_{x_1x_1} = \frac{1}{4}mr^2 + \frac{1}{3}ml^2$ $I_{zz} = \frac{1}{2}mr^2$
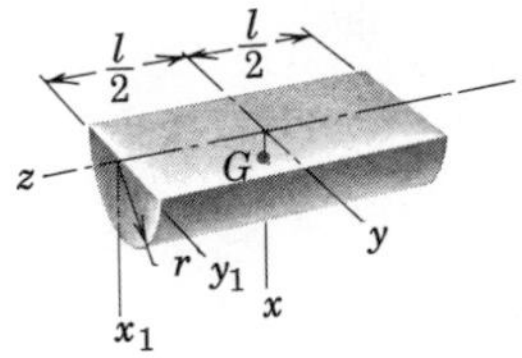 Semicylinder	$\bar{x} = \dfrac{4r}{3\pi}$	$I_{xx} = I_{yy}$ $\quad = \frac{1}{4}mr^2 + \frac{1}{12}ml^2$ $I_{x_1x_1} = I_{y_1y_1}$ $\quad = \frac{1}{4}mr^2 + \frac{1}{3}ml^2$ $I_{zz} = \frac{1}{2}mr^2$ $\bar{I}_{zz} = \left(\dfrac{1}{2} - \dfrac{16}{9\pi^2}\right)mr^2$
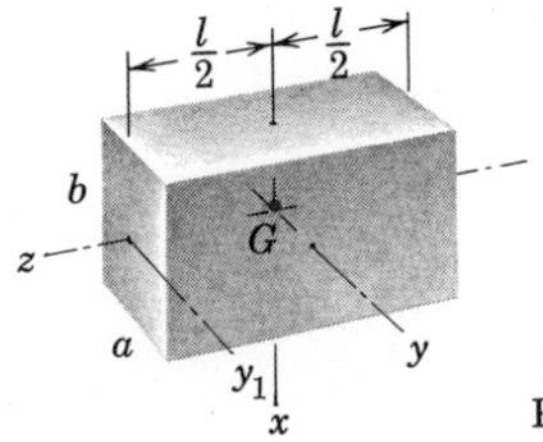 Rectangular Parallelepiped	—	$I_{xx} = \frac{1}{12}m(a^2 + l^2)$ $I_{yy} = \frac{1}{12}m(b^2 + l^2)$ $I_{zz} = \frac{1}{12}m(a^2 + b^2)$ $I_{y_1y_1} = \frac{1}{12}mb^2 + \frac{1}{3}ml^2$

Table C/4   *Continued*

(m = mass of body shown)

BODY	MASS CENTER	MOMENTS OF INERTIA
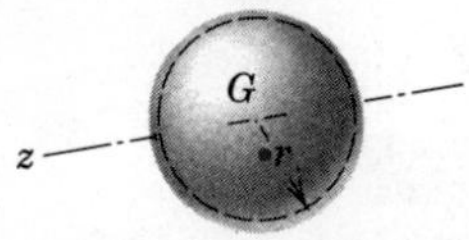 Spherical Shell	—	$I_{zz} = \frac{2}{3}mr^2$
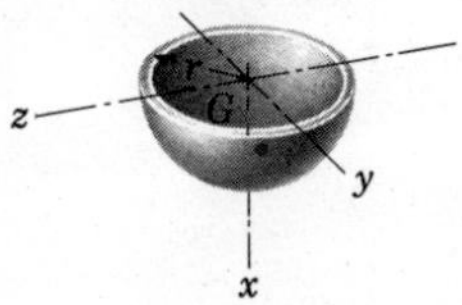 Hemispherical Shell	$\bar{x} = \frac{r}{2}$	$I_{xx} = I_{yy} = I_{zz} = \frac{2}{3}mr^2$ $\bar{I}_{yy} = \bar{I}_{zz} = \frac{5}{12}mr^2$
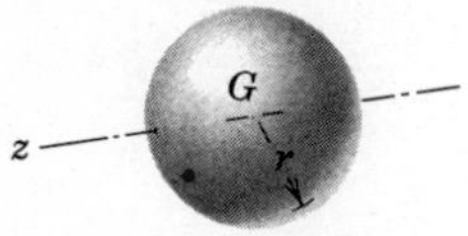 Sphere	—	$I_{zz} = \frac{2}{5}mr^2$
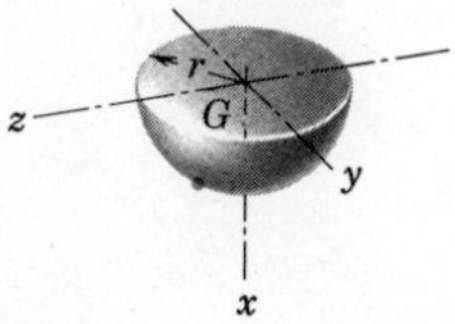 Hemisphere	$\bar{x} = \frac{3r}{8}$	$I_{xx} = I_{yy} = I_{zz} = \frac{2}{5}mr^2$ $\bar{I}_{yy} = \bar{I}_{zz} = \frac{83}{320}mr^2$
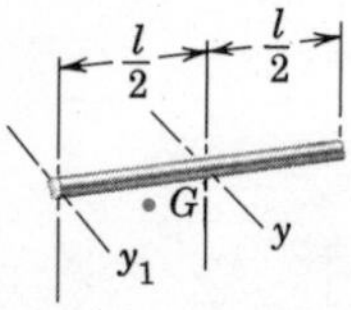 Uniform Slender Rod	—	$I_{yy} = \frac{1}{12}ml^2$ $I_{y_1 y_1} = \frac{1}{3}ml^2$

Table C/4   *Continued*
(m = mass of body shown)

BODY	MASS CENTER	MOMENTS OF INERTIA
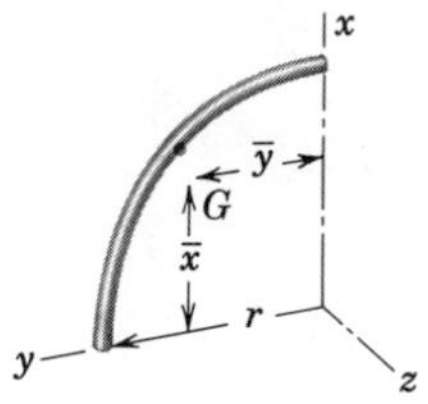 Quarter Circular Rod	$\bar{x} = \bar{y}$ $= \dfrac{2r}{\pi}$	$I_{xx} = I_{yy} = \frac{1}{2}mr^2$ $I_{zz} = mr^2$
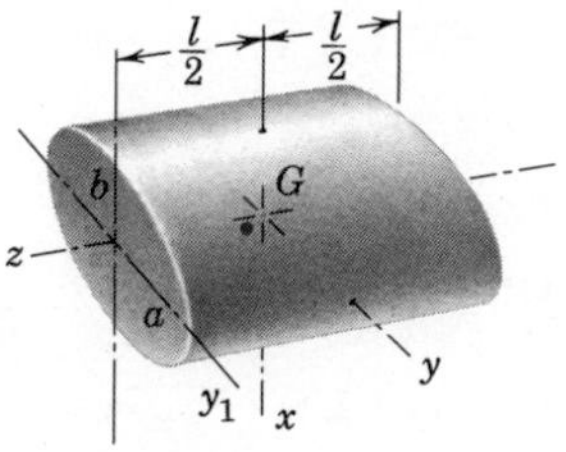 Elliptical Cylinder	—	$I_{xx} = \frac{1}{4}ma^2 + \frac{1}{12}ml^2$ $I_{yy} = \frac{1}{4}mb^2 + \frac{1}{12}ml^2$ $I_{zz} = \frac{1}{4}m(a^2 + b^2)$ $I_{y_1 y_1} = \frac{1}{4}mb^2 + \frac{1}{3}ml^2$
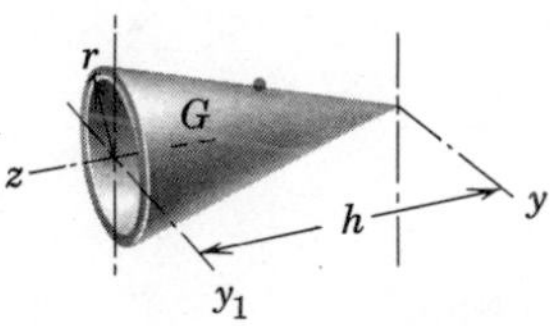 Conical Shell	$\bar{z} = \dfrac{2h}{3}$	$I_{yy} = \frac{1}{4}mr^2 + \frac{1}{2}mh^2$ $I_{y_1 y_1} = \frac{1}{4}mr^2 + \frac{1}{6}mh^2$ $I_{zz} = \frac{1}{2}mr^2$ $\bar{I}_{yy} = \frac{1}{4}mr^2 + \frac{1}{18}mh^2$
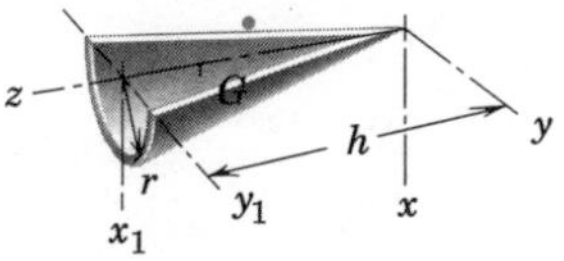 Half Conical Shell	$\bar{x} = \dfrac{4r}{3\pi}$ $\bar{z} = \dfrac{2h}{3}$	$I_{xx} = I_{yy}$ $\quad = \frac{1}{4}mr^2 + \frac{1}{2}mh^2$ $I_{x_1 x_1} = I_{y_1 y_1}$ $\quad = \frac{1}{4}mr^2 + \frac{1}{6}mh^2$ $I_{zz} = \frac{1}{2}mr^2$ $\bar{I}_{zz} = \left(\dfrac{1}{2} - \dfrac{16}{9\pi^2}\right)mr^2$
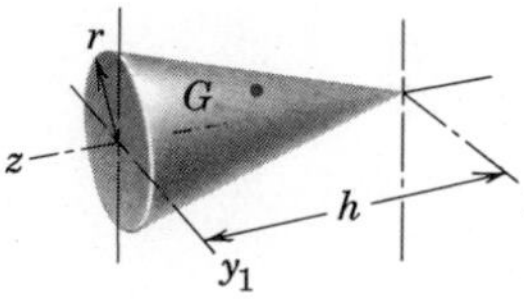 Right Circular Cone	$\bar{z} = \dfrac{3h}{4}$	$I_{yy} = \frac{3}{20}mr^2 + \frac{3}{5}mh^2$ $I_{y_1 y_1} = \frac{3}{20}mr^2 + \frac{1}{10}mh^2$ $I_{zz} = \frac{3}{10}mr^2$ $\bar{I}_{yy} = \frac{3}{20}mr^2 + \frac{3}{80}mh^2$

Table C/4   *Continued*

(m = mass of body shown)

BODY	MASS CENTER	MOMENTS OF INERTIA
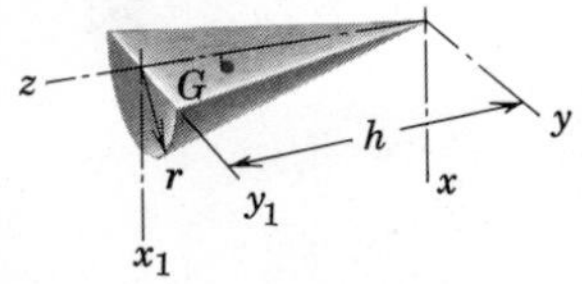 Half Cone	$\bar{x} = \dfrac{r}{\pi}$ $\bar{z} = \dfrac{3h}{4}$	$I_{xx} = I_{yy}$ $\qquad = \frac{3}{20}mr^2 + \frac{3}{5}mh^2$ $I_{x_1 x_1} = I_{y_1 y_1}$ $\qquad = \frac{3}{20}mr^2 + \frac{1}{10}mh^2$ $I_{zz} = \frac{3}{10}mr^2$ $\bar{I}_{zz} = \left(\dfrac{3}{10} - \dfrac{1}{\pi^2}\right)mr^2$
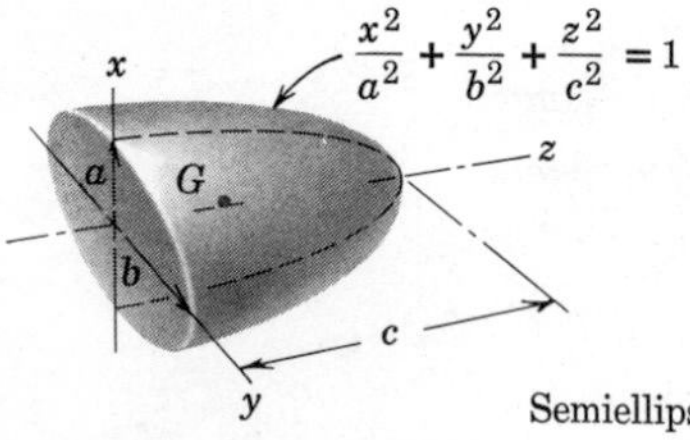 $\dfrac{x^2}{a^2} + \dfrac{y^2}{b^2} + \dfrac{z^2}{c^2} = 1$ Semiellipsoid	$\bar{z} = \dfrac{3c}{8}$	$I_{xx} = \frac{1}{5}m(b^2 + c^2)$ $I_{yy} = \frac{1}{5}m(a^2 + c^2)$ $I_{zz} = \frac{1}{5}m(a^2 + b^2)$ $\bar{I}_{xx} = \frac{1}{5}m(b^2 + \frac{19}{64}c^2)$ $\bar{I}_{yy} = \frac{1}{5}m(a^2 + \frac{19}{64}c^2)$
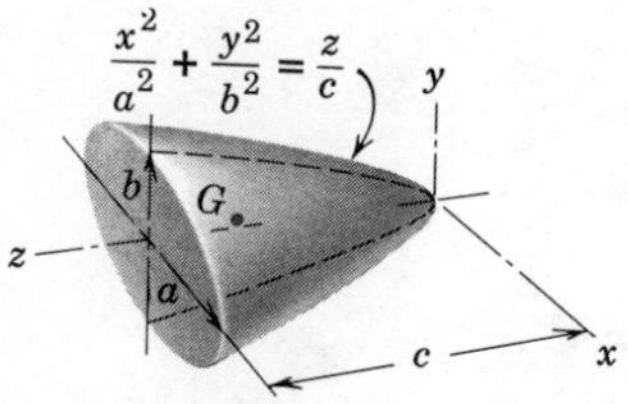 $\dfrac{x^2}{a^2} + \dfrac{y^2}{b^2} = \dfrac{z}{c}$ Elliptic Paraboloid	$\bar{z} = \dfrac{2c}{3}$	$I_{xx} = \frac{1}{6}mb^2 + \frac{1}{2}mc^2$ $I_{yy} = \frac{1}{6}ma^2 + \frac{1}{2}mc^2$ $I_{zz} = \frac{1}{6}m(a^2 + b^2)$ $\bar{I}_{xx} = \frac{1}{6}m(b^2 + \frac{1}{3}c^2)$ $\bar{I}_{yy} = \frac{1}{6}m(a^2 + \frac{1}{3}c^2)$
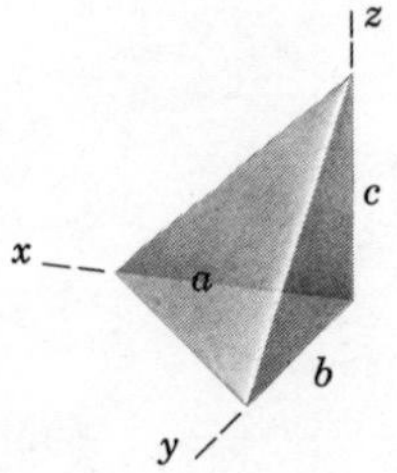 Rectangular Tetrahedron	$\bar{x} = \dfrac{a}{4}$ $\bar{y} = \dfrac{b}{4}$ $\bar{z} = \dfrac{c}{4}$	$I_{xx} = \frac{1}{10}m(b^2 + c^2)$ $I_{yy} = \frac{1}{10}m(a^2 + c^2)$ $I_{zz} = \frac{1}{10}m(a^2 + b^2)$ $\bar{I}_{xx} = \frac{3}{80}m(b^2 + c^2)$ $\bar{I}_{yy} = \frac{3}{80}m(a^2 + c^2)$ $\bar{I}_{zz} = \frac{3}{80}m(a^2 + b^2)$
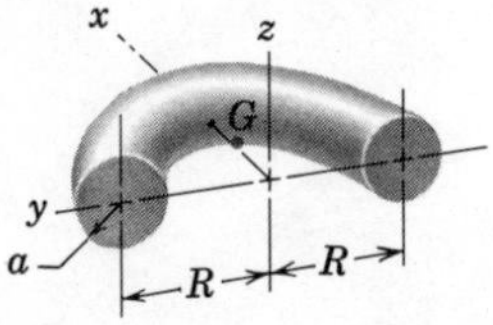 Half Torus	$\bar{x} =$ $\dfrac{a^2 + 4R^2}{2\pi R}$	$I_{xx} = I_{yy} = \frac{1}{2}mR^2 + \frac{5}{8}ma^2$ $I_{zz} = mR^2 + \frac{3}{4}ma^2$

INDEX

CONVERSION CHARTS BETWEEN SI AND U.S. CUSTOMARY UNITS

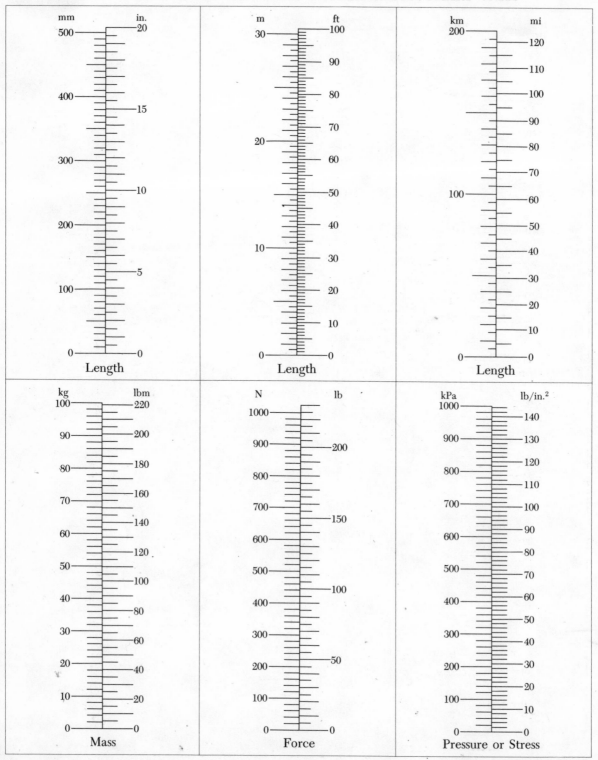